CAMBRIDGE LIBRARY COLLECTION

Books of enduring scholarly value

History

The books reissued in this series include accounts of historical events and movements by eye-witnesses and contemporaries, as well as landmark studies that assembled significant source materials or developed new historiographical methods. The series includes work in social, political and military history on a wide range of periods and regions, giving modern scholars ready access to influential publications of the past.

History of the Royal Society

The Royal Society has been dedicated to scientific inquiry since the seventeenth century. In 1811, Thomas Thomson (1773–1852), a pioneering chemistry teacher who was elected a fellow of the society in the same year, undertook the project of writing a history of the organisation's illustrious past. In this book, published in 1812, Thomson explains how the group began in 1645, initiated by men who met once a week to discuss natural philosophy and mathematics. They were eventually granted a royal charter by Charles II in 1662. The society grew in number and prestige, and began publishing research in its *Philosophical Transactions* in 1665. Thomson's work focuses particularly on the development of the group's many scientific areas of interest and summarises various papers it published. He also includes a full list of the fellowship, from the society's foundation to 1812, and a copy of the society's original charter.

Cambridge University Press has long been a pioneer in the reissuing of out-of-print titles from its own backlist, producing digital reprints of books that are still sought after by scholars and students but could not be reprinted economically using traditional technology. The Cambridge Library Collection extends this activity to a wider range of books which are still of importance to researchers and professionals, either for the source material they contain, or as landmarks in the history of their academic discipline.

Drawing from the world-renowned collections in the Cambridge University Library, and guided by the advice of experts in each subject area, Cambridge University Press is using state-of-the-art scanning machines in its own Printing House to capture the content of each book selected for inclusion. The files are processed to give a consistently clear, crisp image, and the books finished to the high quality standard for which the Press is recognised around the world. The latest print-on-demand technology ensures that the books will remain available indefinitely, and that orders for single or multiple copies can quickly be supplied.

The Cambridge Library Collection will bring back to life books of enduring scholarly value (including out-of-copyright works originally issued by other publishers) across a wide range of disciplines in the humanities and social sciences and in science and technology.

History of the Royal Society

*From its Institution
to the End of the Eighteenth Century*

THOMAS THOMSON

CAMBRIDGE
UNIVERSITY PRESS

CAMBRIDGE UNIVERSITY PRESS

Cambridge, New York, Melbourne, Madrid, Cape Town, Singapore,
São Paolo, Delhi, Dubai, Tokyo, Mexico City

Published in the United States of America by Cambridge University Press, New York

www.cambridge.org
Information on this title: www.cambridge.org/9781108028158

© in this compilation Cambridge University Press 2011

This edition first published 1812
This digitally printed version 2011

ISBN 978-1-108-02815-8 Paperback

HISTORY

OF

THE ROYAL SOCIETY,

FROM

ITS INSTITUTION

TO

THE END OF THE EIGHTEENTH CENTURY.

═══

BY THOMAS THOMSON, M.D. F.R.S. L. & E.

MEMBER OF THE GEOLOGICAL SOCIETY, OF THE WERNERIAN SOCIETY, AND OF THE IMPERIAL
CHIRURGO-MEDICAL ACADEMY OF PETERSBURGH.

═══

LONDON:

PRINTED FOR ROBERT BALDWIN, 47, PATERNOSTER-ROW.

━━

1812.

C. Baldwin, Printer, New Bridge-Street, London.

TO THE

RIGHT HON. SIR JOSEPH BANKS, Bart. President,

AND TO THE

COUNCIL AND FELLOWS OF THE ROYAL SOCIETY,

THIS ATTEMPT TO ELUCIDATE

THE

PHILOSOPHICAL TRANSACTIONS,

IS INSCRIBED,

WITH MUCH RESPECT,

BY THEIR MOST OBEDIENT SERVANT,

THE AUTHOR.

PREFACE.

THE following Work was projected by the Proprietors of the New Abridgement of the Philosophical Transactions, and was intended to consist partly of Biographical Sketches of some of the most eminent Fellows of the Royal Society, and partly of an arranged view of the whole contents of the Philosophical Transactions, from the commencement of the work, in 1665, down to the end of the year 1800, when the Abridgement terminated. The plan being laid before the author some years ago, he readily agreed to undertake the execution of it, conceiving that such a work would be of considerable utility; but in attempting to collect the materials, several difficulties occurred rather of a formidable nature. The first, and not the least fatiguing part of the task, was to peruse the whole of the Philosophical Transactions, and to arrange all the papers under distinct heads, according to the sciences to which they respectively belonged. This took up a much longer portion of time than the author expected, or, indeed, would choose to specify. On proceeding to draw up an account of the papers belonging to each of the sciences, in the order that appeared most convenient, it was immediately perceived that the essays, for the most part, were of so insulated a nature, that it would be impossible to give any interest or connection to the work, if the subjects were strictly confined, as was originally intended, merely to the details to be found in the Transactions. To remedy this defect as much as possible, and to give the reader a greater interest in the sciences as he proceeded, it was thought necessary to begin the history of every science as nearly as possible at its origin, and to give a rapid sketch of its progress till the period of the establishment of the Royal Society. This, it was thought, would put it in the power of every one to judge with more accuracy how far the labours of the Royal Society had contributed to the increase of knowledge, and the advancement of the sciences.

On attempting to give an analysis of the papers in the Transactions, it was perceived that there were some classes of them which could not with propriety be abridged, on account of the nature of the subjects of which they treated. This in particular was the case with the papers on botany. In some branches of science the author was overwhelmed by such a multitude of valuable papers, quite unconnected with each other, that it was impossible, without swelling the work much beyond the length that would have been tolerated, to notice them all. This is the case particularly in the sciences of medicine, mathematics, and chemistry. The only resource left in these sciences was to select those topics which appeared of the most importance; and the author is not without hopes that the selection, which he has made, will meet with the approbation of the reader.

In several branches of science, where the papers in the Transactions are either very few, or of comparatively trifling value, it was thought requisite to introduce the subject with a short outline of the principles of the science. This method was resorted to, because it was found impossible, in any other way, to give such a connection to the parts as would render the subject treated of in a sufficient degree interesting to the reader. Indeed the original plan was to have given a pretty full outline of each of the sciences which occupy a place in the Philosophical Transactions; such, for example, as the section in the first Book which treats of the physiology of plants. But on putting the work to the press, it was soon found that these outlines would increase the size much more than their utility would warrant. On that account several of the longest of them, such as mathematics and chemistry, were omitted. The reader can have no great cause to regret these omissions, as there are such a multiplicity of books both in our own, and in the French language, which supply them. Instead of these outlines, it was thought more entertaining, as well as useful, to substitute historical details which are not so readily to be met with.

Throughout the whole work the references have been made to the Philosophical Transactions rather than to the Abridgement, because it was the object of the author to make it independent of the Abridgement. These references however will serve equally well for the possessors of either work; because in each page of the Abridgement are printed the date and number of the volume where the corresponding paper is found in the Original; and the page of the Original where the paper commences is affixed to the title of each paper in the Abridgement.

CONTENTS.

CONTENTS.

INDEX.

AN

ACCOUNT

OF THE

ROYAL SOCIETY OF LONDON.

————

THERE is nothing which has contributed so much to the rapid progress which Philosophy has made since the revival of letters in Modern Europe, as the exertions of those numerous scientific societies and academies, which have successively been formed in most European countries. Italy, in which polite literature and the arts first made their appearance, has the honour of having given origin to these meritorious associations. England was not slow in imitating the example of the Italian States. The Royal Society was established for the express purpose of advancing experimental philosophy, and is beyond dispute the most magnificent and liberal establishment of the kind which has ever been formed. Its foundation was laid during the time of the civil wars, and was owing to the accidental association of several learned men, who took no part in the disturbances which agitated Great Britain.

About the year 1645, several ingenious men,* who resided in London, and were interested in the progress of mathematics and natural philosophy, agreed to meet once a week to discourse upon subjects connected with these sciences.† The meetings were held sometimes in Dr. Goddard's lodgings, in Wood-street, because he kept in his house an operator for grinding glasses for telescopes; sometimes in Cheapside, and sometimes in Gresham College. In the years 1648 and 1649, several of the gentlemen who attended these meetings being appointed to situations in the University of Oxford, they instituted a similar Society in that city, in conjunction with several eminent men already established there. Among the primitive members of this Oxonian Society, we find the following celebrated names: Dr. Wilkins, Dr. Wallis, Dr. Goddard, Dr. Seth Ward, Dr. Bathurst, Dr. William Petty, Dr. Willis. The meetings were for some time held in Dr. Petty's lodgings; and, when that gentleman went to Ire-

Origin of the Royal Society.

* The most remarkable of these men were, Dr. John Wilkins, Dr. Wallis, Dr. Goddard, Dr. Ent, Dr. Glisson, Dr. Merret, Mr. Foster, and Mr. Haake, a German, who first suggested these weekly meetings. Birch's *Hist. of the Royal Society*, Vol. I. p. 1.

† Wallis's Account of his own Life. *Preface to Mr. Hearne's edition of Langtoft's Chronicle* Vol. I. p. 161.

B

land in 1652, the Society met first in Dr. Wilkins' apartments, and afterwards in those of Mr. Boyle.

The greatest part of these Oxford gentlemen coming to London in 1659, held their meetings twice a week at Gresham College; on Wednesday, after the Astronomical Lecture of Mr. Christopher Wren, and on Thursday, when Mr. Lawrence Rooke lectured on Geometry. Here they were joined by several new associates; among others, by Lord Brouncker, William Brereton, Esq., Sir Paul Neile, John Evelyn, Esq., Thomas Henshaw, Esq., Henry Slingesby, Esq., Dr. Timothy Clarke, Dr. Ent, William Balle, Esq , Abraham Hill, Esq., Dr. William Croune. These meetings were continued till the Members were scattered by the disasters of 1659, after the resignation of Richard Cromwell, when their place of meeting was converted into quarters for soldiers. But after the restoration of King Charles II. in 1660, these meetings were revived, and still more numerously attended On the 28th of November, 1660, a number of gentlemen* met in Mr. Rooke's apartment, Gresham College, and agreed to constitute themselves into a Society, for the promotion of all kinds of experimental philosophy. A set of regulations was drawn up, and a weekly contribution of a shilling was collected from each of the Members, in order to defray the expenses of their experimental investigations. At first the number of Members was limited to 55, but it was afterwards extended, and, finally, admission was left open to every proper candidate. A President, a Secretary, and a Register, were elected out of their body, and an Amanuensis and Operator hired to execute the purposes of the Society.

Such was the origin of the Royal Society of London, which from its very commencement attracted the attention of all the philosophers in Europe, and which was publicly encouraged by Charles II., and many of the principal English nobility. On the 15th July, 1662, a Royal Charter† was granted by Charles II., constituting them a body corporate, under the name of the Royal Society, appointing Lord Viscount Brouncker the first President, Dr. Wilkins and Mr. Oldenburg, the first Secretaries; William Balle, Esq. Treasurer; and twenty-one of the most respectable Members the first Council.‡ This Council

Receive a charter.

---

* The following are their names: Lord Viscount Brouncker, Mr. Boyle, Mr. Bruce, Sir Robert Moray, Sir Paul Neile, Dr. Wilkins, Dr. Goddard, Dr. Petty, Mr. Balle, Mr. Rooke, Mr. Wren, and Mr. Hill. Birch's *Hist. of the Royal Society*, Vol. I. p. 3.

† This Charter will be found in the Appendix, No. I.

‡ Their names were as follows:

Lord Viscount Brouncker	Mr. Henry Slingesby	Mr. William Balle
Sir Robert Moray	Sir William Petty	Mr. Matthew Wren
Mr. Robert Boyle	Dr. Timothy Clarke	Mr. John Evelyn
Mr. William Brereton	Mr. William Erskine	Mr. Thomas Henshaw
Sir Kenelm Digby	Dr. John Wilkins	Mr. Dudley Palmer
Sir Paul Neile	Dr. George Ent	Mr. Abraham Hill
Sir Gilbert Talbot	Dr. Jonathan Goddard	Mr. Henry Oldenburg

was empowered by the charter to nominate, during a period of two months, what persons they thought proper as Members of the Society.* After the elapse of this period all subsequent elections were to be conducted by the Society in a body, and every candidate, in order to be admitted, required the votes of two thirds of the Members present.

The Society first held their meetings at Gresham College; but, about the be- Place of meet-ing. ginning of the year 1667, Mr. Henry Howard, afterwards Duke of Norfolk, made them a present of the Arundel Library, which had been purchased by his grandfather, the Earl of Arundel, during an embassy to Vienna. It had formerly been part of the library of Mathew Corvinus, King of Hungary, erected by him at Buda, in 1485, and after his death, in 1490, it came into the possession of the famous Bilibaldus Pirckeimerus, of Nuremberg, who died in

---

* In consequence of this power, they declared the following persons Fellows of the Royal Society.

James, Lord Annesley	Francis Glisson, M. D.	Henry Powle, Esq.
John Alleyn, Esq.	John Graunt, Esq.	Thomas Povey, Esq.
Elias Asmole, Esq.	Christopher, Lord Hatton	Henry Proby, Esq.
John Austen, Esq.	Charles Howard, Esq.	Philip Packer, Esq.
John Aubrey, Esq.	William Hoare, M. D.	William Quatremaire, M. D.
George, Duke of Buckingham	Sir Robert Harley	Edward, Earl of Sandwich
George, Lord Berkeley	Nathaniel Henshaw, M. D.	Sir James Shaen
Robert, Lord Bruce	James Hays, Esq.	Charles Scarburgh, Esq.
Richard Boyle, Esq.	William Holder, D. D.	Thomas Stanley, Esq.
Thomas Baynes, M. D.	Theodore Haake, Esq.	George Smyth, M. D.
Isaac Barrow, B. D.	William Hammond, Esq.	Alexander Stanhope, Esq.
Peter Balle, M. D.	John Hoskyns, Esq.	Robert Southwell, Esq.
John Brooks, Esq.	Robert Hooke, M. A.	William Schroter, Esq.
David Bruce, M. D.	Richard Jones, Esq.	Thomas Sprat, M. A.
George Bate, M. D.	Alexander, Earl of Kincardin	Christopher Terne, M. D.
William, Lord Cavendish	Sir Andrew King	Samuel Tuke, Esq.
Walter Charleton, M. D.	John, Lord Lucas	Cornelius Vermuyden, Esq.
Edward Cotton, D. D.	James Long, Esq.	Sir Cyril Wyche
Daniel Colwall, Esq.	Anthony Lowthe, Esq.	Sir Peter Wyche
John Clayton, Esq.	John, Lord Visc. Massareene	John Wallis, D. D.
Thomas Cox, M. D.	Sir Anthony Morgan	Edmond Waller, Esq.
William Croone, M. D.	Christopher Merret, M. D.	Joseph Williamson, Esq.
John, Earl of Crawford and	James, Earl of Northampton	Francis Willughby, Esq.
Lindsay	Sir Thomas Nott	William Winde, Esq.
Henry, Marquis of Dorchester	William Neile, Esq.	John Winthrop, Esq.
William, Earl of Devonshire	Jasper Needham, M. D.	Thomas Wren, M. D.
Sir John Denham	Sir William Persal	Christopher Wren, LL. D.
Mr John Dryden	Sir Richard Powle	Edmond Wylde, Esq.
Seth Ward, Bishop of Exeter	Sir Robert Paston	Daniel Whistler, M. D.
Andrew Ellis, Esq.	Sir Peter Pett	Sir Edward Bysshe
Sir Francis Feane	Walter Pope, M. D.	Sir John Birkenhead
Sir John Finsh	John Pell, D. D.	Mons. Christian Huyghens
Mons. Le Febure	Peter Pett, Esq.	Mons. Samuel Sorbiere

1530. It contains, besides a good number of printed books, many rare and valuable manuscripts. At the same time, Mr. Howard gave the Society convenient apartments in Arundel House, to which they removed their meetings, because Gresham College had been rendered unfit for that purpose in consequence of the fire of London. In 1673, they were invited back to Gresham College, by a deputation of the Professors and of the Mercer's Company: and they were induced to accept the offer because their apparatus and collection of curiosities were deposited there, and because Mr. Hooke, their operator, resided in that building. A grant of Chelsea College had been given them by Charles II., and they formed the project of converting it into a house proper for their meetings. Lord Henry Howard had likewise made them a present of a piece of ground near Arundel House, upon which they resolved to build convenient apartments by subscription. But neither of these designs was put into execution. They at last purchased a very convenient house in Crane Court, Fleet-street, in which they continued to hold their meetings, till the British Government, about 30 years ago, furnished them with apartments in Somerset House, where their meetings have ever since been held, and their library and apparatus deposited.

Favourable era for science.   The period of the institution of the Royal Society was peculiarly favourable for the objects in view. The spirit of inquiry, at first feeble, which animated some individuals at the time of the revival of learning, had from numerous causes gathered strength and spread itself over Europe. The Greek literary men, who had taken refuge in Italy after the destruction of the Eastern Empire by the Turks, brought with them a knowledge of their own language and some of the finest models of writing that ever have been produced. These books contributed to form the taste and enlighten the understandings of the poets and historians of Italy, whose labours were fortunately appreciated and encouraged by several of the principal men in that comparatively civilized country. The progress of the mathematical sciences in Italy, France, Germany, and England; the introduction of the new method of philosophizing by Lord Bacon; and the happy illustration of that method, by the successful exertions of Galileo and some of his contemporaries and successors, awakened an enthusiastic ardour in the minds of literary men: the vast field of science lay exposed before them, the true path of investigation had been discovered, and all were eager to enter upon it. In that infant period of science every step was a discovery; every judicious experiment led the fortunate philosopher to eminence. But all were conscious of the immense space which was to be explored, of the little progress that could be made by a single individual, and of the necessity of mutual co-operation, and the division of scientific labour. Hence associations for the purpose of experiment were naturally suggested, as the only means of ensuring speedy and complete success. They furnished the money necessary for convenient apparatus without pressing too severely upon individuals, and enabled the Members to make considerable progress in investigations without dedicating more of their time to such subjects than they could conveniently spare.

At that period, the diffusion of new discoveries was peculiarly difficult, and was chiefly accomplished by epistolary correspondence between philosophers carried on in the Latin language. From the vast multitude of epistles left by Erasmus, and some other eminent men of the same period, we may form some notion of the great portion of time which must have been taken up in this kind of correspondence. Even after every possible exertion, knowledge could be very imperfectly diffused by such means. Hence another important purpose of these associations was to publish, periodically, all the discoveries which came to their knowledge. It is to these Societies, therefore, that we are indebted for the present facility with which knowledge of every kind is diffused over the world: the memoirs and transactions, and the monthly publications which issue from the press in such numbers in most European countries, render it now an easy matter to make oneself acquainted with every improvement in any branch of science almost as soon as it takes place. Formerly, a philosopher could not well appear before the world, unless he had a complete treatise to publish But at present, through these periodical channels, every idea and new fact may be easily and advantageously communicated. The experimenter runs but little risk of losing his labour by investigating what has been already ascertained; while, by the multiplicity of co-operators, emulation and industry are more likely to be maintained.

The period of the incorporation of the Royal Society was peculiarly favourable for the progress of science in Britain. The country had for many years before been engrossed in a civil war, in which the energy and abilities of the various partizans were called into the completest action. The parliament, which from its nature and constitution was likely to contain the men of the greatest talents in the nation, had gained the ascendancy, and in their turn had been obliged to yield to the military usurpation of Oliver Cromwell. The Protector by his energy and abilities had crushed all opposition, and had made the country respected and dreaded by the neighbouring states After his death the kingdom threatened to fall into complete anarchy, when the sudden restoration of King Charles II. healed all divisions, and put an end to revolutionary violence. Then was a favourable time to draw the attention of the rich and the well-informed to the improvement of science, and to direct the effervescence of over-active minds to the advancement of knowledge, instead of political speculations.

The number of eminent men who at that period existed in England, and were disposed to associate for these important purposes, was uncommonly great, and far surpassed what any other country in Europe could produce. The mathematical sciences in particular and astronomy could boast of a great number of very eminent cultivators. Indeed this was the age of British mathematics.

Philosophical Transactions.

The Philosophical Transactions were the result of the labours of the Royal Society. At first they were published by the Secretary in small numbers consisting of a sheet or two, either monthly or seldomer, according as materials were supplied. Many of the papers read before the Society were inserted in these Transactions; but the editor took the liberty of putting into them likewise all the scientific information which he could pick up from any other channel. An analysis of books of science was frequently inserted, and indeed contributed materially to the diffusion of knowledge, and tended very much to bring works of merit into speedy celebrity.

Society proceed by experiment.

From the nature and constitution of the Royal Society the objects of their attention were necessarily unlimited. The physical sciences however, or those which are promoted by experiment, were their declared objects; and experiment was the method which they professed to follow in accomplishing their purpose. Accordingly at first, and indeed for a considerable time after the establishment of the Society, experiments were exhibited before them at every meeting; and a person was appointed with a salary to contrive such experiments as were proper to be shown, and to have every thing ready for the exhibition of them against the meeting of the Society. This practice, in the infancy of experimental investigations, was probably attended with advantages, by putting the spectators upon the right way of investigation, and inducing them to draw legitimate consequences from their observations. But it was not the best way of advancing the sciences. Experiments cannot be advantageously made by a multitude or even before a multitude. Accordingly we find but few discoveries of importance made during these exhibitions, even when the experimenters were eminently successful in their private labours at home. It is only necessary to mention the names of Mr. Boyle, Dr. Hooke, and Dr. Grew, who frequently exhibited experiments before the Society, to be convinced of the truth of this. These experiments however were for the most part but repetitions of experiments already made in private, and exhibited again with perfect propriety for the satisfaction of the Society. In this way the knowledge of the subject was spread rapidly, and fundamental points better established than they otherwise might have been.

The Society, as might have been expected from the nature of such institutions, underwent various periods of vigour and decay; though upon the whole it has been progressive, and has continued to increase both in reputation and in numbers. As the value of the Transactions was in some measure affected by the abilities of the Secretary who had the superintendance of the publication, it may be worth while to lay before our readers the names of the successive Secretaries with the different portions of the Transactions which they respectively edited.

The Secretary, who began the publication of the Transactions, was Mr. Henry Oldenburg. The first Number was published in 1665; and he continued the publication till the period of his death in 1677, and published in all 136 numbers, making very nearly the whole of the first twelve volumes of the Transactions. He was succeeded in office by Dr. Nehemiah Grew, who was chosen Secretary on the 30th of November 1677; he published Number 137, and continued editor for two years. The last number which he published was 142. Dr. Grew was succeeded by Dr. Hooke. But the publication of the Transactions was intermitted for three years. In 1681, Dr. Hooke began to publish what he called *Philosophical Collections;* which have been always considered as constituting a portion of the Transactions, though under a different name. Dr. Hooke continued his publication of the *Collections* during the years 1681 and 1682, and published in all 7 numbers. The discontinuance of the Transactions seems to have been owing to their very limited sale, and to the small profit which accrued to the Secretary who acted as editor. But in the year 1683, Dr. Robert Plot, who had, the preceding year, been chosen Secretary in place of Dr. Hooke, undertook to revive the publication, on condition that the Society would bind themselves to purchase 60 copies of each number. Dr. Plot was editor during the years 1683 and 1684, and published the 13th and 14th volumes, comprehending all the numbers between 143 and 166 inclusive. The 15th volume, comprehending the numbers between 167 and 178 inclusive, was edited by Mr. William Musgrave, who was Secretary during the year 1685.

This volume probably did not give much satisfaction to the Members of the Royal Society. There seems also to have been a deficiency of materials; for it appears, from the registers of the Society, that Dr. Edmund Halley, who had been appointed Clerk to the Society in 1686, offered, on condition that the publication should be renewed, to furnish one fourth of the whole out of his own private stock. It would appear that Dr. Halley was editor of the 16th volume, comprehending the numbers between 179 and 191 inclusive, and published during the years 1686 and 1687. After the publication of this volume there was an interval of three years without the appearance of any thing more, owing obviously to the deficiency of materials. The publication was again revived in 1691, and though Dr. Halley was not the ostensible editor, he appears to have been actively concerned in superintending the publication till the period of his voyage to the southern hemisphere in 1698. For there is an order of council, passed about that time, on record, enjoining Dr. Tyson, Mr. Hart, Dr. Sloane, Mr. Waller, and Dr. Hooke, to assist Dr. Halley in drawing up the Transactions. It is impossible to say how much each of these individuals contributed

Margin notes:
Editors of the Transactions.
1 Oldenburg.
2 Grew.
3 Hooke.
4 Plot.
5 Musgrave.
6 Halley

to the labour of editing; though there is reason to believe that the greatest share of the drudgery fell upon Halley. Volumes 17 and 18, consisting of the numbers between 192 and 214 inclusive, and published during the years 1691,

7 Waller. 1692, 1693, and 1694, were ostensibly edited by Mr. Waller, who had been elected Secretary on the 30th of November 1687. The ostensible editor of all the volumes from the 19th to the 28th inclusive, comprehending from number 215 to number 337, and published in succession between the years 1695 and

8 Sloane. 1713, was Sir Hans Sloane, who had been chosen Secretary on the 30th of
9 Halley. November 1693. Dr. Halley, chosen Secretary in 1713, was the editor of the 29th and 30th volumes, comprehending all the numbers between 338 and 363 inclusive, and published between the years 1714 and 1719, at the rate of two numbers a year. The 31st, 32d, 33d, and 34th volumes, comprehending all the numbers between 364 and 398, and published between the years 1720 and

10 Jurin. 1727, at the rate of two numbers annually, were edited by Dr. James Jurin,
11 Rutty. chosen Secretary at the end of the year 1721. Dr. Rutty, chosen Secretary in 1727, was the editor of the 35th volume, consisting of eight numbers, namely those between 399 and 406 inclusive, and published in the year 1728. Dr.

12 Mortimer. Cromwell Mortimer, elected Secretary in 1730, was the editor of the 36th volume, and continued editor for a period of 20 years, during which he published 11 volumes, comprehending all the numbers between 407 and 496 inclusive. The 46th volume was published in 1750, and it was the last volume that was published in numbers.

13 A Committee of the Society. In the year 1750, a Committee was appointed to superintend the publication; and the Transactions ever since that period have been published in half volumes. For the first 12 years only one half volume was published annually; but from the year 1762, two half volumes or a complete volume have always made their appearance every year.*

---

* This volume, and all those published since, have been each prefaced by the following
ADVERTISEMENT.
" The Committee appointed by the Royal Society to direct the publication of the Philosophical Transactions, take this opportunity to acquaint the public, that it fully appears, as well from the council books and journals of the Society, as from the repeated declarations which have been made in several former Transactions, that the printing of them was always, from time to time, the single act of the respective secretaries, till this present 47th volume. And this information was thought the more necessary, not only as it has been the common opinion that they were published by the authority, and under the direction of the Society itself, but also because several authors, both at home and abroad, have in their writings called them the Transactions of the Royal Society. Whereas in truth the Society, as a body, never did interest themselves any further in their publication, than by occasionally recommending the revival of them to some of their secretaries, when, from the particular circumstances of their affairs, the Transactions had happened for any length of time to be intermitted. And this seems principally to have been done with a view to satisfy the public, that their usual meetings were then continued for the improvement of knowledge, and benefit of mankind, the great ends of their first institution by the royal charters, and which they have ever since steadily pursued.

The preceding details, though of considerable importance, have from unavoidable necessity been so circumstantially minute as to border upon tediousness. On that account we shall exhibit the whole under the form of a table, which will enable the reader by a single glance of the eye to catch all the details.

Year.	Number.	Volume.	Editor.
1665	1— 7		
6	8— 20	1	Mr. Henry Oldenburg.
7	21— 22		
8	23— 40	2	Idem.
9	41— 44	3	Id.
1670	45— 54	4	Id.
	55— 56		
1	57— 68	5	Id.
2	69— 80	6	Id.
	81— 91	7	Id.
3	92—100	8	Id.
4	101—109	9	Id.
5	110—120	10	Id.
6	121—122		
	123—132	11	Id.
7	133—137	12	Id. and Dr. Grew.
8	138—142		Dr. Grew.
Collections.			
1681	1— 3		
2	4— 7		Dr. Hooke.

" But the Society being of late years greatly enlarged, and their communications more numerous, it was thought advisable, that a Committee of their Members should be appointed to reconsider the papers read before them, and select out of them such as they should judge most proper for publication in the future Transactions; which was accordingly done on the 26th of March 1752. And the grounds of their choice are, and will continue to be, the importance or singularity of the subjects, or the advantageous manner of treating them; without pretending to answer for the certainty of the facts, or propriety of the reasonings contained in the several papers so published, which must still rest on the credit or judgment of their respective authors.

" It is likewise necessary on this occasion to remark, that it is an established rule of the Society, to which they will always adhere, never to give their opinion, as a body, on any subject, either of nature or art, that comes before them. And therefore the thanks which are frequently proposed from the chair, to be given to the authors of such papers as are read at their accustomed meetings, or to the persons through whose hands they receive them, are to be considered in no other light than as a matter of civility, in return for the respect shown to the Society by those communications. The like also is to be said with regard to the several projects, inventions, and curiosities of various kinds, which are often exhibited to the Society; the authors whereof, or those who exhibit them, frequently take the liberty to report, and even to certify in the public news-papers, that they have met with the highest applause and approbation. And therefore it is hoped that no regard will hereafter be paid to such reports, and public notices; which in some instances have been too lightly credited, to the dishonour of the Society."

C

Year.	Number.	Volume.	Editor.
1683	143—154	13	Dr. Robert Plot.
4	155—166	14	Idem.
5	167—178	15	Mr. William Musgrave.
6	179—185	16	Dr. Edmond Halley.
7	186—191		
1691	192—194		
2	195	17	Mr. Richard Waller.
3	196—206		
4	207—214	18	Idem.
5	215—218		
6	219—223	19	Sir Hans Sloane.
7	224—235		
8	236—247	20	Idem.
9	248—259	21	Id.
1700	260—267	22	Id.
1	268—276		
2	277—282	23	Id.
3	283—288		
4	289—294	24	Id.
5	295—304		
6	305—308	25	Id.
7	309—312		
8	313—318	26	Id.
9	319—324		
1710	325—328		
1	329—332	27	Id.
2	333—336		
3	337	28	Id.
4	338—342		
5	343—346	29	Dr. Edmond Halley.
6	347—350		
7	351—354		
8	355—358	30	Idem.
9	359—363		
1720	364—366	31	Dr. James Jurin.
1	367—369		
2	370—374	32	Idem.
3	375—380		
4	381—385	33	Id.
5	386—391		
6	392—396	34	Id.
7	397—398		
8	399—406	35	Dr. William Rutty.
9	407—411	36	Dr. Cromwell Mortimer
1730	412—416		
1	417—421	37	Idem.
2	422—426		

Year.	Number.	Volume.	Editor.
1733	427—430 }	38	Dr. Cromwell Mortimer.
4	431—435 }		
5	436—439 }	39	Idem.
6	440—444 }		
7	445—446 }	40	Id.
8	447—451 }		
9	452—455 }		
1740	456—459 }	41	Id.
1	460—461 }		
2	462—467 }	42	Id.
3	468—471 }		
4	472—474 }	43	Id.
5	475—477 }		
6	478—481 }	44	Id.
7	482—484 }		
8	485—490	45	Id.
9	491—493 }	46	Id.
1750	494—496 }		

Year.	Part.	Volume.	Year.	Part.	Volume.
1751	1 }	47	1776	1, 2	66
2	2 }		7	1, 2	67
3	1 }	48	8	1, 2	68
4	2 }		9	1, 2	69
5	1 }	49	1780	1, 2	70
6	2 }		1	1, 2	71
7	1 }	50	2	1, 2	72
8	2 }		3	1, 2	73
9	1 }	51	4	1, 2	74
1760	2 }		5	1, 2	75
1	1 }	52	6	1, 2	76
2	2 }		7	1, 2	77
3	1, 2	53	8	1, 2	78
4	1, 2	54	9	1, 2	79
5	1, 2	55	1790	1, 2	80
6	1, 2	56	1	1, 2	81
7	1, 2	57	2	1, 2	82
8	1, 2	58	3	1, 2	83
9	1, 2	59	4	1, 2	84
1770	1, 2	60	5	1, 2	85
1	1, 2	61	6	1, 2	86
2	1, 2	62	7	1, 2	87
3	1, 2	63	8	1, 2	88
4	1, 2	64	9	1, 2	89
5	1, 2	65	1800	1, 2	90

Transactions
vary in value.

The nature of the papers read before the Society, and of course the value of the Transactions published, were considerably influenced by the President.   If any person will take the trouble to examine the volumes published during the presidency of Martin Folkes, Esq., and to compare them with the rest of the work, he will find a much greater proportion of trifling and puerile papers than are any where else to be found.   It was during this period that Sir John Hill published his *Review of the Works of the Royal Society of London,* in which he endeavours, with all the humour and all the knowledge he was master of, to throw ridicule upon the labours of that illustrious body.   He was induced, it is said, to take this step in consequence of being disappointed in an attempt which he had made to be elected a Fellow   The story- is by no means improbable, though he himself formally disavows its truth *   It cannot be denied that he has selected and exposed a variety of trifling and absurd papers.   But his own humour is coarse and poor, and in more instances than one the statements contained in the papers which he attempts to ridicule are much more accurate than his own.   He affirms that, at the time when he wrote, the Society was entirely under the management of Mr. Henry Baker, a man of acknowledged abilities, but whose knowledge and pursuits were too circumscribed to qualify him for superintending such a Society with advantage.

Immediately on the election of Lord Macclesfield, who succeeded Martin Folkes as President, in 1752, a very striking improvement is observable in the value of the Transactions.   Many excellent papers made their appearance in the Transactions during the presidentship of the Earl of Morton.   The unfortunate dispute between Mr. Wilson and the other Electricians of the Royal Society, about the relative goodness of pointed or knobbed conductors, occupies too great a proportion of the Transactions during the presidentship of Sir John Pringle; yet the volumes of that period contain many memorable papers: as, for example, Dr. Maskelyne's experiments at Schehallien, to determine the density of the earth, with Dr. Hutton's deductions from them ; the experiments of Sir George Shuckburgh Evelyn, and of General Roy, in order to establish correct formulas for measuring heights by the barometer ; the report of the Committee appointed by the Royal Society to determine the proper method of graduating thermometers; the experiments of Mr. Hutchins to ascertain the freezing point of Mercury, and many others.   Whoever will examine the Transactions with care, will easily satisfy himself that by far the most valuable volumes of that work are the 32 which have been published during the presidentship of Sir Joseph Banks, and, fortunately for the progress of science, he has enjoyed that situation for a much longer period than any of his predecessors.

We shall here insert, for the satisfaction of our readers, a list of all the suc-

---

* See the preface to Sir John Hill's book, p. 6.

cessive Presidents of the Royal Society, with the time of their election and the number of years which each held the office.

Presidents' Names.	Date of Election.	Years in Office.	
William Lord Viscount Brouncker	April 22, 1663	14	Presidents of the Society.
Sir Joseph Williamson, Kt.	Nov. 30, 1677	3	
Sir Christopher Wren, Kt.	Nov. 30, 1680	2	
Sir John Hoskins, Bart.	Nov. 30, 1682	1	
Sir Cyril Wyche, Bart.	Nov. 30, 1683	1	
Samuel Pepys, Esq.	Dec. 1, 1684	2	
John, Earl of Carbery	Nov. 30, 1686	3	
Thomas, Earl of Pembroke and Montgomery	Nov. 30, 1689	1	
Sir Robert Southwell, Kt.	Dec. 1, 1690	5	
Charles Montague, Esq. (afterwards Earl of Halifax.)	Nov. 30, 1695	3	
John, Lord Somers.	Nov. 30, 1698	5	
Sir Isaac Newton, Kt.	Nov. 30, 1703	24	
Sir Hans Sloane, Bart.	Nov. 30, 1727	14	
Martin Folkes, Esq	Nov. 30, 1744	11	
George, Earl of Macclesfield	Nov. 30, 1752	12	
James, Earl of Morton.	Nov. 30, 1764	4	
James Burrow, Esq	Sept. 1768		
James West, Esq.	Nov. 30, 1768	4	
James Burrow, Esq.	July 1772		
Sir John Pringle, Bart.	Nov. 30, 1772	6	
Sir Joseph Banks, Bart.	Nov. 30, 1778	Still continues.	

We shall subjoin a list of the Secretaries of the Royal Society, with the date of their elections; from which, the length of time that they continued respectively in office can easily be ascertained. There are always two Secretaries at the same time; and whenever either of them resigns or is displaced, another is always elected in his room. The list therefore will consist of two columns, one for each Secretary.

First Secretary.	Date of Election.	Second Secretary.	Date of Election.	
John Wilkins, D.D.	April 22, 1663	Henry Oldenburg, Esq.	April 22, 1663	Secretaries.
Thomas Henshaw, Esq.	Nov. 30, 1668			
John Evelyn, Esq.	Nov. 30, 1672			
Abraham Hill, Esq.	Nov. 30, 1673			
Thomas Henshaw, Esq.	Nov. 30, 1675			

First Secretary.	Date of Election.	Second Secretary.	Date of Election.
		Nehemiah Grew, M.D. . . Nov. 30, 1677	
Thomas Gale, D.D. . . . . Nov. 30, 1679		Mr. Robert Hooke	
Francis Aston, Esq. . . . . Nov. 30, 1681			

Dec. 9, 1685, both Secretaries resigned.

First Secretary.	Date of Election.	Second Secretary.	Date of Election.
Sir John Hoskyns, Bart. . Dec. 16, 1685		Thomas Gale, D.D. . . . . Dec. 16, 1685	
Richard Waller, Esq. . . . Nov. 30, 1687			
		Hans Sloane, M.D. . . . . Nov. 30, 1693	
John Harris, D.D. . . . . . Nov. 30, 1709			
Richard Waller, Esq. . . . Nov. 30, 1710			
		Edmond Halley, LL.D. . Nov. 30, 1713	
Brook Taylor, LL.D. . . . Jan. 13, 1714			
John Machin, Ast. Prof. Gresh. . . . . . } . . Dec. 1, 1718			
		James Jurin, M.D. . . . . Nov. 30, 1721	
		William Rutty, M.D. . . Nov. 30, 1727	
		Cromwell Mortimer, M.D. . . . . } . . Nov. 30, 1730	
Peter Daval, Esq. Nov. 30, 1747			
		Thomas Birch, D.D. . . . Jan. 21, 1752	
Charles Morton, M.D. . . Nov. 30, 1759			
		Mathew Maty, M.D. . . . Nov. 30, 1765	
Samuel Horsley, LL.B. . . Nov. 30, 1773			
		Joseph Planta, Esq. . . . . Nov. 30, 1776	
Paul Henry Maty, M.A. . Nov. 30, 1778			
Charles Blagden, M.D. . . May 5, 1784			
Edward Whitaker Gray, M.D. . . . . } . . Nov. 30, 1797		Will. Hyde Wollaston, M.D. . . . . } . . Nov. 30, 1806	
Humphry Davy, Esq. . . Jan. 22, 1807			

The following is a list of the Foreign Secretaries of the Royal Society.

Date of Election.	Names.
April 11, 1723	Philip Henry Zollman.
April 18, 1728. . . .	Dr. Dillenius and Dr. Schuchzer. } Appointed by the Council in room of Mr. Zollman (he being obliged to attend the Congress at Soissons) till his return.
Aug. 29, 1748. . . .	Thomas Stack, M.D.
Nov. 20, 1751. . . .	James Parsons, M.D.
March 4, 1762. . . .	Matthew Maty, M.D.

Foreign Secretaries.

Date of Election.	Names.
Dec. 11, 1766....	John Bevis, M.D.
Feb. 13, 1772....	Paul Henry Maty, M.A.
June 30, 1774....	Mr. Joseph Planta.
Jan 14, 1779....	Mr. Charles Hutton.
June 17, 1784....	Rev. Charles Peter Layard, M.A.
Mar. 22, 1804....	Thomas Young, M.D.

That we may lay a complete list of all the office-bearers of the Society before our readers, we shall subjoin the Treasurers and the Clerks.

Treasurers.	Date of Election.	Years in Office.	
William Ball, Esq.	April 22, 1663		Treasurers.
Abraham Hill, Esq.	Nov. 30, 1663	2	
Daniel Colwal, Esq.	Nov. 30, 1665	14	
Abraham Hill, Esq.	Dec. 1, 1679	21	
Alexander Pitfield, Esq.	Nov. 30, 1700	28	
Roger Gale, Esq.	Nov. 30, 1728	8	
James West, Esq.	Nov. 30, 1736	32	
Samuel Wegg, Esq.	Nov. 30, 1768	34	
William Marsden, Esq.	Nov. 30, 1802	Still continues.	

Clerks.	Date of Election.	Years in Office.	
Mr. Wicks	May 13, 1663	23	Clerks.
Dr. Edmond Halley	Jan. 27, 1686		
Mr. Henry Hunt.			
Mr. Alban Thomas.	Dec. 7, 1713	9	
Mr. Francis Hauksbee	May 9, 1723	39	
Mr. Emanuel Mendez da Costa.	Feb. 3, 1763	5	
Mr. John Robertson.	Jan. 7, 1768	9	
Mr. John Robertson, Jun.	Jan. 9, 1777	8	
Mr. George Gilpin.	March 3, 1785	21	
Mr. Stephen Lee.	Nov. 30, 1810	Still continues.	

The only account of a Literary Society which can be at all valuable or interesting, is a detail of the efforts which they have made to increase the stock of knowledge, and to promote the various branches of science to which they have directed their attention The result of these efforts is contained in the Transactions of the Royal Society. Our object therefore will be to take a comprehensive view of the contents of these volumes, and to lay before our readers the additions which the various sciences have received from the labours of this

7

illustrious Society. Now as almost every man of science in the British empire, who has flourished since the original establishment of this Society, has been enrolled among its Fellows, our work will contain in fact a history of the progress of the sciences in Great Britain during the last 150 years. By comparing this progress with the present state of each science, we shall discover at a glance what portion of each originated in Britain, and what portion on the Continent. This comparative view cannot but be highly gratifying to a British reader. We are far from wishing to depreciate the merits of the illustrious philosophers on the Continent: they have been numerous and highly respectable. But owing no doubt to the superior advantages attending a free government, a much greater number of discoveries than ought to have fallen to our share, if we attend only to the comparative population of the different countries, have originated in Britain.

All the subjects treated of in the Philosophical Transactions may be arranged under the five following divisions.

Divisions.

I. NATURAL HISTORY.
II. MATHEMATICS.
III. MECHANICAL PHILOSOPHY.
IV. CHEMISTRY.
V. MISCELLANEOUS ARTICLES.

We shall consider each of these divisions in succession in the five following books; taking the liberty of passing over some of the subjects more slightly than others; either when they occupy only an inferior department in the Philosophical Transactions, or when they are of such a nature as not to be well susceptible of compression and generalization.

BOOK I.

OF NATURAL HISTORY.

NATURAL History, as far at least as it constitutes a department in the Philosophical Transactions, may be divided into four distinct heads; namely,

1 Botany.   3 Mineralogy.
2 Zoology.   4 Geography and Topography.

These will occupy our attention successively in the four following Chapters

CHAP. I.

OF BOTANY.

BOTANY, or that part of Natural History which treats of plants, is one of the most delightful branches of human knowledge, and has always commanded a very numerous body of enthusiastic cultivators, ever since it assumed the form of a science. If we take it in its most extensive signification, it consists, in reality, of three distinct parts, which have been usually cultivated by different individuals. To the first part belongs the collection of plants, their description, Division of and their artificial arrangement into a system in such a manner that every Botany. species may be sufficiently distinguished from all the others, and that it may be in the power of every person who understands, and is in the possession of the system, to discover by means of it the name of every plant with which he may happen to meet. This first part of botany is an essential preliminary to the other two; for it is obvious that no communication whatever can be handed down respecting plants, till we are in possession of the means of discovering the plants, concerning which the writer treats. The second branch of botany may be called the anatomy and physiology of plants. It is exceedingly curious and important. It originated in Britain, and indeed took its rise from the Royal Society; and almost all the additions to it, even to our own times, have been made by the Members of that illustrious body. The third branch of botany relates to the agriculture and œconomical uses of plants.

D

We shall take a general view of these three departments of botany, as far as the labours of the Royal Society are connected with them in the three following sections.

Sect. I.—*Of the Description and Arrangement of Plants.*

It is to this branch of the science that the term *Botany,* in common language, is confined. It has always been considered as highly interesting, ·and has been followed by numerous and enthusiastic cultivators. It must be considered as entirely modern. The ancients, indeed, have left us descriptions, though they are imperfect ones, of many plants; but no systematic arrangement seems ever to have been attempted or conceived by them. Hence the obscurity and uncertainty which overhang all that they have left us on the subject.

Origin of Botany.

The Greek and Roman writers, who treat professedly of plants, unless Aristotle be included, (and the botanical works ascribed to him are unquestionably spurious,) are only three; namely, Theophrastus, Dioscorides, and

Theophrastus. Pliny. Theophrastus, a native of Lesbos, the disciple and successor of Aristotle, born about 370 years before the Christian era, was one of the most celebrated of all the ancients. He wrote two treatises on plants, both of which have come down to us almost entire. The first, entitled *History of Plants;* the second, *On the Causes of Plants.* In these works he mentions and even endeavours to describe about 500 species. But his descriptions are so imperfect, and his allusions so doubtful, that, in spite of the unwearied exertions of his commentators, many of the species to which he alludes remain ambiguous * As a commencement of the study, however, the books have their value, and certainly deserve to be examined as literary curiosities.

Dioscorides.

Dioscorides, a Greek writer, who published a work on Materia Medica, lived, there is every reason to believe, during the reign of Nero. Though he was rather a collector of receipts than a botanist, he was long considered, it is difficult to conceive for what reason, as the prince of botanical writers. Every thing of any value respecting plants was supposed to exist in his works, and therefore indefatigable pains were bestowed by modern commentators, in order to determine the species to which he alludes. But these labours have, if possible, been attended with less success than those bestowed upon his predecessor, Theophrastus.†

Pliny.

Pliny, the elder, may be considered as almost the only naturalist who

---

* Sprengel has bestowed uncommon pains in ascertaining the plants mentioned by Theophrastus. In his *Historia Rei Herbariæ*, Tom. I. p. 71, he has given a table of no fewer than 355 species, with their Linnæan names.

† Sprengel, in his *Hist. Rei Herbariæ*, Tom. I. p. 156, has given a very copious list of the plants mentioned by Dioscorides, with their Linnæan names.

appeared among the Romans. He lived during the reigns of Vespasian and Titus, and fell a sacrifice to his curiosity during an eruption of Mount Vesuvius, in the 56th year of his age and the 79th of the Christian era. His Natural History is one of the most interesting monuments which antiquity has left us. It was compiled, as he himself informs us, from the works of 2500 preceding writers, and includes every branch of natural knowledge with which the ancients were acquainted. His work is divided into 37 books, and 16 of these are devoted to plants. He mentions above 600 species; but in many places is far from accurate, often translating from Theophrastus and Dioscorides, without understanding the correct meaning of these authors, as has been sufficiently proved by Hermolaus Barbarus, Nicolaus Leonicenus, and some others of his early commentators. Like his predecessors, he treats of those plants only which were employed in medicine, or applied to the various purposes of domestic œconomy.*

The Arabians added but little to the botanical knowledge which they derived Arabians. from the Greeks. That nation in fact was never more than half civilized, and their learned men were nearly destitute of invention, and leaned entirely on the Grecian improvements, which they contented themselves with translating and transfusing into their own language.

No progress in botany can be expected during the dark ages of Europe. The cultivation of the sciences was first revived in Italy, where knowledge began to be valued in consequence of the commerce of the different states and the enlightened patronage of the Italian princes. The cultivation of the Greek language was encouraged, and the classic stores of the ancients investigated with care. The first step towards the revival of botany, was to study the works of Theophrastus, Dioscorides, and Pliny, and to determine, as far as possible, the plants which these writers describe. These endeavours produced a whole herd of commentators and translators, who acquired a splendid reputation Commenta-during the 16th century.† But it was soon found that without an examination tors.

* Sprengel has given the Linnæan names of a great number of the plants mentioned by Pliny. *Hist. Rei Herb.* Tom. I. p. 200.

† Some of the most distinguished of these, were

1. Georgius Valla, a native of Placentia. He published, in 1501, an epitome of all the Greek writers on medicine and materia medica.

2. Hermolaus Barbarus, a noble Venetian, who died in 1493. He published, in 1492, a corrected edition of Pliny. The performance was too hasty; yet he corrected many of Pliny's errors, though he overlooked the most weighty.

3. Marcellus Vergilius, a Florentine. He was the first that undertook to correct Dioscorides. He published a corrected edition of that writer with a commentary, at Florence, in 1518.

4. Nicolaus Leonicenus, a Venetian of prodigious learning. He was long a professor at Ferrara and lived till he was above 100. He elucidated the works of Serapion and Pliny.

5. Joannes Manardes, of Ferrara. He was born in 1462, and died in 1536. He illustrated the works of Mesues, and corrected Marcellus Vergilius's edition of Dioscorides

of the plants themselves, it would be vain to expect to make out the descriptions of the ancients. This consideration induced various medical men to make collections of the plants which grew in the countries where they resided, and to publish catalogues containing the result of their labours. This soon informed botanists that every thing respecting plants was not contained in the writings of the ancients; that, in truth, the ancients had attempted to describe, or enumerate, only a very small number of plants, while the great body had escaped their attention.*

But these publications, though they tended to increase the knowledge of plants, were far from being entitled to the name of scientific. They were mere catalogues, without any attempt at arrangement. The real founder of botany, as a science, was Conrad Gesner, who was born at Zurich, in Switzerland, in 1516. He was destined for the church; but marrying at an early age, he was reduced to the most abject poverty, and obliged to turn himself to the study of medicine. He afterwards practised in his native city, and died of the plague in 1565, in the 49th year of his age.

He first suggested the idea of a methodical arrangement of plants, according to classes, orders, genera, and species: an idea which all true botanists have followed ever since, and to which the very existence of botany, as a science, is to be ascribed. He pitched upon the fruit as the part of the plant from which the characters were to be drawn.

Andrew Cæsalpinus, born at Arezzo, in the district of Florence, in 1519, attempted an arrangement of plants, founded also upon the fruit. His knowledge of plants was too limited to enable him to accomplish his plan with success. But his descriptions are excellent, and the new views which he suggested were of the highest importance.

The number of botanical writers, who had by this time appeared in various parts of Europe, was so great, and the names applied to the same plants so numerous and discordant, that the subject was thrown into the greatest confusion; and, unless some person had arisen willing to undertake the herculean labour of arranging the chaotic mass, the science itself was in danger of

Gesner.

Cæsalpinus.

---

* The first publishers of botanical catalogues, were

1. Symphorianus Campegius, of Lyons, physician to the Duke of Lorrain. He published an account of various French plants in 1538.

2. Otho Brunfels, a physician, in Bern, who died in 1534. He delineated the plants of his country in wood. His Herbarium was published in 1537.

3. Hieronimus Tragus, the Friend of Brunfels, born in 1498, died in 1554. His book on the history of plants was published by Conrad Gesner, in 1532.

4. Leon Fuchsius, a physician, at Ingolstadt, born in 1501, died in 1565. He published a history of plants

5. Antonius Musa Brasavolus, a Venetian. He published at Lyons, in 1537, an excellent book, entitled *Examen omnium simplicium medicamentorum.*

destruction. This was in a great measure accomplished by the two illustrious brothers, John and Casper Bauhin, born in Switzerland, and pupils of Gesner. John devoted the whole of his life to the composition of a work, which he left complete in manuscript, and which was first published in 1650. This work was nothing less than an *Universal History of Plants.* Casper Bauhin undertook, and fortunately completed, a much more important and original work which was first printed in 1623, under the name of *Pinax Theatri Botanici.* This book was meant as an index to all the botanical knowledge then in the world; and, as its author informs us, it was the labour of 40 years. In this work about 6000 plants are arranged in twelve books, with some slight attention to system. Each plant is distinguished by a kind of descriptive name, and under it are placed the names given by every preceding author. This work produced a pause in the progress of botany : the learned thought it sufficient if they knew and called the plants by the names which Bauhin had given. Such nearly was the state of botany, as far as the arrangement and description of plants are concerned, at the period when the Royal Society was founded.

[marginal note: The Bauhins.]

The progress which it has made since that period is prodigious, and perhaps unexampled in any other department of science. Nearly 30,000 species of plants have been examined and described, and so admirably arranged, that it is easy to find the place of any one in the system, and even to ascertain its name and relations if it has been already examined and classified. It would be in vain to look for any considerable part of this progress in the Transactions of the Royal Society. That work was never considered as the proper place for registering the labours of botanists, which were usually of too great length and too expensive to be inserted in a periodical work. But a considerable proportion of the most eminent improvers of botany were Members of the Royal Society, and botany stands indebted to Great-Britain for a very great number of the new species which have been added to her treasures. On that account we think it requisite to give a short view of these improvements, that our readers may perceive what part of the progress has been the work of British botanists, and what part has been accomplished by the botanists on the Continent.

[marginal note: Progress since the origin of the Society.]

The botanical papers in the Philosophical Transactions amount to 100 * But of these there are 38 which cannot be considered as of any value. Only 24 new species are described Several of the most eminent botanists, though Fellows of the Society, did not contribute a single paper. It is only necessary to mention the name of Dillenius to be satisfied of this. Since the commencement of the Linnæan Society in 1788, no botanical papers have been admitted

[marginal note: Botanical papers in the Transactions.]

* Among these we do not reckon the numerous lists of new plants annually published for a series of years by the Society of Apothecaries, in obedience to the last will of Sir Hans Sloane, who bequeathed to them his botanic garden on that condition.

into the Transactions of the Royal Society. If it came within our plan to take a view of the ten volumes of Linnæan Transactions which have been already published, we should have it in our power to present our readers with a rich and exquisite feast of botanical knowledge.

One of the declared objects of the Royal Society at its original establishment was botany. Botanical committees were established; directions given to those Members who went abroad to send accounts of rare plants; correspondents were established in America, Asia, and Africa, for the same purpose; and persons were even employed to traverse England in order to collect plants and other natural curiosities. It seems to have been in some measure owing to these liberal and enlightened exertions, and to the subsequent labours of Ray, that the study of botany became extremely fashionable in England, and produced that cluster of eminent men whose illustrious labours Linnæus dignifies with the name of the golden age of botany. Gerard, Johnson, and Parkinson, lived before this period, and do not therefore come within our view. Their writings were of considerable importance at the time in which they appeared. The *Theatrum Botanicum* of Parkinson is the most laboured and original work, though he acquired less reputation than either of the other two. The first British Flora was

How.   published by How in 1650, under the title of *Phytologia Britannica, Natales exhibens indigenarum Stirpium sponte emergentium.* A meritorious book, considering the period and the opportunities of How. It contained 1,220 species; many of which however were not indigenous, and many would now be rejected as varieties.

Merret.   Dr. Christopher Merret, a physician in considerable practice in London, and an early and active Member of the Royal Society, deserves to be mentioned on account of his *Pinax Rerum Naturalium Britannicarum, continens Vegetabilia, Animalia, et Fossilia, in hac Insula reperta,* published in 1667. The greater part of this book is occupied with the plants. He employed Willisel, as he informs us, for five years to travel through England and collect plants for him. He describes 1,400 species; but several of these are not indigenous, and a still greater number are only varieties. His knowledge of the subject was too limited to prevent him from falling into numerous mistakes. The plants are arranged alphabetically, and only the synonyms of Gerard and Parkinson are given. But the most illustrious botanists of these times, as far as description of new species is concerned, were Plukenet, Sloane, Petiver, and the Sherards. Morison and Ray indeed, as the revivers of systematic botany, occupy a superior place, and are entitled to be considered separately.

Plukenet.   Dr. Leonard Plukenet does not appear to have been a Member of the Royal Society, and but few memorials of his life remain. He was born in 1642, appears to have been educated at Cambridge, and settled in London, but never got into extensive practice. He lived at Old Palace-yard, Westminster, where he appears to have cultivated a small botanical garden.   His zeal for botany was excessive; he had correspondents in all parts of the world, and spared no expense to procure new and rare plants.   He had access to the gardens in the neighbourhood of London, and to that at Hampton Court, which was at that time in a flourishing state.   Such was his zeal and industry, that his herbarium at last consisted of 8,000 plants, a much greater number than any botanist had possessed since the days of Bauhin.   The result of all this industry was communicated to the public in four works, containing about 500 new species, and still valuable to the botanist on account of the numerous figures which they contain.   These figures amount to upwards of 2,740, of very unequal merit, but all upon too small a scale.   His works were; 1. Phytographia, or engravings of the more important and rarer plants, published in four parts successively, and consisting of 328 plates.   2. Almagestum Botanicum; *sive Phytographiæ Plukenetianæ Onomasticon, methodo synthetico digestum; exhibens Stirpium exoticarum, rariorum, novarumque nomina, quæ descriptionis locum supplere possint.*   3. Almagesti Botanici Mantissa, *Plantarum novissime delectarum ultra millenarium numerum complectens.*   This may be considered as a continuation and amplification of the preceding works; many new figures are given, and much curious critical information occurs.   4. Amaltheum Botanicum; *seu Stirpium Indicarum alterum Copiæ Cornu, millenas ad minimam et bis centum diversas Species novas et indictas nominatim comprehendens; quarum sexcenæ et insuper selectis Iconibus æneisque Tabulis illustrantur.*   Plukenet began to publish in 1691, and his last work was printed in 1705.   He probably died soon after.

Mr. James Petiver was a contemporary of Plukenet.   He was an apothecary Petiver. in Aldersgate-street, London, where he resided during his whole life.   In course of time he became apothecary to the Charterhouse, and obtained a considerable share of practice in his profession.   He was a very active Member of the Royal Society, and contributed a good many papers to the Transactions, the greater number of which however are on zoological subjects.   He collected a museum of natural history with infinite industry, which must have been of very considerable value, as Sir Hans Sloane offered to purchase it from him for 4,000*l.* which were refused.   His publications, not reckoning a variety of minor ones of inferior importance, nor his detached papers published in the Transactions, were the following: 1. Musei Petiveriani *Centuriæ decem.*   This book contains the names of various rare animals, minerals, and plants.   Several cryptogamous plants make their first appearance in this work; in the investigation of which plants Petiver was very successful 2. Gazophylacii Naturæ et Artis *Decades decem.*   This book was of very considerable value, as it contained engravings and descriptions of many new plants, chiefly American ferns and plants from the Cape of Good Hope.   3. In the third volume of Mr. Ray's History of

Plants there is an account of the rare plants from China, Madras, and Africa, by Petiver, comprehending more than a thousand species. 4. *A Catalogue of Plants found on the Mountains about Geneva, the Jura, La Dole, Saleve; with others growing in the fields, &c. as observed by Gesner, the Bauhines, Chabræus, and Ray.* 5. PTERIGRAPHIA AMERICANA: *Icones continens plusquam* cccc *Filicum variarum Specierum.* Tab. 20. 1712. Fol. 6. *A Catalogue of Mr. Ray's English Herbal, with Figures.* This work Mr. Petiver did not live to complete; two parts only were published, containing 72 plates and figures of 864 species of plants. This work is of considerable importance in enabling us to ascertain the particular varieties described by Ray. The figures, though only outlines and small, are very neat and distinct. Petiver had paid particular attention to English botany, and was the first discoverer of many British plants. The second part of this work was published in 1715, and the author probably died soon after.

Sloane.　　Dr. Sloane, afterwards Sir Hans Sloane, acquired, during his life time, a higher reputation than any botanist that Britain has produced, if we except Ray. Indeed his merit must be admitted to have been very great. He was born in the North of Ireland in the year 1660, of parents originally of Scotch extraction. He studied the preliminary branches of medicine in London for four years, and afterwards went to France for further improvement. At Paris he studied under Tournefort and Du Verney; and he is supposed to have taken his degrees in medicine at Montpelier. He returned to London in 1684, and became the favourite of Dr. Sydenham, who took him into his house, and zealously promoted his interest. In 1684, he was elected a Fellow of the Royal Society, and in 1687 became a Member of the College of Physicians. Though these early advancements held out a flattering prospect of success in his profession, his passion for natural history induced him, in 1687, to relinquish his situation, and to accompany the Duke of Albemarle to Jamaica as his physician. They touched at Madeira, Barbadoes, Nevis, and St. Christopher's; but soon after their arrival in Jamaica the Duke died, and this contracted the term of Dr. Sloane's stay in the island to 15 months. His industry during that time must have been indefatigable, if we attend to the immense harvest which he reaped. The task indeed at that period was not so arduous as it has since become. Dr. Sloane was well acquainted with all the natural knowledge of his age, and he was the first naturalist that ever went to the West-Indies for the express purpose of prosecuting his science. Hence every thing was new, and the whole animal and vegetable kingdoms in a great measure at his disposal. On his return he settled in London as a physician, and soon acquired celebrity and immense practice.

In 1696, he published the *Prodromus* to his history of Jamaica plants, under the title of *Catalogus Plantarum quæ in Insula Jamaica sponte proveniunt, vel vulgo coluntur; cum earundem Synonymis et Locis natalibus; adjectis aliis qui-busdam quæ in Insulis Madeiræ, Barbados, Nieves, et Sancti Christophori nas-cuntur: seu Prodromi Historiæ Naturalis Jamaicæ Pars prima. Præter Indicem valde copiosum Nominum et Synonymorum.* This work gives an account of about 800 species of plants, many of them new, arranged according to the method of Ray. The descriptions and figures are so accurate that little difficulty was found by succeeding botanists in determining the plants to which Sloane al-ludes. Prodigious learning and industry are displayed in settling the synonyms of all the plants mentioned by preceding botanists; in order to prevent the pos-sibility of describing as new those vegetables which were already known. The first volume of his great work, the History of Jamaica, was published in 1707, and the second in 1725. The title of this work is as follows. *A Voyage to the Islands, Madeira, Barbadoes, Nevis, St. Christopher's, and Jamaica; with the Natural History of the Herbs and Trees, Four-footed Beasts, Fishes, Birds, In-sects, Reptiles, &c. To which is prefixed an Introduction, wherein is an Account of the Inhabitants, Air, Water, Diseases, Trade, &c. of that Place; with some Relations concerning the neighbouring Continent and Islands of America.* This work displays the erudition and industry of the author in a most conspicuous point of view. It is curious, that out of 800 plants described by Sloane, 100 are ferns, and most of the rest arborescent vegetables. It is now well understood, that a much greater proportion of the vegetables which occur in the torrid zone are trees, than of those which occur in the temperate climates. The plates amount to 274.

This voyage of Sloane raised a corresponding emulation in France, and induced Louis XIV. to send Plumier to the Caribbee Islands, and Tournefort to the Levant; voyages, both of them productive of considerable advantage to botany. In 1708, Sloane was elected a Foreign Member of the Academy of Sciences, at Paris; in 1719 he was made President of the College of Physicians; and in 1727, on the death of Sir Isaac Newton, he became President of the Royal Society. He was Governor of almost every Hospital in London: to each he presented 100*l.* during his life-time, besides a legacy which he be-queathed them at his death. The papers which he published in the Philoso-phical Transactions, of which he was for some years editor, amount to 35. Several of these are upon medical subjects; several on subjects connected with mineralogy and zoology; but by far the greatest part are botanical, and of very considerable value: containing the first accurate account of various plants, important either in medicine or domestic œconomy; as for example, the myr-tus pimento, or Jamaica pepper; the Winterania canella; the drimys Winteri, or Winter's bark tree; the protea argentea, or silver pine; the coffæa arabica, or coffee shrub, &c. Sir Hans Sloane began very early to form a museum of natural curiosities. His voyage to Jamaica procured him a considerable num-

E

ber of articles, which were prodigiously increased by the indefatigable industry of the collector, especially by the purchase of Courten's and Pettiver's museums. The museum of Sloane at last consisted of above 30,000 articles; besides 200 volumes of dried plants, and an excellent library. It now constitutes the principal part of the British Museum, and is a lasting monument of the liberality and zeal of the collector. In 1741, at the age of 80, he resigned his different situations in London, and retired to Chelsea, where he passed his time in serenity and in the constant exercise of benevolence, till the period of his death, Jan. 11, 1752.

Sherard.

William Sherard, LL. D., was born in 1659, in Leicestershire, educated at Merchant Taylors' School, and afterwards at St. John's College, Oxford. He travelled on the continent, first with Lord Viscount Townsend, and afterwards with a grandson of the first Duke of Bedford. His passion for botany was great, and his progress uncommon, in the investigation both of English and foreign plants, as appears from the obligations which Mr. Ray acknowledges to have lain under to him. About the year 1702, he was appointed Consul at Smyrna, where he resided about 16 years, cultivating a botanic garden and forming his *Herbarium,* which afterwards became unrivalled; comprehending about 12,000 species of plants. After his return to Britain with a handsome fortune, he made another tour to the continent in 1721. It was by his means, chiefly, that the manuscripts of Vaillant were purchased and published by Boerhaave. On his return, he brought over with him the celebrated Dillenius, whom he encouraged to persevere in his investigation of the cryptogamous plants, and whom he kept with him to the period of his death. He bequeathed 3,000*l.* to the University of Oxford, to establish a botanical professorship, on condition that Dillenius should be the first professor: he bequeathed also his botanical library, and his collection and manuscripts. His great work was his Pinax, which however he did not live to complete; and Dillenius, though he undertook to finish it, never found sufficient leisure. The work, in consequence, was never published. The brother of Dr. Sherard, first an apothecary and afterwards a physician, had also a strong passion for botany, and uncommon knowledge of the subject. He cultivated an excellent botanical garden at Eltham, which has been immortalized by the pen of Dillenius. Both brothers were Fellows of the Royal Society. William inserted many papers into the Transactions, but few of them were upon botanical subjects

Dillenius.

John Jacob Dillenius, born at Darmstadt, in 1687, and brought to England by Sherard, where he spent the remainder of his life, is one of the greatest names that botany has to boast of. He was educated at Giessen, a city of Upper Hesse, and first established his reputation as a scientific botanist by his Catalogue of plants growing in the neighbourhood of Giessen, published in 1719. This was the most complete work of the kind which had hitherto appeared, and was particularly distinguished by the uncommon number of Musci and Fungi which it contained, a class of plants hitherto almost entirely neglected, except in England. It was this circumstance that drew the attention of Dr. Sherard, induced him to enter into a correspondence with Dillenius, and, finally, to prevail upon him to accompany him to England. This happened in 1721, and soon after he published a new and very much improved edition of Ray's *Synopsis*.

In August, 1728, his friend and patron, Dr. Sherard, died, and in consequence of his will, he was soon after established as Botanical Professor at Oxford. In 1732 he published his *Hortus Elthamensis*, a work enthusiastically praised by Linnæus, who calls it the most perfect that the world had yet seen. It contains engravings of 417 rare plants, all drawn and etched with his own hand, and with the most minute precision. But his great work, a work which will not easily be surpassed, is his Historia Muscorum, published in 1741. In this work he first subjected mosses, lichens, and algæ, to a rigid investigation, and arranged them into distinct genera and species. It may be interesting to see the general outline of his classification. It was as follows:

He divided the plants of which he treats into two general classes.

I. Musci capitulis floridis destituti.
  A. Non peltati aut tuberculati: *Confervæ.*
  B. Peltis et tuberculis prædati: *Lichenoides* (our *lichens*) divided into foliosos et crustaceos.
II. Musci capitulis floridis donati.
  A. Durioribus.
    a. policoccis: *Marchantia.*
    b. monococcis: *Lichenastrum* (our *jungermania*).
  B. Mollioribus aut pulposis.
    a. prorsus nudis: *Bryum androginum.*
    b. membranis seu capsulis circumdatis.
    α. calyptratis.
      1. e foliorum alis: *Hypnum.*
      2. e summis cauliculis.
      aa. calyptra villosa: *Polytrichum.*
      bb. ——— lævi: *Bryum.*
    β. non calyptratis et sessilibus.
      1. dispersis et solitariis.
      aa. in summis cauliculis: *Sphagnum.*
      bb. in foliorum alis: *Selago.*
      2. aggregatis in spicam: *Lycopodium.*

This work comprehends an account of about 1000 plants, most of them new and of exceedingly difficult investigation. It contained 85 plates, all engraved by Dillenius himself, and with a minute accuracy that renders it easy for us to determine every plant to which he alludes. Posterity will hardly believe that 250 copies of this work were considered as a sufficiently large impression, and that a guinea was reckoned a sufficient price for a copy. So little taste for natural history existed at that time in England. The subject of mosses has been since carried a good deal farther by Hedwig, who has subdivided them into numerous new genera, and made many interesting discoveries respecting their parts of fructification. But the path of investigation was first opened by Dillenius, who had no predecessor nor coadjutor, and the progress which he made in it is really astonishing. He died of an apoplexy in March, 1747, six years after the publication of his great work.

Other Botanists.

It would be tedious to attempt a detailed account of the labours of all the botanists who flourished in England during the early period of the last century. Doody, Bobart, Martyn, Llhwyd, Robinson, Catesby, are distinguished names; while Miller, Collinson, Ellis, Watson, Solander, &c. became conspicuous after the adoption of the Linnæan arrangement by British botanical writers. Let us rather turn our attention to another and superior department of botany, we mean systematic arrangement, and the accurate distribution of plants into classes, orders, genera, and species.

Systematic Botany.

This department had originated with Gesner; but its value had been but faintly seen, and the labours of the Bauhins, important as they were, rather contributed to draw the attention of botanists from classical arrangement. The revivers of systematic botany were Morison and Ray, whose example was imitated, with various success, by several distinguished botanists on the continent.

Morison.

Robert Morison* was born at Aberdeen, in 1620. He was educated at that University, and originally intended for the church. In the civil war which distracted Britain during his youth, he attached himself to the royal cause, and received a dangerous wound at the battle of Brigg, near Aberdeen. On his recovery he went to Paris, applied himself assiduously to the study of natural history, and became so much distinguished for his botanical knowledge that he was recommended by M. Robins, the French King's botanist, to the patronage of the Duke of Orleans, uncle to Louis XIV., who appointed him intendant of his fine garden at Blois, with a handsome salary. This situation he held from 1650, till the death of the Duke in 1660, when King Charles II. invited him over to England, made him King's Physician and Royal Professor of Botany, and gave him a salary of 200*l.* a year. In 1669, he was elected Botanical Professor at Oxford, and lectured in that University till the period of his death, in 1683. He was bruised by the pole of a carriage, while crossing a street in London, and died in consequence of the hurt which he received. If we consider the celebrity which Morison acquired, and the respectable situa-

* There is a Life of Morison, by Bobart, prefixed to the continuation of his History of Plants, published 19 years after his death.

tions which he occupied, it is rather surprising that he never was a Member of the Royal Society. An analytical account, however, of some of his publications is to be found in an early number of the Transactions.

The works published by Morison were three in number. The first was a new edition of the *Hortus Blesensis*, which had been originally published in 1653. Morison greatly increased the catalogue of plants when he re-published it in 1669. He professes to give an account of no fewer than 260 new species; but many of these were either varieties, or they had been already known. Several new plants, however, both from the South of Europe, and natives of France or Britain, are first pointed out by him in this work. His physiological opinions, as far as he delivers them, are, in general, accurate and of considerable value; especially his dissertation, to prove that all plants spring from seeds, and to refute the notion of equivocal generation, at that time adopted by many philosophers. In the year 1672, he published his *Plantarum Umbelliferarum Distributio Nova*, as a specimen of the new methodical arrangement which he meditated. This book possesses very considerable merit, and is admitted to be a very ingenious and even complete distribution of umbellated plants. In 1680 he published a volume of his great work, which he entitled *Plantarum Historiæ Universalis Oxoniensis Pars Secunda.* According to the method universally adopted at that time, he divided all plants into two sets; namely, *trees* and *shrubs*, and *herbs*. The *trees* and *shrubs*, which constituted the first part of his history, were never published. He alleged, as his reason for beginning with the *herbs*, that they were the most numerous and difficult set, and that he began with them that he might not leave the most important part of his work unfinished. He divides all herbaceous plants into fifteen classes, His method. under the following titles :

1 Scandentes.	7 Papposæ lactescentes.	12 Multisiliquæ et Multicapsu-lares.
2 Leguminosæ.	8 Culmiferæ.	
3 Siliquosæ.	9 Umbelliferæ.	13 Bacciferæ.
4 Tricapsulares hexapetalæ.	10 Tricoccæ purgatrices.	14 Capillares.
5 Tricapsulares aliæ.	11 Galeatæ et Verticillatæ.	15 Anomalæ.
6 Corymbiferæ.		

As the method of Morison was not followed by succeeding writers, it is unnecessary to enter into any examination of it. The titles of his classes are sufficient to shew that his method is not uniformly founded on the fruit; but on the fruit and habit conjointly. Of these 15 classes, Morison has given a description of only five. After an interval of 19 years the next four were published by Jacob Bobart, who had been appointed by the University of Oxford, for that purpose. He introduced a considerable number of new plants, discovered after Morison's death; and the figures, which are numerous, are many of them good.

Ray

Ray was a man of much greater powers, and was of much more service to botany in general than Morison, or indeed any other British writer of his age. He may be considered as the person who introduced and established systematic botany in England; and he prosecuted the science with such ardour, and gave it such popularity, that no fewer than 500 new species of British plants are introduced and described in the last editions of his Synopsis, which had been unknown when he published the first edition of that work. All the plants described in the first edition of his Catalogue of English Plants, amount to 1050 species. These were afterwards increased by his own exertions, and those of his friends, to 1550 species. At present we are acquainted with about 2800 species of British plants, rather more than one half of which belong to the cryptogamic class, which was but little attended to before the time of Mr. Ray; and this, in some measure, accounts for the great deficiency in his catalogue. The species of British fungi, at present known, amount to near 600.

John Wray, or, as he always spelt his name after the year 1669, Ray,* was born at Black Notley, near Braintree, in Essex, Nov. 29th, 1628. Though his father was only a common blacksmith, he resolved to give his son a liberal education, and accordingly sent him to the grammar-school at Braintree. He was entered at Catherine Hall, Cambridge, in 1644, and two years after he removed to Trinity College, where the politer sciences were more cultivated. In 1649 he was chosen minor Fellow of Trinity; in 1651 he was made Greek Lecturer; in 1653 Mathematical Lecturer; and in 1655 Humanity Reader. While in these situations, he was tutor to various persons of distinction, with whom he contracted friendships. Mr. Francis Willughby was one of these, with whom, in consequence of the similarity of their tastes and pursuits, he ever after kept up the greatest intimacy and friendship.

In December, 1660, Mr. Ray was ordained both deacon and priest, by Dr Sanderson, bishop of Lincoln; and he continued a Fellow of Trinity College, to which situation he had been chosen some time before, till the passing of the Bartholomew Act, in September, 1662, which, as he did not subscribe, necessarily superseded him.

It was about this period that he undertook different journeys through England, chiefly with a view to ascertain its vegetable and animal productions. The first was undertaken in the summer of 1658, and lasted from the 9th of August to the 18th of September. He travelled alone, and traversed the midland counties of England, and the northern parts of Wales. The second journey was undertaken in 1661, and lasted from July the 26th, to August the 30th He was accompanied by Mr. Willughby and some other gentlemen, and

---

* Dr. Scott published a Life of Ray, in 1760, from materials collected by Dr. Derham. It con tains the fullest account of him which we have seen.

travelled through Durham and Northumberland, to Edinburgh; thence he proceeded to Glasgow; and returned through Cumberland and Westmorland. In his third, and most extensive journey, he was likewise accompanied by Mr. Willughby. It was made in 1662, and lasted from the 8th of May to the 18th of July. He traversed the midland counties, entered Wales by Cheshire, passed through the whole length of the principality; and proceeded through Gloucester and Somersetshire, to the Land's End. From this he returned by the southern counties.

Being freed by the loss of his fellowship from the business of a college life, he was induced, in 1663, to accompany Mr. Willughby, Mr. Skipton, and Mr. Nathaniel Bacon, to the continent. He was absent three years; and, during that time, visited France, Holland, Germany, Switzerland, Italy, Sicily, and Malta   The consequence of this journey was a prodigious increase of natural knowledge. After his return, he passed the year 1666, in reading the publications which had appeared in England during his absence, and in arranging Mr. Willughby's museum of animal and mineral productions, and in completing his own catalogue of English vegetables.

In the summer of 1667, he made a fourth journey through England, accompanied by his friend Mr. Willughby. They set out on the 25th of June, and travelled through the counties of Worcester, Gloucester, and Somerset, to the Land's End; and returned through Hampshire to London, on the 13th of September. He made a fifth journey alone in 1668, into Yorkshire and Westmorland. In 1667, he was chosen a Fellow of the Royal Society, and contributed several papers to the Transactions, but most of them upon zoological or physiological subjects.

In the year 1671, Mr. Ray lost his much valued friend Mr. Francis Willughby, in the 37th year of his age; who appointed him one of his executors; confided to him the education of his two sons; and left him an annuity of 70*l.* In 1673, he married the daughter of Mr. John Oakeley, of Launton, in Oxfordshire. He resided for several years at Middleton-hall, the seat of Mr. Willughby; but after the sons of that gentleman were removed from his care, he retired, first to Sutton Cofield, and afterwards to Falkborne-hall, near Black Notley; at which last place he built a house, and finally settled June 24, 1678. He died on the 17th of January, 1705.

The writings of this indefatigable naturalist are so numerous and various His writings. that we shall not attempt a complete enumeration of them. Those on zoological subjects will come under our review hereafter; and his theological writings, though important and popular, do not, strictly speaking, come within our subject. His first botanical work was his *Catalogus Plantarum circa Cantabrigiam nascentium*, first published in 1660, and two appendixes, containing additional plants, subsequently discovered, were afterwards added. In

1670, he published his *Catalogus Plantarum Angliæ et Insularum adjacentium, tum indigenas tum in agris passim cultas complectens.* This book was long resorted to as classical, by the English botanist. His travels, published in 1673, deserve to be mentioned on account of the great quantity of botanical information which they contain. He gave them the following title: *Observations, Topographical, Moral, and Physiological, made in a Journey through part of the Low Countries, Germany, Italy, and France.* But the work which does him the most credit, by raising him to the rank of a systematic botanist, was first published in 1682; though he had turned his attention to the subject much earlier, and the tables which he drew up for Bishop Wilkins's work, entitled, *Real or Universal Character,* published in 1668, contain, in fact, the outlines of his system. This work was entitled, *Methodus Plantarum nova Brevitatis et Perspicuitatis Causâ synopticè in Tabulis exibita: cum Notis Generum tum summorum tum subalternorum characteristicis. Observationibus nonnullis de Seminibus Plantarum et Indice copioso.* This work was nothing less than an elaborate attempt to reduce all plants under natural classes. The plan was gradually improved in subsequent editions. We shall exhibit the outline of his method in its most improved state.

His method.   All plants were arranged under two sets, namely, Herbaceous and Trees. The first of these was subdivided into 25 classes: the second into eight. The names of these classes were as follow:

HERBS.

1 Submarinæ.	10 Herbæ semine nudo solitario,	18 Multisiliquæ.
2 Fungi.	flore simplice perfecto.	19 Vasculiferæ, monopetalæ, et
3 Musci.	11 Umbelliferæ.	dipetalæ.
4 Capillares.	12 Stellatæ.	20 Siliquosæ et siliculosæ.
5 Apetalæ.	13 Asperifoliæ.	21 Papilionaceæ.
6 Plenipetalæ lactescentes.	14 Verticellatæ.	22 Pentapetalæ.
7 Discoideæ.	15 Polyspermæ.	23 Floriferæ graminifoleæ.
8 Corymbiferæ.	16 Pomiferæ.	24 Stamineæ graminifoleæ.
9 Capitatæ.	17 Bacciferæ.	25 Anomalæ.

TREES AND SHRUBS.

26 Arundinaceæ.	30 Fructu sicco, non siliquoso nec umbil icato.
27 Flore a fructu remoto; seu apetalæ.	31 Siliquosæ non papilionaceæ.
28 Fructu umbilicato; seu pomiferæ et bacciferæ.	32 Siliquosæ papilionaceæ.
29 Fructu non umbilicato: seu pruniferæ.	33 Anomalæ.

From these titles, the reader will perceive that, though Mr. Ray made the *fruit* the basis of his classification, he was not solely guided by it; but attended, likewise, to the habit of the plant; and, in fact, was too much influenced by the figure of the leaves. The method, though elaborate, is perplexed, and exceedingly difficult to follow. The last botanical work of Ray, which we shall mention, was the most elaborate of all his performances, and perhaps the most laborious work ever undertaken and performed by any single botanist. It was no less than an attempt to arrange all known plants under his system, and to give a complete account of every thing known respecting each. This immense task he executed in his *Historia Plantarum Generalis*; consisting of three very thick folio volumes. The first volume appeared in 1686; the second, in 1688; and the third, intended as a supplement, and comprehending all the plants discovered in the interim, was published in 1704. He gives an account of no fewer than 18,600 plants; but the last volume, as was unavoidable from the nature of the subject, abounds in repetitions.

Systematic Botany, thus revived in Britain, by Morison and Ray, soon drew the universal attention of all true botanists; and the methods of the British philosophers were soon followed on the continent by other methods, some of which possessed superior merit. Of the systems which immediately followed that of Ray, by far the best were those of Tournefort and Rivinus, both founded on the flower. Neither of these philosophers was a Member of the Royal Society. Tournefort possessed uncommon merit as a botanist: his method, considering the time in which it appeared, was admirable. It soon superseded almost all other systems; and, indeed, is still followed by a sect of botanists in France.

Joseph Pitton Tournefort* was born in the south of France, in 1656. He was first intended for the church, but his passion for botany was irresistible, and prevented him from pursuing his theological studies. He travelled in search of plants through France, Spain, and the Pyrenæan Mountains. His travels were afterwards extended to Portugal, the Netherlands, and England. He was superintendant of the French King's garden, at Paris; and in 1700 he was sent to Greece and Asia, at the expence of the King, his master, to collect plants and make observations. The result of his observations was afterwards published in his *Travels*, an excellent work, which has been translated into English. While crossing a street in Paris, in 1708, he was struck in the breast by the axle of a carriage: this brought on a hydrothorax, and occasioned his death.

His great work was his *Institutiones Rei Herbariæ*, in three quarto volumes; without doubt the best botanical book before the time of Linnæus. We shall

---

* Besides Fontenelle's *eloge* of Tournefort, which contains but little, there is a Life of him by Lauthier, prefixed to his Voyage into the Levant.

F

His method. present our readers with the outline of his system.   He divided plants into 22
classes, which he distinguished by the following names :

        1 Simplices monopetali campaniformes.
        2 ———————————— infundibuliformes.
        3 ———————————— anomali (arum, aristolochia, &c.)
        4 ———————————— labiati.
        5 Simplices polypetali cruciformes.
        6 ——————————— rosacei.
        7 ——————————— umbellati.
        8 ——————————— caryophyllei.
        9 ——————————— liliacei.
       10 ——————————— papilionacei.
       11 ——————————— anomali (fumaria, aconitum, aquilegia, &c.)
       12 Compositi flosculi (carduus, &c.)
       13 ————— semiflosculi (leontodon, &c.)
       14 ————— radiati (aster, senecio, &c.)
       15 Flores apetali.
       16 Stirpes imperfecti, filices, musci, &c.
       17 Herbæ et suffrutices.
       18 Arbores et frutices flore apetalo cum fructu conjuncto.
       19 Arbores améntaceo flore.
       20 ———— monopetalæ.
       21 ———— flore rosaceo.
       22 ———— flore papilionaceo.

This classification is by no means difficult; and, were it not for the imperfect
characters of a few of the classes, might certainly be followed; though it will
not bear a comparison with the Linnæan method.

Augustus Rivinus. Augustus Quirinus Rivinus was professor at Leipsic.   He was born in 1652,
and died in 1725.   His system was founded upon the number of petals, and
upon their regularity or irregularity.   He published it in 1690, in a tract
entitled, *Introductio .Generalis in Rem Herbariam.*   Though his method
is inferior to that of Tournefort, yet his work contains many admirable observations on the division of plants into genera, and on the way of describing and naming them; observations of which Linnæus afterwards availed
himself.   He first laid aside the division of plants into herbs and trees, as a
useless incumbrance.   On this subject, Ray entered into a controversy with
him.   But the subsequent conduct of botanists has decided in favour of
Rivinus.

Linnæus. We come now to Linnæus, the most extraordinary botanist that ever existed.
That a young Swedish peasant should have taken upon him to publish a new
code of botanical laws, totally different from those formerly received, and that
these laws, in a few years, should have been embraced and implicitly obeyed

by almost all the botanists in Europe, is a phenomenon, we will venture to say, without a parallel in the history of human nature.

Charles Linnæus,* afterwards enobled and created Knight of the Polar Star, was born on the third of May, 1707, at Rashult, a village in the province of Smaland, in Sweden. His father was pastor of the village. He was a lover of gardening and flowers, and cultivated an extensive garden with considerable taste. Young Charles was his constant associate, and made the garden the place of his amusement and recreation. Thus he acquired a taste for plants at a very early age; and his father, to encourage it, gave him a piece of ground of his own to cultivate according to his fancy. He made excursions into the neighbouring fields and woods, and filled it with all the wild plants which he could collect. Thus early was his passion for botany formed, and thus early did he acquire the seeds of his knowledge of the science.

His father destined him for the pulpit, and initiated him himself in the rudiments of Latin, geography, religion, &c. At the age of 10 he sent him to the Latin school, in the adjacent town of Wexicoe. At Wexicoe every spare moment of his time was occupied in collecting and examining plants; and, in consequence, his progress in those studies that fitted him for the clerical profession was very small. In 1724, when he went into the higher college of Wexicoe, some of the professors complained of his little progress to his father, and supplied him with bad testimonials. His father, mortified at the disappointment of his hopes, formed the resolution of binding his son apprentice to a shoe-maker. John Rothman, a physician at Wexicoe, and professor of medicine in the college of that city, had noticed the enthusiastic ardour with which Linnæus cultivated botany, the surprising progress which he had made, and the great genius which he possessed for the study. He stated these particulars to the father of Linnæus, and urged him to give his son liberty to follow his inclinations. He offered to take him into his own family, and to initiate him himself in the rudiments of medicine. These offers, together with the earnest solicitations of Linnæus himself, at last struck the balance, and induced his parents, though with reluctance and little satisfaction, to allow him to enter upon the career of medicine. In Rothman's house he continued three years, and made surprising progress in his favourite pursuits. Here he became acquainted with Tournefort's Institutiones Rei Herbariæ, which first suggested to him a plan, and turned his mind to the systematic arrangement of plants.

It was now requisite for the completion of his medical studies, that he Goes to the University.

* The name Linné, or Linnæus, was adopted by the father of Linnæus, from a large Linden tree, which grew in the neighbourhood of his native place.

The best Life of Linnæus was published in German, by Stoever: there is a translation of it into English, published in 1794, by Mr. Trapp, who must have been a foreigner, as the translation betrays a want of knowledge of the English idiom.

should go to a university; and he made choice of the University of Lund,
because Humærus, one of the professors, was one of his relations; and upon
his friendship, he, in a great measure, depended for subsistence. Unfortunately, before our young philosopher reached the University of Lund, in 1727,
Humærus was dead   But the uncommon diligence, knowledge, and sagacity
of Linnæus, drew the attention of Stobæus, Professor of Physic and Botany
who, aware of the forlorn situation of Linnæus, generously took him into his
own house and supported him during his residence at Lund. Here he had
access to the best botanical books, and to a collection of natural curiosities.
Here too he began to form a herbarium, and to accustom himself to make
descriptions of plants. His ardour for knowledge induced him, in about a year,
His distress. to leave Lund and go to Upsal, the oldest and most celebrated University in
Sweden. Here he was soon reduced to a state of the greatest distress. His
money was exhausted, and his clothes worn out. For some time he supported
himself on the compassion of his fellow students; picked up a scanty meal
where he could procure it; dressed himself in their cast clothes; and was even
reduced to the necessity of mending their old shoes with the bark of trees, to
enable him to prosecute his favourite excursions in quest of plants. While he
was in this situation, Olaus Celsius, First Professor of Divinity, at Upsal, and
the most eminent botanist that Sweden could boast of, happened to meet
Linnæus in the botanic garden. He was astonished at the exactness and extent
of his botanical knowledge, which appeared the more extraordinary the more
closely he examined it. Celsius was, at that time, engaged in preparing his
Hierobotanicon, a work, in which the plants mentioned in the Scriptures were
to be determined and described. It struck him, that Linnæus might be of
considerable service to him in completing this work. On that account, he
invited him to his house, and gave him board and lodging for nothing. Here
he had free access to a very complete botanical library; here he first met with
the work of Vaillant, which pointed out to him the importance of the sexual
organs of plants; and here he first conceived the idea of a new arrangement
of plants, founded on the sexual system; an idea which he afterwards realised
with such complete success. In 1730 he wrote a small treatise on the sexual
system, which, coming into the hands of Olaus Rudbeck, at that time professor
of botany, at Upsal, and an old man, induced him to invite Linnæus into his
house, and to make him his substitute as a lecturer on botany.

His journey
to Lapland.   In the year 1731, a plan was formed by the Academy of Sciences at Upsal,
to send some person to explore the vast and dreary region of Lapland, and to
ascertain the various natural productions which it contained. Celsius and
Rudbeck united in proposing Linnæus as the best fitted for executing this
difficult task, and their choice was unanimously approved. Linnæus embraced
this offer with transport. The smallness of the sum devoted to defray his

expences, which did not exceed seven pounds ten shillings, did not discourage him; neither was he deterred by the dismal country through which he had to travel, covered with snow the greater part of the year, and almost destitute of inhabitants and cultivation. He set out on horseback in the beginning of May, and reached the northern extremity of the gulph of Bothnia, about the end of that month. Here he set out on foot and alone, and with infinite fatigue and difficulty traversed the whole of Lapland, and ascertained its vegetable, animal, and mineral productions. He then bent his course to the mountains which separate Sweden and Norway. These he examined with particular care, and even travelled as far as the Frozen Ocean. Thence he passed through Finland, crossed the gulf, and arrived at Upsal about the end of September. The fruit of this journey was his Flora Laponica, published soon after by the Swedish Academy, and containing above 100 new species of plants, arranged according to his own new system. The journal of his Lapland journey was never published by him. It was written in Swedish, and is in the possession of Dr. Smith, President of the Linnæan Society, of London; who purchased it, together with all the other manuscripts and collections of Linnæus; and has lately favoured the British public with an English translation of it.

After his return to Upsal, Linnæus began to give public lectures on botany and mineralogy. The novelty of his matter, and the ardour with which he inculcated his views, drew general attention, and raised his reputation to the highest pitch. This excited the envy of his competitors. By the laws of the University, no man was entitled to give public lectures at Upsal, till he had taken his degrees. Linnæus had obtained no academical honours, and therefore stood within the meaning of the statute. Dr. Rosen, at that time Lecturer on Medicine, accused him before the Senate of the University, and insisted that the statutes should be put in execution. The Senate, notwithstanding the exertions of the patrons and friends of Linnæus, were under the necessity of enforcing the statutes, and of prohibiting Linnæus from continuing his lectures. This drove the sanguine temper of our young naturalist almost to madness. He drew his sword upon Rosen, at the door of the Senate House, and attempted to stab him. Fortunately he was disarmed by the by-standers; and by the interest of Celsius, this daring outrage, instead of being pursued by expulsion, which would have blasted Linnæus's hopes for ever, was passed off with a reprimand. For some time Linnæus retained his resentment, and persisted in his intention of stabbing Rosen, as soon as he could find him. But by degrees his passion cooled, his resentment subsided, and he listened to milder and more prudent suggestions.

While he was hesitating what plan to follow in order to support himself, his former pupils formed the project of making excursions to improve their mineralogical knowledge, and they offered to put Linnæus at their head. The

Forbid to lecture at Upsal.

offer was accepted, and the excursions begun.   Fahlun, the capital of Dale-
carlia, famous for its copper mines, was one of their principal objects.   Here
Linnæus was introduced to Baron Reuterholm, the governor of the province,
a zealous naturalist, and particularly fond of mineralogy.   He was struck with
the knowledge which Linnæus displayed, and proposed to place his two sons
under his care, in a journey which he projected for the sake of their improvement in mining.   This journey was accordingly made, and after his return to
Fahlun, Linnæus delivered a course of lectures on the method of assaying
metals, and upon other subjects connected with mineralogy.   Here he got
acquainted with Dr. Moræus, the most eminent medical man at Fahlun; fell
in love with his eldest daughter; gained her affections; and solicited her hand
from Dr. Moræus.   The father hesitated.   He was fond of Linnæus, but disliked his dependent situation.   He therefore told him that his daughter should
remain unmarried for three years; and, that if his circumstances were mended
by that period, he should have no hesitation in giving a favourable decision.

Moræus disliked the precarious situation of Linnæus; urged him to lay aside
his botanical pursuits, which were too uncertain to be trusted to, and apply
himself seriously to the practice of medicine.   To put this advice in practice,
it was necessary to obtain a Doctor's degree; and want of money had hitherto
prevented Linnæus from even thinking of such honours.   Now, however, love
stepped in to his assistance, and smoothed the way to his advancement.   It
was the custom with the Swedish medical men to leave their own country and
obtain their medical degrees at a foreign University, and the Universities in
Holland were usually preferred.   Linnæus resolved to imitate this general
example.   Miss Moræus supplied him with a hundred dollars, out of her
savings.   This sum, together with what he had laid up from his lectures and
expeditions, amounted to about 50l.

Goes to Hol-
land.

Having spent the winter months in visiting his friends, Linnæus set out for
Holland, in April, 1735.   He spent about a month at Hamburgh, where he
got acquainted with the eminent men at that place, and inspected their
collections of natural curiosities.   From this city, he proceeded to Harderwyk,
in Guelderland, where he defended a thesis on intermittent fevers, and obtained
the degree of Doctor of Medicine.   From Harderwyk he proceeded to Leyden,
anxious to get acquain ed with the literary characters, who, at that time, distinguished Holland.   Boerhaave, the most celebrated physician of the age, at
that period the glory of Leyden, was an old man, rich, and so much pestered
by the visits of strangers, that it was difficult to obtain access to him.   Linnæus
attempted it in vain.   By the advice of Gronovius, with whom he had become
acquainted, he was induced to publish the outline of his *Systema Naturæ*,
on 14 folio pages.   He sent a copy of this work to Boerhaave, with a letter.
Boerhaave was so much struck with the knowledge and genius displayed in this

publication, that he longed earnestly to get acquainted with the author, and appointed him to meet him at his country house about a mile from Leyden. This house was surrounded by an excellent botanical garden. Here Linnæus had an opportunity of displaying his unrivalled botanical knowledge. Boerhaave exhorted him to remain in Holland, where he could not fail to make his fortune. Linnæus stated that his pecuniary difficulties were such as to preclude the possibility of following his advice. Next day he set out for Amsterdam, and Boerhaave gave him a letter of introduction to Burmann at that time Professor of Botany at Amsterdam. Burmann received him kindly, invited him back, showed him his plants, and was so delighted with his botanical knowledge that he took him into his own house, in order to assist him in drawing up a description of the plants of Ceylon, about which he was at that time employed.

Dr. George Cliffort, then Burgomaster of Amsterdam, and one of the Directors of the Dutch East-India Company, was a zealous patron of natural history in general, and of botany in particular; and had expended vast sums in gratifying this passion. His botanical garden at Hartecamp contained an immense collection of plants from all parts of the world. But they remained undescribed and unarranged. Boerhaave, who had formed the highest and most just opinions of Linnæus's uncommon knowledge and abilities, recommended him to Cliffort as a family physician, and as a man well qualified to arrange and describe his plants. The bargain was immediately struck. Cliffort allowed Linnæus free board and lodging, together with a salary of 1,000 florins a year. At Hartecamp he found a paradise, and seized the opportunity of executing that botanical reform for which he had already made adequate preparations.

His labours, during the years 1736 and 1737, were prodigious, and appear almost too vast for human industry to accomplish. The matter certainly must have been previously collected, and his facility in writing the Latin language must have been very great. His first work was his *Fundamenta Botanica*, which consisted only of 35 duodecimo pages. The theory of botany was reduced to 365 aphorisms, and he displayed in these the basis of his new system. This work was afterwards greatly enlarged, and republished under the title of *Philosophia Botanica*. His next work was his *Bibliotheca Botanica*, in 153 duodecimo pages. In this book he gave an account of above 1,000 preceding botanical writers, and extracted from them a system of botanical researches which he divided into 16 classes. His third work was an account of the *Musa Paradisica*, or Banana Tree, which had blossomed in Cliffort's garden, and enabled him to draw up an accurate description of that very rare plant.

His publications.

In the summer of 1736, he paid a visit to England, induced partly by his desire to get acquainted with the celebrated botanists of that country, and partly to procure plants for the botanical garden of Cliffort. Boerhaave gave

Visits England.

him a very flattering letter to Sir Hans Sloane, at that time in the meridian of his glory. But his reception was cooler than he had reason to expect. Sloane, who had adhered to the system of Ray, was too old to adopt another, and naturally looked with no very friendly eye upon every innovator. From London Linnæus went to Oxford, to visit Dillenius. He found him in company with Sherard, and was treated by him with coldness. He even observed to Sherard when Linnæus entered, that he was the young man who was *confounding* all botany. Though Linnæus was unacquainted with the English language, yet he understood this remark, and next day, when he called to take leave of Dillenius, he requested him to explain to him the meaning of his assertion. Dillenius was thunderstruck, and endeavoured at first to explain away what he had said; but upon being hard pressed by Linnæus, he carried him to his library, and showed him the proof sheets of his *Genera Plantarum*, which Gronovius had sent over without the knowledge of Linnæus. Every page was marked in various places with the letters N. B. which signified, as Dillenius explained them, the false genera to be found in his book. Linnæus defended the accuracy of his genera. They went into the garden to decide by inspection. A blossom of *blitum* was taken up and examined, which according to the general opinion contained three stamina, but according to Linnæus only one. Every blossom was found to agree with the description of Linnæus, and it was so with all the other plants which they examined. This produced a complete change in the conduct of Dillenius. He showed him the herbarium of Sherard, and granted him all the plants which he wanted for Cliffort's garden. But Dillenius, like Sloane, was too old and too celebrated to change his system. He seems even from his letters to Haller to have viewed the rising reputation of the young Swede, and the botanical reform which he was accomplishing, with a degree of jealousy not free from malignity.

In 1737, Linnæus published his *Genera Plantarum*, containing a great part of his proposed reform. Here he described about 935 genera, and upwards of 8,000 species of plants, most of which he had himself examined with scrupulous attention. His next work was his *Flora Laponica*, arranged according to his new system, and dedicated to the Swedish Academy, who had already published an abstract of it in their memoirs. The third and most expensive work which distinguished the year 1737, was his description of Cliffort's botanical garden, entitled *Hortus Cliffortianus*, in 501 folio pages, embellished with 32 plates, engraved by Ehret. His fourth work of 1737, was his *Critica Botanica*, consisting of a commentary on his *Fundamenta Botanica*. Here he examined the old genera of botanists, pointed out the errors, both in nomenclature and arrangement, rejected what was erroneous, and retained only those genera which were conformable to his own regulations. His next work was a description of Cliffort's orchard, entitled *Viridarium Cliffortianum*.

He now left Hartecamp with a resolution to return to Sweden. But on a visit to Van Royen, Botanical Professor at Leyden, who had quarrelled with Boerhaave, he was prevailed upon to remain and assist him in publishing a new arrangement of the Leyden garden. This garden had been arranged by Boer haave. Van Royen proposed to adopt the Linnæan method, but Linnæus out of gratitude to his patron and benefactor, would not listen to this, but suggested a new arrangement, which might be published as Van Royen's own. This plan was adopted, and the arrangement which in fact belonged to Linnæus was published accordingly.

While Linnæus lived with Van Royen he published two other works, the first the Ichthyology of Artedi, his early and particular friend, who had been drowned at Leyden about three years before. The second was his *Classes Plantarum*, in which he gave a critical view of the 16 universal, and 13 partial systems of arrangement which had been introduced into botany before his own time, pointed out the defects and excellencies of each, and finally compared them with his own system. These different publications had raised the reputation of Linnæus to the highest pitch, and his new system began already to be adopted by various distinguished botanists in different countries.

During the whole of this period he had corresponded with his intended bride through the intervention of a common friend for whom he had procured a professor's chair. This friend unfortunately conceived a passion for Miss Moræus, and endeavoured to supplant Linnæus, by representing to her father that Linnæus had remained abroad much longer than his stipulated time, and that there was no prospect of his ever returning to Sweden. Luckily another friend of Linnæus got intimation of the plan, warned Linnæus of his danger, and assured Moræus that his intended son-in-law would infallibly return to his own country, and very speedily too. This intelligence threw Linnæus into a sort of melancholy; he was seized with a fever, and after his recovery resolved to leave Holland without delay. Being so near France he could not resist the temptation of paying a short visit to Paris, and of getting personally acquainted with those botanists with whom he had previously corresponded by letter. Here he was received with politeness and attention; though the French, from their predilection for Tournefort and Vaillant, were but little disposed to adopt his systematic views.

After a residence of about a month in Paris he went on board a ship at Rouen, and in five days reached Helsingburg, in Scania, whence he set out for Stockholm. Here he attempted the practice of medicine, relying upon the reputation which he had acquired to pave his way to employment. For some time he met with nothing but disappointment. Every body was disposed to view his reputation with jealousy and malignity. His wretched antagonist Siegesbeck was every where cried up as having completely overcome him. In

Returns to Sweden.

G

this situation Haller offered to give up in his favour the Botanical Professorship of Gottingen. But before he received Haller's letter, fortune began to smile upon him, and to promise him fortunate days in Sweden. The cure of a fashionable distemper struck the young Swedes with surprize, and recommended him to practice in similar cases. His cures were successful, and his advice became the fashion. He rose in reputation every day, and was called to the lady of an Aulic counsellor troubled with a cough. He prescribed a medicine which she carried about with her for constant use One day, while at a card party with Queen Ulrica Eleonora, this lady made use of her medicine. The Queen inquired what it was, and was informed that it was a remedy for the cough which always procured speedy relief. Her Majesty was troubled with a cough at that very time. She sent for Linnæus, who prescribed the remedy, and the disease vanished. This fortunate accident brought Linnæus into the first practice at Stockholm.

His great and constant patron was Count Charles Gustavus Tessin. By his influence Linnæus was appointed Physician to the Admiralty, and Botanist to the King. Having now a settled income, he married Miss Moræus, five years after he had procured her father's conditional consent. The Stockholm Academy was established about this time, and Linnæus was appointed the First President. In 1741, he made a tour through the islands of Æland and Gothland, at the expence and by orders of the Diet, in order to ascertain their mineral and vege-

Becomes a Professor at Upsal.

table productions. The same year he was appointed Professor of Physic and Anatomy, in the University of Upsal, in room of Roberg who had retired with the whole of his salary. His old antagonist Rosen had the year before been appointed Professor of Botany. The emoluments of both Professorships were equal. They accordingly agreed, with the approbation of the Senate, to exchange their respective chairs, that each might teach the science for which he was best adapted. Thus Linnæus got the botanical chair at Upsal, the original object of his ambition, which he continued to fill for 37 years with greater glory than any literary man in Sweden had ever obtained. His exertions, during the whole of this period, were uncommonly great. He became the legislator of botany. His pupils carried his reputation to all parts of the world ; and, actuated by that enthusiasm which he so successfully inspired, they carried their researches to the remotest countries, and scrupled not to sacrifice their lives to the improvement of their favourite science. Thus botanical knowledge increased with a rapidity altogether unexampled. The number of genera was almost doubled before the death of Linnæus, and since that period the rate of increase has rather accelerated than diminished. South America, New Holland, and New Zealand, have been recently explored, and have added prodigiously to the number of botanical species. But this is a field into which we cannot with propriety enter, as it is scarcely, if at all, connected with the labours of the Royal Society.　Here therefore we shall close our account of the arrangement and description of plants, remarking only that the Linnæan arrangement, which has been almost universally embraced, though upon the whole decidedly superior to every other, is notwithstanding of very unequal merit, some parts being much more perfect than others.　To the cryptogamous plants Linnæus paid much less attention than to the rest.　Hence his arrangement of them has been found inadequate and inexact, and it has been thought necessary to alter it very materially.　Hedwig is the person to whom we are most indebted for his labours in this department.　With a patient industry which admirably fitted him for the task which he undertook, he devoted his life chiefly to the investigation of the musci, and has thrown them into new genera much more satisfactory and much more easily investigated than the genera of Linnæus.　The lichens, fungi, and fuci, have been successfully investigated by various distinguished botanists of Great-Britain, most of whom are still alive, and actively prosecuting their meritorious investigations.　Of the other Linnæan classes perhaps the most defective are the *tetradynamia* and *syngynesia*, in both of which a variety of very exceptionable genera might be pointed out; genera founded entirely upon the arbitrary will of Linnæus, and of course accompanied by descriptions so imperfect, that a beginner finds it extremely difficult to make out the plants alluded to.　This imperfection seems owing to the too great attention which Linnæus paid to the flower, and his total disregard of the seed in constituting the genera.　Whereas in these two classes the seed is in fact of much more importance than the flower.　It would be easy to point out some other parts of the Linnæan classification by no means adequate to the purpose for which it was intended; but it would not be so easy in these cases to apply the proper remedy.

Section 2.—*Of the Anatomy and Physiology of Plants.*

This branch of botany possesses more dignity than the preceding.　It opens to our view some of those beautiful contrivances which nature follows to accomplish her purposes, and which irresistibly lead the enlightened observer to the knowledge of the existence and benevolence of an all-powerful intelligent being.　This branch of knowledge, important as it is, is entirely modern, and owes its origin to the Royal Society, by the Members of which the investigation was begun, and by whom almost every fact relating to the subject hitherto established has been ascertained.　These investigations however are not all to be found in the Transactions of the Royal Society.　The most important have been published in separate treatises, either at the expence or at least with the concurrence of the Royal Society.　The papers in the Transactions relating to this subject amount to 60.　But of these there are 22 which can hardly be considered as of any value.　Subsequent investigations and more complete discocoveries have rendered several of the remaining 38 of comparatively little importance, except in a historical point of view. Some of the most curious papers, on the subject of vegetable functions, have been printed in those volumes of the Transactions, which have made their appearance since the year 1800. These, of course, cannot, with propriety, come under our review.

Grew's work. The first person who began the anatomical examination of plants, was Dr. Nehemiah Grew. He was born in Coventry, about the year 1628. His father, who was a nonconformist, sent him abroad at an early age; and he took the degree of Doctor of Medicine in a Foreign University. After his return to England, he settled in London, became a Fellow of the Medical College, and practised medicine, with some degree of success, till the period of his death, in 1711. He began to turn his attention to the anatomy of plants, as early as the year 1664. In 1670, he put an essay on the subject, which constitutes the first book of his *Anatomy of Plants*, into the hands of his brother-in-law, Dr. Henry Sampson, who shewed it to Mr. Henry Oldenburg, at that time Secretary to the Royal Society. By Mr. Oldenburg, it was given to Dr. Wilkins, Bishop of Chester; by whom the manuscript was read to the Royal Society. That learned body highly approved of it, and ordered it to be printed in 1671. At the suggestion of Dr. Wilkins, Dr. Grew was appointed Curator to the Royal Society, for the Anatomy of Plants. This occasioned the drawing up of the 2d, 3d, and 4th parts of his Anatomy of Plants; and of the various lectures on the same subject, which constitute a part of that work. The whole of these papers were written between the years 1670 and 1676; and read at intervals, during various meetings of the Royal Society. The whole were collected in 1682, and published in a folio volume, by the orders of the Society. It constitutes the book well known by the name of *Grew's Anatomy of Plants;* a book which contains a great deal of valuable and important matter; and which has always been in high estimation, and referred to as a classical work on the subject.

Malpighi wrote on the subject. In the year 1671, after the first part of Dr. Grew's labours were printed, Malpighi, the celebrated Italian Anatomist, and a Member of the Royal Society, sent over a manuscript treatise on the the same subject. It was without figures, and much more concise than the essay of Dr. Grew. The second part of Malpighi's treatise was received in 1674; after which, the work was printed by the Royal Society. Thus, Malpighi began his investigation without any previous knowledge of what had been already done by Grew. His work, as might have been expected from so great a master of anatomical knowledge, is excellent; though the plan is more circumscribed, and the quantity of knowledge communicated upon the whole, is less than is to be found in the work of Grew. There is a very remarkable agreement in the sentiments of these two writers respecting most of the subjects which they describe. This agreement is the more striking, because it holds, not merely in their descriptions, which might have been expected, supposing both correct, but likewise in physiological opinions, and deductions from their observations.

Dr. Stephen Hales, one of the most celebrated names of the last century, the Hales. philosopher who laid open the passage to pneumatic chemistry, which has added so much to our knowledge of nature, and who investigated the functions of animals and vegetables with much patient industry and uncommon success, began his experiments on plants soon after the death of Dr. Grew. He was born on the 7th September, 1677, in the county of Kent. He was descended of an ancient family, his father being a Baronet. He was educated at Cambridge, and immediately distinguished himself by his uncommon industry and skill. He went into orders as early as the age of twenty-five, and was successfully placed in various livings, where he discharged his duty with the most conscientious fidelity. His discoveries were all laid before the Royal Society, as soon as made, and his works were published by order of that learned body. The first volume of his *Statical Essays*, which contains almost all his experiments on plants, was published in 1727 ;* the second, in 1733. He modified some of the opinions of Grew and Ray, relating to the supposed circulation of the sap ; and such was the weight of his reputation, that his sentiments seem to have been implicitly adopted by all his contemporaries ; though the recent discoveries of modern experimenters have shown us that his opinions, on the subject of the motion of the sap, were not quite correct. Besides his *Statical Essays*, he was the author of several other celebrated performances ; the most remarkable of which were his *Essay against the Use of Spirits ;* his experiments on *Freshening Sea Water, and preserving Meat during long Sea Voyages ;* and his *Ventilator ;* for purifying the air of ships, and other confined places. He died on the 4th January, 1761, aged nearly 80 years.

Duhamel de Monceau, a French philosopher, of a high and well-merited Duhamel. reputation, took up the subject where Hales laid it down. In his book, entitled, *Physique des Arbres*, first published in 1757, he relates many ingenious experiments, confirming and extending the opinions of Hales. He investigated also the texture of trees ; traced the position of the sap vessels ; and endeavoured to give precision and extension to the doctrines of Grew and Malpighi, on these subjects. He ventures, in some cases, to call in question the accuracy of the conclusions drawn by these first observers, and has advanced new opinions of his own ; all of them extremely ingenious, and supported by well-contrived experiments, some of which have been confirmed by subsequent enquirers ; while the accuracy of others has been called in question, and, in some cases, destroyed.

* Part of it was read to the Royal Society, in 1719, the year after his admission into that learned body.

Hedwig, a German writer, of great authority, but whose reputation depends chiefly upon his discoveries relating to the mosses, and his new arrangement of that difficult order of plants, followed the same career with Duhamel, and with considerable success. But, perhaps, the greatest improvement in the anatomy of plants, since the time of Grew, was made by Gærtner, a German botanist, who devoted his life to the investigation of the *seeds* of plants.

He was born at Calw, in Suabia, on the 12th of March, 1732. His father had been physician to the Duke of Wirtemberg, but died soon after the birth of his son. Young Gærtner was at first destined for the church, and afterwards for the bar, but he displayed a strong repugnance for both of these professions; while he was actuated by a violent passion for natural history and philosophy. At 18 he went to Gottingen, and studied under Haller, who inspired him with an enthusiastic fondness for anatomy, physiology, and botany. At the age of 20, he travelled through Italy, France, and England; and on his return to Germany, took a Doctor's degree in medicine; and afterwards devoted two years to the study of mathematics, optics, and mechanics. In 1759, he went to Holland; attended the botanical course of Van Royen, at Leyden; and devoted himself entirely to natural history. From Holland, he went to England, and spent above a year in London, and the sea coast, chiefly engaged in the study of ichthyology. He returned to Germany, in 1761, and fixed his residence at Tubingen, where he was, soon after, appointed Professor of Anatomy. In 1768, he was elected a Member of the Imperial Academy of Sciences, at Petersburg; and Professor of Botany and Natural History, in that city. But finding that the whole of his time was taken up in the discharge of the duties attached to this situation, he quitted it in 1770, and returned to Calw, his native city, where he married, and resolved to devote the rest of his life to his favourite object, of giving an anatomical account of the fruits and seeds of plants. To obtain a sufficient number of fruits, for his purpose, he found it necessary to apply to the assistance of others, whose situation had given them an opportunity of making considerable collections of these substances. The collection of Sir Joseph Bankes was the richest in the universe. To that distinguished patron of the sciences he applied in the first place, and met with a reception which gratified his utmost wishes. Sir Joseph made him a present of all his duplicates; allowed him to examine, and even to dissect his whole collection; and procured him liberal supplies from the garden of Kew. On his return to Holland, he met with Thunberg, who had just arrived from his celebrated voyage to Japan, and who was richly furnished with new and interesting fruits and seeds. From this distinguished and liberal botanist, he also received an abundant supply. On his arrival at Calw, he was seized with a disease in his eyes, which threatened to frustrate all his plans. For the space of 20 months he was obliged to confine himself in a dark room. At last, the

7

disease went off of its own accord, without the assistance of any medicine. He resumed his labours with the most indefatigable assiduity; and finished the first volume, containing engravings and dissections of 500 species of fruits; all made with his own hand. After a full and deliberate revisal, the volume was published in 1788. His incessant industry had, by this time, destroyed his health; this made him exert himself the more strenuously, that he might finish the second volume before his labours should be arrested by the hand of death. It was sent to the press accordingly, in 1791. It contains a greater number of seeds and fruits than the first volume, and they are arranged in a much better and much more luminous manner. With this volume he intended to terminate his literary career; but the celebrity which he had acquired was such, that new species of seeds came pouring in upon him from all quarters He was unable to resist the temptation which these new treasures presented to him, and resolved to publish a third volume, by way of supplement to his two former. With this view, he dissected and made drawings of many new seeds; and the evening before his death, was occupied in finishing the description and drawing of the seeds of the *Halleria Lucida.* He died on the 14th July, 1791.*

Gærtner dissected the seeds of no fewer than 1050 species of plants; and published the result of his labours in two large quarto volumes, entitled, *De Fructibus et Seminibus Plantarum.* This book, which enjoys a very high and deserved authority among botanists, contains by far the clearest and best account of the structure of the seeds of plants which has ever appeared; it contains also an arrangement of plants, founded upon this structure; an arrangement, however, which had been long before anticipated by Ray. It contains also very good engravings of the 1050 species of seeds, which Gærtner had examined.

These are the most eminent among those philosophers who have turned their attention to the physiology of plants. There are many others, indeed, who have contributed towards this important investigation. Among these, Bonnet and Senebier hold a distinguished place. Mr. Knight has gained a high and deserved reputation, by his numerous papers on this subject; and has added very much to the precision of our knowledge. His experiments are highly ingenious, and often satisfactory. Unfortunately, the most important of his papers have been published in the *Transactions,* since the year 1800; and therefore they do not come under our review. Mirbel is another modern writer, who has gained reputation by his labours on the anatomy of plants. He has examined the structure of some of the smaller plants; for example, the grasses, with more precision than had previously been done; and he has endeavoured

---

* Annales de Mus. d'Hist. Nat. Tom. I. p. 207.

to add to the knowledge which we before possessed, respecting the structure of plants in general.

Chemical
physiologists.Much curious matter respecting the functions of plants has been brought to light by the investigations of those philosophers, to whom we are chiefly indebted for the improvements of pneumatic chemistry. The subject was begun by Scheele and Priestley. Henry, Percival, Senebier, and Ingenhousz, contributed a multitude of facts. Cruikshanks, Gough, Woodhouse, and Hassenfratz, added many curious circumstances; and, of late, Saussure has contributed a considerable mass of experiments; while Mr. Ellis has set himself to rectify what he conceives to be the mistakes with respect to this subject, into which chemical writers have fallen.

We shall now endeavour to lay before our readers as short and perspicuous a view of the whole subject as possible; and we shall notice, as we proceed, such of the papers in the Philosophical Transactions, as seem to deserve particular attention.

1. Plants, as far, at least, as our observation goes, can be produced only two ways, namely, by *buds,* and by *seeds.*

Plants propa-
gated by buds;The *bud* is the simplest mode, as it does not require the intervention of sexual organs. Buds* always produce plants exactly similar to their parents. There are some plants which, as far as we know, are propagated only by buds. These plants are considered as imperfect. There are numerous species of them, and they all grow with very great rapidity. The fungi, lichens, confervæ, and ulvæ, are plants of this kind. They produce *buds* in abundance, but no seeds. We are not ignorant of the attempts made by Hedwig, and various other learned and laborious botanists, not only to point out the seeds of all of these tribes of plants, but even to demonstrate the presence of sexual organs in them. But the irreconcileable discordancy between the opinions of these various philosophers, and the great uncertainty which every one of them acknowledges, destroy all confidence in their opinions. Besides, the analogy of nature is favourable to the notion that there are tribes of vegetables propagated only by *buds.* For we know that various animals are propagated only that way, and why should not the same law extend to the vegetable kingdom also.

and seeds.But the greater number of plants furnish seeds, from which, when sown in a proper soil, plants similar to the parents are produced. Now, plants pro-
Sexual organs
of plants.ducing seeds have sexual organs; consisting of male and female. These organs constitute the essential part of the flower, which always precedes the perfect seed The male organs consist of three parts; namely, the *filaments,* the *antheræ,* and the *pollen.* The filaments are certain threads to which the

* Buds are of various kinds. For a particular description we refer to Gærtner *De Fructibus et Seminibus Plantarum.* Vol. I. p. 3.

antheræ are fixed.   They are not essential; for, in some flowers, they cannot be observed.   But the antheræ and pollen are essential organs, and therefore never absent.   The antheræ are oblong bags, usually of a yellow colour; placed on the summit of the filaments.   At the time of fecundation, they become covered with a light dust, called the *pollen;* which performs the office of fecundating the female organs by falling on them.   The female organs are four in number; namely, the *stigma,* the *style,* the *ovarium,* and the *seeds.*   The stigma is nothing else than the summit of the style upon which the pollen is deposited. The style is a pillar terminating in the *ovarium,* or vessel in which the seeds are afterwards found.

How the fecundation is performed, we remain entirely ignorant.   We fear Fecundation. the process is inscrutable, and altogether beyond our faculties.   We know only that the style is not sensibly hollow, even when examined by the most powerful microscopes.   The pollen therefore cannot, as some have supposed, make its way through the style to the ovarium   Nothing can pass that way, except in a liquid form, and in quantities so minute as to be insensible to our organs.* Indeed there are some antheræ which, instead of pollen, exude a liquid; a fact which confirms this position.   Fecundation seems to be effected by the union of something from the male and female organs; neither of which, alone, is capable of constituting a complete seed: but what it is which unites, and how the union is brought about, we are entirely ignorant.

Some plants seem to have only the female organs; or, to speak more correctly, the male and female organs are united together in the same part of the plant, so that no distinct male organs can be found.   Such plants do not produce perfect flowers: indeed, in most of them, no flowers whatever can be observed; but only seeds.   Fuci, musci, and filices, are plants of this kind. They have all perfect seeds; but we cannot detect any thing similar to a flower.

The greater number of plants are furnished with both male and female organs   Sometimes, most frequently indeed, both of these exist together in the same flower.   Sometimes the male organs are confined to one flower, and the female to another; these two kinds of flowers often grow together on the same plant; but not unfrequently male flowers only are found upon one plant, and female flowers upon another.   The plant with female flowers alone pro-

---

* There are two papers in the Philosophical Transactions, which attempt to prove that the pollen, or at least grains which it contains, actually pass through the stigma, to the ovarium.   The first, written in 1703, by Mr. Morland, (Phil. Trans. XXIII. 1474;) the second, published in 1765, by Sir F. H. Eyles Stiles, (Ibid. LV. 261.)   This last is an elaborate dissection, accompanied by numerous engravings.   But succeeding observers have been unable to observe the passages described and figured by the author.   He was probably misled by his microscope.

H

produces seeds ; and the seeds are capable of vegetating, only when the female flowers have been impregnated by pollen from the male flowers.

Discovery of the sexes of plants.

This curious and important fact of the sexual organs of plants, and the manner in which impregnation is accomplished, was totally unknown to the ancients. They talk indeed frequently of male and female trees and plants ; but it was some difference between the size of the plants, the smoothness of the leaves, or the shape of the seeds, and no knowledge of the male and female organs, that induced them to impose these denominations. The person who first suggested the real use of the sexual organs, and to whom, of course, the discovery is due, was Sir Thomas Millington, Savilian Professor at Oxford, soon after the establishment of the Royal Society. He mentioned his opinion to Grew, who embraced it, and detailed, at some length, his reasons for adopting it.* In 1681, Bobart, keeper of the botanic garden at Oxford, who probably was acquainted with the opinion of Sir Thomas Millington, made some experiments on the *lichnis dioica*, and convinced himself that the antheræ are necessary for obtaining perfect seeds.† Ray, soon after, adopted the same opinion,‡ and even defended it at considerable length, against the objections of his contemporaries.§ One of the principal opposers of this opinion, when it was first propagated, was Leeuwenhoek, who acquired a very great celebrity by his microscopical observations ; a celebrity long ago entirely vanished. He endeavoured to prove, that the plant existed ready formed in the seed ; an opinion, at that time, pretty generally embraced ; but long since laid aside as absurd and utterly improbable. Reaumur endeavoured to extend the sexual system to the fuci ;‖ but his opinion was refuted by Gmelin. Bradley ascertained the effect of impregnating the stigmata with various kinds of pollen ; and showed that, by this means, hybrids might be obtained ; the taste of fruit changed ; and various other curious effects produced.¶ The same topics were handled, with much success, in a paper by Dr. Blair, published in the Philosophical Transactions, for 1721.** He refutes the opinion advanced by Morland, respecting the passage of the pollen through the style to the embryo, and gives several examples of superfœtation, or of a seed partaking of the properties of three distinct plants, in consequence of the female flower being at once impregnated by two different species of pollen. This subject has been very lately resumed by Mr. Knight,†† the most distinguished of modern physiologists. But he does not seem to have been aware of what had been done on the same

---

* Grew's Anatomy of Plants, p. 171.   † Blair's Botanical Essays, p. 243.

‡ Hist. Plant. I. 17. This book was published in 1686.   § See his preface *ad Syl. Exot.*

‖ Mem. Paris, 1711, p. 371 ; and 1712, p. 26.

¶ A New Improvement of Planting and Gardening. London, 1717.   ** Vol. XXXI. p. 216.

†† Phil. Trans. 1799. Vol. LXXXIX. p. 195.

subject by his predecessor Blair. He succeeded in obtaining several new varieties of peas by cutting off the antheræ from the flowers of certain plants before impregnation, and then introducing the pollen of another kind of pea into the blossoms. He obtained also unequivocal examples of seeds partaking of the qualities of three different plants at the same time. The female flower having been at once impregnated with the pollen of two distinct male flowers.

Geoffroy had published experiments on the Zea Mays in the Memoirs of the French Academy, from which he concluded that even when all the male flowers of this monœcious plant were removed before impregnation, still the seeds continued to grow till they reached their full size. But when planted in the ground they refused to germinate, and of course were imperfect. These assertions induced Mr. Logan to repeat the experiments of Geoffroy with every possible precaution. The result was published in 1736.* When the impregnation was entirely prevented, the seeds in every case withered away without coming to maturity.

Linnæus founded his arrangement of plants upon the sexual system. This arrangement was made fully known to the public in 1736, and after a feeble opposition from some captious botanists it was universally adopted, and with it the opinions respecting the uses of the male and female organs of plants.†

II After the impregnation has taken place, the *ovarium*, or seed vessel, gradually increases in bulk, and the seeds by slow degrees reach maturity. Now a vegetable seed, not reckoning the integuments, which are usually two, consists of the following parts: 1. The *albumen*, so called first by Grew from the resemblance which it bears to the white of an egg. The name was afterwards adopted by Gærtner, and is now universally used by writers on vegetable physiology. The albumen is originally a glary liquid, but becomes white and solid when the seed reaches maturity. It commonly surrounds the other ingredients of the seed, and in many cases, in tne seeds of wheat (*triticum*) for example, it constitutes almost the whole bulk of the seed. This organ seems destined to supply the embryo with food. During a certain period of the growth of the seed it can in most cases be detected, but not unfrequently no traces of it can be observed in the mature seed. Of 1,050 kinds of seeds described by Gærtner, 418 want the albumen, 600 have it, and 32 are doubtful. The seeds of grasses have usually a large albumen, while those of papilionaceous plants, as peas,

Marginal note: Structure of seeds.

---

* Phil. Trans. XXXIX. 192.

† There is a very curious paper in the Philosophical Transactions connected with the seeds of plants, and therefore deserving of mention here. It is by Mr. Barrel, and is printed in the Phil. Trans. 1727. Vol. XXXIV. p. 215. The subject is the growth of the parasytical plant called the *misseltoe (viscum album.)* The seed sticks by a clammy covering to the bark of some tree. The root issues first, proceeds some way, then forms an arch, and adheres by its extremity to the bark. The other extremity now lets go its hold, and gradually becomes the stem of the plant.

beans, &c. are most commonly destitute of it. 2. The *vitellus,* so called by Gærtner from some supposed analogy with the yoke of an egg. It is a solid body placed between the albumen and the embryo, and usually connected with both. It is constantly small, is not always to be seen, and its use has not been very clearly made out. Probably, like the albumen, it is connected with the nourishment of the embryo during its subsequent developement.* 3. The *embryo,* which consists of three parts, namely the *cotyledons,* the *radicle,* and the *plumula.* It is the most important part of the seed, and is never wanting.

The cotyledons vary in number. In some seeds you cannot perceive them at all. But all such seeds are exceedingly small, and on that account their form eludes our examination, even when we are assisted by the best microscopes. Plants yielding such seeds are called *acotyledinous.* The filices, musci, and fuci, belong to this class of plants. Some seeds have only one cotyledon, most seeds have two, and some have three, six, or ten. Of 1,050 genera of seeds described by Gærtner, six are without cotyledons, 123 are monocotyledonous, 920 dicotyledonous, and five or six polycotyledonous. The polycotyledonous genera are pinus, rhizophora, canarium, lepidium, cupressus? and hernandia?† The number of cotyledons in the seeds of plants constitutes the foundation of a very beautiful and useful arrangement of plants by Jussieu, an arrangement indispensable to the general botanist. It is in fact an elaborate attempt at a natural classification of plants; and though it may be considered as only in its infancy, it possesses unquestionable merit and utility. Gærtner has also given us an arrangement of plants according to the cotyledons of the seeds. The cotyledons are usually converted into seed leaves when the seed germinates. In many cases, as in the bean, they give the seed its shape, and constitute almost the whole of its bulk.

In bicotyledonous plants the radicle projects between the two cotyledons. It is a small knob just visible to the naked eye. It is always present in every seed, and is converted by germination into the root of the young plants. The *plumula,* which constitutes the future stem, is also a small knob, and in bicotyledonous seeds it lies concealed between the two cotyledons, and can only be distinguished when they are separated from each other. In many seeds it is not possible to observe the plumula previous to the commencement of germination.

---

* Dr. Smith, in an essay on the Vitellus, published in the 9th volume of the Linnæan Transactions, p. 204, endeavours to prove that the vitellus is only a part of the cotyledons, and ought not to be listinguished from them. Perhaps Gærtner's description of the vitellus is not very accurate, but that in various seeds there is a substance distinct from the albumen, and unconnected with the cotyledons we believe to be indisputable. This substance Gærtner no doubt would have called the vitellus. As to its use, that is another and a very difficult question.

† Mr. Salisbury has shown in a paper, published in the Linnæan Transactions, that the natural order of plants called *orchideæ* have no cotyledons at all.

The number of seeds which plants produce annually is quite enormous, and Number of seeds. can only be accounted for by considering it as an example of the wonderful care taken by the Author of Nature that no species should be lost. To give an example in a plant by no means comparable either in size or fertility to several others that might be mentioned—Mr. Hobson found that a thriving plant of the common mallow (malva sylvatica) yielded in one summer no fewer than 200,000 seeds.*

It is commonly believed that if seeds be kept even for a small number of years they lose the power of germinating altogether. But there is a great diversity between different seeds in this respect. Some, as the acorn, seem to lose their power of germinating almost immediately, while others may be preserved for almost any length of time without injury, provided they be kept in a dry place, or totally excluded from the action of the air. Mr. Ellis pointed out a successful method of bringing seeds without injury from warm climates to England. It consisted in simply surrounding the seed with a covering of tallow or wax. Secretary Hæreus, of Stockholm, found that melon-seeds kept for 41 years grew remarkably well when sown.† In Zurich, in Switzerland, they were formerly in the habit of keeping corn in granaries for 80 years.‡ Some seeds taken from Tournefort's collection, and of course at least 100 years old, were lately sown at Paris, and produced plants as readily as if they had been quite recent. But the most extraordinary example of the length of time that some seeds retain their vegetating power was exhibited in London after the fire which laid waste so great a part of that city in 1666. Several plants of *sinapis arvensis* and *trifolium repens*,§ speedily grew up in the site of several of the demolished houses. Mr. Ray examined the soil, and found that it contained the seeds of these plants. These seeds must have lain buried under the houses of London for several hundred years without losing their faculty of vegetating.

The most curious paper on seeds, in the Philosophical Transactions, is an account of the seeds of fern by Mr. Miles.‖ It is in fact rather a description of the curious mechanism of the seed vessels, than the seed of ferns which are too small to be examined. This mechanism was discovered by Dr Hooke as early as 1669. Swammerdam describes it with great accuracy, and exhibited it to the Botanical Professor of Leyden as early as 1673. The account given by Grew of this mechanism in his Anatomy of Plants is very inaccurate.

* Phil. Trans. 1743. Vol. XLII. p. 320.
† Phil. Trans. Vol. XLII. p. 115.
‡ Phil. Trans. 1667. Vol. II. p. 464.
§ Stating this from recollection (for I cannot at present remember where I met with the anecdote) it is possible I may not have given the proper species. The plants, as far as I remember, were *wild mustard* and *white clover*.
‖ Phil. Trans. Vol. XLI. p. 770.

Structure of
plants.

III. The opinion of physiologists respecting the ultimate structure of plants is divided. Grew and Malpighi considered them as composed of extremely minute fibres interlacing each other in all directions. This opinion was embraced by Duhamel, by Hedwig, and by almost all the German physiologists. Mirbel, on the other hand, has advanced that the ultimate structure of plants is vesicular and similar to the texture of a honeycomb The walls which constitute the cells are transparent, and two contiguous cells have only one wall between them. Though these two opinions have been each sustained with great obstinacy, and though they have lately occasioned a violent controversy between Mirbel and the German Physiologists Sprængel, Link, Rudolphi, Bernhardi, and Treviranus,* they do not appear to us to differ so much from one another as the opposite parties suppose; and if Mirbel will compare some of the figures given by Duhamel with his own, he will find the difference not quite so great as he at present supposes. We fear that a discussion of this kind is not likely to advance our knowledge of the vegetable kingdom. The ultimate structure both of animals, plants, and minerals, is beyond the scrutiny of our imperfect senses. To search for it is to run after a phantom, and to subject ourselves to the infallible certainty of being imposed upon by optical and other deceptions, which it is not in the power even of the most patient and cautious man to guard against.

The pith.

Trees, and indeed all plants, are composed essentially of three distinct parts, the *pith*, the *wood*, and the *bark*. The pith is placed in the centre, and is at first succulent, but dries up, and apparently becomes useless as the plant advances in age. It is composed of vesicles as has been fully ascertained by Grew and Malpighi, and by all succeeding anatomists.

The wood.

The wood surrounds the pith. It consists of two distinct substances. The internal cylinder is harder and more compact, and is called the *perfect wood*, the external cylinder is of a softer and looser texture, and is called the *alburnum*. Each of them is composed of concentric circles, one of which is formed every year. It is always the outermost of these circles that is formed last, and the circles of the alburnum gradually change into perfect wood. In the bicotyledonous trees, there are horizontal rays which proceed from the pith to the circumference of the stem, and these rays are composed of the same materials with the pith. Hence they have been called medullary rays. By the carpenters of this country these rays are distinguished by the name of the *silver grain* of the wood. In monocotyledonous trees these medullary rays are not to be observed. It is remarkable that these trees do not send out lateral branches, and that when their top is cut off they uniformly die.

* The reader who wishes to make himself acquainted with this dispute, and with the opinions of Mirbel, respecting the structure of plants, may consult the *Exposition de la Theorie de l'Organisation Vegetale*, published by Mirbel in 1809, in answer to the papers of the German physiologists who gained the prize proposed by the Royal Society of Gottingen in 1804.

The bark constitutes the external covering of the stem.   It consists of three The bark. parts, namely, the *epidermis,* or outer skin, which may be rubbed off and is renewed again; many monocotyledonous plants have no distinct *epidermis.* The *parenchyma,* or green pulpy matter which lies within the epidermis.   It constitutes a considerable part of the bark; is composed of vesicles, and seems to be an organ of great importance in the vegetable economy.   The *cortex liber,* or *true bark;* which lies innermost, and is composed of a number of layers like the leaf of a book.   The most generally received opinion is, that these layers are gradually converted into wood; while new layers are formed in their place.   The experiments of Duhamel, successfully repeated by Mirbel, were considered as demonstrating the truth of this opinion, but Mr. Knight has lately started some doubts on the subject.

Vegetables, like animals, are furnished with a variety of vessels destined to Vessels of plants. convey their various liquids to the proper places, and to contribute to the development of the various parts of the plant.   Great attention has been paid to these organs by anatomists; because the knowledge of their structure and distribution was conceived to be the best means of throwing light upon the œconomy of the vegetable kingdom.   They were first discovered by Grew and Malpighi; some additional knowledge was gained by the observations of Duhamel and Hedwig; but Mirbel is the physiologist who has studied the subject with the greatest industry.   The vessels of vegetables, as far as we are acquainted with them, are of five different kinds.   1. *Tracheæ.* They are the largest vessels in plants, and may often be discovered by the naked eye.   They were first observed by Grew and Malpighi, who gave them the name of tracheæ, because they found them full of air, and thought them air vessels.   Their structure is very remarkable; they are composed of a fibre, twisted round like a cork-screw, so that by laying hold of the extremities of this fibre, it may be drawn out to a considerable length.   Hedwig conceived that this fibre was a hollow tube, and that it conve ed sap; but his opinion has not been confirmed by the observations of succeeding physiologists.   The tracheæ do not seem to be attached to the vegetable in which they are situated, except by their extremities.   They are never found in the pith or bark, but only in the wood; and, according to Mirbel, chiefly in a zone immediately surrounding the pith. 2. *False tracheæ.* These vessels had been previously observed by others, but the attention of physiologists was first particularly called to them by Mirbel, who assigned them the name by which they are distinguished.   They resemble tracheæ, in appearance, but are not composed of a fibre, twisted like a cork-screw, and cannot therefore be drawn out in length by laying hold of their extremities.   According to Mirbel, they are full of horizontal slits, each of which is surrounded by a glandular protuberance.   Others deny the existence of these slits, though the protuberance is obvious enough.   The fact is, that the slits, if they exist, are so small, that it is very difficult to perceive them, and still more difficult to avoid being misled by optical deceptions. These false tracheæ, like the true, are found only in the wood. 3. *Porous tubes.* These are vessels which are in the form of long tubes, and have no resemblance to the tracheæ. According to Mirbel, to whom we owe the most exact account of them, they are full of pores, usually placed in horizontal rows, and each pore is surrounded by a glandular protuberance. These porous tubes, like the two last sets of vessels, are confined to the wood. According to Mirbel, and his opinion is very likely to be accurate, those three vessels are all destined for the same purpose, and often run into one another; so that if we trace a vessel for some length, we find one part of it a tracheæ, another a false tracheæ, and a third a porous tube. 4. *Necklace tubes, (vaisseaux en chapelet.)* These vessels are first mentioned by Mirbel. According to him, they consist of a number of vesicles attached to each other by their extremities, and thus constituting a long tube; having a diaphragm perforated with numerous pores at the junction of each pair of cells; at which place, also, the diameter of the tube is somewhat contracted. These tubes are found in similar situations with the preceding, and doubtless answer the same purpose. 5. *Proper vessels.* These are found in the bark; and sometimes, according to Mirbel, in the pith, but never in the wood. Hedwig endeavoured to prove, that the structure of these vessels was similar to that of the tracheæ; but his opinions have been refuted by Knight and Mirbel. They are simple tubes, without any pores, sometimes single and large, sometimes numerous, small, and grouped together. They contain the peculiar juice. Mirbel supposes them rather reservoirs than proper vessels; but the rapidity with which the whole of their contents are discharged when they are wounded or cut, is rather hostile to that notion.

The leaves.   The composition of the roots, the branches, and even the foot-stalks of the leaves, seems to be similar to that of the stem. The leaves themselves are composed of abundance of vessels, diverging from the end of the foot-stalk, which gives the leaf its shape. These vessels are double, lying one over the other, as was first shown by Nicholls;[*] who found that the skeletons of leaves are always double. The rest of the leaf consists of the parenchyma of the bark, covered with the epidermis. Neither the cortical layers, nor the wood, can be found in the leaves.

Food of plants.   IV. As vegetables are continually increasing in bulk, and forming new parts, it is obvious that they require food as well as animals; and that this food must be digested and prepared, before it can be assimilated to the vegetable organs. As plants have their roots fixed in the earth, and their branches in

---

* Phil. Trans. Vol. XXXVI. p. 371.

the atmosphere, and as they will not vegetate well when deprived of either of these elements, it is obvious that both must contribute, somehow, to their nourishment. The effect of manure, in promoting the vegetation of plants, and the well-known fact that soils are speedily exhausted if they be made to bear crops, without being supplied with manure, is a proof that something else besides mere air and earth contributes to the food of vegetables. A very great number of experiments and observations have been made to elucidate this subject; but the success of philosophers has not corresponded with their industry. Indeed the subject is as difficult as it is important, and much laborious investigation is still wanting before we can establish a satisfactory theory. The following may be considered as a pretty accurate summary of the present state of our knowledge.

Plants will not vegetate without a supply of water; but pure water alone is not sufficient to produce a healthy and perfect vegetation. It is requisite that the water should hold in solution a certain substance, which exists in abundance in decayed animal and vegetable matter. This substance is similar, in many of its properties, to vegetable *extractive*. Water, in order to answer well the purposes of vegetation, ought not to contain above a certain proportion of this matter; not more than five per cent. at most seems requisite. Accordingly, too much manure is found just as hurtful to vegetation as too little. Particular salts seem useful to the vegetation of certain plants, probably acting as stimulants. Thus, gypsum promotes the vegetation of clover; nitre, of nettles and sun-flower; nitrate of soda, of barley; common salt, of the salsolas, and other plants, which vegetate upon the sea coast.

The food of plants is absorbed by the extremities of the roots. As these extremities are very small, and do not appear perforated even when examined by the best microscopes, it is obvious that this food can enter only in a liquid state; it must therefore be previously in solution by water. From the extremities of the roots, this liquid matter is conveyed to the very summit of the plant. It has been ascertained that it passes entirely through the wood; and, according to Knight, chiefly through the alburnum; and that none of it is to be found either in the bark or pith. It passes up through the tracheæ, false tracheæ, and porous tubes; and is known by the name of *sap*. It flows, in Ascent of the the greatest quantity, in spring, just before the leaves expand. At that period, sap. if the extremity of a branch of several plants, as the vine, be cut off, or if the stem be perforated as far as the perfect wood, as in the birch or the maple, the sap flows out with the more or less velocity, and the plant is then said to bleed. From the experiments of Hales, we learn that the sap ascends in plants with a very considerable force, a force which he found capable of balancing a column of mercury 38 inches high:* even when it has proceeded to the

---

* Hales' Vegetable Statics. Vol. I. p. 114.

distance of 44 feet from the root, he still found the sap to balance a column of 31 feet of water.*

Matter thrown off by transpiration. This sap passes on to the surface of the leaves of the plant, where a considerable portion of it is exhaled, as was first shewn by the experiments of Dr. Woodward, and afterwards by those of Dr. Hales. Dr. Woodward's experiments have acquired great celebrity, and are constantly referred to by all writers on vegetable physiology. They consisted in putting sprigs of vegetables into the mouths of phials, filled with water, allowing them to vegetate for some time, and then determining the quantity of water which they have imbibed, and the quantity of weight which they have gained. The difference obviously indicates the quantity of moisture exhaled by the plant. He found that plants acquired the least increase of weight, when they grew in pure water; and the greatest when they grew in water mixed with garden mould. The following table exhibits a view of Dr. Woodward's experiments.†

Phial.	Kinds of plants and water.	Weight of plants.		Gained in 77 days.	Expence of water.	Proportion of increase of plants to expence of water
		When put in.	When taken out.			
A	Common spear mint, set in spring water.	gr. 27	gr. 42	gr. 15	gr. 2558	1:170$\frac{8}{13}$
B	Ditto, set in rain water............	28$\frac{1}{4}$	45$\frac{1}{4}$	17$\frac{1}{2}$	3004	1:171$\frac{23}{35}$
C	Ditto, in Thames water............	28	54	26	2493	1: 95$\frac{33}{26}$
D	Common night shade, in spring water...	49	106	57	3708	1: 65$\frac{3}{7}$
E	Lathyris, in spring water............	98	101$\frac{1}{2}$	3$\frac{1}{2}$	2501	1:714$\frac{4}{7}$
	Plants, all spear mint.			Gained in 56 days.		
H	Hyde Park conduit water............	127	255	128	14190	
I	The same......................	110	249	139	13140	
K	The same, mixed with an ounce and half of garden mould............	76	244	168	10731	
L	Ditto, with ditto	92	376	284	14950	
M	Ditto, distilled..................	114	155	41	8803	
N	Residue of ditto, left in the still.......	81	175	94	4344	

Dr. Hales' experiments were of a similar nature. He found that a cabbage transmitted daily a quantity of moisture, equal to about half its weight; and

* Hales' Vegetable Statics. Vol. I. p. 117.   † Phil. Trans. 1699. Vol. XXI. p. 193.

that a sun-flower, three feet high, transmitted in a day 1lb. 14 oz., avoirdupois, of moisture.*　He found that the transpiration was nearly confined to the day, very little taking place during the night ;† and that it was much promoted by heat, and stopped by rain and frost.‡　He showed that the quantity of transpiration, in the same plant, is proportional to the surface of the leaves, and that when the leaves are taken off, the transpiration nearly ceases.§　These conclusions are confirmed by the subsequent observations of Millar, Guettard, and Senebier.

Thus it appears that a liquid is absorbed by the roots of plants, that it ascends through the wood in proper vessels to the leaves, where a considerable portion of it is exhaled ; that it ascends with a very considerable force ; and that the ascent is somehow connected with the leaves, soon ceasing when they are removed.　Now it becomes a problem, of considerable importance, to ascertain the power which occasions this motion.　It obviously depends upon the life of the plant, because, when the plant dies, the ascent of the sap entirely ceases ; but in what way does the living plant exert a force which causes the sap to ascend ?　We are afraid that no very satisfactory answer can be given to this question.　Most philosophers have endeavoured to account for the rise upon mechanical principles ; but not one of the solutions proposed, even the most recent, can be admitted.　Grew accounted for the rise by the *levity* of the sap ; an opinion too absurd to merit an examination.　Malpighi supposed (as is undoubtedly the case) that plants contained in their vesicles abundance of air, and that the sap was forced up by the dilation of this air, occasioned by heat.　If the sap vessels were air tight, and furnished with valves, this solution would possess considerable plausibility ; but as neither of these states seems to exist, it is plain that no dilatation of the air could force the sap upwards. Capillary attraction has been a favourite mode of resolving the difficulty ; but if this were the cause of the ascent of the sap, it is plain that trees could never bleed ; because the attraction of the uppermost film of the vessels for the sap, being as great as that of the lowest, would just balance the attraction, and prevent the sap from leaving the vessels at all.　Mirbel accounts for the ascent of the sap by the *suction* of the leaves and bark ; but this explanation is just as inadmissible as the others.　*Suction* means nothing more than the production of a vacuum, by withdrawing the air or liquid sucked.　Suppose such a vacuum produced in the leaves, which could only happen if they were hard and air tight, neither of which properties ever belongs to them ; the consequence would be, that the external air, acting upon the sap in the vessels, would force it up into the leaves.　Hence it is obvious, if suction be the cause

Marginal note: Why does the sap ascend?

---

* Vegetable Statics.　Vol. I. p. 5, 15.　　　† Ibid.
‡ Ibid. p. 27, 48.　　　　　　　　　§ Ibid. p. 30.

of the ascent of the sap, that the force of that ascent can never be greater than the atmospherical pressure.   But Hales found the force of the sap at least one fifth part greater than the pressure of the atmosphere; hence we are quite certain that the cause of the ascent of the sap is not suction.   Besides all this, it is obvious that Mirbel's account of the structure of the sap vessels, which he says are full of pores, and freely communicate, almost every where, with the cells of the wood, is quite incompatible with the rise of sap by suction; unless he be prepared to affirm that air cannot penetrate by the external bark, and that the suction is applied, not to the sap vessels alone, but to the whole wood. Now these positions are obviously ill founded.   Mr. Knight seems inclined to conceive the sap vessels as supplied with valves, and to account for the ascent of the sap by the expansion of the included air.   But even if we were to admit the presence of valves, the contrary of which seems to be clearly demonstrated, still the ascent of the sap would be unaccountable that way, if the sap vessels be full of pores, as Mirbel has endeavoured to shew, and as Knight himself seems disposed to admit.   Even supposing the vessels not porous and perfectly air tight, still the expansion of air would not occasion the sap to ascend with a force balancing 38 inches of mercury; unless the air were to expand so much that its bulk were doubled, and one-fifth more.   Now this would require a temperature equivalent at least to 608° of Fahrenheit's thermometer: a degree of heat which certainly never takes place, and which would destroy vegetation altogether.   This supposition then, even supposing we concede to Mr. Knight every thing which he thinks proper to demand, is absolutely inadmissible.   The ascent of the sap, indeed, might be accounted for, if we were to suppose the vessels furnished with valves, destitute of pores, and an indefinite quantity of elastic fluid to be generated within the plant.   But these suppositions we allow to be contrary to the fact, and therefore inadmissible.

The only remaining opinion is that of Saussure, senior, who explained the ascent of the sap by the contraction of the vessels.   We confess this appears to us the most likely mode of explaining the phenomenon; though we do not see how the anatomical structure of the vessels of plants can be reconciled with it: for the texture of the sap vessels, and their connexion with the neighbouring parts of the wood, are such, that we cannot very well see how contraction is possible; and even if we were to concede the possibility of contracting, still if the sap vessels be full of pores, and if they communicate freely with the cellular texture of the wood, the contraction of the vessels would not occasion the Cannot be ex- ascent, but the extravasation of the sap.   Upon the whole, the safest concluplained. sion that we can draw, in the present state of our knowledge, is, that the ascent of the sap is owing to some power exerted by the living vegetable; but what that power is we are still entirely ignorant.   Our knowledge of the ascent of the sap in vegetables is, at present, nearly in the same state as that of anato-

3

mists with respect to the animal fluids, before Harvey's great discovery of the circulation of the blood.

From the experiments of Hales, it appears that there is a free lateral communication of the sap between the parts of plants; so that a tree may be nourished by absorbing moisture by the extremity of a branch. A tree will even live though its roots be taken out of the earth, provided its branches inosculate with trees upon each side of it. From the experiments of Bonnet, it appears, that even the under surface of the leaves is capable of absorbing enough of moisture to keep a whole branch alive a considerable time. From the experiments of Duhamel, it follows, that the sap moves equally well either from the roots to the summits of the branches, or, in the contrary direction, from the summits of the branches to the extremities of the roots. But Knight has lately shown, that the conclusions drawn from these experiments require to be somewhat modified; for he found a marked superiority in the facility with which the sap flows in its natural direction.

Hales and Senebier contrived to collect portions of the liquid, which exhales from the leaves of plants, by inclosing a living branch within a large glass vessel. It possessed nearly the properties of pure water; but contained in it some vegetable matter: for, when kept for some time, it acquired a putrid smell. Senebier found in the liquid, collected in this manner from the vine, traces of carbonate, and sulphate of lime, with some mucilaginous and resinous matter.

This enormous quantity of matter, almost purely aqueous, transpired by the leaves of plants, must make a very considerable difference in the nature of the resident sap retained by these organs, or at least in the proportion of matter held in solution in it. Other changes go on in the leaves at the same time; for these organs are the instruments employed by plants to digest their food, and convert it into the various liquids requisite for the purposes of vegetation. They are, in some measure, equivalent to the lungs and stomach of animals. In what way the digestion is performed we have no distinct notion: but it has been ascertained, that during the day the leaves of plants absorb carbonic acid, decompose it, emit part of the oxygen, in the form of gas, retain the carbon, and a portion of the oxygen, and emit also a portion of azotic gas, in such a quantity that the bulk of the atmosphere surrounding them remains nearly unaltered. The absorption of carbonic acid, and the emission of oxygen gas, were first observed by Priestley. Both were confirmed by the subsequent experiments of Percival, Henry, and Ingenhousz. Senebier ascertained that the emission of oxygen was the consequence of the absorption of carbonic acid; and Saussure, junior, determined by experiment that the proportion of oxygen gas, in the atmosphere surrounding a plant vegetating in air containing carbonic acid and exposed to the light, was increased, while that of the carbonic acid was diminished; that the whole oxygen in the carbonic acid gas absorbed

Functions of the leaves.

was not to be found in the surrounding atmosphere, and that the deficiency was made up by a quantity of azotic gas. There seems to be some reason to distrust the correctness of Saussure's conclusions respecting the azotic gas, emitted by plants. In all likelihood he mistook for azotic gas some of the numerous combustible gases which are formed of hydrogen and carbon.

It appears also, from the experiments of Ingenhousz and Saussure, junior, that plants have the property of absorbing oxygen gas, and that this absorption takes place chiefly in the night. Sometimes the oxygen gas is barely absorbed; but in other cases, it is either partly or wholly converted into carbonic acid gas. Fleshy leaved plants absorb the least oxygen gas. Plants growing in marshy situations likewise absorb little. Evergreen plants absorb rather more; but the most of all is absorbed by those plants which lose their leaves during the winter. The greatest quantity of absorption, observed by Saussure, amounted to eight times the bulk of the plant; and the least, to rather more than half the bulk of the plant, in a night's time.

It appears also, from the experiments of Bonnet, that the leaves of plants have the property of absorbing moisture during the night. In trees, this absorption is performed by the under surface of the leaf, while the transpiration is confined to the upper surface.

These are all the changes upon the atmosphere surrounding plants which have been detected. It is probable that plants, by this means, absorb a considerable portion of the carbon which they contain. Hence probably the reason why the proportion of carbonic acid gas in the atmosphere never increases; notwithstanding the immense quantity which is constantly thrown into it by the respiration of animals, and the processes of combustion. We cannot account for the double process of absorbing and emitting carbonic acid and oxygen. Some philosophers, on that account solely, have refused to admit the reality of those processes, and endeavoured to prove that the changes produced upon air by plants, are precisely similar to those produced by animals; that is to say, consist entirely in converting a portion of the surrounding oxygen into carbonic acid gas. But the experiments of Senebier, Ingenhousz, and Saussure, junior, are too exact to be rejected, at least without repetition. We are rather surprised, that no British physiologist has thought it worth his while to subject the whole doctrine to a rigid experimental investigation. Much curious information would be the infallible consequence, of great importance to the progress of vegetable physiology, and which might also throw considerable light on some of the processes of agriculture.

By these and other unknown processes which take place in the leaves, the sap is converted into that state which fits it for the purposes of vegetation. From the leaves in this new state it is sent back to all the parts of the plant by means of vessels situated in the bark, and which proceed only towards the root and never in a contrary direction. From the observations of Mr. Knight it would not be surprizing if these vessels were furnished with a kind of imperfect valves He has shown at least that they are much better fitted to carry the *true* True sap. *sap* (as the liquid which they contain has been called) towards the root than in the contrary direction. Mirbel is inclined to consider these vessels rather as reservoirs than as true vessels, and he conceives that the liquid which they contain is not intended for the nourishment of the plant but for some other purpose. The nourishing sap he considers as the liquid which appears in spring between the wood and the bark, and to which, after Grew, he has given the name of *cambium*. We think the opinion of Grew on this subject, and which appears to have been adopted by Duhamel, to be more probable. Grew considers the cambium as the first rudiments of the new layer of wood, and as consisting of a congeries of vessels so extremely fine that the least violence offered to them reduces them to the state of a liquid pulp. Mr. Knight has rendered it very probable that a portion of the true sap is lodged towards the end of the year in the alburnum, to serve as food for the developement of the leaves against the next season. It is not unlikely that a portion of this deposited sap may be destined likewise for forming the new layer of wood, and may therefore be in reality the source of the cambium, which only makes its appearance at a particular season.

V. The motion of the fluids in vegetables, though not similar to the circulation of the blood in animals, is no less real, and the force of the ascending sap has been shown by Dr. Hales considerably to exceed the force of the circulation of the blood in a horse. As this motion is confined to the living vegetable, and ceases entirely after death, it is difficult to avoid ascribing it to the same cause which occasions the motion of blood in animals, we mean *irritability*. By irri- Irritability. tability is meant the property of contracting when certain stimulants are applied. Now there are not wanting proofs that vegetables possess irritability as well as animals. We shall mention one, because it is contained in the Philosophical Transactions, though it would not be difficult to produce others if requisite. Dr. Smith found that when the inside of the stamina of the berberis communis, or common barberry bush, are touched ever so slightly, they immediately start inwards till the antheræ strike against the stigma. But no motion is produced by touching any other part of the stamen or even the antheræ.*

SECT. III.—*Of Agriculture and the Œconomical Uses of Plants.*

This very important subject is obviously too extensive to admit of our considering it in a general view. It cannot be doubted that agriculture has been more an object of study, and has been brought to a much greater degree of

* Phil. Trans. 1788, Vol. LXXVIII. p. 158.

perfection in Great-Britain than in any other part of the world. In some counties of England these improvements are rather of an old date, and the attention bestowed on the subject for many years past has been attended with very
beneficial consequences. In the art of rearing cattle, and in the method of
ameliorating the breed both of cattle and horses, some counties of England still
continue unrivalled; but in the art of cultivating the ground, and raising crops
of corn, Norfolk seems to be the county that originated the first improvements.
From that county they were carried to Roxburgh and Berwickshire, in Scotland; and by the enlightened farmers of these provinces and of Haddington-
shire, immense improvements were made both in œconomizing labour, and in
improving the management of land and the rotation of crops. It is in these
three counties that farming, strictly so called, has acquired the greatest degree
of perfection. These improvements are gradually making their way into the
other southern counties of Scotland, and they have been long understood in
that part of England which borders on the Tweed. But in the south western
counties of England, and even in the inland counties as far north as Yorkshire,
the system of farming is still exceedingly defective. The farmers in general are
destitute of sufficient capital, and almost always of a good education; the
fields are too small and too much surrounded with trees; the plowing utensils
are ill contrived, and the number of horses and men employed above all calculation too great. The tythes and poor rates undoubtedly impede the improvement of agriculture in England; but the system which prevails in many places
of not giving the farmers leases of their farms obviously forms an insuperable
check to all improvement. For no man will choose to lay out money in improving a farm while he is conscious that he may be turned out before he has
reaped any advantage from his improvements. But it would be improper, in a
work of this nature, to dwell longer on a subject of this kind. We refer the
reader who wishes for full information on the subject to the Agricultural Reports for the different counties of Great-Britain, published by the Board of
Agriculture, and to the works of Mr. Arthur Young and Mr. Marshall, and,
above all, to the System of *Agriculture*, printed in the Edinburgh Encyclopædia, and since published in a separate state.

Agriculture forms but a very subordinate branch in the Philosophical Transactions. The number of papers on the subject in that work amount to 40, and
more than the half of these are of a very old date. There are 15 of these
papers which at present can scarcely be considered as of any value. I shall
take a short view of the most interesting of the remaining 25.

I. The first paper that deserves notice is a review or rather an outline of
Evelyn's *Sylva* and *Pomona*, the first book ever written and published by the
express orders of the Royal Society.* This work is universally known, and is

* Phil. Trans. 1669. Vol. IV. p. 1069.

one of the most valuable books on the œconomy of forest and fruit trees that has ever appeared. A new and splendid edition of it, enriched with numerous notes, was published in 1786 by the late Dr. Hunter of York. Mr. John Evelyn was one of the original Members of the Royal Society, and one of its greatest ornaments. He was born in 1620, and died in 1706. He was very active in promoting the restoration, and on that account was honoured with particular marks of attention both by Charles II. and James II.

II. The second paper which deserves notice is a letter written by Mr. Evelyn Drill plough. to the President of the Royal Society, giving an account of a drill plough invented by an Austrian, and used in Spain at the time when the letter was written.* Though this plough is far inferior to the drill ploughs at present in use in this country, yet the fact deserves notice both in a historical point of view, and because it serves in some measure to show the comparative progress of agriculture at that period in Spain and in England.

III. In the year 1675, the method of farming in Cornwall was to take four Farming in crops of corn, and then allow the land to lie six or seven years in grass.† At Cornwall. that time they were in the habit of manuring their fields with a kind of sea sand formed by the trituration of sea shells. That sand was found to answer best which was covered by the sea. They laid 300 sacks of this sand on the acre, when the land lay very near the sea; and gradually diminished the quantity even to 20 or 30 sacks as the distance increased. The effect, according to Dr. Cox, the author of the paper, was to increase the size of the ear and diminish the length of the straw, and he mentions a good crop of barley which he had seen where the ear was as long as all the rest of the straw ‡ The same mode of manuring land was practised in Devonshire. They were accustomed also to apply quicklime as a manure in that county; and some farmers (but the practice is condemned by Dr. Bury as injurious) were in the habit of paring the surface of poor land and burning it. This practice was so common that it was known by the name of *Devonshiring*.§ Sea shells were used as a manure for the turf lands of Londonderry and Donegal, in Ireland, and were found to render the fields very productive. Quicklime was also employed for the same purpose, but was not found to answer quite so well.‖ This method of rendering mossy soils productive has been lately practised in Scotland, and considered by the introducers of it as a new discovery. They were not aware that it had been practised nearly a century before in Ireland.

Hardly any improvements make their way more slowly from one country to another than improvements in agriculture. This must be ascribed entirely to the

---

* Phil. Trans. 1670. Vol. V p. 1056.　　　　　† Phil. Trans. 1675. Vol. IX. p. 293.

‡ This method of manuring land in Cornwall is still continued. Probably the agricultural state of that county is not much altered since the beginning of the last century.

§ Phil. Trans. 1708. Vol. XXVI. p. 142.　　　　‖ Phil. Trans. 1708. Vol. XXVI. p. 53.

ignorance and violent prejudices of the farmers. Though the value of sea shells as a manure had been known in Cornwall at least as early as 1675; and though the process followed had been published in the Transactions, yet in 1744 the use of them was so little understood in Suffolk, that a farmer who found it out by accident soon realized an ample fortune by the discovery.*

Farming in the Highlands of Scotland.   IV. In the year 1675, Sir George Mackenzie gives some account of the mode of husbandry practised in the Highlands of Scotland. Oats and barley appear to have been the only grain raised. Sea weed was employed as a manure, which was found to answer much better for barley than oats. The increase he says often amounts to 18 times the quantity sown.†

Saffron.   V. The culture of saffron was introduced into England from the Low Countries, and is still practised in several counties, especially in the neighbourhood of Cambridge. Saffron consists of the dried styles and stigmata of the flowers of the *crocus sativus.* Some of the cakes consist of the parts of the flowers without any addition; but towards the end of the crop, when the flowers become small, it is customary to sprinkle the cake with a quantity of beer. On account of the high price of saffron, and the small quantity produced, *safflower* has been introduced by apothecaries as a substitute, and is commonly sold instead of it. There are two different papers in the Philosophical Transactions giving an account of the method of cultivating the plant, and preparing the saffron. The first by the Hon. Charles Howard, afterwards Duke of Norfolk ;‡ and the second, which is more circumstantial and complete, by Dr. James Douglass.§

Indian corn.   VI When North America was first planted by the English, they observed that the Indians cultivated a sort of grain peculiar to that country, and now well known by the name of *Zea Mays* or *Indian corn.* It grows to the length of six or eight feet, and the grains are arranged in several rows upon the head which has a considerable size. This corn has a sweetish taste, and makes very palatable bread. It is much cultivated by the Americans, and is even employed for making beer, and for various purposes of domestic œconomy. There is in the Philosophical Transactions a very minute and accurate account of the mode of cultivating Mays by Mr. Winthrop.‖ The mode practised by the Americans at that period to convert it into malt is curious. They spread it on garden soil, covered it slightly with earth, and allowed it to remain till the surface was covered over with a green vegetation. It was then dug up, cleaned and dried, and used as malt. Mr. Winthrop assures us that the usual mode of malting barley had been tried with maize but without success. The Americans had

---

* Phil. Trans. 1744. Vol. XLIII. p. 191.   † Phil. Trans. 1675. Vol. X. p. 396.
‡ Phil. Trans. 1678. Vol. XII. p. 945.   § Phil. Trans. 1728. Vol. XXXV. p. 566.
‖ Phil. Trans. 1678. Vol. XII. p. 1065.

another method of preparing beer from mays which is no less curious. The mays was ground and baked into bread. The bread was broken into lumps which were infused in hot water; this water was then fermented and constituted good beer.

VII. In the Transactions for 1685, we have the weight of a cubic foot of a variety of substances, as determined by the experiments of the Philosophical Society of Oxford. From this table we may deduce the following as at that time the weight of a bushel of wheat, barley, and oats, raised in England.

Weight of corn in 1685.

> Wheat............59·89lb. avoirdupois.
> Barley...........51·18.
> Oats ....36·71.*

The weight of the bushel of the same grain at present is considerably greater. For example, good Norfolk barley usually weighs 53lb. per bushel. I have even weighed barley from Essex, and from the Carse of Gowery, in Scotland, heavier than 54lb. per bushel. From this increase of weight it seems reasonable to conclude that the quality of grain has considerably improved in Great-Britain since the year 1685.

VIII. Coffee has been used in Eastern nations from time immemorial; but its introduction into England, as we learn from a curious paper in the Transactions by Mr. Houghton,† is but of very recent date. In the year 1652, Mr. Daniel Edwards, a Smyrna merchant, brought over with him into England a Greek servant named Pasqua, who made his coffee which he drank two or three dishes at a time, twice or thrice a day. This gentleman seems to have been one of the first persons that made use of coffee in England; though Dr. Harvey, the discoverer of the circulation of the blood, is said likewise to have frequently drunk it. It gradually made its way into private houses, which induced Mr. Edwards to set up Pasqua as a coffeeman. He got a shed in the church-yard of St. Michael, Cornhill, and thus opened the first coffee-house in England. In the year 1699, the annual consumption of coffee, in Great-Britain, amounted to about 100 tons, and it sold at the rate of 14l. per ton.

Coffee.

IX. The great profit that results from plantations of trees, and the annual increase of their value, was never better demonstrated than by the following table drawn up by Mr. Marsham of Stratton, from his own observations.‡

Growth of trees.

---

* Phil. Trans. 1685. Vol. XV. p. 926.   † Phil. Trans. 1699. Vol. XXI. p. 311.
‡ Phil. Trans. 1759. Vol. LI. p. 7.

Measure of trees taken in April 1743, *before they began to shoot ; and again in autumn* 1758, *after the year's growth was completed. The measure taken at five feet from the earth.*

	Circumference in Spring 1743.			Ditto in Autumn 1758.			Increase in 16 years.			Content in 1743.			Content in 1758.			Solid increase in 16 years.		
	Feet.	Inches.	8ths of inch.	Feet.	Inches.	8ths of inch.	Feet.	Inches.	8ths of inch.	Cubic feet.	Quarters.	Inches.	Cubic feet.	Quarters.	Inches.	Cubic feet.	Quarters.	Inches.
1. Ash, planted since 1647.	9	10	4	11	1	0	1	2	4	60	1	318	76	3	43	16	1	57
2. Oak past thriving, but sound......	9	4	4	10	1	0	0	8	4	54	1	336	63	2	80	9	0	176
3. Oak about 80 years old..........	6	3	3	7	8	3	1	5	0	24	1	284	36	2	408	12	1	124
4. Scotch fir, seed in 1698........	5	4	6	6	6	0	1	1	2	17	3	48	26	1	270	8	2	222
5. Oak, planted about 60 years......	5	11	1	7	2	3	1	3	2	21	3	224	32	0	174	10	0	382
6. Spanish chesnut near 60 years old.	4	4	0	5	6	3	1	2	3	11	2	408	18	3	270	7	0	294
7. Do. 45 years old................	2	9	6	4	4	4	1	6	6	4	2	391	11	2	408	7	0	17
8. Oak, planted in 1720............	2	11	2	5	1	2	2	2	0	5	1	116	16	0	260	10	3	144
9. Scotch fir, planted 1734; 2 feet high	1	11	6	4	0	0	2	0	2	2	2	0	10	0	0	7	2	0
10. Pinaster, planted in 1734........	2	5	1	4	3	1	1	10	0	3	2	260	11	1	68	7	2	240
11. Oak, set an acorn in 1719........	1	7	0	2	8	2	1	1	2	1	2	116	4	1	336	2	3	220
12. Oak, planted in 1720............	2	9	5	4	9	4	1	11	7	4	2	391	14	0	176	9	1	217
										213	0	300	322	0	333	109	0	33

Mode of increasing fruit.
X. Mr. Fitzgerald cut off a portion of bark at the bottom of a fruit tree branch, and immediately bound it on again. The consequence was, that the branch bore much fruit, while the bark grew again to the wood as at first. He recommends this as a good method of increasing the quantity of fruit upon trees.*

Produce of a wheat seed.
XI. Mr. Millar by repeated divisions obtained from a single seed of wheat 500 plants. These yielded 21,109 ears, or about 576,840 grains. The whole amounted to three pecks and three quarters of clean corn, and weighed 47lb. 7oz. This is perhaps the greatest instance of increase upon record; yet it appears from the observations of Dr. Watson who relates it, that the plants would have admitted of still greater divisions, and that of course the increase might have been still greater.†

Indigenous trees.
XII. Mr. Daines Barrington published an elaborate paper, to shew that the Spanish chesnut and lime trees are not indigenous in England. He is of the same opinion with respect to the elm, the yew, and several other of our common trees.‡ Dr. Ducarel adopted the opposite opinion; and has proved, in a satisfactory manner, that the Spanish chesnut was common in England during the time of the Saxons.§ The subject is not very susceptible of decision.

* Phil. Trans. 1762. Vol. LII. p. 71.   † Phil. Trans. 1768. Vol. LVIII. p. 203.
‡ Phil. Trans. 1769. Vol. LIX. p. 23.   § Phil. Trans. 1771. Vol. LXI. p. 136.

Mr. Barrington's opinion would appear very probable; and the time when these trees, and probably many other useful plants, were transplanted into England, was, no doubt, when the Romans had possession of the country, and, of course, anterior to the period of the Saxons.

XIII. Dr. Hales, in his Vegetable Statics, had recommended washing the *Washing trees* stems of trees, as a means of promoting their growth. This advice was followed by Mr. Marsham, who published two papers in the Transactions, giving an account of the results.* In all the trials which he made, and they were pretty numerous, the washed trees vegetated faster, and increased more in diameter, in a given time, than the unwashed trees. It is difficult to explain the reason of this advantage, resulting from washing the stems of trees, unless we are to ascribe it to the removal of the lichens and mosses, which may, perhaps, withdraw a considerable portion of the juices of the tree for their own nourishment.

XIV. We shall conclude our extracts from the Philosophical Transactions, *Grafting of* on the subject of the agriculture and œconomy of plants, with a curious paper *fruit trees.* by Mr. Knight, on the grafting of fruit trees. From his observations, it appears that most of the old varieties of apple and pear tree, are worn out with old age. When a young tree is engrafted, from a branch of an old tree, the branch retains all the defects of the old tree; exhibits the same symptoms of old age; and is subject to the same diseases. On the contrary, if the graft be taken from a very young tree, (as a tree raised lately from seed,) it possesses all the health of the young plant; vegetates with vigour, but will not bear fruit till it has acquired the age at which the tree, from which it was taken, becomes capable of bearing. Grafts from old trees bear fruit immediately, but are as unhealthy as the parent trees. Grafts from young trees are healthy and vigorous, but do not bear fruit till the tree, from which they were taken, has reached its thirteenth or fourteenth year.†

CHAP. II.

OF ZOOLOGY.

This department of Natural History must always have possessed attractions for philosophers; as man, the chief and sovereign of terrestrial animals, belongs to it. Even the politician and the man of business cannot view it with

* Phil. Trans. 1777. Vol. LXVII. p. 12; and 1781. Vol. LXXI. p. 449.
† Phil. Trans. 1795. Vol. LXXXV. p. 290

indifference, as no small share of our power, and almost the whole of our comforts, depend upon the use which we make of the inferior animals. Hence it becomes a curious and highly useful inquiry to ascertain the dispositions, powers, and propensities of those animals, which assist us in our labour, or contribute to our food.

Division of zoology.

It consists of various departments, which it will be requisite to consider separately. The first department is occupied with the *arrangement and description* of animals. It is to this branch of the subject, that the term *Zoology* is usually confined; but we must here take it in a more extended sense, that we may include under it several branches of knowledge which cannot be conveniently placed any where else. Zoologists have found, that the only method of arranging animals systematically, which can be followed with advantage, is founded upon the knowledge of their structure. Now this knowledge can be acquired only by *anatomy*. Hence anatomy is intimately, and indeed inseparably, connected with the arrangement of animals. But anatomy has been also made subservient to the purposes of surgery and medicine. On that account, human anatomy has been exalted into a state of much greater comparative importance than the anatomy of the inferior animals. The term *anatomy*, without any epithet, is usually applied to human anatomy; while the anatomy of the inferior animals, is called *comparative* anatomy. The *functions* of the different animals, and especially of man, have been studied with great care; chiefly, it must be acknowledged, by the physicians, and with a medical object in view; and this study has been raised to the dignity of a science, and distinguished by the name of *physiology*. It naturally comes to be considered after anatomy, human and comparative, upon which it is in a great measure dependent. Animals, especially man, are liable to various accidents and diseases, which derange the functions and injure the health. Much attention and study has been bestowed upon the method of remedying these accidents, and curing these diseases, especially when they occur in man; and this study has given birth to two sciences, or rather professions, we mean *surgery* and *medicine*. These, as far as they are possessed of any principles, are dependent for them upon anatomy and physiology; and therefore naturally come to be considered after these two branches of knowledge.

All of these various departments occupy a very conspicuous place in the Philosophical Transactions, and come, therefore, necessarily under our review. We shall accordingly divide this chapter into five sections; treating respectively, 1. Of the arrangement and description of animals. 2. Of anatomy. 3. Of comparative anatomy. 4. Of physiology. 5. Of surgery and medicine.

SECT. I.—*Of the Arrangement and Description of Animals.*

Aristotle.

The earliest writer, who turned his attention to the animal kingdom was

6

Aristotle, the tutor of Alexander the Great, and one of the most eminent philosophers that Greece ever produced. Alexander was at considerable expense in procuring animals for his examination. Aristotle's history of animals is divided into nine books, and consists chiefly of generals. He treats, in the first place, of the anatomy of animals; then, of their generation at great length; and maintains a doctrine which, in consequence of his authority, long continued prevalent in the schools, we mean the doctrine of equivocal generation. The last part of his work is occupied with the dispositions of animals. He notices many curious particulars, which were copied by Ælian, Gesner, and Aldrovandus; and, from these authors, made their way into the writings of Buffon; who has, in consequence, gained credit for many original observations, which do not belong to him. Aristotle divides animals into viviparous and oviparous. The first division comprehends quadrupeds; the second, birds, fishes, and insects   Every body now knows that this division is defective, for some quadrupeds, (lizards for example,) are oviparous, while there are various fishes and insects that are viviparous.

Pliny, in his Natural History, has not neglected animals. He has devoted *Pliny.* four books to the consideration of them. The 8th book of his work treats of land animals, or quadrupeds; the 9th, of fishes; the 10th, of birds; and the 11th, of insects. He has less arrangement than Aristotle; and his zoology must be admitted to be inferior to that of the Grecian philosopher. Indeed, it is a mere sketch; but a sketch executed by the hand of a master.

Ælian, a Greek writer, of the second century of the Christian era, published *Ælian.* a work on animals, in 17 books. An edition of this book, in two quarto volumes, was edited by Gronovius, and published in London in 1744. It contains little or no original matter, and is, in every respect, except in bulk, inferior to the writings of his predecessors.

These are the only ancient writers, deserving of notice, who wrote upon zoology. The science, for many ages, slumbered in oblivion, and can scarcely be said to have drawn the attention of mankind for a moment, before the auspicious period of the revival of letters. Soon after this period two very laborious writers on zoology appeared, namely, Aldrovandus and Gesner. Ulysses Aldrovandus was an Italian, and published his zoology at a very ad- *Aldrovandus.* vanced age. It consists of six very thick folios. He treats, first, of quadrupeds, then of birds, fishes, whales, insects, and serpents. He has given wooden figures of many of the animals which he describes; and several of them are tolerably exact representations. The book is an immense mass of erudition, ill-digested, and introduced without discrimination; and contains a prodigious quantity of matter totally unconnected with the subjects of which he treats. Gesner, like Aldrovandus, arranged his voluminous history according to the Aristotelian method.

Johnston, a Dutch zoologist, published his work on zoology, in 1657. He can be considered in no other light than as a copyist of Gesner and Aldrovandus; but as he left out the extraneous matter with which these authors are loaded, he, in point of fact, considerably improved the subject.

Such nearly was the state of zoology, when the Royal Society was constituted; for it would be tedious to enumerate all the writers who attempted, with various success, to illustrate particular branches of this extensive subject. The Aristotelian arrangement had not been altered; but a great improvement had been introduced, by enumerating and describing every particular animal, and by giving figures of them, which contributed materially to remove doubt and ambiguity. Soon after the establishment of the Royal Society, a new and more brilliant æra in the science of zoology commenced. This was owing chiefly, if not entirely, to the exertions of two illustrious Fellows of the Society, Willughby and Ray; and to the generous liberality of the Society, and some of the most eminent Members, who agreed to defray the expense of the publication of Willughby's works.

Francis Willughby, Esq., was a gentleman of Warwickshire, born in 1634, and the intimate friend, companion, and patron of Mr. Ray, with whom he had contracted an acquaintance at Cambridge; an acquaintance soon warmed into friendship by the congeniality of their tastes and their pursuits. With Mr. Ray, he made repeated journies through the different counties of England, in order to become acquainted with its plants and animals; and with Mr. Ray, Mr. Shippon, and Mr. Nathaniel Bacon, he made a three years' tour on the continent, with the same objects in view. He had made extraordinary collections of birds and fishes, and in these he had been greatly assisted by Mr. Ray. But, before he had time to complete his arrangements, far less to commit the result of his labours to the press, he was snatched away by a sudden death, in the thirty-seventh year of his age. He left to Mr. Ray the charge of educating his two infant sons, with an annuity of 70 pounds; which constituted, ever after, the chief part of Mr. Ray's income. In the year 1676, Mr. Ray published the Ornithology of his friend; to which he himself had given arrangement, and which he had enriched with many valuable materials drawn from his own stores. It was published under the following title: Ornithologiæ Libri tres: *in quibus Aves omnes hactenus cognitæ in Methodum Naturis suis convenientem redactæ accuratè describuntur. Iconibus elegantissimis et vivarum Avium simillimis Æri incisis illustrantur. Totum Opus recognovit, digessit, supplevit Johanes Raius.* Mr. Ray translated this book into English, and published it two years after with large additions. The figures were engraved at the expense of Mrs. Willughby, and were wholly inadequate to the merit of the work.

In 1686, the Ichthiology of Mr. Willughby was likewise published by Mr.

5

Ray. It had been left in a still more imperfect state than the ornithology. Mr. Ray wrote the whole of the first and second books himself, and probably added a great deal of new materials to the two others. The work was printed at Oxford, at the expence of the Royal Society, the Members of which contributed to furnish the plates; indeed, by far the greatest part of them were engraved at the expence of Sir Robert Southwell, at that time President of the Royal Society. The book came out under the following title: *Francisci Willughbeii, Armig.* DE HISTORIA PISCIUM *Libri quatuor, jussu et sumptu S. Reg. Lond. editi. Totum Opus recognovit, cooptavit, supplevit, Librum etiam primum et secundum integros adjecit J. Raius.*

Ray contributed more materially to the improvement of zoology, than any other person of the age in which he lived. His classification was greatly superior to that of any preceding writer. He divided animals into *sanguinea* and *exsanguinea*. The *sanguinea* were subdivided into those that *breathe with lungs,* and those that have *gills.* The former are separated into those whose heart has *two ventricles,* and those which have only *one.* The latter division contains *reptiles;* the former *viviparous quadrupeds, whales, and birds.* The *animalia bronchiis respirantia* include all fishes properly so called; the whale, and all the exsanguinea being excluded. The *animalia exsanguinea* are divided into *greater* and *less:* the latter division contains *insects;* the former is again subdivided into three genera; the first of which includes the *mollusca;* the second, the *crustacea;* and the third, the *testacea,* or *shell-fish.* Upon this arrangement it is not necessary to make any observations. The subsequent classification of Linnæus was merely an extension and improvement of the method of Ray. The first of Mr. Ray's zoological works, if we except the ornithology and ichthyology of Willughby, which he edited, was his *Synopsis Quadrupedum;* published in 1693. Soon after he drew up his *Synopsis Avium et Piscium;* as we learn from his letters, and from the preface to those works, by Dr. Derham; but the manuscripts lay unpublished in the hands of the bookseller, till they were purchased by Mr. Innys, and prepared for the press by Dr. Derham; who added the figures, and made some additions. They were published in 1713, after Mr. Ray's death. The last work in which Mr. Ray engaged, was his *Methodus Insectorum.* He had been accumulating materials for it during many years, and was at last induced to engage in it at the age of seventy-five, by the persuasion of Dr. Derham. He confined his object to British insects. Mr. Ray did not live to finish this work. It was published, after his death, by Dr. Derham, in 1710.

Linnæus, the great reformer of botanical classification, did not neglect zoology. His powers of arrangement were so great, and his industry so indefatigable, that this extensive branch of natural history lies under almost as many obligations to him as botany does. His *Systema Naturæ* was first pub-

Ray.

Linnæus.

L

lished in 1785; and he continued to improve it during the whole period of his life. The Linnæan classification of animals is generally considered as the most unexceptionable, and has been much more universally followed than any other, notwithstanding the great number of systems which have made their appearance, during the last century, in various parts of Europe, especially in France; where, from the influence and celebrity of Buffon, less justice was done to Linnæus than in other countries. Linnæus divided animals into six classes; namely, *mammalia, birds, amphibia, fishes, worms, insects.* His classification of birds, and of insects, constitutes the best part of his system; and it is, at the same time, the most original.

Pennant.   Pennant, one of the most celebrated and valuable of British zoological writers, followed chiefly the classification of Ray. He does not treat of insects and worms, but confines himself to the larger animals. This is the case, also, with Buffon, who owes his celebrity chiefly to his style. He neglected and despised artificial classification, and bestowed much of his time in endeavouring to establish certain favourite systematic opinions; all of which, perhaps without exception, are erroneous and absurd.

Royal Society.   Though zoology has always constituted a favourite pursuit with the Members of the Royal Society, and though it is indebted to that learned body for a very considerable part of the progress which it has made, yet, from the nature of the Philosophical Transactions, it is not possible that they should contain any thing like a complete system of zoology; or an artificial classification of animals. All that we have a right to expect, is a variety of detached papers on zoological subjects; we find in them, reckoning from the commencement of the work to the year 1800, no fewer than 290 papers, connected with the arrangement and description of animals. These papers, as might be expected, are of various merits; many of them excellent; others reaching only mediocrity; and there are 125 which seem, in the present advanced state of the science, scarcely to be entitled to any attention whatever. It is curious to observe, that a large proportion of these trifling papers were published when Martin Folkes, Esq., was President of the Society; and not a few of them were written by Mr. Arderon, an inhabitant of Norwich, who seemed to possess abilities and industry, but to be totally illiterate, as far as systematic zoology was concerned. No fewer than 301 species of animals are described in the zoological papers, published in the Transactions; several of which are new, but certainly not the whole; and it is difficult to say how many.

But we shall have it in our power to judge with more accuracy of the zoological papers, in the Philosophical Transactions, if we take a view of the different classes of animals in succession; and we may adopt the Linnæan classification, as more usually followed in this country.

Mammalia.   I. MAMMALIA. The species of mammalia hitherto discovered and described

3

amount to about 250. The most celebrated arrangements of these animals are those of Ray, Klein, Linnæus, Brisson, and Daubenton. Buffon follows no arrangement whatever. Pennant has adopted the arrangement of Ray. Klein was the zoological antagonist of Linnæus, and has written successfully on most parts of the science ; though his system, as a whole, appears to be considerably inferior to that of his celebrated antagonist. The French naturalists object to Linnæus the new meanings which he has given to the names of his orders, and the forcible union of animals of very different natures under the same order ; but they do not attend to the object of an artificial classification, which is merely to facilitate the learner in discovering the name of every animal. Every artificial arrangement is, in fact, intended to supply the place of a lexicon of natural history, and of course that arrangement deserves the preference which can be consulted with the greatest facility.

The species of mammalia described in the Transactions amount to 36. The following is a list of the names of these animals, with the names of the authors to whom we are indebted for the account of them :

Species described in the Transactions.

Mus lemmus, or sable mouse. Sir Paul Rycaut.	Mustela lutreola, or lesser otter.
Simia mormor, or mantegor......Dr. Tyson.	Mr. John Reinhold Forster.
Cervus alces, or moose deer.....Mr. Dudley.	Mustela martes, or pine marten. ......Ditto.
Sciurus volans, or flying squirrel. ..M. Klein.	Mustela erminea, or stoat and ermine..Ditto.
Lemur volans, or flying macauco........Ditto.	Mustela nivalis, or common weasel.....Ditto.
Monodon monoceros, or sea unicorn.	Shunk.........................Ditto.
Dr. Steigertahl.	Hystrix dorsata, or Canada porcupine..Ditto.
Phoca barbata, or great seal.....Dr. Parsons.	Castor fiber, or beaver...............Ditto.
Asiatic rhinoceros..................Ditto.	Castor zibethicus, or musk beaver.....Ditto.
Simia jacchus, a small monkey........Ditto.	Lepus timidus, or Alpine hare........Ditto.
Felis Caracal.....................Ditto.	American hare. ....Ditto.
African rhinoceros.................Ditto.	Quebec marmot...................Ditto.
Simia lar, or tailless monkey...Mr. De Visme.	Sciurus vulgaris, or common squirrel...Ditto.
Calmelopardalis..............Mr. Carteret.	Great flying squirrel................Ditto.
Physeter catodon, or spermaceti whale.	Field mouse. ....Ditto
James Robinson, Esq.	Mus sylvaticus, or field mouse........Ditto.
Antelope picta, or nyl ghau.	Mus terrestris, or short tailed mouse. ..Ditto.
Dr. William Hunter.	Sorex araneus, or fetid shrew........Ditto.
Canis lagopus, or arctic fox.	Shrew.........................Ditto.
Mr. John Reinhold Forster.	Tyger cat, of the Cape of Good Hope ..Ditto.

The animals described by Forster were from Hudson's Bay. The greater number, if not the whole of them, had been previously described by Mr. Pennant.

II. BIRDS. The species of birds are much more numerous than those of the quadrupeds ; about 1800 have been described and classified. Belon was one of the first of the moderns who treated of these animals. His history of birds was published in 1555, and forms a single folio volume. Willughby's work, published in 1686 by Ray, established a new era in this branch of natural history

Birds.

A great many classifications have been since proposed, the best of which is that of Linnæus founded on the bill and feet. Barrerre, Klein, Mæhring, Salerne, and Brisson, also published classifications with more or less success. Edwards published a work on the subject in 1748, with figures, which at the time were excellent. Much valuable matter was published by Sloane in his Account of Jamaica, and by Catesby in his work on Carolina. But the most complete work on the subject which has appeared in this country is the work of Dr. Latham.

Species described in the Transactions.

There are 69 species of birds described in the Philosophical Transactions. The following is a list of them, together with the authors to whom we are indebted for describing them:

Phænicopterus ruber, or flamingo.
            Dr. Douglass.
Tetrao umbellus, or American pheasant.
            Mr. Edwards.
Otis tetrax, or little bustard. . . . . Ditto.
Tringa lobata. . . . . Ditto.
Chinese pheasant. . . . . Ditto.
Aptenodyta patochonica, or Patagonian penguin. . . . . Mr. Pennant.
Vultur serpentarius. . . . . Mr. Edwards.
Columba cristata of Latham. . . Dr. Badenach.
Falco columbarius, or pigeon hawk. . . Forster.
Falco spadiceus. Ditto.
Falco sacer, or partridge hawk. . . . . Ditto.
Strix brachiotos, or short-eared owl. . . . Ditto.
Strix nyctea, or snowy owl. . . . . Ditto.
Strix funerea, or Canada owl. . . . . Ditto.
Strix paperina, or little owl. . . . . Ditto.
Strix nebulosa, or grey owl. . : . . . . Ditto.
Lanius excubitor, or great butcher bird. Ditto.
Corvus canadensis, or cinereous crow. . . Ditto.
Corvus pica, or magpie. . . . . Ditto.
Picus auratus, or gold wing woodpecker. Ditto.
Picus villosus, or hairy woodpecker. . . . Ditto.
Picus tridactylus, or three-toed woodpecker.
              Ditto.
Tetrao canadensis, or spotted grous. . . . Ditto.
Tetrao lagopus, or white grous. . . . . . . Ditto.
Tetrao togatus, or shoulderknot grous. . Ditto.
Tetrao phasianellus, or long tailed grous. Ditto.
Columba migratoria, or migratory pigeon.
              Ditto.
Alauda alpestris, or shore lark. . . . . . . Ditto.
Turdus migratorius, or American fieldfare.
              Ditto.
Turdus merula? uncertain. . . . . Ditto.
Loxia curvirostra, or cross bill. . . . . Ditto.

Loxia enucleator, or pine grosbeak. . . Forster.
Emberiza nivalis, or bunting. . . . . Ditto.
Emberiza leucophrys. . . . . Ditto.
Fringilla laponica, or finch. . . . . Ditto.
Fringilla linaria, lesser red-headed linnet. Ditto.
Fringilla montana, mountain sparrow. . . Ditto.
Fringilla hudsonia. . . . . Ditto.
Muscicapa striata. . . . . Ditto.
Motacilla calendula, ruby-crowned wren. Ditto.
Porus atricapillus, black cap titmouse. . . Ditto.
Porus hudsonicus, Hudson Bay titmouse. Ditto.
Ardea canadensis, Canada crane. . . . . . Ditto.
Ardea americana, hooping crane. . . . . . Ditto.
Ardea stellarius, bittern. . . . . Ditto.
Scolopax totanus, spotted woodcock. . . Ditto.
Scolopax laponica, red godwit. . . . . Ditto.
Scolopax borealis, Eskimaux curlew. . . Ditto.
Tringa interpres, turnstone. . . . . Ditto.
Tringa helvetica. . . . . Ditto.
Anas marila, scaup duck. . . . . Ditto.
Anas nivalis, snow goose. . . . . Ditto.
Anas canadensis, Canada goose. . . . . . . Ditto
Anas albeola, red duck. . . . . Ditto.
Anas clangula, golden eye. . . . . Ditto.
Anas perspicillata, black duck. . . . . Ditto.
Anas glacialis, swallow tail. . . . . Ditto.
Anas crocea, teal. . . . . Ditto.
Anas histrionica, harlequin duck. . . . . . Ditto.
Anas boschas, mallard drake. . . . . Ditto.
Peliconus onocrotalos. . . . . Ditto.
Colymbus glacialis, northern diver. . . . . Ditto.
Colymbus auritus, eared grebe. . . . . Ditto.
Larus parasiticus, arctic gull. . . . . Ditto.
Sterna hirundo, common tern. . . . . Ditto.
Cuculus indicator. . . . . Sparman.
Meleagris gallo-pavo, the turkey. . . . Pennant.
Motacilla arundinacea, reed wren. . . Lightfoot.

The birds described by Forster consisted of a collection sent from Hudson's Bay by Mr. Graham. Most of them were well known previously; though some few were new species. It is remarkable enough that the motacilla arundinacea, described by Lightfoot, was an unknown species, although the bird haunts the Colne, in the neighbourhood of Uxbridge, not many miles from London. Pennant's Dissertation on the Turkey is curious. He demonstrates that it was unknown to the ancients, that it is a native of America, that it was brought first from Mexico to Spain, and that it was introduced into England as early as 1524. By 1583 it had become a common article of food.

III. AMPHIBIA. The amphibious animals are divided into two orders; the *Amphibia.* first consisting of those which have feet; the second of those which are destitute of limbs, and consist of the serpent tribe. As the name imports, this class of animals is capable of living both on land and in water. They are not nearly so numerous as the other classes of animals, though it is difficult to form an estimate of the number of species which they comprehend.

There are only eight species described in the Transactions. The following is *Species described in the* a list of them, together with the names of the authors to whom we are obliged *Transactions.* for the descriptions:

Rattle snake. .... Dudley.	New species of chamælion. .... Parsons.
Lacerta gangetica, crocodile of the Ganges.	Manis pentadactyla. .... Hampe.
Edwards.	Testudo ferox, soft-shelled turtle. .. Pennant.
Siren lacertina..................... Ellis.	Testudo coriacea. .... Ditto.
Coluber cerastes, horned viper.........Ditto.	

The most remarkable of these animals is the siren lacertina. Specimens of it *Siren lacer-* were sent from Charlestown, in Carolina, by Dr. Garden, to Mr. Ellis. It was *tina.* at first suspected to be the larva of some species of lizard; but upon examining all the larvæ of these animals in the British Museum, Mr. Ellis found that they all had four feet, whereas this animal has only two. Ellis sent a specimen of the animal to Linnæus at Upsal, who hesitated whether to consider it as a new genus, or as the larva of some unknown lizard. It is now admitted by naturalists to constitute a new genus, and the name Siren proposed by Linnæus has been adopted. Several additional species of this genus have been since discovered.

There are many remarkable stories in the Philosophical Transactions about *Fascinating* the fascinating power of serpents, some of which it is difficult to disbelieve, *power of ser* especially one related by Dr. Sprengel, which he himself witnessed in Italy. A *pents.* viper-catcher in Italy had a collection of about 60 living vipers in a room for sale. Among these was a pregnant female viper, for it is remarkable that these animals, when in confinement, feed only when in the pregnant state. A live mouse was thrown into the room among the serpents. As soon as the pregnant female and the mouse exchanged looks, the mouse began to run round the viper in a circle, uttering faint cries. At last he ran into the open mouth of the viper, and was swallowed; the viper all the while remaining motionless in the middle of the room. The Hon Paul Dudley relates another story of a rattle snake in North America. A squirrel was seen to leap in a restless manner from one bough to another, at last it came down the tree and ran behind a log. On going behind the log to see what was become of the animal, a rattlesnake was observed which had just swallowed the squirrel. It is difficult to determine what to make of these, and many other similar stories that are to be found in books of natural history, relating to the fascinating power of serpents. The most probable account of this supposed power which we have seen is in a paper by Sir Hans Sloane in the Philosophical Transactions. According to him the serpent always begins by biting his prey. He then retires to a little distance, and waits the event. The animal is instantly enfeebled by the effects of the poison, and in no long period of time drops down dead. The serpent then advances, and licking the animal all over to lubricate the skin, swallows him entire. This account is very plausible and satisfactory; but cannot be admitted as a complete explanation, unless we deny the truth of the numerous stories concerning the fascinating power of serpents, some of which indeed do not seem to be well authenticated; but others depend upon evidence that it would be difficult to reject.

Fishes.

IV. FISHES. Fishes are much more numerous than the amphibious animals; but much fewer than the birds. From the nature of the element in which they live, it is not likely that all the species are already known to naturalists Indeed, even upon the coasts of Great Britain, various new species have been discovered by Mr. Montague, and described in the Linnæan and Wernerian Transactions within these few years.

The ichthyology of Willughby and Ray constitutes the first great step made by the moderns, in the knowledge and description of these animals. The work of Artedi, though posthumous, and of consequence incomplete, was notwithstanding greatly superior to every thing that had preceded it. Peter

Artedi.

Artedi was born in Angermannia, one of the provinces of Sweden, in 1705. His father was a clergyman, and originally intended his son for the church; but his love of natural history was invincible, and induced him to turn his attention to medicine, as a profession that would permit him to cultivate his favourite study. When Linnæus went to Upsal, in 1727, he inquired who was the young man of greatest knowledge, and who devoted the most of his time to study. All were unanimous in repeating the name of Peter Artedi. Linnæus went to see him, and formed with him an intimacy and friendship which became every day stronger. The ardour of both was equal; and the emulation produced by their mutual discoveries contributed still more to increase their industry Artedi early devoted the greatest part of his attention to the study of fishes ; and Linnæus assures us, that his knowledge of that class of animals greatly exceeded his own. He went to London, in 1734, as a place where he might have better opportunities of increasing his knowledge of fishes, than at Upsal. Linnæus met him at Leyden, in 1735, with his finances exhausted, and preparing to return to his own country. He recommended him to the celebrated apothecary Seba, at Amsterdam, to assist him in describing his cabinet of natural history. Linnæus and he indulged in a long friendly conversation, on the 27th of September, 1737 : Linnæus showed him his *Fundamenta Botanica* ; and Artedi communicated his *Philosophia Ichthyologica* ; a work which had occupied him for many years. On his return home at night, he stumbled in the dark and fell into a canal, and was drowned. Linnæus took charge of his MSS., and published his work on fishes, in 1738. Artedi was the first who divided fishes into orders and genera, and who established accurate principles for the description and arrangement of the class. His orders were five, and his genera fifty-two.

Though Linnæus did not adopt the arrangement of Artedi, he was obviously indebted to him for much of the precision which he introduced into this class of animals. He divides fishes into five orders ; and the division depends chiefly upon the absence, or position when present, of the ventral fins. His genera amount to 61.

One of the latest systems of ichthyology, which has made its appearance, is Lacepede. the work of Lacepede, a French naturalist of considerable celebrity. The first volume of his work was published in 1798. Four others, which complete the book, have appeared since. It is in quarto, and embellished with plates of almost all the species described. Lacepede has an arrangement of his own, according to which he has drawn up his book.

The species of fish described in the Transactions, amount to 19 The following is a list of their names, with the the authors to whom we are indebted for the descriptions : Species described in the Transactions.

Callionimus lyra, gemmeous dragonet. Tyson.	Accipenser sturio, sturgeon. . . . . Forster.
Lophius piscatorius, sea devil. . . . . . . Parsons.	Gadus lota, burbot. . . . . Ditto.
Zeus luna, king fish. . . . . Mortimer.	Salmo lavaretus, gwiniad. . . . . Ditto.
Silurus anguillaris, silvor. . . . . Russel.	Cyprinus catostomus, sucker. . . . . Ditto.†
Silurus cous, zakzuk. . . . : . . . . Ditto.	Raja torpedo, electric ray. . . . . Walsh.
Ophidium mastacem, balus . . . . Ditto.	Squalus glaucus, blue shark. . . . . Watson.
Lophius conubicus, long angler. . . . . Ferguson.	Exocætus volitans, flying fish. . . . . Brown.
Chætodon rostratus. . . . . Schlosser.	Ophidium barbatum, bearded ophidium.
Siæna jaculatrix. . . . . Pallas.	Broussonet.
Chætodon, ————.* . . . . Tyson.	Tetrodon electricus, electric tetrodon. Paterson.

* Phil. Trans. Vol. LXI. p. 247.
† Phil. Trans. Vol. LXIII. p. 149. The four fishes described by Forster were from Hudson's Bay.

Journies of
eels.

The numerous stories circulated respecting the journies of eels on dry land, during the night time, are well known to every reader. Dr. Plot, in his History of Staffordshire, gives an account of their passage across meadows. These stories have been probably exaggerated, and have not always obtained credit. This makes an account, contained in the Philosophical Transactions, of some importance. Mr. Arderon in that paper states, that he saw a number of eels sliding up the flood-gates of the water-works, at Norwich, though six feet above the surface of the water, and perfectly dry and smooth. They ascended with the greatest facility, and made their way into the water above the gates.*

Chætodon
rostratus.

We must not omit to mention a curious account, given in the Transactions, of a method employed by the chætodon rostratus, a fish which frequents the sea-shores, in the East-Indies, to catch its prey. When it observes a fly sitting on the plants that grow in shallow water, it swims on to the distance of five or six feet, and then, with a surprising dexterity, it ejects out of its tubular mouth a single drop of water, which never fails striking the fly into the sea, where it soon becomes its prey. Hommel, a Dutch governor in the East-Indies, made a number of experiments to satisfy himself that the fish actually possessed this property, and saw it repeatedly dart a drop of water at a fly which he had fixed upon a stick within its reach † The chætodon rostratus is not the only species of fish that possesses this curious property; Pallas describes another, the siæna jaculatrix; which likewise secures flies by a similar contrivance.‡

Fecundity of
fish.

The prodigious fecundity of fish has long been well known and admired. Petit found 342,144 eggs, in a carp; and Leewenhoeck estimated the eggs of a cod at 9,384,000. But the most complete information on this subject, any where to be met with, is in a paper published in the Transactions, by Harmer. His method was to weigh the roe of the fish with accuracy, and then to count the eggs contained in a certain number of grains. From this, it was easy to determine the number contained in the whole roe. The following table exhibits the result of his researches on this subject.§

* Phil. Trans. 1747. Vol. XLIV. p. 395.   † Phil. Trans. 1764. Vol. LIV. p. 89.
‡ Phil. Trans. 1766. Vol. LVI. p. 186.   § Phil. Trans. 1767. Vol. LVII. p. 280.

Names of Fish.	Weight of Ditto.			Weight of the Roe.	Number of Eggs in Ditto.
	Avoirdupois.				
Carp, ...No. 1	16 oz.	12	dr.	1,265 gr.	101,200
2	25	8		2,571	203,109
Cod-fish, ...	....			12,540	3,686,760
Flounder, ...No. 1	2	14		182½	133,407
2	3	8½		152	225,568
3	6	12		598	351,026
4	24	4		2,200	1,357,400
Herring, ...No. 1	4	3		367	32,663
2	5	0		236½	21,285
3	3	13		259	23,569
4	5	10		480	36,960
5	4	6½		366	29,646
6	4	8		420½	27,753
7	5	1		490½	32,863
Lobster, ...No. 1	14	8		....	7,227
2	36	0		1,671	21,699
Mackerel, ...No. 1	20	0		1,027	454,961
2	20	0		949	430,846
3	18	0		1,223½	546,681
Perch, ...No. 1	8	9		765½	28,323
2	5	10		502	20,582
Pickerel, ...No. 1	56	4		5,100½	49,304
2	....			3,248	80,388
3	48	10½		3,184	33,´32
Prawn, ...No. 1	0	127	gr.	....	3,806
2	0	94½		....	3,479
3	0	100½		....	3,579
Roach, ...No. 1	2	0	dr.	114	9,604
2	6	8		671	43,615
3	3	8		346½	29,799
4	2	2		153	9,486
5	10	6½		361	81,586
6	9	10¼		417	113,841
7	3	8		213½	45,475
Shrimp, (with light-coloured spawn.) No. 1	0	17½	gr.	3	3,057
2	0	39		7	6,807
3	....			....	4,601
Ditto, (with dark-coloured spawn,) ..No. 1	0	31		5	4,090
2	0	22		4	2,849
Smelt, ...No. 1	2	0	dr.	149½	38,278
2	0	289½	gr.	50	14,411
3	1	14	dr.	157½	29,925
4	1	12		145½	30,991
5	1	7		149	24,287
6	1	5		136	23,800
Soal, ...No. 1	14	8		542½	100,362
2	5	0		179½	38,772
Tench, ...No. 1	40	0		....	383,252
2	28	8		533½	280,087
3	8	14¾		224	83,104
4	9	8		284½	108,963
5	12	8		366	138,348
6	27	9½		1,969	350,482
7	14	15		866	138,560

M

Worms.　　V. Worms　This class of animals, as Linnæus left it, is exceedingly heterogeneous, and has since, with great propriety, been split into various subdivisions
by succeeding naturalists.　Linnæus divided it into four orders; namely, 1. Intestina, including naked simple animals, without limbs.　2. Mollusca, naked
simple animals, furnished with limbs.　3. Testacea, shells, the calcareous
covering of mollusca.　4. Zoophita, composite animals, efflorescing like vegetables.　5. Infusoria, minute animalcules, seldom visible to the naked eye.
These orders are obviously independent of each other, and might each, without
Intestina.　impropriety, constitute a distinct class.　The first order, though denominated
intestina, consists only in part of worms which inhabit the intestines of other
animals; several of the genera are found only in the earth, or the ocean.

Species de-　　The species belonging to this order, described in the Transactions, amount to
scribed in the　eight.　The following is a list of their names, together with the authors to
Transactions.　whom we are indebted for an account of them.

Three species of tænia, two in the dog, one in the mouse.....................Lister.	Gordius aquaticus.* ....Lister.
Tænia globosa. ....Tyson.	Gordius medinensis, guinea worm......Ditto.
Worm in the human tongue...........Dent.	Planaria clavata. ....Garcin.
	Fasciola hepatica ....Bidloo.

Mollusca.　　The mollusca constitute a very numerous class of animals, concerning the
œconomy of which many curious and interesting observations have been made.
It is to be regretted that these observations have been too little attended to by
systematic writers, and are scarcely to be found in any of the books which
they have written on the subject.　The number of species described in the
Transactions, amounts only to 10.　The following is a list of them, together
with the names of the authors to whom we are indebted for the descriptions:

Asterias caput medusæ, star-fish. ..Winthrop.	Actinia gemmacea.‡ ....Gaertner.
Aphrodita aculeata, sea-mouse....Molineaux.	Actinia mesembryanthemum.‡ ....Ditto.
American cuttle fish.† ....Baker.	Holothuria pentactes.‡ ....Ditto.
Actinia cereus.‡ ....Gaertner.	Ascidia pedunculata. ....Russel.
Actinia bellis.‡ ....Ditto.	Actinia sociata, clustered animal flower..Ellis.

Shells.　　The testacea, or shells, constitute the covering or habitation of many of the
mollusca.　They have been too little studied with reference to the animals
which inhabit them.　Instead of classing shells in a particular order by themselves, it would have been better, had it been practicable, to have described
them along with the animals that form them.　But this has not hitherto been
done; and indeed zoologists are entirely unacquainted with the nature of the

---

* Phil. Trans. 1672. Vol. VII. p. 4064.　Lister found this animal in a species of beetle.　By Gmelin it is constituted a peculiar genus, under the name of Filoria Scarabæi.

† Phil. Trans. 1758. Vol. L p.777.　　　　　　‡ Phil. Trans. 1762. Vol. LII. p. 75.

1

animal that inhabits many of the shells. Shells constituting a beautiful set of bodies, easily collected and preserved, were very early formed into cabinets. The study of them was long ridiculed by the wits of the age, as an abuse of time and waste of money; but its importance has lately become strikingly evident to all those who make the structure of the earth, and the various changes which it has undergone, an object of their attention. For shells are found in abundance in a great variety of rocks and positions; they constitute the medals of the ancient world; and from an accurate acquaintance with their different species, and with the nature of the animals that inhabited them, many curious and important deductions respecting the formation and changes of the crust of the earth may be drawn.

The first important works on conchology were by Bonanni and Lister. Dr. Lister Lister's book formed a new æra in the science, and contributed chiefly to give celebrity to this excellent naturalist. He practised medicine for many years at York; where he cultivated entomology and conchology with much ardour and success. His circumstances must have been at least easy, as he invited Mr Ray to come and live with him, when that admirable philosopher was, in some measure, turned loose by the death of his friend and patron Mr. Willughby Dr. Lister afterwards settled in London, and was an early and active Member of the Royal Society. His *Historia sive Synopsis Methodica Conchyliorum* was published in 1685. It contains very accurate figures of all the shells known in his time, amounting to upwards of 1,000. It deserves to be recorded that they were all drawn by his two daughters, Susanna and Anne Lister The copper-plates of this work became the property of .the University of Oxford, and a second edition was published at Oxford in 1770, under the care of Huddesford. This edition wants two or three of the plates in the original work; but to make up for this deficiency, two or three new plates have been added. The figures of shells which it contains amount to 1,153, besides a number of anatomical drawings and representations of fossil bodies. This work of Lister, notwithstanding the progress which the study has since made, still retains its value, and is indispensable to the student of conchology.

Gualtieri's *Index Testarum Conchyliarum*, published at Florence in 1742, is Gualtieri. remarkable for the accuracy of its specific distinctions. But the most important work on shells which has hitherto been published, a work which it will be very difficult to surpass, is a Danish one in eleven quarto volumes, begun in 1769, and finished in 1795. The first three volumes were drawn up by Martini, Martini. the last eight by Chemnitz. This work contains figures of above 3,800 shells, all elegantly coloured.

The arrangement of shells began at a very early period. Aristotle divided Arrangement of shells. them into univalve, bivalve, and turbinated, and he imposed the names of various genera still retained by modern naturalists; for example, *lepas, solen, pinna,* *nerita.** Hardly any improvement in the arrangement was made from the time of Aristotle to that of Linnæus. This celebrated naturalist at first established eight genera of shells, namely, *cochlea, nautilus, cypræa, haliotis, patella, dentalium, concha, lepas.* But he afterwards found the necessity of taking several of these genera to pieces, so that in consequence he increased the number of genera to 35. Several valuable works on the subject have lately appeared in Britain. Donovan's and Montague's books deserve to be mentioned with applause, and much merit is due to the catalogue of British shells published by Racket and Maton, in the Linnæan Transactions, and the catalogue of Scotch shells by Captain Laskey in the Wernerian Transactions.

The number of testacea, described in the Philosophical Transactions, amount to 60; though it must be acknowledged that several are included in that number that properly belong to the mollusca, as they consist of animals not provided with any shell. The difficulty of determining their specific names has induced me to allow them to remain where they were placed by Ellis in his description of barnacles, among which he has thought proper to class them. The following is a list of these testacea, together with the names of the authors to whom we are indebted for the descriptions:

Species described in the Transactions.	Turbo bidens. } .... Anonymous.[†] Turbo perversus.}	Pholas conoides.................Parsons.	
	Twenty-seven snail shells.[‡] .... Lister.	Twenty species of barnacles.§ .... Ellis.	
	Murex canaliculatus. .... Witzen.	Nautilus lacustris...............Lightfoot.	
	Ostrea Virginica. ....Ditto.	Helix fontana........ ....Ditto.	
	Buccinum lapillus, purple fish. .... Cole.	Helix spinulosa.....................Ditto.	
	Lepas diadema, pediculus ceti...... Sibbald.	Turbo helicinus ....Ditto.	
	Teredo navalis. ....Baster.	Patella oblonga.‖ ....Ditto.	

Zoophites. The zoophites are a class of animals most of which inhabit the sea, and have so striking a resemblance to plants, that for many ages they were classed by naturalists in the vegetable kingdom. The discovery of their real nature was first made by Peyssonnel, a French physician, who had been many years settled at Guadaloupe. He sent a manuscript Treatise on Coral to the Royal Society, a full account of which, drawn up by Dr. Watson, was published in the Transactions for the year 1751. The subject was much more completely and accurately investigated by Mr. John Ellis, a merchant in London, who died in **Ellis.** 1776, one of the greatest naturalists who adorned the last century, so prolific in discoveries concerning subjects connected with natural history. It does not

---

* See Aristotle, Book IV. Chap. IV.
† Phil. Trans. 1669. Vol. IV. p. 1011.   ‡ Phil. Trans. 1674. Vol. IX. p. 96. They are chiefly helices, buccina, buliæ, with two mytili. Figures are given so that the species may be readily determined.
§ Phil. Trans. 1758. Vol. L. p. 845. Most of the barnacles described belong to the genus lepas, though not the whole.   ‖ Phil. Trans. 1786. Vol. LXXVI. p. 160.

appear that Mr. Ellis was assisted in his investigations by the Treatise of Peyssonnel. His mind was first turned to the subject by a collection of corralines sent him from Anglesey, which he arranged upon paper so as to form a kind of natural landscape. Dr. Hales, in one of his visits to Mr. Ellis, was struck with the beauty of the picture, and requested him to make some similar ones for the Princess Dowager of Wales, to contribute to the instruction and amusement of the young princesses. The request was complied with : Mr. Ellis, by means of his friends who resided on the sea coast, made a new collection of corralines, and he was induced to examine them with a microscope, and to attempt to arrange them according to their appearances. These observations inducing him to suspect that several of his corralines were animals, he drew up a paper on the subject, which was read at a meeting of the Royal Society in 1752. His opinions were still further confirmed by several other Members of the Society ; but being conscious of a deficiency of accurate knowledge, he withdrew his paper, and formed the resolution of making a set of observations on corralines in situ. For this purpose he went, in August 1752, to the Isle of Sheppy, accompanied by Mr. Brooking, a painter, and the observations which he made still further confirmed him in his opinions. In 1754, he prevailed on Ehret, the celebrated engraver, to accompany him to Brighthelmstone, where they made drawings, and formed a collection of zoophites. In 1755, he published ·the result of all his investigations under the title of an *Essay towards a Natural History of Corralines*. This is one of the most accurate books ever published, whether we consider the plates, the descriptions, or the observations which demonstrate the animal nature of the zoophites. His opinions on the subject were opposed by Job Baster, a Dutch physician and naturalist, who published various dissertations in the Philosophical Transactions in order to prove that corralines were of a vegetable nature. But his arguments were victoriously refuted by Ellis, whose opinions on the subject were almost immediately assented to by naturalists in general, and have been only further confirmed by every subsequent examination of the subject.

In 1786, a posthumous work, entitled, *Natural History of many curious and uncommon Zoophites*, was published by a daughter of Mr. Ellis at the request of Sir Joseph Banks. The plates had been completed under the inspection of Mr. Ellis himself, and the descriptions had been drawn up by Dr. Solander. In this work the zoophites are divided into 16 genera ; namely, *actinia, hydra, flustra, cellaria, tubularia, sertularia, pennatula, gorgonia, antipathes, isis, corralina, millepora, madrepora, alcyonium, spongia.* It is scarcely necessary to observe that this is the best systematic account of the zoophites which has yet appeared. The subject has also been investigated by Bohadsh and Muller, and above all by Pallas, whose work on zoophites is classical.

The species of zoophites described in the Transactions amount to 18. The following is a list of them, together with the names of the authors to whom we are indebted for the descriptions:

Hydra viridis, common polypus.(a)	Sertularia pumila.(b) .... Ditto.
Leuwenhoeck.	Sertularia scruposa.(b) .... Ditto.
Sertularia neritina. .... Ellis.	Alcyonium Schlosseri. .... Schlosser.
Pennatula encrinus . ... Ditto.	Isis asteria. .... Ellis.
Sertularia geniculata.(b) .... Ditto.	Pennatula phosphorea. .... Ditto.
Sertularia antennina.(b) .... Ditto.	Pennatula purpurea.(c)............ Ditto.
Sertularia volubilis.(b) .... Ditto.	Three species of sponge (d) from the Mediter-
Sertularia rosacea.(b) .... Ditto.	ranean.................... Strange.

The *infusoria* consist of animalcules almost always too minute to be visible to the naked eye, and observed in water either fresh or stagnant. The number of them is very considerable; but from their extreme minuteness it is exceedingly difficult to determine their structure with any tolerable accuracy. Leuwenhoeck, one of the first describers of these little animals, calculated that some of the species which he observed were so small that a common-sized drop of water might contain a million of them. The following is a list of those described in the Transactions, together with the names of the authors to whom we are indebted for the descriptions:

Vibrio aceti.(e) .... Leuwenhoeck.	Vorticella convallaria.(h) .... Ditto.
Eight species described by .... Ditto:(f)	Vorticella anastatica.(i) .... Brady.
Volvox globator.(g) .... Ditto.	Vorticella socialis.(i)............... Ditto.

Such are the species of vermes described in the Philosophical Transactions. But that voluminous work contains many particulars respecting the œconomy and animal nature of various species which deserve to be mentioned. The hydatids, or small watery vesicles, which occur so frequently in the diseased organs of the human body and of the inferior animals, especially the liver, were long considered as nothing else than small bladders filled with water. Dr. Tyson first showed them to be a species of worm, and to be very tenacious of

(a) Phil. Trans. 1703. Vol. XXIII. p. 1304. This animal was afterwards particularly described by Trembley. Phil. Trans. 1742. Vol. XLII. p. 283. Reaumur observed it about the same time.

(b) Phil. Trans. 1754. Vol. XLVIII. p. 627. 'It was in this paper that Ellis first demonstrated the animal nature of the sertularia.

(c) Phil. Trans. 1763. Vol. LIII. p. 419. In this paper Ellis describes three other pennatulas from Bohadsch.

(d) Ibid. 1770. Vol. LX. p. 179. Ellis first explained the nature of sponge in a satisfactory manner in a paper published in 1765. Phil. Trans. Vol. LV. p. 280.

(e) Phil. Trans. 1676. Vol. XI. p. 653.

(f) Ibid. 1677. Vol. XII. p. 821. His description is so imperfect that it is not easy to determine the animals he alludes to.

(g) Phil. Trans. 1700. Vol. XXII. p. 509.

(h) Phil. Trans. 1703. Vol. XXIII. p. 1304. He describes also other species of vorticella in the same paper.

(i) Ibid. 1755. Vol. XLIX. p. 248.

life.* They are now considered as tæniæ, and constitute two species of that very extensive genus ; namely, the *tænia visceralis* and *tænia cellulosa*.

The manner in which the young of the murex canaliculatus, a species of shell fish, are produced is curious, and was first pointed out by Mr. Witzen.† It holds with all the species of the *murex* and the *buccinum*. At a certain season of the year a kind of matrix or sack, about 12 inches long, issues from the shell. This long body, resembling a sausage, is filled with a great number of round cells, each containing its little shell fish. These gradually increase in size till at last their weight breaking them off from the common matrix, they fall to the bottom of the sea and shift for themselves. *Production of certain shell fish.*

The celebrated *purple dye* of the Tyrians, so highly prized by the ancients, is well known to have been obtained from a certain species of shell fish found upon the coast of Phenicia. The use of this dye was superseded by the introduction of cochineal, which yields a much more beautiful though not so permanent a colour. The animals inhabiting various species of *murex* and *buccinum* furnish this colouring matter. Among these none is more remarkable than the *buccinum lapillus,* a shell fish to be found in abundance on the shores of Britain. The nature of this colouring matter has not yet been ascertained, though from the suite of changes which take place when it is applied to cloth it would appear to have some analogy to indigo, or to consist of a base capable of uniting with various doses of oxygen, and of varying its colour accordingly. There is a curious paper on the subject by Mr. William Cole of Bristol, in an early volume of the Transactions.‡ A person living on the sea coast in Ireland had made considerable profit by marking fine linen for ladies and gentlemen of a delicate durable purple colour. Mr. Cole being informed that the colouring matter was obtained from a shell fish, examined all the shell fish on the coast, and at last in the buccinum lapillus found a liquid which answered the purpose. When the shell is removed, and the animal exposed, a white vessel appears near the head of the fish, and the liquor in this vessel possesses the tinging property. When applied to linen it appears at first of a pleasant green colour, and the writing being exposed to the light the green gradually increases in intensity, and at last changes to blue ; the blue soon acquires a tint of red, and at last becomes a fine purple of great intensity ; and beyond this no further change is produced by the action of the light of the sun. These changes go on with greater or less celerity, according to the intensity of the light to which the linen is exposed, and accordingly take place much more rapidly in summer than in winter. *Tyrian purple dye.*

About the middle of the last century, the peculiar nature of the hydra viridis, *Polypus.*

* Phil. Trans. 1691. Vol. XVII. p. 506.   † Phil. Trans. 1693. Vol. XVII. p. 870..
‡ Phil. Trans. 1695. Vol. XV. p. 1278.

or fresh-water polypus, an animal originally discovered by Leuwenhoeck, and described by him in an early volume of the Transactions, drew the attention of naturalists. It was found that it could be propagated indefinitely by cutting it in pieces, every piece spredily becoming a complete animal. Trembley, a philosopher of Geneva, but at that time residing in Holland, seems to have been the first that ascertained the point by correct experiments. His paper was no sooner published in the Philosophical Transactions, than a multitude of experimenters set themselves to verify the observations which he had made. Of these, the most eminent was Reaumur, whose curious experiments on the subject were published in the Memoires of the French Academy. Martin Folkes also, at that time President of the Royal Society, and who valued himself upon his knowledge of zoology, published a set of observations on the same animal. Dicquemare extended his observations to the *actinia*; the different species of which he distinguished by the name of the sea anemonies; and showed that they could be propagated by cutting them in pieces, as well as the polypus; and that when a piece was cut from them they soon recovered again the lost member.*

Insects.

VI. Insects. Insects constitute by far the most numerous class in the animal kingdom, not fewer than 18,000 species being already described and classified. The first book on the subject was the joint production of several learned men, among whom was Conrade Gesner. It was published in 1634,

Mouffet's treatise.

by Dr. Mouffet, an English Physician. The subject was cultivated with great success about the beginning of the last century, by Goedart and Swammerdam. To this last writer we are indebted for a great part of the anatomical knowledge of the structure of these animals, which we at present possess. His book is written in a dull prolix style; but is fraught with much curious and important information. Reaumur turned his attention to insects soon after the beginning of the last century; and the voluminous book, which he published on the subject, is by far the most valuable of all his numerous works. He has laid open to us as it were a new world, and has traced the surprising changes which these animals undergo, with much persevering industry and success. It is to this book chiefly that Reaumur is indebted for the celebrity which he has acquired, and which will last as long as Natural History continues to be cultivated or respected. De Geer's book on insects is the counterpart of Reaumur's, and hardly less extensive or accurate.

Linnæus.

Linnæus was the first person who succeeded in arranging this numerous class of animals into genera and species. His classification of them is admirable,

Fabricius.

and has been most generally followed. Fabricius indeed, who was his pupil,

* Phil. Trans. 1773. Vol. LXIII. p. 361; and in two subsequent dissertations, published in Vol. LXV. p. 207; and Vol. LXVII. p. 56.

and who devoted the whole of his life to the study of these animals, contrived a new classification; but, by the generality of entomologists, it is considered as inferior in value to the Linnæan. Fabricius, however, must be admitted to have described a much greater number of insects than any other naturalist. He was a Dane; born at Tundern, in 1742. He was, during the greatest part of his life, Professor of Natural History, at Kiell; and died of a dropsy, in 1807, in the sixty-fifth year of his age. Except by Dr. Lister, this branch of zoology has not, till lately, been much cultivated in Britain; but we at present possess several persons very well acquainted with it; and some of the works on British insects, which have lately appeared in this country, are of very considerable value.

The number of insects described in the Philosophical Transactions, amount to 60 species. The following table exhibits a list of their names, together with the authors to whom we are indebted for the descriptions:

Species described in the Transactions.

Apis willughbiella. (a) King and Willughby.	Coccus polonicus. (l)............Breyne.
Thirty three species of spiders. (b).... Lister.	Monoculus apus. (m)..............Klein.
Three species of ants. (c)...... ......King.	Ephemera vulgata. (n)...........Collinson.
Lampyris italica, flying glow worm. (d) Waller.	Phalena ecconymella. (o).........Skelton.
	Sphex, from Pennsylvania (p)......Bartram.
Phalena granella, wolf. (e).... Leuwenhoeck.	Coccus cacti coccineliferi, cochineal. (q) Ellis.
Musca putris, cheese maggot. (e) ....Ditto.	Cicada septendecim. (r).........Collinson.
Scarabæus melolontha, cockchafer. (f) Molyneux.	Coccus polonicus, Polish cochineal. (s) Wolfe.
	Cancer stagnalis. (t) ....King.
Ptinus fatidicus, death watch. (g) .... Allen.	Aphis, of the rose tree. (u)......Richardson.
Termes pulsatorium, death watch. (h) Derham.	Different species of the monoculi. (v) Muller.
Acarus, which occasions the itch. (i) Bonomo.	Termites, of Africa. (w)........Smeathman.
Dermestes polygraphus. (j) ....Dudley.	Coccus lacca, lac insect. (x) ....Ker.
Vespa, common wasp. (k)..........Derham.	Tenthredo, that destroys turnips. (y) Marshall.

The œconomy and transformations of insects present some of the most curious and interesting particulars to which the naturalist can direct his atten-

(a) Phil. Tans. 1670. Vol. V, p. 2098, and 2100.
(b) Ibid. 1671. Vol. VI, p. 2170. They are all English insects, but the last four in the list are not spiders.
(c) Ibid. 1667. Vol. II. p. 425.　　　　　(d) Phil. Trans. 1685. Vol. XV. p. 841.
(e) Ibid. 1694. Vol. XVIII. p. 194.　　　(f) Ibid. 1697. Vol. XIX. p. 741.
(g) Ibid. 1698. Vol. XX. p. 376.　　　　(h) Ibid. 1701. Vol. XXII. p. 832; and Vol. XXIV. p. 1586.
(i) Ibid. 1703. Vol. XXIII. p. 1296.　　(j) Ibid. 1705. Vol. XXIV. p. 1859.
(k) Ibid. 1724. Vol. XXXIII. p. 53.　　(l) Ibid. 1732. Vol. XXXVII. p. 444.
(m) Ibid. 1738. Vol. XL. p. 150.　　　　(n) Ibid. 1746. Vol. XLIV. p. 329.
(o) Ibid. 1748. Vol. XLV. p. 281.　　　(p) Ibid. 1749. Vol. XLVI. p. 278.
(q) Ibid. 1762. Vol. LII. p. 661.　　　　(r) Ibid. 1764. Vol. LIV. p. 65.
(s) Ibid. 1764. Vol. LIV. p. 91. LVI. p. 184.　(t) Ibid. 1767. Vol. LVII. p. 72.
(u) Ibid. 1771. Vol. LXI. p. 182.　　　(v) Ibid. 1771. Vol. LXI. p. 230.
(w) Ibid. 1781. Vol. LXXI. p. 139.　　(x) Ibid. 1781. Vol. LXXI. p. 371
(y) Ibid. 1783. Vol. LXXIII. p. 217.

tion   The work of Reaumur is full of these accounts, and owes to them a great part of its interest.   Several very curious details likewise are to be found in the Philosophical Transactions; and it will be worth while to notice the most remarkable of these in this place.

Apis willughbiella.

The way in which the apis willughbiella of Kirby is hatched was first discovered by Dr. Edmond King; and was farther elucidated by Mr. Willughby. The parent bee forms long cells in the wood of old willows, and in these borows, which are always in the direction of the grain of the wood, she deposites her eggs.   Each egg is enveloped in pieces of rose leaves, rolled up in the form of a cartridge, and containing along with it a quantity of food, sufficient to support the young animal while in the state of a maggot.   The maggots, after consuming the greatest part of their food, surround themselves with a kind of theca, where they undergo their transformation, and at last issue out in the form of full-grown bees.*

Ichneumon fly.

The ichneumon fly deposites its eggs in the body of different caterpillars. Sometimes the quantity of maggots produced from them is so great, that the caterpillar shrinks up and dies; sometimes it changes to an aurelia, but never becomes a butterfly, always dying before-hand; while the ichneumon flies make their way out of the body of the animal.†

Ants.

The œconomy and industry of ants has been celebrated in all ages.   The opinion, indeed, entertained by the ancients, that they laid up food for the winter, is not true; because these animals remain in a torpid state during the cold weather, and of course have no occasion for food during that period Their spawn is a white substance, similar to pounded sugar or the crumb of bread.   It contains in it white specks, which, when viewed with a microscope, resemble the eggs of fowls.   These specks are gradually changed into maggots, as small as mites: these grow to a certain size, and are then surrounded with a white bag.   In that state they are of the size of a grain of wheat, and are commonly, though improperly, called ants' eggs.   The little animal, in these cases, is gradually changed into a perfect ant, which, at a certain period, breaks from its prison and mingles with its fellows.‡

Spiders.

The spider tribe are commonly viewed with dislike by the generality of mankind; partly, no doubt, on account of their appearance; but chiefly on account of their voracity, and the deceitful manner in which they entrap other insects. But their œconomy and manners are exceedingly striking, and have attracted the admiration of all those naturalists who have devoted any part of their time to the examination of these industrious animals.   Dr. Lister was one of the earliest Members of the Royal Society, who attended to the œconomy of the

* Phil. Trans. Vol. V. p. 2098.   † Phil. Trans. Vol. VI. p. 2279.
‡ Phil. Trans. Vol. II. p. 425.

spider tribe; and there are various papers by him in the Transactions, on the subject. Willughby and Ray likewise endeavoured to elucidate several particulars respecting them; but it is to Leuwenhoeck that we are indebted for the most minute and curious details respecting them. Their number is very considerable; a catalogue of no fewer than 33 British species having been drawn up by Dr. Lister. They are each furnished with eight immoveable eyes, placed in various directions to enable them to see on all sides at once, so as the better to observe and secure their prey. They have likewise eight very long legs, besides two shorter ones near the mouth, which answer the purpose of hands, and which are armed with two rows of teeth, and with a long hollow moveable fang, through which they infuse the poison into the wound which they inflict. They deposit an incredible number of very small eggs, which they envelope round with a very strong web, and which are gradually hatched by the heat of the weather. The young spiders are at first exceedingly small, so as to be barely visible to the naked eye; but they very rapidly increase in size, and begin to spin. The instruments of spinning are in the posterior part of their body, and no fewer than eight different places, at which threads issue, may be discerned. Each of these threads, according to Leuwenhoeck, when it issues at first, consists of a great number of exceedingly fine threads, which speedily unite and form one large one. The animal has the power of darting out a very long thread, not attached to any thing, by means of which it is wafted about in the air, where it often remains a considerable time, and where it is doubtless employed in catching its prey. In spinning its web, it makes the extremity of the thread adhere to any substance that it chooses, by pressing the part of the body from which the thread is to proceed against it. It then moves off to a distance, and by that means the thread is drawn out. Spiders are exceedingly voracious, so that they attack and destroy one another whenever they meet; and, in case of a rencounter, never separate till one of them is lifeless. They can live without food for a very long period; but always eat with the greatest voracity when it is in their power. They not only suck out the juices of those insects which they catch, but also devour their solid parts. Leuwenhoeck even affirms that wings and legs of flies may be sometimes observed, unaltered, in their excrements.*

It is well known that certain species of winged insects sometimes appear in Scarabæus such numbers, as to destroy all vegetation for many miles round those places melolontha. which they haunt. The locust is a well-known example of this kind. There is a species of beetle, the *scarabæus melolontha*, which occasionally proves nearly as destructive. Dr. Molyneux relates an example of this which occurred in Ireland, in 1688. Vast swarms of these insects were brought into Galway,

* Phil. Trans. Vol. XXII. p. 867.

by a south-west wind; during the day they hung in clusters from the boughs of trees, but about sun set they took wing, and darkened the air with their vast numbers. It was difficult to make one's way through them; for they dashed themselves against the face, and occasioned a considerable pain. They fed upon the leaves of trees, and the noise they made in eating was like the sawing of timber. For two or three miles round every tree was completely stript of its leaves, so that the county bore the appearance of the depth of winter. They made their way also into the gardens, and devoured the blossoms and leaves of the fruit trees, and some of the more delicate were entirely killed. Smoke was found to be a preservative. The burning of heath and ferns kept them out of gardens, or drove them out if they had already taken possession. Towards autumn they entirely disappeared, probably lodging under ground in a dormant state; for next season immense quantities of them were found in some places in holes under ground. They deposited their eggs below the surface of the earth, which soon were hatched and became white caterpillars, which fed upon the roots of the corn, and thus destroyed the future harvest. Cold weather proved very injurious to them, and destroyed them by millions.*

Fleas.
     Cestone has given an account of the transformation of common fleas, which is precisely similar to what happens to winged insects. According to him, these animals deposite eggs of the size of nits upon animals, or articles of furniture. From these eggs, in process of time, a worm is generated, which moves about very nimbly, and when it has acquired a certain size spins itself a theca, in which it remains enveloped till it is converted into a perfect flea. In this state it bursts its prison and escapes.†

Crab.
     There is no tribe of insects whose œconomy is more singular or more unaccountable than the crab. These animals associate together in considerable numbers, and if carried off and dropt into the sea, at the distance of several miles, they find their way back again to their old haunts. They cast their crust once a year, and when deprived of it look like a mishapen mass of jelly. The surface of this jelly gradually hardens into a new crust, of larger dimensions than the old; for it is in this way that the creature increases in size. If any of its legs be wounded, it has the property of throwing off the whole leg at the joint, a quantity of mucilaginous matter issues out from the extremity, which immediately stops the bleeding and is gradually hardened into a new limb. If this mucous matter be forcibly removed, the animal speedily bleeds to death ‡

May fly.
     Many insects continue a very considerable period in their caterpillar and aurelia state; but after they have assumed the appearance of a winged insect,

---

* Phil. Trans. Vol. XIX. p. 741.   † Phil. Trans. Vol. XXI. p. 42.
‡ Collinson. Phil. Trans. Vol. XLIV. p. 70.

very speedily die.　One of the most remarkable of these is the *ephemera vulgata*, or May fly; of which a very entertaining account is given in the Transactions, by Mr. Collinson.　For the greatest part of the year the nymph lies quiet at the bottom or sides of rivers; but about the end of May it rises to the surface, splits open, and out starts a beautiful fly, with three long remarkable hairs at its tail.　These flies attach themselves to the trees that vegetate on the banks of the river, and in a day or two they cast their skin and leave it behind them The males are now much smaller than the females.　They fly under the trees at some small distance from the river, whither the females repair, and fly briskly about for some hours together, striking against each other in the air　When impregnated, the females return to the river, and drop their eggs into it successively for two or three days.　When this task is finished, their strength is exhausted, and they either drop into the river or become the prey of those fish and birds which feed upon them.　The males never repair to the river at all, but drop down and die under the trees, where they originally took up their habitation.　During the whole of this period they never eat, and are not even provided with an apparatus for that purpose; the sole purpose of their assuming the winged state being to copulate and deposite their eggs.*

There is a set of small insects, called aphides, which inhabit peculiar trees, Aphides. and almost every tree has a species, and sometimes two or three species of its own; so that this tribe of insects is exceedingly numerous; and it is very singular and extraordinary in another respect.　Dr. Richardson has given in the Transactions a particular account of the largest species of aphis which inhabits the rose tree.　Eggs are deposited by the animal on the branches of the plant in autumn, and as near the bud as possible.　These eggs are covered with a hard crust, and are not injured by the coldness of the winter.　As soon as the mild temperature of spring comes on, the eggs are hatched, and the animals produced are all females.　These females, without any intercourse with the male sex, soon produce a new offspring; this offspring is not an egg, but a female animal like its mother.　This female produces young like its mother, all likewise females; and in the course of the summer no less than ten generations succeed one another.　The first nine of these are all females and viviparous; but the tenth consists of a mixture of males and females.　The males have wings, but the females are destitute of these organs.　They copulate during autumn, and the female deposites eggs, which remain on the plant all the winter, and are hatched next year; so that, in these singular creatures, one copulation produces no less than ten generations, without the necessity of a new impregnation.　The fecundity is now exhausted, males and females are produced;

* Phil. Trans. Vol. XLIV. p. 329.

the impregnation is again repeated, and proves efficacious as before for the production of ten successive generations.*

The number of insects which inhabit the warmer regions of the earth, is much greater than we, who inhabit the colder climates, have any conception of. The great purpose which these little animals answer in the economy of nature, is to consume and destroy dead animal and vegetable matter, in order to prevent the baneful effects which would result from an accumulation of putrid substances. As the rapidity of putrefaction is much greater in the torrid than in the temperate zones, it is obviously necessary that those substances which are liable to undergo the putrid change should be more speedily removed out of the way. Hence the reason of a great increase of the insect tribe, whose task it is to perform this great and important office. Of all these insects, there is none **Termites.** which performs more work in this way than the *termites*, or *white ants*, as they are improperly termed ; nor is there any set of insects whose economy is more singular or perfect. They exist together in kingdoms, and build cities, which are extremely populous. There are a considerable number of species, some of which build their nests or cities on trees, others on the ground ; but perhaps the most singular species of all is the *termes bellicosus* ; of which an excellent account is given in the Philosophical Transactions, by Mr. Smeathman. They are found both in India and Africa ; but the species which Smeathman observed inhabited the middle regions of Africa. They build their habitations on the surface of the ground, in the form of a cone rounded on the top, of the height of four or five feet, and of vast extent, considering the size of the animals. The walls are built of clay, and so very strong as to bear the weight of the heaviest animals, without being crushed. Wild bulls often station themselves on their summits ; and they are usually pitched upon by Europeans as stations, when they wish to observe any objects at a distance. They contain a vast number of compartments ; that in the centre always holds the king and queen, and is increased in size as necessity requires. Other surrounding compartments are destined for the attendants of the queen, which are very numerous, scarcely fewer than 100,000. Other compartments are used as nurseries for hatching the eggs and rearing the young, till they become capable of joining the rest of the society. Others are destined for holding provisions, of which they have always a plentiful store provided. The society consists of three different kinds of animals ; or, to speak more accurately, of the termites in three different states. The first state is that of *labourers*.; they consist of small animals, without wings or visible eyes, and without any distinction of sex. These are by far the most numerous, and upon them the whole labour of the community

---

* Phil. Trans. Vol. LXI. n. 182.

falls. They build their cities and procure the provisions, and feed and nourish the young. The *soldiers* constitute the second state of the animal. They exist in every colony, in considerable numbers, though they are much fewer than the labourers. They are about the size of 15 labourers, have a very large head armed with a kind of awl-shaped forceps, with which they penetrate and wound those animals which they attack. Their business is to defend the community, which they do with the greatest obstinacy and bravery. They seem to exercise a kind of control over the labourers. The third state is that of the king and queen. This is the perfect state of the animal. In this state they are furnished with four wings and two bright eyes, and the distinction of sex is very obvious. Whenever they undergo this change, they immediately fly away, in order to establish new colonies. Their wings last them only a few hours, and are then broken off or rendered useless. In this state they are attacked by an infinite number of enemies; birds, fish, insects, and even man, feed upon them with voracity, so that very few escape with life. A few pairs, however, fall in with some of the labourers of their own species. These little animals immediately pitch upon them for their king and queen, and without loss of time inclose them in a wall of clay, round which they raise their whole city, with wonderful celerity. The holes leading to the royal cell are only large enough to admit the labourers, so that the royal pair cannot get out; but are ever after fed and protected by the labouring insects. The abdomen of the queen gradually increases prodigiously in size; so that she becomes as bulky as 20 or 30,000 labourers, and she lays an infinite number of eggs, laying at the rate of 60 in a minute, or 80,000 and upwards in 24 hours. These eggs are immediately removed by her attendants, and carried to the nurseries. These animals never venture out in the open air, where they are no match for their numerous enemies, but construct large galleries under ground, by means of which they make their way to their favourite articles of food. Decayed wood, or wood that is not vegetating, is the article which they chiefly select, and scarcely any wood is capable of resisting their destructive ravages. When they make their way into a stake or beam of wood, they hollow it completely within, but leave the outside entire, so that a board or shelf, which shall appear to the eye entire, shall be no heavier than if it were a hollow bag of paper. When the beams have a weight to support, the animals are aware of this, and fill up the interstices as they eat them out, with moistened clay, so that in a very short time the wood is converted into a beam of clay. Boxes, fallen trees, stakes in a hedge, nothing can resist their destructive ravages; and when once they make their way their numbers are so immense, that they complete the destruction in a very short period of time. When their cities are broken open the soldiers march out with impetuosity, and attack every thing that comes in their way. Wherever they strike they immediately draw blood, and they not unfrequently beat off the bare-legged negroes altogether. As soon as the assailant has withdrawn, the labourers issue forth in prodigious numbers, and very speedily erect a new wall, and repair the damage which their habitation has sustained.*

Coccus.

The *coccus* constitutes a genus of insects, of a very peculiar nature. The female fixes herself immoveable upon the branch of some tree, and, at a certain period, the young animals burst their way through the body of their parent. These animals furnish coloured juices, which have been always highly prized as dye stuffs. The cochineal, which is a species of this animal, a native of Mexico, and inhabiting a kind of cactus, furnishes the finest red colour known, and has, in some measure, superseded the use of some other species, formerly cultivated in Europe, for a similar purpose. But perhaps there is no species of this singular animal of more importance than the coccus lacca, which inhabits the uncultivated mountains on both sides of the Ganges, and is found on four species of trees; namely, the ficus religiosa, ficus indica, plaso, and rhamnus jujuba. But upon the last it is not so common. An account of this insect was published in the Philosophical Transactions, by Mr. James Kerr. The insect attaches itself to the branches of the tree which it inhabits, and is gradually surrounded by a clear viscid juice, which soon hardens and incloses the animal in a cell. This matter constitutes what is called *gum lac*, a species of resin, which probably exudes with little alteration from the fig-trees upon which the animal is fixed. It is produced in great abundance by these little insects, and is an article of great importance in India, and even in Europe, being applied to a great variety of uses. The insect itself contains a red juice, which is employed in India for dying red, and gives to lac a considerable part of its value. The lac, before it is imported into Europe, has been deprived of its colouring matter by the Indian dyers.†

Section II.—*Of Anatomy.*

The various diseases and accidents, both internal and external, to which mankind are liable, would necessarily turn their attention to medicine at a very early period. The art of healing would speedily become a profession, and those who devoted themselves to it would almost immediately discover the importance of being acquainted with the structure of the human body, and endeavour, by every means in their power, to make themselves acquainted with that structure. In general diseases indeed, such as fever, palsy, epilepsy, &c., the seat of the disease was not so apparent, and the malady might therefore be ascribed to the malevolent interposition of some deity, or the preternatural power of some enemy, and might accordingly be administered to by sacrifices, charms, and

---

* Phil. Trans. Vol. LXXI. p. 139.   † Phil. Trans. Vol. LXXI. p. 374.

incantations. But when any of the organs were disordered, or when any wound was inflicted, the seat of the malady was obvious to every one, and every rational method of cure must be dependent in some measure upon a knowledge of the structure of that part where the disease was situated. The importance of this kind of knowledge was so obvious, that we find marks of it in the most ancient writings upon record. Innumerable examples, for instance, occur in Homer of his knowledge of the internal structure of the human body. A spear is thrust through the right side of Pharcelus under the bone to the bladder;* Apizaon's liver is wounded by Euripulus's spear piercing under the precordia;† the acetabulum coccendicis of Æneas was broken, and the nerves near it, so that he must have remained for ever lame had not his cure been undertaken by his goddess mother. ‡

But such was the veneration which the ancients entertained for their dead, and such the guilt which they ascribed to the violation of them, that few opportunities would occur to ancient physicians of dissecting human bodies, and these few they would hardly venture to indulge. The only knowledge of the human body which they could acquire would be derived from the practice of embalming, and from occasional wounds which would expose some of the internal cavities of the body. They would be under the necessity of confining their dissections to brutes, and deducing from their observations upon them inferences respecting the structure of the human body. These prejudices were never overcome by the ancients. Accordingly we are not to look among them for any accurate anatomical knowledge The three most celebrated medical men among the ancients, any of whose writings have come down to us, were Hippocrates, Celsus, and Galen. *Anatomy of the ancients.*

Hippocrates, who lived during the Peloponnesian war, was the father of medicine, and in point of reputation stands first in the list of medical men. As to anatomy his works contain very little. There is not even evidence sufficient to convince us that he had attempted to dissect brutes, far less human bodies. Though the well known story of his interview with Democrates of Abdera, whom he was employed to cure of madness, is a demonstration that Democrates, at least in those days, was sensible of the great importance of anatomy, and had applied himself to determine the structure of the inferior animals. Celsus, who flourished during the Augustan age, has left us a system of medicine written with great elegance and purity of style. But the anatomical part is not more perfect in him than in Hippocrates. Indeed the few anatomical observations which he gives are evidently translated from the writings of the Grecian physicians his predecessors. Galen is the writer that contains by far the most anatomy of all the ancients. The reputation that he gained was enor- *Hippocrates.* *Celsus.* *Galen.*

* Iliad, Book V. v. 65.   † Iliad, Book XI. v. 574.   ‡ Iliad, Book V. v. 304.

O

mous, and for many ages he ruled over medicine with despotic sway. His writings constituted the oracles of the schools, and to venture to call any of his opinions in question was deemed unlawful and heretical.

He has given a much more complete anatomical account of the human body than any of his predecessors, or even successors, for a thousand years after. There can be no doubt that he dissected the bodies of the inferior animals. But Vesalius, the first of the moderns who ventured to call in question his infallibility, affirmed that he never had dissected a human subject, and that all his notions respecting the structure of the human body were drawn from his examination of the lower animals. This assertion occasioned a violent controversy, which did not soon subside. The physicians of the times started up in arms, and in formidable numbers, to defend their oracle. But Vesalius, supported by his knowledge of human anatomy, was invincible, though single, and victoriously repelled all their attacks. At present when the dispute has been for many years at an end, and the reputation of Galen brought to its true level, it is the opinion of some of the most learned and laborious anatomists, and some of the best judges, that the accusation of Vesalius is well founded, and that Galen never did dissect a human body. This in particular was the opinion of Haller If this judgment respecting Galen is to be adopted, and for our part we think it very likely to be correct, then it follows that the anatomy of the human body was an art with which the ancients were unacquainted, and that all their knowledge of the science was derived from the dissection of brutes.

It is not necessary to mention any of the Grecian physicians who lived after Galen, nor the Arabian physicians who flourished during the age of the Caliphs, because neither of them added any thing to the anatomical knowledge which they derived from the writings of Galen. Nor need we notice the schoolmen who disputed about medical subjects in the different universities of Europe during the dark ages. Mundinus indeed, who lived in the 13th century, and wrote a treatise on the three cavities of the human body, gives proof that he was a practical anatomist, and has made some additions to the anatomy of Galen. His book is a barbarous jargon, half Latin, half Arabic; but it gained him a high reputation, and was for a considerable time taught in all the universities of Italy.

Mondinus

The destruction of the Grecian empire, and the capture of Constantinople by the Turks, drove the literary men of that country for refuge into Italy; and with them they brought the ancient classics of their country, and diffused a knowledge of them, together with their language, over Europe. Hippocrates and Galen thus came to be known in their original dress; for before that period they had been studied only from miserable Arabic translations. The editing of these ancient fathers of medicine, and studying and commenting upon them, occupied the medical world for a considerable time; but the human mind had now received an irresistible impulse which it was impossible to confine.   Anatomy came to be taught in the universities as a practical art: human subjects were publicly dissected, and the different parts of the human body accurately explained and described.   It was impossible in that age, when the works of Galen were in the hands of every student, to avoid comparing his descriptions with the parts of the human subject actually before the eyes of the readers. Thus his deficiencies and mistakes were gradually detected, and whispered among the practical anatomists of Europe.   His dominion over medicine and anatomy was tottering on its basis when a young man of 28 years boldly resolved to hurl him from his throne.

This was Andrew Vesalius of Brussels, who published his anatomy in the Vesalius year 1543, a work which occupies a most conspicuous place in the annals of science, which freed medicine from the trammels of authority, which laid the foundation of genuine anatomy, and even contained a bold and pretty full outline of the science.   This book exhibited a regular and minute description of the human body, illustrated by excellent engravings, and a resolute exposure of the mistakes of Galen, whose ignorance Vesalius is at great pains to point out. This conduct raised against him a host of antagonists, and among others Sylvius, who had been Vesalius's master.   Vesalius it is said was appointed Professor of Anatomy at Padua, and was the first anatomist that ever received a salary.

Thus he laid the foundation of the celebrated Paduan school, which for a Paduan period of nearly 200 years was constantly filled by anatomists of celebrity, to school. which medical students resorted from all parts of the world, and where almost every one who made a figure, either in medicine or anatomy, received his education.   The ancient prejudice against dissecting dead bodies still continuing in France, Germany, and Britain; Italy was the only country in which it was practised.   To Italy therefore every student of medicine, who wished to acquire a knowledge of anatomy, was obliged to resort.   It would be tedious to detail all the great anatomists that Italy produced during this period.   Perhaps the two most celebrated and valuable of all were Falopius and Eustachius; the first, Professor of Anatomy at Padua; the second, at Rome.   Both lived in the time of Vesalius.   Falopius published his anatomical work in 1561, and intended it Falopius. obviously as a supplement to the book of Vesalius.   For he passes over slightly those parts which Vesalius had treated fully and accurately, and dwells on those parts which the former had either omitted or inaccurately described. Eustachius Eustachius. published his Opuscula Anatomica in 1563; but his great work which he had promised, and for which he had engraven 46 copper-plates, was never published by him.   The copper-plates, after having been lost for many years, were afterwards found at Rome in the beginning of the last century, and published in 1714 by Lancisi, who added a short explanatory text, because Eustachius's own account of his plates could not be found.

Fabricius ab Aquapendente, also Professor of Anatomy at Padua, deserves likewise to be mentioned, because he was the master of Harvey, and because his particular pursuits and discoveries obviously bent the original mind of Harvey into a particular channel, and thus paved the way for the greatest physiological discovery which has ever been made—a discovery to which we are indebted for almost all the knowledge of the animal œconomy that we possess,

and which raised Harvey to a degree of celebrity among medical philosophers second only to that of Hippocrates himself. It is hardly necessary to say that the discovery alluded to is that of the *circulation of the blood*. Harvey was a demonstrator of anatomy in London, and taught his doctrine as early as 1616; but he did not publish his demonstration till 1628. It was entitled, *Exercitatio Anatomica de Motu Cordis et Sanguinis in Animalibus*. His doctrine was immediately attacked by a host of antagonists, the most violent of whom was Riolanus, who at that time taught anatomy at Paris. Harvey refuted all the arguments of his opponents with great modesty and success in his second and third Exercitatio, published in 1649. Harvey died in 1657, only three years before the establishment of the Royal Society, at the advanced age of 80. Long before his death he had the satisfaction of seeing his discovery universally admitted, and the arguments of his antagonists tacitly given up. Harvey, besides his discovery of the circulation of the blood, was the author also of another great work scarcely less remarkable for solidity of reasoning and ingenuity of observation, namely, his *Exercitationes de Generatione Animalium, quibus accedunt quædam de Partu, de Membranis ac Humoribus, de Conceptione, &c.* In this work he overturned a doctrine first broached by Aristotle, and obstinately maintained ever after, namely, the doctrine of equivocal generation, and substituted in its stead another opinion now universally admitted to be true, an opinion which he expressed by the phrase *omnia ex ovo*.

About the time when the Royal Society was established, anatomy, which hitherto had been cultivated chiefly in Italy, began to be attended to in the more northern regions of Europe. A crowd of anatomists made their appearance in Britain, France, Holland, Denmark, who soon acquired celebrity, and made the most rapid and unlooked for advances in their favourite science. The Royal Society at its original establishment considered anatomy as one of the most important objects to which they could direct their attention. Dissections were frequently exhibited before them, and they even procured an order from the crown to have the dead bodies of malefactors delivered to them by the sheriffs of London for that purpose. Some of the most distinguished names in anatomy, who contributed essentially to the progress of the science, were active Members of the Society, and were in the habit of laying the results of their discoveries before that illustrious body. But in a periodical work like the Transactions, published so frequently, and in such small portions at a time, it was not possible, nor would it have been proper, to insert a complete system of anatomy, or even a complete account of any of the complicated organs of which it was composed. Short notices of discoveries and improvements were all that could be expected, and all that was intended, sufficient to ascertain the dates and secure to the inventors all the credit which they deserved, and at the same time to inform the anatomical world what advances their favourite science was making by the labours of the Royal Society. Before proceeding to notice the anatomical papers published in the Philosophical Transactions, it will be worth while to take a short survey of the different branches of anatomy in order to point out the anatomists to whom each is chiefly indebted for its present state of improvement.

I. The knowledge of the bones constitutes the foundation of all anatomy, The bones. because it is by their means alone that the position of the soft parts can be accurately marked out. This knowledge is likewise the simplest and most easily attained. It was in some measure known to the ancients, who could not avoid a certain degree of acquaintance with human bones. Whole skeletons must have occasionally presented themselves to their view. There is even a story told by some anatomists, though I do not know upon what authority it is founded, that Hippocrates himself was in possession of a skeleton cast in brass, and that he consecrated it to the oracle of Apollo at Delphos. Two circumstances respecting bones have occupied the attention of anatomists. 1. An accurate description and delineation of them. 2. An investigation of their structure and use. The first of these, though in appearance simple, is not so in reality. Accordingly the description of the bones acquired its present state of simplicity and perfection only by slow degrees. Vesalius has given us a description of the bones, and every anatomist since his time has done the same thing; yet it was not till the beginning of the last century that the task was completed. Cheselden gave excellent plates of the bones in his Anatomy first published in 1713. Dr. Monro's Treatise on the Bones, published at Edin- Monro burgh in 1726, is by far the best account which ever has appeared. It may be considered as the completion of the subject, since it would be exceedingly difficult and perfectly unnecessary to introduce any improvement upon the descriptions which Dr. Monro has given. Winslow deserves likewise to be mentioned. Winslow. His Anatomy, published in 1732, contains an excellent description of the bones, though not equal to that previously given by Monro. Winslow was a Dane who had changed his religion and settled in France. He became a Member of the Academy of Sciences, and a demonstrator on anatomy at Paris. According to Haller, who had been his pupil, his abilities were not so remarkable as his industry But by dint of assiduity he became an excellent anatomist. He was the author of a great number of anatomical dissertations on various subjects, which appeared successively in the Memoires of the French Academy

But his System of Anatomy, in four volumes, was his great work upon which his celebrity is chiefly founded. He died about the year 1760, leaving a widow and family in rather destitute circumstances. He had a son a captain of a trading vessel, who was on a voyage to China at the time of his father's death, and who, on his return to France, made application to the British Government, by means of the Royal Society, for leave to sail with impunity, notwithstanding the war and the ruined state of the French navy.

With respect to the structure and growth of the bones, the best account of the subject which has yet appeared is by Clopton Havers, one of the early Members of the Royal Society, and an account of his treatise is published in one of the numbers of the Philosphical Transactions.* The subject was after-wards resumed by Duhamel, and by Haller, one of the most indefatigable anatomists, and perhaps the most industrious literary character that ever ex-isted. He was a native of Switzerland; and, when only a boy, was initiated in anatomy, by the lectures of Duverney. He then went to Leyden, at that time, by the unrivaled talents of Boerhaave as a teacher, raised to the first rank among Medical Universities. Here he profited by the instructions of Boer-haave, and the anatomical talents of Albinus, who filled the anatomical chair. From Leyden he passed over to London, where he lived in the house of Dr. Douglass, assisted at his dissections, and profited by his knowledge. His next remove was to Paris, where he put himself under the tuition of Le Dran and Winslow. Here he became so notorious, for the great number of dead bodies which he procured for his private dissections, that the indignation of the popu-lace was raised, and information against him was lodged with the police, which obliged him to flee from Paris with all speed, and return to his own country. Mieg, the Professor of Anatomy at Basil, at that time labouring under ill health, employed Haller, first as his dissector, and afterwards to give lectures in his room. Soon afterwards he was chosen Professor of Anatomy at Berne. His various anatomical and botanical labours having by this time raised Haller to a high degree of reputation, he was invited in consequence to Gottinguen, where a new University was established, and where particular care was taken to procure the best qualified persons to fill each of the professorial chairs. Here he continued for some time to exercise his double duty of Pro-fessor and President of the Gottinguen Academy, till want of health, as he informs us, obliged him to throw up his situations and return to his own coun-try. He pubilshed, during a period of 50 years, an infinite number of anato-mical dissertations upon almost every subject. But the greatest and most valu-able of all his works is his *Elementa Physiologiæ Corporis Humani*, in eight large quarto volumes; the first of which was published in 1757; the last,

Haller.

* Phil. Trans. 1691. Vol. XVII. p. 544.

1

in 1766. It is a work of inconceivable labour, and indicates an acquaintance with books so extensive as almost to appear beyond the power of the longest life to acquire.

II. After the bones, the muscles, by means of which the various parts of the The Muscles body are put in motion, and which in some measure fill up and give shape to the human figure, obviously claim the next attention of the anatomist. They are very numerous and complicated, and of course could not have been fully known to the ancients; though it must be admitted that they were acquainted with a considerable number of them. Vesalius is the first anatomist who Vesalius. attempted a minute description of them; and his attempt, considering the difficulty of the subject, was attended with the happiest success. It was his antagonist Sylvius who first imposed names upon a great many of the muscles, and these names they still continue to bear; for before that time they were merely denoted by numbers, and by the part of the body to which they were connected; and as each anatomist used numbers peculiar to himself, it was exceedingly difficult to understand his descriptions without a thorough knowledge of the subject. Vesalius's description of the muscles, improved indeed somewhat by his rivals and successors, continued the best account of them till the anatomy of Winslow was published, who far excelled his predecessors in minute accuracy. The plates of the muscles, published by Mr. William Cowper, a little before the commencement of the last century, are excellent. Cowper. But this ingenious surgeon and anatomist was guilty of a literary theft, which remains as a slur upon his reputation. Bidloo, Professor of Anatomy in the University of Leyden, published a valuable work, under the title of *Anatomia Corporis Humani,* accompanied with excellent plates, where all the parts are delineated as large as the life. Cowper purchased 300 copies of these plates, and published them in London in 1698, with an English explanation drawn up by himself, in which he had considerably improved upon Bidloo, and with a few additional figures. In this publication no notice was taken of Bidloo, except once or twice in the preface, in which he is censured. Bidloo complained of this theft to the Royal Society, and immediately wrote a very severe pamphlet, entitled *Gulielmus Cowperus citatus coram Tribunali.* Cowper, instead of acknowledging the impropriety of his conduct, published a virulent pamphlet, entitled *Vindiciæ;* in which he endeavours to shew that Bidloo did not understand his own figures, from which he infers that they were not really Bidloo's figures; that they had been engraved by Swammerdam, and purchased by Bidloo from Swammerdam's widow. A malicious charge, which some subsequent writers have been malevolent enough to propagate and defend.

Dr. James Douglas deserves also to be mentioned as an excellent writer on Douglas the muscles. He published his *Myographia,* or a comparative description of the muscles in a man and in a quadruped, in 1707; and it was confessedly the best account of the subject which had appeared when he wrote. But the person who may be considered as having completed the description of the muscles as far as structure, origin, and insertion are concerned, is Albinus, the Prince of Anatomists. Bernard Siegfried Albinus, when scarcely 20 years of age, was raised to the Anatomical Chair at Leyden, by the discerning and disinterested recommendation of Boerhaave; and he filled it for 50 years; during the whole of which period he devoted his undisturbed attention to anatomical pursuits. Those intervals which were left him free from the care of the young men under his charge, he dedicated to a careful examination of the origin and the insertion of every muscle, and he employed the most skilful artists to draw representations of the whole. His plates upon almost every subject of anatomy are admirable, both for execution and exactness. Indeed it would be a difficult and a useless task to endeavour to excel him in either. He died about the year 1770.

Albinus.

There is another subject intimately connected with the muscles, and necessary for completing our knowledge of them, which has been greatly and unaccountably neglected by anatomists in general; I mean the way in which they combine in producing the motions of the human body. Anatomists content themselves with describing the action of each muscle separately, or in conjunction with its fellow; but this is far from being sufficient to give an idea of the motions of the human body. Winslow indeed, in his anatomy, has attempted something better, and has given us some excellent observations on the way in which the complicated motions of the human body are produced. But by far the best account of the subject is given by Dr. Barclay, of Edinburgh, in a work which he lately published on the muscles. In this excellent work, which fills up a blank unaccountably left in anatomy, we have a very complete and satisfactory account of the way in which the various muscles combine, and contribute their parts, to produce the different complicated motions of the human body.

Muscular motion.

III. The blood-vessels which pervade every part of the human system, and distribute the blood to every organ of the body, are too important to have been neglected by anatomists or physicians. But the ancients had no correct ideas respecting either the nature or distribution of these vessels. According to them the arteries were filled with air, and not with blood, and the chief seat of the blood, from which it was distributed to other parts of the body, was not the heart, but the liver. Though considerable advances were made in the knowledge of the blood-vessels by modern anatomists; though they had ascertained that the arteries as well as the veins were filled with blood; though Fabricius ab Aquapendente had discovered the valves in the veins, which might have led him to a knowledge of the real motion of the blood; and though what is called the lesser circulation, or the circulation of the blood through the lungs, was

The blood-vessels.

known to Columbus and Cæsalpinus, and is mentioned by Servetus; yet it was our immortal Harvey who first removed the veil, and explained the subject with luminous simplicity, by his discovery of the circulation. This discovery was not the result of any new observations respecting the structure of the heart and blood-vessels, but of accurate consequences deduced from the structure already known, and especially from the valves of the veins discovered by his master Fabricius, which probably led Harvey at first to turn his attention to this important subject. He observed that the structure of the heart was muscular, and that its motions were performed by the contractions of these muscles. The auricles contract first, and thereby propel the blood into the ventricles. The ventricles contract next, and as the valves prevent the possibility of the blood returning to the auricles, it is thrown into the arteries. Here it would constantly accumulate and speedily burst the vessels, unless it were carried by them to all parts of the body, and afterwards conveyed by them to the extreme vessels of the veins. In the veins, as is obvious from the structure of their valves, the motion of the blood is from the small veins to the trunks, and thence to the heart, by which organ it is again delivered to the arteries. Thus the circulation perpetually goes on as long as life continues. *Circulation.*

The attempts to deny the reality of this great discovery, and, when that was impossible, to rob Harvey of the honour of it, were innumerable. Aristotle and Galen, among the ancients; and Servetus, Cæsalpinus, and Columbus, among the moderns, were quoted as philosophers who were acquainted with the circulation, and taught it in their writings. Another story was circulated, and even countenanced by Bartholine. According to it the discovery was made by Father Paul, the Venetian, who communicated it to Fabricius ab Aquapendente. But these philosophers, apprehensive of the fate of Galileo, if they should venture to propagate a doctrine so contrary to the received opinions, carefully concealed it, or revealed it only to their favourite pupils. Fabricius communicated it to Harvey, who, being under no restraint in England, did not hesitate to communicate it to the public. This ridiculous story was treated at the time with more attention than it deserved; and Dr. Clarke was at the trouble to refute it in the Philosophical Transactions, and to shew how it originated. Harvey, it seems, and Father Paul were intimate friends, and the knowledge of the circulation which Father Paul possessed was derived from Dr. Harvey, by means of the Venetian Envoy then in England.*

Harvey's doctrine of the circulation was zealously defended by Sir George Ent, one of the early Members of the Royal Society, a physician of considerable eminence, and a very zealous promoter of the objects of the Society. He had a dispute with Thurston respecting the use of the lungs, which scarcely

* Phil. Trans. Vol. II. p. 672.

P

deserves to be noticed.   According to Thurston the blood underwent a *commi-nution* in the lungs; an opinion which Ent successfully refuted.

Lower.   The most complete Treatise on the Heart hitherto published is by Dr. Lower, who was one of the most distinguished anatomists of the 17th century.   He first explained the reason of the difference of colour between venous and arterial blood.   It is well known that arterial blood is of a beautiful scarlet colour, whereas venous blood is dark and almost brown.   Lower showed that the colour of the arterial blood was owing to the action of the air, and that venous blood when exposed to the air immediately acquires the florid colour of the arterial blood.   It was Lower and Dr. King that first performed the celebrated experiment of the transfusion of the blood of one animal into another, which at that time was considered as likely to lead to the most important medical discoveries, and which occasioned a violent dispute between the English and French anatomists with respect to priority of invention.   This experiment has long lost its celebrity, and the hopes entertained that it might be the means of producing extraordinary cures naturally terminated with the humoral doctrines of pathology which are now nearly exploded from medicine.

Anatomical injections.   The knowledge of the distribution of the blood vessels was very much promoted by the method of filling them with injections of coloured liquid wax, which, afterwards hardening, rendered it easy to trace their most minute ramifications.   This art was discovered in Holland, and is believed to have been first practised by Swammerdam.   Ruysch was indebted to it for a considerable share of his celebrity.   It was alleged that he had originally learned it from Swammerdam; but this he himself refused to admit.   It is not unlikely that he might have known that Swammerdam employed some artificial injection, and from this hint he may have gone to work and hit upon a method which answered the purpose.   Senac published an excellent Treatise on the Structure of the Heart, its Action and Diseases, in 1749.

The brain and nerves.   IV. The brain and nerves constitute one of the most important but the most mysterious part of the human body.   They form as it were the link that unite together mind and body.   All the motions of the body, every sensation, and even the faculty of thinking itself, depends upon them.   Hence they claim, and have obtained, the most minute and accurate examination of anatomists.   Though it must be confessed that the physiological improvements resulting from this investigation are scarcely adequate to the labour bestowed upon it.   We are Willis.   indebted to Dr. Thomas Willis, one of the original Members of the Royal Society, for the first accurate anatomy of the brain and nerves.   He was appointed Sedleian Professor of Natural Philosophy at Oxford in 1660, and some years afterwards he removed to London, where he distinguished himself by a variety of medical and anatomical publications.   His anatomy of the brain was by far the best of his works.   He was assisted in his dissections by Dr. Lower, and in his drawings by Sir Christopher Wren. Though his anatomical descriptions be good, yet his physiological opinions must be acknowledged to be altogether extravagant and absurd. For example, he lodges common sense in the corpus striatum of the brain, imagination in the corpus callosum, and memory in the cineritious matter which encompasses the medullary.

Raymond Vieussens, Professor of Anatomy at Montpelier, is the next author after Willis that deserves to be mentioned. He published his *Neurographia Universalis* in 1685. In this work he greatly improved the description of the brain and nerves, though he has fallen into mistakes in some points. His obligations to his predecessor Willis are sufficiently obvious. It is to Vieussens that we are indebted for imposing the names upon a great many parts of the brain which are still adhered to by anatomists.* It must be acknowledge that Vieussens was not very happy in the terms which he imposed; for scarcely any thing is to be met with more whimsical or absurd than most of the names by which the different parts of the brain are distinguished. Fortunately these terms are all in a learned language, and of course their absurdity is not nearly so obvious as it would be if we translated them into our own tongue. *[Vieussens.]*

The enumeration of the nerves given by Willis is still adhered to by anatomists. But since the time of Willis and Vieussens, many splendid sets of plates, illustrating every thing respecting the brain and nerves, have been published. Perhaps the finest of these in point of execution at least, if not the most accurate, are the superb plates painted and published by Vicq. D'Azyre, not many years before the commencement of the French revolution.

V. Besides the blood vessels, there is another set of vessels of no less importance in the human system, by no means so obvious, namely, the *absorbents*. These vessels are of two different kinds, and were discovered at different times. The first kind consists of a set of vessels originating in the alimentary canal, the use of which is to take up the chyle as it forms, and carry it to a large vessel called the *thoracic duct*, which conveys it into the trunk of a vein very near the heart. As the colour of the chyle is similar to that of milk, these vessels have received the name of lacteals. These vessels are obscurely hinted at by Hippocrates and Galen, and are supposed also to have been known to Erasistratus. But they had been overlooked and forgotten by anatomists when they were discovered a second time by Casper Asellius of Cremona, an anatomist of great merit and modesty, who died in 1626. His account of the lacteals was a *[The absorbents.]* *[Lacteals.]*

---

* The extreme fondness for coining new technical terms, by which the French have always been distinguished, is very remarkable. In anatomy, chemistry, mineralogy, and even in mathematics, they have been great inventors of words, and nothing in general can be more unfortunate than the terms which they have contrived. Is this propensity to be ascribed to the extreme poverty of the French language, and the rigid laws laid down by the academy, which, in some measure, preclude the introduction of all new words except technical terms?

posthumous work, not published till 1628. It deserves to be recorded as a proof that the account of these vessels, found in the writings of Hippocrates and Galen, are not very obvious, that Dr. Harvey never would admit the reality of this discovery of Asellius.

Thoracic duct.

The thoracic duct was first discovered by Pecquet, an anatomist of Paris, who published an account of it in 1651. Next year John Van Horne, a Dutch Professor, published an account of the same organ, and laid claim to at least a share of the merit of the discovery, appealing to his pupils for the truth of his assertion, that he had publicly exhibited the thoracic duct before any account of it by Pecquet had appeared. But the public did not allow of the claim of Van Horne, and gave the whole merit of the discovery to Pecquet.

Lymphatics.

The other set of absorbent vessels arise from all the cavities of the body, are filled with a liquid almost as limpid as water, and are so thin and transparent that it is no wonder they remained long unknown. Their discovery was made in the year 1651 or 1652, and two claimants appeared, each alleging that the original discovery was made by himself. The first of these claimants was Thomas

Bartholine.

Bartholine, a Dane, and one of the most active and illustrious philosophers of his age. He had inspired the Danes with an ardour for the sciences, and had taken every means in his power to encourage and reward every appearance of

Rudbeck.

merit. The other claimant was Olaus Rudbeck, a Swede, a Professor at Upsal, and not inferior in celebrity to Bartholine himself. As far as it is possible to determine from the claims urged, and the concessions made by each of these antagonists, both of them were entitled to the merit of the discovery; but the priority of time seems to belong to Rudbeck, for he observed the lymphatic vessels as early as the month of March, while they were not seen by Bartholine till near the end of the year. Dr. Timothy Clark, in a dissertation which he published in the Philosophical Transactions, endeavours to deprive both of these philosophers of the merit of the discovery But he does not appear to have been accurately informed of the date at which the foreign anatomists first observed the lymphatics. For he affirms that our countryman Jolif, while examining the spermatic vessels in 1652, accidentally observed the lymphatic vessels swelled by his attempt to render the blood vessels of the part more turgid.* Now this claim, urged 16 years after the period of the alleged observation, allowing it all the validity that Dr. Clark could desire, does not invalidate the claims of Rudbeck and Bartholine, since their observations had been at least as early if not earlier than the supposed discovery of Jolif.

Many observations on the lymphatics were made by Dr. Hunter and Dr. Monro, who entered into a very hot dispute about priority of discovery. Respecting the merits of this dispute it is not necessary to decide. But certainly

* Phil. Trans. 1668. Vol. II. p. 672.

the claims of Dr. Hunter were preposterous in the extreme, when he boasted that his discovery was more important than Harvey's far famed one of the circulation.   An excellent description of the lymphatics was published by Mr. William Hewson in 1774.   But the completest account of the subject which has yet appeared was drawn up by Mr. Cruikshanks.   Masgagni has published by far the most splendid and accurate plates of the absorbent system that have yet appeared.

VI. Such is an account of the improvements made by the moderns in the most important parts of anatomy.   The bones, blood vessels, brain and nerves, and lymphatics, owe much of their present advanced state to the labours of British anatomists; but the muscles have been more indebted to foreigners. With respect to the other branches of anatomy, it would be tedious to enter into Other a minute detail.   We may mention however in general, that the first account of branches of anatomy. the glands was given by Wharton, that for the best description of the organs of generation we are indebted to De Graaf, Van Horne, and Swammerdam. Duverney gave a full description of the ear; Malpighi of the tongue; Leuwenhoeck, Zinn, and J. Hunter, of the eye; Steno, and Caspar Bartholine, junior, described the salivary vessels; Albinus, and John Hunter, gave an accurate account of the teeth; Monro gave a full account of the bursæ mucosæ; and Kirkringius, Hunter, and others, gave an account of the growth of the fœtus. In short, the subject is so extensive, and the number of anatomists who have written upon particular parts of the human body so great, that even a list of the bare names of the authors of merit would swell to a size incompatible with the limits of the present work.*

The anatomical papers which occur in the Philosophical Transactions amount to 118.   A considerable number of these relate to morbid anatomy, which, although they may be of considerable importance in a medical point of view, either as explaining the symptoms, or as suggesting a particular mode of treatment, yet they cannot with propriety be noticed here; because the details into which it would be requisite to enter, in order to point out the importance of the case, and the inferences which might be deduced from it, could be admitted only with propriety into a medical treatise.   A considerable number of these papers, especially in the early volumes of the Transactions, are reviews and analyses of anatomical works published at the time, or historical details of anatomical controversies which have long ago lost all their interest. These circumstances reduce greatly the number of anatomical papers which it

---

* The readers who wish to become acquainted with the writings and merits of anatomists, will obtain satisfaction by consulting Haller's Bibliotheca Anatomica; in two thick quartos. He brings down the subject to 1776, which is sufficiently low.

is proper to notice here. The following papers I consider as the most important anatomical dissertations contained in the Transactions.

1. Description of the epiploon, and the formation of fat. By Malpighi. Phil. Trans Vol. II. p. 533.

2. De viscerum structurâ exercitatio anatomica. Marcelli Malpighi. Phil. Trans. Vol. III. p. 888. This is an account of a book; but so full, and well drawn up, that it deserves to be read with attention.

3. Dissection of a man killed by thunder. Phil. Trans. Vol. I. p. 222. This dissection contains nothing particular; but as it is communicated by the celebrated Dr. Wallis, and was performed by Willis, Mallington, and Lower, it is too curious to be omitted without notice.

4. Anatomical observations on the structure of the nose. By Duverney. Phil. Trans. Vol. XII. p. 976.

5. Account of a body, after being long buried, almost wholly converted into hair. Philosophical Collections, No. II. p. 10. This is an account of a woman that had been buried at Nuremberg, and her coffin had been covered by two others. Forty-three years after her interment the coffin was opened, and the body was observed to retain its original shape, but was wholly covered with a very thick growth of brown hair, long and much curled. On touching the body it fell to powder, and nothing remained but the hair, part of which was exhibited to the Royal Society.

6. Anatomical observations on an abscess in the liver; also a great number of stones in the gall bag, and bilious vessels; an unusual conformation of the emulgents and pelvis; a strange conjunction of both kidneys; and great dilatation of the vena cava. By Edward Tyson, A.M. and M.S. Oxon. Phil. Trans. Vol. XII. p. 1035. This paper possesses considerable value, and was one of the first that threw light upon the symptoms of the liver complaint.

7. Four ureters in a child, and on the glandulæ renales. By Dr. Edward Tyson. Phil. Trans. Vol. XII. p. 1039.

8. Casp. Bartholini Thom. Fil. de ductu salivali hactenus non descripto, observatio anatomica. Phil. Trans. Vol. XIV. p. 749.

9. A letter from Mr. Anthony Leuwenhoeck, F.R.S. dated April 14, 1684, containing observations on the crystalline humour of the eye, &c. Phil. Trans. Vol. XIV. p. 790. This paper contains the first account of the fibrous structure of the crystalline humour.

10. Osteologia nova, or some observations on the bones, &c. communicated to the Royal Society in several discourses read at their meetings. By Clopton Havers, M.D. and R.S. Soc. This is an account of Havers's well known book on the Structure and Growth of the Bones.

11. An extract of a letter from Bernard Connor, M.D. to Sir Charles Walgrave, giving an account of an extraordinary human skeleton, having the vertebræ of the back, the ribs, and several bones, down to the os sacrum, all firmly united into one solid bone without jointing or cartilage. Phil. Trans. Vol. XIX. p. 21.

12. Account of two glands, and their excretory ducts lately discovered in human bodies. By Mr. William Cowper, F.R.S. Phil. Trans. Vol. XXI. p. 364. These are two glands situated near the prostate, and serving to secrete a mucilaginous matter to lubricate the urethra, and defend it from the action of the urine.

13. The human alantois fully discovered. By Rich. Hale, M.D. Phil. Trans. Vol. XXII. p. 835.

14. An account of several schemes of arteries and veins dissected from adult human bodies, and given to the repository of the Royal Society. By John Evelyn, Esq. F.R.S. To which are subjoined a description of the extremities of those vessels, and the manner the blood is seen by the microscope to pass from the arteries to the veins in quadrupeds when living, with some surgical observations and figures after the life. By William Cowper, F.R.S. Phil. Trans. Vol. XXIII p. 1177. This is a very

curious paper, and contains several important particulars respecting the anastomoses of arteries, and the advantages to be derived from this kind of knowledge in cases of aneurism and wounded arteries.

15. Observations on the glands in the human spleen, and on a fracture in the upper part of the thigh-bone. By J. Douglass, M.D. et R.S.S. Phil. Trans. 1716. Vol. XXIX. p. 499.

16. An account of the external maxillary and other salivary glands: also of the insertions of all the lymphatics, both above and below the subclavians, into the veins. By Richard Hale, M.D. and F.R.S. Phil. Trans. 1720. Vol. XXXI. p. 5. This is an elaborate paper, containing a good deal of historical information respecting the maxillary glands.

17. Concerning the difference in the height of a human body, between morning and night. By the Rev. Mr. Wasse; Rector of Aynho, in Northamptonshire. Phil. Trans. 1724. Vol. XXXIII. p. 87. Mr. Wasse found that the height of an ordinary sized man was diminished an inch, after labour. The truth of this curious fact is now fully established. Every person is shorter at night than in the morning when he rises out of bed. This is owing to the distances between the bones of the spine being diminished by their weight and pressure upon each other.

18. Dissection of two eyes, which had been affected with cataract; communicated in a letter from ———, to Samuel Molyneux, Esq.; Secretary to the Prince of Wales, and F.R.S. Phil. Trans. 1724. Vol. XXXIII. p. 149. This paper shows what is now well known, that, after couching, the lens of the eye is absorbed and disappears.

19. Account of Margaret Cutting, a young woman at Wickham market, in Suffolk, who speaks readily and intelligibly, though she has lost her tongue. By Mr. Henry Baker, F.R.S. Phil. Trans. 1742. Vol. XLII. p. 143. This is the famous case of Margaret Cutting, concerning which it is not necessary to make any remarks. Mr. Baker's account not proving satisfactory to many of the Members of the Society, the woman was brought to London, examined in a meeting of the Society, and an accurate account of her case, differing but little from the previous statement of Mr. Baker, was afterwards published by Dr. Parsons. Phil. Trans. 1747. Vol. XLIV. p. 621.

20. Of the structure and diseases of articulating cartilages. By William Hunter, Surgeon. Phil. Trans. 1743. Vol. XLII. p. 514. This is a very entertaining as well as accurate account of some of the most curious parts of the animal œconomy.

21. On the use of the ganglions of the nerves. By James Johnstone, M.D. Phil. Trans. 1764. Vol. LIV. p. 177.

22. A description of the lymphatics of the urethra and neck of the bladder. By Henry Watson, Surgeon to the Westminster Hospital, and F.R.S. Phil. Trans. 1769. Vol. LIX. p. 392.

23. Of a remarkable transposition of the viscera. By Matthew Baillie, M.D. Phil. Trans. 1788. Vol. LXXVIII. p. 350.

24. Of a particular change of structure in the human ovarium. By Matthew Baillie, M.D. Phil. Trans. 1789. Vol. LXXIX. p. 71. In this paper, Dr. Baillie endeavours to shew that the change of the ovarium into a fatty matter, intermixed with hair and bones, which is occasionally perceived, is not the consequence of the sexual intercourse, but a change which the part assumes of itself.

25. Account of two instances of uncommon formation in the viscera of the human body. By Mr. John Abernethy, Assistant-Surgeon to St. Bartholomew's Hospital. Phil. Trans. 1793. Vol. LXXXIII. p. 59.

26. Description of an extraordinary production of human generation, with observations. By John Clarke, M.D. Phil Trans. 1793. Vol. LXXXIII. p. 154.

27 Observations on the foramina thebesii of the heart. By Mr. John Abernethy, F.R.S. Phil. Trans. 1798. Vol. LXXXVIII. p. 103. Mr. Abernethy endeavours to shew, that the use of these foramina (which have been the occasion of so much hesitation among anatomists,) is to prevent the heart from being too much distended in cases of diseases of the lungs.

28. Of the orifice in the retina of the human eye, discovered by Prof. Soemmering: with proofs of this appearance being extended to the eyes of other animals. By Everard Home, Esq., F.R.S.

Phil. Trans. 1798. Vol. LXXXVIII. p. 332.   Mr. Home shews that this orifice is not peculiar to the human eye, as he observed it also in the eye of the monkey.

29. On a very unusual formation of the human heart.   By Mr. J. Wilson, Surgeon.   Phil. Trans. 1798. Vol. LXXXVIII. p. 346.

30. Of a tumour found in the substance of the human placenta.   By John Clarke, M.D.   Phil. Trans. 1798. Vol. LXXXVIII. p. 361.

31. Additions to a paper on the subject of a child with a double head.   By Everard Home, Esq., F.R.S.   Phil. Trans. 1799. Vol. LXXXIX. p. 28.   The original paper, to which this is an addition, was published in Phil. Trans. 1790. Vol. LXXX. p. 296.

32. The Croonian lecture, on the structure and uses of the membrana tympani of the ear.   By Everard Home, Esq., F.R.S.   Phil. Trans. 1800. Vol. XC. p. 1..

These are the most interesting and important of the anatomical papers, which have been published in the Philosophical Transactions.   Perhaps, in strict propriety, some of them ought to have been reserved for a following section, when we take under our consideration the physiological labours of the Royal Society.   It is impossible to terminate these observations on anatomy, without mentioning the imperfect state of its nomenclature, which must strike every person in the least degree conversant with the subject.   We do not allude so much to the names of the bones, muscles, blood-vessels, and nerves; which, though awkward and inconvenient in some respects, yet answer the purpose sufficiently well, but to the general terms which are used to express the position of the several organs with respect to each other, and with respect to the body in general.   These general terms, indeed, we make a shift to understand when the human body only is treated of; but, in many cases of comparative anatomy, they are little else than an absolute jargon, which Oëdipus himself could not unravel.   Anatomists ought immediately to adopt the new general anatomical terms proposed by Dr. Barclay, in his Essay on the Anatomical Nomenclature, which are elegant and precise, and infinitely more simple and more easily remembered than the terms which are in common use.

Anatomical nomenclature.

Sect. III.—*Of Comparative Anatomy.*

This branch of anatomy is, in reality, the oldest, for it was to it alone that the ancient physicians turned their attention, and from it they drew inferences with respect to the anatomy of the human body.   Vesalius has shewn that Galen drew most of his knowledge of anatomy from the dissection of apes.   There is a good deal of comparative anatomy to be found in the first part of Aristotle's Treatise on Animals.   The science has been cultivated by the moderns with much assiduity, and has been of more importance in pointing out the use of the different organs than human anatomy itself.   Because particular organs being wanting, or imperfectly developed in certain sets of animals, an opportunity is afforded of determining what effect these deficiencies occasion in the animal

Importance.

œconomy in general. Another reason has induced even those naturalists, whose immediate object was not directed towards physiology, to pay particular attention to comparative anatomy. It has been found that an accurate arrangement of animals into classes, orders, genera, and species, can only be accomplished by an accurate knowledge of the structure and arrangement of their organs. For the arrrangement of Linnæus, though in fact somewhat dependent upon structure, has been found, in some parts, inadequate and incongruous. His class of vermes in particular contains a heterogeneous mixture of animals, which have been since subdivided into four classes by separating the mollusca, the zoophites, the intestinal worms, and the infusoria, from each other. On this account, comparative anatomy has been cultivated by the greater number of zoologists, and indeed has received from them a great and valuable increase of facts.

Swammerdam was one of the earliest and most diligent anatomists who turned Swammerdam. their chief attention to comparative anatomy; and his work, of which a translation has been published in English, contains much curious information concerning the anatomy of insects and worms. In 1681, Gerard Blasius published Blasius. a large quarto volume, expressly devoted to comparative anatomy, in which he gives the anatomy of the different classes of animals much in detail, and accompanied by figures. In 1746, a thin octavo volume on comparative anatomy was published in London. It was a posthumous work of Dr. Monro, Monro. and was possessed of great merit. After giving a view of comparative anatomy in general, and pointing out its importance, he gives the anatomy of the dog, the cow, fowls in general, the cock, carnivorous fowls, and of fishes.

But it would be a tedious and indeed a useless task to attempt an enumeration of the writers on comparative anatomy, who have made their appearance since the revival of learning in modern Europe. But some of the most celebrated names, are Steno, Collins, Duverney, Petit, Lyonnet, Haller, Monro, Hunter, Home, Geoffroy, Vicq D'Azyr, Camper, Comparetti, Scarpa, Blumenbach, Poli, Kielmeyer, Harwood, Barthez, &c. One of the most laborious comparative anatomists of the last century was Daubenton. It is well known that Daubenton. he was employed along with Buffon in publishing the Natural History of Animals; and that all the dissections, which constitute the most valuable part of the work, were executed by him. Vicq D'Azyr likewise published some excellent dissertations on comparative anatomy, chiefly in the Memoires of the French Academy. Britain, and the Royal Society in particular, ever since its foundation, has always possessed a succession of eminent and celebrated comparative anatomists, who have enriched the Transactions with numerous important papers on the subject; though it must be acknowledged that the number of papers on comparative anatomy is greater in the Memoires of the

Q

French Academy, than in our national publication. This was owing to the pains taken during the reign of Louis XIV., to furnish the Academy with Papers in the
Transactions. proper animals, and the number of anatomists who received a salary, and of course devoted themselves to anatomical subjects. Of late years, both the number and the value of the papers on comparative anatomy, published in the Philosophical Transactions, have greatly increased. The most eminent of those persons, who have inserted papers on comparative anatomy into the Philosophical Transactions, are Tyson, Swammerdam, Brown, Leuwenhoeck, Cowper, Blair, Price, Klein, Hewson, John Hunter, Camper, Abernethy, Home, Carlisle. The papers published on the subject, reckoning down to the year 1800, amount to 90. But of these about 20, consisting chiefly of microscopical observations by Leuwenhoeck, or of concretions and monstrosities found among the lower animals, cannot be considered of much value.* I shall proceed to give a catalogue of the more important of the remaining papers; but, from the nature of the subject, it will not be possible to enter into any details.

1. Some observations on vipers. By Signor Redi. Phil. Trans. 1666. Vol. I. p. 160. In this paper, Redi gives the first true account of the situation of the poison of vipers; namely, two small bags placed on each side of the mouth, from which a tube passes and goes through the hollow fangs of the animal. Thus a portion of the poison, when the animal bites, is squeezed through the fangs and lodged in the wound. The subject, in consequence of a controversy excited about the truth of this opinion, is continued in three subsequent papers of the Transactions; namely, Vol. V. p. 2034; Vol. VI. p. 3006; and Vol. VII. p. 5060.

2. Dissection of a large fish, the *chimæra monstrosa,* or sea fox; and of a lion in the King's library at Paris, in 1657. Phil. Trans. 1667. Vol. II. p. 532. This is merely an abridgement of the observations published by the dissectors at Paris.

3. Anatomical description of a chamelion *(lacerta camelion),* beaver *(castor fiber),* dromedary, bear, gazelle. Phil. Trans. 1669. Vol. IV p. 987. These animals, like the preceding, were dissected in the King's library, at Paris. The account contained in the Transactions is merely a review of the essay published on the occasion by the dissectors.

4. Th. Bartholini dissertatio de cygni anatome, nunc aucta a Casp. Bartholino. Phil. Trans. 1669 Vol. IV. p. 1017. The book here reviewed is chiefly occupied with an account of the curious structure of the wind-pipe of the swan, which is of a great length, and bent in a curious manner.

5. Extract of a letter from Signor Malpighi to the editor, concerning the structure of the lungs of frogs, tortoises, &c., and the more perfect animals; also on the texture of the spleen, &c. Phil.

---

* Two elementary systems of comparative anatomy have made their appearance, of very considerable value, and deserving the perusal of all who wish to possess general views on the subject. The first by Cuvier, Lecturer on Comparative Anatomy, in Paris, who has gained a high and well-earned reputation by his papers on comparative anatomy, which have been both numerous and important. It was published in 1800, in four octavo volumes, under the title of Lectures on Comparative Anatomy; and in 1802 was translated into English, by Mr. Ross, under the inspection of Mr. Macartney, Lecturer on Comparative Anatomy, in London. The second work was published by Blumenbach, a well-known German Naturalist, who has likewise acquired a high reputation. It is in one volume, and of course contains fewer details than Cuvier's; but it is more entertaining and better arranged. It was translated into English by Mr. Lawrence. Both of these translations are faithful, and do credit to the translators.

Trans. 1671. Vol. VI. p. 2149. In this paper, the celebrated author endeavours to shew that the lungs of frogs, &c., are covered with a muscle.

6. Account of the dissection of a porpoise. By Mr. John Ray. Phil. Trans. 1671. Vol. VI. p. 2274. This is a very particular account of the anatomy of that animal; and from the great size of the brain Mr. Ray infers the superior sagacity of the animal; not being aware that intellect is not proportional to the absolute size of the brain, but to the size of the brain compared with that of the nerves. There is another dissection of the porpoise, published in 1681, in the second number of the Philosophical Collections. The dissection was made at Gresham College, by Dr. Tyson.

7. Some observations made by a microscope contrived by M. Leuwenhoeck, in Holland, lately communicated by Dr. Regnerus de Graaf. Phil. Trans. 1673. Vol. VIII. p. 6037. This is remarkable, as being the first of Leuwenhoeck's papers published in the Transactions. It contains some curious observations on the structure of the bee and louse. Figures connected with this paper are given at p. 6116 of the same volume.

8. Extract of two letters from Dr. Swammerdam, concerning some animals that, having lungs, are yet found to be without the arterious vein, *(pulmonary artery,)* together with some other curious particulars. Dated Amsterdam, Jan. 24, 1673. Phil. Trans. Vol. VIII. p. 6040. This letter relates to frogs, and some other amphibious animals, which are well known to want the pulmonary artery, and on that account can remain much longer without breathing than man, quadrupeds, or birds.

9. Memoires pour servir a l'histoire naturelle des animaux; to which is joined another tract totally different, entituled, La Mesure de la Terre. Paris, 1671. Phil. Trans. 1676. Vol. XI. p. 591. A concise and accurate account of the contents of the French book is given in this paper. The dissections related in it are those of a lion, chat pard, sea fox (*squalus vulpes,* a kind of shark,) lynx, otter, civet cat, elk, coati mondi of Brazil. The book was afterwards translated by Mr. Richard Waller, and an account of the translation is given in Phil. Trans. 1687. Vol. XVI. p. 371.

10. Observationi intorno alle torpedini, fatte da Stephano Lorenzini. Fiorentino, 1681. Philosophical Collections, No. I. p. 42. This is remarkable for being one of the first accounts of the torpedo. A much more complete account of this wonderful animal is given in a subsequent volume of the Transactions, when the animal was dissected by Mr. John Hunter.

11. An account of the dissection of an ostrich. By Edward Brown, F.R.S., and of the College of Physicians, 1682. Philosophical Collections, No. V. p. 147.

12. Vipera caudisona Americana; or the anatomy of the rattle-snake; dissected at the repository of the Royal Society, in January 1683. By Edward Tyson, M.D. Coll. Med. Lond. Cand. et R.S. Soc. Phil. Trans. 1683. Vol. XIII. p. 25. This is a very minute account, with figures of the structure of this singular and inactive animal. The intention of the rattle has never been explained in a satisfactory manner.

13. The lumbricus latus *(tænia solium)*; or a discourse read before the Royal Society concerning the jointed worm. By Eward Tyson, M.D. Col. Med. Lond. nec non Reg. Societ. Soc. Phil. Trans. 1683. Vol. XIII. p. 113. This is a very elaborate account of a very singular genus of worms, the tænia, which are found in the intestines of various animals, particularly in the human intestines. The anatomy of this extraordinary animal has been lately still farther elucidated by Mr. Carlisle, in an early volume of the Linnæan Transactions.

14. Lumbricus teres *(ascaris lumbricoides)*; or some anatomical observations on the round worm bred in human bodies. By Edward Tyson, M.D., &c. Phil. Trans. Vol. XIII. p. 154. This is also a valuable paper. The animal described exists often in the human intestines, and produces many distressing symptoms.

15. Tajaçu, seu aper Mexicanus moschiferus *(sus tajaçu)*; or the anatomy of the Mexican musk hog, &c. By Edward Tyson, F.R.S. Phil. Trans. Vol. XIII. p. 359. This is the only species of hog found in the New World. It is smaller than the common hog. Dr. Tyson, in this paper, ac-

cording to Buffon, has fallen into very material error.   He has affirmed that the animal has three stomachs, whereas it has in reality but one stomach, parted a little, like that of the tapir, by two strictures or contractions.

16. An observation concerning a blemish in a horse's eye, not hitherto discovered by any author. By Dr. Rich. Lower.   Phil. Trans. 1668. Vol. II. p. 613.   This paper gives an account of certain spongy excrescences that grow from the uvea of horses' eyes.

17. A letter of Malpighi to Jacobus Sponius, giving a minute account of the uterus of the cow. Phil. Trans. 1684. Vol. XIV. p. 630.

18. Observations on the dissection of a rat.   By Mr. R. W., S.R.S.   Phil. Trans. 1693. Vol. XVII. p. 594.

19. Anatomical observations on the heads of fowls.   By the late Allen Moullen, M.D. S.R.S. Phil. Trans. 1693. Vol. XVII. p. 711.

20. Observations on the dissection of a paroquet.   By Mr. Richard Waller.   Phil. Trans. 1694. Vol. XVIII. p. 153.

21. On the structure of the internal parts of fish.   By Dr. Charles Preston.   Phil. Trans. 1697. Vol. XIX. p. 419.

22. The dissection of the scallop.   By Dr. Martin Lister, F.R.S.   Phil. Trans. 1697. Vol. XIX. p. 567.

23. The anatomical history of the leech.   By M. Paupart.   Phil. Trans. Vol. XIX. p. 722.

24. Concerning the eggs of snails. By Mr. Anth. Van Leuwenhoeck. Phil Trans. Vol. XIX. p. 790.

25. Carigueya, seu marcupiale Americanum; or, the anatomy of an opossum, (didelphis marsupialis;) dissected at Gresham College.   By Edward Tyson, M.D. F.R.S.   Phil. Trans. 1698. Vol. XX. p. 105.

26. Abstract of letters sent to Sir C. H. relating to some microscopical observations.   Communicated by Sir C. H. to the editor.   Phil. Trans. 1703. Vol. XXIII. p 1357.   The animals described in this paper are the louse, a mite, a muscle, the lepas balanus, various larvæ, ticks, and some infusoria.   The observations are not of much value.

27. The anatomy of those parts of a male opossum that differ from the female.   By William Cower, F.R.S.   Phil. Trans. 1704. Vol. XXIV. p. 1576.   The structure of the male organs of generation in the opossum is quite peculiar.   It is very particularly described in this paper.

28. Microscopical observations on the structure of the spleen, and proboscis of fleas.   By Mr. Anthony Van Leuwenhoeck.   Phil. Trans. 1706. Vol. XXV. p. 2305.

29. The anatomy and osteology of an elephant; being an exact description of all the bones of the elephant which died near Dundee, April 27, 1706, with their several dimensions.   By Mr. Patrick Blair, surgeon, &c.   Phil. Trans. 1710. Vol. XXVII. p. 53.   This is a most surprizing paper, and contains a most minute and accurate account of the anatomy of the elephant, especially of the osteology of that animal.   If we consider that Dr. Blair made all his observations on one animal, we must admit that his exertions must have been uncommon, and his address great to have made his account so minute as it is.

30. Anatomical description of the heart of the land tortoise from America.   By Mr. Paul Bussiere, surgeon, F.R.S. Phil. Trans. Vol. XXVII. p. 170.   The heart of this animal, as is the case with the whole tribe of amphibia, has only one ventricle.

31. A description of that curious natural machine the wood-pecker's tongue, &c.   By Richard Waller, Esq. late Secretary to the Royal Society.   Phil. Trans. 1716. Vol. XXIX. p. 509.   This animal bores holes in sound oak and beech trees with its beak, and builds its nest in them.   Its tongue is long and pointed, it suddenly darts it out three or four inches beyond the point of its bill, and draws it in again very speedily with the insect spitted upon its point.

32. An account of the coati mondi of Brazil.   By Dr. George Mackenzie.   Phil. Trans. 1723. Vol. XXXII. p. 317.   In this paper Dr. Mackenzie compares his dissection with that of the same animal by the French Academicians, and points out many differences, which he supposes owing to the

3

difference in the sex of the animals examined, his being a female, while that of the French was a male.

33. Some observations on an ostrich dissected by order of Sir Hans Sloane, Bart. By Mr. John Ranby, surgeon. Phil. Trans. Vol. XXXIII. p. 223. Some additions are made to this paper in Vol. XXXVI. p. 275. There is besides another dissection of an ostrich, by Mr. George Warren, surgeon, in Cambridge, in Vol. XXXIV. p. 113.

34. Anatomy of the mus Alpinus, or marmot. By J. Jas. Scheuczer, of Zurich, M.D. F.R.S. Phil. Trans. 1727. Vol. XXIV. p. 237.

35. The dissection of the poisonous apparatus of a rattle-snake, made by the direction of Sir Hans Sloane, Bart. With an account of the quick effects of its poison. By John Ranby, surgeon, F.R.S. Phil. Trans. Vol. XXXV. p. 377. In this paper the structure of the fangs, and the way that the poison is conveyed to the wound inflicted, are sufficiently explained.

36. An anatomical description of worms found in the kidneys of wolves. By Mr. James Theodorus Klein, Secretary of the City of Dantzic, F.R.S. Phil. Trans. 1730. Vol. XXXVI. p. 269.

37. An account of the hermaphrodite lobster presented to the Royal Society, May 7, by Mr. Fisher of Newgate Market, examined and dissected, pursuant to an order of the Society. By F. Nicholls, M.D. F.R.S. Phil. Trans. Vol. XXXVI. p. 290.

38. The dissection of a female beaver, and an account of castor found in her. By C. Mortimer, M.D. R.S.S. Phil. Trans. 1733. Vol. XXXVIII. p. 172. In this paper Dr. Mortimer gives an excellent historical account of every thing that had been previously observed respecting the anatomy of the beaver. His own dissection is defective, because the animal was previously so much torn by a dog that several of the parts were destroyed.

39. An account of some peculiar advantages in the structure of the asperæ arteriæ, or wind-pipes of several birds, and in the land tortoise. By Dr. Parsons, F.R.S. Phil. Trans. 1766. Vol. LVI. p. 204. This is a curious paper, giving an account of the tortuosities of the wind-pipe (with figures) in several birds, particularly water fowl.

40. An account of the lymphatic system in birds. By Mr. Wm. Hewson, Reader in Anatomy. Phil. Trans. 1768. Vol. LVIII. p. 217. This is one of the papers which conferred celebrity upon Mr. Hewson. It contains the first account of the lymphatics of birds ever published.

41. An account of the gymnotus electricus. By John Hunter, F.R.S. Phil. Trans. 1775. Vol. LXV. p. 395. This is a minute account of the electrical organs of the torpedo. Mr. Walsh having made experiments on the electric power of this surprizing fish was anxious to get an exact account of the organs which possessed the electric property, he accordingly procured a torpedo, and it was at his request that Mr. Hunter dissected it. Mr. Hunter never having enjoyed the benefit of a liberal education, his papers generally required to be corrected, or even new modelled, before they were committed to the press. The present paper was drawn up by Dr. Bancroft from J. Hunter's dissection, at the request of Mr. Walsh.

42. On the organs of speech of the orang outang. By Peter Camper, M.D. late Professor of Anatomy, &c. in the University of Groningen and F.R.S. Phil. Trans. 1779. Vol. LXIX. p. 139. This is a curious paper. Camper shows, contrary to Dr. Tyson's assertion in a former paper in the Transactions, that these organs in the orang outang are very different from the human, and thence deduces that the animal is not capable of articulate speech like man.

43. A microscopic description of the eyes of the monoculus polyphemus Linnæi. By Mr. William André, surgeon. Phil. Trans. 1782. Vol. LXXII. p. 440.

44. A description of the teeth of the anarrhichas lupus Linnæi, and of those of the chætodon nigricans of the same author; to which is added, an attempt to prove that the teeth of cartilaginous fishes are perpetually renewed. By Mr. William André, surgeon. Phil. Trans. 1784. Vol. LXXIV. p. 274.

45. Observations on the structure and economy of whales. By John Hunter, Esq. F.R.S. Phil. Trans. 1787. Vol. LXXVII. p. 371. This is a very long and elaborate paper, and contains the fullest information respecting the anatomy of the whale tribe yet offered to the public.

46. Some particulars in the anatomy of a whale. By Mr. John Abernethy. Phil. Trans. 1796. Vol. LXXXVI. p. 27. This paper contains several curious particulars, from which the author deduces some important physiological conclusions.

47. A description of the anatomy of the sea otter from a dissection made Nov. 15, 1795. By E. Home, Esq. F.R.S. and Mr. Arch. Menzies.

48. The dissection of an hermaphrodite dog. With observations on hermaphrodites in general. By Everard Home, Esq. F.R.S. Phil. Trans. 1799. Vol. LXXXIX. p. 157. This is a very curious paper, containing a great collection of very important facts, and some general deductions from them.

49. On the structure of the teeth of graminivorous quadrupeds; particularly those of the elephant and sus Ethiopicus. By E. Home, Esq. F.R.S. This paper contains much valuable information respecting the structure and growth of teeth; but not very susceptible of abridgement; as is indeed the case with most papers on comparative anatomy.

50. Account of a peculiarity in the distribution of the arteries sent to the limbs of slow moving animals; with some other similar facts. By Mr. Anthony Carlisle, surgeon. Phil. Trans. 1800. Vol. XC. p. 98. Mr. Carlisle's papers are all uncommonly ingenious and valuable. It is much to be regretted that he has favoured the world with so few.

51. Some observations on the head of the ornithorhynchus paradoxus. By E. Home, Esq. F.R.S. Phil. Trans. Vol. XC. p. 432. The ornithorhynchus is a singular animal from New Holland, having attached to its mouth a beak very similar in appearance to the bill of a duck. In this paper we have a curious account of the structure of the mouth.

From the preceding catalogue, we see that the Philosophical Transactions contain an anatomical account of the following animals:

Viper.	Squalus vulpes, or sea-fox.	Opossum.
Chimera monstrosa, or sea-fox.	Lynx.	Flea.
	Otter.	Elephant.
Lion.	Civet cat.	Land tortoise.
Chamelion.	Elk.	Woodpecker.
Beaver.	Coati mondi.	Ostrich.
Dromedary.	Torpedo.	Marmot.
Bear.	Ratle-snake.	Worms, from the kidneys of wolves.
Gazelle.	Tænia solium.	
Swan.	Ascaris lumbricoides.	Lobster.
Frog.	Sus Tajaçu.	Orang outang.
Porpoise.	Rat.	Whales.
Bee.	Paroquet.	Sea otter.
Louse.	Scallop.	Hermaphrodite dog.
Chatpard	Leech.	Ornithorhynchus paradoxus.

This is a pretty extensive list; though small when compared to the vast number of inferior animals which constitute the object of comparative anatomy. To complete the list of the labours on comparative anatomy, performed by British anatomists, it is proper to mention Dr. Monro's work on fishes, which has added considerably to our knowledge of these animals. Mr. Carlisle, in a paper lately published in the Transactions, and which unfortunately does not come under our review, has communicated some very valuable information respecting the muscular system of these animals.

Section IV —*Of Physiology.*

The term *Physiology,* originally synonimous with *Natural Philosophy,* has, like many other Greek words, been very much restricted in its signification by modern philosophers. It is now applied exclusively to that science which treats of the *properties* and *functions* of *living bodies.* The first treatise on physiology, at least which has come down to us, is Galen's book *De Usu Partium.* Galen. Though possessed of considerable merit, considering the period in which it was produced, it was necessarily very imperfect, in consequence of the small progress which had been made in anatomy. The anatomists were in fact the great Improvers of improvers of physiology, and it is to their discoveries that we are indebted for physiology. the greater part of the knowledge which we possess of the functions of living bodies. The discovery of the circulation of the blood, and of the absorbent system, contributed in no small degree to account for various phenomena and diseases formerly inexplicable. For our part, however, we must acknowledge that the science of *physiology,* notwithstanding the vast multitude of writers who have attempted to elucidate it, and notwithstanding the numerous theories which have been advanced, and the supercilious confidence with which they have been maintained, appears to us to have hardly made any real advances even in the most modern times. We allow, indeed, that a great deal Present state of knowledge has been gained of the mechanical structure of living bodies, and of the science. that various plausible chemical explanations have been offered of some of the functions; respiration, for example, and nutrition. But the action of the nervous system; the manner in which it occasions motion, sensation, and perception; the connexion between organization and thinking; animal heat; the nature of generation; and, in short, upon what the phenomena of life depend; are problems, just as far from solution at present, as they were in the days of Hippocrates and Galen. The opinions, even of the most recent writers on the subject, are too absurd to merit the smallest attention. Indeed we are apprehensive that the subject itself is beyond the reach of the human faculties, and that nothing better than chimæras and vain imaginations can ever be produced on it.

The science of physiology has been chiefly cultivated by medical men, and the practice of medicine, or the method employed by physicians in attempting to cure diseases, has been almost entirely regulated by the physiological system which happened to be in vogue. Now it is curious to remark, that the principles of physiology have uniformly depended upon the particular science that happened to be in fashion. When the opinions of Aristotle and Galen reigned paramount in the schools, all the functions of the living body were explained by having recourse to *occult qualities* and *occult faculties*. When Paracelsus drew the attention of Europe to chemistry, that science was considered as paramount to account for all the animal functions; and the physiologists of the time explained every thing by means of fermentation, sublimation, distillation, filtration, concoction, and other similar processes, familiar to the chemists of the age. When Newton and his contemporaries laid the foundation of mechanics, upon the rigid principles of mathematical demonstration, physiologists embraced with eagerness the fashionable doctrines; the human body was converted into a hydraulical machine; the force of the heart and the velocity of the blood were rigidly ascertained; and every thing was accounted for by the size, and shape, and motion, of the different particles of matter of which the body was composed. When mechanical philosophy began to lose its novelty, it was in some measure supplanted as a fashionable study by a peculiar species of metaphysics, which was prosecuted with much ardour for a time, till it at length terminated in universal scepticism. During the progress of this enticing science, physiologists laid hold of its notions and doctrines, and two opposite systems were produced, the more ancient explaining every thing by the action of a *living principle*, and the more modern by a principle somewhat indefinite, to which they gave the name of *irritability*. The recent discoveries in pneumatic chemistry having again brought that bewitching science into fashion, a new race of physiologists has arisen, who ascribe every thing to chemical principles; and ring changes upon the words *oxygen, hydrogen, carbon,* and *azote,* by means of which, in their opinion, every function in the living body may be sufficiently and satisfactorily explained.

It would carry us too far, and would be spending time to very little purpose, to take a particular view of the physiological systems which distinguished the last century, and each of which for a time possessed its admirers, and gave celebrity and success to the physician who first broached it. Boerhaave, Hoffman, Stahl, Cullen, Brown, Darwin, are the most distinguished names. The system of Boerhaave has lost all its defenders; but the opinions of Stahl, of Hoffman, as modified and altered by Cullen, and of Brown, are still adhered to by numerous sects. Darwin's system has not been so fortunate. His Zoonomia was published at an unlucky period; his opinions deviated too far from those of his contemporaries; and his knowledge of chemistry, upon which his theories chiefly depended, was too confined and inaccurate to attract much respect or confidence. The short-lived celebrity of his botanic garden, and the extravagant hypotheses which he advanced in various departments of the science, contributed likewise to injure the success of his system. The physiology of Haller is by far the most important work on the subject which has hitherto appeared, and indeed will not be easily surpassed by succeeding physiologists. I consider it as the most stupendous monument of industry which the eighteenth century produced. Instead of indulging, like most of his contemporaries and prede-cessors, in constructing an ingenious hypothesis to account for the functions of living bodies, Haller undertook the gigantic task of collecting all the *facts* relative to the subject, which had been ascertained. This he accomplished in his Elements of Physiology, which consists of eight pretty thick quarto volumes.

Let us imitate the example of this illustrious philosopher, and instead of attempting an outline of the science, which would be of very little value, let let us endeavour to collect and arrange all the physiological facts which have made their appearance in the Philosophical Transactions. The physiological *Papers on the* papers, which have been published in that voluminous work, amount to no *subject in the Philosophical* fewer than 220; and contain a vast collection of curious and important facts. *Transactions.* They are of very various importance; and, indeed, no fewer than 81 of these papers in the present state of our knowledge, may, without impropriety, be overlooked, as of no manner of consequence. Every thing contained in the *Topics touch-* remaining 139 papers may be arranged under the eight following heads; under *ed upon in that work.* which may be placed all the physiological topics touched upon in the Philo-sophical Transactions.

1 The circulation of the blood.　　5 Vision.
2 Respiration.　　　　　　　　　　6 Organs of motion.
3 Action of the skin.　　　　　　　7 Nourishment and digestion.
4 Nervous system.　　　　　　　　8 Generation.

Let us take a view of each of these topics in succession.

I. CIRCULATION. We have already detailed the great discovery of the circu- *The circula-* lation of the blood, by Harvey, at sufficient length; and mentioned the illiberal *tion.* attempts of his contemporaries and of some of his successors to rob him of the merit of the discovery, by ascribing it to ancient writers, who had formed no conception whatever on the subject. Three particular points occupied physi-ologists after this discovery; namely, to determine the quantity of blood in man and other animals; to ascertain the nature and properties of blood; and to calculate the force and velocity with which it flows. Dr. Allen Moulin *Quantity of* found the quantity of blood in various animals as follows: *blood in ani-* *crals.*

	Weight of animal.				Weight of blood.		Proportion of blood to the animal
Sheep ..	118 lb.	0 oz.	0 dr.	0 gr.	5·25 lb.	0 gr.	$\frac{1}{13}$
Lamb ..	30½	0	0	0	1· 5	0	$\frac{1}{20}$
Duck ..	2	14	0	50	1· 5 oz.	53	$\frac{3}{18}$
Rabbit ..	0	10	7	50	2 dr.	57	$\frac{1}{30}$

In the heart of a dog, whose blood he had coagulated by an injection into
R

the vessels, he found six oz. of blood ; and in that of another dog treated in the same manner, he found a still greater quantity.   Now, on the supposition that the blood in a man bears the same proportion to his weight as in the lamb, it follows, that a man weighing 160 lb. contains about 8 lb. of blood. Supposing only four oz. of blood to enter the heart at each diastole, and supposing the pulse to beat at the rate of 75 times in a minute, then it will follow that the whole blood in a man circulates 140 times in an hour.*   Dr. James Keill published a book on secretion in the year 1709, in which the different secretions, and indeed all the diseases and disorders of the human body are accounted for, by the attraction of the particles of matter for each other, and by the different degrees of velocity with which the blood moves in different organs. A pretty detailed account of this book is given in the Philosophical Transactions.†   Dr. Keill differs entirely from Dr. Moulin, in his opinion respecting the quantity of blood in animals.   The mode taken to determine the point by Moulin was to let the animal bleed to death, and then to weigh the quantity of blood which flowed out.   Dr. Keill demonstrates that the whole blood never flows out in these cases, and that the quantity obtained depends upon the size of the vessel punctured, being always greater the smaller the size of the wounded vessel is.   All this is very correct : but when Dr. Keill estimates the quantity of blood in a man weighing 160 lb. at 100 lb., it is impossible to avoid rejecting his calculation as altogether extravagant.   A controversy upon this subject took place between Dr. Keill and Dr. Jurin, who opposed his conclusions ; but as this controversy depended upon the application of the laws of dead matter to the living body, it is quite unnecessary to enter upon it here. The subject has long lost the whole of its interest, it being now universally admitted that the living body is regulated by laws quite different from those which regulate dead matter.   In fact, we have no better means of determining the velocity of the circulation than those proposed by Dr. Moulin, and they must be allowed to be very inadequate means.

Force with which the blood flows.

With respect to the force with which the blood flows in different animals, there is nothing of any importance to be found in the Philosophical Transactions ; but there is a work on the subject which contains much curious and valuable matter, we mean the second volume of Hales' Statical Essays.   As this work was read to the Royal Society, and as it was published at the particular desire of the Society, it would be improper to omit stating the results which this illustrious philosopher obtained.

Dr. Hales having laid open the crural and carotid arteries of various animals, as horses, sheep, and dogs, fixed into them a long glass tube, by means of a

---

* Phil. Trans. 1687. Vol. XVI. p. 433.   † Phil. Trans. Vol. XXVI. p. 324.

brass pipe, and observed how high the blood was impelled in each by the action of the heart and arteries. The following table exhibits the results obtained.

Animals.	Weight of animals.	Height the blood rose in the tube.		Pulses in a minute.	Blood that flowed out.	Velocity of blood in the aorta.*	Blood thrown into the aorta per minute.
		Feet.	Inches.		lb.		lb.
Horse, 1st.	....	8	3	....	42·2	....	....
2d.	....	9	8	....	40	....	....
3d.	825 lb.	9	6	36	31·82	86·85	13·75
Ox......	1600	....		38	....	76·95	18·14
Sheep....	91	6	5½	65	....	174·5	4·593
Doe....	....	4	2	....	....	....	....
Dog, 1st.	52	6	8	97	....	144·77	4·34
2d.	24	2	8	....	....	130·9	3·7
3d.	18	4	8	....	....	130·0	2·3
4th.	12	3	3	....	....	120 0	1·85

If the experiment had been made on man, Dr. Hales conceives that the blood in the tube would have stood at the height of seven feet six inches. From these data, he calculates the force with which the heart contracts, which he shows to be equal to the weight of 51·5 lb.; and in the third horse to 113·22 lb.; in the sheep, it amounted to 36·56 lb.; and in the first dog to 33·61 lb. He ascertained that the blood moves eleven times as fast in the large arteries as in the capillary vessels, and that the motion through the capillary vessels of the lungs is much more rapid and free than through those of the body. The force necessary to burst the aorta, he found to be 50 times greater than what it had to sustain. The veins are weaker, but the force to which they are exposed is proportionably less These are the most important conclusions to be found in the Hydrostatics of Hales. His reasoning about animal heat is too inaccurate to deserve recapitulation. He has drawn a number of practical inferences from his experiments, which are of some importance in a medical point of view; and advantage has been taken of them to improve certain surgical operations.

A great many observations were made by physicians and physiologists during Nature of the last century, to determine the nature of blood; because most of the physi- blood. ological opinions, then in vogue, were connected with the motion and size of the particles of that liquid. Leuwenhoeck examined it before the microscope, and found that its red colour depended upon the presence of a number of round globules, which floated in it. He afterwards thought that he ascertained that each of these globules was composed of eight smaller globules; and

* Estimated in inches per minute.

each of these of eight still smaller.* These fanciful observations were embraced with great eagerness by the humoral pathologists of the times. Secretion was accounted for by the breaking down of these globules; inflammation, by a globule of too large a size having made its way into a capillary vessel; and the effect of various medicines, by attenuating or breaking down the globules of the blood into globules of a still smaller size. These opinions continued prevalent till the year 1761, when Father Di Torre, of Naples, having examined blood with very powerful microscopes, pronounced the supposed globules to be hollow articulated rings.† These new opinions being detailed in the Transactions, by Sir F. H. Eyles Stiles, occasioned a controversy, which terminated in destroying the authority both of Leuwenhoeck and Di Torre, and in convincing physiologists that the supposed structure of the globules of the blood, observed both by the one and the other, was nothing more than an optical deception.

When blood, after being drawn from an animal, is allowed to remain for some time in a vessel, it coagulates, and gradually separates into two parts namely, a fluid of a greenish yellow colour, called the *serum*, and a solid part retaining the red colour of the blood, and called the *crassamentum*. From some experiments on the specific gravity of blood and serum, made by Mr. Boyle, it was inferred that the serum was specifically heavier than the crassamentum, and of course than the whole blood. The discordancy of this opinion with the common observation, that the crassamentum always sinks to the bottom of the serum unless buoyed up by air, induced Dr. Jurin to make an accurate set of experiments to determine the point. They were published in the Philosophical Transactions for 1719.‡ He found from an average of numerous experiments, all nearly agreeing with each other, that the specific gravity of human blood is 1 054, of serum 1·030, and of the moist crassamentum 1 084. He endeavoured to determine the proportion which the crassamentum bears to the serum, by distilling off the liquid part of blood by means of a low heat; and he concluded from the results which he obtained that $\frac{12}{17}$ths of the blood is serum, and $\frac{5}{17}$ths crassamentum. But it is unnecessary to observe that his method was faulty, and entitled to no confidence whatever.

In the year 1770, Mr. Hewson inserted two valuable papers in the Transactions, upon the blood.§ His object was to determine the circumstances which occasion the coagulation of the blood; and, if possible, account for a phenomenon which has always proved a stumbling-block to physiologists But his endeavours were not crowned with success; the matter still remains obscure. We have no better explanation than that proposed by Mr. John Hunter, that

---

* Phil. Trans. 1674. Vol. IX. p. 23 and 121.   † Phil. Trans. 1765. Vol. LV. p. 246.
‡ Vol. XXX. p. 1000.   § Vol. LX. p. 368 and 384.

it depends upon the *vitality* of the blood. But how this supposed vitality acts, is unscrutable. It is well known, that when animals are killed by lightning the blood does not coagulate at all. In these cases, according to Mr. Hunter, the vitality is destroyed instantaneously, and of course has not time to act. But when blood is drawn out of a vein, it retains its vitality for a certain time. Every person must be sensible that the term *vitality*, as applied to the blood by Mr. Hunter, is merely the name of an *occult principle*. Hewson, in 1773, published another paper on the blood, in which he endeavoured to rectify the opinions of Leuwenhoeck, and Di Torre, respecting the globules of that liquid.* The only other paper respecting the blood that requires to be mentioned, is one by Dr. Darwin, in which he shows, by experiments, that blood contains no elastic fluid, unless it has been exposed to the air.† Though the opinion advanced in this paper is probably well founded, the experiments are not calculated to demonstrate its truth.

Towards the end of the 17th century, much attention was paid by philoso- Transfusion phers to the art of transfusing blood from one animal to another. The subject of blood. was first started in consequence of a set of experiments made by Sir Christopher Wren, while Savilian Professor at Oxford. He injected liquids of various kinds into the blood-vessels of dogs, in order to determine the effect which each liquid would produce upon the animal. Soon after, Dr. Lower exhibited at Oxford the transfusion of blood from one dog to another, by means of quills inserted in the carotid artery of one dog and into the jugular vein of another. Soon after, Dr. King suggested the possibility of transfusing the blood from the vein instead of the artery. About a year after, the experiment was made at Paris, by Dr. Denys, upon a man who had the blood of an animal transfused into him, while his own blood was allowed to flow out. The experiment was attended with no inconvenience whatever to the person upon whom it was made It was conceived upon the principles of the humoral pathology then in vogue, that almost all diseases were occasioned by the diseased state of the fluids, especially of the blood: hence it was believed that by this happy expedient, diseases might be at once got rid of by drawing off the diseased blood, and introducing healthy blood from some sound animal. There are no less than 11 papers upon the subject in the earlier volumes of the Philosophical Transactions. A controversy was even begun between the French and English philosophers on the subject, each claiming for his own country the merit of having first started the happy idea. But it was at last agreed upon that the idea belonged to Dr. Lower, and that the British physiologists first tried the experiment upon the lower animals, and the French upon man. The expected advantages resulting

---

Phil. Trans. Vol. LXIII. p. 303.   † Phil. Trans. 1774. Vol. LXIV. p. 344.

from this practice have been long known to be visionary. It is not therefore worth while to enter into the minute details of the experiments.

Respiration.

II. RESPIRATION. This function is known to be in some degree necessary to all animals, though it is performed in different ways. In man and land animals it is performed by lungs, and the whole blood of the body is made to circulate through these organs as regularly as through the system itself. In amphibious animals the circulation through the lungs can be interrupted for a time at the pleasure of the animal, which can thus suspend its respiration for a certain period. In fishes the blood is made to circulate through the gills, as it does through the lungs of land animals; the air held in solution in water comes in contact with these organs, and is as necessary for the fish as the inhalation of atmospheric air is to land animals. In insects and worms the air is drawn in by means of numerous pores arranged along their bodies, and when their pores are stopped up so as to prevent the ingress and egress of the aerial fluid, the animal dies. We are ignorant of the respiratory organs of zoophites and infusoria. Some of the intestinal worms ought from their situation to be independent of respiration altogether. But we are hardly possessed of the means of determining the point with accuracy.

Mechanical part.

The mechanical part of respiration was ascertained at an early period. By the action of the diaphragm, and the intercostal muscles, &c. the thorax is elevated and its capacity increased, a vacuum would be the consequence; but as the lungs communicate with the external air, that fluid is forced in through the nostrils, and fills the cavity. It is forced out again by the contraction of the thorax: and this alternate inspiration and expiration continues during the life of the animal. Haller, in a paper published in the Transactions for 1750,* showed that the intercostal muscles are active during respiration, concerning which anatomists had formerly been in doubt. He showed also that there was no air between the pleura and the ribs, as had been supposed by some; but that the lungs filled the whole cavity of the thorax, whether in a state of contraction or dilatation. Lower, as early as 1667, showed that the diaphragm is the chief organ of respiration; and that if the nerves of a dog which go to the diaphragm be cut, the animal breathes exactly like a broken-winded horse.†

Effect of air.

In the year 1667, Fracassati wrote a paper‡ to show, contrary to the opinion then entertained, that the black colour of blood, at the bottom of a dish filled with the liquid, was not owing to a mixture of the melancholy humour, but to its not being exposed to the air; for he found that when it was turned up, and placed in contact with the atmosphere, it became of a florid red. Lower soon after showed that the scarlet colour of the arterial blood is owing to its being

---

* Vol. XLVI. p. 325.   † Phil. Trans. Vol. II. p. 544.   ‡ Phil. Trans. Vol. II. p. 493.

exposed to the influence of air in the lungs. In the year 1668, Mayow published his celebrated tract on Respiration, in which he endeavoured to explain that function, and certainly anticipated some of the most fashionable modern theories on the subject. According to him, air contains certain subtile *nitrous* particles, which are absorbed by the blood in the lungs; and when air is deprived of them, it becomes unfit for respiration. The blood thus impregnated possesses the property of stimulating the heart, and thus in fact continues the circulation upon which all the vital functions depend. Hence the reason of the sudden death which ensues when respiration is suspended.* That so little attention was paid to this theory when published was partly owing to the many absurd opinions with which it was mixed, and partly to the hypothesis of Stahl, which very speedily attracted the sole attention of chemists and physiologists.

But perhaps the most important papers on respiration, which appear in the early volumes of the Transactions, are those of Mr. Boyle. They consist of experiments demonstrating the necessity of air for all animals. The animals were enclosed in the exhausted receiver of an air pump, and the phenomena that took place were observed. These experiments labour under a considerable defect. The degree of exhaustion produced is not noted, so that we do not know how rare the atmosphere was to which the animals were exposed. There is reason to believe that Mr. Boyle's air pump was very imperfect; of course the exhaustion would be proportionably incomplete.

Ducks lost all signs of life in about two minutes; vipers continued alive about two hours and a half; frogs were killed in appearance in about three hours, but being left long enough in the open air again recovered; kittens, the day after being kittened, were found just as unable to live without air as full grown animals; fishes lived much longer than land animals, but were likewise killed by being kept under an exhausted receiver, except oysters and craw-fish, which resisted the utmost efforts of the air pump. Sloe worms and leeches were also found to live long in an exhausted receiver, but insects and caterpillars were soon killed. Mr. Boyle likewise showed the unfavourable effects of too rarefied an atmosphere upon animals exposed to it.† About the same period Dr. Hook exhibited an experiment to the Royal Society, at that time deemed of considerable importance. He removed the ribs and diaphragm of a dog, and kept him alive for an hour by blowing into his lungs with a pair of bellows. The experiment succeeded equally when the lungs were pricked so as to let out the air; and they were kept constantly distended, though without any motion, by blowing a continued stream of air into them. Thus showing that it was the renewal of the air, and not the motion of the lungs, that constituted the essential part of respiration.‡

* Phil. Trans. Vol. III. p. 833.　　　　† Phil. Trans. 1670. Vol. V. p. 2011 and 2026.
‡ Ibid. 1667. Vol. II. p. 539.

These are the only papers contained in the Transactions relating to respiration, except a curious dissertation by Dr. Parsons on Amphibious Animals, in which he explains the peculiarities in their structure that enable them to suspend their respiration for some time at pleasure.* There is also a paper, by Mr. John Hunter, on certain receptacles of air in birds, which communicate with the lungs, and are lodged both in the fleshy parts and in the hollow bones of those animals.† But these receptacles are rather subservient to the purposes of flying, by rendering the animals as light as possible, than of respiration.

Death by drowning, being obviously connected with respiration, the analysis of a book, on the subject, written by Conrad Becker, which is given in an early volume of the Transactions,‡ deserves to be mentioned here. It was the universal opinion at that period that drowning was occasioned by water getting into the lungs and filling them. It was supposed likewise that much water got into the stomach and intestines. Becker endeavoured to show that both opinions were ill founded. He mentions several cases of drowned persons examined by himself, in which no water whatever was found either in the lungs or alimentary canal. It is now well known that Becker's opinion is correct. A little water indeed occasionally makes its way into the lungs, but the death of the animal, which is kept immersed under water, is owing entirely to the absence of air, and the consequent interruption of respiration.

Animal heat.   Animal heat has always been considered as connected with respiration, because those animals that do not breathe are cold blooded; and when animals breathe hard their heat is at the same time increased. But in what manner heat is produced by breathing remained for many ages altogether inconceivable. The ancients supposed that the use of respiration was to moderate and diminish the heat of animals, which otherwise had a tendency to run into excess. After the discovery of the circulation of the blood, animal heat was conceived to be produced by the friction of the blood against the vessels. Even Dr. Hales goes into this notion; though he supposes that the electricity which, in his opinion, is excited by the circulation, contributes to the evolution of heat. Dr. Cromwell Mortimer, in a paper published in the Transactions for 1745,§ shows that heat is never produced by the friction of liquids against solids; and therefore considers the theory of animal heat at that time received as erroneous. He himself proposes another. According to him, phosphorus is distributed in considerable quantity all over the animal body. By respiration air is brought in contact with this very combustible body. The consequence is a constant and gradual combustion of phosphorus in every part of the body, and this combustion in his opinion occasions the heat which we perceive in the breathing animals. This opinion was too absurd to produce many converts, and has been long

---

* Phil. Trans 1766. Vol. LVI. p. 193.   † Phil. Trans. 1774. Vol. LXIV. p. 205.
‡ Ibid. 1705. Vol. XXIV. p. 2152.   § Vol. XLIII. p. 473.

since forgotten. The theory by which animal heat is at present explained, and it must be allowed to be a very plausible one, was first proposed by Dr. Crawford. Arterial blood according to him has a greater capacity for heat than venous blood. Venous blood is changed into arterial blood in the lungs, and at the same time absorbs a quantity of heat given out by that portion of the air inspired which is changed into carbonic acid gas. During the circulation the arterial blood is gradually converted into venous blood; of course its capacity for heat is constantly diminishing, and therefore it is giving out heat during the whole circulation. The truth of this theory depends entirely upon the different capacities of venous and arterial blood for heat. This Dr. Crawford laboured with great assiduity to establish. But the experiments are of so very delicate a nature that it is difficult to receive them with perfect confidence.*

III. ACTION OF THE SKIN. It is well known that the skin has the property of throwing out a certain vapour from the body, known by the name of *perspiration*, and in some cases likewise a liquid matter, known by the name of *sweat*. These fluids are considerable in quantity, as has been ascertained by the statical physicians; and the state of the function has considerable influence upon the health of the animal. The pores by which this matter is discharged are very visible, especially upon the palms of the hand and the soles of the feet. The palms of the hand, especially at the extremities of the fingers, may be seen disposed in regular ridges. The pores are set in these ridges, and may be easily distinguished by means of a moderate sized glass.† *[margin: Action of the skin.]*

One of the medical men to whom we are obliged for some of the most valuable statical experiments, was Dr. John Lining, of Charlestown, in South Carolina. They were continued for a whole year. He ascertained his weight in the morning and evening; the weight of the food which he swallowed, and the weight of the urine and alvine excretions ejected. The result of these very troublesome experiments was published in 1743, in the Philosophical Transactions.‡ The tables are curious; but too long to be inserted here, and they do not easily admit of abridgment.

It will be proper to mention here a very elaborate and ingenious essay by Dr. John Mitchell, to account for the difference of colour in people of different *[margin: Colour of negroes]*

* There is a very interesting paper in the Philosophical Transactions (1792. Vol. LXXXII. p. 199.) by Dr. Currie, of Liverpool, on the changes in animal heat occasioned by immersion in cold salt, and fresh water, and by passing repeatedly from the water into air, and vice versa. These experiments have not yet received any attention from chemists and physiologists. They appear to me to subvert every theory of animal heat hitherto proposed, not even excepting the theory of Dr. Crawford, ingenious and plausible as it is. Some very curious experiments by Mr. Brodie, published in the Transactions for 1811, are equally incompatible with the present chemical theory of respiration. But there is an apparent discordance between them and those of Dr. Currie, which ought to be cleared up before any attempts can be made to construct a new theory of animal heat.

† Grew. Phil. Trans. 1684. Vol. XIV. p. 566. ‡ Vol. XLII. p. 491.

climates, published in the Transactions for 1744.* The following is the outline of this very elaborate composition :—the colour of white people proceeds
from the colour which the epidermis transmits; that is, from the colour of the
parts under the epidermis, rather than from any colour of its own The skins
of negroes are of a thicker substance, and denser texture, than those of white
people, and transmit no colour through them. The part of the skin which
appears black in negroes is the corpus reticulare cutis, and external lamella of
the epidermis : and all other parts are of the same colour in them with those of
white people, except the fibres which pass between those two parts. The colour
of negroes does not proceed from any black humour or fluid parts contained in
their skins; for there is none such in any part of their bodies more than in
white people. The epidermis, especially its external lamella, is divided into
two parts, by its pores and scales, two hundred times less than the particles of
bodies on which their colours depend. These are the different data which he
establishes, and he deduces from them a pretty satisfactory explanation of the
black colour of negroes. But his reasoning is extended to a very great length,
and could not be abridged without stripping it of much of its plausibility, and
of course, without doing it injustice.

Living bodies capable of bearing a high temperature. Connected with the action of the skin, and no doubt depending upon it, is
the power which the human body has of bearing a very high temperature, with-
out any change in the temperature of the body itself. Accurate experiments
upon this subject (for it is needless to mention the inaccurate trials of Boerhaave) were first made by M. Tillet at Paris, in 1764. The subject was resumed by Dr. Fordyce in 1775, and we have two curious papers by Sir Charles
Blagden in the Transactions, giving an account of the phenomena which were
observed. Dr. Fordyce heated a room by means of flues running along the
floor, and by means of boiling water, to the temperature of about 120°, he then
stripped himself to the shirt, and went into the room; the temperature of his
body continued at about 100°, but his pulse increased considerably in rapidity,
beating 126 times in a minute. Streams of water condensed upon his body,
and ran down to the floor. This was the vapour of the hot water by means of
which the room was heated, condensed by the action of the comparatively cold
surface of his body. Dr. Fordyce invited the Hon. Capt. Phipps, Sir Joseph
Banks, Dr. Solander, and Sir Charles Blagden, to repeat and verify these experiments. These gentlemen attended two several days, and during the last of
these days there were also present Lord Seaforth, Sir George Home, Mr. Dun-
das, and Mr. Nooth. The room was heated by means of a'large iron cocle
placed in the centre and heated red hot The thermometer, in some of their
trials, rose to 260° This degree of heat they found they could bear for a con-

* Vol. XLIII. p. 102.

siderable time without any great inconvenience, both when dressed and when naked. The temperature of the body continued at 100°, the pulse was greatly quickened, increasing to 144 beats in a minute, which was rather more than double its ordinary velocity. The perspiration, as might have been expected, was very violent, and no doubt, together with the imperfect conducting power of air, occasioned that equability of temperature which the body preserved. To prove that the heat of the air was really as great as they stated, and that there was no error in their thermometer, they introduced eggs and beef steaks, which were perfectly roasted in a short time. Water introduced remained stationary at 140°, the evaporation preventing any further increase; but when this evaporation was stopped by covering its surface with oil or melted wax, the water was soon heated so as to boil briskly.* Experiments attended with nearly a similar result were made at the same time by Dr. Dobson, at Liverpool.†

IV. NERVOUS SYSTEM. The brain is somehow or other the seat of intelli- gence; every disorder in it is attended with a corresponding disorder of the intellectual faculties; the exercise of the senses, and even the motion of every muscle, depend upon the nerves which pass into these organs, and are destroyed entirely by injuring or cuting the nerves which belong to the organs. These are facts which are perfectly well understood; but in what way the brain and nerves exert this amazing power remains still a secret. Descartes started the idea of a peculiar fluid, secreted in the brain, and moving along the nerves, which produced the effect. The opinion, though not serving in the least to elucidate the powers of the nerves, was adopted with eagerness; and the exist- ence of a nervous fluid was long a favourite opinion of physiologists. And though no vacuity can be perceived in the nerves, even when viewed by the best microscopes, we are not sure that the notion even at present is perfectly exploded. It certainly existed in Italy not many years ago, and some of our own young physiologists were fond of supporting it.

But although we know nothing of the way the brain and nerves exert their power, several curious and important particulars respecting them are delivered in the Philosophical Transactions. Dr. Haighton, by a set of well contrived and decisive experiments, has demonstrated that when the nerves are separated by cutting them in two, they again unite in process of time, and the nervous energy passes through them as at first. He divided the eighth pair of nerves in a dog on both sides. These nerves supply the stomach; the animal became disordered and died in two days. When the nerve on one side only was cut, the dog was but little affected, but when the remaining nerve was cut three days afterwards, the dog died in a few days. When the nerve of one side was

Marginal notes: Nervous sys- tem. Nervous flui[d] Nerves unite again when cut.

* Phil. Trans. Vol. LXV. p. 111 and 484.   † Phil. Trans. Vol. LXV. p. 463.

cut, and an interval of six weeks suffered to elapse before cutting the nerve on the other side, the dog suffered comparatively little and recovered completely. This was not owing to any other nerves becoming larger and supplying the place of the eighth pair in the stomach; for, when the dog who had thus recovered had the eighth pair cut again, on both sides at the same time, he died in two days, as was the case with the first dog. Hence it follows, that his recovery in the former case was owing to the re-union of the nerve which had been cut first, before the other nerve was separated.*

Structure of nerves. A great variety of opinions, respecting the structure of nerves, has been entertained by those physiologists who have examined them with microscopes According to some, they are composed of globules; according to others, of hollow cylinders; according to a third party, of fibres. Animal substances are less fitted for microscopical observations than almost any other; and wherever great magnifying powers are employed, the observer is almost certain to be deceived by his glasses. The knowledge of this fact induced Mr. Home, to whom we are indebted for a paper upon the subject in the Philosophical Transactions,† to use single microscopes, which did not magnify more than 23 times. The result of his observations, assisted by Mr. Ramsden and Mr. Clift, was, that the optic nerve of the cat, where it enters the eye, is transparent; and that the optic nerve of the horse, and of the cat, is partly transparent, partly opaque. The opaque portion consists of a number of cylinders, and the transparent consists of a gelatinous matter, which joins them together. These cylindrical bodies are least numerous where the nerve issues from the brain, and they gradually increase in number as it proceeds towards the eye, so as to increase from about 40 to nearly 500.

Mr. Cruikshanks made a number of experiments on dogs, pretty similar in their nature to those stated above, as performed by Dr. Haighton. The result was similar. The nerves after being cut, again united, and the parts to which they were sent performed their functions as usual. This happened even when a portion of the nerves was cut out and removed. He found that when the spinal marrow was cut through, below the place where the phrenic nerve passes off, the animal lives for some time, continuing to breathe by means of the diaphragm. But when the phrenic nerve is divided, and then the spinal marrow cut in two at the same place, instant death takes place.‡

Sleep. Sleep is somehow connected with the nervous system; though no satisfactory explanation of it can be given. The different theories of Haller, Brown and Darwin, though each had its zealous defenders and supporters, are not entitled to any consideration whatever, as they are inconsistent with fact, and inade-

* Phil. Trans. 1795. Vol. LXXXV. p. 190.   † Phil. Trans. 1799. Vol. LXXXIX. p. 1.
‡ Phil. Trans. 1795. Vol. LXXXV. p. 177.

quate to explain the phenomena. The time of sleeping, and even the length of time devoted to sleep, depend much upon custom. A sailor can sleep at any time, and awake at any time; but most other persons sleep only during the night. In general not more than one third of the 24 hours is devoted to sleep; but there are some instances on record of a state of sleep continuing for a preternatural length of time, though every effort was used to rouse the sleepy person. One of the most remarkable of these cases is related by Dr. Oliver, Long continued sleep. in an early volume of the Philosophical Transactions.* Samuel Chilton, of Tinsbury, near Bath, a labourer, about 25 years of age, of a robust habit of body, not fat but fleshy, and having dark brown hair, fell asleep on the 13th of May, 1694, and continued asleep for a month; when he awoke as usual and went about his ordinary avocations. Food was placed beside him, which he eat regularly every day, and had the usual evacuations. He fell asleep again in 1696, and continued asleep for 17 weeks, during the last six of which he eat nothing. He fell into a third sleepy fit in 1697, which continued for six months. During these fits he was subjected to many experiments, by the curious who went to see him, and were anxious to discover whether his sleep was real or pretended Volatile alkali was poured up his nose, his nostrils were stuffed with hellebore, pins were thrust into his flesh, he was blooded; but all these trials he bore without appearing to have the least sensibility of pain.

The state of torpor, into which many animals fall during winter, is obviously Torpor. analogous to sleep. But it differs from sleep in being occasioned solely by temperature. Hybernating animals always assume their torpid state whenever the thermometer sinks to a certain point. Almost all animals seem to be susceptible of this state, at least to a certain extent, not even excepting man. For the apparent death produced by cold is probably nothing else but a species of torpor, out of which the animal, in most cases, might be roused, if the requisite caution in applying the heat were attended to; for death, in most cases, seems to be produced not by the cold, but by the incautious application of heat, which bursts the vessels and destroys the texture of the body. It is well known that if any part of the body be frost-bitten, an incautious application of heat infallibly produces mortification, and destroys the part. There is a remarkable example, in the Transactions, of a woman almost naked lying buried for six days under the snow, and yet recovering.† In this case it is scarcely possible to avoid supposing that the woman must have been in a state of torpor, otherwise she would certainly have endeavoured to have made her way home.

The action of poisons upon the animal body seems to depend chiefly upon Poisons. the effect which they produce upon the nervous system. Some of the mineral poisons indeed seem to corrode and destroy the texture of the solid parts of the

---

* Phil. Trans. 1705. Vol. XXIV. p. 2177.   † Phil. Trans. 1713. Vol. XXVIII. p. 265.

body; and some of the animal and vegetable poisons probably produce their fatal effects, by causing some change in the blood. But these effects alone would not account for the suddenness with which the fatal symptoms appear, while a variety of poisonous drugs, opium for example, seem to act upon the nervous system alone. Though very little light has been hitherto thrown upon the action of poisons; yet it is of importance to register the effects which are produced, as it is by means of these effects alone that any valuable inferences can be drawn. There are two papers upon the subject in the Philosophical Transactions: the first giving an account of a variety of experiments, made at Montpelier, in 1679, by Mr. Courten,* chiefly by injecting various liquors in the jugular and crural veins of dogs. The following are the results :

LIQUORS INJECTED.	EFFECTS.
1 oz. of emetic wine	Violent evacuations. The dog died in convulsions.
1½ dr. of sal ammoniac	Instantaneous death in convulsions.
1 dr. of salt of tartar	Instant death in convulsions.
1 oz. of human urine	No bad symptoms.
Decoction of 2 dr. of white hellebore	Instant death.
Vinegar	No bad symptoms.
2 dr. of sugar in 1 oz. of water	No bad symptoms.
1½ dr. of muriatic acid	Instant death.
1 oz. of camphorated alcohol	Instant death.
50 gr. opium in 1 oz. water	Gradual convulsions and death.
1½ dr. opium in 1½ oz. water	Sleep, apoplexy, death.
1½ dr. common salt in 1½ oz. of water	Thirst, but no other bad symptom.
½ oz. olive oil, warm	Apoplexy. Death.
1 oz. olive oil	Instant death.
10 dr. of alcohol	Gradual death, without any symptom of pain.
3 dr. of alcohol	Drunkenness, but gradually recovered.
5 oz. of white wine	Drunkenness, but recovered.
1 oz. of decoction of tobacco	Violent convulsions and death.
10 drops oil of sage, mixed with ½ dr. of sugar.	No bad symptom.
1 dr. white vitriol	Instant death.
30 gr. salt of urine	Violent convulsions, but recovered.
2 dr. of senna, in water	Violent vomiting, great weakness, but recovered.

Two ounces of the juice of Dutch nightshade, of hemlock, of wolfsbane, swallowed by a dog, did no harm; two drams of white hellebore produced violent evacuations, but the dog recovered. A dog bit by a viper was much revived by being made to swallow a solution of volatile alkali in water, and finally recovered. Fifteen grains of the dried root of napellus, or monkshood, given to a dog, produced violent evacuations and much weakness, but the animal finally recovered.

The other paper on poisons in the Transactions was published in 1751, by Herissant,† a physician in Paris. It gives an account of a set of experiments on the effects of the poison of lamas and of ticunas; poisons extracted by the

* Phil. Trans. Vol. XXVII. p. 485.   † Phil. Trans. Vol. XLVII. p. 75.

natives of South America, from certain plants. With them they anoint the points of their arrows, and the animals wounded, though never so slightly, drop down dead almost immediately. M. de la Condamine, who went with the other French academicians to measure a degree of the meridian in Peru, brought with him a quantity of this poison to Paris, and supplied Herissant, who was anxious to make some experiments on it, with a considerable quantity. It had nearly the appearance and smell of treacle. Herissant was directed to dissolve it in water, and boil it down to a proper consistency. The instruments to be employed in wounding the animals were then to be dipt into this extract, and allowed to remain till it dried upon them. Herissant found that the vapour which proceeded from this poison while boiling, disordered both himself and a young man who was with him so much that their lives were endangered. But Fontana, who afterwards repeated the experiments of Herissant, did not find any injurious effects from being exposed to this vapour. Hence there is reason to suspect that the imagination had some effect in occasioning the symptoms of which Herissant complained. A single drop of this poison, put into a wound, occasioned instant palsy and death within a minute But, when the poison was swallowed by animals, it did not prove fatal if taken only in moderate quantities. Sugar and salt, which had been proposed as remedies, were tried without effect. The only successful mode of treatment was the immediate application of a red hot iron to the wound, or the amputation of the wounded limb immediately after the infliction of the wound. All quadrupeds and birds, tried by Herissant, were speedily killed when this poison was infused into a wound ; but amphibious animals, fishes, worms, and insects, though they suffered some inconvenience, uniformly recovered.

There is a circumstance which is not the least singular, of the many obstruse points connected with the nervous system. Bones, tendons, and ligaments in their healthy state, are found to have no sensibility whatever ; so that they may be pricked with pointed instruments, or burnt with corrosive liquids, without the animal giving any symptoms whatever of feeling pain. Haller made many experiments to determine this point. And in the year 1755, Dr. Brocklesby made a number of experiments on the tendons and ligaments, all demonstrating their want of sensibility.* But when these parts of the body are affected with disease, they become most exquisitely sensible. ^{Sensibility of diseased bones and tendons.}

V. Vision. This is the only one of the Senses, respecting which any physiological papers occur in the Philosophical Transactions. There are two papers relative to vision; both of considerable celebrity, and of considerable importance. ^{Vision.}

Mariotte, a French mathematician of considerable eminence, found, that

---

* Phil. Trans. Vol. XLIX. p. 240.

when the rays of light from an object fell upon the optic nerve, the object became invisible. He fastened upon a wall about the height of his eye a small round paper, to serve as a fixed object of vision. He placed another paper on the side towards his right eye, a little lower than the first paper and about two feet from it. He then placed himself directly opposite to the first paper, with his left eye shut, and drawing back by little and little he found when he was about ten feet from the wall that the second paper totally disappeared.* He conceived that it disappeared because the rays from it fell upon the optic nerve. This conclusion has been generally acquiesced in, and indeed is very probable; though it is possible that the deficiency of vision is owing to the rays falling upon the hole in the retina, near the optic nerve, discovered by Soemmering in the human eyes, and by Home in the eye of the ape. From this discovery, Mariotte drew as a conclusion that the retina was not the medium of vision, as had been hitherto supposed, but the choroid coat. This opinion occasioned a controversy between Mariotte and Pequet, which is continued in several papers published successively in the Philosophical Transactions. Pequet's opinion is now universally adopted; indeed every thing known, respecting the nervous system, leads irresistibly to that conclusion. A nerve is absolutely necessary for the exercise of all the senses, and the retina is an expansion, according to the common opinion, of the optic nerve.

Blind boy restored to sight.   The second paper on vision contained in the Transactions is of great celebrity, and throws a great deal of light on the nature of sight. Accordingly, it is perpetually referred to by metaphysical writers. It is a paper by Cheselden, giving an account of the observations made by a young gentleman, who had been born blind, and who was couched when between 13 and 14 years of age.† Before the operation he could distinguish certain colours, as black, white, and scarlet, when they were placed in a strong light, but he was incapable of perceiving the shape of any thing. When he first saw, he thought that all objects touched his eyes, as what he felt did his skin; and only learned very slowly and gradually to judge of distances. He thought no objects so agreeable as those which were smooth and regular; though he could form no judgment of their shape, or guess what it was in any object that was pleasing to him. At first he thought that pictures were nothing else than party-coloured planes, or surfaces diversified with variety of paint; but, about two months after he was couched, he discovered all at once that they represented solids. He was very much confounded at the discovery, expecting that the pictures would feel like the things which they represented, and was amazed when he found those parts, which, by their light and shadow appeared round and uneven, felt only flat like the rest, and asked which was the lying sense, feeling or seeing? Being shown

* Phil. Trans. 1668. Vol. II. p. 668.   † Phil. Trans. 1728. Vol. XXXV. p. 447.

his father's picture in a locket, at his mother's watch, and told what it was, he acknowledged a likeness, but was vastly surprised; asking, how it could be that a large face could be expressed in so little room; saying, it should have seemed as impossible to him as to put a bushel of any thing into a pint. At first he could bear but very little light, and the things which he saw he thought extremely large; but on seeing things larger, those first seen he conceived less; never being able to imagine any lines beyond the bounds which he saw. The room which he was in he said he knew to be but part of the house, yet ye could not conceive that the whole house could look larger. Being afterwards couched in the other eye, he said that objects at first appeared large to this eye, but not so large as they did at first to the other; and looking on the same object with both eyes, he thought it looked about twice as large as with the first couched eye only, but not double, that they could anywise discover.

There is a third paper relating to vision, in the Philosophical Transactions, giving an account of a number of scales lying over each other in that part of the sclerotic coat of the eyes of birds which is next the cornea.* The author of the paper, Mr. Pierce Smith, considers the observation as new, and claims the discovery as his own. His account however, in both particulars, is inaccurate. Malpighi had long ago made the observation, and published an account of it; so that there was nothing new in the observation of Mr. Smith. But ignorance is not the only charge that may be brought against the author of this paper. The observation which he claims in it was not his own originally. He derived his knowledge of this particularity in the eyes of birds from a preparation of Dr. Barclay, at that time a Student of Medicine, in Edinburgh; and the dissections which he describes were subsequent to his examination of this preparation. As Mr. Smith's paper contains in fact nothing new, the injury which he did to Dr. Barclay by publishing his dissection as his own, was in reality trifling. But his obvious intention was to acquire credit by appropriating to himself what he conceived to be the discovery of another, and a fellow student. On that account his paper deserves to be stigmatized.

VI. ORGANS OF MOTION. Dr. William Croone, one of the original Members of the Royal Society, and a philosopher and physician of considerable eminence, left behind him, at his death in 1684, the plan of an annual lecture on muscular motion before the Royal Society. This plan was put in execution by his widow. The first lecture was read in 1738, by Dr. Alexander Stuart, Physician to the Queen;† and has been continued ever since. These annual lectures, for a considerable number of years, have been regularly published in the Philosophical Transactions. They have been drawn up by the most eminent

Organs of motion.

---

* Phil. Trans. 1795. Vol. LXXXV. p. 263.
† Birche's History of the Royal Society, Vol. IV. p. 340.

physiologists, who were Members of the Society, and contain a great collection of very curious and important facts, respecting the muscles and their motions. But few of the lecturers have attempted to investigate, far less to explain, the nature of muscular motion.   Indeed the subject appears to be altogether beyond the reach of our faculties.   We know that the motion of many of the muscles is obedient to the will; that when a muscle produces motion it becomes shorter, and at the same time swells out and becomes stiff.   The degree of contraction produced is not great, and particular pains have been taken to make it as small as possible.   When a muscle acts constantly, as the heart and diaphragm, the contraction which it undergoes is never great.   When a muscle contracts very forcibly, it feels fatigued, and unless it be relieved by speedy relaxation, it is apt to lose its tone, and cannot be exerted again without pain: rest, however, gradually restores it to its wonted energy.   As to the cause of muscular motion. no less than six different hypothetical explanations have been offered; but all of them so absurd, or so hypothetical, that barely to mention them is sufficient.

Hypothetical causes of muscular motion.   1. The action of a peculiar fluid secreted in the brain, passing along the nerves and therefore called the *nervous fluid*, was long a favourite hypothesis of physiologists.   It was the opinion of Dr. Stuart, who read the first Croonian lecture on muscular motion.   But this opinion, which in fact explained nothing, and the truth of which it was impossible to prove, has been long laid aside by the greater number of medical men.   2. Dr. Hartley endeavoured to prove, that the intention of the will was communicated from the brain to the muscle, by means of a vibration produced in the nerves.   This opinion was adopted by Dr. Priestley, who published a kind of abstract of the worst part of Dr. Hartley's Treatise on Man; and endeavoured, by means of it, to demonstrate the principles of materialism.   This opinion is exceedingly absurd, and never had many advocates.   Supposing we were to concede all that those who maintain it demand, namely, a vibration in the nerves; we are as far as ever from understanding how this vibration produces muscular motion.   But the fact is, that nerves are the worst substances possible to be thrown into vibrations, and seem scarcely susceptible of it.   They are soft substances, destitute of elasticity, and of every requisite for vibration.   But though neither of these hypotheses seems capable of being maintained, we must admit that the power which the muscles have to produce motion, is somehow derived from the brain; because, whenever the nerves going to a muscle are cut or tied, all muscular motion is destroyed.   3. Another hypothetical explanation of muscular motion is, that it is owing to a quantity of electric matter, transmitted from the brain to the muscles by the intervention of the nerves.   This hypothesis is as little satisfactory as the two preceding; because, if it were granted, the motion of the muscle would still be as inexplicable as ever; and because the transmission of electric matter along the nerves in animal bodies is inconsistent with the laws which electricity follows in its passage through bodies. To obviate this difficulty, some physiologists have supposed that the electric matter belonging to animal bodies follows peculiar laws of its own. When this opinion is explained, it means that a nervous fluid is secreted in the brain, having properties analogous to the electric fluid; but differing in the way in which it passes through bodies. This opinion, having nothing whatever to support it, is unworthy of examination.

A fourth opinion was proposed by Mayow in his Treatise on Respiration. According to him, the spiritu-sulphureous particles of the blood, and the nitro-aërial particles imbibed from the air by respiration, are thrown together into every muscle previous to action. These two substances, when they come in contact, occasion an effervescence, which swells up the muscle and occasions motion. This opinion, not being intelligible, is not capable of discussion. 5. Dr. Morton, in a paper published in the Philosophical Transactions about the year 1751, ascribed muscular motion to a sudden heat generated in the blood by the will. This would not serve to explain the phenomena, though it were admitted and it is well known that no such sudden generation of heat in muscles takes place in muscular motion. The cold blooded animals, as fish, possess muscular motion as perfectly as quadrupeds, yet a very moderate heat, when applied to them, is found sufficient to destroy life. 6. The last explanation of muscular motion, with which I am acquainted, was proposed by Dr. George Fordyce in a Croonian lecture on muscular motion, which he read before the Royal Society in the year 1788.* All motions, according to him, are produced by attractions or repulsions. There are various kinds of attractions known to philosophers, as the attraction of gravitation, of electricity, of magnetism. Now muscular motion is produced by a particular species of attraction between the parts of a muscle, which makes them approach nearer each other. As this attraction is confined to the living state of the muscle, Dr. Fordyce distinguishes it by the name of the *attraction of life*. This attraction, according to him, is inscrutable, like the ultimate attractions which regulate the motions of inanimate matter; but we may determine the laws according to which it acts. And this, according to him, should be the sole object of our inquiries, when we turn our attention to muscular motion. This explanation of Dr Fordyce, we are afraid, is nothing else than the introduction of a new word, of which we do not know the meaning. An attraction confined to animal matter, dependent upon the will, and capable of being suspended at pleasure, has no analogy to gravitation, or any other attraction belonging to inanimate matter, which has been dignified with a name. Even if its existence were taken for granted, we should have no notions of the nature of muscular motion, unless we knew how it is connected with the brain and nerves, and how it is influenced by the will. Muscular motion

* Phil. Trans. Vol. LXXVIII. p. 23.

in fact resolves itself into the nervous energy, which is obviously inscrutable in the present imperfect state of our faculties.

The greater number of physiologists have not attempted to explain the nature of muscular motion; but have satisfied themselves with endeavouring to point out some of its laws, or with noticing some peculiarities belonging to the *Experiments* muscles of certain parts of the body. One of the most interesting topics made *on the eye.* choice of is the *eye*, a subject intended by Mr. John Hunter for a lecture on muscular motion. He unfortunately died before his experiments were completed; but the subject was continued by Mr. Everard Home in several repeated lectures, with a great deal of ingenuity and success. He was assisted in his experiments by Mr. Ramsden and Sir Henry Englefield. Mr. Hunter had conceived that the crystalline lens of the eye is muscular, and that the adjustment of the eye to different distances is performed by the action of these muscles. Leuwenhoeck had long before observed that the lens was fibrous. Mr. Hunter had found in the eye of the cuttle-fish a lens in which this fibrous structure was very conspicuous. These fibres Mr. Hunter supposed acted as muscles. But when Mr. Home tried the experiment on an eye deprived of the lens by extraction, it was found to possess the power of adjustment as completely as a perfect eye. Hence it was obvious that Mr. Hunter's opinion was inaccurate. Upon examining the subject experimentally, it was found that the cornea changed its shape in order to adjust the eye; and Mr. Home concluded that the adjustment was accomplished partly by the action of the cornea, partly by the elongation of the axis of vision, and partly by a change of place in the lens itself.

Specific gra- It may not be improper to notice here a curious paper published in the Trans-
vity of men. actions for 1757 by Mr. Robertson.* His object was to determine the specific gravity of living men. For this purpose he constructed a vessel into which (when containing water) men might be immersed, and he determined the specific gravity by the bulk of water displaced, or, which is the same thing, by the rise of the water on the side of the vessel. Ten trials were made in this way on ten labouring men belonging to the ordinary of Portsmouth yard. They were all thin and varied in size from 6 feet 2 inches to five feet 3¼ inches. The following table exhibits the specific gravities of each of these, arranged according to the height of each; the specific gravity of rain water being reckoned 1·000.

	Height. Feet. Inch.	Weight. lb.	Sp. Gravity.
1.	6.. 2	161	1·0012
2.	5.. 10¾	147	0·9096
3.	5.. 9¼	156	0·9961
4.	5.. 6¾	140	0·8111

* Phil. Trans. Vol. L. p. 30.

	Height.		Weight.	Sp. Gravity.
	Feet.	Inch.	lb.	
5	5	5⅞	158	0·8977
6	5	5½	158	0·8600
7	5	4⅔	140	0·8230
8	5	3⅓	132	0·8429
9	5	4⅛	121	0·7986
10	5	3¼	146	0·9972

From this table it appears, that all the persons tried were specifically lighter than water, except one. Hence it is obvious that most persons naturally swim in water, and that drowning might often be avoided if the person that falls into the water were not deprived of his presence of mind.

VII. NOURISHMENT. This function is not so inscrutable, so completely be- Nourishment yond our faculties, as the nervous influence or muscular motion ; though we must acknowledge that hitherto the phenomena of digestion and assimilation have not been explained in a satisfactory manner. Animals, during one period of their life, are increasing in size, and during the whole of it their organs are undergoing almost perpetual changes. Hence the necessity of food to repair the waste and increase the bulk. Now almost every substance belonging to the animal and vegetable kingdom constitutes the food of animals, since it may be truly said that there is hardly any one of these substances upon which some animal or other does not feed. The quantity of food necessary varies in a surprising and inexplicable manner. Some animals, spiders for example, can live for months or even years without food, though they eat voraciously whenever they can procure food. Some animals, as bears, marmots, bats, &c. spend the winter in a dormant state. They sleep during the whole winter season, and of course take no food ; though during summer, when they are in a state of activity, they require a supply as well as any other quadruped. Even in man himself, though in general he requires frequent supplies of food during the whole of his life, examples are not wanting of life being prolonged, and even a certain degree of health and strength continuing for years without any supply of food whatever, or with a supply totally inadequate to the usual demands of nature.

Two remarkable examples of abstinence from food occur in the Philosophical Abstinence from food. Transactions. Both of the individuals were natives of the Highlands of Scotland, and both cases are very well authenticated. John Ferguson was employed in taking care of cattle in the Highlands. In the year 1724, having overheated himself by running on the mountains, he drank excessively from a spring of cold water, fell asleep on the spot and awaked next day in a fever. He recovered, but lost all relish for food, and for 18 years afterwards he took no other nourishment than pure water with now and then, during a certain period of the year, a draught of clarified whey.   During the whole of this time he continued in his employment, and enjoyed health and a certain portion of strength.*   The other example is of a more recent date.   A young woman, in Rosshire, fell ill, took to her bed, lost the faculty of speech, and the use of her eye-lids; her jaws became locked, and she refused all sustenance.   In this state she had continued for 13 years, when the account of her case was drawn up.   Her parents often attempted to convey nourishment into her mouth, first by forcing open her jaws, and afterwards through the hole left by two of her teeth which fell out; but she did not appear to swallow any of it.   At first she drank occasionally some water, but afterwards gave it up entirely.   She did not continue always in bed, but after some years got up and employed herself in spinning wool.†

Digestion.   The mechanism of digestion (if the expression may be tolerated) in man, and the larger animals, is tolerably well understood.   The food is taken into the mouth, where it is ground between the teeth and mixed with the saliva.   It is then swallowed and conveyed into the stomach, where it remains for some time, and is converted into a kind of pulp, known by the name of *chyme* This chyme passes gradually into the intestinal canal, where, by the peristaltic motion of the bowels and the agency of the bile, the pancreatic juice, and perhaps other means not yet discovered, it is separated into two distinct substances; a liquid matter, similar to milk in appearance, called *chyle*; and a solid matter, or at least a matter of greater consistence, which is protruded along the canal, and gradually thrown out of the body as excrementitious.   The chyle is absorbed by the lacteals, passes into the thoracic duct, from which it is conveyed into the left subclavian vein, mixes with the blood, and is gradually converted into that important liquid.   The blood circulates through the whole body, and furnishes materials to all the organs to supply their waste, and continue their functions.   From it too all the different liquids of the body are secreted.   Thus, digestion serves to increase the quantity of the blood, from which, as from a store-house, every thing necessary for the supply of the animal is drawn.‡

In some tribes of animals, as birds, trituration is not performed in the mouth but in the stomach, which, for that purpose, is made powerfully muscular. And these animals are in the habit of swallowing small pebbles, to assist the trituration.   It would appear that some species of fish, the gillaroo trout for example, found in some rivers in Ireland, have stomachs of the same nature.§ But most fish swallow their food entire, and it is gradually converted into chyme by the digestive powers of the stomach, without any trituration.   Some

* Phil. Trans. 1742. Vol. XLII. p. 240.   † Phil. Trans. 1777. Vol. LXVII. p. 1.
‡ Cowper.   Phil. Trans. 1696. Vol. XIX. p. 231.
§ Barrington.   Phil. Trans. 1774. Vol. LXIV. p. 116.

animals, as oxen, swallow their food entire, and afterwards bring it up to the mouth, chew it at their leisure, and swallow it again. This is called chewing the cud. Examples of men possessing this power are on record. There is an account, in the Philosophical Transactions, of a man at Bristol who regularly chewed the cud. Dr. Sloane, who gives the account, mentions several other instances of a similar nature.*

Assimilation is not confined merely to the muscles and soft parts of the body; even the bones undergo as rapid a change as any other organs. Mizaldus, in a book published in 1566, entitled *Memorabilium, Utilium, ac Jucundorum Centuriæ Novem*, mentions, that when animals are made to swallow *madder*, their bones are died red. This fact was afterwards discovered a second time by Mr Belchier, who published an account of it in the Transactions for 1736. He observed, that hogs which had fed on bran boiled with an infusion of madder, had their bones dyed red.† Duhamel repeated the experiments of Belchier upon different fowls and pigeons, and found that three days' use of the food was sufficient to give a red colour to the bones of the animals.‡ If the animals give up the use of madder, their bones speedily recover their whiteness. These experiments are sufficient to demonstrate the rapidity with which the substance of the bones is changed. There is an affinity between the phosphate of lime and the colouring matter of madder. Hence that matter combines with the phosphate of lime, which always constitutes part of the food. The compound being deposited in the bones obviously occasions the red colour.

Effect of madder on the bones.

Physiologists have not been slow in endeavouring to explain the process of digestion; and a great variety of hypotheses have made their way into medical books. Most of these hypotheses have been restricted to explain what happens in the stomach. The subsequent processes in the intestinal canal and blood-vessels, though much more curious and complicated, have attracted less attention. Most of the hypothetical explanations of digestion are discussed at some length in the Transactions. A bare mention of them here will be sufficient, as they have been for the most part rejected and exploded. The old doctrine of Aristotle and Galen supposed that the food was converted into chyme in the stomach, solely by the action of heat, as meat or vegetables may be boiled to a pulp, by the continued action of heat and water. This hypothesis, which conceived the stomach to be precisely similar to a stew pot, has received the name of *elixation*, from modern physiologists; and has been long since laid aside as utterly inconsistent with the phenomena of digestion, which takes place in cold as well as warm-blooded animals. Another hypothesis advanced at an early period, and continued down almost to our own times, was, that food is digested

Attempts to explain digestion.

---

* Phil. Trans. 1691. Vol. XVII. p. 525. † Phil. Trans. Vol. XXXIX. p. 287.
‡ Phil. Trans. 1740. Vol. XLI. p. 390.

in the stomach by the process of *fermentation*; and many methods were devised for bringing on this fermentation. Some supposed that the food itself fermented, without any additional ingredient; others, that a portion of the digested food, still remaining in the stomach, was mixed with the newly swallowed food, and thus brought on a fermentation, just as leaven does when mixed with dough. Clopton Havers conceived the saliva to be the ferment According to him, saliva consists of two distinct liquids; one similar in its nature to oil of vitriol, and the other to oil of turpentine. These two liquids being mixed with the food in the mouth, begin to act on each other, and produce a gentle fermentation, which reduces the food to chyme.* Dr. Harvey ascribed digestion to trituration alone; but the most prevalent opinion has always been, that the change is produced by the chemical action of some substance upon the food, though there has been very little agreement as to the nature of the liquid which performs this important office. According to Tilingius, the substance which acts is a nitrous salt; according to Willis, it is an acid and sulphur; according to Mayow, it is the nitro-aërial principle; according to Diemerbroeck and Sylvius, it is the saliva. Leigh endeavoured to show that it was a menstruum, secreted for the purpose in the stomach;† and this opinion is at present generally received. The liquid which produces this important effect is called the *gastric juice*. Many exertions have been made to procure it, and ascertain its properties, but these exertions have not been crowned with success. It has been ascertained, however, that it acts as a solvent; that it is neither acid nor alkaline; that it is not capable of dissolving all substances indifferently; that the husk, or outer coat, of the different species of corn resists its action. Hence the necessity of trituration, in order to enable this juice to perform its functions with effect. From the observations of Mr. John Hunter, it appears that this juice, after the death of the animal, is capable of acting upon the stomach itself, and of reducing it to a pulpy consistence, similar to digested food.‡

Action of bile. The bile, which is thrown into the intestinal canal, not far from its origin, and to the formation of which the liver, the spleen, with a very curious set of vessels, seem to be entirely devoted, is, without doubt, a very important liquid. Though the way in which it acts, and how it contributes to the formation or the separation of chyle, is not understood. Dr. Stuart, from a curious case, of which he has given an account in the Transactions,§ has concluded that it is essential to life, and that if it be not thrown into the intestines, no chyle is absorbed by the lacteals. A Mr. Menzies, of the Guards, was wounded with a sword in the body; he lived about six days, hardly slept at all during the

---

* Phil. Trans. 1699. Vol. XXI. p. 233.   † Phil. Trans. 1684. Vol. XIV. p. 694.
‡ Ibid. 1772. Vol. LXII. p. 447.   § Ibid. 1730. Vol. XXXVI. p. 341.

time, had no stool, though every method to procure evacuation was tried. On examining the body, after death, the gall bladder was found punctured, and the bile had made its way into the cavity of the abdomen; every other organ was sound. Dr. Stuart inferred that the peristaltic motion of the bowels, in this case, was stopped. From this, he concluded that this motion is occasioned by the stimulus of the bile; and all his other inferences followed of course. But this seems to be going much farther than the single case produced by Dr. Stuart will warrant. One cannot see why a wound in the gall bladder should prevent any bile whatever from making its way into the intestines, as there is another channel through which it may enter. It is much more probable that death, in this case, was occasioned by some other cause than the mere absence of bile.

VIII. GENERATION. This, as it is one of the most important, so it is one Generation. of the most inexplicable functions of animals. It is well known that animals are of two sexes, male and female, and that the concurrence of both is necessary for the propagation of the species. The females only produce young; but the fecundating power of the male is an essential requisite. Some animals indeed, as snails, unite, in one individual, the organs of both sexes; yet copulation is necessary in them as well as in other animals. They mutually impregnate each other, and both the individuals produce young. In fowls, one copulation is capable of fecundating at least 20 eggs, laid in as many successive days. And what is still more extraordinary and unaccountable, in a genus of small insects, called *aphides*, living upon trees and bushes, one copulation is sufficient to impregnate ten successive generations of these animals. The first nine are all females; but the tenth is a mixture of males and females, which copulate together, and impregnate ten successive generations, as before.

The doctrine of equivocal generation; or, that animals are produced from Harvey's putrefaction and fermentation, without the necessity of any parent, was taught doctrine. by the ancients, especially by Aristotle; and was long universally received in the schools and universities of Europe. This doctrine was first formally refuted by Harvey, who proved that all animals spring from eggs, previously deposited by their parents. This opinion was rapidly propagated through Europe, and speedily overturned the Aristotelian doctrine. One of the ablest supporters was Redi, an Italian philosopher of considerable celebrity, who demonstrated, by numerous and satisfactory experiments, that animals never breed in putrid animal or vegetable substances; except when flies and other similar insects have access to them, and deposite their eggs in them. Redi, however, was not quite free from absurd opinions on the subject of generation: for he conceived that gall insects, which make their appearance on oaks and other trees; and the worms, occasionally found in the livers of sheep, were produced by the

U

energy of the plants and animals in which they are found ;* whereas, it is now well known that they also originate from parent animals, as well as all others.

Experiments of Needham and Buffon.

There is only one set of experiments which seem still to give some countenance to the doctrine of equivocal generation. Needham and Buffon found successively that animal and vegetable infusions, in boiling water, though put into a phial while scalding hot, and kept ever after close corked, generated animalcula in a few days. These experiments were afterwards repeated and confirmed by Mr. Wright.† Now it is conceived, that the boiling water would have destroyed any ova of these animalcula, previously deposited upon the animal or vegetable substances employed. And hence it has been inferred, that the animalcula in question have been produced by equivocal generation. But we have no right to suppose that boiling water is capable of destroying the ova of all animals, because it is capable of destroying a great many. The heat of boiling water will not prevent the seeds of many vegetables from germinating. It is likely that the ova of the animalcula in question are capable of resisting its action.

Hypothesis of Leuwenhoeck.

The opinion respecting generation, anciently received, was, that there was a female semen as well as a male; that the two were mixed together in the uterus, and that the foetus was produced by the mixture. But Leuwenhoeck, having observed a vast number of animalcula in the male semen of different animals, was induced, in consequence, to propagate a different opinion; which very speedily acquired celebrity; and was even pretty generally acceded to According to him, one of these animalcula constitutes the rudiments of the future foetus. It is merely lodged in a convenient nidus, in the uterus, and all that the female has to do is to supply it with proper nourishment. Dr. Garden, of Aberdeen, made an addition to this theory. According to him, during the period of copulation, one or more ova make their way from the female ovaria to the uterus; such of the animalcula of the male semen, as were fortunate enough to get into these ova, found a proper nidus, were nourished, and gradually increasing in size, became foetuses; while all the others, amounting by Leuwenhoeck's calculation to several millions, speedily perished for want of nourishment.‡

Abandoned.

Subsequent physiologists having examined the semen of various animals, found some in which no traces of animalcula could be found. These were as capable of impregnating the female, as those in which animalcula abounded. This was sufficient to overturn the hypothesis of Leuwenhoeck. Accordingly it was abandoned, and consigned to ridicule as speedily as it had been adopted.

---

* Phil. Trans. 1670. Vol. V. p. 1175.   † Phil. Trans. 1755. Vol. XLIX. p. 553.
‡ Ibid. 1691. Vol. XVI. p. 474.

Indeed there is reason to believe, that Leuwenhoeck had sometimes mistaken small crystals of phosphate of lime, which abound in the semen, for animalcula. The opinion of Harvey, that the fœtus proceeds from an ovum detached from the female ovarium, and somehow fecundated by the influence of the male semen, came to be universally admitted. This opinion was supported with considerable keenness by Kirkringius.* But De Graff is the anatomist who threw the greatest light on the subject. He ascertained, by numerous experiments on rabbits, that the ovaries are the seat of conception; that one or more of their vesicles become changed; that they are enlarged, lose their transparency, and become opaque and reddish coloured; that the number of vesicles thus altered corresponds with the number of fœtuses; that these changed vesicles at a certain period after copulation discharge a substance, which, being laid hold of by the fimbriated extremity of the falopian tube, and conveyed into the uterus, soon assumes a visible vesicular form, and is called an *ovum;* and that this ovum gradually evolves different organs, and becomes a fœtus. Considerable additional light has been thrown upon this obscure subject Experiments by a curious paper of Dr. Haighton, published in the Philosophical Transac- of Haighton. tions for 1797.† He has demonstrated by very decisive experiments upon rabbits that the male semen does not pass into the falopian tube or ovaries; but merely stimulates the vagina. One or more ova pass in consequence of this stimulus into the uterus, and become the fœtus. These experiments render the process of generation if possible still more obscure than it was. The male semen certainly answers some other purposes than merely acting as a stimulus to the parts, otherwise we could not see why it should be so absolutely necessary in all cases as it is. Fowls often lay eggs without impregnation; but unless they have had connection with a male these eggs are incapable of being hatched. Toads, frogs, and fish, exclude the ova out of the body without any previous connection with the male, and the male afterwards sprinkles his semen on the excluded ova, without which they are incapable of being converted into animals. Here the semen is applied not to the female organs, but to the semen, and therefore cannot be intended merely to stimulate these organs.

Mr. Cruikshanks has demonstrated by a set of well conducted experiments on rabbits, that the ovum is formed in the ovarium, that it comes out of it after conception; that it passes along the falopian tube, and that it takes some days in making its way to the uterus.‡ Female animals have two ovaries, one on each side. Mr. John Hunter extirpated one of these organs from a sow, and compared her breeding powers with another sow of the same age, and treated in every respect the same as the first sow; but retaining both her ovaries. The

* Phil. Trans. 1672. Vol. VII. p. 4018.   † Vol. LXXXVII. p. 159.
‡ Phil. Trans. 1797. Vol. LXXXVII. p. 197.

spayed sow continued to breed till she was six years of age ; the perfect sow till she was eight : the spayed sow had 76 pigs ; the perfect sow 162.* From this experiment it would appear that a determinate number of ova exist at first in each ovarium, and that when these are exhausted the animal ceases to breed. The loss of an ovarium must of course have a considerable influence both on the time of breeding, and upon the quantity of young produced.

Nourishment of the fœtus.

Needham published a work giving an anatomical account of the vessels, liquors, coverings, nourishment, &c. of the fœtus, from which he acquired a high reputation. Of this work there is a good analysis in the Philosophical Transactions.† There can be no doubt that the fœtus is nourished, at least chiefly, by means of the umbilical canal. Hence the reason why respiration is unnecessary, and why the fœtus can live and grow without some of its most essential organs, as is well known sometimes to happen. The fœtus in utero is surrounded by a liquor called the *liquor amnii,* and it has been supposed by some that this liquor makes its way into the mouth of the fœtus, and serves it for nourishment. Dr. Fleming mentions a fact which at first sight seems to prove the truth of this opinion, at least with regard to some animals. Examining the meconium of a still born calf, he found it full of white hairs similar to the hairs with which the skin of the animal was covered.‡ It is well known that upon calves there is always a great deal of very loose hair Hence he supposed that this loose hair had mixed with the liquor amnii, been swallowed by the calf, and thus made its way to the meconium. But that this mode of nourishment is only partial, if it exists at all. is obvious from a multitude of facts which might be produced ; fœtuses have come into the world entirely destitute of a head, and of course incapable of swallowing. Mr. Brady gives an example of a living puppy whelped without any mouth whatever,§ of course incapable of swallowing or of being nourished by the liquor amnii.

Children cry in the womb.

The fœtus in utero is surrounded with a liquid, and consequently excluded from air. Voice, it is well known, depends upon the action of air on the throat. Accordingly fish and all animals that live constantly under water are absolutely dumb. It is on that account very difficult to explain a fact which sometimes happens, namely, a child crying in the womb. Yet there are examples of this upon record, supported by evidence which cannot be set aside. Dr. Derham relates an instance which came under his own examination. The child cried almost every day for six weeks before delivery, and so loud that it could be heard in the next room.‖ Chickens sometimes cry before they burst their shell ; but this is not so surprising, as these animals are at that time obviously surrounded

* Phil. Trans. 1787. Vol. LXXVII. p. 233.
† Phil. Trans. 1667. Vol. II. p. 503.
‡ Phil. Trans. 1755. Vol. XLIX. p. 254.
§ Phil. Trans. 1705. Vol. XXIV. p. 2176.
‖ Phil. Trans. 1709. Vol. XXVI. p. 485.

with air; probably even breathing has begun in them some little time before they make their way out of their shell.

In certain animals, as man, and most quadrupeds, the fœtus is attached to the uterus, and nourished by the mother by means of the navel string. In others, as in birds and fishes, the ovum is excluded at once, and the fœtus is nourished by means of the food deposited in the egg without any connection with the parents whatever. The animals of the opossum tribe seem to be inter- mediate in this respect between quadrupeds and birds. Mr. Home has given us Organs of the kanguroo. a curious account of the generative organs of the kanguroo, an animal belong- ing to this tribe. The embryo in these animals does not appear to be attached to the uterus. When not more than 21 grains in weight, it is excluded and re- ceived into the false belly of the mother; the mouth of the fœtus attaches itself to the nipple of the mother, and the young animal continues in that situation for nine months, when it is expelled.*

The number of young produced at once differs according to the animal. Number of young pro- duced at once. Some fish produce several thousands at a time. Dogs, cats, swine, &c. produce from four to 20 at a time. But man and the large quadrupeds are usually li- mited to one. In the human race twins are by no means uncommon. They occur at an average about once in every 80 births. Three at a time is much more uncommon, though it sometimes happens. An instance lately occurred in Berwickshire, in which all the children survived. Four at a birth is a very un- common occurrence; but there is a well authenticated case of that kind in the Philosophical Transactions. It happened in Lancashire, and the children were delivered by Mr. Hull, a surgeon, in the neighbourhood. Two of them were still born, and two were alive but died soon after. Dr. Garthshore has attached to the case a number of curious examples of numerous births recorded by medi- cal authors. One of them an example of no fewer than eight children at a birth, one of whom grew up to manhood, and was alive when the account was drawn up.†

Physiologists have laid it down as a maxim that when animals are capable of Whether dif- ferent species can breed. breeding together, and of producing young perfect in all their organs, and ca- pable of propagating, they belong to the same species; but when the young animal is incapable of propagating, then the parents belong to different species. Thus the horse and the ass belong to different species, because the mule which they produce is incapable of propagating. Relying upon these data, Mr. John Hunter has published a variety of documents showing that wolves, jackals, and dogs, breed together, and produce an offspring capable of propagating. Hence he concludes that these animals belong to the same species. The wolf he con- ceives to be the original, the dog to be the wolf tamed, and the jackal the dog

Phil. Trans. 1795. Vol. LXXXV. p. 221.   † Phil. Trans. 1787. Vol. LXXVII. p. 311.

6

run wild.* But this mode of reasoning, though strenuously insisted upon by Mr. Buffon, and though it became popular in consequence, seems entirely hypothetical and incapable of proof. This capacity of propagating perfect animals certainly demonstrates a similarity in the organs of generation of the animals which breed; but this is not sufficient to demonstrate identity of species. It is conceivable that a perfect animal might be bred between man and the ourang outang; but this surely would be insufficient to demonstrate identity of species.

Why females are more numerous than males. There is only one other paper in the Transactions connected with this subject which we shall touch upon. It is well known that in most European countries the number of male births exceed that of females; yet if a census be taken of the inhabitants of a country, the number of females is always found to exceed that of males. This is commonly accounted for by the more exposed life which males lead. Hence it is conceived that they are liable to many more casualties than the females. Wars, emigrations, the sea, are mentioned, and these are conceived fully to account for the difference. But Dr. Clarke has shown that other causes intervene besides these. For from a table of births and causalties in the Dublin Lying-in-Hospital, for a series of years, he has shown that there is a greater proportion of still-born males than females, and that more male children die during their infancy than females. He accounts for this by showing that male children are larger than female, and in particular that the head of a male fœtus has a greater circumference than that of a female. Hence he conceives that male children are more likely to be stunted of nourishment before delivery than female, and from the greater size of the head they are much more likely to be injured during delivery than females.† The last we conceive to be the true reason of the greater mortality of male than female children.

Section V —*Of Medicine and Surgery.*

Medicine must have been, in some measure, coeval with mankind. For such is the frailty of our nature, and so numerous are the accidents to which we are exposed, that diseases and wounds must have speedily made their appearance, and the sufferers must have cast about on every side for assistance and relief. Surgery was probably the first department of medical knowledge which mankind attempted to acquire. Diseases were naturally enough ascribed to the direct agency of the offended deities, whose resentment they attempted to remove by sacrifices and prayers. But a wound received in war, or the bite of an enraged or poisonous animal, did not require any such explanation. They were injuries obviously inflicted by their fellow creatures. Accordingly we find that the medical men celebrated for their skill by Homer were in fact only surgeons. When a pestilential fever thinned the Grecian ranks, Podalirius and Machaon were not called. The disease was ascribed to the resentment of Apollo, and his anger was deprecated, and the disease removed, by restoring the captive daughter of his favourite priest. But when any of the Grecian heroes was wounded in battle these medical men immediately attended, and displayed their skill in extracting the dart, stopping the blood, and healing the wound.

India, Chaldea, and Egypt, being the most ancient and the first civilized kingdoms, medicine no doubt made its first appearance among them. Apollo and Esculapius, to whom the origin of the art is ascribed by the Greeks, appear to have been both Egyptians. But what progress the science had made in these countries is altogether unknown. It is from the Greeks that we have derived all our arts and sciences, and that people have given us no information respecting the state of medicine in Egypt. It was probably very low, entirely unconnected with philosophy and science, and depending upon the delusive practices of magic and superstition. *Origin of medicine.*

Even in Greece, which obviously derived its first medical knowledge from the Egyptians, the art continued long in a very low state. And Hippocrates is acknowledged by the universal voice of antiquity to have been the first who connected medicine with philosophy, and who established his practice upon rational and scientific principles. He has therefore been universally considered as the true and legitimate founder of medicine, and has always enjoyed a higher reputation than any other medical man whatever. He is the first physician whose writings still exist. With him therefore commences the science as far as our knowledge of the subject extends. In all probability he borrowed much of his knowledge of the symptoms of the diseases, and of the effect of various medicines, from his predecessors; but as we have no information concerning the knowledge of his predecessors, the whole of the science, as far as he has delivered it in his writings, has been ascribed to him. *Hippocrates.*

Hippocrates was born in the Isle of Cos, in the first year of the 80th Olympiad, about 30 years before the commencement of the Peloponnesian war. On the father's side he was of the family of the Asclepiades, or the descendants of Esculapius, a family in which the knowledge of the medical art had descended without interruption from the illustrious founder. The Asclepiades were all physicians, and the knowledge of diseases was confined to their family. Thus he was at an early period initiated in all the medical knowledge of the times. He added to this the study of philosophy, which he considered as essential to the physician, a keen, constant, and accurate observation, the knowledge of anatomy as far as it was in his power to attain it without dissecting human subjects, and an industry and sagacity upon which medical celebrity, in all ages and countries, in a great measure depends. A considerable part of the works of Hippocrates still remains, from which a pretty accurate estimate may be formed of his knowledge and practice.

His description of diseases is very accurate and complete, and has not been surpassed by any succeeding writer.   His knowledge of symptoms was wonderfully precise, and his prognosis, founded on that knowledge though in some cases it appears to us absurd, was upon the whole so exact that it must have been, in a great measure, drawn from the observations of his predecessors.   For it is impossible for one man, however busy and acute, to make so many observations during the course of a short life time as his aphorisms obviously require. His philosophy, founded upon the Pythagorean, led him to put much confidence in the value of numbers.   Upon this was founded his faith in odd numbers, and his opinion respecting the periods and revolutions of diseases, and certain days on which all the important changes took place.   These memorable days were called *critical*, and he lays down his regulations respecting them with much pains and precision.   Though the philosophy upon which these critical days were founded has been long banished from medicine, the belief in critical days is by no means laid aside.   We find it still taught by some of the most recent and most celebrated medical writers.   It is an opinion however that can neither be supported by reasoning nor observation; and would not have been allowed to affect medical opinions for so long a period, had it not been for the unbounded reputation and real merit of the original introducer of it.   For it may be said with truth, that no man is able permanently to injure a science by the introduction of absurdities, excepting one who has contributed essentially to its progress and celebrity

His physiology. His physiological opinions deserve to be mentioned, because they have had considerable influence upon the opinions of his successors, and because his language still continues to be used by modern physiologists; even when they are unacquainted with, or do not believe, the philosophy of Hippocrates.   He conceived that there was implanted in man a living principle, or soul, which was only a part of φυσις, or *nature*, concerning which he speaks in the highest terms.   " Nature," says he, " is alone sufficient to animals for all things.   It knows of itself all that is necessary for them, without requiring any instruction or information."   It governs the body by means of certain δυναμεις, or faculties, which are subordinate to it, act from necessity. and perform the various functions of the body, such as hearing, seeing, digesting, &c.

His practice. The medical practice of Hippocrates was regulated by these opinions: Nature, in all cases, performed the cure ; the business of the physician was only to look on, and registe her operations.   Accordingly, in his treatment of acute diseases, he seldom administered any powerful remedies, at least at first, but trusting entirely to the influence of nature, satisfied himself with regulating the diet and the dress of the patient. The body, according to Hippocrates, contained four humours; namely, *blood, phlegm, yellow and black bile*. Disease was occasioned by the undue accumulation of some one of these humours in particular organs, or by the unnatural separation of these organs from each other. The object of the physician was, to get rid of the peccant humour, and for this purpose various evacuant remedies were applied: bleeding, purging, blisters, or, at least sinapisms, diaphoretics, and diuretics. Certain medicines were conceived to have the property of driving off certain humours, and they were applied accordingly, when the disease was ascribed to the redundancy or peccant nature of the particular humour. Bleeding was frequently carried by Hippocrates, till it produced fainting; and all his purgatives, such as helebore, scammony, verdigris, were of the most drastic nature. The use of leeches appears to have been unknown to him; but cupping-glasses, pretty much as they are used at present, are very particularly described by him.

The surgery of Hippocrates, (for he practised at once all the different parts His surgery. of the medical art,) was as violent as his medical treatment was mild. The actual cautery was very frequently applied on five or six parts of the body at once, and the ulcers produced by this cruel remedy were kept open for a considerable time by corrosive dressings. It is not necessary to describe his mode of purging the head by violent sternutatories; nor of evacuating matter from the lungs by exciting violent coughing.*

Hippocrates having taught the art of medicine to all persons without distinction, who chose to become his pupils, medicine was no longer, as formerly, confined to a particular family, but spread itself, like philosophy, over all the states of Greece. Thessalus and Draco, the two sons, and Polybius, the son-in-law of Hippocrates, exercised the profession with eclat after the death of their father; especially Polybius, who acquired a high reputation, and intro- Polybius. duced, it is supposed, considerable changes into the practice of medicine, as taught by his father-in-law. A considerable period elapsed after the death of these men, before any physician of great eminence made his appearance. Diocles indeed, who seems to have flourished not many years after the death of that father of medicine, acquired such celebrity, that he was distinguished by the Athenians by the name of the second Hippocrates; but as none of his writings remain, we know little of his method of practice, or of his opinions. Herophilus, who was a physician in Alexandria, during the reign of Ptolemy Herophilus & Soter; and Erasistratus, physician to Antiochus, King of Syria, made great Erasistratus. improvements in anatomy, or rather indeed laid the foundation of the science.

* The reader will find a very detailed account of the practice of Hippocrates, with a list of the diseases that he describes, and all the medicines which he employed, in Le Clerc's Histoire de la Medicine, p. 112—253.

Being permitted by their respective sovereigns to inspect the bodies of crimi
nals, a practice before that time considered as impious, and never allowed,
they acquired, in consequence, a high reputation, and became in some measure the founders of sects, which bore their name and followed their opinions.
But as their writings have not come down to our times, all our knowledge of
their opinions and practice is derived from Galen.

Division of medicine.   About this period, or a little after, as we are informed by Celsus, medicine,
which had hitherto been all practised by one individual, came to be divided
into three distinct professions, namely, *dietetics, pharmaceutics,* and *surgery.*
Those who practised dietetics confined themselves to regulating the food of
their patients; upon a strict attention to which, they made the whole of medicine depend. Those who practised pharmaceutics administered internal medicines; while the practitioners of surgery confined their attention to
wounds, fractures, dislocations, ulcers, and other external diseases, to which
surgeons confine themselves at present. It is hardly possible that dietetics and
pharmaceutics should have been practised exclusively by different individuals;
for, in many cases, physicians would find it necessary both to regulate the food
of their patients, and to administer some medicines, in order to obviate certain
symptoms.

Soon after this period the Grecian physicians separated into two sects, as we
are informed by Celsus, who regulated their practice on quite different principles. These were the *empirics* and the *dogmatists:* Both considered Hippocrates as their founder, and both seemed to do so with considerable justice.
The empirics. The empirics discarded philosophy from medicine, and made the art depend
upon experience alone. According to them, the physician has nothing to do
with the causes of diseases; he ought only to make himself acquainted with the
symptoms, and to determine by experience what is the proper method of cure.
The dogma tists. The dogmatists, on the other hand, considered philosophy and medicine as inseparably connected. They began always by investigating the cause of every
disease; and this cause, once known, led them, they conceived, to the only
rational and efficacious mode of cure.

Celsus gives us, in his preface, the reasoning by which these two sects supported their different opinions, at considerable length; and, in the conclusion
which he draws, he holds a medium between both; acknowledging that the
practice of medicine has been chiefly promoted by experience, but affirming
that reasoning, and even an investigation of the causes of disease, as far as
possible, ought not to be neglected by medical men. By this sensible decision
every person, we presume, will be disposed to abide; for though it must be
granted that reasoning goes but a very little way in guiding the practical physician; though it will hardly be denied that absurd attempts at philosophizing,
and inaccurate physiological theories, have much more frequently led physicians

astray, and deranged their practice, than contributed to the advancement of the medical art; though the improvements in anatomy, in chemistry, and in science in general, have not contributed so much as might have been expected to the improvement of medicine; yet we presume, that there is no person so hardy or inconsiderate as to maintain that the science has made no progress since the days of Hippocrates; or that no part of this progress has been owing to the reasoning and investigations of medical men. Various important and indisputable improvements might be mentioned, sufficient to rescue the profession from such severe stigmas; as for example, the use of mercury in the venereal disease; the cool regimen in fevers; the use of bark in intermittents; the method of preventing the spreading of contagion, and the diffusion of pestilential diseases. Even in those maladies which medicine can do but little to assuage, the progress of knowledge has made itself conspicuous, and has at least contributed to alleviate the afflictions, and to sooth the distresses, of those whom it cannot completely relieve.

These two sects continued to divide the medical world of Greece for several centuries. Meanwhile Rome was advancing with gigantic steps to the empire of the world. That city increased enormously in opulence and extent, and held out a tempting bait to the needy literary men of Greece, who flocked to it in great numbers. Among these was Asclepiades, of Praso, who came to Asclepiades. Rome, as Pliny informs us, during the Mithridatic war. He at first attempted to distinguish himself by teaching rhetoric, but was unsuccessful. He then thought of commencing physician; and, having studied the art for some time, considered his abilities as sufficient to raise him to distinction. Archagathus, a Grecian physician, who had settled some time before at Rome, had been driven from the city in consequence of the severity of his remedies. This was sufficient to induce Asclepiades to adopt a contrary method According to him, patients ought to be cured safely, quickly, and agreeably.* His medical project succeeded beyond his hopes; he speedily rose to the highest reputation, overturned the method of practice established by his predecessors, and raised himself to the rank of a second Esculapius. His philosophy appears to have been that of Epicurus and Democritus; and his remedies, according to Pliny, were reducible to five kinds; namely, *abstinence from animal food, abstinence from wine, friction, walking,* and *riding,* or *gestation.* His practice, as it has been transmitted to us by the writings of Celsus, Galen, and Cælius Aurelianus, hardly corresponds with the pretensions of Asclepiades, and appears scarcely less severe than that of his predecessors.

Asclepiades seems rather entitled to be considered as an expert quack, than

* Tutè, celeriter, et jucunde. Id votum est (says Celsus, Lib. III. Cap. 4.); sed fere periculosa esse nimia. et festinatio, et voluptas solet.

as a skilful physician. This, indeed, is the character which Pliny ascribes to him; but he had the merit of setting free his successors from the trammels of antiquity, to which the medical art had been before that period bound His successors, though in some measure the disciples of Asclepiades, aspired to the honour of thinking for themselves, and accordingly almost every subsequent physician of eminence introduced some innovation. The most celebrated of them all v as Themison who founded the sect of the *methodists*, who continued to flourish with considerable lustre for several centuries.

Themison.

Themison was of Laodicea, and, from what Celsus says of him, appears to have lived and speculated about the beginning of the reign of Augustus. His object was to point out an easy method of conveying the knowledge of medicine to others. All diseases, according to him, might be reduced under three classes; those which proceed from *tension*, those which proceed from *relaxation*, and those which proceed from a mixture of the two.* Some diseases are *acute*, others *chronic;* some go on increasing for a certain time, become stationary, and then diminish in the same way as they increased. The treatment must differ according to the stage of the disease, and according to its acute or chronic nature. Such was the small number of heads under which Themison reduced all diseases; and according to him, every disease that belongs to the same class, whatever be the organ afflicted, and whatever be the temperament of the patient, is to be treated exactly in the same way. It was only necessary after this to know to what class a disease belonged, and to be acquainted with the remedies belonging to the class, to be able to treat it according to rule. Thus, bleeding and purging were medicines appropriated to the diseases of tension; cold water and refrigerants, to the diseases of relaxation. It deserves to be mentioned, that Themison is the first physician who is recorded to have employed *leeches;* but, as Cælius Aurelianus, from whom we derive this information, does not mention the application of leeches as a new remedy, they were probably employed by others before the time of Themison, though no notice is any where taken of the time of their introduction.

The pneumatists.

There is another sect of physicians known by the name of *pneumatists*, which originated with Atheneus, a native of Cilicia, who appears to have lived during the reign of Trajan. He believed, with the stoics, that a certain intelligent principle, or πνευμα, pervaded every living creature;† and diseases were

---

* The Greek terms Themison used, were σίεγμον, σίεγιωσις, τασις, ρωσις, συναγωγη, πυκιωσις, which I translate *tension;* and ρωδις, ρυσις, χαλασις, αίονια, χυσις, αραιωσις, which I translate *relaxation*.

† Though this doctrine was only introduced into medicine by Atheneus, it had been long very generally received by the philosophers and poets. Thus Virgil says,

> Principio cœlum, ac terras, camposque liquentes,
> Lucentemque globum lunæ, Titaniaque astra,
> Spiritus intus alit: totamque, infusa per artus,
> Mens agitat molem, et magno se corpore miscet.

occasioned by every thing which incommoded or injured this living principle, or πνευμα. The only writer belonging to this sect, whose works remain, is Aritæus, of Capadocia. His description of diseases possesses great merit; but his practice is not inferior, in point of violence, to that of any physician of antiquity. In various cases, the disease itself seems to have been more tolerable than the cure.

Celsus, who lived during the Augustan age, and who seems from his book Celsus. to have been a physician, though the contrary has been maintained by some, has left us a system of medicine, written with much elegance of style, and abounding in the soundest sense and the most accurate observations. It is by far the most unexceptionable system which antiquity has left us. Celsus does not appear to have attached himself to any sect, but to have adopted the eclectic principles, and to have culled from all what was the most valuable in each. He has given us also a good deal of historical information. Several of the preceding observations have been taken from him.

But the most celebrated of all the ancient physicians, Hippocrates excepted, Galen. is Galen, who was born in Pergamus, a city of Asia Minor, in the year 131 of our æra, and during the reign of the Emperor Adrian. He flourished during the reign of the Antonines; which may be considered as the golden age of the Roman empire. Galen was possessed of an independent fortune. His father, who was himself a philosopher, initiated him early in the learning of the times, and his own ardour was quite insatiable. He attached himself to medicine as his favourite study, and such was his industry and abilities that he collected into one body all the knowledge of his predecessors; added many new facts to their store, and thus completed, it may be said, the whole body of ancient medicine. Hippocrates was the standard whom he professed to follow. The medical sects which had risen since the time of that illustrious father of medicine, he considered as injurious innovations, and attacked them accordingly. His writings are so numerous and so various, that it would be impossible to give any idea of his opinions, without going greatly beyond the limits which have been assigned to this work. We shall therefore satisfy ourselves with referring to Le Clerc, who has given a very full and accurate account of the medical opinions of Galen.*

Galen was the last of the Greek physicians that deserve to be mentioned. His successors, confounded at the vastness of his knowledge, made no attempts to increase it, but satisfied themselves with consulting the ample stores which he had provided. Science in Greece lost its lustre, and made no progress. The writings of her ancient sages were locked up in libraries or monasteries, and hardly consulted. Superstition laid her iron hand upon the nation, and crushed

---

* Hist. de la Medicine. Part III.

her beyond the possibility of recovery. Meanwhile, the sciences began to raise
The Arabians. their heads in another and an unexpected quarter of the world. After the Mahometans had extended their empire from the banks of the Ganges to the Pillars of Hercules, and the Atlantic Ocean, their Caliphs, seated on the borders
of the Euphrates, began to excite a spirit of inquiry, and a taste for the
sciences, among their courtiers and subjects. The works of the ancient Greeks
were sought for with avidity, and translated into Arabic.

Among the sciences which the Arabians cultivated, medicine holds a distinguished place. They cultivated it with much success; and, different from what
happened in the other sciences, the Arabian physicians considerably out-
stripped their Grecian masters in the mildness and efficacy of the new remedies
which they administered. This was partly owing to the country in which they
lived, and partly to the knowledge of chemistry, which they cultivated with
some success. We are indebted to them for several of the mildest and best purgatives in the materia medica; such as *manna, senna, rhubarb;* we owe to
them our knowledge of distillation, and of various compound medicines, manufactured by means of the still. They likewise introduced *sugar,* which soon
superseded the use of honey, which had been formerly employed by the Grecian
physicians.

As to the theory of medicine, the Arabians made no additions or changes in
it, but satisfied themselves with copying their Grecian masters. But some
diseases are described by them which appear to have been unknown to the
Greeks and Romans. The *small-pox* and *measles* may be mentioned as examples. These dreadful diseases seem to have been formerly confined to the
eastern and southern regions of the earth, and to have made their way into
Europe, and the cultivated parts of Asia, in consequence of the conquests of
the Saracens. The small-pox has now made its way to every inhabited part of
the globe, and, in its ravages, proves more destructive to the human race than
any other disease whatever.

The most celebrated of the Arabian physicians were Rhazes, Avicenna, and
Mesué, who lived between the tenth and twelfth centuries. It is in their
writings that we find the first details concerning chemical medicines. Avicenna,
for example, describes rose water; and Mesue gives us the process for making
a distilled oil, which he calls *oil of bricks.* We may therefore date the period
at which chemical medicines were introduced into pharmacy, about the year
1000 of the Christian æra.

Medicine in Europe during the dark ages. It was from the Arabians chiefly, in consequence of their literary establish-
ments in Spain, that the rude nations in Europe, during the middle ages,
acquired any information. Hence their physicians were the great authorities
who decided all disputes, and directed the practice of the medical men in
Europe. Avicenna, who was the most celebrated of the Arabians, ruled with

despotic sway for several centuries, in all our schools; and to doubt or decline his authority was considered as the most unpardonable of all heresies. Galen, known in Europe through the impure channel of an Arabian translation, enjoyed an equal authority. The medicine of Europe then, during the dark ages, was all drawn from the writings of Galen and Avicenna. Their opinions, on every subject, were embraced without appeal; their descriptions of diseases were alone current; and their medicines only were administered.

Anatomy had the merit of first shaking the throne of this redoubted duumvirate, *The dominion* and of setting loose the opinions of medical men, and leaving them to be guided *of Galen and* by reason and experience. A spirit of inquiry had arisen in Europe, excited, or *Avicenna overthrown.* at least prodigiously increased, by the new religious opinions of Luther and his followers. The conquest of Constantinople by the Turks had driven the literary men from that capital into Italy, and they brought with them a knowledge of their own language, and of the inestimable writings of their ancestors. These excellent works supplied the ravenous appetites of the European literary men with the most substantial food. The Greek physicians were studied, translated, and commented upon. The writings and opinions of Hippocrates became known; Galen was studied in his own language, and free from many of the gross faults with which the Arabians had loaded him. This itself was a reformation, but it was not sufficient. The anatomical knowledge of Galen was still looked up to as perfect, and his opinions acceded to in consequence, without discussion. At last Vesalius, of Brussels, a man of independent fortune, and so enthusiastic an admirer of anatomy that he used to rob the gibbets, and dissect the bodies in his bed-room, affirmed that Galen's knowledge of anatomy was derived from the dissection of quadrupeds, and not of human bodies; that it was inaccurate in many particulars; and, to prove the truth of his assertions, published a System of Anatomy of his own, at the age of twenty-seven. This produced a violent controversy: the admirers of Galen and Avicenna took the alarm, examined the pretensions of the young anatomist, sifted his performance, and loaded it with abuse. But the anatomy of Vesalius, though not entirely free from those very faults of which he had accused Galen, stood this ordeal without injury, and was speedily acknowledged infinitely superior to every preceding system of anatomy. This successful work deprived Galen of much of his authority; and, notwithstanding several spirited attempts to revive his lost credit, his pathology and physiology were gradually discarded from the schools of medicine.

Chemistry was the first science that attempted to take possession of the vacant *Chemical* throne. The Arabian physicians had introduced chemical medicines, and the *medicine* practice of chemistry into pharmacy; and Raymond Lully, Arnoldus de Villa Nova, Basil Valantine, and a few other European alchymists, had made some important additions to the list of chemical medicines; but the person who first

attempted to render chemistry paramount in medicine, and to explain the whole
of physiology upon chemical principles, was Aureolus Philippus Theophrastus
Paracelsus Bombastus ab Hohenheim, one of the most extraordinary and unaccountable men that ever existed. He was born in Einsidlen, in Switzerland,
in the year 1493.   His father is supposed to have been a medical man, and he
instructed him, as Paracelsus informs us himself, in all the medical knowledge
of antiquity.   He learned the practice of chemistry from Sigismond Fugerus,
a German of some celebrity, who took him as a pupil.   He travelled through
most of the countries of Europe, and consulted indifferently, as he informs us
himself, physicians, old women, conjurors, and chemists.   After this extraordinary education, he commenced the practice of medicine, at a very early age,
and with the most unbounded applause.   He was invited by the magistrates to
Basil, and delivered medical lectures in that city for about two years.   At the
end of that time he quarreled with the magistrates about a medical fee, left the
town, rambled over Germany in a state of continual dissipation, and at last
died at Saltzburg, in the forty-eighth year of his age.   His works, printed in
two large folios, consist of such a monstrous tissue of absurdities, that it is next
to impossible to develope his opinions.   His practice was wholly chemical, and
his medicines almost all drawn from the laboratory.   He was a believer in magic,
and anxious to pass for a profound magician.   He affirmed that he had received
a letter from Galen, dated in the infernal regions; and that he had held a dis-
putation with Avicenna concerning his potable gold.   He believed in the existence of a universal medicine, which he called a *quintessence*, extracted from all
things, and capable of curing all diseases.   He affirms that he was in possession
of this grand arcanum, and gives various examples of the prodigies which it
had performed in his hands.   But he allowed that it was difficult to obtain it,
and on that account points out various medicines for particular diseases, more
easily procurable.   One method of discovering these remedies was, by observing
the signature of each plant; thus, the euphrasia officinalis, having a mark on
its blossom similar to the eye, indicates, by that signature, that it is a remedy
provided by nature for sore eyes.   In like manner, the leaf of the scrofularia
officinalis, having some resemblance to the texture of the lungs, indicated its
fitness for curing the diseases of that important organ.   He affirmed that the
human body was a microcosm, or little world; and that parts bearing a strong
analogy to all the parts of the great world might be traced in it.   He made
certain organs of the body connected with the planets and with the metals, and
hence conceived that the diseases of each would be cured by the metals which
they represent.   Thus, the heart was connected with the sun, and with gold;
the brain, with the moon, and with silver; the kidneys and testacles, with
Venus, and copper; and the spleen, with Saturn, and lead.

He affirmed that all things were produced from an original *prime matter,* which was itself destitute of form, and of consequence invisible: that the seeds of all things were created at the beginning, but remain invisible, and wrapped up in this prime matter, till called out and brought to light by the agency of a divine being whom he calls Archæus. The elements of things, he affirmed, were not *fire, air, water,* and *earth,* as had been taught by the Greeks; but *salt, sulphur,* and *mercury,* a set of principles obviously borrowed from his predecessors the alchymists. Anatomy, as commonly taught, or the figure and position of the different organs of the body, he considered as a useless piece of knowledge. But the real and important anatomy was the chemical analysis of the different organs of the body, in order to determine the proportion of salt, sulphur, and mercury, which each of them contain, and the changes of these proportions produced by age or disease. He describes, at some length, the particular diseases occasioned by the prevalence of any of these principles in particular organs. Such seem to be the most important principles scattered through the writings of Paracelsus. But his writings abound so much in barbarous terms of his own coining, not accompanied by any explanation, that in various places they are not to be understood. In all probability he did not himself understand what he dictated; but was induced to publish unintelligible jargon, in order to maintain and extend the impressions in his favour, which had already produced such important effects in those countries which he frequented. He openly laid claim to the monarchy of medicine, and the most extraordinary stories were propagated of the wonderful cures which he performed, little short, many of them, of miraculous. But Libavius, who was himself a chemist of eminence, and well qualified to form an accurate judgment of his conduct, has given us a very different account. According to him, he seldom succeeded in curing his patients: many were deceived by false hopes, and many were destroyed by his remedies.

The monarchy, which Paracelsus flattered himself that he had established, scarcely outlived the founder. His disciples were lost in the mazes of his absurdities and enigmas. Though Crollius, and a few others, propped up the credit of the Paracelsine doctrines for some years, and though Van Helmont, destitute as he was of imposture, and possessed of original genius, lent them a favourable hearing, they contained so many inherent contradictions and incompatibilities, that they were gradually abandoned by all the world, before the short-lived influence of chemical theory had time to produce any great or permanent effect upon the reasoning and language of medical men.

Soon after the æra of Paracelsus and his reveries, mathematics came to be cultivated with much ardour and success in almost every part of southern Europe. During the whole of the 17th century mathematical discoveries of importance followed each other in quick succession, till the science at length engrossed the exclusive attention of literary men. Physicians, who have never

Mathematical medicine.

Y

been behind their contemporaries in knowledge, were naturally induced to study this all-powerful science, which lent its aid to every other, and successfully developed the principles of all. The consequence of this soon became conspicuous in their practice and reasoning. The human body was converted into a complicated machine, all the movements of which were regulated by the most rigid mathematical laws. Borelli calculated the force of the muscles, and the degree of contraction which they underwent in moving the bones. The discovery of the circulation by Harvey enabled them to calculate the force of the heart and arteries, the velocity of the blood, and the quantity of it which exists in different animals. The blood vessels were converted into hydraulical engines, the laws of which were explained according to the principles of hydrostatics. Diseases were accounted for by the globules of the blood, and other fluids, being forced into vessels of too small a diameter; and the salutary effect of different medicines was readily explained by the power which they had to diminish the size of the globules of the blood. Every thing, in short, was accounted for on mathematical principles; and treatises of medicine were filled with all the technical terms of mathematicians, and every proposition underwent a systematic demonstration.

The celebrated Pitcairn, first a Professor at Leyden, and afterwards at Edinburgh, was one of the most conspicuous of these mathematical physicians. The doctrine was also embraced by Boerhaave, and supported by all his eloquence, and by the far-famed celebrity of his name. Dr. Keill, Dr. Jurin, Dr. Friend, and even Dr. Hales, were likewise conspicuous supporters of mathematical medicine, and contributed by their writings to give it eclat, or to support its credit.

But these erroneous notions, however specious, or however propped up by celebrated names, could not always maintain their credit. The dissimilarity between the living body and a mechanical engine was too great, and the impossibility of accounting for all the functions on these principles was too glaring to escape, for any length of time, the penetration or the attention of the medical world. Two celebrated physicians in Germany, even during the period of their greatest celebrity, had refused their assent to the mathematical theories of medicine. These were Hoffman and Stahl; two men who left a very high reputation behind them, and who still retain many adherents in various parts of Europe They saw the great influence of the brain, spinal marrow, and nerves, upon all the functions and actions of the living body, and the impossibility of explaining this influence by any suppositions which reduced the living body to a mere mechanical machine. Stahl had recourse to a *living principle, or soul,* by means of which he accounted for all the functions of the living body; and he founded upon this physiological explanation a system of medical doctrines possessed of much ingenuity, but at the same time abundantly intricate and ob-

Living principle introduced.

scure. Hoffman had recourse to a principle of irritability residing in certain Doctrine of irritability. solids, a principle connected with, and depending on, the nerves, and which enabled him to remove the seat of disease from the liquids, where it had been uniformly placed by all preceding medical men, and to consider it as residing in the solids. The theory of Hoffman was embraced, with some limitations by the celebrated Dr. Cullen, who taught it for many years in the University of Edinburgh. His theory of fever, of which he has given an outline in his work entitled, *First Lines of the Practice of Medicine*, is obviously a modification of the doctrine of Hoffman.

Neither ought the doctrine of Brown, which attracted so much attention, Brunonian system. and acquired so much celebrity in consequence of its apparent simplicity, to be considered in any other light than as a scion sprouting from the doctrine of Hoffman. John Brown, the author of this system, resided long in Edinburgh, and supported himself by teaching medical pupils. He was at first patronized by Dr. Cullen, but afterwards quarrelled with him. He was a man of undoubted genius, but involved himself in various imprudences, which prevented his success in life, and kept him in continual poverty. His system, like that of the Methodists of old, has for its object to facilitate the method of practising medicine, and certainly exceeds all others in that respect. It reduces all diseases to two classes, and of course two methods of treatment will answer for all maladies whatever. The different organs of the body, according to the Brunonian doctrine, in order to perform the functions for which they were destined, require to be in a state of *excitement*, and this excitement is produced by *stimulants*. Health, and even life itself, depends upon the different organs being in their proper degree of excitement. When the stimulants are too powerful or too long continued, they at first produce an excessive degree of excitement, but this state gradually exhausts itself, and the organ is left in a state of debility, either with too little excitement or with none at all. According to this very convenient doctrine all medicines are reduced to those which produce excitement, and those which diminish it. Or, according to the most orthodox class of Brunonians, all medicines are reduced to *stimulants* alone.

The only other medical doctrine which has appeared in this country is the System of Darwin. Zoonomia of Dr. Darwin, a most elaborate work which is not very susceptible of abridgment. I consider it as merely the Brunonian doctrine under a disguised form, introduced with a greater parade of learning and science. The great number of new terms which it contains makes it a very irksome task to peruse it. Great pains are taken, throughout, to discard every appearance of a living principle, and to establish medical practice on the broad and vulgar basis of materialism.

Such is a short sketch of the history of medicine down to our own time From a view of the whole we are entitled to conclude, that medicine, if we consider it as a science, has made very little progress since the days of Hippocrates. New theories indeed, and new modes of explanation, have successively appeared in abundance; but they have been all equally visionary and absurd, and have vanished without leaving any traces behind them. If we attend to the opinions and writings of the medical men of the present day, we find a propensity among some to restore the exploded chemical doctrines of Paracelsus, amended, of course, and improved. Others lean pretty strongly to the humoral doctrines of the ancients, obviously in consequence of the strong bias which the popular systems of Hoffman and Cullen gave medicine, to direct the whole of its attention to the solids. On the other hand, if we view medicine as an art, it has undergone great and progressive improvements. The materia medica has been gradually almost entirely changed; and the present mode of treating patients, though perhaps not less efficacious than that of the ancients, is infinitely superior in every thing that concerns the feelings of the patient. The improvements in surgery are not less evident and important.

Medical papers in the Transactions.

The medical papers in the Philosophical Transactions amount to no fewer than 478, and constitute no small proportion of the whole work. Of these 201 belong to the department of medicine, and 277 to that of surgery. They are of very various merit, as must be expected in such a collection. Indeed there are no fewer than 283 which, in the present state of these arts, may be safely considered as possessed of very little value. The remaining 195 are upon the most miscellaneous subjects. All of them deserve the closest examination of medical men. But, as they are neither capable of classification nor abridgment, a full account of them could not be given here without occupying a much greater space than would be consistent either with the nature of this work, or with the patience of our readers. I shall therefore satisfy myself with noticing a few of the most prominent articles.

Sydenham.

I One of the most eminent physicians that England has produced was Dr. Sydenham, the patron and friend of Sir Hans Sloane. An analysis of his work on fevers occurs in an early volume of the Transactions.* The great merit of this celebrated physician consists in the accurate descriptions which he has left us of several diseases which first became conspicuous in his time. His account of the small-pox, and of his medical treatment of that disease, is admirable, and contributed in no small degree to establish his celebrity. He was the first person who introduced the cooling regimen in fevers, a method of treatment frequently attended with the happiest effects. Though it must be acknowledged, that he did not sufficiently distinguish between the typhus and the inflammatory fever, and on that account he sometimes carried his bleedings to an excess. He contributed also essentially to introduce the Peruvian bark as a cure for intermittents.

* Vol. I. p. 210.

II. Cantharides, as an internal remedy, were employed by Hippocrates, Cantharides though the application of them externally in blisters was not introduced till long after. Hippocrates seems to have considered them as efficacious in cases of disuria.   There is a very remarkable instance of their efficacy in that very troublesome disease, recorded in the Philosophical Transactions by Mr. Yonge.* A lady, 54 years of age, long tormented by fits of the stone, and afterwards afflicted with dropsy, soon after being cured of the latter disease fell into a total suppression of urine, which continued for many days, and baffled all remedies. Mr. Yonge at last gave her five cantharides without head, wings, or legs, weighing four grains and a half.   These were made up into two pills, with an equal quantity of camphor.   Next day about noon the flood came, and continued for 48 hours, during which she voided an immense quantity of urine, and her disease went off without any bad symptoms whatever.

III. The small-pox is so dreadful a distemper, and so universally diffused over Small-pox. the surface of the earth, that every thing respecting its medical treatment claims particular attention, and rises to a degree of importance which no other disease can command. The cow-pox, so happily substituted for it by Dr. Jenner, would indeed eradicate this pestilential distemper in a very few years, if it could be universally adopted.   But this, we fear, is not likely speedily to take place. Even in the streets of London, the city which may be considered as the centre and source from which vaccination originated, small-pox patients are to be met with in numbers.   Much more likely are they to abound in distant countries, where the new substitute can only make its way by slow and imperceptible steps. There is an excellent paper on the treatment of the small-pox, by the celebrated Dr. Huxham, of Plymouth, in an early volume of the Philosophical Transactions.†   This paper contains many valuable observations on the medical treatment of the small-pox.   Dr Huxham particularly inculcates the necessity of keeping the bowels open ; and gives some striking examples of the good effects arising from that practice.   Indeed the propriety, and even the necessity, of this practice is now very generally known.   The disease very seldom ends fatally, when the bowels have been kept uniformly and regularly open.

IV. The efficacy of cold water in fevers has been inculcated by the oldest Cold water physicians.   It was a remedy even employed by Hippocrates.   It was not, how- in fevers. ever, used by the Greek physicians in enormous quantities, or to the exclusion of all other remedies ; but a new method of using it was introduced into Naples about the year 1729 ; an account of which is given in the Philosophical Transactions, by Dr. Cyrillus, of Naples.‡   The patient was kept without food, and made to drink from one to two pints of water, cooled with snow, every two

* Phil. Trans. 1702. Vol. XXIII. p. 1210.   † Phil. Trans. 1725. Vol. XXXIII. p. 379.
‡ Ibid. Vol. XXXVI. p. 142.

hours, day and night, except when asleep. This practice was persisted in for three days, or even longer if the fever required it. Some cases are stated, in which it was persisted in for ten days. The use of food, during the administration of the water, was considered as injurious. In general, the water kept the bowels open; but if it did not, it was thought requisite to keep them open either by administering injections or almond oil. When sweats broke out, it was requisite to check them; and if nothing else would do, the administration of the cold water was to be laid aside. This method, though strongly recommended by Dr. Cyrillus, and though warmly panegyrized in the year 1722, by Dr. Hancock, was never much practised in this country. But the external use of cold water in fevers, which seems to depend upon the same principle, has been warmly recommended by Dr. Wright, and by the late Dr Currie, of Liverpool; whose publication on the subject attracted considerable notice, and gave the practice currency and celebrity.

Spontaneous combustion.

V. There are several very remarkable examples recorded of persons taking fire, and being burned to ashes, without any visible cause. A very striking instance is given in the Philosophical Transactions for 1745,* from an Italian treatise, by Joseph Bianchini, a Prebend in the city of Verona, where the accident happened. " The Countess Cornelia Bandi, in the sixty-second year of her age, was all day as well as she used to be, but at night was observed, when at supper dull and heavy. She retired, was put to bed, where she passed three hours or more in familiar discourses with her maid, and in some prayers; at last falling asleep, the door was shut. In the morning, the maid taking notice that her mistress did not awake at the usual hour, went into the bedchamber and called her; but not being answered, doubting some ill accident, opened the window, and saw the corpse of her mistress in the deplorable condition following:

" Four feet distance from the bed there was an heap of ashes, two legs untouched, from the foot to the knee, with their stockings on: between them was the lady's head; whose brains, half of the back part of the scull, and the whole chin, were burnt to ashes; among which were found three fingers blackened. All the rest was ashes, which had this peculiar quality, that they left in the hand, when taken up, a greasy and stinking moisture. The air in the room was also observed cumbered with soot floating in it: a small oil lamp on the floor was covered with ashes, but no oil in it: two candles in candlesticks, on a table, stood upright; the cotton was left in both, but the tallow was gone and vanished; somewhat of moisture was about the feet of the candlesticks: the bed received no damage: the blankets and sheets were only raised on one side, as when a person rises up from it, or goes in: the whole furniture, as well as the bed, was spread over with moist and ash-coloured soot, which had penetrated into the chest of drawers, even to foul the linen; nay, the soot was also gone into a neighbouring kitchen, and hung on the walls, moveables, and utensils of it. From the pantry a piece of bread, covered with that soot and grown black, was given to the dogs, which refused to eat it. In the room above it was noticed, that from the lower part of the windows trickled down a greasy, loathsome, yellowish liquor; and thereabout they smelt a stink without knowing of what, and saw the soot fly around."

Such is the account of this extraordinary phenomenon, as given in the Philosophical Transactions. Several other examples of similar deaths are recorded in the same paper; and in the same volume of the Transactions, there is an account of a woman at Ipswich, who perished in the same extraordinary manner, during the spring of 1744. Several similar cases have occurred since that period, both in Britain and on the Continent. It is remarkable that all the cases recorded happened in the night, and when no spectator whatever was present. The sufferers had, usually, either a lamp or a candle. But all the circumstances render it impossible for us to conclude that they were destroyed in consequence of their clothes catching fire, for most of them were undressed; and portions of their clothes always escaped entirely, which could not very well have happened if they had caught fire first. The phenomenon itself is still unexplained. All the attempts to account for it are extremely nugatory, and deserve no attention whatever.

VI. The dreadful consequences of confined and stagnant air, in places *Jail distemper.* crowded with inhabitants, are now universally known, and carefully guarded against, both in prisons and ships. But prisons were formerly so small, crowded, and ill contrived, as to be exceedingly injurious to the health of the miserable wretches who were doomed to languish in them. The consequence was a malignant and most infectious fever, well known by the name of the *jail distemper.* In the year 1750, the Lord Mayor of London, two of the Judges, and an Alderman on the Bench, in consequence of the state of the felons tried at the Old Bailey Sessions, were seized with this fatal distemper, and died. This induced the magistrates of London to resolve upon attempting to render Newgate more healthy; and they consulted Dr. Hales and Sir John Pringle, about the method which they should follow. Dr. Hales recommended the use of his *ventilator;* a machine contrived to pump out the air of any place, and thus occasion a perpetual renovation of it. The machine was accordingly erected, and the salutary effects became speedily apparent. The deaths in Newgate were reduced from seven or eight a week, to about two in a month. Eleven men were employed in erecting the ventilator, and of these no fewer than seven were seized with the jail distemper. A very interesting account of the cases of these men, and of the mode of treatment, was drawn up by Sir

John Pringle, and published in the Philosophical Transactions.* From this account, the very infectious nature of the fever is obvious. One of the men died, and the whole of his family, one after another, were seized with the disease. They all recovered except his mother-in-law, to whom it proved fatal. It was the enlightened exertions of Dr. Hales and Sir John Pringle, that first turned the attention of mankind, in a forcible manner, to the importance of ventilation, which led to the subsequent improvements introduced into our ships, and our prisons, by Cooke and Howard.

Dropsy.
VII. There are few diseases less under the control of the physician than dropsies. They most frequently proceed from an exhausted constitution, which nothing in the power of the physician is able to restore and invigorate. Every successful mode of treating this disease, therefore, claims the particular attention of medical men. There is a very simple method of treatment described in the Philosophical Transactions, by Dr. Oliver,† which proved successful in three different cases; and which, from its simplicity and gentle stimulating nature, seems to promise as fairly as any other to restore the lost tone of the dropsical parts. This method is nothing else than *friction*. The diseased part is rubbed every day for an hour with sallad oil, by means of a warm hand In a few days the quantity of urine increases, and the parts resume their usual tone.

Blisters in fever with cough.
VIII. The next medical paper which we shall notice, is by the celebrated Dr. Whytt, of Edinburgh. He points out the great advantage of blisters, in cases of fever attended with cough, occasioned by an accumulation of pituitous matter in the lungs, when the pulse is so small, and the strength so exhausted, that the patient cannot well bear bleeding. In such cases, blisters always lower the pulse in a remarkable degree; and when sufficiently large, and often enough repeated, seldom fail to remove the disease.‡

Effect of camphor.
IX. A remarkable experiment made by Mr. Alexander, a surgeon, in Edinburgh, on the effect which camphor produces when swallowed as a medicine, is described in the Philosophical Transactions;§ and deserves to be noticed as a warning to medical men, respecting the quantity of that substance which may be swallowed at a time with impunity. Disputes existing at the time whether camphor increased or diminished the heat of the body; Mr. Alexander resolved to make an experiment on himself, in order to resolve the point. He swallowed one scruple of camphor, and found that it produced no sensible change on the heat of the body; but it lowered the pulse a few beats per minute. Next day he swallowed two scruples of camphor: the effects at first were the same as on the preceding day; but he gradually became giddy and confused, and at last wholly lost his recollection; and was seized with convulsions. His pulse

---

* Phil. Trans. 1753. Vol. XLVIII. p. 42.   † Phil. Trans. 1756. Vol. XLIX. p. 46
‡ Ibid. 1758. Vol. V. p. 569.   § Ibid. 1767. Vol. LVII. p. 65.

rose to 100 in a minute. Dr. Monro being sent for, and the cause of the disorder accidentally discovered by means of a written paper on the table, he was made to vomit, and gradually recovered his senses. But, for a long time, his intellect continued confused and unsteady. He was seized with a violent head-ache, but after a night's sleep he found himself quite recovered. From this case, it would appear that camphor, in small doses, acts as a sedative; while in larger doses it is a powerful and dangerous stimulant.

X. The remarkable state of health which the crews of Captain Cooke's ships enjoyed, during his long and memorable voyages of discovery, when contrasted with the dreadful ravages committed by the scurvy upon the crews of former ships, when long out at sea, is a convincing proof of the great importance of a proper attention to the food of the seamen, and to the cleanliness and ventilation of the vessel. Captain Cooke has given an account of the method which he employed, in a paper published in the Philosophical Transactions;* for which he was rewarded with the Copleyan Medal. His men were furnished with a certain quantity of fresh wort every day; they had a stock of sour crout; sugar was substituted for oil; and wheat flour for oatmeal. The men were always plentifully supplied with fresh water, and with fresh animal and vegetable food, wherever it was to be procured. They were at three watches, and when wet had always dry clothes to put on. Great care was taken to keep the ship clean; and fires were frequently burnt in the well and between decks, in order to produce a proper ventilation. Captain Cooke considers the practice of sprinkling vinegar upon the decks as of very little use; and there can be no doubt whatever, from the experiments which have been made upon the subject, that this liquor, though more expensive, is not in the least more efficacious as a detergent than mere water. It might be laid aside in the navy without impropriety, and would occasion a considerable saving. _[margin: Cooke's method of preserving the health of his crew.]_

XI. The remaining papers which we mean to notice are more closely connected with surgery than with medicine. Indeed, the number of important surgical papers in the Transactions is much greater than of medical, strictly so called. One of the most dreadful diseases to which mankind are subject, is the stone. The only cure is the extraction of the calculus, and this is generally a radical cure; though the operation sometimes terminates fatally. There is reason to believe, that the calculi first form in the kidney; that they pass along the ureter, while only of a small size; and being deposited in the bladder, gradually increase in bulk. Sometimes, however, they remain in the kidney, and occasion a most excruciating and hopeless disease; because the kidneys lie so deep, and are so situated, that it is extremely difficult to cut into them, or to perform the operation called *nephrotomy*. There is one successful case, however, _[margin: Case of nephrotomy.]_

* Phil. Trans. 1776. Vol. LXVI. p. 402.

of the performance of this desperate operation, recorded in the Philosophical Transactions; and as it is almost singular in its nature, we could not with propriety pass it over in silence. Mr. Hobson, the English Consul at Venice, having been long afflicted with the stone in the kidney, was at length attacked with a fit of such duration and violence, that it reduced him almost to desperation; and finding no relief from any means that had been used, he determined to apply himself to Dominic de Marchetti, a famed and experienced practitioner at Padua; intreating him to cut the stone out of his kidney; being fully persuaded that no other method remained capable of relieving him. Marchetti represented the extreme difficulty of the operation; that he had never attempted it; and that it would in all probability destroy him. But Mr. Hobson persisting, he was at length prevailed upon to undertake it. He began with his knife cutting gradually upon the region of the kidney, till the blood disturbed the operation, so that he could not finish it at that attempt; wherefore, dressing the wound till the next day, he then resumed and completed the operation; and, taking out of the kidney two or three small stones, he dressed up the wound. From this instant Mr Hobson was freed from the severity of his pain, and in a reasonable time was able to walk about his chamber, having been in no danger either from flux of blood or fever. The wound refused to heal, became fistulous, and a small quantity of urine was constantly discharged from it. Some time after, his wife, while dressing it, took out another small stone, of the figure and magnitude of a date stone; after the removal of which he felt no more pain whatever. He recovered his health perfectly, merely covered the place with a linen cloth to receive the urinous discharge, and could ride 40 miles a day without inconvenience.*

Trachæa cut through and healed.

XII. There is a very remarkable case recorded in the Philosophical Transactions, of Nicholas Hobb, a man aged about sixty-three, who was set upon by ruffians, knocked down, had his trachæa cut somewhat below the pomum Adami, together with several large blood-vessels; and, after being robbed, was left for dead. After some time he recovered so much sense and strength as to thrust his neck-cloth into the wound, and to crawl to his own house, which was at no great distance. Mr. Keen, a surgeon in the neighbourhood, was called: he found the two parts of the trachæa at a great distance from each other; however by raising his feet above his head he contrived to bring them near each other, and sewed up the enormous wound. The stitches repeatedly gave way, but were as often renewed, and in process of time new flesh granulated; pieces of bone from the osseous cartilages of the trachæa were discharged, and the wound finally closed. The man was as well as ever, in most respects; only the epiglottis did not exactly cover the rima of the larynx as at first, so that he was obliged

* Phil. Trans. 1696. Vol. XIX. p. 333.

to use some precautions in swallowing liquids, and even solid food, to prevent them from making their way into the trachæa.*

Dr. Musgrave, on the authority of this case, recommends the propriety of laryngotomy in all cases where diseases in the throat threaten suffocation. There can be no doubt of the propriety of the recommendation; and, at present, surgeons perform the operation without hesitation wherever it is necessary.

XIII. The method of inoculating for the small-pox, now so universally known, was first introduced at Constantinople by the Circassians, Georgians, and other Asiatic nations, about the year 1670. People were at first shy in imitating the example of these men; but, perceiving the good effects which resulted from the practice, it graduall made its way into general use. The first account of this practice was published in the Philosophical Transactions for 1714, in two different papers. The first was a letter from Emanuel Timoni,† an Italian, who at that time practised medicine at Constantinople. The second account was written by Jacob Pylarini,‡ a native of Cephalonia, who practised medicine successively in Constantinople, Smyrna, and Moscow, and was likewise Venetian Consul at Smyrna. His account is more minute than that of Timoni, and he seems to have had an active hand in bringing the practice into repute in Turkey. Inoculation began to be practised in London soon after the publication of these papers. But it made its way exceedingly slowly in consequence of the violent prejudices which it had to combat. In the year 1722, Dr. Nettleton began to try it at Halifax, in Yorkshire, and he published a detail of his proceedings in the Philosophical Transactions.§ It had been likewise introduced into New England.‖ The whole number inoculated in England by the year 1722 amounted, according to Dr. Jurin, to 182. Of these two died. Dr. Jurin, in an admirable paper which he published on inoculation, demonstrated that, in the small-pox taken the natural way, the deaths amounted to rather more than one in 14.¶ Thus the superiority of inoculation became conspicuous at the very outset. For, even by the enemies of that method, the deaths were only estimated at one in 91. They are in fact somewhat lower than that ratio. From various papers published in the Transactions** by Dr. Williams and Mr Wright, it appears that inoculation had been practised in Wales long before the method was made known from Constantinople. The Welch called their process *buying the small-pox*. It is remarkable that this phrase has been applied to inoculation in the most distant and uncivilized countries, indicating the antiquity of the practice, and the probability that it has originated from some common inventor.

Inoculation for the small-pox.

* Phil. Trans. 1699. Vol. XXI. p. 398.　　† Vol. XXIX. p. 72.　　‡ Ibid. p. 393.
§ Vol. XXXII. p. 35.　　‖ Vol. XXXII. p. 33.
¶ Phil. Trans. 1722. Vol. XXXII. p. 213.　　** Vol. XXXII. p. 262, 267.

Sir Hans Sloane interested himself greatly in promoting the practice of inoculation; and he drew up an historical detail of the proceedings in London, which was afterwards published.* It was in consequence of a letter which he wrote to Dr. Sherard, at that time in Smyrna, that the account of inoculation by Dr. Pylarini was sent over and inserted in the Philosophical Transactions. This notice lay dormant till Mr. Wortley Montague, then Ambassador at Constantinople, and Lady Mary, his wife, inoculated their son, and brought him in safety to England. Upon this Queen Caroline, at that time Princess of Wales, begged the lives of six condemned criminals, who had never had the small-pox. They were inoculated, and all took the disease except one woman. To make further trial, Queen Caroline procured half a dozen of the charity children belonging to St. James's parish, who were inoculated; and all of them, except one (who had had the disease before, but concealed it for the sake of the reward) went through it with the symptoms of a favourable kind of the distemper. Queen Caroline afterwards consulted Sir Hans Sloane about inoculating her own family. Sir Hans approved of the process, but refused to advise her Majesty to put it in practice in her own family, as not being certain of the consequences that might follow, and on account of the great importance of the persons experimented on to the public. This opinion determined both Queen Caroline and King George I. The operation was performed, and the children went through the disease favourably. Sir Hans Sloane adds that out of 200 cases of inoculation that he had seen, only one terminated fatally.

There is a curious historical account of the introduction and success of inoculation in New England, drawn up by Mr. Gale, and published in the Philosophical Transactions for 1765.† From this memoir it appears, that the natural small-pox was particularly fatal in that colony. The number of deaths averaged one in seven. When inoculation was introduced, the deaths amounted to one in 30. By improvements and proper precautions they were reduced to one in 80, and finally, by the administration of mercury, to one in 800. In consequence of the careless way in which persons ill of the inoculated small-pox were accustomed to mix with society, and thus spread the infection, the magistrates of New England were induced to prohibit the practice of inoculation altogether, under severe penalties, and forbidding those who went to the other states in order to be inoculated, from returning to their own homes within a shorter period than 20 days.

XIV. It is a very remarkable circumstance in the history of medicine, and of mankind, that some of the most formidable diseases of modern times are not mentioned by the ancient physicians, and appear to have been unknown to the Greeks and Romans. The small-pox and the measles, for example, are never

Phil. Trans. 1755. Vol. XLIX. p. 516.   † Vol. LV. p. 193.

6

mentioned by any of the Greek physicians, and first make their appearance in the writings of the Arabians. The first medical notice of the venereal disease ~~Venereal disease.~~ is of a much later date. Physicians almost uniformly agree to refer its first appearance to the year 1494. The story told by Fallopius, the first writer on the subject, is this. Charles, King of France, having made an inroad into Italy, took possession of the Milanese and Tuscany, and invaded Naples, at that time in the possession of the Spaniards. The Spanish troops, not being a match for the large French army which was brought against them, had recourse to stratagem to accomplish their destruction. After poisoning the wells, and mixing the French bread with plaster, they sent into their camp several prostitutes infected with the venereal disease. This stratagem succeeded. The whole French army was infected with the malady, which was spread in consequence over Italy and Europe with prodigious rapidity, and filled the world with terror and despair. According to Peter Martyr (as quoted by Fallopius) this disease was communicated to the soldiers and sailors who accompanied Columbus on his memorable expedition to the West Indies in 1492. It was spread over the whole of the West India islands, and was communicated by the Indian women to those Spaniards whom they admitted to their embraces. Some soldiers from Columbus's ship, having gone directly to Naples, had communicated the disorder, and made it known to the Spanish and Neapolitan physicians just before the French invasion.

Such is the account of the origin of this formidable disease given by the Italian physicians. But we must confess that it is loaded with several circumstances which it is not easy to believe. There can be no doubt whatever that the disease became first publicly known about the year 1494, and that this notoriety was occasioned by the general prevalence of it in the French army lying in the Neapolitan dominions. But that it did not previously exist in Europe, and that it was imported from the West India islands, are not so clear. In the first place we have no evidence whatever that the disease existed in the West India islands and America. It does not appear to exist at present among the Indian tribes, and has never been observed in any of the more recently discovered islands in the South Sea. There does indeed exist a disease in Otaheite, bearing some resemblance to it, mistaken for it by most of our European voyagers, and thought to have been imported from Europe to these delightful regions. But it would appear from recent and more accurate examination, that this disease differs entirely from our venereal disease both in its symptoms and termination. Even if it had existed in the West India islands, the short period of a year, which is all that elapsed between the discovery of America by Columbus and the appearance of the venereal disease in Italy, appears too limited to account for so general a diffusion of the malady, not only over Italy but every other country of Europe. In the year 1497, it was so prevalent in Edinburgh, as to have produced the greatest consternation in the king and his council. There is a proclamation on the subject still existing in the records of the town council of Edinburgh. It went by the name of the *grandgor ;* and all those that were infected with it were sent to Inchkeith, with their physicians, to remain there till cured, in order effectually to prevent the further spreading of the infection.*

Mr. Beckett has published three elaborate papers in the Philosophical Transactions to prove that the venereal disease existed in England long before the year 1494.† The evidence which he adduces, though not absolutely demonstrative, is so strong as to produce conviction in any person who will weigh it with attention. There can be no doubt that a disease existed in the sexual organs of both sexes at a very early period. Some of the descriptions of this disease seem to identify it with gonorrhœa, while others describe symptoms similar to those which exist in syphilis, such as chancres and bubos. Hence it is not unlikely, that both states of the disease existed in England at an early

Supposed the same with the leprosy.
period. Mr. Beckett conceives that what was called at that time the *leprosy* was nothing else than the venereal disease. He has shown at least that it was different from what we now call the leprosy, that it was cured by mercury, and that it disappeared when the venereal disease became known by its real name. Mr. Beckett brings his documents from the registers of the stews, which were by authority allowed to be kept on the Bankside, in Southwark, under the jurisdiction of the Bishop of Winchester. Among other regulations drawn up as early as the year 1162, is this: *no stew-holder to keep any woman that hath the perilous infirmity of burning.* In a book in the custody of the Bishop of Winchester, supposed to have been written about 1430, there is found this regulation: *that no stew-holder keep noo woman wythin his hous that hath any sycknesse of brenning, but that she be putte out, upon the peyne of makeit a fine unto the lord of a hundred shylyngs.* John Arden, Esq. who was surgeon to Richard II. and afterwards to Henry IV. defines *burning* to be a certain inward heat, and excoriation of the urethra, which is a tolerably correct description of the gonorrhœa. He gives us an account of the mode of cure which was by injecting a mixture of milk and oil of violets, some sugar, ptisane, and oil of almonds being likewise added. Mr. Beckett shows that the word *burning* was given by English medical writers to the venereal disease even as late as the year 1530.

Mr. Beckett further shows, by quotations from ancient English writers, that this disease arose from the impure embraces of infected persons, that it degenerated into an ulceration and putrefaction of the genital organs, which termi-

* Phil. Trans. 1743. Vol. XLII. p. 420.
† Phil. Trans. 1718. Vol. XXX. p. 839. Vol. XXXI. p. 47, and p. 108.

nated in death. He quotes a curious passage from Dr. Thomas Gascoigne, a clergyman, who gives an account of the symptoms, and affirms, that John of Gaunt died of this disease ; and that he showed the putrefied state of his genitals to King Richard II., who visited him while he lay on his death-bed. Mr. Beckett likewise shows, that the Italian medical writers, who lived during the siege of Naples, in 1494, were by no means unanimous in considering the disease as new, far less as imported from the new world. Upon the whole, it is exceedingly probable that the disease existed previously under a milder form ; that, in consequence of some unknown accident, it assumed a more virulent aspect about the period of the siege of Naples. This virulence drew the attention of the medical world more particularly to it, and induced them to give it a new name, and to invent a new mode of treating it. For many years past it has lost this peculiar virulence, except in a few rare constitutions ; and it will not be surprising, if it should again sink into the same insignificance as it was in before the siege of Naples. Mercury has been considered as an indispensable medicine. Perhaps it might be so formerly ; but there is reason to doubt its necessity at present, at least in every case. Nor can there be much hesitation in affirming, that the violent courses of mercury, formerly administered in the slightest cases of venereal infection, were much more destructive to the constitution than the disease itself. M. Christien, a physician in Montpellier, has lately announced the application of various preparations of gold, as a certain cure for the venereal disease. His experiments have been repeated with success, in various parts of France. This shows us that mercury is not absolutely necessary. Probably, in all the cases tried by Christien and his followers, the disease would have disappeared of itself, without the administration of any medicine whatever.

XV. Though lithotomy must be admitted as a radical cure for the stone, and though, in general, it may be performed without impropriety, or without much risk to the patient, yet there are cases when, from old age and other circumstances, no surgeon would venture to recommend so formidable an operation. In such cases, Mr. Douglas recommends the formation of an artificial fistula, in the perinæum, which he affirms would remove all the disagreeable symptoms, by enabling the patient, at pleasure, to introduce an oiled probe, and thus push back the stone whenever it occasions a suppression of urine. He affirms, that the fistula may even be gradually enlarged, so much as to admit of the extraction of the stone, by means of a forceps.* It is rather surprising that so simple a process, recommended by so eminent a surgeon, and skilful anatomist, should not yet have come into general use. *Palliative treatment for the stone.*

XVI. Mortifications of different parts of the body, from an internal cause, *Putrefaction cured by bark.*

* Phil. Trans. 1727. Vol. XXXV. p. 318.

were long considered as incurable; and accordingly, whenever they appeared, the case was given up by surgeons as hopeless. About the year 1730, Mr. Rushworth, a surgeon, in Northampton, discovered that their progress was stopped by the copious administration of Peruvian bark. The truth of the discovery was confirmed by the numerous observations of Mr. Sergeant Amyand; and, in 1732, Mr. Douglas published an account of the method of treatment in this disease; accompanied by ·a variety of cases confirming the efficacy of bark, as a cure for mortification. An account of this treatise, by Dr. Douglas, is published in the Philosophical Transactions.* This remedy is now universally employed, and considered as a specific; except in one case of mortification, to which chimney-sweepers are liable, and which Mr. Pott discovered might be cured by the administration of opium.

Ascites cured by injections.   XVII. There is an account in the Philosophical Transactions of a very ingenious attempt by Mr. Warwick, a surgeon at Truro, to cure ascites, by injecting warm astringent liquors into the cavity of the abdomen, after the dropsical liquid had been drawn off by tapping.† His experiment was made upon a woman near fifty years of age, who had been previously tapped, and 36 pints of liquid drawn from the abdomen. The cavity filled again with great rapidity, and he was called a second time to perform the same operation. After having drawn off part of the serum collected, he injected a mixture of warm claret and Bristol water, and repeated the injection twice. A syncope came on, but the woman gradually recovered her senses, and in a few days her health was restored, as completely as ever it had been; and though she was afterwards afflicted with a tertian fever, yet her dropsy did not return. There can be no doubt that such injections operate as complete cures in cases of partial dropsies. Thus the hydrocele is usually cured by a similar injection. But, in a confirmed ascites, the constitution is usually so much broken down, the cavity is so large, and the organs which it contains of so much importance, and so necessary for life, that such a practice could not be expected to be frequently attended with success. Mr. Warwick's case, however, and some others that have been since recorded, show us that the disease is not always incurable, and that it is, at least, worth while to try the efficacy of such injections.

Improvement in lithotomy.   XVIII. The gradual improvements in the operation of lithotomy constitute a very curious part of surgical history. The present improved state of the lateral operation, which is now almost the only one practised, except in very peculiar circumstances, was the gradual result of the successive improvements of various ingenious surgeons. One improvement of some importance, introduced by Dr. Mudge, a very ingenious surgeon and physician at Plymouth, is described in the Philosophical Transactions.‡ After cutting into the bladder,

---

* Phil. Trans. Vol. XXXVII. p. 429.   † Phil. Trans. 1744. Vol. XLIII. p. 12.
‡ Ibid. 1749. Vol. XLVI. p. 24.

and introducing the forceps, it was customary for operators to extract the stone by force. If the calculus was large, this could not be done without forcibly tearing the bladder; and Mr. Mudge shows that most of the violent and fatal symptoms which follow the operation, are owing to this laceration, which is much more injurious to the parts than cutting them. An attempt to cut the bladder, upon the forceps, while in a flaccid state, could hardly succeed; and if it did, the rectum was almost sure to be cut at the same time. Mr. Mudge remedied this defect, and put it in the power of the surgeon to dilate the wound of the bladder, at pleasure, by leaving a staff in the bladder, on which there was a groove cut; along his groove a proper knife was slid, and thus the bladder dilated at pleasure.

XIX. The cataract is a well-known disease of the eye, in which the crys- Extraction of talline lens becomes opaque, and obstructs vision. The cure is the removal of the lens of the eye. the lens; for the eye can perform its functions without this part of it, with tolerable accuracy. The method commonly practised is called *couching*. The lens is forced out of its place into the back part of the eye, where it is speedily absorbed and carried off. M. Daviel, a surgeon in Paris, about the year 1745, contrived a new mode of operating. He cut a considerable opening in the cornea, and through this opening, by gently pressing the eye, he forced out the lens, and thus removed it at once out of the eye. He employed three different instruments in succession; and, in consequence of the discharge of the aqueous humour, and the flaccidity of the cornea, it was difficult, and required a very steady hand, to dilate the cut in the cornea sufficiently, without cutting the iris, which would injure the eye materially. There is a description of Daviel's process, published in the Philosophical Transactions, by Dr. Hope.* The subject was taken up by Mr. Sharp, a celebrated surgeon in London, who improved the process, by employing a knife capable of performing the whole cutting at once, without employing successive instruments, which greatly facilitated the process. In two papers, which he published on the subject,† he has given an account of his method in detail; and has likewise related the cases of a variety of patients whom he successfully treated in this way. Notwithstanding the apparent superiority of this operation, the method of couching still continues to keep its ground; owing, probably, to its greater facility and seeming mildness; though in fact, it occasions greater uneasiness than the process of extracting the lens.

XX. Deafness is one of the diseases the least under the control of the Deafness physician. Unless it proceed from some obvious disorder in the external cured. meatus, or from swelled tonsils, it is usually given up as a hopeless case by

---

* Phil. Trans. 1751. Vol. XLVII. p. 530.
† Phil. Trans. 1753. Vol. XLVIII. p. 161 and p. 322.

medical men; and the patient left to shift for himself. Every additional means suggested to alleviate this distressing disease acquires, on that account, considerable importance. Hence it will be proper to mention a method suggested by Mr. Wathen, which he tried in various instances with success. He conceived that deafness was sometimes occasioned by the obstruction of the Eustachian tube, which prevented the external air from making its way into the internal cavity of the tympanum. He proposed, therefore, to try injections of warm liquids into that tube, by means of a silver tube introduced through the nose. In six cases tried, this injection was attended with success.*

Regeneration of bones. XXI. The power which the bones have to regenerate in the human body after being partly exterminated, deserves the serious attention of the surgeon. In all probability, fewer amputations would take place if this renovating power in nature were sufficiently attended to by surgeons. Various cases are related in the Transactions, of the renovation of bone after being cut out, and the restoration of the limb to its original functions, or at least to a considerable part of them. It may be worth while to mention some of the most remarkable of these examples: M. Le Cat, a surgeon, in Rouen, who has published a great many papers in the Philosophical Transactions, cut three inches, ten lines, out of the bone of the shoulder of a soldier The arm was kept at its usual length by machinery. In a short time, a callous formed; and the arm recovered its energy completely.† Mr. White, a celebrated surgeon, of Manchester, recommended the excision of the head of the os humeri, in certain cases; and affirmed, that the arm healed after the operation, and recovered its energy. He gives some cases, in his Surgical Essays, confirming the truth of this opinion. In the Philosophical Transactions there is a case given by Mr. Bent, surgeon, of the extirpation of the head of the os humeri, with success. The patient was able, after the cure, to lift the arm six inches from the side, and to perform every function that required no higher an elevation.‡

Wounds of the intestines healed. XXI. Wounds of the intestines are generally, and with reason, dreaded by surgeons; and, for the most part, they terminate fatally. There are not wanting instances, however, when such wounds have healed of themselves, without any disastrous symptoms intervening. Several examples of this kind occur in the Philosophical Transactions. We shall notice one related by Mr. Nourse, surgeon at Oxford.§ James Langford, a lad of twenty-one, was stabbed with a knife in the left side of the belly; the wound was about three inches long; great part of the intestines were protruded; and it was found necessary to enlarge the wound in order to return them into the abdomen. The colon was wounded near that part which terminates in the rectum. For a considerable

---

* Phil. Trans. 1755. Vol. XLIX. p. 213.   † Phil. Trans. 1766. Vol. LIV. p. 270.
‡ Phil. Trans. 1774. Vol. LXIV. p. 353.   § Phil. Trans. 1776. Vol. LXVI. p. 426.

time, fæces were discharged by the wound in the belly, and blood by the anus; but, by degrees, all the bad symptoms were removed, and the wound healed in about three weeks, without any bad symptom whatever.

Such are a few of the most striking medical papers to be found in the Philosophical Transactions. It would be easy to extend our observations, on this important subject, almost to any length; but we have already, we fear, rather exceeded the bounds prescribed to us in this work. We, therefore, take leave of this interesting subject, inviting all medical men to consult this valuable repository of important medical knowledge.

CHAP. III.

OF MINERALOGY.

The term *mineral* is at present applied to all the different substances of which this earth, which we inhabit, is composed; and *Mineralogy* is the science which treat of these minerals. Now, there are three different points in which minerals may be viewed: we may consider them as distinct substances, and arrange them into a system by dividing them into genera and species; we may consider the way in which they are deposited in the earth, the order in which they lie, and their connexion with each other; or, we may consider the best way of extracting from the earth such of them as can be applied to valuable purposes. The first of these views of the subject is, in this country, Division of usually distinguished by the name of *Mineralogy*; but, in Germany, it is called the subject. *Oryctognosy*; the second, is called *Geognosy*, or *Geology*; the third, *Mining*. We shall consider each of these branches of the science, as far as the papers in the Philosophical Transactions on these subjects are concerned, in the three following sections.

Section I.—*Of Oryctognosy.*

This branch of the science is quite recent; its real origin can hardly be dated Little culti-further back than about 40 years ago. The ancients have left us little that is vated by the valuable on the subject. Theophrastus wrote a Treatise on Stones, which was ancients. translated into English by Sir John Hill, and first brought him into the notice of the public. This translation contains many valuable notes, most of which have been pillaged by more recent mineralogical writers, without acknowledgement. Pliny, in the last two books of his Natural History, has likewise given us a treatise on stones; but it is very difficult to determine many of the species to which he refers.* The Arabian philosophers turned some of their attention to this subject; and Avicenna has left us a Treatise on Minerals. After the revival of learning in Europe, mineralogy acquired a portion of the attention of naturalists, as well as botany and zoology; but much less progress was made in it, partly on account of the difficulty of the subject, and partly on account of the infant state of chemistry, on which mineralogy is obliged to depend for a good deal of her exact information, as far as the division of minerals into species is concerned

Linnæus. Linnæus's mineralogy was far inferior to his arrangement of the two other kingdoms of nature; though he first brought into view the importance of crystallization, which has since been laid hold of by the French mineralogists with such happy success in determining the species of miner ls. It is hardly worth while to notice the various systems of mineralogy which made their appearance in various countries in succession, after the publication of the Linnæan system.

Cronstedt. But the first system of sterling value was that of Cronstedt; who formed his arrangement according to the composition of the various stones. His method was adopted by Bergman, in his Sciagraphia; by Werner, and by almost all succeeding writers on the subject.

The manner of describing minerals, and the technical language by which the description is conveyed, were invented by Werner, of Freyberg. who published
External characters. his treatise on the *external characters*, in 1773. His mode of describing minerals has been universally adopted. As to the arrangement of minerals into species, two different methods are followed by the German and French schools.
Wernerian method. According to Werner, the species of minerals are merely artificial associations, for the conveniency of description. Accordingly, he has made them to depend
Haüyan method. upon a certain agreement in all the external characters. Haüy, on the other hand, conceives that minerals have been divided into species by nature herself, as well as animals and vegetables. Identity of form constitutes, according to him, the specific character. Accordingly, all those minerals which have the same primitive form (with certain exceptions) are placed under the same species. This method of Haüy recommends itself at once by its simplicity, and the exact limits which it enables us to assign to every species; whereas the Wernerians are obliged to admit, that different species pass, by imperceptible shades, into each other, and that many minerals exist which cannot be referred to any well-defined species, but lie intermediate between two. This is a necessary consequence of the opinion which they entertain that nature has not

---

* The late Dr. Walker, of the University of Edinburgh, understood Pliny's account of stones, better than most persons. His new Mineralogical Nomenclature consists chiefly of words taken from Pliny, and used in the same sense with that writer. In that point of view it deserves some notice.

divided minerals into species; but that species are merely artificial associations of minerals, contrived by the mineralogist for the convenience of arrangement and description.

After considering the subject with as much attention as was in my power, I have convinced myself that both parties carried their opinions too far, and that both methods, when properly circumscribed, might be made to harmonize in the construction of a better system of mineralogy than we possess at present. That no well defined species exist in nature, is an assertion which, if made without limitation, cannot be admitted. What can be better defined, or more constant in its characters and composition, than diamond, fluate of lime, tinstone, and many others which will readily occur to the recollection of every mineralogist. On the other hand, to make species in every case depend upon form, is to exclude from the mineral kingdom a great number of very important substances which never assume a crystallized form. Chalk, for example, fullers'-earth, and porcelain clay, according to the rigid rules laid down by the followers of Haüy, could not be admitted as mineral species, though they are of much more importance to society, and therefore better entitled to the attention of mineralogists, than many substances found in a well defined crystallized state. Let therefore the Haüyan method of determining species be introduced where it can; but let it not be introduced to the exclusion of the Wernerian, which enables us to distinguish those minerals that lie beyond the limits of the system of Haüy.

How far coincidence in form is sufficient to characterize crystallized minerals, is a point which does not yet seem to me to be completely determined. I am disposed to believe that it will prove sufficient, but am not quite certain. It is not impossible, that the number of primitive forms may be too small for the purpose, and that the number of subdivisions which it will be found necessary to introduce into the species according to that plan may be inconvenient. Identity of form may perhaps rather characterize the genus than the species. What would somewhat incline one to this opinion is, the great number of coincidences of the primitive forms of different species, which Haüy has lately pointed out in his *Tableau Comparatif*, and the consequent diminution of the number of his species. Thus he has united augite, sahlite, coccolite, and diopside, under one species; schillerstone, smaragdite, and bronzite, under another; zoizite and pistazite, under a third; and topaz and schorlous beryl, under a fourth.

The species of minerals at present known, and described by Werner, amount to about 300. But Haüy has reduced them to a much smaller number And there can be no hesitation in admitting, that several of the Wernerian species might, without impropriety, be united together. As for example, zircon and hyacinth, corundum and adamantine spar, beryl and emerald, &c. His subdivisions were made previous to the knowledge of the identity of the composi tion of these respective minerals, and he has not thought proper to alter it since.

Papers in the Transactions.

From the newness of this branch of mineralogy, we are not to expect much valuable information on the subject in the Philosophical Transactions. Indeed by far the best papers on the subject have made their appearance very recently, and are contained in the volumes published since the commencement of the present century, and which, of course, do not come under our review. There are 38 mineralogical papers contained in the Philosophical Transactions; but of these no fewer than 25 may be fairly neglected as of very trifling value. In the remaining 13 papers we are not to look for accurate descriptions of minerals; but the facts which they contain are useful in a historical point of view. One gives an account of the formation of cloth and paper from amianthus, discovered in Italy.* Another mentions the existence of a great quantity of native sub-carbonate of soda in the neighbourhood of Smyrna, which the inhabitants of the country employed in making soap.† Another paper contains a

Fullers'-earth in Bedford-shire.

description of the pits from which fullers'-earth is dug in Bedfordshire, by Mr. Holloway. These pits lie near Woburn, in a ridge of sand hills which run from Cambridge to Oxford, at the distance of about eight miles from the Chiltern hills. From the surface to the depth of about 14 yards, there occurs nothing but sand, through the middle of which, at about six yards from the surface, there runs a thin bed of red-coloured sandstone. The fullers'-earth lies under the sand, and is about eight feet thick. The first foot is so much mixed with sand as to be useless The upper part of the fullers'-earth is coloured yellow but it becomes lighter coloured as we descend deeper. Under the fullers'-earth is a bed of white rough stone about two feet thick, and below this occurs sand again; but no fullers'-earth occurs deeper. These beds are nearly horizontal, and extend a considerable way ‡

Discovery of diamonds in Brazils.

The high value attached to diamonds depends not so much upon their beauty and hardness, as upon their great scarcity, and the labour and expense necessary in procuring them. Hitherto they have been observed only in the torrid zone and Brazil is the only country in America where they have been found There is a paper in the Philosophical Transactions giving an historical account of their discovery in that country by Dr. De Castro Sarmento. Near the capital of the county Do Serro do Frio flows the river Do Milho Verde, where they used to dig for gold, or rather to extract it from the alluvial soil. The miners, during their search for gold, found several diamonds, which they were induced to lay aside in consequence of their particular shape and great beauty, though they were ignorant what they were. At last, in the year 1728, a miner

* Phil. Trans. 1671. Vol. VI. p. 2167.   † Phil. Trans. 1696. Vol. XIX. p. 228.
‡ Phil. Trans. 1723. Vol. XXXII. p. 419.

came to the country, who, suspecting these stones to be diamonds, made some experiments on the subject, and satisfied himself that his conjecture was well founded.　He set himself in consequence to search for diamonds in the alluvial soil of the country, and the other miners followed his example.　Diamonds were even found among the sand of the river though in less abundance.　Ever since that period the searching for diamonds in that country has been continued with good success.*　The specific gravity of diamonds was first accurately determined by Mr. Ellicot, who employed a balance for the purpose that turned with the 200th part of a grain.†　The following table exhibits the result of his trials :

Diamonds weighed.	Weight in grain.	Sp. Gravity.
1 A Brazil diamond, fine water, rough coat	92·425	3·518
2 A ditto, fine water, rough coat	88·21	3·521
3 Ditto, fine bright coat	10·025	3·511
4 Ditto, fine bright coat	9·560	3·501
5 An East India diamond, pale blue	26·485	3·512
6 Ditto, bright yellow	23·33	3·524
7 Ditto, very fine water, bright coat	20·66	3·525
8 Ditto, very bad water, honey-comb coat	20·38	3·519
9 Ditto, very hard bluish coat	22·5	3·515
10 Ditto, very soft, good water	22·615	3·525
11 Ditto, a large red foul in it	25·48	3·514
12 Ditto, soft, bad water	29·525	3·521
13 Ditto, soft, brown coat	26·535	3·516
14 Ditto, very deep green coat	25·25	3·521
Mean sp. gr. of the Brazil diamonds		3·513
Mean ditto of East India diamonds		3·519
Mean of both		3·517

Specific gravity of diamonds.

Our accurate knowledge of the precious stones cannot be dated further back than the publication of Romé de Lisle's Crystallography, and has been very much improved of late years.　Hence, a paper published in the Philosophical Transactions for 1747, by Mr. Dingley, giving a catalogue of the precious stones, and mentioning those employed by the ancients for engraving on, is an object of some curiosity.　As he gives no other description of the stones which he mentions, except their colour, his account is involved in some obscurity. The following is his catalogue :

Catalogue of the precious stones

Beryl, red and yellow.‡　　　　　　Jacinth, deep tawny red.
Chalcedony.　　　　　　　　　　　Chrysolite, light grass green.
Plasma, green with white spots.　　　Crystal, or oriental pebble, silver white.

* Phil. Trans. 1731. Vol. XXXVII. p. 199.　　　† Phil. Trans. 1745. Vol. XLIII. p. 468
‡ Probably jasper, cornelian, &c

Garnet, deep red claret.
Amethyst, purple.
Diamond, white.
Ruby, red or crimson.
Emerald, deep green.
Aqua marina, bluish sea green.

Topaz, ripe citron yellow.
Sapphire, deep sky blue, or silver white.
Cornelian, red or white.
Opal, white and changeable.
Vermillion stone, more tawny than jacinth.

All these are more or less transparent. **The following are opaque:**

Cat's eye, brown.
Red jasper, red ochre.
Jet, black.
Agates.
Blood stone, green, veined with red and white.
    Heliotrope?
Onyx, white and black.

Sardonyx, brown and white.
Agate onyx, two kinds of white, opaque and
    transparent.
Alabaster, white and yellow.
Toad's eye, black.
Turquoise, yellowish blue inclining to green.
Lapis lazuli, deep blue.

The ancients engraved usually on the beryl (jasper), sometimes on chalcedony, plasma, and jacinth. Rarely on the chrysolite, crystal, garnet, and amethyst. They also engraved on several of the opaque stones.*

Tourmaline, the lyncurium of the ancients. About the year 1757, in consequence of a dissertation on the subject published by Æpinus, and another by Wilke, the electrical properties of the tourmaline came to be accurately known. At that time, the stone itself was considered as very rare, and as a great curiosity, though it is now known to be a very common constituent of primitive mountains, and occurs abundantly in different parts of Great Britain. In the year 1759, Sir William Watson wrote a dissertation to prove that the *lyncurium* of the ancients, a stone described by Theophrastus, and noticed by Pliny, was nothing else than the tourmaline; and the evidence which he adduces from the properties of the stone, and the circumstances respecting these properties noticed by Theophrastus, leaves little doubt that his conjecture is well founded.†

Native tin. Tin is one of the metals which exhibits the fewest varieties of ores. Indeed almost all the tin used by artists is obtained from the ore called *tinstone*, which is an oxide of tin more or less mixed with foreign matter. Hitherto mineralogists have not admitted native tin into their systems. Yet, from a paper on the subject published by Dr. Borlase in the Philosophical Transactions, there can be no doubt that *native tin,* though it be very scarce and only in small quantities, yet does now and then occur. In that paper he describes three specimens of native tin found in Cornwall, in the centre of very large masses of tin ore. His description is so indistinct and unsatisfactory, that it is difficult to form any notion of the constituents of the specimens. However, from the experiments of

* Phil. Trans. Vol. XLIV. p. 502.   † Phil. Trans. Vol. LI. p. 394.

Γ Costa on the supposed tin, there can be no doubt that it was really that
metal. In the specimens, the tin was surrounded by a white substance, which
Borlase considered as quartz; but which, he says, turned out on examination to
be arsenic. A fourth specimen of native tin was afterwards found in the centre
of a transparent crystal of tin diamonds, as they are called in Cornwall. With
respect to this last specimen, there could be no doubt of its being natural.
Hence it puts the existence of native tin beyond dispute.*

Native lead is equally uncommon with native tin, and has not yet been ad- Native lead.
mitted into a mineralogical system. It deserves therefore to be noticed, that it
was found by Dr. Morris, though in no great quantity, interspersed in lead ore
(probably galena) from Monmouthshire.† This, as far as I know, is the only
example on record, where this very rare mineral was observed.

These are the only mineralogical papers occurring in the Transactions that
seem entitled to notice; and it must be confessed, that they are not of very great
importance.

SECTION II.—*Of Geognosy.*

The term *Geology* has been applied in two different senses. Naturalists have Meaning of
been always fond of speculating about the original formation of the earth, and the term.
about the changes which it has undergone since its original creation. Accordingly, various fanciful theories have been constructed in succession, either
founded entirely upon the imagination of the constructor, or partly upon imagination, and partly upon the account of the creation contained in the Old
Testament. These whimsical hypotheses have been dignified with the title of
theories of the earth, and the term *geology* has been very frequently applied to
them. These theories appeared in great numbers during the last century.
Burnet, Whiston, Woodward, Ray, Lazaro Moro, Buffon, Whitehurst, &c.
are the authors of the most celebrated. It is remarkable, that most of these are
to be found in the Philosophical Transactions. At present, the geological world
is divided between two opposite hypotheses respecting the original formation of
the earth. The first, at the head of which stands Werner (and with him almost all mineralogists agree in sentiment), conceives the surface of the earth to
have been gradually formed by successive depositions from an ocean which originally covered the whole surface; but which, in process of time, by some unknown and unaccountable means, withdrew and disappeared. The second, at
the head of which is Dr. Hutton (and with him coincide Lazaro Moro, Dr.
Hooke, Buffon, and a few other mineralogists), conceives, that the surface of the
earth has been formed by the operation of fire, which first melted the stony

* Phil. Trans. 1766. Vol. LVI. p. 35 and 305.—Vol. LIX. p. 47.
† Phil. Trans. 1773. Vol. LXIII. p. 20.

masses, and afterwards raised them to their present elevation. These two sects have been distinguished by the fanciful names of Neptunists and Vulcanists.

Present mean- ing.
The second meaning which has been given to the term *geology* is, an account of all the stony masses which compose the crust of the earth; the order in which they lie with respect to each other, and with respect to the substances which they contain. Geology, as thus explained, was cultivated with some success by the late Mr. Whitehurst; but, it is to the learned labours of Werner, of Freyberg, that we are indebted for almost all the progress which it has hitherto made. To distinguish this important branch of science from the absurd speculations about the formation of the earth, Werner has given it the name of *geognosy;* a name which we have adopted.

Papers in the Transactions on the subject.
The Philosophical Transactions, though they contain no regular treatise on geognosy, yet exhibit a very great mass of important information on the subject; information, indeed, which it is often difficult to appreciate, on account of the imperfection of the mineralogical nomenclature which the authors of various papers were obliged to employ, as no better existed when they wrote. The papers on the subject amount to no fewer than 251, of very various value; and I reckon no fewer of them than 85 to be trifling: the remaining 166, however, contain a great collection of facts, of which geognosts have not hitherto availed themselves sufficiently. I shall endeavour to point out the most important facts which they contain, as briefly as possible. Instead of noticing every paper individually, it will be more entertaining, as well as more concise, to arrange the subject under a small number of distinct heads.

Petrifactions.
I. PETRIFACTIONS. There is no fact connected with geognosy more interesting, or more extraordinary, than the existence of a vast quantity of animal and vegetable substances in the mineral kingdom. These substances occur in three states; sometimes they are a little altered; sometimes they are converted into stone; and sometimes only the impressions of them, or the moulds in which they have been inclosed, remain. The term petrifaction, strictly speaking, applies only to the second of these states; but it has been usual, for the sake of conveniency, to extend it to all the three. The number of papers on this subject, in the Philosophical Transactions, amount to 42. The following are the facts which they contain:

Vegetable.
1. *Vegetable petrifactions.* Wood occurs in great abundance in many parts of England, buried at various depths under the surface, and very little altered either in its texture or properties. Lincolnshire, and a considerable tract of country on the banks of the Humber and the Thames, contain abundance of trees, at no great depth below the surface. In the isle of Axholme, lying partly in Lincoln and partly in Yorkshire, oak, fir, and other trees, are frequently found under ground, in the moor. The roots are still found as they grew, in firm earth, under the moor; but the trees had been cut down. The

size of many of these trees is immense: some firs have been taken up 36 yards long. The quantity of these trees was so great, that, during the 17th century, many cart loads of them were taken up every year.* Oaks, and fir trees, are also dug up at Youle, about 12 miles below York; where the Dun runs into the Humber. They are at some depth, and lie in a bog, which is covered with sand, and the sand with soil.† The famous levels of Hatfield chase, in Yorkshire, consisting of 180,000 acres, were, during the 17th century, drained at a vast expence, by Sir Cornelius Vermuiden, a Dutchman. Even as low as the bottom of the Ouse, and in the whole marsh, abundance of firs, oaks, birch, beech, yew, thorn, willow, ash, &c., were found.‡

De la Pryme endeavours to account for the existence of so much wood all over that tract of country, by supposing, that when the Romans were employed in the conquest of Britain, the whole country was covered with forests, in which the Britons lurked, and from which they were accustomed to make frequent incursions upon the Romans. To put an end to these troublesome attacks, the Romans cut down all the woods, and left the trees as they fell. Soil and vegetable matter gradually accumulated over these trees, and preserved them from destruction.§ This conjecture must be allowed to possess plausibility. The trees must be admitted to be destroyed on purpose, as many of them exhibit the marks of the hatchet, and many of them show evident traces of fire; besides, various remains of Roman cutting instruments, and other utensils, are found intermixed with the prostrate trees.

About the year 1705, there was an inundation of the Thames, at Dagenham and Havering marshes, in Essex, which made an excavation nearly 20 feet deep, and laid open a great number of trees, mostly alder (as was supposed from their texture), buried under a soil, obviously composed of the mud of the Thames. Trees, in a similar situation, have been discovered all along the Essex banks of the river. Dr. Derham, to whom we owe the account of these trees supposes, with considerable plausibility, that they had been overturned by some inundation of the Thames, and afterwards covered by the repeated overflowings of the river, which took place every tide, before its waters were confined by means of artificial banks.‖

It is a common opinion among geologists, that pit coal is of vegetable origin, and that it has been brought to its present state by means of some chemical process, with which we are still unacquainted. There is one circumstance which gives this opinion, though at first sight it may appear extravagant, consider-

---

* Phil. Trans. 1671. Vol. V. p. 2050.

† Dr. Richardson. Phil. Trans. 1697. Vol. XIX. p. 526.

‡ Rev. Abraham de la Pryme. Phil. Trans. 1701. Vol. XXII. p. 980.

§ Phil. Trans. 1701. Vol. XXII. p. 980.   ‖ Phil. Trans. 1712. Vol. XXVII. p. 478.

able plausibility, we mean, the existence of vast depositions of matter, half way as it were between perfect wood and p rfect pit coal ; betraying obviously its vegetable nature, and yet so nearly approximating to pit coal in several respects, that it has been generally distinguished by the name of coal: One of the most remarkable of these depositions exists in Devonshire, about 13 miles south-west Bovey coal. of Exeter, and is well known under the name of Bovey coal. It has been very well described by Dr. Mills, in the Philosophical Transactions ; and its vegetable nature has been ascertained by Mr. Hatchett, in a set of experiments which, unfortunately, do not come under our review.

The beds of coal are 70 feet thick, but there are beds of clay interspersed On the north side they come within a foot of the surface, and dip south at the rate of about 20 inches per fathom. The deepest beds are the blackest and heaviest, and have the closest resemblance to coal. The upper resemble wood strongly, and are considered as wood by the people who dig them. They are brown, and become exceedingly friable when dry. They burn with a flame similar to wood. They have exactly the appearance of wood which has been rendered quite soft by some unknown cause, and, while in this state, has been crushed flat by the weight of the incumbent earth. This is the case, not only with Bovey coal, but it holds also with all the beds of wood coal, hitherto observed in every part of the earth. Dr. Mills and Dr. Miller have endeavoured, by subjecting Bovey coal to destructive distillation, and comparing the effects with the destructive distillation of wood, to prove, that it has not been formed from the vegetable kingdom, but that it is purely and originally a mineral.* But their arguments are not only inconclusive, but inconsistent with many circumstances connected with the coal. Mr. Hatchett found in it both *extractive* and *resin ;* substances peculiar to the vegetable kingdom.

Petrified wood at Lough Neagh. In the neighbourhood of Lough Neagh, in Ireland, very large masses of petrified wood are found, all of them siliceous, and of course exceedingly hard. The grain of the wood is still very distinct ; and, in some specimens, pieces of wood still unchanged may be detected. The wood is of various kinds : holly, ash, and oak, are mentioned by different writers ; but it is no easy matter to determine, with accuracy, the species of wood from a petrified specimen. On the petrified wood of this lake, there are no fewer than four papers in the Philosophical Transactions ;† the best of which is that by Mr. Simon. In these papers, an attempt is made to prove that Lough Neagh possesses the property of petrifying wood, when left for some time in it ; and this property is even ascribed to the sand, which extends to some distance on the sides of the

* Phil. Trans. 1760. Vol. LI. p. 534 and 941.

† Molineaux. Vol. XIV. p. 552. (1684.)—Smyth. Vol. XV. p. 1108. (1685.)—Simon. Vol. XLIV. p. 305. (1746.)—Bishop Berkeley. Vol. XLIV. p. 325. (1746.)

Lough. But there is no foundation whatever for this opinion. The petrifactions are of a very old date. The waters of the Lough in no respect differ from other waters.

Besides the petrified wood from Lough Neagh, concerning the reality of which there can be no doubt, there is an account, in the Transactions, of a great variety of petrified fruits, dug up in the isle of Sheppey. The account is by Dr. Parsons; and is so imperfect that it is difficult to make any thing out of it. There can be little doubt, however, that these supposed petrifactions are entirely fanciful; and nothing else than accidental imitations of the fruits in question. The following is a list of the supposed petrified fruits, described and figured by Parsons.*

Fig.	Water melon.	Horse chesnut.
Myrobalon.	Plum stone.	Yellow myrobalon.
Phaseolus.	Cherry stone.	Halmo coco.
Seed of an American gourd.	Euonomus.	Walnut.
Coffee berries.	Soap-tree berry.	Ear of corn, or of
Beans.	Germen huræ.	grass.
Underground pea.	Mango stone.	Cocculus indicus.
Acorn.	Long bean.	

2. *Shells and Zoophites.* These are by far the most abundant of all the animal remains which occur in the mineral kingdom. They occur most frequently in lime-stone rocks, though they are not entirely confined to them. I have seen them pretty often in slate clay, and bituminous shale. These substances occur in the mineral kingdom in two states: sometimes they are unaltered; sometimes they are petrified, or converted into stone. The stony matter is most commonly carbonate of lime; but very frequently also it is pure silica. This, in particular, is the case with most of the numerous petrifactions of echini, and other similar animals, which occur so abundantly in chalk. Some mineralogists go so far as to suppose, that all the flints which occur in such quantities in chalk, and in lime-stone, are petrifactions. But this has not been proved, nor even rendered probable; though it must be acknowledged, that petrifactions very frequently occur in flints. Some kinds of lime-stone are wholly made up of shells; and, in the South Seas, the coral accumulates in such quantities as to form islands of considerable size. The papers in the Transactions, respecting the depositions of shells and coral, in various parts of the kingdom, are not of very great value, though the facts are worth enumerating.

Dr. Lister gives a description of *trochitæ* and *entrochi*, or *St. Cuthbert's beads*, as they are usually called in this country, hich occur abundantly in

the scars, at Broughton and Stock, small villages in Craven.*   They consist of
lime-stone, and are obviously remains of different species of the *isis*, especially
the *entrocha*, which is usually found petrified.   They occur also in abundance
in Holy Island, upon the coast of Northumberland, where there is a bed of
lime-stone, that is thickly planted with them.   By the sailors on that island,
they are considered as a preservative from drowning, and therefore sought after
with eagerness.

Lister also gives an account of *glossopetræ*, or *shark's teeth ;* which occur in
considerable quantity in the isle of Sheppey, and in a quarry near Malton.†
These, indeed, ought rather to be reckoned among bones than shells; but, in
mineralogical collections (probably on account of their size), they are usually
put along with shells.

To the same indefatigable conchologist, we owe an account of the *astroites*,
or *star stone ;* also a species of zoophite, consisting of lime-stone; which he
found deposited in beds of clay, in the Yorkshire wolds.‡

At Reading, in Berkshire, there is a bed of oyster-shells, lying over chalk,
a circumstance by no means common.)   This bed is two feet thick, and is
covered by the following beds: 1. clay ; 2. fullers'-earth ; 3. sand ; 4. red clay
for bricks, which constitutes the bed at the surface of the earth.   The oyster-
shells are very brittle, and consist entirely of carbonate of lime.   I presume,
though Dr. Brewer, to whom we owe the account, makes no mention of the
circumstance, that all the animal membrane, which fresh oyster-shells contain,
had disappeared.   This is usually the case with shells, when found in similar
situations, and is the cause of their brittleness.§

At Broughton, in Lincolnshire, there are two quarries, containing abundance of fresh-water shells.   These shells occur in a blue stone, which De la
Pryme. to whom we owe the account, conceives to have been formerly in the
state of clay, and to have become gradually indurated. The shells are pectinites,
echini, conchites, some pieces of coral, and shells of the fresh-water muscle,
(probably the margatifera).‖

About a mile from Reculver, in Kent, there occurs a bed of shells, consisting entirely of white conchites, and very brittle.   This bed is twelve feet thick,
lies in a greenish sand, and contains here and there in it pieces of wood.¶

The cliff at Harwich, at the side of the entrance of the river, is about 50 feet

---

* Phil. Trans. 1673. Vol. VIII. p. 6181.   † Phil. Trans. 1675. Vol. IX. p. 221.
‡ Phil. Trans. 1675. Vol. IX. p. 274.   § Phil. Trans. 1700. Vol. XXII. p. 484.
‖ Phil. Trans. 1700. Vol. XXII. p. 677.   The names of the shells, given in the text, are Lister's.
Not having Dr. Lister's book at hand, it is not possible to give the Linnæan synonimes.   Nor,
indeed, could they be given with safety, as De la Pryme enumerates only genera; and does not
venture to give specific names.
¶ Stephen Gray.   Phil. Trans. 1701. Vol. XXII. p. 762.

high, and consists of sand. In this sand are found a great variety of shells.
Mr. Dale has given us a catalogue of them, to the number of 28. The following are the names of genera which he found, according to the method of Lister,
at that time generally followed : 10 species of buccina, 2 of cochleæ, 1 nerita,
1 turbo, 1 pecten, 1 auricularia, 7 pectunculi, 4 conchæ, 1 trigonella.*

On digging a moorish pasture, in Northamptonshire, abundance of snail and
river shells were found. The digging was not continued farther than three feet
but the proportion of shells always increased the lower down they got. The
experiment was tried over a considerable extent of ground, with the same effect.
The shells found, distinguished by their Listerian names, were, buccinum exi-
guum, cochlea umbilicata, cochlea citrina, and common striped snail shell
The river shells were, periwinkle of three wreaths, and a periwinkle of five.
Fresh shells, of the same species, are still found in the neighbouring peat soil.†
The preceding account is by no means satisfactory ; it is probable, that the place
examined had been, at some preceding period, overflowed with water, and the
shells left mixed with the mud and sand, after the retreat of the water.

There is a species of petrifaction distinguished by the name of *belemnites ;*
usually cylindrical, or conical ; sometimes containing a hollow nucleus, divided
into compartments ; sometimes not. They are often of considerable length,
consist of carbonate of lime, and are found imbedded in chalk, sand-stone,
sand, and clay. Considerable doubts, respecting their nature, were entertained
by mineralogists. Da Costa endeavours to prove, that they are not petrifactions,
but merely minerals, which have accidentally assumed a particular shape.‡
But Mr. Baker, junior, described two belemnites, from a chalk-pit, near Norfolk ; having, the one, an oyster-shell, and the other, two of those vermiculi
commonly found on sea shells, attached to it. Hence he infers, with considerable probability, that the belemnites themselves are marine productions.§ Mr.
Platt examined the subject with much attention, and has shown, in a very con-
vincing manner, that the belemnites are real marine petrifactions. He has
rendered it probable, that they constitute a specie of nautilus ; and, on that
supposition, has explained their formation in a satisfactory manner.∥ The
belemnites occur very frequently in the coarser kinds of marble ; and may be
often seen, of considerable size, in old marble chimney-pieces. To the belemnites may be referred another similar petrifaction, the *orthoceratites,* which
occurs likewise in marble ;¶ and which is found at Kelwick, near Fulham.**

Much curious information, respecting the occurrence of shells in the mineral

* Phil. Trans. 1704. Vol. XXIV p. 1568.   † Morton. Phil. Trans. 1706. Vol. XXVIII. p. 2210
‡ Phil. Trans. 1747. Vol. XLIV. p. 397.   § Phil. Trans. 1748. Vol. XLV. p. 598.
∥ Phil. Trans. 1764. Vol. LIV. p. 38.   ¶ Wright. Phil. Trans. 1755. Vol. XLIX. p. 670
** Himsel. Phil. Trans. 1758. Vol. L. p. 692.

kingdom, has been lately given to the public, by La Marke and by Cuvier, and Brogniart.* These two last philosophers have given a most interesting description of the structure of the country round Peris; and have drawn some curious inferences from the alternate occurrence of fresh and salt water shells, in different beds. Mr. Parkinson has given a similar account of the soil round London ;† a subject, not so interesting as the country round Paris, from the different nature of the beds; but still possessed of considerable interest and importance.

Bones.

3. *Bones.* The bones of animals, both sea and land, occur also in considerable quantity in the mineral kingdom; though the quantity is by no means comparable to that of shells and zoophites. Bones are seldom petrified; for they still retain their phosphate of lime, which constitutes, in some measure, their distinguishing character. But they have undergone a great change; for most commonly (though not always) they are destitute of their gelatinous and cartilaginous constituents, and are precisely in the same state as if they had been exposed to a red heat in the open air. The facts respecting mineral bones, detailed in the Transactions, possess considerable importance.

The head and horns of an unknown species of stag, much larger than any European species, is found abundantly under ground in Ireland, usually in a kind of marl. In one of these skeletons, measured by Molineux, the length, in a straight line, from the tip of one horn to that of another, was 12 feet. The horns were both palmated. Molineux conceived, that these skeletons belonged to the American moose deer.‡ But Pennant has shown, that it differs in various particulars from the head and horns of the moose; and that it belongs to a species of deer unknown in the live state, and probably extinct.

The bones of elephants have been found, at various times, scattered over most parts of Europe; a fact which has puzzled naturalists considerably, because the climate is at present too cold for these animals; and their remains, of course, seem to indicate some change in the nature of the earth, or, at least, in the climates. Much light has been thrown upon this obscure subject, by the anatomical knowledge of the Hunters, and, of late, by the indefatigable labours of Cuvier. These philosophers have demonstrated, that various bones, considered at first as belonging to elephants, certainly belong to an animal totally different from any species of elephant at present known. Cuvier has constituted this animal a peculiar genus, and more species of it than one seem formerly to have existed; though the whole genus appears to have been extinct, from a very remote period. Still, however, it can scarcely be doubted, that the bones of

---

* See the Annales de Museum d'Hist. Nat. passim.
† Transactions of the Geological Society of London. Vol. I.
‡ Phil. Trans. 1697. Vol. XIX. p. 489.

real elephants are occasionally dug up in various parts of Europe. The follow-ing examples are recorded in the Philosophical Transactions:

The skeleton of an enormous elephant, which is conceived to have been 24 feet high, was found in white sand at the bottom of a mountain in Thuringia.* Four elephants' teeth were dug up in Ireland. They were found about four feet under ground, near a small brook which divides the counties of Cavan and Monaghan.† The tusk of an elephant was dug up at the end of Gray's-Inn-lane, London, and another in Northamptonshire, together with a grinder of the same enormous animal.‡ The bones of an elephant were found under ground in the Isle of Sheppey on the sea coast. They speedily fell to powder when exposed to the air, but their size and shape had been previously ascer-tained.§ Sir Hans Sloane endeavours to prove, and his proofs are pretty con-vincing, that the bones of supposed giants, found in various parts of the world, and described by different authors, were in reality elephants' bones.‖

Near the river Ohio, in America, at a place called the Great Buffalo lick, a great number of bones belonging to some enormous animals were discovered by Croghan.¶ These were at first supposed to be elephants' bones. But Mr. John Hunter, having examined the teeth, pronounced them to belong to a carnivorous animal. Dr. Hunter published a dissertation on the subject, founded chiefly on his brother's observations,** and since that time they have been admitted to be-long to some enormous unknown animal, and the name *Mammoth*, applied to the bones of a large animal found in Siberia, has been applied to them. Cuvier has shown that the American and Siberian Mammoth belong to different species. Some years ago a complete skeleton of the American Mammoth was exhibited in London by Mr. Peale.

A vast collection of fossil bones have been found in the rock of Gibraltar. They were conceived to be human bones; but, Dr. Hunter having examined them, demonstrated that they belonged to a quadruped.††

A stag's head and horns were found at Matlock, Derbyshire, on a rock which, from the description of it given by Mr. Barker, seems to be calcareous tuffa.‡‡

In the mountain of St. Peter, near Maestricht, were found in the year 1770, abundance of fossil bones, which Camper showed to belong to a kind of physe-

---

* Tentzel. Phil. Trans. 1697. Vol. XIX. p. 757.
† Neville and Molyneux. Phil. Trans. 1715. Vol. XXIX. p. 367 and 370.
‡ Sloane. Phil. Trans. 1728. Vol. XXXV. p. 457.   § Jacob. Phil. Trans. 1754. Vol. XLVIII. p. 662,
‖ Phil. Trans. 1728. Vol. XXXV. p. 497.   ¶ Phil. Trans. 1767. Vol. LVII. p. 464.
** Phil. Trans. 1768. Vol. LVIII. p. 34.   †† Phil. Trans. 1770. Vol. LX. p. 414.
‡‡ Phil. Trans. 1785. Vol. LXXV. p. 353.

ter.* A splendid work was afterwards published on these bones by Fojas de St. Fonde.

Mosses.

II. Mosses. The great abundance of mosses in the colder parts of the earth, and the rapidity with which in certain circumstances they are renewed, while in the torrid zone they are totally wanting, are circumstances claiming the close attention of geologists. Pit-coal, in like manner, abounds in cold climates; but (as far as I know) has never been found in the torrid zone. Hence the opinion entertained by some that mosses by length of time, and by being exposed to considerable pressure, are converted into pit-coal, has at least some little plausibility in its favour. That peat is entirely of vegetable origin cannot be doubted; that it consists chiefly of *sphagnum palustre*, and other similar plants which delight in moisture, is probable, and that it is formed by the action of water on dead vegetable matter appears abundantly evident. But an accurate explanation of the suite of changes which take place during the formation of peat has not been given. A precise chemical examination of peat, and a comparison of its different varieties with those of pit-coal, would be a valuable addition to geological knowledge.† In the mean time it cannot be doubted that mosses act a very important part in nature, and that they afford us the means of detecting many curious changes which have taken place in process of time on the surface of the earth.

Mosses always occur on plains, though these are frequently situated at some considerable height on the side of a mountain. They are always in the neighbourhood of ground higher than themselves, and are obviously the receptacle of water collected from neighbouring acclivities. They frequently contain many large trees, chiefly oak and fir, and this even in countries where trees will no longer grow, as the Orkneys and Western Islands of Scotland.‡ There is in an early number of the Philosophical Transactions a very valuable paper on the origin of mosses, by the Earl of Cromartie. And this nobleman, who was about the age of 80 when he wrote, had the singular good fortune to wit-

* Phil. Trans. 1786. Vol. LXXVI. p. 443.

† I have made some experiments on the subject, but never could find any of the constituents of peat, mentioned by Dr. Rennie in his work on *peat*, in any of the varieties of that combustible which I was able to procure. I have examined peat from Lancashire, from Perthshire, from the neighbourhood of Stirling, from the neighbourhood of Edinburgh, and from a moss between Stirling and Glasgow. Dry peat is tasteless and insoluble in water. But. if you steep it in water for several months it tinges the liquid reddish brown similar to moss water. In water thus tinged I never could detect any thing by re-agents, except some slight traces of vegetable *extractive*.

‡ The reason why wood does not grow in these places is not any change of the climate for the worse, but an increase of the number of sheep and cattle, which prevent trees from propagating themselves by seeds, the only way in which forests can be continued in a cold climate. Young trees brought from a distance and planted are sure to die.

ness with his own eyes all the different steps of the process within a period of rather less than 50 years.

In the year 1651, when his Lordship was 19 years old, going from a place Origin of peat. called Achadiscald to Gonnazd, in the parish of Lochbrun, he passed by a very high hill, which rose in a constant acclivity from the sea. At less than half a mile up from the sea there is a plain about half a mile round, and from it the hill rises in a constant steepness for more than a mile in ascent. This little plain was at that time all covered over with a firm standing wood, which was so very old that not only the trees had no green leaves, but the bark was quite thrown off; which the old countrymen, who were with his Lordship, said was the universal manner in which fir woods terminated, and that in 20 or 30 years after the trees would commonly cast themselves up from the roots, and so lie in heaps till the people cut and carried them away. About 15 years afterwards his Lordship had occasion to come the same way, and observed that there was not a tree, nor even a single root of all the old wood remaining; but instead of them, the whole bounds where the wood had stood was all over a flat green ground covered over with a plain green moss. He was told that nobody had been at the trouble to carry away the trees; but that being all overturned from their roots by the winds, the moisture from the high grounds stagnated among them, and they had in consequence been covered over by the green moss. His Lordship was informed that nobody could pass over it because the scurf of the *fog* would not support them; but he thought proper to make the experiment, sunk in consequence up to the arm pits, and was drawn out by his attendants. Before the year 1699, the whole piece of ground was turned into a common moss, and the country people were digging peats out of it. At first they were soft and spongy, but gradually improved, and at the time when his Lordship wrote (1711) they were good.*

This is the most valuable paper on mosses that I have ever met with. His Lordship's observations and conclusions are confirmed by Sir Hans Sloane, who gives a description of the mosses in Ireland, and shows them to be similar in every respect to the mosses in Scotland.†

The only other paper on mosses in the Philosophical Transactions which deserves Irruption of particular notice is written by the late Dr. Walker, Professor of Natural History the Solway moss. in the University of Edinburgh. It gives us an account of the irruption of the Solway moss, which took place on the 16th of December, 1772. The Solway flow contains 1,300 acres of very deep and tender moss, which before this accident were impassable even in summer to a foot passenger. It was mostly of the quag kind, which is a sort of moss covered at top with a turf of heath and coarse aquatic grasses; but so soft and watery below, that if a pole is once thrust

Phil. Trans. 1711. Vol. XXVII. p. 296.   † Phil. Trans. 1711. Vol. XXVII. p. 302.

through the turf it can easily be pushed, though perhaps 15 or 20 feet long, to the bottom. The surface of the flow was at different places between 50 and 80 feet higher than the fine fertile plain between it and the river Esk. About the middle of the flow were the deepest quags, and there the moss was elevated higher above the plain than in any part of the neighbourhood. From this to the farm called the Gap, upon the plain, there was a broad gully, though not very deep, through which a brook used to run. In consequence of an uncommon fall of rain the moss being quite overcharged burst at these quags, about eleven o'clock at night; and, finding a descent at hand, poured its contents through the gully into the plain. It surprized the inhabitants of twelve farms in their beds. Nobody was lost, but many persons saved their lives with great difficulty. By the next morning 200 acres were entirely overwhelmed, and in a few days it overspread about 400 acres. The land covered was all enclosed with hedges, and bore excellent crops of wheat and turnips. The greatest part of the plain was covered with 15 feet of moss, and in some hollows the moss could not be less than 30 feet deep. The ridge of the flow itself subsided in consequence of this inundation about 25 feet.[*]

The existence of petrifactions in such abundance, incorporated with different rocks, is a demonstrative proof that this earth which we inhabit has undergone great changes since its original formation. Now what are the sources of these changes? There are three sources of changes which immediately present themselves to our view as soon as we consider the subject; namely, earthquakes, volcanoes, and inundations; and to one or other of these sources, or to all of them united, geologists have been in the habit of ascribing these changes. Let us consider each of them in succession, as far as the information contained in the Philosophical Transactions on the subject will enable us to go.

Earthquakes.   III. EARTHQUAKES. If we consider the violent effects which earthquakes produce, especially in hilly countries, where their effects are most perceptible, we cannot hesitate to admit that considerable changes in the position of some of the strata at the surface of the earth may have originated from them. The number of papers on earthquakes in the Philosophical Transactions amount to about 130. But, the greater number of these containing hardly any other information than the existence of an earthquake in a particular spot at a given period, I did not include them among the number of geognostical papers, because very little geognostical information can be deduced from them.

With the exception of a few descriptions of earthquakes contained in the earlier volumes of the Transactions, almost the whole of these papers refer to the period between 1750 and 1756 inclusive, a period remarkable for the frequency and the violence of the earthquakes which were felt in almost every part

---

[*] Phil. Trans. 1772. Vol. LXII. p. 123.

of the civilized world.   No fewer than 55 of them relate to the earthquakes which were felt all over the south-east of England during the year 1750, and which some of the writers distinguish emphatically by the name of the year of earthquakes.   There are 45 papers which give an account of the earthquake which destroyed Lisbon on the 1st of November 1755.   This appears, from all these descriptions, to have been one of the most extensive earthquakes ever felt, having extended itself every way not less than 4,000 miles.

As to the phenomena of earthquakes, they scarcely admit of a general description.   It has been supposed by some that they only take place in very dry seasons; but this does not appear to be a well established fact.   To the same place they seem always to come in the same direction.   Thus every earthquake felt at Lisbon has come from the north-west.   The agitation of the earth lasts only for a very short time, but it is often repeated two or three times within a very small interval.   The effect has by most persons been compared to an undulation passing along the surface of the earth, and different persons affirm that they have traced the progress of this undulation for a considerable way by the nodding of the trees in succession.   The earthquake is always attended by a loud noise.

As to the cause of earthquakes we are not in possession of any very satisfactory theory.   Dr. Stukely, who has inserted a paper on the subject in the Philosophical Transactions, afterwards published separately, has endeavoured to account for them by means of the Franklinian theory of electricity.   But his reasoning is so loose and unsatisfactory, that it is not entitled to a serious examination.   Neither is the hypothesis of Dr. Hales more plausible or better supported.   He conceives the phenomenon to be occasioned by a sudden vacuum formed in the atmosphere.   In dry seasons he thought that sulphureous exhalations rose in abundance from the earth.   These, he conceived, might make their way into the atmosphere enveloped in clouds; but the moment atmospherical air was admitted to them a condensation would take place, which might be so extensive as to occasion an earthquake.   This reasoning was founded on an experiment which Dr. Hales had made many years before.   By mixing together nitric acid and iron pyrites, an effervescence was produced, and a particular air generated, which remained without any alteration while confined over water; but, when mixed with atmospherical air, assumed a fine reddish brown colour, and diminished in bulk.   The air which he had thus formed he conceived to be the same with the supposed sulphureous exhalations which rise from the earth. The mistake of Dr. Hales was afterwards clearly shown by the discoveries of Dr. Priestley and Mr. Lavoisier.

The air which Dr. Hales had generated is the same substance which is now distinguished by the name of *nitrous gas*.   It is produced not from the pyrites, but from the decomposition of the nitric acid.   When mixed with atmospherical air, it unites with the oxygen of that air, and immediately forms again nitric

Cause of
earthquakes.

acid.   Hence the red fumes and the diminution of bulk.   For the nitric acid, as
soon as formed, is absorbed by the water over which the experiment is made.
But we have no proof that nitrous gas is emitted from the earth: on the contrary, we have every reason to believe that it is not. But even supposing nitrous
gas by some unknown means to be introduced into the atmosphere in considerable quantity, and to be suddenly mixed with the atmospherical air, the consequence would, no doubt, be a violent tempest, the air rushing from all quarters
to fill up the vacuum; but no reason can be adduced why such a vacuum
should occasion any motion in the earth, far less a motion so extensive as to be
felt over 4,000 miles of the earth's surface.

We must conceive the cause of earthquakes to be lodged within the earth's
surface.   And certainly no cause capable of producing the effect presents itself
to us with so much facility as the sudden generation of an enormous quantity
of elastic fluid within the bowels of the earth.   Mr. Michell has written a
very elaborate and ingenious dissertation on this subject,* in which he endeavours to convince his readers, that earthquakes are occasioned by the sudden inundation of water upon the fires which he conceives to exist in considerable
numbers at various depths under the surface of the earth.   The water thus
heated being converted into steam, he supposes, forces its way between the different strata of the earth which he conceives to be horizontal, and, heaving
them up in succession, may occasion an earthquake over a considerable extent
of country.   When the internal fire lies deep under the surface, the earthquake
will move with greater velocity; when it is more superficial, it will move over a
smaller tract of country, and with less velocity.   Where the strata raised by
this steam are covered by a smaller number of incumbent strata, the earthquake
will be more violent. This he conceives to be the case with mountainous countries; and if the strata be very thin, the fire may even break out and constitute a
volcano.   This frequently happens on the summit of mountains.   And the irruption of a volcano always puts an end to the earthquake for the time.   Of
this Mr. Michell brings several instances from Peru.   The internal fire which
occasioned the earthquake of Lisbon he conceives to exist under the Atlantic
Ocean, in a direction north-west from Lisbon, and at the distance of about a
degree.   As to the depth, there are no data to determine that point, but he
guesses it to be intermediate between a mile and a half and three miles.

Though this hypothesis of Mr. Michell is attended with many difficulties,
which it is not possible at present to explain, yet it must be allowed to be the
most plausible solution of earthquakes hitherto proposed.   Werner conceives
all volcanoes to be situated in a coal country; and explains them, by supposing
the coal, at an immense depth, by some means or other to be set on fire.   But

* Phil. Trans. 1760. Vol. LI. p. 566.

there are several circumstances which militate against this very simple and ingenious explanation. We have never heard of any coal occurring in tropical countries; yet it is well known that volcanoes are very common in these countries. The account which Humboldt gives of the composition of several volcanic mountains, in South America, is utterly irreconcileable with their focus being in coal, unless we admit that coal may exist below granite; a phenomenon which has never been observed. But, if coal be not the combustible substance, we can form no idea of what the matter can be which supplies so many internal fires with fuel for so many ages.

An idea has been thrown out that the combustible matter may be potassium, sodium, and the metallic bases of the earths; which would take fire by mere contact with water, and therefore might be conceived to burn any length of time, without the assistance of external air. That such a combustion is possible, cannot be denied; but it is scarcely reconcileable with the nature of the substances thrown out of volcanoes. We never find great quantities of potash, or soda, among the lavas of volcanoes, nor of perfect glass; as we ought to do, on the supposition that the bases of the alkalies, or alkaline earths, supplied the materials of the combustion. Hence there does not appear any reason to believe that these substances supply volcanoes with fewel. Nothing remains, therefore, but the bases of silica and alumina; which are still hypothetical, and concerning which, it would, consequently, be absurd to speculate. Besides, the quantity of sulphureous vapours, of sulphureous acid, and of sulphur, which are emitted from volcanoes, announces the presence of that combustible substance.

But, if sulphur, iron, and charcoal, (as is most likely) be the food which supplies volcanoes with the means of burning, it is impossible to explain the existence of subterraneous fire at great depths under the earth, and quite unconnected with the external air. That they do in some cases exist, in such situations, must be admitted; but every attempt hitherto made to explain how, must be allowed to be unsatisfactory. Upon the whole, I feel rather inclined to adopt the opinion of Mr. Michell, respecting the cause of earthquakes; though that opinion still leaves much to account for.

Dr. Hutton, of Edinburgh, and his disciples Mr Playfair and Sir James Hall, have got rid of these difficulties, by conceiving the whole central part of this globe to be a mass of *solid fire*. But this supposition, independent of its inexplicability, leads to consequences which are inconsistent with the phenomena of nature; and of course overturn it. To reconcile this red hot nucleus with the phenomena of earthquakes, it is obvious that it must extend to within a few miles of the surface of the earth. We are not sufficiently acquainted with the conducting powers of the earth to know what absolute effect such a central mass would have upon the temperature of the surface of the globe; but we are certain that it would have considerable effect. It is true, that we may conceive the waste of heat, from the surface, just to balance the quantity proceeding from the centre.   On such a supposition, the heat of the surface indeed would not increase, but it would certainly remain stationary.   But this is not the case, we find the heat of the surface to vary, and to be entirely regulated by the sun, indicating irresistibly that the temperature is derived from that luminary, and not from the centre of the earth.   Neither are the polar regions the hottest, as they ought to be on the supposition of a central fire; since they are much nearer the centre than the equatorial regions; but the coldest, because the sun's rays strike them most obliquely.   Neither do we find the temperature of the earth increase, the deeper we dig in mines, as it must do, if the heat proceeded from a central fire.   On the contrary, we speedily come to a certain temperature, (always the mean temperature of the country in which the mine is situated) which never varies, how deep soever we proceed.   These, and various other phenomena which will readily occur to the reader, appear to me so utterly irreconcilable with the hypothesis of a central fire, as to oblige every person who weighs them with sufficient care to abandon that hypothesis entirely.

Volcanoes.   IV. VOLCANOES. Volcanoes constitute, without doubt, the most striking and formidable geognostic phenomenon which nature has presented to our view. They are not, indeed, so destructive to the lives of the human race as earthquakes; but they present the eye with something much more terrific. Their number is very considerable, nearly two hundred having been reckoned by different writers.   There is an immense range of them running from north to south, on the continent of America, and occupying the summits of many of the Andes, and the mountains of Mexico and California.   There is a considerable number along the east coast of Asia, and in the Indian islands.   Iceland alone contains eight volcanoes.   The two volcanoes with which we are best acquainted, are those of Ætna and Vesuvius.   Ætna has been burning as far back as the records of history go.   We have an account of an eruption during the expedition of the Argonauts; which took place at least 12 centuries before the commencement of the Christian æra.   The following dates of remarkable eruptions of this volcano are given in a paper published in an early volume of the Philosophical Transactions:

> 476 years before Christ.   Mentioned by Thucydides.
> 40 years after Christ.   During the reign of Caligula.
> 812 years after Christ.   During the time of Charlemagne.
> 1284
> 1329
> 1444
> 1536
> 1633
> 1650*

* Phil. Trans. 1669. Vol. IV. p. 967.

The greatest eruption of Ætna, which has taken place for many years, happened in 1669; when, by the torrent of lava, the town of Catanea was destroyed. This eruption was particularly described by Borelli; and an abstract of his account is given in the Philosophical Transactions.* Two subsequent eruptions, namely, those of 1755 and 1766, are described in subsequent volumes of the Transactions.

Vesuvius is a much smaller mountain than Ætna; but its eruptions are more frequent, and not less formidable. The first account of it, as an active volcano, is that dreadful eruption in which Pliny the elder lost his life; which happened during the reign of Titus, about the year 72 of the Christian æra. Though no previous mention of Vesuvius, as a volcano, is any where to be found on record, yet it is not supposed that the mountain began only to be a volcano at the time of the eruption which destroyed Pliny. Pliny the younger, who gives a very circumstantial account of the whole affair, in two letters to Tacitus, never once hints at any such thing; nor was the least surprise expressed at the eruption, which, undoubtedly, he would have done, had it been the first that ever happened. But, it is probable, that the mountain had lain quiet for a considerable number of years before; and this, in some measure, accounts for the consternation that seized all those who lived in its vicinity. In the Philosophical Transactions, there is an account of no fewer than nine eruptions of this mountain, which took place during the last century. The dates of these eruptions are as follows: 1707;† 1737;‡ 1751;§ 1754;‖ 1760;¶ 1766;** 1767;** 1779;†† 1784.‡‡

The last four of these eruptions are described by Sir William Hamilton; who, during a period of about 40 years, attended so assiduously to the operations of that mountain. We owe to him, likewise, a description of Mount Ætna.

We have, in the Philosophical Transactions, an account, likewise, of the Peak of Teneriffe, which is a volcano; but not so frequent in its eruptions as Vesuvius, or even-Ætna.§§ We have also an interesting account of a new volcano, which burst out in the Archipelago, near the island of St. Erini, on the 8th of May, 1707; raising up, at the same time, a new island, out of the bosom of the deep.‖‖

The number of papers on volcanoes, in the Philosophical Transactions, amounts to 27. But, a student of geognosy would peruse these papers in vain

---

* Phil. Trans. Vol. IV. p. 1028.　　† Phil. Trans. 1713. Vol. XXVIII. p. 22.
‡ Ibid. 1739. Vol. XLI. p. 252.　　§ Ibid. 1751. Vol. XLVII. p. 409.
‖ Ibid. 1755. Vol. XLIX. p. 24.　　¶ Ibid. 1761. Vol. LII. p. 39.
** Ibid. 1767. Vol. LVII. p. 188.　　†† Ibid. 1780. Vol. LXX. p. 42.
‡‡ Ibid. 1786. Vol. LXXVI. p. 360.　　§§ Ibid. 1715. Vol. XXIX. p. 317.
‖‖ Phil. Trans. 1708. Vol. XXVI. p. 67. The account is drawn up by Dr. William Sherard, at that time Consul at Smyrna.

if he were in search of accurate and satisfactory details respecting the nature of the rocks of which volcanic hills are composed. The most valuable book on the subject, which I have seen, is Dolomieu's account of the Lipari islands. He had attended to the phenomena of volcanoes, with great attention, for many years; but, from the papers which he published on the subject in the Journal de Physique, there is reason to suspect that his knowledge of rocks was not very precise. He gives us a list of a prodigious variety of minerals, which he affirms to exist among lavas, and to have been vomited out of volcanoes. But the list is so extensive, and so far exceeds what has been observed by others, that it is by no means improbable that he confounded lava with various rocks, in the neighbourhood of volcanoes indeed, but never vomited from the volcanic crater in the form of lava, nor interspersed in lava.

We are neither acquainted with the nature of the rocks of which Ætna or Vesuvius is composed: indeed, the task is extremely difficult; for these mountains are so surrounded with lava on all sides, that the rock itself, of which the mountain consisted before the volcano commenced, may perhaps be entirely concealed from view. The prodigious extent of Ætna, about a hundred miles in circumference, and the vast number of volcanic hills attached to its sides, at least 44 in number, render the geognostic examination of the rocks, of which it was originally composed, almost impossible. Sir William Hamilton and several other writers on volcanoes suppose, that, previous to the commencement of the volcano, no mountain whatever existed; and that the whole mountain has been formed by successive eruptions from the crater. But this opinion is both improbable in itself, and destitute of evidence. Werner conceives, that volcanic hills are always composed of green stone, basalt, and the other rocks which constitute the independent coal formation, and the floetz trap. But the observations of Humboldt, on the volcanoes of America, will not admit of such conclusions; unless, indeed, we conceive *granite* and *porphyry* to belong to these formations; a supposition not very improbable, if we attend to the late observations of Von Buch and Professor Jameson; the former of whom in Norway, and the latter in Scotland, have found *granite* among secondary rocks.

Geologists are still deplorably ignorant of every thing relating to volcanic rocks. As a proof of this, it may be stated with truth, that the best treatise on volcanoes, which has hitherto appeared, was written by Bergmann, and published before the year 1780. Even *lava* itself, or the melted matter which issues from volcanoes, has been frequently confounded with *basalt* and *green stone*. At one time it was the fashion to consider all hills composed of these two last rocks; and, in short, all *floetz trap* hills as extinct volcanoes. Thus, Fojas de St. Fond, in his travels through Scotland, finds every where abundance of extinct volcanoes; and Mr. Raspe inserted a paper in the Philosophical Transactions, describing similar hills in Hessia, under the same appellation.* In like manner, the hills of Auvergne, in France, have been considered as extinct volcanoes by the French mineralogists; even D'Aubaisson, who was educated in an opposite school, has come to the same conclusion. Notwithstanding this, there can be little doubt that these hills stand in the same predicament with the fltoez trap hills of Scotland and Germany. These opinions, respecting the volcanic nature of basalt and green stone, are now pretty generally laid aside. The followers of Dr. Hutton, indeed, contend that they have been melted by heat, but do not suppose that they have ever belonged to a volcanic mountain.

Considerable doubts are entertained by some, whether pumice be a volcanic substance or not; and these doubts are founded on the supposition that pumice, though often observed in the neighbourhood of volcanoes, has never been seen mixed with lava, or actually flowing from a volcano. If the evidence of Tournefort be considered as sufficient, there can be no doubt that pumice is occasionally thrown out of volcanoes; for he describes various examples of it in his Voyage to the Levant. Pumice has been repeatedly observed floating on the surface of the sea, in immense quantity. To give one example, Mr. Dove, the Captain of an Indiaman, observed it floating in the Atlantic Ocean, in immense abundance, over a tract of not less than 317 miles in length. The pumice was first observed in south latitude 35° 36', west longitude 4° 9', and the shoals of it continued for several days.† Now it is impossible to account for such appearances on any other supposition than that the pumice has been thrown from the bottom of the sea by some volcanic force; and, if that be admitted, it will follow that pumice, at least sometimes, though perhaps not always, is volcanic.

There is a phenomenon, in some measure connected with volcanoes, which deserves to be mentioned. Coal pits, in some cases, take fire, and continue burning for a considerable length of time. There is one example of this accident, related in two different papers in the Philosophical Transactions. About the year 1648, a coal mine at Benwell, a village near Newcastle-upon-Tyne, was accidentally kindled by a candle: at first, the fire was so feeble, that the reward of half-a-crown, which was asked by a person who offered to extinguish it, was refused. But it gradually increased, and had continued burning for thirty years when the account, in the Transactions, was drawn up; and, it was not conceived, that it ever could be extinguished till the fuel was burnt out.‡ Other examples, of a similar kind, have happened in Scotland and in Germany.

Burning coal pits.

---

* Phil. Trans. 1771. Vol. LXI. p. 580.   † Phil. Trans. 1728. Vol. XXXV. p. 444.
‡ Phil. Trans. 1675. Vol. XI. p. 762; and Ibid. 1746. Vol. XLIV. p. 221.

Hot springs.   Connected with volcanoes, and probably depending upon the same cause, are the hot springs, which issue out of the earth in such numbers, in different countries.   No satisfactory explanation of the temperature of these springs, and, above all, of their wonderful equability in this respect, for a very long series of years, has ever been offered.   They are very frequently connected with volcanoes, and appear in great numbers in volcanic countries.   In such cases, we naturally ascribe the temperature of the spring to the heat of the volcano; but, when they occur at a considerable distance from volcanic countries, such an explanation cannot be applied.   Thus, the hot spring at Bath has continued at a temperature higher than that of the air for a period not less than 2000 years; yet it is so far from any volcano, that we cannot, without a very violent and improbable extension of volcanic fires, ascribe it to their energy.   There are various decompositions of mineral bodies, which generate considerable heat. These decompositions are usually brought about by means of water; or, to speak more properly, water is itself the substance which is decomposed, and which generates heat by its decomposition.   Thus, for example, there are varieties of pyrites, which are converted into sulphate of iron, by the contact of water, and such a change is accompanied by an evolution of heat.   Were we to suppose the Bath spring to flow through a bed of such pyrites, its heat might be occasioned by such a decomposition.   Such, probably, is the way in which those mineral springs, that contain sulphureted hydrogen gas, receive their impregnation.   But we are pretty certain, that such a supposition will not apply to Bath water: first, because it does not contain the notable quantity of sulphate of iron, which would be necessary upon such a supposition; and, secondly, because instead of sulphureted hydrogen gas, which would infallibly result from such a decomposition of pyrites, there is an evolution of azotic gas.   This evolution of azotic gas, however, is a decisive proof that the heat of Bath waters is owing to some decomposition or other, which takes place within the surface of the earth; though, from our imperfect acquaintance with the nature of the mineral strata, through which the water flows, we cannot give any satisfactory information about what that decomposition actually is.

I shall here give the small number of facts, concerning hot springs, contained in the Philosophical Transactions; and mention also, as somewhat connected with the subject, some trials made by Mr. Douglas, to determine the temperature of the sea at great depths, in high latitudes.

Temperature of Bath water.   1. The following table contains the temperature of Bath water, taken at different places.   I have repeated some of these trials, and found them correct. The table is drawn up by Mr. John Howard.*

* Phil. Trans. 1767. Vol. LVII. p. 201.

King's bath pump ....113°	Pump in the bath ....113°
Hot bath pump 114	Cross bath, coolest part 89
Cross bath pump 108	Ditto, warmest part..................... 90
Hot bath, coolest part................ 96	Cross bath pump 107
Ditto, warmest part.....:............ 97	Pump in the Market-place, Bath 54
Pump in the hot bath 113	Springs on Claverton 47
King's bath, coolest part............. 99	St. James's spring water.... 43
Ditto, hottest part 101	Springs on Lansdown 45
Queen's bath, coolest part.......... . 97	Old Well-house, Bristol................. 67
Ditto, warmest....................... 98	New Well, ditto........................ 76

These temperatures were taken in the months of November and December, 1765. The scale was Fahrenheit's, and the thermometer constructed by Bird.

2. In the island of St. Miguel, one of the Azores, which exhibits, according to Mr. Masson, very obvious marks of having abounded in volcanoes, there are a considerable number of hot springs of various temperature ; some boiling hot, others cooler, and some so low that they are used as baths, and have acquired great celebrity for the wonderful cures they have performed. Mr. Masson informs us, that these springs are surrounded with abundance of native sulphur, which he affirms is exhaled by them in abundance. A circumstance which renders it probable, that the heat of these springs depends upon the decomposition of pyrites.* *Hot springs in St. Miguel.*

3. The following table exhibits the result of the trials made by Charles Douglas, Esq. Captain of his Majesty's ship the Emerald, in the year 1769, off the coasts of Norway and Lapland, to determine the temperature of the sea at considerable depths :

Temperature of the sea.

Date.	Latitude.	Temperature.			Depth.
		Open air.	Surface of sea.	Bottom.	
May 12	70° 40′	27°	36°	39°	87 fathoms.
17	Ditto.	38	37	89	90
22	70° 32′	40	37	39	80
June 29	70° 54′	47	44	40	98
July 7	70 45	46	46	44	70
8	68 43	46	47	52	260
8	68 43	46	47	46	100
9	65 25	48	48	48	210
9	65 25	48	48	46	100
10	64 40	52	52	46	141
10	64 40	52	52	45	75

The experiments were made by letting down the thermometer enclosed in a tin cylinder filled with water, and letting it remain at the bottom for half an hour. It was sunk by means of the deep sea sounding lead.†

* Masson. Phil. Trans. 1778. Vol. LXVIII. p. 601.   † Phil. Trans. 1770. Vol. LX. p. 39.

Deluge.

V. DELUGE. Not only the account of a deluge related in the Book of Genesis, and the traditions to the same effect preserved by all ancient nations; but the abundant remains of sea shells and coral, found at great distances from the sea, at great heights, and intermixed with various rocks, have induced mineralogists, without exception, to agree that at some former period the whole of this earth was covered with the sea. Various hypothetical explanations of the way in which this supposed deluge took place have been from time to time published, and several of these are to be found in the Philosophical Transactions. It is not necessary to take notice of the old hypothesis of Burnet, who conceived that the antediluvian world consisted of a thin smooth crust spread over the whole sea, and that this crust breaking occasioned the deluge, and the present uneven surface of the earth; nor of Whiston, who ascribed the deluge to the effect of the tail of a comet; because these opinions have many years ago lost all their supporters. Nor is any attention at present paid to the hypothesis of Buffon, who conceived the earth to have been splintered from the sun by the blow of a comet, and accounted for the deluge by suppositions equally arbitrary, and inconsistent with the phenomena.

Woodward's theory.

Dr. Woodward was the first writer on the subject, who acquired a splendid reputation by his theory. He was a native of Derbyshire, and born in the year 1665. After having spent some years at the grammar school in the country, he was sent to London, and bound apprentice to a linen draper. But this situation not according with the philosophical turn of his mind, he soon left it; and turning his attention to science and medicine made a very considerable progress in both in a few years. In 1692, he was appointed to the vacant physical professorship in Gresham College, and in 1695, the degree of Doctor of Medicine was conferred upon him by Archbishop Tenison, and he was afterwards admitted to the same degree by the University of Cambridge. His *Natural History of the Earth*, first published in 1695, and afterwards considerably augmented and improved, was the work to which he stood indebted for his reputation. It engaged him in a good deal of controversy with some foreign writers; but his authority was such, that his opinions, though not always correct, generally prevailed. He seems to have been a man rather of a violent temper. In the year 1710, when a Member of the Council of the Royal Society along with Sir Hans Sloane, in a dispute with that eminent physician, he expressed himself in terms which were considered insulting by the other Members of Council. On that account, he was required to make an apology to Sir Hans, to which refusing his consent, he was in consequence expelled the Council. He died in 1728. Besides the Natural History of the Earth, he wrote *His Method of Fossils, The State of Physic and Diseases*, and several papers in the Philosophical Transactions.*

* See Ward's Lives of the Gresham Professors.

4

In his Natural History of the Earth, after refuting the hypotheses of his predecessors, he proceeds to show, that the present state of the earth is the consequence of the universal deluge; that the waters took up and dissolved all the minerals and rocks, and gradually deposited them along with the sea shells; and he affirms that all rocks lie in the order of their specific gravity. It is not worth while to examine into his theory with any minuteness, as it has long lost all its authority. Several of the positions, however, which he laid down continue still to find a place in every theory which has succeeded him. He first showed, that an ordinary deluge would not account for the appearances which we see in the present strata of the earth, and that we must conceive the rocks themselves to have been held in solution and gradually deposited.*

Dr. Halley supposed, that the position of the poles of the earth was changed Halley's by the blow of a comet. This he conceives would account for the deluge, and theory. occasion such a dreadful agitation in the waters as would be sufficient to produce all the effects which have taken place.† This opinion is still a favourite one with astronomers, and has been illustrated at some length by La Place in his popular System of Astronomy. But Dr. Halley himself seems afterwards rather to suppose, that the intermixture of sea shells with the strata of the earth took place at a period much anterior to the Mosaic deluge.‡ This opinion is now universally adopted by modern geologists. Dr. Halley's opinions, in the paper above referred to, bear a striking resemblance to the notions entertained on the subject by the late Dr. Hutton of Edinburgh, which have been so much admired by Mr. Playfair for their originality and philosophical nature. I mean, that the earth was doomed at various periods to be buried in the sea and raised up again. And this, he conceived, might be necessary for the purposes of vegetation.

In the forty-ninth volume of the Philosophical Transactions there is an elaborate paper by Dr. Edward Wright, in which he endeavours to refute Buffon's notion, that the earth had been for a long period covered by the sea, and that the mountains had been formed by currents in the ocean. He shows very clearly, that Buffon was but little acquainted with mountains, and that the account which he gave of them was, in most respects, inaccurate. Dr. Wright himself has recourse to the universal deluge to explain all the phenomena; but he does not attempt to show how they proceed from it.§

Mr. Edward King, a very ingenious but whimsical man, advanced a new King's theory of the deluge in 1767. According to him, the sea merely changed its theory. place at the flood. The bottom of the ancient sea rose up while the surface of

---

* For a particular account of Woodward's Theory, see Phil. Trans. 1695. Vol. XIX. p. 115.
† Phil. Trans. 1724. Vol. XXXIII. p. 118.   ‡ Ibid. Vol. XXXIII. p. 123.
§ Phil. Trans. 1755. Vol. XLIX. p. 672.

the ancient dry land subsided. In consequence what was formerly sea became land, and what was formerly land became sea.* This opinion has been adopted by Mr. Deluc in his late geological publications. There can be no doubt, that all the phenomena may be accounted for on such a supposition. But as we have no means of determining either its truth or its falsehood, it is beyond the province of philosophical investigation. Such suppositions have nothing whatever to do with science. Geology has been much injured by them. Its real object, as far as it deserves cultivation, is not to speculate about the original formation of the earth, but to ascertain with precision the relative situation of all its strata. And this might be done with precision, without indulging in fanciful speculations no less inconsistent with true taste than with a philosophical spirit.

Saltness of the sea.

Connected, in some degree, with this subject, is an ingenious proposal of Dr. Halley for determining the age of the earth. All lakes, according to him, that have no outlet, are more or less salt; while those from which a river runs are of necessity as fresh as the water which flows into them. This is the case with the Caspian sea, with the Dead sea, with the lake of Mexico, and with the lake Titicaca in Peru; though the last two are much less salt than the two former. Now Dr. Halley conceives that the saltness of these lakes is owing to this circumstance. The river water which runs into them, and by means of which they are kept up, contains in it a small quantity of salt, which it has imbibed from the soil over which it flows. This water, when it constitutes the lake, is continually rising from the surface of it in the state of vapour. But this vapour containing no salt in it, we see obviously, that the proportion of salt in the lake must continually increase, though at a very slow rate, salt being always flowing into it, and none ever leaving it. Now Dr. Halley thinks that the ocean is precisely in the same circumstance as these lakes, since it may be considered as an enormous lake into which water is continually flowing, and from which none escapes except by evaporation. Hence, to determine the age of the earth, we have only to ascertain the rate at which the saltness of the sea increases. Thus, if we find that in the year 1800 it contains a certain portion of salt more than it did in the year 1700, we have only to find what proportion that increase bears to the whole salt in the sea, from which we can at once deduce the time that has elapsed since the period when the sea first began to give out water by evaporation.†

It will be admitted at once, that this is a very ingenious and plausible speculation; but it will not bear a rigid examination. We have no evidence whatever, that the sea was not salt at its original formation. Indeed there is a presumption in favour of that opinion; because many of the animals which it

---

* Phil. Trans. 1767. Vol. LVII. p. 44.   † Phil. Trans. 1715. Vol. XXIX. p. 296

contains cannot live in fresh water. Hence we must either admit that the sea remained for many ages uninhabited, or that it was salt at its first formation. But, granting that the sea was originally fresh, it would not follow that it became salt by evaporation, unless we were certain that the vapour which rises from the sea is absolutely destitute of salt. But we have evidence that this is not the case. Margraaff found salt in rain water, which must have been originally raised by evaporation, either from the sea or the land; and, if we suppose the latter, the supposition makes more strongly for the reality of the vapour from sea water containing some salt. But even if this point were given up, still there is another consideration which would make it difficult or impossible to deduce any conclusions from the rate at which the saltness of the sea increases. It is true that salt is mixed, to a certain amount, with almost every mineral in nature, as follows from the galvanic experiments of Mr. Davy. But the proportion of it is very various in different places. Sometimes, as in Cheshire and in Poland, we find it deposited in prodigious quantities, so as to form beds of enormous thickness. In other cases it is loosely scattered, but in very inconsiderable quantities, in rocks and the soil. While in other cases, it is so intimately mixed, that it cannot be separated by any other method, with which we are acquainted, than the galvanic energy. Water flowing through minerals of such different natures must dissolve very different proportions of salt. In beds of the first kind, it may, in certain circumstances, become saturated with salt, and will always dissolve so much as to be entitled to the name of a salt spring. In beds of the second kind, it will dissolve only a very minute quantity of salt; and, in those of the third kind, if any such exist of any considerable extent, it will dissolve none at all. Hence waters, when they began to flow into the sea and into lakes, would contain very different proportions of salt, according to the nature of the country through which they flowed; and hence different lakes, and different parts of the sea, would possess different degrees of saltness, and would increase in saltness at very different rates. Finally, this increase of saltness in the sea, if it takes place at all, must do so with inconceivable slowness; for the specific gravity of sea water has never been observed to increase since the first time that it ever was taken, which is more than a century ago.

VI. HEIGHT OF HILLS. The ridges of mountains which traverse so very considerable a portion of the earth's surface, claim the particular attention of the geognost. They give every country its particular aspect. They give rise to all the rivers, and determine the course which they are under the necessity of following, in order to reach the sea. They exhibit to our view the rocks, of which the earth is composed; and, what is equally important and unexpected, the highest mountains are usually composed of those rocks which lie lowest, in the order of strata, and which, accordingly, in level countries, lie at the greatest depth below the surface.   It is a point of considerable importance to determine the height of mountains, not only on this account, but because this height has considerable influence upon the climate; and because, if the heights of all mountains were once accurately settled, we should be enabled to determine how far the commonly received opinion, that the mountains are annually wasted by the weather, and of course diminishing in height, is well founded.   Much pains have been taken to determine the height of mountains; and the method of measuring these heights by the barometer is, upon the whole, the most accurate. The labours by which that method was established, and the ultimate formulæ at last pitched upon, will come under our observation in a subsequent part of this work.   At present, that we may not omit any thing of importance, we shall give a table of the heights of different mountains in France; as determined geometrically, during the trigonometrical survey of that kingdom, between the years 1669 and 1703.   We copy the table from a paper by Dr. Scheuchzer, on the height of mountains.   His own measurements by the the barometer, being inaccurate, do not merit transcription.*

	French feet.
Mont Claret, in Provence	1662
La Massane, in Roussillon	2382
Bugarach, a mountain in Languedoc	3888

Mountains in Auvergne.

Le Puy de Domme, near Clermont	4860
La Courlande	5028
La Coste	5106
Le Puy de Violent	5118
Le Cantal	5904
Le Mont d'Or	6180

Mountain in Avignon.

Le Mont Ventoux	6216

Pyrenean Mountains.

S. Barthelemy dans le Paix de Foix	7110
La Montagne de Mousset	7548
Le Conigou	8640

To these heights we shall add that of the Peak of Teneriffe, as determined geometrically by Dr. Thomas Heberden.   He found it to be 15396 English feet above the level of the sea.†

Changes on the earth's surface.   VII. CHANGES. A favourite object of geologists has always been to speculate on the supposed changes to which this earth has been subjected, since the period of its original creation.   They have frequently supported their hypothetical reasoning, by producing certain changes, the existence of which is

* Phil. Trans. 1728. Vol. XXXV. p. 577.   † Phil. Trans. 1751. Vol. XLVII. p. 353.

actually supported by historical documents. Hence it becomes an object of considerable importance to register, correctly, all the changes of that kind upon record; I shall therefore state, in this place, the small number of such facts contained in the Philosophical Transactions. Some of the facts, already noticed in a preceding part of this section, might, without impropriety, be introduced under this head; as the flow of the Solway moss; the subterraneous forests in Lincoln and Yorkshire; and the bones of quadrupeds, found at considerable depths under ground. But these, having been already stated at sufficient length, need not be again repeated. The remaining papers, respecting changes in the surface of the earth, are only six in number, and relate to three changes, all, in some measure, hypothetical.

1. The first three of these papers relate to a supposed isthmus, which formerly joined Great Britain and France, and which was conceived to have been broken down by the sea, before the commencement of any accurate historical records respecting these islands. Two of them were written by Dr. Wallis, within two years of his death, when he had reached his eighty-fifth year; and afford a surprizing proof of the activity of mind, and the perfect command of his faculties, which that extraordinary man retained at so advanced an age.* The third paper, on the same subject, was written by Dr. William Musgrave.† This opinion was first broached by Camden, supported by Sumner, and opposed by Vossius. Dr. Wallis supports it with much learning, and points out the effects of the rupture with much ingenuity. He conceives, that the tradition mentioned by Plato, of the destruction of an island in the Atlantic Ocean, related to the rupture of this isthmus; and Dr. Musgrave quotes the well-known passage in Virgil:

—— Penitus toto *divisos* orbe Britannos.

as a proof that Virgil was aware of such a rupture, and alluded to it. When men have recourse to such proofs as these in support of an opinion, it affords clear evidence that they have nothing better to produce.

2. Few things are more curious or difficult to explain than the prodigious quantity of coral formed in the sea, especially in the tropical regions. Coral is the produce of different species of vermes, and it consists chiefly of carbonate of lime. Now it is difficult to conceive where these animals procure such prodigious quantities of this substance. Sea water, indeed, contains traces of sulphate of lime, but no other calcareous salt, as far as we know. Hence it would appear, that these creatures must either decompose sulphate of lime, though the quantity of that salt contained in sea water seems inadequate to supply their wants, or, they must form carbonate of lime from the constituents of sea

Supposed isthmus between Britain and France.

Coral islands.

* Phil. Trans. 1701. Vol. XXII. p. 967 and 1030.   † Phil. Trans. 1717. Vol. XXX. p. 589

water, in a way totally above our comprehension.   Be that as it may, there is one consequence of this copious formation of coral in the tropical regions of considerable importance to navigation, which has been clearly pointed out by Mr. Dalrymple, and is now pretty well understood.   The winds and waves accumulate these corals in large banks, which, entangling the sand, gradually rise above the surface of the waves, and form islands.   These, in process of time, probably by the agency of birds, become covered with vegetation, and frequently loaded with timber.   The bottom of these islands is nothing else than a coral bank; the surface is a black soil, formed of a mixture of sand and decayed vegetable matter; the whole island is flat, long, and narrow, and extends usually in its greatest length from north to south, because, almost all the winds between the tropics blow either from the east or the west.   The sides of these islands frequently constitute a perpendicular wall; and the sea, at a little distance from them, is of unfathomable depth.*

Supposed formation of marble.   3. The last paper in the Transactions which we shall notice, on this subject, is by Mr. Raspe; and we notice it, chiefly, to point out the mistake into which he is fallen.   The Abbé Vegni traced the hot mineral waters of St. Philip, situated at Radicofani, in Tuscany, on the road from Florence to Rome, to a small hill, composed of white marble, from which they flowed in several rivulets.   He found that these waters deposited a great quantity of shining white tophus, with which, not only the sides of the channels, along which they flowed, became encrusted, but likewise all kinds of hard bodies that were thrown into them; and this, in such manner, that when the said tophus was dexterously broken off, it retained exactly the form and shape of the bodies on which it had been deposited.   This tophus Mr. Raspe considers as exactly the same with white marble; and hence infers, that all white marble has been deposited from springs in this manner.†   But the tophus in question is quite different in its characters from *granular lime-stone*, to which true white marble belongs.   It is the mineral well known under the name of *calctuff* white, susceptible of a fine polish, and, in many cases, very beautiful.   It is often cut into slabs, and used for ornamental purposes; and, as it is composed of the same constituents with granular lime-stone, it may, and often is, called *white marble;* but, when the two minerals are compared, their difference is at once conspicuous.   Granular lime-stone is a congeries of minute crystals; and, in some cases, as in the parian marble, these crystals are of such a size that their form can be distinctly seen with the naked eye.   But the grain of calctuff is extremely fine, and no appearance of crystallization can be observed in it.   Its texture is similar to that of chalk, only more compact.   Hence it is obvious, that the mode in which calctuff is formed can never be extended to the formation of *granular*

* Phil. Trans. 1767. Vol. LVII. p. 394.   † Phil. Trans. 1770. Vol. LX. p. 47.

lime-stone. Sir James Hall has shown, that chalk, shells, and, probably, likewise calctuff, when violently heated in close vessels, undergo a species of fusion, and assume the appearance of granular lime-stone. Bucholz has ascertained, that the same thing takes place even in open vessels, provided the heat be applied suddenly. How far these experiments will account for the existence of granular lime-stone in such abundance, and the contrast between it and chalk, shells, and calctuff, as Sir James Hall seems to think they will, is a very different question, and could not be discussed without the introduction of a variety of topics foreign to our present subject.

VIII. DESCRIPTIONS. The only way of really improving geognosy, is by an accurate examination of the whole surface of the earth; and describing, with fidelity, the rocks of which it is composed; and the relative situation of these rocks. Considerable progress has of late years been made in such a survey. But, it could not be attempted, with any precision, till the science of mineralogy had made considerable progress; till we were acquainted with the names of the different rocks which constitute the earth's surface; and able to distinguish and describe these rock by their characters. Some valuable papers, on this subject, have appeared in those volumes of the Transactions which have been published since the commencement of the present century. But the small number of descriptions which occur in the preceding volumes are of comparatively little consequence. Yet, as descriptions constitute the only true means of improving the science, it would be improper to omit noticing those contained in the Transactions, even though of inferior value. Descriptions.

1. There is nothing which has contributed more essentially to the prosperity of Great Britain, or, which has tended more to produce and foster the important manufactures, by which this industrious and enterprizing island is distinguished, than the mines of coal, which are scattered in such profusion under its surface. We find every manufacturing town set down in the midst of a coal country: Bristol, Birmingham, Wolverhampton, Sheffield, Newcastle, and Glasgow, afford striking instances, and many more will readily occur to the recollection of every one who is acquainted with Great Britain. An accurate account of the different coal fields, in this island, would be a very valuable addition to our geognostic knowledge. Several excellent tracts on the subject have been published at different periods, especially by Mr. Williams, in his *Mineral Kingdom;* but, unfortunately, the terms employed in these tracts, being the usual ones employed by the miners, are too vague and too little understood to convey any exact information. There is a paper in an early volume of the Transactions, giving an account of the coal mines of Mendip, in Somersetshire. They occur in that county, as every where else in the low country, and are not to be found in the hills. The beds of coal are not horizontal, but sloping, and they dip to the south-east at the rate of about 22 inches per Mendip coal-
mines.

fathom.   Hence they would speedily sink so deep that it would not be possible to work them, were it not that the beds are here and there intersected by perpendicular dykes, or veins, of a different kind of mineral, (sometimes clay, but usually green stone); and, upon the other side of this vein, they are all found considerably raised up.   There are seven different beds of coal at Mendip, lying at regular distances below each other, and separated by beds of a different kind of matter.   From the names which Mr. Strachey applies to these beds, it is impossible to make out the nature of every mineral which occurs.   Sand stone, slate slay, and bituminous shale, seem to be the principal.   The deepest bed of coal, which is ten inches thick, and is not considered as worth working, lies about 38 fathoms, or 228 feet below the surface of the earth.[*]

Giant's Causeway.
2. The cliff on the northern coast of Ireland, usually distinguished by the name of the Giant's Causeway, has attracted a great deal of the attention of philosophers.   There are no fewer than six descriptions of it in the Philosophical Transactions; two of them by the celebrated oriental traveller Dr. Pococke.[†]   But none of them convey any idea of the nature of the stone.   The circumstance which has excited general attention is, the existence of a great number of perpendicular pillars, usually hexagonal, composed of single stones placed above one another, and jointed into each other very neatly.   These pillars, from the specimen which I have seen, appear to be *green stone*; and the *black stone*, mentioned by Pococke, in which I have seen several cornua ammonis, is very analogous to *flinty slate*, if not the very same.   This columnar structure is very commonly observed in green stone, basalt, and porphyry slate, and occurs abundantly in many parts of the world.   Staffa, Arthurseat, near Edinburgh, many hills in Fifeshire, Italy, and Germany, may be mentioned as familiar examples.   This columnar structure led to the supposition, that these hills had been formed by fusion, and hence they were conceived to be extinct volcanoes.   A violent controversy was the result.   It was impossible to bring decisive evidence in favour of either opinion; but the greater number of geologists have acceded to the opinion of Werner and his disciples, that these rocks exhibit no proofs of ever having been in a state of fusion.

Strata of Boston.
3. In the year 1783, the Magistrates of Boston, in Lincolnshire, employed a person to sink a well in the market-place of that town   Accordingly, after encountering a great many difficulties, a hole was bored, to the depth of 478 feet 8¼ inches below the surface.   But the attempt was not attended with success, no sufficient supply of water having been procured, and what did rise being too salt to be employed for the purposes of domestic economy.   In the Philosophical Transactions, there is an account of the different beds of earth

* Phil. Trans. 1719. Vol. XXX. p. 968.
† Phil. Trans. 1748. Vol. XLV. p. 124; and 1752. Vol. XLVIII. p. 226.

through which they bored.   As an experiment of this kind is seldom made, the description, though not expressed in language that can be always understood with precision, yet deserves to be mentioned. The following table exhibits the different beds passed through, as far as the paper in the Transactions enables us to ascertain them.*

Beds.	Depth from surface in feet.	Thickness of beds in feet.
Light coloured blue clay	36	36
Sand and gravel.	37½	1½
Blue clay.	48	10½
Dark green stone.	48½	½
Blue clay.	75	26½
Light coloured stone		½
Dark blue clay.	114	38½
Stone.		8 inches.
Gravel.		½
Dark clay, like black lead	174	63
Chalkly clay, with flints		3 inches.
Dark coloured clay	210	36
Light coloured clay.		½
Dark coloured clay	342	31½
Shells and white coloured earth.		½ inch.
Light coloured earth		
Dark coloured clay	447	5
Dark earth, with chalk and gravel	449	2. 10 inches.
Dark coloured earth.	454	5
Ditto, with chalk and gravel	456	2
Dark coloured earth.	457	1
Light coloured earth	462	5
Dark coloured earth	470	8
Ditto, with chalk and gravel		7 inches.
Rag stone.		1
Dark coloured earth.	472	1
Dark coloured silt, with chalk.	475	3
Ditto, without chalk.	478 8½ inches.	3 8½ inches.

4. Caverns are very commonly met with both in lime-stone and sand-stone Caverns. rocks.   Three remarkable caverns are described in the Philosophical Transactions, but the description of the first two is very unsatisfactory, as nothing is said about the nature of the rocks in which they occur. The third cavern is mentioned to occur in lime-stone. The first cavern is Okey Hole; which Okey Hole. occurs in the Mendip hills, about a mile from Wells, in Somersetshire. It extends in length about 200 yards; and varies in height, from eight fathoms to a few feet.† The second cavern described is the celebrated one in Derbyshire,

* Phil. Trans. 1787. Vol. LXXVII. p. 50.   † Beaumont. Philos. Collec. 1681. No. II. p. 1.

Elden Hole. known by the name of Elden Hole. Mr. Lloyd descended into it, and found it 186 feet deep. At the bottom, it consisted of a very spacious cavern, divided into two compartments, the first, like an oven ; the second, like the dome of a glass-house furnace. From the stalactites, which Mr. Lloyd describes as hanging from the roof of this cavern, we may reasonably conclude that it occurs in a lime-stone rock.*

Dunmore Park. The third cavern is that of Dunmore Park, in Kilkenny, in Ireland. The mouth of it is at the top of a small lime-stone hill ; and its depth is 90 feet. Its bottom and top are covered with the finest specimens of stalactites. These stalactites are of the same nature with calctuff, and admit of a fine polish.†

Iron found in Siberia. 5. We shall terminate this part of our subject, with noticing the account of an enormous mass of iron, found by Pallas, on the ridge of a mountain in Siberia, in the neighbourhood of the river Jenisca. This enormous mass, which was discovered by the Russian miners, and first observed by Pallas, weighed above 1680 Russian pounds. It was cavernous, and the vesicles were filled with a yellowish green vitreous substance, similar, in appearance, to chrysolite. On the outside it was rusty. The first accounts of this singular discovery were communicated to the Society in a letter from M. de Steplin, Counsellor of State to the Russian Empress.‡ Afterwards, Pallas himself communicated a more exact description, both of the mass itself, and of the situation where it was found.§ This mass has been subjected to a chemical analysis, by Klaproth, and other eminent chemists, and found to be an alloy of iron and nickel. Proust found the native iron, subsequently discovered in Peru, to be, likewise, an alloy of iron and nickel. The same holds with iron discovered in the south of Africa. Now, as the iron which exists in meteoric stones is an alloy of the same kind, and as this alloy has never been observed in any ore of iron, and as no rational account can be given of the formation of these enormous masses of iron by human means, it has been conjectured with considerable plausibility, and the conjecture has been generally acceded to, that these masses, like meteoric stones, have fallen from the atmosphere.

Such, as far as I am able to judge, is a pretty complete account of all the geognostic information of value, to be found in the Philosophical Transactions. As a description of the structure of the earth, or, as an account of the different rocks of which mountains are composed, and the relative position of these rocks, the Philosophical Transactions, along with all other publications of the period, must be admitted to be exceedingly defective. Some important observations on these subjects will be found in Whitehurst's Theory of the Earth ; and some valuable facts have been communicated by Saussure and Deluc. But,

* Phil. Trans. 1771. Vol. LXI. p. 250.   † Walker. Phil. Trans. 1773. Vol. LXIII. p. 16.
‡ Phil. Trans. 1774. Vol. LXIV. p. 461.   § †Phil. Trans. 1776. Vol. LXVI. p. 528.

the philosopher to whom we are indebted for laying the real foundation of geognosy, for establishing its fundamental principles, and pointing out the true method of improving it, by accurate observations, was Werner, the celebrated Professor of Mineralogy, at Freyberg, in Saxony. He has not, himself, indeed, published any thing on the subject; but has satisfied himself with teaching it in his Annual Courses of Lectures. As hardly any sources of getting information respecting this important science exist in this country,* it will probably be expected, that I should give a short sketch of the Wernerian geognosy in this place. This I shall endeavour to do; but, as my limits are confined, it will be impossible to enter into any details.

The rocks, of which this globe is composed, as far as we have been able to Sketch of the Wernerian geognosy penetrate below the surface, amount to 36; and they all occupy a determinate position with respect to each other. They extend round the whole earth, and inclose the central nucleus like the coats of an onion. Not that they are every where spherical, or uninterrupted: partly owing to inequalities in the central nucleus, over which they are deposited, and partly to other causes, they rise higher in one place and sink lower in another, sometimes slowly, and sometimes abruptly; and they are entirely wanting in many paticular spots, having either never been deposited, or having been removed and carried away by some unknown cause. The position of the different rocks, being thus constant, has been pitched upon as the basis of the classification of them. They have been divided into five classes; and the term, *formation,* has been applied to them, Formations. from the supposition that each class had been formed about the same time. Those rocks which lie lowest down, or nearest the central nucleus, belong to the first, or primitive formation; and those which lie highest up, or immediately at the surface, belong to the fourth, or latest formation; for the fifth formation includes only the volcanic matters, and of course is confined to particular spots. The names of the formations are as follow: 1. Primitive.— 2. Transition.—3. Floetz.—4. Alluvial.—5. Volcanic. We are not to suppose, however, that the rocks belonging to the primitive formation are always at a great depth below the surface. On the contrary, they frequently constitute mountains; and the highest mountains, on the surface of the earth, are com posed of them. In these cases, we must suppose the subsequent formations either never to have been deposited, or to have been removed and carried off by some unknown means. In like manner, the transition and the floetz formations often constitute mountains, and appear at the surface; and this must be accounted for in the same way.

The primitive formation consists of five classes of rocks, which follow each Primitive formations.

---

* I am acquainted with no English treatise on the subject, except Professor Jameson's excellent outlines of it, which constitute the third volume of his Mineralogy. A work, to which I refer my readers, who wish for accurate information respecting the structure and position of rocks.

other in succession, in the order of their names, beginning with the lowest. These are as follow: 1. Granite.—2. Gneiss.—3. Mica-slate.—4. Clay-slate.— 5. Porphyry and Sienite. Alternating with gneiss, mica-slate, and clay-slate, there occur beds of several other rocks, which, being of no great extent compared with them, and being frequently repeated; have been termed *subordinate* formations. These are primitive lime-stone, primitive trap,* quartz, flinty slate, and gypsum. Along with porphyry and sienite, occur serpentine and granite. Primitive rocks contain no petrifactions. They constitute the highest mountains on the face of the earth. They are evidently chemical compounds, and contain no minerals, which show themselves to have been mechanical depositions. They must have been formed before the earth was inhabited.

Transition formations.

The transition formation consists, likewise, of five classes of rocks, only one of which, namely, *grey wacke*, is peculiar to it, and characterizes it. These rocks are as follow: 1. Transition lime-stone.—2. Grey wacke, and grey wacke slate.—3. Transition trap.—4. Transition flinty slate.—5. Transition gypsum. I believe Professor Jameson has likewise discovered porphyry and granite, among transition rocks. These rocks contain petrifactions of animals and vegetables; but they are of the lowest order, both of animals and vegetables, and generally consist of species which can be no longer found in a recent state. Hence these rocks must have been formed after the earth contained both animals and vegetables. But we have no proofs that the inhabitants of the earth, at the formation of these rocks, were of the same species as at present; far less, that any of the large animals existed.

Floetz formations.

Our knowledge of the rocks which constitute the floetz formation is not so far advanced as that of the two preceding. They occur in a level country, and are usually covered by soil. Hence the investigation is much more difficult, and will be brought to perfection very slowly. We are acquainted with the following floetz formations:

1. *Sand-stone.* Of this, there are various formations; three at least have been ascertained. The lowest of all is red, and distinguished by the name of *old red sand-stone.*

2. *Lime-stone.* Of this, also, there are at least three formations, if we include chalk, the position of which is not very well understood.

3. *Floetz gypsum.* Of this, there are two formations, one of which is distinguished by alternating with rock salt.

4. *Floetz trap.* This consists of green stone, and was first observed by Professor Jameson, in his examination of Dumfriesshire.

5. *Independent coal.* This formation, besides coal, which characterizes it,

---

* By *trap*, mineralogists understand those rocks which are distinguished by a notable quantity of *hornblende*. They consist of hornblende, hornblende-slate, and green-stone, which is a mixture of hornblende and felspar.

1

contains a variety of other rocks, in beds, chiefly sand-stone, green-stone, clay iron-stone, lime-stone, shale.

6. *Newest floetz trap.* This formation generally caps the hills in those countries where it occurs. It consists of basalt, wacke, grey-stone, porphyry-slate, green-stone, trap tuff, clay-stone, sand-stone, &c. The floetz formations abound in petrifactions and in mechanical depositions. Most of the metallic ores occur in the primitive and transition formations.

The alluvial formation consists of the loose soil, gravel, sand, moss, &c., Alluvial formation which cover the surface of the earth, and do not require to be particularly described. Neither is it necessary to mention, at any length, the volcanic formations, which of course consist of the ashes and lavas vomited out of volcanoes.

Such is a sketch of the structure of the earth, according to the Wernerian geognosy; a sketch which will be allowed to be imperfect; because the whole surface has not yet been examined by persons qualified for the task. But, imperfect as it is, its principles furnish us with the means of making accurate observations, and the great outline has been amply confirmed by every recent observation, upon the accuracy of which any reliance can be put. I may refer to the observations of Humboldt, on the Andes, and of Mr. Farey, upon the structure of a considerable part of England. Mr. Farey's observations are the more remarkable, because he does not appear to be acquainted with the Wernerian geognosy. They certainly do him credit as an observer; but are, unfortunately, too circumstantially minute for common use. General principles, stripped of all useless details, are much more attractive, and much more easily understood and remembered, than a multitude of minute observations, which require an unusual portion of attention in the reader, in order to be able to draw the proper deductions from them.

Section III.—*Of Mining.*

The art of detecting the ores of metals in the earth, and of working them with advantage, constitutes one of the most important objects of political œconomy. It was carried to a considerable extent by the Grecian states, especially the Athenians; and was likewise pretty successfully practised during the early period of the Roman empire. Of the modern nations of Europe, the Germans have paid the greatest attention to mining, and the French the least. Britain possesses many important mining districts, especially Cornwall, Anglesey, Northumberland, Cumberland, and Derbyshire. The mines in Cornwall and Derbyshire British mining districts. have been worked since the period of the Roman conquest of Britain, and probably long before that period; for the Phœnicians were in the habit of drawing tin from Spain and Cornwall, in the remotest ages of antiquity. It is a very remarkable circumstance, and a striking proof of the want of education among our miners, that, notwithstanding the importance of our mining districts, and the vast quantity of iron, copper, lead, tin, and zinc, every year extracted from the bowels of the earth, in Great Britain, no systematic treatise on mining has ever made its appearance. We have some treatises on the subject, translated from the Spanish and German, all of them of a very old date, and, consequently, imperfect; but no general account has yet appeared of the practices followed in this country, in extracting the ore from the mine, and obtaining from it the metal which it contains. We have, indeed, some valuable works giving an account of the practices followed in particular districts, especially Cornwall and Derbyshire; and some important information on the mode of smelting, &c., in Watson's *Chemical Essays;* the most elegant chemical work which has hitherto appeared in any language, a work not appreciated in this country according to its value.

The papers on mining, contained in the Philosophical Transactions, amount to 29; and more than the half of these were published before the year 1672. Of these there are 15 which, in the present advanced state of our knowledge of the subject, cannot be considered as of any value. The rest consist, either in descriptions of particular mines, or in details of certain processes followed in certain districts. I shall proceed to give an account of the most important statements which they contain.

Blasting by gunpowder.   1. The method of blasting rocks by gunpowder is now so familiar to miners, that little attention is paid to the importance of it. But the methods, practised before its introduction, were so imperfect, that the use of gunpowder may be considered as constituting an important æra in mining. In a very early number of the Philosophical Transactions, we have a description of the method of blasting rocks by gunpowder, by Sir Robert Moray, where the invention is ascribed to M. Du Son.*

Mines in Transylvania and Hungary.   2. Transylvania and Hungary have long been celebrated as mining countries, and supply annually a considerable quantity of gold. Dr. Brown, a physician of London, of considerable eminence, who travelled on the continent in 1668, communicated a description of these mines to the Royal Society, which was published in the Philosophical Transactions. He describes a salt mine of great note, near Eperies, in Upper Hungary, no less than 1080 feet deep. The ground was not rocky but clay; and the salt, being ground to powder, was used without any refining. The gold mines of Chremnitz, he describes as of the greatest value, and as having been wrought for 900 years. His description of the ore is so imperfect, as to be of no value; but he gives an exact detail of the mode of extracting the gold by amalgamation, a process lately so much improved by Baron Born and his coadjutors. There was a mine of mercury about seven miles from Cremnitz, but it was not wrought. Dr. Brown makes

* Phil. Trans. 1665. Vol. I. p. 82.

some curious observations on the filling up of the passages formerly dug out in the old mines.   By the trickling down of moisture along the sides of these passages they gradually become narrower, and approach each other.*

3. The most extensive salt mines known are those in Poland, not far from Cracovia, at a small town called Wilizka.   There is an imperfect description of them in the Philosophical Transactions.   According to that account, they are 1,200 feet deep; and about 1,000 miners are constantly employed in them. The horses kept under ground become speedily blind, in consequence of the sharpness of the salt; and the hoofs of one of them, who had been long in the mine, were as long again as usual.† *Salt mines in Poland.*

Salt had been prepared in Cheshire for many years by boiling down the salt springs which are common in that country; but in 1670, a rock of salt was discovered, from which they expected to be able to manufacture the article with greater advantage.‡ *In Cheshire.*

4. There is a pretty long paper, published in the Transactions for 1671, giving an account of the method of working the tin mines in Cornwall, and of preparing and smelting the ore, and obtaining the tin.   As the terms used are all provincial words, peculiar to the Cornish miners, without any explanation, the paper is not very intelligible to ordinary readers.§   The publication of a dictionary, explaining all the words used by the miners in the different mining provinces of Great Britain, would be a work of considerable value, and would greatly facilitate the attempts that may be made to improve the state of our mines.   Something of the kind indeed has been attempted with respect to Cornwall, but at a period when the science of mineralogy was not sufficiently advanced to admit of an explanation of every term. *Tin mines in Cornwall.*

5. It is very remarkable that diamonds, by far the most beautiful and most highly valued of all the precious stones, have hitherto been found only in the torrid zone, in India, and Brazil.   The mines, as they are called, consist of nothing else than the alluvial soil, no doubt originally washed down from mountainous districts.   In this soil, diamonds are found scattered very unequally; sometimes occuring in great abundance, sometimes very sparingly.   In some places the stones are all small, never exceeding a few grains in weight; in other places they are occasionally found as heavy as nine ounces.   Diamonds have never yet been observed in their original position; from which circumstance one would be apt to suppose that this repository must be some of the newest rocks; as these are the rocks which from their situation are most liable to be worn away.   In the Philosophical Transactions there is a particular ac-

* Phil. Trans. 1670. Vol. V. p. 1189.   † Phil. Trans. 1670. Vol. V. p. 1099.
‡ Phil. Trans. 1670. Vol. V. p. 2015.   § Phil. Trans. 1671. Vol. VI. p. 2096.   See also
a paper on the same subject by Merret, Phil. Trans. 1678. Vol. XII. p. 949.

Diamond
mines in
India.

count of the different places in India, on both sides of the Ganges, where diamonds occur, the manner of digging and washing the soil, and the colour of the different soils thus wrought. The paper was communicated to the Royal Society by the Earl Marshal of England. The description of the soil and rocks is so imperfect as to convey no information. But there is one curious particular mentioned, that deserves to be noticed; we mean, the mode pursued by the Indians to dig through rocks. They kindled a fire on the rock, and when it was very hot they poured water on it. By this means the rock cracked and split into shivers for a certain depth; these fragments were removed, the fire again kindled, and water poured upon the hot rock as before. Thus they proceeded till they reached the requisite depth.*

Calamine.

6. Calamine is an oxide of zinc, which has been employed from time immemorial in the making of brass. There is in the Transactions an account of the method followed by the miners in digging this ore out of the earth, and preparing it for sale.† Calamine occurs frequently in beds, and seems, in some cases, to exist in great abundance. The Mendip hills, in Somersetshire, were famous for their calamine mines; though, I believe, the ore is now pretty much exhausted in that quarter. From the description given of the calamine mines in Somersetshire, by Mr. Pooley, it is plain that he considered the calamine as occuring in veins. It is always mixed with some galena or sulphuret of lead. It is dug out of the earth, and being broken into small pieces is exposed to the action of a current of water, which washes away the light earthy matter, and leaves the calamine. The whole is then thrown into deep wooden vessels full of water, and agitated for a considerable time. The galena sinks to the bottom, the calamine is deposited in the middle, and the earthy matter on the surface. The calamine, thus separated from its impurities, is ground to powder, and is then fit for sale.‡

7. Dr. Nichols, Professor of Anatomy in Oxford, seems to have examined the structure of metallic veins with more accuracy than any other British writer of the early part of the last century. He published some observations on the mines of Devonshire and Cornwall, which possess some value.§ He has given us an explanation of several of the Cornish words used by the miners, of which the following may serve as a specimen:

Mining terms.

      Load, a vein.
      Living load, a vein containing metallic ore.

---

* Phil. Trans. 1677. Vol. XII. p. 907.   † Pooley. Phil. Trans. 1693. Vol. XVII. p. 672.

‡ The practice followed at present, in Somersetshire, differs in several particulars from the account given in the text; though the object in view, and even the means employed, are the very same.

§ Phil. Trans. 1728. Vol. XXXV. p. 402 and 480.

Dead load, a vein destitute of ore.

Flooking, a cross vein of stone; or dyke, as called in the North of
England.

He has some observations on the crystals of tinstone, and on its specific gravity compared with that of tin, and on its colour, much more accurate than
any to be met with in chemical books of the same period.

8. In the year 1751, there was a very rich copper mine wrought at the river Precipitation
Arklow, in the county of Wicklow, in Ireland. From this mine there ran a of copper by iron.
stream of blue-coloured water, of so deleterious a nature as to destroy all the
fish in the river Arklow. One of the workmen, leaving an iron shovel in this
stream, found it some days after encrusted with copper. This induced Mr.
Johnston, one of the proprietors of the mine, to make a set of experiments on
the subject, from which he concluded, that the blue water contained an acid
holding copper in solution, that iron had a stronger affinity for the acid than
copper, that the consequence of this affinity was the precipitation of the copper, and the solution of the iron when pieces of that metal were put into the
blue water. These ideas induced the miners to dig a great many pits for the
reception of this water, and to put bars of iron into them. By this means they
obtained a very great quantity of copper, much purer and more valuable than
the copper which they obtained from the ore itself by smelting.*

9. The superiority of Swedish iron over that of other countries, for the Swedish iron
making of steel, is well known. Hitherto the British steel makers have not been
able to employ British iron in their processes; they have found it too brittle to
bear cementation. Attempts are at present making by some very spirited steel
makers at Sheffield; and, from the products already obtained, good hopes are
entertained of ultimate success. This superiority of the Swedish iron depends
upon the great purity of the ore from which the iron is smelted. One of the
most remarkable of these mines, if the name can with propriety be applied to
it, is Taberg, a mountain of considerable size, composed entirely of pure iron
ore, and occurring in a large tract of sand over which it seems to have been
deposited. This mountain has been wrought for nearly these three hundred
years, and yet its size is scarcely diminished.†

* Phil. Trans. 1751. Vol. XLVII. p. 500 and Vol. XLVIII. p. 181.
† Ascanius. Phil. Trans. 1755. Vol. XLIX. p. 30.

CHAP. IV.

OF GEOGRAPHY AND TOPOGRAPHY.

WE are sensible that the title of this chapter, as a branch of natural history, will appear objectionable to many persons. There can be no doubt, that the topics which we mean to comprehend under it might have easily been arranged under other branches of science. But if we had followed that plan, which we acknowledge to be the usual one, the different topics would have still occupied a place by themselves, so that nothing would have been gained for those sciences under which they would have been placed. Thus for example, had we arranged the facts respecting the latitude and longitude of places, which we mean to give in this chapter, under the head of astronomy, they must still have continued as distinct from astronomy, strictly so called, as they do at present, and have constituted a section apart. The fact is, that the geographical position of all the different spots of the earth constitutes a very important part of the natural history of the earth, as the description of the manners and customs of different nations do of the natural history of man. We might therefore have placed the former under *geognosy*, and the latter under *zoology*. But, upon the whole, it appeared better to place them under a particular department by themselves.

The papers relating to geography and topography, in the Philosophical Transactions, amount to 67, and contain several particulars of considerable importance. They may be arranged under four distinct heads, namely: 1. The latitudes and longitudes of different places as determined by astronomical observations. 2. The observations detailed in the Transactions on the construction of maps. 3. The papers on hydrology. 4. The papers on topography.

Latitudes and longitudes of places.

1. LATITUDES AND LONGITUDES OF PLACES. The following table exhibits a view of the different latitudes and longitudes determined in the Philosophical Transactions:

Places.	Latitudes.	Longitudes.	Observers.
Greenwich	51° 28′ 39½″ N.	0° 0′ 0″	Bradley.(a)
Ballasore, East Indies	21 20 0	86 20 0 E	Halley and Harry.(b)
Moscow, Russia	55 34 0	38 45 0	(c)
Yereslaw, Russia	57 44 0	0 0 0	(c)
Wologda, Russia	59 19 0	0 0 0	(c)

(a) Bradley and Maskelyne. Phil. Trans. 1787. Vol. LXXVII. p. 151.
(b) By the occultation of the bull's eye by the moon, 1682. Phil. Collect. No. V. p. 124.
(c) Phil. Trans. 1690. Vol. XVII. p. 453.

Places.	Latitudes.	Longitudes.	Observers.
Woslak, Russia	61° 15′ 0″ N.		(a)
Arkangel, Russia	64 30 0		(a)
Constantinople, Turkey	41 6 0	28° 53′ 49″ E.	Greaves.(b)
Rhodes, Turkey	37 50 0		Ditto.(c)
Pekin, China	39 55 30	116 12 35	Cassini.(d)
Canton, China	0 0 0	113 5 5	Ditto.(d)
Lisbon, Portugal	38 42 30	9 7½ 0 W.	Bradley.(e)
New York, America	0 0 0	74 4 0	Ditto.(e)
Buenos Ayres, America	34 35 0 S.	58 0 0	Halley.(f)
Magdalen islands, America	47 41 0 N.	61 38 0	Holland.(g)
Island of Entry, ibid.	47 17 0	61 20 0	Ditto.(g)
Bird's island, ibid.	47 55 0		Ditto.(g)
Bryon island, ibid.	47 52 0		Ditto.(g)
Island battery, ibid.	45 54 0		Ditto.(g)
Cape North, ibid.	47 2 0		Ditto.(g)
St. Paul's island, ibid.	47 11 0		Ditto.(g)
Dartmouth harbour, ibid.	46 13 0		Ditto.(g)
Conway harbour, ibid.	46 20 0		Ditto.(g)
Niganishe, ibid.	46 44 0		Ditto.(g)
Namur	50 28 32	4 44 50 E.	Pigot.(h)
Luxemburg	49 37 6	6 11 50	Ditto.(h)
La Heese	51 23 2	4 47 20	Ditto.(h)
Ostende	51 15 10	2 58 20	Ditto.(h)
Tournai	50 36 57		Ditto.(h)
Cork	51 54 0	8 29 15 W.	Longfield.(i)
Madras	13 4 54		Steevens.(k)
Brinhill, near the Severn	51 14 56½	3 5 58	Pigot.(l)
Quantock hill, ibid.	51 8 48½	3 21 57	Ditto.(l)
Landmark tour, ibid.	51 5 5	3 26 52	Ditto.(l)
Watchet hill, ibid.	51 6 56	3 30 1	Ditto.(l)
St. Hilary's church, ibid.	51 26 44½	3 34 23	Ditto.(l)
Llanmace church, ibid.	51 24 39	3 37 24	Ditto.(l)
Minehead, ibid.	51 12 42½	3 37 52	Ditto.(l)
Frampton house, ibid.	51 25 0	3 38 2	Ditto.(l)
Llantwit church, ibid.	51 24 13	3 38 38	Ditto.(l)
Llangwynewar hill, ibid.	51 28 28½	3 39 22	Ditto.(l)
St. Donat's castle, ibid.	51 23 49	3 41 15	Ditto.(l)
Huston point, ibid.	51 13 29½	3 44 14	Ditto.(l)
Leemouth, ibid.	51 13 54	3 57 7	Ditto.(l)
Hangman hill, ibid.	51 12 22	4 10 35	Ditto.(l)
Stockholm		18 5 15 E.	Wargentin.(m)

(a) Phil. Trans. 1690. Vol. XVII. p. 453.

(b) The Greeks made it 43° 5′, the Arabians 45°. The real latitude is 41° 1′ 24″. Phil. Trans. 1685. Vol. XV. p. 1295.　　(c) Ibid.　　(d) Ibid. 1698. Vol. XX. p. 53 and 371.　　(e) Ibid. 1726. Vol. XXXIV. p. 85 and 92.

(f) From an observation of Pere Feuillée. Ibid. 1721. Vol. XXXII. p. 2.

(g) He was sent out for the purpose by the Lords Commissioners of Trade and Plantations. Phil. Trans. 1768. Vol. LVIII. p. 46.　　(h) Phil. Trans. 1776. Vol. LVI. p. 182.　　(i) Ibid. 1779. Vol. LXIX. p. 163.

(k) Ibid. p. 182.　　(l) Phil. Trans. 1790. Vol. LXXX. p. 385.　　(m) Phil. Trans. 1777. Vol. LXVII. p. 162.

Places.	Latitudes.	Longitudes.	Observers.
Paris	48° 50′ 14″ N	2° 20′ 5″ E	Dalby.(a)
Landscrone, Sweden	55 52 23	0 15 16	Bugge.(b)
Hoeen church, ibid	55 54 38	0 5 56	Ditto.(b)
Kullen lighthouse, ibid	56 18 3	0 7 58 W	Ditto.(b)
Frankeklint, Langlande	55 9 44	1 38 47	Ditto.(b)
Kongsberg, Moen	54 58 3	0 4 12	Ditto.(b)
Sproeisland, Belt	55 19 56	1 37 0	Ditto.(b)
Copenhagen, Zealand	55 41 4		Ditto.(b)
Roeskilde, ibid	55 38 25	0 29 48	Ditto.(b)
Holbeck church, ibid	55 43 2	2 51 26	Ditto.(b)
Kallundborg church, ibid	55 40 54	1 29 12	Ditto.(b)
Korsor lighthouse, ibid	55 20 22	1 27 0	Ditto.(b)
Nestved church, ibid	55 13 55	0 49 12	Ditto.(b)
Wordingborg tower, ibid	55 0 32	0 40 4	Ditto.(b)
Ringsted church, ibid	55 26 51	0 47 20	Ditto.(b)
Skagen, N. Jutland	57 43 44	1 57 55	Ditto.(b)
Hioring, ibid	57 27 44	2 35 17	Ditto.(b)
Fladstrand, ibid	57 27 3	2 2 15	Ditto.(b)
Sœbye, ibid	57 20 2	2 2 36	Ditto.(b)
Aalborg, ibid	57 2 57	2 39 4	Ditto.(b)
Nibe, ibid	56 59 4	2 55 54	Ditto.(b)
Greenaae, ibid	56 24 57	1 41 49	Ditto.(b)
Rander's, ibid	56 27 48	2 32 3	Ditto.(b)
Viborg, ibid	57 27 11	3 9 25	Ditto.(b)
Aarhuus, ibid	56 9 35	2 21 40	Ditto.(b)
Ribe, ibid	55 19 57	3 48 25	Ditto.(b)
Hudersleben, S. Jutland	55 15 15	3 4 56	Ditto.(b)
Norborg, ibid	55 3 53	2 49 53	Ditto.(b)
Apenrade, ibid	55 2 57	3 9 7	Ditto.(b)
Tondern, S. Jutland	54 56 30	3 41 53	Ditto.(b)
Sonderborg, ibid	54 54 59	2 47 1	Ditto.(b)
Flensborg, ibid	54 47 18	3 8 5	Ditto.(b)
Husum church, ibid	54 28 29	3 31 3	Ditto.(b)
Gluckstadt, Holstein	53 47 44	3 8 43	Ditto.(b)
Hesseloe, Cattegat	56 11 46	0 51 44	Ditto.(b)
Anholt lighthouse	56 44 20	0 55 24	Ditto.(b)

Size of England.

Such are all the latitudes and longitudes of places to be found in the Philosophical Transactions, or at least the greatest number of them. There is one other paper on a subject somewhat connected with this, which it may be proper to notice. Dr. Grew, from a very simple calculation, estimates the number of acres, in England, at 46,080,000. He then compares the size of England and Holland, and observes, that if England were as populous in proportion to its

(a) Phil. Trans. 1791. Vol. LXXXI. p. 286.
(b) The longitudes are from the observatory of Copenhagen. Phil. Trans. 1794. Vol. LXXXIV. p. 43.

2

size as Holland, it would contain 110 millions of inhabitants.* But Dr. Grew's estimate is obviously too great. He makes the length of England, from Berwick to the southernmost point, 400 miles, whereas it certainly does not exceed 350. The breadth of England, where greatest, he makes 360 miles, whereas it cannot be reckoned more than 300, if it amounts to that quantity.

II. Construction of Maps. The construction of maps belongs to geometry, Construction depending upon the various methods of projecting a sphere upon a plane sur-of maps. face. The method most commonly used for maps is what is called the stereographic projection, in which the eye is usually conceived to be in the surface of the sphere, opposite to that which is delineated. In maps of this construction, the meridians cut the circles of latitude perpendicularly, which they ought to do; but the central parts of a map of this kind, when it represents a considerable tract of country, for example, a hemisphere, are contracted greatly below their proper size. Rectilinear maps were very early adopted, in which the meridians were described parallel to each other, and the degrees of longitude and latitude every where equal: the rhumbs were consequently right lines, and hereby it was thought that the courses or bearings of places would be more easily determined. But such maps were soon found insufficient for nautical purposes, as giving results totally erroneous. The first step to improve them was made by Gerard Mercator, who published a map about the year Mercator's 1550, in which the degrees of latitude were increased from the equator to the projection. pole. But he did not show upon what principle he had constructed these maps. About the year 1590, Mr. Edward Wright discovered the true principles on which such charts should be constructed, and communicated his discovery to one Iodocus Hondius, an engraver, who, contrary to his engagement, published it as his own. This induced Mr. Wright, in the year 1599, to communicate his method of construction to the public, in his book, entitled *Correction of Errors in Navigation*. In the preface to this work, he explains the share which Mercator had in this important invention, and gives an account, likewise, of the treacherous conduct of Hondius. This invention of Wright is perfect in its kind, and is of the utmost importance to navigators.

The greater number of maps seem to be laid down according to no fixed principle of projection, and are often very erroneous, as far as the distances between places is concerned. On this account Mr. Murdoch, in a paper which he published in the Philosophical Transactions, proposed a new method of projecting maps, in which these errors are the least possible. It consisted in finding a conical surface, equal to the spherical surface, which was to be given, and then, in spreading out this conical surface into a plane. He gives the rules for finding the conical surface, and for constructing maps accurately,

* Phil. Trans. 1711. Vol. XXVII. p. 266.

according to his principles.* There can be no doubt, that maps constructed on this principle would, in many cases, be preferable to the common ones; though I do not recollect ever to have seen any map actually made according to the rules laid down by Murdoch.

Rate at which
camels travel.

It may be worth while to mention here an ingenious method proposed by Major Rennel, for determining distances in those parts of Africa and Asia, where more accurate methods cannot be had recourse to. In these countries long journies are performed by means of camels, between places at a very great distance from each other. Now these caravans always travel the same number of hours every day, and Major Rennel has shown, that camels always travel at the rate of $2\frac{1}{2}$ miles in the hour. Hence, knowing the number of hours travelled, or even the number of days, it is easy to determine the distance. And if the bearings have been taken with a pocket compass, the position of the places on the globe may be determined with considerable accuracy.†

Hydrology.

III. HYDROLOGY. The papers on this subject, in the Philosophical Transactions, are only two: the first giving an account of a remarkable lake in Carniola; the other, of the falls of Niagara.

Zirchnitzer
Sea.

In Carniola there is a lake called the Zirchnitzer Sea, two German miles long and one broad. In the month of June it sinks under ground, through a great number of holes, which empty themselves in succession. At this period, the inhabitants in the neighbourhood are employed in fishing, and a vast quantity of fish are caught at the different holes. The fish in the lake are of three kinds, namely, the eel pout, the tench, and the pike, all of which grow to a large size. The lake also contains many crabs; but they do not constitute an agreeable article of food. While the lake is dry, the soil produces a fine crop of grass, which is cut down for hay. In some parts of it millet is sown, and a considerable crop obtained. In the month of September, the water rushes back again through the holes with great impetuosity, and the lake is restored to its original size. This extraordinary alternation in this lake is accounted for in the following manner: the country is hilly, and the lake is surrounded with rising grounds. It has no visible exit, yet seven rivulets empty themselves into it. By subterraneous channels, it communicates with two lakes concealed underground, the one situated below, the other above its own level. Into the first it empties itself by means of the holes in its bottom. From the second, it receives a supply equal to its waste, which prevents it from sinking underground during the winter. From the lowest lake a considerable river runs. In the summer, the uppermost lake not being fed as usual by rain, becomes smaller, and ceases to supply the Zirchnitzer Sea with water. The waste of this lake, therefore, being greater than the supply, it is drained in consequence, and dis-

Phil. Trans. 1758. Vol. L. p. 553.   † Phil. Trans. 1791. Vol. LXXXI. p. 129.

appears. When the uppermost lake is restored to its usual size, it affords the proper quantity of water ; hence the lowest lake swells, and at last forces part of its contents through the holes into the open air, and thus restores the Zirchnitzer Sea to its original size.*

The falls of Niagara were measured in the year 1722, by order of the gover- Falls of Niagara. nor of Canada, by means of a plumb-line, and were found to be 156 French feet, which amounts to 176¼ English feet.†

IV. TOPOGRAPHY. The countries and people, of which an account is given Topography. in the Philosophical Transactions, are, Hudson's Bay, the North American Indians, the Patagonians, the Falkland Islands, and Thibet.

1. Mr. Wales was sent out to Hudson's Bay, by the Royal Society, as a Hudson's Bay. proper station for observing the transit of Venus over the sun's disk, in 1769. He continued in that country for 13 months; and, after his return, a very entertaining account of the climate, soil, and mode of living, with some observations on the inhabitants, was published by him in the Philosophical Transactions. The place where he wintered was at the Factory, on Churchill river, in north latitude 58° 55¼′, west longitude 94° 50¾′. The following are the most important observations contained in this paper: on his approaching the coast the vessel was visited by some Eskimaux, in their boat, whom he praises highly for the excellence of their dispositions, and their skill in all the arts which they practised. The sea at that time was full of islands of ice, and he had an opportunity of ascertaining that the green-coloured ice, contrary to the opinion at that time received, was fresh, and not salt; provided sufficient care was taken to wash it from the salt water in which it was swimming, before examining it. The islands of ice moved with considerable velocity, and, as they went to the south, they were not long able to resist the united action of the solar rays and the heat of the sea, but speedily melted into water.

The atmosphere was usually hazy, and Mr. Wales observed with surprise, that the islands of ice, when seen at a distance, appeared considerably elevated, though when they got near them they were found to be scarcely raised above the level of the sea. When they approached the coast, the land was seen distinctly from the deck, but from the mast-head it was quite invisible. Mr. Wales was so surprised at this fact, that he went to the mast-head, and verified it himself. The country all round the factory for several miles is rocky, or composed of loose sand; the soil poor, and utterly unfit for every kind of corn except oats. There were many firs about 20 feet high, but no other trees,

* Brown.   Phil. Trans. 1669. Vol. IV. p. 1080 ; and J. Weichard Valvasar. Ibid. 1687. Vol. XVI. p. 411.
† Paul Dudley.   Phil. Trans. 1722. Vol. XXXII. p. 69.

though the country was covered with bushes, consisting of dwarf willows and birches. There were many gooseberry and black currant bushes.

The country abounds with a prodigious number of birds, geese, ducks, partridges, &c., which come in succession, and supply the inhabitants with the greatest part of their food. During summer there are three insects which are remarkably troublesome; these are the *moschetto;* a kind of small fly called *sand fly*, which comes in such prodigious swarms as to fill the whole air, and to be exceedingly troublesome; and a very large fly, like the English *flesh fly*, which, when it lights on any part of the body, always draws blood.

The ground was covered with snow by the middle of October. In January the cold became very intense; the beams of their houses were continually cracking with a very loud noise, in consequence of the expansion of the frost, and the rocks split with a prodigious explosion. The wood of their beds was covered with a thick coat of ice, and brandy exposed to the open air soon became solid. The first thaw appeared on the 19th March, for a few minutes at noon, and on the 26th of March it thawed in reality. By the 23d of April, the ground was bare of snow in many places, and spring was begun by the end of that month. The month of May was of a pleasant temperature, neither too hot nor too cold. On the 16th of June the ice in the river broke up and went to sea, which enabled the inhabitants to begin their fishing pursuits. Great numbers of salmon were caught, and a fish which Mr. Wales calls *képling*, and which he describes as the most delicate of all fish, far exceeding in flavour every other. On the 22d of January, the coldest day, the thermometer stood at 45° below zero, and on the third of July, which was the hottest day, it stood at 80°; so that the range of temperature in this climate amounts to 125°. In England, the range of temperature rarely exceeds 80°, and has scarcely ever been observed more than 100°.

The inhabitants of the country are all thin, and rather tall; they are copper coloured, with wide mouths, thick lips, and lank black hair. They appear rather of a melancholic temperament, as is the case with most rude nations; but they are friendly, good natured, honest, and hospitable. They lead a wandering life, and depend, for their subsistence, on hunting. They believe in two superior beings, a good and a bad, and pay a sort of worship to both.

Mr. Wales observed several marble rocks in the country, and specimens of copper and copper ore. He mentions, also, different kinds of pyrites and micas. The atmosphere is never clear for 24 hours together; there is a perpetual haze round the horizon, which gives occasion to two red circles of light, which make their appearance before sun-rise. The aurora borealis is not to be compared, in point of brightness, with what is often seen in England.*

* Phil. Trans. 1770. Vol. LX. p. 100.

2. There are two papers in the Philosophical Transactions, on the North North American Indians. American Indians; one by Sir William Johnson, who was for a considerable time governor of one of the British settlements in that country;* another by Mr. M'Causland, surgeon to the 18th regiment, who probably spent some years in that country, during the American war.† The accounts communicated in these papers are not of much importance, neither of the writers having had an opportunity of viewing these people near enough to form an estimate of them sufficiently exact and minute. The following are the most important particulars:

It was conceived by many European writers, and brought as a proof that the Indians in America belong to a different species from the Europeans, that they were destitute of beards. But this opinion is not well founded. They are in the habit, indeed, of pulling out the hairs of their chins, as fast as they grow, and are all provided with instruments for the purpose, and this gives them the appearance of a smooth chin; but those among them who have left off that practice, require to shave as much as the natives of Europe. The order of their government, and their civil polity, is much less perfect than it was formerly; owing to the proximity of Europeans, the diminution of their numbers, the introduction of spirituous liquors, and a certain corruption of manners which they have imbibed from the lower orders of their civilized neighbours. When they want to convey their thoughts to those who are at a distance, they do not use hieroglyphics, but draw as accurate a picture of the thing as they can. Thus, if they wish to communicate a warlike expedition, they draw a few trees, to represent the country; or a canoe, if the expedition is by water, and as many warriors as go upon it, or at least a certain number. Each nation is divided into several tribes, which have their peculiar designation and mark, as the turtle, the bear, the wolf, the snake, the deer, &c. Every nation has a *sachem*, or chief, who is most commonly chosen by election, and in consequence of superior merit; though in some nations the office is hereditary. But his power is very limited; and the small portion which he possesses, he is under the necessity of using with great caution and prudence. Witchcraft and theft are held in great detestation by them, and are usually punished with death; but murder is left to the relations of the deceased, to avenge as they think proper; and they are commonly satisfied with something less than a capital punishment, because the standing maxim of every individual is, to diminish the numbers of the community as little as possible.

They possess indefatigable patience, and of course are capable of learning our arts with great success; but the universal opinion of them all being, that

---

* Phil. Trans. 1773. Vol. LXIII. p. 142.　　　　† Phil. Trans. 1786. Vol. LXXVI. p. 229.

every thing is contemptible except war, they are in consequence prevented from improving in civilization; because every such improvement is looked upon as a degradation. Their language is very emphatic, and abounds with the boldest figures, like that of the Asiatic nations. The Indians on the river St. Lawrence, in New England, and on the Ohio, understand one another, and of course speak nearly the same language; but the six nations, though surrounded by these, speak a totally different language. The letters M and P are wanting in it, and they do not understand, nor are capable of conversing, with the other American Indians.

Patagonians.   3. The Patagonians are a tribe of Indians who inhabit the extremity of the Southern Continent of America. This country was visited by several of our navigators, who represented the inhabitants as of a gigantic stature. According to Carteret, they are never below six feet, and vary from six feet to six feet seven inches.* Succeeding accounts have taken away considerably from this surprizing stature, and have produced a belief, that the inhabitants seen by Carteret had, by some accident, been of an uncommon size.

Falkland islands.   4. The Falkland islands, or, as the French call them, the Maloine islands, are situated between the latitude 52° 26′ and 51° 6′ south, and longitude 56° and 60° 30′ west. They are not very far from the coast of South America, and are very numerous, forming a mass of broken high lands, or very low sedgy keys and sunken rocks. There was a project formed at one time in this country to establish a colony on these islands. In consequence of this idea, they were very carefully surveyed, and an account of them published in the Philosophical Transactions, by Mr. Clayton, of the navy.† The following are the particulars contained in that paper:

The eastmost island is the largest. There was a settlement formed on it by M. Bougainville, in 1764; and this settlement was purchased by the Crown of Spain, but soon abandoned. The island next to this is also large. The third island is called Saunder's island. On it the English settlement was made, chiefly because it afforded the most convenient landing place. No tree is to be found on these islands, but they contain a variety of shrubs, grapes, and other plants, which Mr. Clayton describes; but, as he was unacquainted with botany, his descriptions are quite insufficient to convey any adequate notion of the plants. The south and west winds blow over these islands for two thirds of the year, and are often boisterous. The north and north-west winds are the most pleasant, but seldom blow long; the north-east winds are moist, foggy, and unwholesome. The winds from east to south are very pernicious, blasting every thing, and even destroying animals, but, fortunately, they never blow

* Phil. Trans. 1770. Vol. LX. p. 20.   † Phil. Trans. 1776. Vol. LXVI. p. 99.

long at a time. The sea abounds with mullets, muscles, limpets, cray-fish, crabs, logger-heads, and with a transparent fish of the size of a herring. The river in the large island furnishes delicious trout.

There is only one quadruped on these islands, a species of fox, very like the English fox, but exceedingly shy. The amphibious animals are the sea lion and the seal, of which there are various species. The birds, are penguins, geese, ducks, widgeons, teals, shelldrakes, snipes, besides many small birds and sea birds. The soil of the island is boggy, barren, and rocky. The summer is usually cold, the thermometer hardly rising higher than 64°. The winter, on the other hand, is warm, for the north wind blows most frequently, and the thermometer seldom sinks below the freezing point. Mr. Clayton describes a singular vegetable, which rises like a mole hill, about six feet round, exsudes a kind of gum like balsam of capaiva, and is covered with short dark-green leaves. It would seem, from this account of it, to be a species of cactus, or some plant analagous to that singular genus.

5. The country, called Thibet, is of an immense size, and appears, from the Thibet. united result of all observations, to be the most elevated spot on the face of the earth. It is divided from Bengal by a ridge of mountains, being a continuation of the vast chain of Caucasus, which, passing along the north of Persia, penetrates between Bengal and Thibet, and loses itself in China. Thibet had scarcely ever been visited by any European since the time of Marco Polo, except a few wandering and ignorant missionaries, till an accident put it in the power of Mr. Hastings, Governor-general of Bengal, to send a deputation to explore the country. On the death of one of the petty Rajahs, on the northern border of Bengal, a dispute arose about the succession, and the inhabitants of Boutan, which constitutes the southern part of Thibet, where called in by one party, while the British took the field to assist the other. Our troops, in an attack which they made upon a town called Gooch Behar, first met with the Boutaners, and nothing could exceed the surprise of both parties at the rencounter. The Boutaners, instead of the timid and naked Indians who fled at the first onset, found themselves opposed by a body of men regularly clothed and disciplined, moving in exact order, officered by men whose complexions and appearances were quite new to them, and employing fire-arms with great steadiness and execution. The British, on the other hand, met with a robust and uncouth race of men, in dresses quite new, armed with bows and arrows, and other weapons, to which they had not been accustomed. The town was taken, and many articles belonging to the Boutaners were sent down to Mr Hastings. The Tayshoo Lama, who at that time governed Thibet, during the infancy of the Delai Lama, interposed his good offices in behalf of the Dah Terriah, or Governor of Boutan, who was dependent upon him, and sent a letter to Mr. Hastings soliciting peace. The request was readily granted, upon

the most favourable terms; and Mr. Hastings took the opportunity of sending an embassy to the Tayshoo Lama, in order to negociate a treaty of alliance and commerce, and to get some accurate information respecting the country of Thibet. Mr. Bogle was selected for the purpose, and dispatched. He penetrated to the capital of Thibet, was absent 15 months, and executed his commission to the entire satisfaction of the Governor-general. Mr. John Stewart, at that time in Bengal, drew up an account of this journey from authentic documents, and this account was published in the Philosophical Transactions.* A subsequent journey to the same country was made by Mr. Saunders, a surgeon, and he communicated a very entertaining account of what he saw to the Royal Society, which was likewise published in the Philosophical Transactions.† It relates chiefly to the vegetable and mineral productions of Thibet, to the climate, and to the diseases to which the natives are liable. The following are the most important particulars contained in these two excellent papers.

The country is divided into two districts, very different from each other in situation, aspect, and climate. The southernmost of these districts is called by the inhabitants of Bengal, Boutan; and by the natives, Docpo. It is under the government of the Dah Terriah, as he is called, a sort of sovereign dependent upon the Delai Lama. Boutan is exceedingly hilly, and there is a constant ascent from the south to the north of it. The hills are covered with wood; through the valleys flow innumerable rivulets; the country is fertile, and the climate delicious. Mr. Saunders describes a very great number of plants which he met with, both Indian and European, for the country is capable of bringing both to a state of perfection. The people are copper-coloured, rather tall, hasty in their tempers, addicted to the use of spirituous liquors, but very honest in their dealings. The farthest north, and most extensive district, is called Thibet, and, by the natives, Pû. It begins at the summit of the ridge where Boutan terminates, and continues almost to the borders of the Russian empire. The whole of it is upon an elevated base, and consists of broad valleys and moderate hills. It is exceedingly bleak and desolate, being destitute of wood, and almost of vegetation. The climate, notwithstanding the position of the country so near the tropic of Cancer, is exceedingly severe. At Chamnanning, which is only 8° north of Calcutta, Mr. Bogle frequently observed the thermometer at 3°; and Mr. Saunders, whose thermometer scale went no lower than 16°, was unable to determine the real temperature. The standing waters were frozen in the middle of April.

The inhabitants of Thibet are a smaller race than their southern neighbours of Boutan, and less robust; their complections are fairer, and their faces Tarta-

---

* Phil. Trans. 1777. Vol. LXVII. p. 465.   † Phil. Trans. 1789. Vol. LXXIX. p. 79.

rian. They are mild and cheerful in their dispositions, and the higher ranks, according to Mr. Bogle, are remarkably polite and easy in their manners. The common people dress in coarse woollen stuffs, the product of the country; but the higher ranks wear European broad cloth, or Chinese silk lined with the finest Siberian furs. Their food is chiefly milk, cheese, and butter, with a coarse species of barley and peas. They have also plenty of animal food, and they procure wheat and rice from Bengal. They have a particular way of preparing their mutton, by exposing the whole carcase, newly killed, to the cold winds of August and September, which speedily dry up the juices, and in this state it will keep all the year round, and is eaten raw.

Thibet is considered as depending upon China, and there are two mandarines and 1,000 Chinese, who reside by way of garrison in Lahissa, the capital of the country, to support the government. But, in reality, the influence of the Chinese extends but a very little way; and the Delai Lama, whose power is founded upon religious veneration, and upon personal affection, rules within the country with uncontrouled power. The Delai Lama constitutes the head of one of the widest spread religions upon earth; he is venerated by the Tartars, and, among some of the remote tribes, is considered as the Deity himself. Even the Emperor of China acknowledges him in his religious capacity, and supports his nuncio, at Pekin, at a great expence. The Lamas, or priests, form a very numerous and powerful body in Thibet, upon whom, in some measure, depends the government of the state. The Delai Lama is both priest and king. When he dies the opinion entertained is, that his soul leaves an old crazy habitation, and takes up its abode in the body of an infant born on the same day. This infant is always a Lama, and is discovered by certain marks known only to the Lamas. During his minority (for he is made Delai Lama as soon as discovered) the Tayshoo Lama, who is second in power, governs the country in his name. There are many monasteries of these Lamas in the country, and they resemble the Roman Catholic convents so much in their appearance, that it is very easy to mistake them for Christians. Indeed the religion has been supposed to be a corruption of Christianity; it is simple, conveys exalted notions of the Deity, and inculcates a pure morality; but it is greatly corrupted.

Polygamy exists in this country under a very peculiar form. It is customary for a family of five or six brothers to have a single wife among them, and they live in this way in the greatest harmony. They neither bury nor burn their dead, but expose them on the sides of mountains to be devoured by wild beasts, or destroyed by the weather. They have a peculiar species of cow, with a fine flowing tail like a mare's, which they hold in great veneration. They venerate also the waters of the Ganges, and requested liberty from Mr. Hastings to build a place of worship on its banks; a request which was very readily granted. The Delai Lama lives at Patelli, a vast palace about seven miles from Lahassa, and situated on the banks of the Barampooter. The rooms are large and highly finished in the Chinese style; but they are unacquainted with the use of stairs and of windows, they go up into their upper apartments by means of ladders, and their windows are merely holes in the ceiling which they can shut against the weather. Lahassa is a large town, and is frequented by numerous caravans of merchants from China, Cassimere, Siberia, &c. The caravans take two years in travelling between Lahassa and Pekin, a distance of 2,000 miles; yet an express arrives from the one capital to the other in three weeks. The trade with Siberia is carried on by caravans to Seling, situated on the Baykal-lake. The inhabitants of Thibet have four staples sufficient to insure them an extensive trade with all their neighbours. These are—1. The cow tails, mentioned above, which are sold high, and used all over India to brush away flies from the face. 2. The fine wool of which the shawls are made. This wool is bought up by the Cassimere merchants, and carried into their own country, where it is manufactured into shawls. 3. Musk, which is obtained from a species of deer, an inhabitant of their mountains. 4. Gold, which is found in considerable quantities in their rivers in the form of dust. There are likewise mines of it in the northern provinces belonging to the Delai Lama, which are farmed out to adventurers at a certain rent. The Tayshoo Lama, whom Mr. Bogle visited, was a very intelligent and excellent man, well acquainted with the policy and power of the different nations of Asia and with Russia, and eager to get acquainted with that of the other European states. Mr. Bogle, at his request, drew up a statement of the most interesting particulars relative to Europe in the language of Hindostan, which was translated into that of Thibet for the use of the Lama.

Mr. Saunders describes Thibet as abounding in valuable minerals; but his description is not sufficiently precise to convey much information. He mentions lime-stone, granite, quartz, flint, and chalk, as having occurred to his observation, and says, that mines of gold, lead, mercury, and copper, seem to abound. But the fuel is so scarce, that they use it only for dressing food, and spend even the cold winter months without fire, trusting for heat to the furs in which they wrap themselves.

Diseases.   The diseases are not numerous. Swellings in the neck are very common. This disease has been usually ascribed to snow water; but Mr. Saunders, with great judgment, hesitates to admit the accuracy of that notion. The venereal disease is likewise common, and is cured by mercury the same as in Europe. The medicine which they employ is thus prepared. They mix together alum, nitre, vermilion, and mercury, in a certain proportion, and put the mixture into the bottom of a large earthen pot. Over this they lute a small inverted pot. A measured quantity of fuel is burnt over the top of the small inverted pot, and the fire continued for about 40 minutes. When the pots are unluted, the mercury has disappeared, and the medicine is fit for use. This medicine consists essentially either of an oxyde of mercury or a nitrate, or more probably of a mixture of both. The alum, assisted by heat, will decompose the nitre, and the nitric acid set at liberty will speedily oxydise the mercury, and combine with it if in sufficient quantity. The vermilion seems useless. They carry salivation to a great height, and continue it for ten or twelve days. When the small-pox appears among them, they shut up the infected to prevent the disease from spreading, and leave them to their fate.

BOOK II.

OF MATHEMATICS.

THE word *Mathematics,* originally Μαθησις, signifies, literally, *instruction;* Meaning of the word. and is usually supposed to have been applied by the Greeks to a science of which they were in some measure the inventors, in consequence of the high estimation in which they held it. This science treats of those *quantities* which Object. are capable of being measured, and which on that account have been called *proper quantities.* They consist of the following three, namely, *extension, duration, number.* All other quantities are *improper,* and can only be introduced into mathematics, by artificially applying to them some proper quantity as a measure. Thus *velocity,* considered by itself, is not susceptible of measurement, and, of course, we cannot determine how much one velocity surpasses another. But if we make the *space* passed over in a given time the measure of velocities, then they become susceptible of accurate comparison, and consequently of mathematical investigation. Some improper quantities are of such a kind that we cannot apply any proper quantity as an artificial measure of them. Beauty, virtue, love, and many others, are of this nature. Such quantities can never become the objects of mathematical investigation. We may, indeed, throw our reasoning respecting them into a mathematical shape, and ring changes upon mathematical formulas; but such reasoning can never promote our knowledge of these improper quantities, or contribute in the least to the advancement of truth.*

The proper subjects of mathematics, then, are *number* and *extension;* for *duration* may be easily joined to either the one or the other of these quantities. Hence the science naturally divides itself into two parts; namely, *arithmetic,* which treats of *number;* and *geometry,* which treats of *extension.* These two parts were, for many ages, considered separately. But, they have been gradually

---

* From not attending to these obvious principles, different metaphysical and moral treatises have appeared, in which the various objects of these sciences have been treated with all the parade of mathematical demonstration. The best account of *quantity,* which we have seen, and it ought to be studied by every metaphysician, is by the celebrated Dr. Reid, in Phil. Trans. 1748. Vol. XLV. p. 505. In the New Abridgement of the Transactions, by Dr. Hutton, &c., this paper is, by some inadvertence, ascribed to Dr. Miles, who was only the transmitter of it to the Society.

6

so united and blended together, that, in the present improved state of the science, they no longer admit of separate investigation. They constitute two streams, rising from distant sources, and running a long and parallel course: at first, only small and inconsiderable, but, fed each during its progress by a thousand rills, they swell into two mighty rivers, which, gradually approaching, at last mix their waters together, and flow united into the vast ocean of truth.

Importance. Mathematics constitute the most splendid monument of the human intellect. They are of vast extent, and, unlike the other sciences, they have never suffered any retrogression. Every mathematical truth, when once developed by that rigid principle of demonstration, which alone the true mathematician admits, constitutes ever after a part of the science, which no man, who understands the subject, will venture to call in question. At times, indeed, the progress of the science has been in a great measure stationary, while at others it has moved with prodigious celerity but at no period has it been retrograde. Mathematics have been divided into two great branches. The first branch, which treats of pure number and extension, has been called *pure mathematics ;* the second branch comprehends most of the sciences belonging to *mechanical philosophy ;* which, by means of certain contrivances, have been subjected to mathematical investigation, and owe most of their improvements to that powerful auxiliary. This branch, which is of prodigious extent, has received the name of *mixed mathematics.* It will occupy our attention in the next book. As to *pure mathematics,* which constitute our present object, they comprehend such a vast mass of important facts, and would require so very copious a developement in order to lay them fully before our readers, that we feel ourselves overwhelmed with superabundance of matter, and cannot even venture upon a subdivision of the subject, without the certainty of over-stepping the just limits of the present work. The papers on pure mathematics, in the Philosophical Transactions, amount to 208 ; and the writers of them are no fewer than 74.* Instead of

---

* The following is a list of all the persons who have published papers on pure mathematics, in the Philosophical Transactions :

Mercator.	James Gregory.	Viviani.	Castilion.
Collins.	Brounker.	Robartes.	Jones.
Petty.	Wren.	Cotes.	T. Simpson.
Wood.	Barrow.	Keill.	Reid.
Wallis.	Sluse.	Maclaurin.	Robertson.
Halley.	Leibnitz.	Machin.	Dodson.
Demoivre.	Ash.	Simson.	Landen.
Thornycroft.	D. Gregory.	Stirling.	Murdoch.
Colson.	Caswell.	Brakenridge.	Pemberton.
Craig.	Newton.	Gersten.	Bayes.
Brook Taylor.	Sault.	Eames.	Waring.
Hobbes.	Ditton.	Stone.	Price.

attempting a regular account of them, which would require a much greater space than we can allow, we shall rather venture upon a short historical sketch of the progress of mathematics, from their original invention, dwelling chiefly upon those improvements which have originated with the Members of the Royal Society ; and noticing, as we proceed, some of the most important papers which have made their appearance in the Philosophical Transactions.

The origin of mathematics, like that of all other sciences, is buried in im- Origin. penetrable obscurity. The Greeks drew their first notions of the science from the Egyptians, who indeed were their masters in almost every species of knowledge. What progress it had made among the Egyptians we have no accurate means of knowing ; but, from the nature of the discoveries, which threw the first Grecian mathematicians into raptures,* we may be certain that they had never passed the first elements of mathematical knowledge. Suidas informs us, that Anaximander, the celebrated successor and disciple of Thales, wrote the first elementary treatise on geometry. This philosopher flourished about 560 years before the commencement of the Christian æra. This period, therefore, as far as regards historical information, may be considered as but little posterior to the origin of mathematics in Greece. Hippocrates, of Chios, who enjoyed Hippocrates. a very high reputation among the ancients, shows us by his researches that, in his time, geometry had advanced to more difficult speculations. He was a merchant, and, if we believe Aristotle, of a degree of simplicity bordering on stupidity, of which the custom-house officers at Byzantium taking advantage, plundered him so unmercifully, that he was reduced to a state bordering upon ruin. He went to Athens, in order if possible to re-establish his affairs, and happening by accident to saunter into a school of philosophers, he was so much captivated by the charms of geometry that he forthwith renounced his mercantile speculations, and devoted himself wholly to the study of mathematics. The discovery which rendered him famous among the ancients, and which has transmitted his name to posterity, is that of a space contained between two arches of circles, and known by the name of the lunula of Hippocrates, which is perfectly quadrible. His reasoning is founded on the 47th proposition of the first book of Euclid, and the quadrature is rather a piece of address, by

Horsefall.	Hutton.	Atwood.	Wildbore.
Horseley.	Maseres.	Wales.	L'Huillier.
Winthrop.	Playfair.	Vince.	Sewel.
Lexel.	Milner.	Landerbeck.	Wood.
Stedman.	Hellins.	Nicholson.	Brougham.
Lyons.	Ludlam.	Morgan.	Wilson.
Glenie.	Stanhope.		

* Pythagoras, for example, is said to have sacrificed a hecatomb, when he discovered that the square of the hypothenuse of a right angled triangle is equal to the squares of the two sides.

removing a portion common to two spaces, and thus leaving a circular space, equal to a rectilineal, than a real quadrature. Attempts have been made to discover the quadrature of the circle by the same process, but they have been unsuccessful.

Plato.

But it is to Plato, and to the Platonic school founded by that illustrious philosopher, that geometry lies under the greatest obligations for its early advancement. The history of Plato is so generally known, that it is only necessary to touch upon it here. He was an Athenian, and is believed to have been born during the 88th olympiad. He turned his first attention to poetry, and composed four dramatic pieces before he had completed his 18th year, intending to stand candidate for the laurel on the olympic stage. But happening by accident, on the day before the contest, to hear an harangue of Socrates, before the Bacchanals, he burnt all his poems that very evening, and entered himself soon after a pupil of that celebrated philosopher. When Socrates was condemned for impiety, Plato, who was at that time a senator, exerted himself in favour of his master, and offered him a sum of money sufficient to procure his discharge; which Socrates refused to accept. Disgusted at the fate of this illustrious philosopher, he left Athens, and travelled for many years in quest of knowledge. He first visited Italy, to learn the doctrines of the Pythagorean school; thence he passed to Cyrene, where Theodorus, at that time, enjoyed a high reputation for his skill in mathematics; from Cyrene he passed into Egypt, where he placed himself under the tuition of the priests, and learned from them all their opinions respecting every branch of science with which they were acquainted; his three voyages to Sicily, and the treatment which he received from the elder and younger Dionysius; his friendship for Dion; his exertions to procure liberty for the Syracusans; the resentment of the elder Dionysius at the freedom of his animadversions; his delivering him into the hand of the Lacedemonian ambassador, who sold him for a slave; the purchase of him by Anniceres, the Cyrenean, who generously restored him his liberty, and sent him back to Athens—All these are historical facts so universally known, that the bare mention of them here is sufficient. He finally settled at Athens, and taught in the academy; a low marshy situation, covered with trees, remarkably unhealthy, which he pitched upon, it is said, as a place well suited to reduce the too great corpulency which he had contracted. He died in the eighty-first year of his age, in the full enjoyment of the highest reputation; being almost adored by all the inhabitants of Greece. His estimation of mathematical knowledge was so great, that he ordered the following inscription to be engraved over the gate of the academy: *Let no one presume to enter here unless he has a taste for geometry and mathematics.* He has left us no writings on mathematical subjects, so that we do not know what his individual discoveries in that science were; though it is universally admitted that he paid very particular attention to it. Almost the whole of his works are in the form of dialogues, and his style was universally admired by all the ancients, who compared it with that of Homer, and considered him as a philosopher what Homer was as a poet. Cicero says, that Jupiter, were he to condescend to speak to men, would make use of the style of Plato; and the encomium of Quintilian is, if possible, still higher.

Three very remarkable mathematical discoveries were made by the Platonic school very soon after the time of Plato himself; namely, the analysis used by the ancients, the conic sections, and geometrical loci. *Discoveries of the Platonic school.*

The method of *analysis* is universally ascribed to Plato himself, and was without doubt one of the most important made by the ancients in any department of mathematics, because it supplied them with an instrument for extending the bounds of mathematical investigations. There are two modes of proceeding in mathematics; by *synthesis* and *analysis*. The first is employed, when we wish only to communicate the knowledge which has been already acquired. We set out from some principle universally admitted, and proceed without interruption from consequence to consequence, deducing one truth from another previously ascertained. The Elements of Euclid, and indeed almost all the mathematical writings of the ancients, exhibit specimens of this method, which have always been admired for their elegance and beauty. Analysis proceeds in a different manner, and is employed when we wish to resolve some problem, or discover the truth or falsehood of some mathematical position. It begins by supposing the truth of the point to be discussed, and deduces the consequences that follow from this assumption, till at last we come to some consequence obviously true or obviously false, from which the truth or falsehood of the topic investigated becomes evident. Then, by tracing our steps backwards from this point, we establish the demonstration. Archimedes and Apollonius exhibit several specimens of this mode of investigation, and Pappus uses it almost perpetually.* There is a peculiar species of analysis used by the ancients known by the name of *porisms*, on which Euclid wrote a treatise unfortunately lost, concerning the nature of which modern mathematicians were long at a loss what opinion to form. At last Dr. Simson restored the lost book of Euclid from some indications given by Pappus, and thus explained the nature of porisms. The subject has been carried further by Mr. Playfair, in the Transactions of the Royal Society of Edinburgh; and Mr. Wallis, Mr. Brougham, and others, have since furnished us with an ample store of curious porisms. There are two papers on the subject in the Philosophical Transactions. The first by Dr. Simson, giving a great number of Euclid's porisms, *Analysis. Porisms.*

---

* Mr. Leslie, in his elementary work on geometry lately published, has given us ample specimens of the analysis of the ancients.

and containing the first details respecting porisms.* The second is a valuable paper by Mr. Brougham, which contains a variety of curious porisms chiefly in the higher geometry.†

Conic sections. The discoverer of the conic sections, those curves which make so conspicuous a figure both in mathematics and in nature, is not known. Some, from some allusions in a commentary by Proclus on Euclid's Elements, have supposed that the inventor was Plato himself; others, from an allusion by Eratosthenes, have ascribed the discovery to Menechmes. But as antiquity has been silent on the subject, we have no right to draw conclusions. It is certain, however, that they were known at least soon after the age of Plato, that they were discovered in the Platonic school, and that considerable progress was made by that school in the investigation of their properties. The conic sections consist of three different curves (supposing the circle to be a species of ellipse); namely, the *parabola*, the *ellipse*, and *hyperbola*, formed by cutting the cone by a plane passing through it in various directions. The planetary bodies, comets, projectiles, move in these curves; which renders it necessary to be acquainted with their properties before we can investigate the laws which regulate the motions of these bodies.

Geometrical loci. The third invention of the Platonic school is that of *geometrical loci*. By *locus*, in geometry, is meant a series of points, each of which resolves a problem, capable by its nature of a variety of solutions. Suppose, for example, it were required to describe a given number of equal triangles upon a certain base. We have only to describe a line parallel to that base, and construct at pleasure triangles upon the base, and having their apex in the parallel. We know, from the first elements of mathematics, that they would be all equal. In this case the parallel drawn is the *locus* of all these equal triangles. Suppose again, that we were required to construct an indefinite number of right-angled triangles upon a given line. We have only to describe upon the line a semicircle. As in that case all the lines drawn from the extremity of the given line to any point in the circumference will form a right angle. Here the semicircle described is the locus of all these triangles. The doctrine of *loci*, on account of its importance, has been treated of, and elucidated, with great care by the moderns. There is one problem on the subject, a problem of great celebrity among the ancients, which has been fully and elegantly resolved by Dr. Pemberton in the Philosophical Transactions; namely, the locus for three and four lines.‡

Problem of doubling a cube. There were two problems of great celebrity among the ancients which took their rise in the Platonic school, and therefore cannot be omitted here. We mean the problem for doubling the cube, and for the trisection of an angle.

---

* Phil. Trans. 1723. Vol. XXXII. p. 330.   † Phil. Trans. 1798. Vol. LXXXVIII. p. 378.
‡ Phil. Trans. 1763. Vol. LIII. p. 496.

Valerius Maximus* gives us the following account of the origin of the first of these problems. A pestilence ravaging Attica, a deputation was sent to Delphi to consult the oracle of Apollo on the means of removing the destructive disease. The oracle in answer enjoined them to double the altar of the god. This command appearing of easy execution, they pulled down the old altar, and, doubling each side, they constructed another, not double, but eight times as large as before. The plague still continuing, a new deputation was sent to inquire the reason. The god informed them that they had not executed his commands. Suspecting some mystery in this duplication, they consulted the mathematicians, who were very much puzzled what answer to return. Plato in particular, who was first consulted, not being able to resolve the problem, saved himself by a piece of finesse, and sent them to Euclid as to a geometrician by trade. This story is obviously a fable; and the reference of Plato to Euclid, who was posterior to him by half a century, shows it to a demonstration. The difficulty of the problem was sufficient to give it charms in the eye of mathematicians, who have been always accustomed to treat with contempt what was of easy solution, and to bend the whole efforts of their mind to the explanation of difficult and arduous problems. It was soon perceived, that this problem might be reduced to the discovery of two mean proportionals between two given lines. This cannot be effected by means of elementary geometry. The employment of a conic section, or of a curve of a higher order, is necessary. The ancients have furnished us with various solutions of the problem, all remarkable for their ingenuity. Plato gave a mechanical solution. Archytas resolved the problem by means of a curve described upon a cylinder. At last Menechmes gave two very elegant solutions, by means of two parabolas. Nicomedes afterwards invented the conchoid, a curve by means of which the problem is solved with much ease and elegance.

The trisection of an angle is a problem scarcely less ancient than the duplication of the cube, and scarcely less celebrated. It involves difficulties of a similar kind, and cannot be resolved by means of the ruler and compass alone. Other lines of more difficult construction must be introduced. The quadratrix, a curve invented by Dinostrates, seems to have been contrived on purpose for the solution of this problem. Dinostrates was the brother of Menechmes, a disciple of Plato, who distinguished himself by his celebrated solution of the problem for doubling the cube, and who appears to have made great improvements in the theory of conic sections. There is a paper in the Transactions upon the construction of the quadratrix.† *[margin: Trisection of an angle.]*

We come now to notice the most extraordinary mathematician among the ancients, to whom antiquity, with one consent, gave the palm of genius and in- *[margin: Archimedes.]*

---

* Lib. VII. c. 13.   † Phil. Trans. 1700. Vol. XXII. p. 445.

vention, we mean Archimedes, a name familiar to every tiro in mathematics or science. Dr. Wallis, speaking of him, employs the following remarkable expressions: *Vir stupendæ sagacitatis, qui prima fundamenta posuit inventionum fer omnium, de quibus promovendis ætas nostra gloriatur.* Archimedes was born at Syracuse, about the year 287, before the commencement of the Christian æra; and, if we believe Plutarch, was a near relation of Hiero king of Syracuse.

His pursuits. His mechanical inventions were as extraordinary as his mathematical speculations; but unfortunately he did not consider it as worth his while to convey any account of them to posterity, so that the greater part of them, to the great injury of mankind, have been unfortunately lost. A splendid edition of the works of Archimedes, with a Latin translation, was published at Oxford in 1792. The edition had been prepared by Torelli, an Italian mathematician, who dying before he was able to put it to the press, the manuscript was purchased by some English gentlemen, and published at Oxford. Archimedes devoted himself, in a peculiar manner, to the mensuration of curve surfaces and solids, a subject at that time new in mathematics; and, in consequence of his perseverance and sagacity, he reaped a rich harvest of brilliant discoveries. In his two books on the *sphere* and *cylinder*, he determines the surface of these bodies, their solidity, and that of their various sections. His discovery, that the sphere is two thirds of the circumscribed cylinder, both in surface and solidity, seems to have given him peculiar delight, as he requested to have these figures engraven on his tomb; a request which was complied with, and by means of which Cicero, many years afterwards, discovered it in the neighbourhood of Syracuse. In his book on the mensuration of the circle, he demonstrates that this figure is equal to a triangle, whose base is equal to the circumference, and its height to the radius. Then, by methods now familiarly known to every tiro in mathematics, he demonstrates that if radius be unity, the circumference is less than 3 $\frac{1}{7}$, and greater than 3$\frac{10}{71}$. His quadrature of the parabola, which he demonstrated equal to $\frac{2}{3}$ of the circumscribed rectangle, and his curious investigations respecting the spiral which retains his name, have been particularly admired by the moderns. His *psammites* makes us, in some measure, acquainted with the system of arithmetic employed by the Greeks, which, though difficult and intricate when compared with ours, nevertheless was tolerably perfect as far as principles went. It would be tedious to run over all his writings; but we must not omit to mention that he laid the foundation of the science of hydrostatics. He was led to turn his thoughts to the subject by the following accident. Hiero, King of Syracuse, had given a quantity of gold to a goldsmith to make into a coronet. The work was executed in an exquisite manner, and was of the requisite weight; but some circumstances led the king to suspect, that the jeweller had abstracted for his own use a portion of the gold, and replaced it with silver. Wishing to ascertain whether his suspicions were well founded, without destroying the coronet, he applied to Archimedes, who for some time was at a loss for a solution. At last the truth struck him all at once while bathing, and he was so transported with the discovery that, if we believe Vitruvius, he ran naked through the streets of Syracuse crying out, Ευρηκα, ευρηκα; *I have found it, I have found it.* His discovery was, that every body, when plunged into water, loses a portion of its weight equal to the weight of a quantity of water equal to its own bulk. Hence he deduced the method of determining the specific gravity of bodies as at present practised; and by comparing the specific gravity of the coronet with that of gold and of silver it was easy, supposing no mutual penetration, to determine whether the coronet was an alloy; and, if it was, what proportion of silver it contained.

To him likewise we are indebted for laying the foundation of the science of *statics,* and for giving us a demonstration of the fundamental property of the lever. The ancients ascribed to Archimedes 40 mechanical inventions, but the names only of a few of them have been transmitted to posterity. The ingenious instrument for raising water, known by the name of Archimedes's screw, and which he is said to have invented in Egypt, is one of them, and is familiarly known to the moderns. The sphere, which he contrived to represent the motions of the stars, is spoken of with rapture by Cicero,[*] and has been celebrated by different poets.[†] The account is without doubt greatly embellished. But the mechanical invention of Archimedes, which has given occasion to the greatest discussion, is the burning mirrors by which he is said to have set on fire the His burning Roman ships during the siege of Syracuse. The story is not mentioned by mirrors. Polybius, who wrote the history of that siege not more than 50 years after. It is passed over in equal silence by Livy and by Plutarch, though they dwell with a sort of complaisance upon the exploits of Archimedes. It is founded entirely upon relations of Zonaras and of Tzetzes, who wrote many ages after the time

---

[*] In his Tusculan Questions, and his Treatise de Nat. Deorum.

[†] The epigram of Claudian is well known. As it is beautiful, and contains the only information on the subject transmitted by the ancients, we shall insert it.

> Jupiter in parvo cum cerneret ethera vitro,
> Risit, et ad superos talia dicta dedit:
> Huccine mortalis progressa potentia curæ;
> Jam meus in fragili luditur orbe labor.
> Jura poli, utrumque fidem, legesque deorum,
> Ecce Syracusius transtulit arte senex.
> Inclusus variis famulatur spiritus astris,
> Et vivum certis motibus urget opus.
> Percurrit proprium mentitus signifer annum,
> Et simulata nova Cynthia mense redit.
> Jamque suum volvens audax industria mundum
> Gaudet, et humana sidera mente regit.
> Quid falso insontem tonitru salmonea miror,
> Æmula naturæ parva reperta manus.

of Archimedes.   But these writers appeal to the authority of Dion, of Diodorus Siculus, Heron, Pappus, and Anthemius, men certainly of sufficient weight to establish the truth of any fact to which they gave their united testimony. The verses of Tzetzes are so remarkable, that we shall here insert a translation of them :

> Cum autem Marcellus removisset illas (naves) ad jactum arcus,
> Educens quod speculum fabricavit senex,
> A distantia autem commemorati speculi,
> Parva ejusmodi specilla cum posuisset angulis quadrupla
> Quæ movebantur scamis, et quibusdam γυγγλιμοις,*
> Medium illud posuit radiorum solis.
> Refractis deinceps in hoc radiis
> Exarsio sublata est formidabilis ignita navibus, &c.
> Dion et Diodorus scribunt historiam ;
> Et cum ipsis multi meminerunt Archimedis ;
> Anthemius quidem imprimis, qui paradoxa scripsit,
> Heron, Philon, Pappusque, ac omnis mechanographus,
> Ex quibus legimus et speculorum incensiones,
> Omnemque aliam descriptionem rerum mechanicarum
> Ponderum tractricem, pneumaticam ac hydroscopia,
> Idque ex senis hujus Archimedis libris.

It is obvious, that it was not possible for Archimedes to burn the ships of the Romans by any simple mirror, either concave or convex. The focus of all such, however enormous we can suppose them, is too short to be of any use for such a purpose.   The above passage of Tzetzes led Kircher to another idea. He conceived that Archimedes might use a number of small plain mirrors, so placed that all of them might reflect the light of the sun to the same point; and Buffon afterwards constructed a burning glass upon this principle, which was capable of melting lead and tin at the distance of 140 French feet, and of kindling wood at a much greater distance.†

But let us hasten to finish the life of this extraordinary mathematician.   The successor of Hiero having unfortunately quarrelled with the Romans, that ambitious people laid hold of the opportunity to make themselves masters of Sicily. After various successes, the celebrated Marcellus, the Roman General, laid siege to Syracuse in the year 212, before the commencement of the Christian æra.   It was upon this occasion that the mechanical genius of Archimedes burst forth in all its lustre.   His countrymen, confounded at the success, and at the name of the Romans, would have made little resistance; but Archimedes roused their courage, and animated them to one of the most vigorous defences recorded in history.   The machines of Archimedes, each more formidable than another, dis-

---

* Hinges.   † Mem. Par. 1746.

concerted the Roman engineers, and filled the breasts of the soldiers with terror and dismay. The appearance of a beam of wood, or a bit of rope, from the walls, set them instantly to flight. Marcellus, despairing of success, converted the siege into a blockade, and waited patiently for a favourable opportunity to surprise the place. The confidence of the Syracusans at length furnished him with one. Occupied in celebrating the festival of Diana, they left their walls defenceless. The Romans, always on the watch, perceived the error of their enemies, applied scaling ladders to the wall, entered the city without opposition, took it, and pillaged it. Marcellus had given orders to spare the life of Archimedes, but they were not obeyed. This extraordinary man, totally absorbed in a mathematical investigation, was insensible of all that had happened; and when a Roman soldier entered his apartment, and ordered him to follow him to the general, he requested him to be gone, and not to derange his figure. The soldier, either provoked at this answer, or tempted by the richness of his apartment, put him to death. Thus perished Archimedes with the liberty of his country.

Archimedes was attached to no school, and of course had no predecessor nor successor in Syracuse. But in his time there existed a celebrated mathematical school at Alexandria, in Egypt, founded by the Ptolemies, the successors of Alexander the Great, where a succession of eminent mathematicians appeared, and where the last embers of Grecian science continued to be fanned and cherished, after they had become cold and extinct in every other part of the Roman empire. One of the first, and not the least celebrated of the mathematicians, whom the Ptolemies collected in Egypt, was Euclid, the author of the *Elements of Geometry;* a book better known, and which has run through a greater number of editions than any other, the Old and New Testament excepted. Hardly any thing is known concerning Euclid. He is said (and Pappus is quoted, though I cannot find the passage,) to have been born in Alexandria, and to have taught mathematics in that city about 300 years before the commencement of the Christian era. His character is drawn by Pappus, in the most advantageous terms. Good-natured and modest; he was peculiarly attentive to all mathematicians, and never attempted to injure their reputation nor diminish their laurels. Ptolemy, King of Egypt, was fond of his conversation, and asked him one day if there was not an easier road to geometry than by his Elements? No, Prince, says he, there is no road made for kings to the mathematics.* His Elements have been translated into almost all languages. They consist of 13 books; but there are usually two others added, which are considered as the workmanship of Hypsicles, of Alexandria. The first six of these books treat of plane figures; the next three of arithmetic; the tenth, of

Margin note: Alexandrian school.

Margin note: Euclid.

---

* Non est regia ad mathematicam via.

2 K

commensurable and incommensurable quantities; the eleventh, twelfth, and thirteenth, of solids. The seventh, eighth, ninth, tenth, and thirteenth, are omitted in the ordinary editions, as of less value. The best English edition of Euclid, is Dr. Simson's, of Glasgow, which has been for many years classical in Great Britain. Most subsequent editions are merely Simson's, with a few changes. Mr. Playfair, indeed, has substituted the investigations of Archimedes for the eleventh and twelfth books of Euclid, which we consider as an improvement. Many attempts have been made by mathematicians to improve the arrangement adopted by Euclid, by introducing a more methodical one; but they have been all unsuccessful, losing in point of demonstration more than they gained in point of arrangement. In France, Euclid, as an elementary book, has been set aside, and other books substituted, much shorter, but greatly inferior in point of rigid demonstration. In this country, however, Euclid, and a just sense of the elegance of the ancient mode of demonstration, is still retained. Besides the Elements, Euclid was the author of many other mathematical works; some of which still remain. The Greek text of Euclid was first published at Basil, in 1533, under the care of J. Grynæus. It was accompanied with the copious commentary on the first book of the Elements, by Proclus. David Gregory published a superb edition of all Euclid's works, in 1703, in Greek and Latin. The edition is in folio, and does honour to the Oxford press.

Eratosthenes.   Eratosthenes is another mathematician of the Alexandrian school, who lived about the same period with Archimedes, and who was the author of various important works on geometry and astronomy, none of which remain. We mention him here on account of a curious method which he contrived for determining the prime numbers, known by the name of the *Sieve of Eratosthenes*. An excellent account of this valuable invention is given by Dr. Horseley, in the Philosophical Transactions, which he has partly divined and partly copied from the very imperfect account of it given by Nicomachus.*

Apollonius.   The next mathematician that claims our attention is Apollonius, the most celebrated of all those that adorned the Alexandrian school, and second only to Archimedes himself. He was born in Pamphilia, about 150 years before the commencement of the Christian era, and flourished chiefly during the reign of Ptolemy Philopater. Pappus gives us a very unfavourable idea of his disposition, representing him as vain, jealous, and revengeful, and ever ready to repress his contemporaries, and prevent them from rivalling him in his own department. His writings were very numerous and profound. But the most celebrated are his Conics, which contain every thing that the ancients knew respecting these curves. The first four books consist of a collection of the facts

---

* Phil. Trans. 1772. Vol. LXII. p. 327.

ascertained on the subject by his predecessors. The last four books contain his own profound investigations on these celebrated curves. An edition of the first four books, with a Latin translation, was published in 1537, by Memmius, a noble Venetian. The last four were supposed to be lost, and efforts were made by different geometricians to supply their place, from the indications given by Pappus. Maurolicus and Viviani were the most remarkable of these. While the latter was at work collecting materials for his intended publication, the celebrated Golius returned from the East, loaded with Arabic manuscripts, among which were the seven first books of the Conics of Apollonius. Golius announced his intention of publishing a translation of this work, so eagerly sought after by mathematicians; but some accident or other prevented him from accomplishing it, and the public thought no more of his manuscript; and it was generally concluded, notwithstanding the notice given by Golius, that the rest of Apollonius was lost for ever. But in the year 1658, Borelli, while examining the library of the Medici, during his passage through Florence, found an Arabian manuscript, the Italian title of which announced the eight. books of Apollonius on Conics. Delighted with this discovery, he ran over the manuscript, and judged, from a comparison of the figures, that it was really the work of the Grecian geometer. He made a religious moronite translate the title of the fifth part, which, according to the known division of Apollonius, treated of maxima & minima. The Duke of Tuscany generously trusted him with the manuscript, which he carried to Rome, and, by the assistance of *Abraham Ecchellensis,* skilled in the Oriental languages, it was translated into Latin, and published in 1661, with learned notes; rendered necessary by the extreme conciseness of the Arabian translator, or rather abridger. This edition contained only the first seven books of Apollonius, and there is every reason to believe that the eighth is entirely lost, as it is wanting in every Arabic manuscript hitherto examined. In 1710, a splendid edition of Apollonius was published at Oxford, under the care of Dr. Halley. The eighth book has been so well restored by that celebrated mathematician, so thoroughly acquainted with the geometry of the ancients, that we have no reason to regret the loss. All the other writings of Apollonius are lost; except his treatise *De Sectione Rationis,* which Dr. Halley published in 1708, from an Arabic manuscript; but most of them have been restored by modern mathematicians, from the account of them given by Pappus.

We must pass over a considerable number of years before we come to any other mathematician that deserves particular notice. The Alexandrian school produced indeed several eminent men, during the interval; but they drew their celebrity rather from astronomy than from geometry. Diophantus, the man of Diophantus. whom we have now to speak, is supposed to have lived during the reign of the Emperor Julian, or about the year 365 of our era. Diophantus is celebrated in mathematics as the inventor of algebra; or at least the first writer on that important branch of analysis, whose works have come down to us. His work is entitled *Arithmeticorum Libri*, and only the first six books remain, together with a book *De Numeris Multangulis*. The algebra of Diophantus was pretty much the same as it continued among the moderns, till the introduction of letters instead of numbers. The unknown quantity, or the quantity sought, he denotes by στι; its square he calls δυναμις, *potentia*, and denotes by δʸ; the cube is called κυβος, and denoted by κ; the biquadrate is denoted by δδʸ; and the fifth power by δκʸ. Algebra, in his time, had advanced as far as the solution of quadratic equations. But what constitutes the chief merit of his algebra, is the address with which he resolves indeterminate problems: a branch of analysis attended with peculiar difficulties.*

With Diophantus, the glory of the Alexandrian school in some measure terminated: most of the geometers who followed contented themselves with writing commentaries on the labours of their predecessors, instead of exerting themselves to extend the bounds of the science. Pappus, indeed, deserves to be excepted from this reproach. His writings contain a great many valuable historical details, with some appearances of original research; and we are indebted to him for most of our knowledge of the writings of those ancient mathematicians whose works have not come down to our times. Diocles, too, the inventor of the cyssoid, deserves to be exempted from the reproach of want of originality. This curve is very much approved of by Sir Isaac Newton, and solves many problems famous among the ancients with peculiar elegance and facility. Diocles must have been posterior to Pappus, as that writer takes no notice of him in his enumeration of the various methods proposed by mathe-

* The reader will find some fine specimens of this kind of analysis in the second volume of Euler's algebra, which is entirely devoted to it. We shall give a specimen of this species of problem in the following epitaph on Diophantus.

> Hic Diophantus habet tumulum, qui tempora vitæ ez
>   Illius mira denotat arte tibi.
> Egit sextantem juvenis, lanugine malas
>   Vestire hinc cœpit parte duodecima.
> Septante uxori post hæc sociatur, et anno
>   Formosus quinto nascitur inde puer.
> Semissem ætatis postquam attigit ille paternæ,
>   Infelix subita morte peremptus obit.
> Quatuor æstates genitor lugere superstes
>   Cogitur; hinc annos illius esse quære.

The meaning of the epitaph is, that Diophantus passed the 6th part of his life in childhood, the 12th in youth; that after a 7th part of his life and five years more passed in a barren marriage he had a son, who died after attaining half the age of his father, who survived him four years. The question is to find a number whose 6th, 12th, 7th with 5, $\frac{1}{2}$ with 4, make up the number. The answer is easy, and is 84.

maticians for doubling the cube.   We may therefore fix the period of his life at about the sixth century of the Christian era.

We may now close our account of Grecian mathematicians.   That unfortunate nation, assailed at once by the double evil of despotism and superstition, gradually lost that philosophic spirit which characterized it for so many ages, and occupied itself with the most miserable objects which bigotry and besotted ignorance could pursue.   Happily for posterity a better spirit animated the Arabians, who, after conquering all their opponents, and carrying the tenets Arabians. of Mahomet from the shores of India to the coast of the Atlantic Ocean, were conquered in their turn by letters and the sciences.   They procured, with great care, the manuscripts of all the ancient Grecian philosophers, especially the mathematicians, and translated them into their own language.   By this means they were preserved from destruction, and when civilization and the sciences again shewed themselves in Europe, we were, by their means, supplied with these invaluable writings, and enabled at once to profit by the accumulated labours of the sages of antiquity." But it is not necessary to dwell on these Arabian philosophers.   As far as mathematics are concerned they added little to the discoveries of their masters, and when they ventured to make changes, they rather injured and disfigured than improved their Grecian models.

Let us return then to Europe, the country in which, owing to some unknown fatality, all the great improvements in the arts and sciences have hitherto originated. the country of activity, of invention, and innovation.   After some ages of barbarism, which the genius of Alfred and of Charlemagne in vain attempted to soften, Europe began in the tenth century to shew some feeble symptoms of improvement.   The first person who claims our notice is the famous Gerbert, Gerbert. afterwards raised to the papacy, and distinguished by the name of Sylvester II He was born in Auvergne, about the commencement of the tenth century, and educated in the monastery of Fleuri, where Abbon, who was Abbot, had instituted a school, which drew to it all the young men of abilities of the order. Gerbert, however, was soon dissatisfied with the information which he could acquire at this school, and passed into Spain, where the Saracens had established two famous academies, at Cordova and Grenada, to which students flocked from the east and from the west.   Here Gerbert soon surpassed his masters in mathematical knowledge.   From some of his writings which have been published, it appears that he was acquainted both with Euclid and Archimedes; but what has given him celebrity, and for which we lie under the greatest obli- Introduction gations to him, he first introduced the Arabian numbers and the Arabian me- of the Arabian thod of notation into Christian Europe.* notation.

---

* Abacum, says William, of Malmesbury, certe primus a Saracenis rapiens regulas, dedit quæ a sudantibus abacistis vix intelliguntur.

With respect to the origin of the nine digits, and the decimal notation at present in use, we know nothing for certain. It is commonly supposed to have been an Indian invention, to have been borrowed from the Indians by the Arabians, and it is certain that it was brought by Gerbert from the Saracens in Spain into France, and from that country very speedily propagated over Europe. The period of the introduction appears to have been between 970 and 980. The preceding account is taken from Dr. Wallis's algebra, who has been at great pains in collecting information on this curious subject. Indeed, that we are indebted to Gerbert for the introduction of the present numbers, is obvious from William, of Malmsbury, and from the letters of Gerbert himself. Kircher ascribes the introduction of them to the tables of Alphonso, King of Castile; but this is undoubtedly too late a date, as these tables were not constructed till the 13th century. There are no less than nine papers on this curious subject in the Philosophical Transactions, chiefly employed in giving an account of old inscriptions in Arabic figures still remaining in England. Dr. Wallis describes a 'mantle-piece with this date A° Do^i M° 133.* Mr. Luffkin mentions a house in Colchester, in which there is a stone with this date engraven on it 1090.† Mr. Cope describes a mantle-piece in Widgell Hall, Hertfordshire, with this date engraven on it M16, which he considers as 1016 ‡ He likewise mentions a gate at Worcester, having engraven on it 97v, in black letter characters, which he considers as 975.§ But the validity of all these dates is denied by Mr. Ward, who published three elaborate papers on the subject in the Philosophical Transactions. The first date he conceives should be read 1233, instead of 1133; the second, he considers as 1490; the third is simply 1000, in Roman characters, and what has been read 16, he conceives to be the letters I. G.; the fourth, he considers as MXV, or 1015, all Roman. Mr. Ward is probably accurate with respect to all these inscriptions except the first, which, there can be little doubt, is really M133, as Dr. Wallis read it. As to the others they are of little consequence, as Mr. Barlow has brought forward another date, so distinct that it is impossible to mistake it. This date is upon the parish church of Romney, Southampton, and is 1011.‖ Thus, there appears sufficient evidence that the Arabic numerals were known in England about the year 1000; which serves, in some degree, to establish the above period, which we have assigned for their introduction into France by Gerbert. But they do not seem to have come into common use for some time; for it would appear, from a passage quoted above, that William of Malmesbury, who wrote more than 200 years after, did not well understand their use: if he had, he would scarcely have represented it as so difficult to be understood.

---

* Phil. Trans. 1683. Vol. XIII. p. 399.   † Phil. Trans. 1699. Vol. XXI. p. 287.
‡ Phil. Trans. 1735. Vol. XXXIX. p. 119.   § Phil. Trans. 1735. Vol. XXXIX. p. 131.
‖ Phil. Trans. 1741. Vol. XLI. p. 652.

But the tenth century found Europe still too barbarous to detain us long. Except this fortunate improvement of Gerbert, hardly any addition was made to our knowledge. We may even pass over the next four centuries, during which, though Europe was slowly imbibing knowledge from the Arabians, and though Italy began to put on the smiling aspect of trade and civilization, her progress was slow and faltering, and all her efforts sadly thwarted by the baneful superstition which still involved mankind. Thus, Roger Bacon was persecuted for his superior knowledge, thrown into prison, and prohibited from writing and studying. But in the 15th century, the real symptoms of improvement appeared, and the *reformation* in religion, which speedily followed, at once unfettered the minds of men, and produced the most splendid and unlooked for effects. Leonardo, of Pisa, travelling into the East for instruction, about the beginning of the 15th century, made himself acquainted with algebra, as it was cultivated among the Arabians, and on his return into Italy informed his countrymen of the science. Here it was cultivated with success; became speedily known to all mathematicians, as is obvious from the work of Regiomontanus, on *triangles;* in which he refers to the rules for resolving a quadratic equation, as to a practice familiarly known. But the two men to whom Mathematics owes most in the 15th century, and who may be considered as in reality the revivers of the sciences in modern Europe, were Purbach and Regiomontanus. *Leonardo of Pisa.*

George Purbach was born in Austria, in 1423, and was a disciple of John de Gmunden, who taught astronomy in the University of Vienna, at the commencement of the 15th century. Purbach devoted himself to astronomical pursuits with peculiar predilection, and travelled through Italy and various parts of Europe, in order to acquire information from those who cultivated the science. He finally settled at Vienna, where he succeeded Gmunden in the University. He dedicated much of his time to astronomical observations, and invented various instruments to facilitate his pursuits. Trigonometry lies under particular obligations to him: he banished from it the sexagesimal notation, and supposed the radius divided into 600,000 parts. He constructed a variety of tables of sines, to facilitate astronomical calculation. He was at considerable pains in improving the vulgar editions of the works of Ptolemy, which were extremely faulty; but being ignorant both of Greek and Arabic, he could correct the blunders only by collating the different editions together, and by his knowledge of astronomy. When Cardinal Bessarion, who was himself a Greek, came to Vienna, as the Nuncio of the Pope, Purbach was easily persuaded to accompany him to Rome, in order to learn the Greek language, and was just preparing to set out when he was suddenly cut off by a disease, in the year 1461, in the flower of his age. *Purbach.*

John Muller was born in the town of Königsberg, in Franconia, and, on

Regiomon-
tanus.

that account, took the name of *Regiomontanus.* The year of his birth was 1436. When only 14 years of age, his zeal for mathematics and astronomy induced him to put himself under the care of Purbach, who was at that time in the height of his reputation. He became his favourite pupil, or rather his intimate friend, continued with him till the time of his death, and ably seconded all his views. When Purbach died, Regiomontanus set out for Italy, in the suite of Cardinal Bessarion, where he learned Greek, and published a new edition of Ptolemy's *Almageste,* translated from the original. He undertook and executed a prodigious number of translations from the Greek, not only of astronomical works, but of books in every department of mathematics. Archimedes, Euclid, and Apollonius, and many others passed through his hands. This did not prevent him from composing many original works of his own, in which trigonometry and various other departments of mathematics were greatly improved. To him, in particular, we owe the first developement of our present system of decimal fractions; which, though very simple, is notwithstanding exceedingly useful and important. He returned to Germany about the year 1471, and fixed himself at Nuremberg, where he employed his time in making astronomical observations. In the year 1475 he returned to Rome, being invited by Pope Sixtus IV., who had formed the project of reforming the calendar, and conceived that Regiomontanus was the most proper person to manage the alteration. For this purpose he made him great promises, and had even named him Bishop of Ratisbon; but Regiomontanus died suddenly, in 1476, before he had well commenced his reform.

Lucas de
Borgo.

Lucas Paccioli, usually called Lucas de Borgo, because he was of the Borgo-San-Sepulchro, in Tuscany, deserves likewise to be mentioned, as a man who contributed essentially to the introduction of mathematics into Italy. He was a Franciscan who travelled into the East, and afterwards taught mathematics at Naples, Venice, and Milan. He published three different works about the commencement of the 16th century; one a kind of system of arithmetic, algebra, and elementary geometry; another, on a line cut in extreme and mean ratio; the third treats of polygons and regular figures.

Commenta-
tors.

The end of the 15th and the beginning of the 16th century was the age of commentators. After the recovery of the previous remains of ancient mathematics, the first step undoubtedly was to publish correct editions of these works, in order to enable mathematicians to understand them, and to profit by them. These commentators, therefore, were a set of useful men, and contributed essentially to the progress of the science. The first place among them is due to Commandin; but it would be tedious and useless to run over a dry catalogue of names long since forgotten, and consigned to oblivion. They constituted a body of pioneers, useful indeed, and necessary to the science, but not of sufficient importance, individually, to claim peculiar mention. Let us rather turn our attention to the improvements which gradually were introduced into mathematics, after the study of them was thus happily revived. The 16th century exhibits several improvements of considerable importance; but the 17th century is the golden age of mathematical science. During that fortunate and busy era, its bounds were enlarged on every side: new branches of mathematics, and new methods of calculation sprung up in every quarter, till at last the fluctionary calculus of Newton made its appearance, and more than doubled the power and the resources of the mathematician. We shall make a few observations on the discoveries of the 16th century; but we shall dwell at greater length upon the new methods of the 17th century, because some of the chief of them were made by the Founders, or Members, of the Royal Society, and because various important documents, relative to these new methods, are to be found in the Transactions.

Nicolo Tartaglia is the mathematician who seems to hold the first rank Tartaglia. among the geometers of the 16th century. He was born at Brescia, of a family in the lowest situation of life. His father was a messenger, and dying while Tartaglia was very young, left his family in extreme misery. Soon after, the French army on its return from Naples, having gained the battle of Fornoul, took Brescia and pillaged it. Tartaglia was wounded during the assault, and on that account assumed the name which he bore. Recovered from this accident, he taught himself to read, and was obliged to get a writing book by trick, pretending that he wanted to take a model of the letters of the alphabet. These first difficulties overcome, he soon acquired all the learning of his time, and established himself in Venice, where he taught mathematics. His great dis- Solution of covery was the method of resolving cubic equations, commonly known under tions. the name of Cardan's rule, on account of a circumstance which deserves to be detailed. Tartaglia having discovered the method of resolving cubic equations, in consequence of a challenge by one Florido, an itinerant mathematician, communicated it to Cardan, who was his particular friend, binding him at the same time not to reveal it. This did not prevent Cardan, some time after, from publishing it as his own. Tartaglia was provoked at this conduct, and complained of it bitterly. He went farther: in order to prove to the world that he was the real inventor of the formula, and that he was a superior mathematician to Cardan, he challenged him to resolve, within a given time, 31 propositions, which he should propose to him, and allowing Cardan to propose as many to him, undertaking to resolve the propositions of Cardan, or at least a greater number of them than Cardan would of his. The challenge was accepted; and it appears that Cardan was scarcely able to resolve any of the questions proposed by Tartaglia, and the few which he did resolve, not till after the appointed time; while, on the contrary, Tartaglia resolved in a few days almost the whole of Cardan's. But both sides, as is usual in these cases, claimed the victory; the quarrel continued, and was terminated only by the death of the combatants.

Cardan.

Cardan himself was a very extraordinary man. He was born in Milan, in 1501, and died at Rome in 1575. He was a physician, and attached himself with ardour to every branch of knowledge. We find him at once aiming at oratory, geometry, algebra, astronomy, morality, philology, and the various branches of natural history. Algebra, however, was the science which he understood best, and which has been the source of his reputation. Though it be true that the method of resolving cubic equations is not his, yet it must be admitted that he added considerably to the elucidation of the nature of cubic equations.

Solution of biquadratics.

The method of solving biquadratic equations was discovered very soon after the discovery of the solution of cubic equations. The discovery is due to Louis Ferrari, born at Milan, and a pupil of Cardan. A certain adventurer, called John Colla, who took pleasure in puzzling the mathematicians of the age, had proposed a certain problem, which involved a biquadratic equation, and about the solution of which the learned were divided. Cardan requested his pupil, Louis Ferrari, with whose genius he was well acquainted, to turn his attention to this sort of equations, and endeavour to find the solution. Ferrari did so, and speedily succeeded. It is with regret that we are obliged to add, that the moral character of Ferrari appears to have been very bad. Cardan, who himself was not a model of morality, assures us that his conduct was so disgraceful, that he was ashamed to allow him to visit him.

Ramus.

The only other mathematician of the 16th century, (except those concerned in the introduction of new methods), whose name we think it necessary to mention, is the celebrated Peter Ramus, a French Protestant, who distinguished himself very much by his zeal for the mathematics, and his contempt for the vain disputations of the schools at that time in vogue. He perished in the disgraceful massacre of St Bartholomew, a day on which so many eminent Frenchmen lost their lives. For though the Hugonots never exceeded in number the sixth part of the population of France, they comprehended within their body all the learning and science, and almost all the talents, of the country.

Let us now proceed to the radical improvements which mathematics underwent, and the new methods introduced, about the end of the 16th, and during the course of the 17th century.

Vieta.

I. The first improvement which we shall notice was made by Vieta, a name deservedly illustrious in the annals of mathematics. He first substituted letters instead of known quantities in algebra; an improvement, which may seem of little importance to those who are ill informed on the subject, but which mathematicians know has contributed essentially to the progress of the science. Vieta was born at Fontenai, in Poitou, about the year 1540. He was *maître des requêtes* at Paris, and, notwithstanding the multiplicity of his occupations, found time to cultivate mathematics, and to outstrip all his contemporaries in his knowledge of that science. According to M. de Thou, he used sometimes to sit whole days at table absorbed in mathematical investigations, and his servants were under the necessity of bringing him food to supply the waste of strength occasioned by this incessant labour. He had violent disputes with Scaliger and Clavius. He refuted the pretended quadrature of the circle of the first; but the calendar which he proposed to substitute for that adopted by Gregory XIII., at the suggestion of Lilius and Clavius, was a very defective one, and its errors were very well poinfed out by Clavius. Vieta was well skilled in the difficult art of explaining ciphers. He died at Paris in the year 1603, at the age of sixty-three. His works are scarcely intelligible to modern mathematicians, in consequence of the vast number of Greek terms with which they abound.

The next improver of algebra was Thomas Harriot, who was born at Oxford Harriot. in the year 1560; he distinguished himself so much by his mathematical knowledge, that Sir Walter Raleigh made choice of him to be his instructor in that science, and carried him out with him to Virginia, when he planted his first colony in that country. Harriot constructed a map of Virginia, and published an account of the country after his return, under the title of *A brief and true Report of the new-found Land of Virginia;* which is to be found at the end of the third volume of *Hakluit's Voyages.* He was afterwards introduced by Sir Walter, to Henry, Earl of Northumberland, who allowed him a handsome pension; as he did to Mr. Hues and Mr. Walter Warner. When the Earl, in 1606, was committed to the Tower for life, these three learned men still continued to associate and converse with him, and were known by the name of the Earl of Northumberland's three Magi. Mr. Harriot removed some time afterwards to Sion College, and he died in 1621, of a cancer in his lip. His *Artis Analyticæ Praxis* was published after his death, by his friend Mr. Walter Warner, and contains his analytical improvements. The great discovery which he made was, that all the higher equations are composed of as many simple equations multiplied together as are equal to the greatest index of the unknown quantity. From this it follows, that every equation has as many roots as are equivalent to the index of its highest power. An equation of the fifth power has five roots; one of the fourth, four; one of the third, three; and so on. Dr. Wallis, in his Algebra, has given an elaborate catalogue of the discoveries of Harriot; but he has been accused of partiality by the French mathematicians, and of an undue eagerness to reduce the discoveries of Descartes as low as possible.

Several important additions to the doctrine of equations have been made by General solution of equa- Descartes, and by Sir Isaac Newton; and the most persevering efforts have tions.

been made to discover a general method of resolving equations of every degree. Euler, Waring, Bezout, and La Grange, have particularly distinguished themselves in these investigations, and probably various other mathematicians, with whose writings I am unacquainted. But all their efforts have been in vain; new and unforeseen difficulties have arisen on every side, and a general resolution of equations seems to be as distant and as hopeless as ever. Whether the problem is in itself insoluble, or whether the want of a solution be owing to the imperfection of our methods and of our modes of investigation, has not yet been determined. Mathematicians, deterred by the numerous difficulties which opposed the general solution of equations, and by the prodigious length of calculations which such investigations required, have turned their attention to a method of approximation; and various very convenient methods have been discovered, which scarcely leave any room to regret, in a practical point of view, that the general solution of equations still continues unknown. Newton was the first inventor of a method of approximation, which he communicated to Dr. Barrow in 1669.* In 1669, Mr. Joseph Raphson, in his *Universal Analysis of Equations,* proposed another method, which was further prosecuted by M. de Lagney, Professor of Mathematics at Paris. In 1694, Dr. Halley published two new methods of accomplishing the same thing.† The methods of these three mathematicians are those still most generally followed. Newton's method continued unknown, except from a small specimen of it given by Dr. Wallis in his Algebra, till the publication of the *Universal Arithmetic.* These three methods bear a close resemblance to each other. Other methods of approximation have been since proposed by Taylor, Thomas Simpson, the Bernoullis, Euler, and La Grange; and the subject has been, in a great measure, exhausted.

Logarithms.

II. The next improvement in mathematics, which we have to mention, is the introduction of logarithms, those numbers so important by diminishing the labour of tedious calculations, and which play so conspicuous a part in the transcendental analysis. For this admirable discovery we are indebted to John

Napier.

Napier, Baron of Merchiston, near Edinburgh. He was descended from an ancient family; and his father, Sir Archibald Napier, was knighted by James VI. and made Master of the Mint, an office which he sustained with honour and dignity. Napier seems to have turned the bent of his genius towards the discovery of methods to facilitate and abridge trigonometrical calculations; and various contrivances were proposed by him in succession, all remarkable for their ingenuity. The last and most memorable of all was his discovery of logarithms. There is a story told by Mr. Wood, but it does not appear en-

* Methods indeed, though imperfect, had been proposed by Vieta, Harriot, and Oughtred.
† Phil. Trans. Vol. XVIII. p. 136.

titled to any attention, that one Dr. Craig, a Scotchman, coming out of Denmark into his own country, called upon John Napier, Baron of Merchiston, near Edinburgh, and told him of a new invention, in Denmark, by Longomontanus, to save tedious multiplications and divisions in astronomical calculations.   Napier being solicitous to know further of him of this matter, he could give no other account of it than that it was by proportional numbers; which hint Napier taking, desired him, at his return, to call upon him again.   Craig, after an interval of some weeks, did so, and Napier then showed him a rude draught of what he called *canon mirabilis logarithmorum*.   Had there been any truth in such a story, we may be sure that Longomontanus and the Danes would not have abstained from laying their claim to so admirable a discovery.   Napier has also been considered as having been anticipated in his invention by Stifels, and by Juste Byrge, two German mathematicians; but these allegations originating from jealousy, or from national partiality, are entitled to no attention whatever, and Napier's claims have for many years been allowed by the universal consent of all mankind.

Logarithms are numbers which follow an arithmetical proportion, while the natural numbers corresponding to them follow the geometrical.   Hence the addition of logarithms corresponds with the multiplication of natural numbers, and their subtraction with the division of natural numbers.   Dividing them by the index of the power is equivalent to the extraction of the root denoted by that number.   Thus, to divide a logarithm by 3 is the same thing as to extract the cube root of a natural number.   Hence the admirable use of a table of logarithms, and the great importance of which they must be in all tedious calculations where it would be necessary to multiply or divide large numbers.   The logarithms which first presented themselves to Napier were those at present known by the name of hyperbolic logarithms.   But it afterwards occurred to him that logarithms, similar to those in our modern tables, in which the logarithm of 1 is 0; that of 10, 1; that of 100, 2; &c., would be more convenient.   But he died, in 1618, before he had time to put his new plan in execution; but not till he had explained its nature to Mr. Henry Briggs, Gresham Professor of Mathematics, who had seen at once all the importance of logarithms, and had early devoted himself to bring them to perfection.

Henry Briggs was born at Worleywood, near Halifax, in Yorkshire, about Briggs. the year 1556.   After being educated at a grammar-school in the country, he was sent to St. John's College, Cambridge, November 5, 1579; admitted a scholar of the house in 1581, and took the degree of Bachelor of Arts; that of Master in 1585; and was chosen Fellow of his College, March 29, 1588. He applied himself chiefly to the study of mathematics; and, in 1592, was made Examiner and Lecturer in that faculty, and soon after Reader of the Physic Lecture founded by Dr. Linacer.   On the settlement of Gresham College he was made first Professor of Geometry there. In 1620, he was appointed Savilian Professor of Geometry in Oxford; in consequence of which appointment he resigned his Gresham Professorship. In this situation he continued till his death. As soon as the Napierian discovery of logarithms was announced, he made two successive journeys into Scotland, to confer with the discoverer himself, and settle plans for the calculation and construction of logarithmic tables. An account of the nature and properties of logarithms was published at Edinburgh, in 1618, by Robert Napier, the son of the great discoverer, under the following title: *Mirifici Logarithmorum Canonis constructio et eorum ad Naturales ipsorum Numeros Habitudines, una cum Appendice de alia eaque prestantiori Logarithmorum Specie condenda,* &c. &c. This book had been written, and was ready for the press, when John Napier, the inventor of logarithms, was prevented from publishing it by his death. The same year Briggs published a table of the logarithms of the first 1,000 natural numbers, under the title of *Logarithmorum Chilias prima.* In 1624, he published, under the title of *Arithmetica Logarithmica,* the logarithms of all numbers from 1 to 20,000, and from 90,000 to 100,000, calculated to 14 decimal places. Briggs

Gunter. was assisted in his calculations by Gunter, who was Professor of Astronomy at Gresham College, and the contriver of the graduated rule which passes under his name. He calculated the logarithms of the sines and tangents, and published a table of them in 1620, entitled, *Canon of Triangles.* Briggs had made considerable progress in a table of sines and tangents, calculated to 100 parts of a degree, (for he wished to introduce the decimal notation into trigonometry) but died, in 1630, before he had completed it. It was finished by

Gellibrand. Henry Gellibrand, who had succeeded Gunter in the astronomical chair, at Gresham College; and he published it in 1633, under the title of *Trigonometria Britannica.*

Kepler. One of the first persons on the Continent, who properly appreciated the importance of logarithms, was Kepler. He published a work on the subject in 1624, in which he simplified the theory considerably, and developed the views

Vlacq. of Napier with great sagacity and simplicity. Adrian Vlacq, a Dutchman, reprinted Brigg's book, at Gauda, in 1628, and the same year he published a French translation of it, in which he filled up the gap left by Briggs, and gave a complete table of the logarithms of all numbers from 1 to 100,000, calculated to 11 decimal places. In 1633, he reprinted Gellibrand's *Trigonometria Britannica,* and the same year he published a work of his own on the subject, entitled, *Trigonometria Artificialis, seu Magnus Canon Logarithmicus, ad radium* 1.00000,00000, *et ad dena scrupula secunda, ab Adriano Gaudano constructus.* Since that period a great number of logarithmic tables have appeared in different countries. We shall mention some of the most remarkable

Sherwin. of these. In 1706, Sherwin published a set of tables, at London, entitled,

Tables of Logarithms, for all numbers from 1 to 102,100; and for the sines and tangents to every ten seconds of each degree in the quadrant; as also for the sines of the first 72 minutes to every single second; with other necessary tables. This book was often reprinted, several editions being edited by Gardiner; the last of which, according to Dr. Hutton, abounds with errors. A set of tables was published by Schulze, at Berlin, in 1778; another by Vega, at Vienna, in 1783. A very splendid set of tables of the logarithms of the sines and tangents, calculated to seconds by Taylor, and edited by Dr. Maskelyne, was published in London, in 1792. A set of tables by Callet, was published at Paris, in 1783, and a new stereotype edition of the same work, remarkable for beauty of execution and for accuracy, was published in 1795. We shall terminate this list, which might be much further extended, by mentioning the logarithmic tables of Dr. Hutton, which are commonly used in this country, and have superseded those of Sherwin. They are equally conspicuous for the beauty of execution and for accuracy, and will not be easily surpassed by any succeeding tables. Dr. Hutton has prefixed to this book a curious and interesting history of logarithms, to which we have been indebted for most of the facts contained in the preceding sketch. It was this history likewise which suggested to Baron Maseres the plan of his *Scriptores Logarithmici,* in which he has collected together all the valuable tracts upon logarithms, published since the original discovery by Napier to the present times.

　III. We now come to the developement of those new methods which appeared in such numbers during the 17th century, and which advanced the science of mathematics to the most unlooked for perfection. The first of these methods, and one of the least important, is that of Father Guldin. It is hinted at by Pappus with considerable clearness, and Guldin has been blamed, and probably with justice, for taking no notice of the claim of that ancient geometrician. Guldin was born at St. Gall, in 1577, and, having abjured the Protestant religion, entered into the society of Jesuits as a temporal coadjutor. His superiors, perceiving that he possessed a great genius for mathematics, sent him to Rome, where he was a Professor of Mathematics for some years, and he afterwards taught the science at Gratz and at Vienna. He published a variety of works, but the most important of them is his *Centrobaryca,* or his Treatise of the *Centre of Gravity.* Every body knows that there is a certain point in every figure called its centre of gravity, and this point is such that if we suppose an axis to pass through it, and to support the figure, the figure will remain at rest in any position that we please to give it. The centre of gravity possesses several other remarkable properties well known to mathematicians. The new method of Guldin consisted in the application of the centre of gravity to measure solids of revolution. This discovery consisted in the following proposition: *Every solid formed by the rotation of a line, or a surface round*

Callet.

Hutton.

Guldin's method.

an immoveable axis, is equal to the product of the generating quantity multi-plied by the course of its centre of gravity. Thus, for example, the rectan-gular cone is formed by the revolution of a right angled triangle round one of its sides. The centre of gravity of such a triangle is distant from the axis by a space equal to one third of the base. Hence, according to Guldin's rule, the cone is equal to the generating triangle multiplied by the third of the circum-ference of the base of the cone. Hence it is obviously the third of the cylin-der, having the same base and height as we know it from other means to be. In this manner Guldin deduced the solidity of all the solids of revolution with great facility. Though he did not succeed in demonstrating the truth of his fundamental proposition. But its truth was afterwards demonstrated by Caval-leri during a dispute between him and Guldin, and he informs us that it had been drawn up many years before by one of his pupils.

Method of in-divisibles.

Cavalleri.

IV. The method of *indivisibles* was of much greater importance than the method of Guldin, and constitutes the real origin of the immense improve-ments afterwards made. Bonaventure Cavalleri was born at Milan, in 1598, and entered when young into the order of Hieronimites. He displayed so much genius during the progress of his studies, that, after he had taken orders, his superiors thought proper to send him to Pisa, that he might improve him-self in that celebrated university. Here he became acquainted with Benedict Castelli, one of the disciples of Galileo, who advised him to study geometry as a means of alleviating his pain, and of affording him amusement during the fits of the gout with which he had been attacked. Cavalleri made such progress in his mathematical studies, and perused the writings of the ancient geometricians with so much rapidity and avidity, that Galileo and his disciples predicted the celebrity which he would one day acquire. This prediction was speedily accomplished by his invention of the method of indivisibles, of which he was in possession before the year 1629. For Magin, Professor of Astronomy in the University of Bologna, dying that year, Cavalleri stood candidate for the chair ; and, as a proof that he deserved it, sent to the magistrates of that city his *Treatise on Indivisibles*, and another Dissertation on *Conic Sections*. His claim was sustained, and he began to execute the duties of his new professorship in 1629.

Cavalleri supposed surface or space to be composed of an infinity of parts which constitute its ultimate elements, and which are obtained by dividing it continually into slices parallel to each other. These last elements he called *indivisibles*, and by determining the ratio in which these indivisibles increased and decreased, he was enabled to measure the surfaces, or solids, which were composed of them. It must be acknowledged that this mode of explaining the foundation of his method was not quite consistent with rigid geometry, and on that account he was attacked with some acrimony by Father Guldin. This at-

3

tack obliged him to explain himself with more precision; and it was easy for him to show that his indivisibles were nothing else than the inscribed and circumscribed parallelograms of Archimedes reduced to their simple ratios, and therefore much more easily managed. The geometry of indivisibles may be divided into two parts. The object of the first part is to compare figures with each other by means of the equality, or constant ratio, which exists between their elements. Thus he shows the ratio of parallelograms, triangles, &c. on the same base and between the same parallels, and reduces the whole to this general proposition: All figures, whose elements increase or decrease in a similar manner from the base to the summit, are to the uniform figure of the same base and height in the same ratio. The second part consists in pointing out the method of determining the ratio of an infinite number of lines or planes, increasing or decreasing according to a certain law with the same number of elements homogeneous with the first, and all equal to each other. Thus, for example, the cone, according to the language of Cavalleri, is composed of circles decreasing from the base to the summit, while the cylinder of the same base and height is composed of an infinite number of circles all equal to each other. In the cone these circles decrease from the base to the apex as the squares of the terms of an arithmetical progression. If therefore we can determine the ratio between an infinite number of terms decreasing according to that law, and an infinite number of terms all equal to the first term of the first progression, we shall have the proportion between the cone and the cylinder of the same base and height. But Cavalleri shows that the one is the third of the other; hence it follows that the cone is the third part of the cylinder. Cavalleri applied this method to a great variety of figures, and solved a vast number of problems which had escaped all the efforts of preceding geometers with the greatest facility. In particular he applied it to the solution of a great many curious and difficult problems, which had been some time before laid before mathematicians by Kepler, and he succeeded without difficulty in solving almost the whole of them. Cavalleri was the author of various other mathematical works, and during the whole of his life-time occupied the first rank among the Italian mathematicians.

V. We must here interrupt our account of the new methods introduced into mathematics, in order to give the history of the discoveries respecting the cycloid, in consequence of their importance, and the great celebrity which they acquired.

The cycloid is a curve formed by a circle rolling upon a straight line, situated The cycloid. in a plane, while a point in the circumference, during its revolution, traces the cycloid upon the plane. Who the discoverer of this curve was, is not very well known. Wallis ascribes it to the Cardinal de Cusa; but this opinion seems to be entirely destitute of foundation. Galileo appears in fact to have been the first person that thought of it; for, in a letter to Torricelli, dated 1639, he says, that he had considered it about 40 years before, and that he thought it, from its pleasing figure, well calculated for the arch of bridges. Father Mersenne, in the year 1628, proposed to Roberval the problem of finding the area of the cycloid; but he was at that time unequal to the task. He therefore made no very strenuous efforts to resolve it, but set himself to study the ancients, especially Archimedes; and, after an interval of six years, the problem being proposed to him again, he succeeded in demonstrating, that the area of the cycloid is equal to three times the area of the generating circle.

Roberval.

Gilles Personnier de Roberval, the author of this discovery, was born in 1602, at Roberval, a village in the diocese of Beauvais. He came to Paris in 1627, got acquainted with the literary men of that capital, and began speedily to figure among the principal geometers of the age. He was a Member of the Academy of Sciences from its original institution, in 1665, and held, during forty years, the situation of Mathematical Professor in the College of Gervais; a professorship founded by Ramus, and which, according to the statutes, was to be exposed to competition every three years. Though an excellent mathematician, he was destitute of the art of communicating his discoveries with neatness and precision. His demonstrations are distinguished by a want of elegance, and by a profusion of lines and figures, which renders it very difficult and very fatiguing to peruse them. There is an anecdote told of him, which the enemies of mathematics have laid hold of, as a proof that they unfit the mind for relishing compositions of taste and elegance. Being present during the acting of a tragedy, and being afterwards asked what he thought of it, he answered the question by asking *What did it prove?* But this reply proves nothing, unless we knew the merits of the tragedy in question. It was probably execrable stuff, and Roberval's answer only a mathematical and whimsical method of announcing his opinion of it. Were a mathematician to go to a London theatre, in the present day, he might put such a question with the greatest propriety, and without any reproach either to his good sense or his taste.

Descartes's observations on the cycloid.

Mersenne, in his *Harmonie Universelle*, published in 1637, gives Roberval's discovery of the area of the cycloid. In 1638 he gave an account of the discovery to Descartes, and, from the answer of that celebrated mathematician sprang the origin of the violent quarrel between him and Roberval, which lasted so many years. Descartes said, that the discovery was a beautiful one, and quite new to himself; but that it was by no means difficult, and perfectly within the range of any person moderately skilled in geometry; and, to prove the truth of his opinion, gave a hasty sketch of the demonstration of the relation between the cycloid and the generating circle, which he developed more at length in a succeeding letter. Roberval, mortified at this decision of Descartes, affirmed that he had been aided in his demonstration by his knowledge of the result; an assertion which might very well be true. Descartes, to demonstrate his superiority over Roberval, determined the tangents of the cycloid, sent the solution to Mersenne, and, at the same time, challenged Roberval and Fermat, with whom he was at that time at variance, to resolve the problem.

Pierre de Fermat, Counsellor in the Parliament of Toulouse, was one of the Fermat. greatest mathematicians of the age in which he lived. His discoveries respecting parabolas and spirals, his method of maxima & minima, and his method of tangents, procured him great and deserved celebrity. Fermat soon resolved the problem respecting the tangents of the cycloid, and sent in his solution to Mersenne; but Roberval was not able to succeed. He sent in five or six solutions one after the other, as appears from the letters of Descartes, none of which was satisfactory. At last, the method of Fermat having come to his knowledge, he resolved the problem, but was not able to find a demonstration of his own solution.

In the year 1639 Mersenne wrote to Galileo, and informed him that certain problems relative to the cycloid at that time occupied the attention of the French mathematicians. Galileo wrote to Cavalleri, and requested him to endeavour to determine the cycloidal area; but Cavalleri, skilful mathematician as he was, could not discover the method of solving the problem, and Galileo died in 1642, without knowing whether it was possible to find the area of the cycloid or not. But Torricelli and Viviani, two of Galileo's disciples, and his Discoveries of companions in his old age, knowing that their master had interested himself in the Italians. the cycloid, set about investigating its properties; and about the year 1643 Torricelli determined the area, and Viviani the tangents, of this figure; and Torricelli inserted these discoveries by way of appendix to his works, which were published that year. This excited the jealousy of Roberval, who wrote him a long pedantic letter, claiming these discoveries for himself, and calling the *immortal gods* to witness that he had discovered them many years before the Italian mathematicians had turned their thoughts to the subject. Torricelli, worn out with the perpetual appeals of Roberval, at last wrote him that it was of very little consequence whether the problems, relative to the cycloid,. were first solved in Italy or France; that he did not call himself the discoverer of them; that the area of the curve was unknown in Italy till after the death of Galileo; and that he did not get his information respecting it from France; but, that he had found out the demonstration himself, and he did not concern himself much whether his word was believed or not, as he knew that it was conformable to the dictates of truth. If Roberval was jealous of that discovery, he would abandon it to whom ever chose to take it up, provided they did not attempt to wrest it from him by violence.

Roberval next examined the solids, formed by the revolution of the cycloid, about its base, and about its axis, and found that the first was to the cylinder of the same base and height, as 5 to 8; and the second, nearly as 11 to $17\frac{7}{9}\frac{9}{3}\frac{1}{3}$. Mersenne having mentioned these particulars to Torricelli, he returned for answer, that he had discovered the same things a few months before.

The theory of the cycloid remained stationary for some years, when it was brought again upon the field by the celebrated Pascal, in the year 1658. Blaise Pascal* was one of the most extraordinary men of the 17th century. His father was a mathematician, and intimately connected with all the celebrated geometers at that time in France. He was anxious to conceal the science from his son till he arrived at a certain age, lest it should turn his attention from other studies, to which it was necessary that he should attend. But it was difficult to live in the house of Pascal without hearing of geometry, because it was the constant subject of conversation. This circumstance, together with the care with which he was excluded from all means of studying its principles, created in the mind of the young philosopher an eager desire to become acquainted with this prohibited science, and he created, in consequence, a kind of geometry for himself. When he was of the age of 12, his father going one day into his apartment, found him occupied with a geometrical figure, and perceived, to his great astonishment, that he had just demonstrated the 32d proposition of Euclid, or the equality of the three angles of a triangle to two right angles. The father of Pascal no longer refused to initiate his son in the principles of a science for which he had displayed so strong a predilection; but put into his hands a Euclid, which he ran over like a romance. He was henceforth admitted to the philosophical conversations of his literary friends, and, at the age of 16, when other young men are only entering upon the study, Pascal was a consummate mathematician, and had composed a treatise on conic sections; in which, from one proposition accompanied with 400 corollaries, he demonstrated most of the properties of these celebrated curves. At an early age, he, in some measure, abandoned the pursuit of mathematics for religion and morality, which he considered as of greater importance. His *Provincial Letters*, written in consequence of the religious disputes which at that time divided France, constitute one of the finest specimens of eloquence that has ever been produced, and continue even at present to be read and admired, though the disputes, from which they originated, have been long since buried in oblivion. He occasionally, however, turned himself back to geometrical pursuits, and the cycloid at last caught his attention. Wishing to withdraw his mind from the violent pain, occasioned by a disease with which he was afflicted, he examined the segments of the cycloid, determined their area, their

* A most interesting Life of Pascal has been published by Bossut.

2

centre of gravity, and the solids, formed by their revolution round their axes. These discoveries would probably have remained unpublished, had not several of his friends, who conceived that it would be for the advantage of religion to show the world the great genius of one of its most zealous defenders, persuaded Pascal to make use of these problems, to try the strength of the different mathematicians of Europe. Accordingly, under the name of A. Dettonville, he *Proposes a prize relative to the cycloid.* addressed a circular letter to geometers, dated June 1658, inviting them to resolve his problems, and promising a prize of 40 pistoles to the first person who resolved them, and 20 to the second. The first of October was the day by which it was necessary to send in the solutions properly authenticated, and M. de Carcavi was the person to whom they were to be sent. These problems were of so difficult a nature, that there were few mathematicians of the age capable of resolving them. Accordingly, only two men sent in solutions, and stood candidates for the prize. Those were the celebrated Dr. Wallis, and Father Lalouere, a Jesuit of Toulouse. There were some other persons, however, who, without laying any claim to the prize, took the opportunity of sending in solutions of some important problems relative to the cycloid. Slusius, for example, Ricci, Huygens, and Sir Christopher Wren; the last of whom gave the method of rectifying the cycloidal arch. Commissioners being appointed to examine the two papers sent to M. de Carcavi, that of Father Lalouere was declared at once to have no title to the prize. He had only fulfilled a part of the task assigned by Pascal, and what he had done was inaccurate. Neither was the prize assigned to Dr. Wallis. The commissioners published a detailed account of the reasons which induced them to decide against him. These were a number of mistakes into which he had fallen, and concerning which, as the original paper of Dr. Wallis was never published, we have no means of judging. Dr. Wallis himself, when he published his tract on the cycloid, denied their validity; though he acknowledged a few inadvertencies into which he had fallen: but these, he said, were of little moment. There can be no doubt that this decision embittered him against the French mathematicians, and contributed to that severity with which he handled them in his algebra, and all his historical writings which appeared after this famous prize question of Pascal. In 1659, Pascal published the solution of all his problems, under the title of letters of *M. Dettonville* to *M. Carcavi*. He had already favoured the world with an historical account of the cycloid, in which the information which he communicated, obviously derived from his friend Roberval, is far from accurate.

VI. One of the greatest improvements in mathematics was made by Descartes, who, by connecting together geometry and algebraical analysis, supplied mathematicians with a convenient instrument for improving the science, and with the means of generalizing, and, in some measure, completing the doctrine

Descartes.   of curves. René Descartes was born at *La Haye,* in Touraine, the 31st of March, 1596. While yet a boy he showed such an inquisitive turn of mind that his father used to call him his *little philosopher.* He was sent to the Jesuit's College, at *La Fleche,* in 1604, and put under the tuition of Father Charlet. Here he made great progress in the learned languages, and in polite literature, and contracted an acquaintance with different gentlemen, who afterwards became eminent in the republic of letters. But having passed through his course of philosophy, with little satisfaction to himself, he left the college in 1612, and began to turn his attention to a military life; but the weakness of his constitution not permitting him to go to the army, he was sent to Paris under a tutor, where he contracted so many literary acquaintances, and his passion for philosophy was so strongly rivetted by their example and exhortations, that he laid aside his amusements, and continued for some time in a very laborious course of study. In 1616 he went to Holland, and entered as a volunteer into the troops of the Prince of Orange. While he lay at garrison, at Breda, he solved a difficult mathematical problem, which, having been affixed in the streets by some unknown person, procured him a great deal of reputation. He left Breda in 1619, and entering into the army of the Duke of Bavaria, made several campaigns in Germany; till at last, becoming disgusted with a military life, he left the army, sold his estate, and spent some time in travelling through different parts of Europe. His love of liberty induced him to settle in Holland, where he wrote most of his books, and where he refused several advantageous offers, made him by the French King, to prevail upon him to return to his own country. He was induced, however, to comply with an invitation made him by Queen Christina, and died at the court of Sweden in 1650. Seventeen years after, his bones were removed to Paris, and a magnificent monument erected to his memory in the chuch of St. Genevieve. It is not our object at present to give a catalogue of his numerous works, far less to enter into any examination of his philosophy, which has been long since overturned. His pursuits were exceedingly various. He was a mathematician, a metaphysician, a moralist, a chemist, an anatomist, and a naturalist. His works were of considerable service at the time when they were published, and contributed essentially to set free the minds of men from the trammels of authority.

His mathematical discoveries.   But it is to his geometry that he is indebted for the only part of his reputation which will never fade. It was he that first pointed out the importance of the negative roots of equations; and he added an important fact to the theory of Harriot, respecting equations, by laying down this rule, that there are as many positive roots of an equation as there are changes in the signs, and as many negative roots as there are signs of the same kind, following each other. This rule holds when there are no imaginary roots, and it even holds when there are, if we conceive the imaginary roots to be both negative and positive at the same time. But the important part of his mathematical labours was his application of algebra to the investigation of curves. This he did, by conceiving every curve to be formed by the motion of a straight line along the axis of the curve, while a point moving along this line, according to a certain law, describes the curve. This line, intercepted between the axis and the curve, is called the *ordinate*; while the part of the axis, intercepted between the ordinate and the apex of the curve, is called the *abscisse*. The ordinate is always denoted by the letter y, and the abscisse by the letter x. The nature of the curve is expressed, by stating the relation between the ordinate and the abscisse. Thus, for example, in the parabola let the parameter be represented by p, the ordinate by y, and the abscisse by x, we have always $y^2 = px$. This is called the equation of the parabola; and, from this equation, the different properties of the parabola may be deduced with great facility. Hence, in order to investigate the properties of a curve, we have only to determine its equation. In the ellipse, if a and b be the two conjugate semi-diameters, the equation is $y^2 = \frac{b^2}{a^2} (2ax - x^2)$. Descartes arranged curves into different orders, according to the highest powers of the letters x and y, in their equations. Those in which these letters are only of the first power, belong to the first order. The straight line alone has such an equation. The curves of the second order are four: the circle, the ellipse, the parabola, and the hyperbola; or, they are only three, if we consider the circle as a species of ellipse The curves of the third order were enumerated by Sir Isaac Newton, and reckoning some few that have been added since his time, they amount to 78. Those of the fourth order have never been reckoned; but they probably amount to nearly 2000. Beyond this, mathematicians have not ventured to go: the task would be endless and useless, as the number of curves is obviously infinite; and whenever the equation of a curve is given, we have it in our power to delineate it, and to examine its properties. Descartes' geometry of curves possesses this further remarkable circumstance in its favour, that it is still the best method of examining curves, being better adapted for that purpose, in several respects, than the fluctionary calculus.

Different orders of curves.

VII. We now come to the discoveries of Dr. Wallis, which claim particular attention; because they constituted the germ from which some of the most important of the Newtonian discoveries originated. Dr. John Wallis was born at Ashford, in Kent, the 23d of November, 1616. His father was a clergyman, and died when his son was only in his sixth year. He was first put under the charge of Mr. James Moffat, a Scotchman, and trained up in his school at Leygreen, near Tenterden, till about the year 1630, when he was sent to Felsteed school, in Essex, under the direction of Mr. Martin Holbech. At

Wallis.

Christmas, 1631, he went home, and understanding that his brother had been learning arithmetic, he requested him to show him what he had been doing. His brother complying with this request, Dr. Wallis set himself to acquire this new piece of knowledge, and, in ten days, made himself master of the four common rules of arithmetic: the rule of three, fellowship, false position, practice, and the reduction of coins. This was his first introduction to mathematics, and, as he informs us himself, all the learning that he ever got.* Under Mr. Holbech he learned Latin, Greek, and Hebrew, and the rudiments of Logic, Music, and French. Thus accomplished, he was sent to the University of Cambridge, and admitted of Emanuel College. He afterwards removed to Queen's College, where he was chosen a Fellow; a situation which he held till his marriage, which, of necessity, vacated it. In 1640 he went into orders, quitted the University, and became Chaplain, first to Sir Richard Darby, and afterwards to Lady Vere. Next year he was chosen one of the Secretaries to the Assembly of Divines, at Westminster. In the year 1642, a chaplain of Sir William Waller, one evening at supper, at Lady Vere's, in London, showed Dr. Wallis an intercepted letter, in cipher, and asked him if he could make any thing of it? Dr. Wallis said, perhaps he might if it was only a new alphabet. He took it with him, and in two hours deciphered it, and sent the chaplain a copy of it next morning. This was his first attempt at deciphering, in which afterwards he became so skilful and so famous.†

In 1645 he married Susanna, daughter of John and Rachael Glyde, of Northam, in Sussex; and soon after settled in London, at a time when philosophical studies of all kinds were greatly impeded by the civil wars. In 1649 he was appointed Savillian Professor of Geometry, in Oxford, and Keeper of the Archives, in that city; situations which he held till his death. On the restoration, he was appointed one of the Chaplains to his Majesty; King Charles II. having a great respect for him, on account of various services which he had done to his father and to himself. He died in 1703, on the 28th of October, being eighty-seven years three months and five days old. His memory was prodigious. He informs us himself that he extracted the square root of 3, while in bed in the night, and found it 1·73205,08075,68877,29353. At the request of a foreigner he took another number, consisting of 53 places of figures, and in the dark of the night extracted its root to 27 places, by the usual process. Both of these numbers he retained in his mind, till he afterwards wrote them down by day-light.‡

Dr. Wallis was the author of a great many books, not only on mathematics, but divinity, and other subjects. His English Grammar, and his Treatise on

---

* Wallis's account of his own life.   Langtoft's Chronicle, Vol. I. p. 146.
† Ibid. p. 158.   ‡ Phil. Trans. 1685. Vol. XV. p. 1269.

the Method of Teaching the Deaf and Dumb to speak, deserve particular attention. His dispute with Hobbes, and his refutation of the absurd geometrical opinions of that paradoxical man, excited much attention at the time, but have now lost their interest, because the reputation of Hobbes has long since vanished. His mathematical works were published in three large volumes, during his own lifetime. The most important of them all is his *Arithmetic of* *His arithmetic of infinites.* *Infinites;* which contains, not only a rich harvest of important discoveries, but constituted the germ from which almost all the discoveries of the age originated. He first expressed the denominators of fractions by negative exponents, guided by this analogy; in the progression x^3, x^2, x^1, x^0, $\frac{1}{x}$, $\frac{1}{x^2}$, $\frac{1}{x^3}$, in which the terms are in geometrical progression, the exponents, till the fractions commence, are obviously in arithmetical progression 3, 2, 1, 0. This progression ought to continue, and therefore $\frac{1}{x}$ is the same thing as x^{-1}; $\frac{1}{x^2}$ the same as x^{-2}, and so on. Hence he wrote the progression this way, x^3, x^2, x^1, x^0, x^{-1}, x^{-2}, x^{-3}. This happy thought gave him the quadrature of all spaces, both plane and solid, whose elements are reciprocally as some power of the abscisse. The equation of all hyperbolas, for example, is $y = \frac{1}{x^m}$ or $y = x^{-m}$. But it was demonstrated that in all curves, whose equation was $y = x^m$, the area of the curve is to the parallelogram of the same base and height, as $1 : m + 1$; therefore, in hyperbolas, this ratio must be $1 : -m + 1$. Suppose $m = \frac{1}{2}$ as it is in one species of hyperbola, then $-m + 1 = \frac{1}{2}$, and we have the ratio of it to the parallelogram of the same base and height $1 : \frac{1}{2}$. In the ordinary hyperbola the ratio is $1 : -1 + 1$, or $1 : 0$, indicating that the asymptotic spaces of the ordinary hyperbola are infinite. Wallis drew from this view of the subject a very ingenious mode of obtaining an expression for the quadrature of the circle. He observed, that we possessed the absolute quadrature of all the figures whose ordinates are expressed by $(1 - x^2)^0$, $(1 - x^2)^1$, $(1 - x^2)^2$, $(1 - x^2)^3$, &c. But the first of these, according to the rules of the arithmetic of infinites, is the equation of a figure equal to unity, or to the parallelogram circumscribed; the second is $\frac{2}{3}$ of it; the third, $\frac{8}{15}$; the fourth, $\frac{48}{105}$; supposing $x = 1$. Thus we have a series of terms 1, $\frac{2}{3}$, $\frac{8}{15}$, $\frac{48}{105}$, &c., each of which expresses the relation to the circumscribed parallelogram of the figure, whose ordinate is expressed by $(1 - x^2)^0$, $(1 - x^2)^1$, $(1 - x^2)^2$, $(1 - x^2)^3$, &c. But the exponents of this last series, 0, 1, 2, 3, &c., are in arithmetical progression. Hence, if we wished to introduce a new term between each of the old, it is obvious that the term which would come between $(1 - x^2)^0$ and $(1 - x^2)^1$ would be $(1 - x^2)^{\frac{1}{2}}$. But this is the expression for the ordinate of the circle. Hence we would have the quadrature of the circle, if in the series 1, $\frac{2}{3}$, $\frac{8}{15}$, $\frac{48}{105}$, &c., it were possible to interpolate a term between 1 and $\frac{2}{3}$. This, however, is a difficult task, and

Wallis investigated it with much genius and address. Not being able to find it in finite terms, he expressed it by the infinite fraction $\dfrac{2 \times 4 \times 4 \times 6 \times 6 \times 8 \times 8, \&c.}{3 \times 3 \times 5 \times 5 \times 7 \times 7 \times 9, \&c.}$

These are but a small number of the important discoveries which Wallis made by the application of his method. He determined the various properties of the cycloid, the cyssoid, the conchoid, the parabola, the spiral, &c. &c. His method gave occasion also to the first rectification of a curve which was done by Mr. William Neil; and Wallis demonstrated that the curve which he rectified was the semicubic parabola.* It gave occasion also to the discovery of

Continued
fractions. continued fractions. Being dissatisfied with the expression which he had found for the quadrature of the circle, he wrote to Lord Brouncker, requesting him to turn his thoughts to the subject. His Lordship did so, and ascertained that if the square be 1 the circle is equal to

$$\cfrac{1}{1 + \cfrac{1}{2 + \cfrac{9}{2 + \cfrac{25}{2 + \cfrac{49}{2 + \&c.}}}}}$$

prolonged to infinity; and if we stop short at any place, the limits are alternately too large and too small. The nature and properties of continued fractions were afterwards investigated by Huygens, and in our own times the theory of them has been completed by La Grange. It is to the arithmetic of infinites of Wallis, likewise, that we owe the celebrated discovery of Mercator respecting

Mercator. the hyperbola. Mercator was born in Holstein, and came over to England about the year 1660, and continued to reside in London till his death. His proper name was Kauffman, which he translated into Latin, as was customary in these times. Wallis had shown that the expression for the ordinate of the hyperbola, taking the origin of the abscisse upon the asymptote at a certain distance from the centre, was $\dfrac{1}{2 + x}$; but this expression did not come within his rules, and he had attempted in vain to bring it under them. The ingenious idea of dividing the numerator of this fraction by its denominator occurred to Mercator, and he obtained in consequence this infinite series $1 - x + x^2 - v^3 + x^4$, &c., the well known series for the hyperbola. Mercator published this discovery in his *Logarithmotechnia*, printed in the year 1668. He gave his work that title because he employs his series in the construction of logarithms, which depend, as every body knows, on the quadrature of the hyperbolic area between its asymptotes. Lord Brouncker had previously discovered a very simple series for the hyperbola; namely, $\dfrac{1}{1\cdot2} + \dfrac{1}{3\cdot4} + \dfrac{1}{5\cdot6} + \dfrac{1}{7\cdot8}$, &c. to infinity. This discovery was made before the year 1657, but was not published till 1668.†

---

* Phil. Trans. 1673. Vol. VIII. p. 6146.   † Phil. Trans. Vol. II. p. 645.

VIII. The next improvement which we have to mention is the method of tangents of Dr. Barrow, which approaches very nearly to the fluctionary calculus. Isaac Barrow was born, in London in 1630, and was sent for two or Barrow. three years to the Charterhouse School, and afterwards to Felstead School, in Essex, where his progress was so conspicuous, that his master appointed him tutor to the son of Lord Viscount Fairfax, at that time at the school. When 14 years of age he was admitted a Pensioner of Peterhouse, in Cambridge, under his uncle Mr. Isaac Barrow, then Fellow of that College. When he was qualified for the university, he was entered a pensioner in Trinity College, the 5th February, 1645. In 1647, he was chosen a scholar of the house, and though he was always a staunch loyalist, yet his merit and his prudence were so great, that he preserved the respect and good opinion of his superiors. He turned his attention to all parts of learning, especially to natural philosophy, and studied the writings of Galileo and Lord Bacon, as infinitely superior to the trifling studies of the schools. In 1648, he was chosen Fellow of his College entirely from his superior merit, for he had no friends to recommend him, and the times were very unfavourable to men of his opinions. This induced him to turn his thoughts to medicine, and he made some progress in anatomy, botany, and medicine; but afterwards he was prevailed upon by the advice of his uncle to resume his theological studies. These studies turned his attention to chronology, which he soon found so intimately connected with astronomy, that he thought it necessary to make himself acquainted with the principles of that science. Astronomy was so dependent upon geometry that he was naturally led to turn his attention to the mathematics, and he perused Euclid, and made great progress in that delightful science. In 1652, he stood candidate for the Greek Professorship which had been held by Dr. Duport; but his principles being known to be favourable to monarchy, another was preferred. This induced him to leave England: he went to France, then to Italy, then to Smyrna, and thence to Constantinople. In 1659, he returned, through Germany and Holland, to his own country. After the restoration, being neglected by Charles II. he expressed his sense of the neglect in the following lines:

> Te magis optârat rediturum, Carole, nemo.
> Te reducem sensit, Carole, nemo minus.

His merit however was such that he could not long remain without preferment. In July 1662, he was elected Professor of Geometry at Gresham College; and in 1663, when Mr. Lucas founded his Mathematical Professorship at Cambridge, Dr. Barrow was so powerfully recommended by Dr. Wilkins, that he was chosen the first Professor. In 1669, he resigned his chair to Sir Isaac Newton, and resolved to dedicate the rest of his life to theology. In 1672, he was appointed Master of Trinity College by a royal mandate, and Charles II. declared that he had given the place to the best scholar in England. In 1675, he was chosen Vice-chancellor of the University. He died of a fever in 1677, when only 47 years of age. He was a man of an athletic size, but remarkable for the mildness of his disposition. One day, while living at the house of a friend, he went out into the garden, and was attacked by the dog, a large mastiff, placed on purpose to guard the house. After some struggling, he got the dog under him, and held him fast. At first he thought of strangling the animal; but, considering that the dog was only discharging his duty, he relented, and thought it better to hold the animal fast till the family rose and came to his assistance. His sermons are perhaps the best in point of matter in the English language, though having been written before the metaphysical part of our language acquired that precision which it possesses at present, it is not always easy to understand them. Charles II. used to call him his unfair preacher, because when he discussed a subject he left nothing to his successors to do. He had a great predilection for poetry, and has left us a variety of poems, chiefly in Latin.

But it is to his geometrical writings that he is chiefly indebted for his celebrity. He was so fond of the science that he inscribed at the beginning of his edition of Apollonius; Θεος γεομετρει, *tu autem, Domine, quantus es geometra.* When his death approached, he is said to have expressed his joy, and congratulated himself, that he was at length going to learn, in the bosom of the Divinity, the solution of a great many intricate problems of geometry and astronomy. His *Lectiones Geometricæ* contain a great number of important researches relating to the dimensions and properties of curve surfaces. But we

His method of tangents.

shall at present notice only his method of tangents, and explain it as far as that can be done without a figure. This method is only that of Fermat simplified and generalized. He conceives the ordinate and the abscissa of the curve to receive each an infinitely small increment. These increments, together with the small portion of the curve intercepted between them, form an infinitesimal triangle, similar to the triangle formed by the ordinate, subtangent, and the tangent of the curve. From the equation of the curve he finds the relation between these two increments, and then says, as the increment of the ordinate is to the increment of the abscissa, so is the ordinate to the subtangent, which solves the problem. To give an example, the equation of the common parabola is $y^2 = px$. Let the increment of x the abscissa be e, and that of y the ordinate a. When these increments take place, the equation becomes $y^2 + 2ay + a^2 = px + pe$. From this equation take away from each side y^2 and px which are equal by the original equation, there remains $2ay + a^2 = pe$. But as a is a quantity infinitely small, a^2 may be thrown away. There remains then $2ay = pe$ which gives us this proportion, $a : e :: p : 2y$, or $a : e :: p : 2\sqrt{px}$.

Hence we have $p : 2\sqrt{px} :: \sqrt{px}$ (the ordinate) : the subtangent, which gives us the subtangent $= \dfrac{2\sqrt{px} \times \sqrt{px}}{p} = 2x$. So that in the parabola the subtangent is equal to twice the abscissa, as is well known to be the case. Every person must see the remarkable similarity between this method and the fluctionary calculus.

IX. We now come to the fluctionary calculus, the greatest improvement hitherto introduced into mathematics; an improvement of such magnitude, that the most arduous investigations of former mathematicians are performed by means of it with such facility that they resemble the amusements of children. For this admirable invention we are indebted to the genius of Sir Isaac Newton, a man by the universal consent of all his successors, and of almost all his contemporaries, placed at the head of mathematics and of science; and allowed to be the most splendid genius that has yet adorned human nature. On that account, and on account of the disputes between him and Leibnitz, about the discovery of the differential calculus, we think it proper to enter into a more minute detail of the particulars of Newton's life, than was requisite with respect to those mathematicians of whom we have already spoken.*

Sir Isaac Newton was born on Christmas-day old style, 1642,† at Wools- Newton. thorpe, in the parish of Colsterworth, in the county of Lincoln, near three months after the death of his father, who was descended from the eldest branch of the family of Sir John Newton, Bart. and was Lord of the Manor of Woolsthorpe. The family came originally from Newtown, in the county of Lancaster, from which probably they took their name. His mother was Hannah Ayscough, of an ancient and honourable family in the county of Lincoln. She was married a second time to the Rev. Barnabas Smith, Rector of North Witham, and had by him a son and two daughters, from whom were descended the four nephews and nieces, who inherited Sir Isaac's personal estate. Sir Isaac went to two little day schools at Skillington and Stoke, till he was 12 years old, when he was sent to the great school at Grantham, under Mr. Stokes, who had the character of being a very good schoolmaster. While at

---

* The first life of Newton that appeared was drawn up by Fontenelle from materials furnished by Newton's nephew, and published in the Memoires of the French Academy. This is the life from which all succeeding biographers of Newton extracted their materials. But like almost all the *Eloges*, published in the *Memoires of the Academy*, it is very inaccurate, and intended rather to display the abilities, and answer the private views of Fontenelle, than to convey accurate information to the readers. Mr. Edmund Turnor has lately favoured the world with the original life of Newton, drawn up by Mr. Conduitt, Newton's nephew, for the information of Fontenelle, and with a most interesting letter of Dr. Stukely on the same subject, from the MSS. in possession of the Earl of Portsmouth. See *Turnor's Collections for the Town and Soke of Grantham*, published in 1806. From these papers we have taken our account.

† When born he was so little, that his mother used to say that he might have been put into a quart mug, and so unlikely to live, that two women, who were sent to Lady Pakenham's, at North Witham, for something for him, did not expect to find him alive at their return.

4

Grantham, he boarded in the house of Mr. Clark, an apothecary, whose brother was at that time usher of the school.

His mechanical inventions when a boy.

Here he soon gave proofs of a surprising genius, and astonished his acquaintances by his mechanical contrivances. Instead of playing among other boys, he always busied himself in making curiosities, and models of wood of different kinds. For this purpose he had got little saws, hatchets, hammers, and all sorts of tools, which he knew how to use with great dexterity. He even went so far as to make a wooden clock. A new windmill was set up about this time near Grantham in the way to Gunnerby. Young Newton's imitating genius was excited, and by frequently prying into the fabric of it, as they were making it, he contrived to make a very perfect model, which was considered at least equal to the workmanship of the original. This sometimes he set upon the house-top where he lodged, and clothing it with sails the wind readily turned it. He put a mouse into this machine which he called his *miller*, and he contrived matters so that the mouse would turn round the mill whenever he thought proper. He used to joke too about the miller eating the corn that was put into the mill. Another of his contrivances was a water clock, which he made out of a box that he begged from the brother of his landlord's wife. It was about four feet in height, and of a proportional breadth. There was a dial plate at top with figures for the hours. The index was turned by a piece of wood which either fell or rose by water dropping. This stood in the room where he lay, and he took care every morning to supply it with its proper quantity of water.

These fancies sometimes engrossed so much of his thoughts, that he was apt to neglect his book, and dull boys were now and then put over him in his form. But this made him redouble his pains to overtake them, and such was his capacity that he could soon do it, and outstrip them when he pleased; and this was taken notice of by his master.* Still nothing could induce him to lay aside his mechanical inventions; but during holidays, and every moment allotted to play, he employed himself in knocking and hammering in his lodging room, pursuing the strong bent of his inclination, not only in things serious but likewise in ludicrous contrivances, calculated to please his school-fellows as well as himself. As for example, paper kites, which he first introduced at Grantham. He took pains to find out their proper proportion and figures, and the proper place for fixing the string to them. He made lanterns of paper crimpled, which he used to go to school by in winter mornings with a candle, and he tied them to the tails of his kites in a dark night, which at first frightened the country people exceedingly, who took his candles for comets. He was no less diligent in

---

* Sir Isaac used to relate that he was very negligent at school and very low in it, till the boy above him gave him a kick on the belly, which put him to great pain. Not content with having thrashed his adversary, Sir Isaac could not rest till he had got before him in the school, and from that time he continued rising till he was the head boy.

observing the motion of the sun, especially in the yard of the house where he lived, against the wall and roof wherein he drove pegs, to mark the hours and half hours made by the shade. These, by some years' observation, he made very exact, so that any body knew what o'clock it was by Isaac's dial, as they usually called it.

His turn for drawing, which he acquired without any assistance, was equally remarkable with his mechanical inventions. He filled his whole room with pictures of his own making; copied partly from prints, and partly from the life. Among others, were portraits of several of the kings, of Dr. Donne, and of Mr. Stokes, his school-master.*

Mrs. Vincent was niece to the wife of Sir Isaac's landlord, at Grantham, and lived with him in the same house. According to her account, he very seldom joined with his school-fellows in their boyish amusements, but chose rather to be at home, even among the girls, and would frequently make little tables, cup-boards, and other utensils, for her and her play-fellows to set their babies and trinkets in. She mentioned, likewise, a cart which he made with four wheels, in which he would sit, and by turning a windlass about, make it carry him round the house wherever he pleased. He is said to have contracted an attachment to Mrs. Vincent, whose maiden name was Storey, and would have married her, but being himself a Fellow of a College, with hardly any other income, and she having little or no fortune of her own, he judged it imprudent to enter into any matrimonial connection. But he continued to visit her as long as he lived, after her marriage, and repeatedly supplied her with money when she wanted it.

During all this time the mother of Sir Isaac lived at North Witham, with her second husband; but, upon his death, she returned to Woolsthorp, and, in order to save expences as much as she could, she recalled her son from school,

---

* He informed his nephew, Mr. Conduitt, that he had a facility in making verses. This is the more remarkable, as he had been heard to express a contempt for poetry. Hence it is probable, that the following lines, which he wrote under the portrait of Charles I., were of his own composition. They were given by Dr. Stukely, from Mrs. Vincent, who repeated them from memory.

> A secret art my soul requires to try,
> If prayers can give me what the wars deny.
> Three crowns distinguished here in order do
> Present their objects to my knowing view.
> Earth's crown, thus at my feet, I can disdain,
> Which heavy is, and, at the best, but vain.
> But now a crown of thorns I gladly greet:
> Sharp is this crown, but not so sharp as sweet.
> The crown of glory that I yonder see,
> Is full of bliss and of eternity.

If Newton wrote these lines, it must be remembered that they were written when he was only a boy at school.

Takes charge of his farm.

in order to make him serviceable at Woolsthorp, in managing the farm and country business. Here he was employed in superintending the tillage, grazing, and harvest; and he was frequently sent on Saturdays to Grantham market, with corn and other commodities to sell, and to carry home what necessaries were proper to be bought at a market-town for a family; but, on account of his youth, his mother used to send a trusty old servant along with him, to put him in the way of business. Their inn was at the Saracen's Head, in Westgate, where, as soon as they had put up their horses, Isaac generally left the man to manage the marketing, and, retiring to Mr. Clarke's garret, where he used to lodge, entertained himself with a parcel of old books, till it was time to go home again; or else he would stop by the way, between home and Grantham, and lie under a hedge studying, till the man went to town and did the business, and called upon him on his way back. When at home, if his mother ordered him into the fields to look after the sheep, the corn, or upon any other rural employment, it went on very heavily under his management. His chief delight was to sit under a tree with a book in his hands, or to busy himself with his knife in cutting wood for models of somewhat or other that struck his fancy; or he would get to a stream and make mill-wheels.*

Goes to Cambridge.

This conduct of her son induced his mother to send him to Grantham school again for nine months; and then to Trinity College, in Cambridge, where he was admitted the fifth of June, 1660. He always informed himself, before hand, of the books which his tutor intended to read, and when he came to the lectures, he found he knew more of them than his tutor himself. The first books which he read for that purpose, were Saunderson's Logic, and Kepler's Optics. A desire to know whether there was any thing in judicial astrology, first put him upon studying mathematics. He discovered the emptiness of that study as soon as he erected a figure; for which purpose he made use of two or three problems in Euclid, which he turned to by means of an index. He did not then read the rest, looking upon it as a book containing only plain and obvious things. This neglect of the ancient mathematicians, we are told by Dr. Pemberton, he afterwards regretted The modern books which he read gave his mind, he conceived, a wrong bias, vitiated his taste, and prevented him from attaining that elegance of demonstration which he admired in the ancients. The first mathematical book that he read was Descartes' Geometry; and he made himself master of it by dint of genius and application, without going through the usual steps, or having the assistance of any person. His next book was the Arithmetic of Infinites, by Dr. Wallis. On these books he wrote comments as he read them, and reaped a rich harvest of discoveries, or,

* The preceding details were collected upon the spot, by Dr. Stukely, from Sir Isaac's school-fellows, and the companions of his boyhood.

more properly, indeed, made almost all his mathematical discoveries as he proceeded in their perusal.

In 1664 he bought a prism, as appears by some of his own accounts of expenses at Cambridge, to try some experiments upon Descartes's doctrine of colours, and soon satisfied himself that the hypothesis of Descartes was destitute of foundation. The further prosecution of the subject satisfied him respecting the real nature of light and colours. He soon after drew up an account of his doctrine, which was published in the Philosophical Transactions, and unfortunately gave origin to a controversy between him and some foreign opticians, which produced an unhappy effect upon his mind, and prevented him from publishing his mathematical discoveries, as he had originally intended. He communicated them, however, to Dr. Barrow, who sent an account of them to Collins and Oldenburg, and by that means they came to be known to the Members of the Royal Society.

He laid the foundation of all his discoveries before he was 24 years of age. *Discovery respecting gravitation.* In the year 1665, when he retired to his own estate on account of the plague, the idea of his system of gravitation first occurred to him, in consequence of seeing an apple fall from a tree. This remarkable apple-tree is still remaining, and is usually shown to strangers as a curiosity. At that time, not being in possession of any accurate measure of the earth's surface, he estimated the force of gravity erroneously, and found, in consequence, that it was not capable alone of retaining the moon in her orbit. This induced him to dismiss his hypothesis, at that time, as erroneous. But afterwards, when Picard had measured a degree of the earth's surface, with tolerable accuracy, he was enabled to make a more precise estimate, and found that the force of gravity exactly accounted for the moon's motion in her orbit. He applied his doctrine to the planets and the whole solar system, and found it to account, in a satisfactory manner, for the whole phenomena of the motions of these bodies.

In 1667 he was elected Fellow of Trinity College, in Cambridge; and, in *Made Professor.* 1669, Dr. Barrow resigned his Mathematical Professorship to him. In 1671 he was elected Fellow of the Royal Society. It is stated, I do not know upon what authority, that at this time he was so poor that he was obliged to apply to the Society for a dispensation from the usual contribution of a shilling a week, which all the Fellows of the Society regularly pay. His estates (for he had two) were worth about 80*l.* a year. I do not know the value of his Fellowship, nor of his Professorship, but both together could not be less, surely, than 100*l.* a year; so that his income, at that time, could not well be less than 200*l.* a year. Upon this, indeed, he had his mother and her family to support; but if we consider the expence of living in 1671, we can hardly pronouce Mr. Newton, at that time, a poor man. In 1675 he had a dispensation from King Charles II. for retaining his Fellowship without taking orders. In 1687 he

was chosen one of the Delegates to represent the University of Cambridge, before the High Commission Court, to answer for their refusing to admit Father Francis Master of Arts, upon the King's mandamus, without his taking the oaths prescribed by the statutes. He was a great instrument in persuading his colleagues to persist in the maintenance of their rights and privileges. So strenuous, indeed, was the defence which he made, that the Crown thought proper to drop its pretensions. In 1688 he was chosen by the University of Cambridge, Member of the Convention Parliament, and sat in it till its dissolution. He was chosen Member of Parliament also for the same University in 1701. In 1696 the Earl of Halifax, at that time Mr. Montague, and Chancellor of the Exchequer, who was a great patron of the learned, wrote him a letter to Cambridge, acquainting him that he had prevailed with the King to
Made Warden
of the Mint. make him Warden of the Mint.* In this post he did signal service in the great re-coinage, which took place soon after. In 1699 he was made Master and Worker of the Mint; in which situation he continued till his death, and behaved himself with an universal character of integrity and disinterestedness. He had frequent opportunities of employing his skill in mathematics and chemistry, particularly in his Table of Assays of Foreign Coins, which is printed at the end of Dr. Arbuthnot's Book of Coins.

In 1701 he made Mr. Whiston his Deputy Professor of Mathematics at Cambridge, and gave him all the salary from that time, though he did not absolutely resign the Professorship till 1703. In the year 1703 he was chosen President of the Royal Society, and continued to fill that honourable situation till the time of his death. In 1705 he was knighted by Queen Anne, at Cambridge.

At the University of Cambridge, he spent the greatest part of his time in his closet, and when he was tired with the severer studies of philosophy, his relief and amusement was going to some other study, as history, chronology, divinity, chemistry; all of which he examined with the greatest attention, as appears by the many papers which he left behind him on those subjects. After his coming to London, all the time that he had to spare from his business, and from the

---

* The letter was as follows:

    Sir,

       I am very glad that at last I can give you a good proof of my friendship, and the esteem the King has of your merits. Mr. Overton, the Warden of the Mint, is made one of the Commissioners of the Customs, and the King has promised me to make Mr. Newton Warden of the Mint. The office is the most proper for you; 'tis the chief office in the Mint, 'tis worth five or six hundred pounds per annum, and has not too much business to require more attendance than you may spare. I desire that you will come up as soon as you can, and will take care of your warrant in the mean time. Let me see you as soon as you come to town, that I may carry you to kiss the King's hand. I believe you may have a lodging near me.   I am, &c.

                                      CHARLES MONTAGUE.

civilities of life, in which he was scrupulously exact and complaisant, was employed in the same way; and he was hardly ever alone without a pen in his hand, and a book before him: and in all the studies which he undertook, he had a perseverance and patience equal to his sagacity and invention. His niece, afterwards married to Mr. Conduitt, who succeeded him as Master of the Mint, lived with him about twenty years, during his residence in London. He always lived in a very handsome, generous manner, though without osten- Mode of living in London. tation or vanity; always hospitable, and, upon proper occasions, he gave splendid entertainments. He was generous and charitable without bounds; and he used to say, that they who gave away nothing till they died, never gave. This, perhaps, was one reason why he never made a will. Scarcely any man of his circumstances ever gave away so much, during his own life time, in alms, in encouraging ingenuity and learning, and to his relations: nor, upon all occasions, showed a greater contempt of his own money, or a more scrupulous frugality of that which belonged to the public, or to any society he was entrusted for. He refused pensions and additional employments that were offered him; he was highly honoured and respected in all reigns, and under all administrations, even by those whom he opposed; for in every situation he showed an inflexible attachment to the cause of liberty, and to the constitution of Great Britain. George the Second, and Queen Caroline, showed him particular marks of their favour and esteem, and often conversed with him for hours together. Queen Caroline in particular, who was so great a patroness of learned men, used to take delight in his company, and was accustomed to congratulate herself that she lived in the same country, and at the same time, with so illustrious a person.

Notwithstanding the extraordinary honours that were paid him, he had so humble an opinion of himself, that he had no relish for the applause which he received. He was so little vain and desirous of glory from any of his works, that he would have let others run away with the glory of those inventions which have done so much honour to human nature, if his friends and countrymen had not been more jealous than he was of his own glory, and the honour of his country. He was exceedingly courteous and affable, even to the lowest, and never despised any man for want of capacity; but always expressed freely his resentment against any immorality or impiety. He not only showed a great and constant regard to religion in general, as well by an exemplary life as in all his writings, but was also a firm believer of revealed religion; as appears by the many papers which he left behind him on the subject.* But his notion of the

* I have heard it affirmed by some of the self-constituted philosophers of the present day, that Sir Isaac Newton believed the Christian Religion merely because he was born in a Christian country; that he never examined it; and that he left behind him a cart load of papers on religious subjects, which Dr. Horsley examined, and declared unfit for publication. These gentlemen do not perceive

Christian Religion was not founded on a narrow bottom, nor his charity and morality so scanty, as to show a coldness to those who thought otherwise than he did in matters indifferent: much less to admit of persecution, of which he always expressed the strongest abhorrence and detestation. He had such a mildness of temper that a melancholy story would often draw tears from him, and he was exceedingly shocked at any act of cruelty to man or beast; mercy to both being the topic that he loved to dwell upon. An innate modesty and simplicity showed itself in all his actions and expressions. His whole life was one continued series of labour, patience, charity, generosity, temperance, piety, goodness, and all other virtues, without a mixture of any known vice whatsoever.

He was blessed with a very happy and vigorous constitution: he was of a middle stature, and rather plump in his latter years: he had a very lively and piercing eye, a comely and gracious aspect, and a fine head of hair, as white as silver, without any baldness. To the time of his last illness he had the bloom and colour of a young man. He never wore spectacles, nor lost more than one tooth till the day of his death. About five years before his death, he was troubled with an incontinence of urine, and sometimes with a stillicidium; both of which continued to afflict him, more or less, according to the motion to which he was exposed. On this account he sold his chariot, and went always in a chair; and he gave up dining abroad, or with much company at home. He eat little flesh, and lived chiefly upon broth, vegetables, and fruit, of which he always eat heartily. In August, 1724, he voided, without any pain, a stone, about the size of a pea, which came away in two pieces; one some days after the other. In January, 1725, he had a violent cough and inflammation of the lungs, upon which, he was persuaded with considerable difficulty to take a house in Kensington, where he had, in his 84th year, a fit of the gout, for the second time, having had a slight attack of it some years before. This fit left him in better health than he had enjoyed for several years. In the winter of 1725, he wanted to resign his situation of Master of the Mint to his nephew, Mr. Conduitt: that gentleman would not permit his resignation, but offered to conduct the whole business in his place; and, for about a year before his death Sir Isaac hardly ever went to the Mint, trusting entirely to the management of his nephew.

His death.   On Tuesday, the last day of February, 1727, he went to town, in order to attend a meeting of the Royal Society. Next day Mr. Conduitt paid him a

that their assertions are inconsistent with each other. No body who has ever read a page of Newton's works would believe that he could write a cart-load of papers on a subject which he never examined. Newton's religious opinions were not orthodox; for example, he did not believe in the Trinity. This gives us the reason why Horsley, the champion of the Trinity, found Newton's papers unfit for publication. But it is much to be regretted that they have never seen the light.

visit, and found him apparently in better health than he had enjoyed for several years. Sir Isaac was sensible of it himself, and told his nephew, smiling, that he had slept the Sunday before from eleven at night till eight in the morning, without waking. But his fatigue in attending the Society, and in making and receiving visits, brought his old complaint violently upon him. Dr. Mead and Mr. Cheselden were carried out to Kensington to see him, by Mr. Conduitt. They immediately said, that his disease was the stone in the bladder; and gave no hopes of his recovery. The stone was probably removed from the place where it lay quiet, by the great motion and fatigue of his last journey to London. From this time he had violent fits of pain, with scarcely any intermission; and though the drops of sweat ran down his face with anguish, he never complained, nor cried out, nor showed the least sign of peevishness or impatience: and, during the short intervals from that violent torture, would smile and talk with his usual cheerfulness. On Wednesday, the 15th March, he was somewhat better, and fallacious hopes were entertained of his recovery. On Saturday, the 18th March, he read the newspapers, and held a pretty long conversation with Dr. Mead, and had all his senses perfect. But that evening at six, and all Sunday, he was insensible, and died on Monday, between one and two o'clock in the morning. Thus, he reached the age of eighty-four years and a few months, and retained all his senses and faculties to the end of his life, strong, vigorous, and lively; and he continued writing and studying many hours every day till the period of his last illness. He died worth 32,000*l*. of personal estate; which was divided between his four nephews and nieces of half blood. The land which he had of his father and mother descended to his heir of the whole blood, John Newton, whose great grandfather was Sir Isaac's uncle. A little before his death he gave away an estate, which he had in Berkshire, to the sons and daughters of Mrs. Conduitt's brother; and an estate at Kensington to Mrs. Conduitt's daughter. From this lady the present Earl of Portsmouth is descended.

Sir Isaac Newton was buried with great magnificence, at the public expence. His funeral. On the 28th of March, he lay in state in the Jerusalem Chamber, and was buried from thence in Westminster Abbey, near the entry into the choir. The spot is one of the most conspicuous in the Abbey, and had been previously refused to different noblemen, who had applied for it. The pall was supported by the Lord High Chancellor, the Dukes of Montrose and Roxborough, and the Earls of Pembroke, Sussex, and Macclesfield, being Fellows of the Royal Society. The Hon. Sir Michael Newton, Knight of the Bath, was chief mourner, and was followed by some other relations, and some eminent persons intimately acquainted with Sir Isaac. The office was performed by the Bishop of Rochester, attended by the Prebend and Choir. A magnificent monument was erected to his memory, with the following inscription:

4

H. S. E.
ISAACUS NEWTON, EQUES AURATUS,
QUI ANIMI VI PROPE DIVINA
PLANETARUM MOTUS, FIGURAS,
COMETARUM SEMITAS OCEANIQUE ÆSTUS,
SUA MATHESI FACEM PRÆFERENTE,
PRIMUS DEMONSTRAVIT.
RADIORUM LUCIS DISSIMILITUDINES,
COLORUMQUE INDE NASCENTIUM PROPRIETATES,
QUAS NEMO ANTEA VEL SUSPICATUS ERAT, PERVESTIGAVIT.
NATURÆ, ANTIQUITATIS, S. SCRIPTURÆ,
SEDULUS, SAGAX, FIDUS INTERPRES,
DEI OPT. MAX. MAJESTATEM PHILOSOPHIA ASSERUIT,
EVANGELII SIMPLICITATEM MORIBUS EXPRESSIT.
SIBI GRATULENTUR MORTALES, TALE TANTUMQUE EXTITISSE
HUMANI GENERIS DECUS.
NATUS XXV. DECEMB. MDCXLII. OBIIT XX. MART.
MDCCXXVI.

His mathematical discoveries. The mathematical discoveries of Sir Isaac Newton were so numerous and so important, that it is no easy task to give an idea of them. As his geometrical studies were conducted, in a great measure, without a master; and as the first books, to which he paid particular attention, were the Geometry of Descartes, and the Arithmetic of Infinites of Dr. Wallis, he never possessed any intimate acquaintance with the methods of the ancient mathematicians; a circumstance which, as we are informed by Dr. Pemberton, he afterwards regretted; but which, probably, contributed to render his invention so fertile and so happy. He made a great many discoveries while perusing the two works above-mentioned; and we have complete evidence that he was in possession of all his inventions before the age of 24. A complete collection of his works was published in 1779, by Dr. Horsley, in five quarto volumes, accompanied by a commentary, which, however, is any thing but complete. It is to be regretted that the mathematical world is yet destitute of a good commentary on the works of this consummate mathematician. Some of his books, indeed, have been fully commented on. Thus the Jesuits' copy of the Principia, if it has any fault, abounds too much with notes; and Stirling's Commentary on Newton's Treatise respecting Lines of the Third Order, is excellent. We have also a very elaborate Commentary on his Universal Arithmetic; and, perhaps his Optics stands in need of no other commentary than the few optical discoveries which have been made since he wrote, and which enable us to rectify one or two of his opinions on that difficult subject.

Newton communicated many of his original discoveries to Dr. Barrow, at that time Professor of Mathematics, at Cambridge; and by him they were made known to various other British mathematicians. He likewise entered into

3

a correspondence with Collins and Oldenburg, and by them was induced to write several long letters to Mr. Leibnitz, in which he gave an historical detail of the way that he was led to some of his most considerable discoveries. All these letters were afterwards published in the *Commercium Epistolicum*. The correspondence also between James Gregory and Collins, published in the same book, throws considerable light upon the order and time of Newton's mathematical discoveries. One of his first discoveries struck him while perusing Wallis's Arithmetic of Infinites, about the year 1663. Wallis had shown the method of finding the quadrature of all curves, the ordinates of which are expressed by $(1 - x^2)^m$, x being the abscissa, supposing m a whole number, either positive or negative or zero; and that when m was respectively 0, 1, 2, 3, 4, &c., the areas corresponding to the abscissa x were respectively x; $x - \frac{1}{3}x^3$; $x - \frac{2}{3}x^3 + \frac{1}{5}x^5$; $x - \frac{3}{3}x^3 + \frac{3}{5}x^5 - \frac{1}{7}x^7$; &c.; and he showed, that if a number could be interpolated between x and $x - \frac{1}{3}x^3$ in the second series, corresponding to the interpolation of $\frac{1}{2}$ in the first series between 0 and 1, that this number would represent the quadrature of the circle. But Wallis could not succeed in making this interpolation; it was left for one of the first steps of Newton, in his mathematical career. Newton arranged the terms of the second series given above, under each other in order, and examined them as follows:

$$x$$
$$x - \tfrac{1}{3}x^3$$
$$x - \tfrac{2}{3}x^3 + \tfrac{1}{5}x^5$$
$$x - \tfrac{3}{3}x^3 + \tfrac{3}{5}x^5 - \tfrac{1}{7}x^7$$
$$x - \tfrac{4}{3}x^3 + \tfrac{6}{5}x^5 - \tfrac{4}{7}x^7 + \tfrac{1}{9}x^9$$

On considering this table, Newton observed, that the first terms are all x; that the signs are alternately positive and negative; that the powers of x increase by the odd numbers; that the co-efficient of the first term is 1; that the co-efficient of all the other terms are fractions; that the denominators of these fractions are always the indices of x, in the respective terms; that the numerators in the second terms are the ordinary numbers; in the third terms, the triangular numbers; in the fourth terms, the pyramidal numbers; &c. These observations made him master of the laws that regulated the whole of the series. Hence he concluded, that having to develope in general $(1 - x^2)^m$, the series of numerators for the respective fractions in the different terms must be 1; m; $\frac{m \cdot m - 1}{1.2}$; $\frac{m \cdot m - 1 \cdot m - 2}{1.2.3}$, &c., for these are the expressions which represent the natural, triangular, and pyramidal numbers. Now this will hold good whether m be a whole number or a fraction. In the case which occasioned the investigation, namely, $(1 - x^2)^{\frac{1}{2}}$; $m = \frac{1}{2}$, and, consequently, the numerators deduced from the preceding formulas are 1, $\frac{1}{2}$, $\frac{1}{8}$, $\frac{1}{16}$, $\frac{5}{128}$, &c. These, multiplied into the terms of the series, namely, $x - \frac{x^3}{3} + \frac{x^5}{5} - \frac{x^7}{7} + \frac{x^9}{9}$, &c., Quadrature of the circle.

give us the following series: $x - \dfrac{x^3}{6} - \dfrac{1}{8.5}x^5 - \dfrac{1}{16.7}x^7 - \dfrac{5}{128.9}x^9$, &c., a series which obviously represents the area of the circular segment, corresponding to the abscissa x. This investigation led him likewise to the discovery of the binomial theorem, so celebrated in algebra, and of so much importance in an infinite number of investigations.

Binomial
theorem.

Newton had already made these discoveries, and many others, when the *Logarithmotechnia* of Mercator was published; which contains only a particular case of the theory just explained. But, from an excess of modesty and of diffidence, he made no attempt to publish his discoveries, expressing his conviction, that mathematicians would discover them all before he was of an age sufficiently mature to appear, with propriety, before the mathematical world. But Dr. Barrow having contracted an acquaintance with him soon after, speedily understood his value, and exhorted him not to conceal so many treasures from men of science: he even prevailed upon him to allow him to transmit to some of his friends in London a paper containing a summary view of some of his discoveries. This paper was afterwards published under the title of *Analysis per Equationes Numero Terminorum Infinitas.* Besides the method of extracting the roots of all equations, and of reducing fractional and irrational expressions into infinite series, it contains the application of all these discoveries to the quadrature, and the rectification of curves; together with different series for the circle and hyperbola. He does not confine himself to geometrical curves, but gives some examples of the quadrature of mechanical curves. He speaks of a method of tangents, of which he was in possession, in which he was not stopped by surd quantities, and which applied equally well to mechanical and geometrical curves. Finally, we find in this extraordinary paper the method of *fluctions* and of *fluents*, explained and demonstrated with sufficient clearness; from which it follows, irresistibly, that before that period he was in possession of that admirable calculus: for the editors of this paper, which was published in the *Commercium Epistolicum*, attest that it was faithfully taken from the copy which Collins had transcribed, from the manuscript sent by Barrow. At the request of Dr. Barrow, he drew up a full account of this method, which was only described in the first tract with great conciseness. This new work he entitled *Methodus Fluxionum, et Serierum Infinitarum.* This last book he meant to publish at the end of an English Translation of the Algebra of Kinckuysens, which he had enriched with notes. But, in consequence of the disagreeable disputes into which he had been dragged, by his discoveries respecting the different refrangibility of the rays of light, he altered his intention, and the treatise, to the great injury of mathematics, and ultimately, likewise, to the diminution of his own peace, lay unpublished till after his death.

Fluctionary
calculus.

About the time that this paper of Newton's was sent to London, or about the year 1668, James Gregory published his *Exercitationes,* a book which contained several important facts connected with the discoveries which Newton had made. In particular there is a new demonstration of Mercator's Series for the Hyperbola. Collins communicated Newton's discoveries to various mathematicians, and among others to Gregory. He first sent him Newton's Series for the Circle, concerning the accuracy of which Gregory at first had his doubts; but he soon discovered his mistake, and by pondering over the subject for about a year, there appears sufficient evidence from his letters in the *Commercium Epis-* *tolicum,* that he divined Newton's method, and consequently had the merit of discovering the fluctionary calculus at least in part. But he declined publishing any thing on the subject, as he states in one of his letters, that he might not interfere with the rights of the original inventor.

Discovered also by Gregory.

James Gregory, having been one of the most eminent mathematicians of his time, and a Fellow of the Royal Society, we must not omit this opportunity of giving a few particulars respecting his life. He was born at Aberdeen in 1639, and made great proficiency in classical learning; but still greater in mathematics and philosophy. He invented his reflecting telescope when only 24 years of age, and came to London, in 1664, in order to get it constructed. But not being able to find any person capable of grinding a speculum in the truly parabolic figure, he was obliged to lay aside his design for the present, on account of the difficulty of putting it in execution. Next year he went to Italy, at that time considered as the great school for mathematics and science in general; and while in that country printed his book, entitled, *Vera Circuli et Hyberbolæ Quadratura,* a book which raised his reputation to a great height; though some of his positions were afterwards attacked by Huygens, and a controversy ensued in which Mr. Gregory was considered by mathematicians as in the right. Mr. Gregory was Professor of Mathematics in Aberdeen, and died of a fever in 1675, when only thirty-six years of age. Besides his work above-mentioned, he published likewise a Treatise on Optics; and the *Commercium Epistolicum* contains many egregious specimens of his analytical skill.

James Gregory.

Newton had made all his discoveries before his future rival Leibnitz had entered upon his mathematical career. Godfrey William Leibnitz was born at Leipsic on the 23d of June, old style, 1646. His father was Professor of Ethics, and Secretary of the University of Leipsic, and dying when his son William was very young, the care of his education devolved upon his mother, who spared no expence to make it as complete as possible. His passion for knowledge was excessive, and at the age of 15 he began to embrace, with incredible ardour, every species of learning. Poetry, history, antiquities, philosophy, mathematics, jurisprudence, and law, engaged him in succession, and with equal eagerness. His poetical career did not go far, and seems to have terminated about the 15th year of his age. After this he studied philosophy and mathematics at Jena and at Leipsic, till the year 1663, when he was made

Leibnitz.

Master of Arts. He next applied his mind with great assiduity to the study of the Greek philosophers, and made many efforts to reconcile the opinions of Plato and Aristotle, and those of Aristotle and Descartes. But finding his endeavours fruitless, he abandoned these barren pursuits for the law, in which he made such progress that, in the year 1667, he was made Doctor of Laws by the University of Altorf, and was offered a Professorship extraordinary in the law, which however he declined accepting.

About this time Baron de Boinebourg, meeting him at an entertainment, was so much pleased with his conversation, that he procured him the patronage of the Elector of Mentz, and, in 1672, he went to Paris to manage some affairs of that Elector. Here he became acquainted with Huygens, and with some other Members of the Academy of Sciences, and began to turn his attention to mathematical investigations, but his knowledge of the subject was still very imperfect. In 1673, he went to London, where he was introduced to Oldenburg and Collins, and by their means was made acquainted with some of the Newtonian discoveries. With these gentlemen he ever after kept up a correspondence, and procured from them a great many important mathematical facts, particularly different series of Newton and Gregory; and two remarkable letters were sent him by Newton himself at the request of Mr. Collins, in which that illustrious philosopher gives a detail of the steps by which he was led to his various discoveries, and announces his possession of the fluctionary calculus, though he does not give any explanation of it. From London he returned again to France, and, in the year 1676, he came a second time to England, and there is complete evidence that, during his residence in London, he had an opportunity of perusing all the different documents afterwards published in the *Commercium Epistolicum,* among which there is a concise account of the fluctionary method of Newton.

From London he returned to Germany by the way of Holland, having attached himself to the service of the Elector of Hanover. Here he was so much engrossed in business that he had no longer leisure to prosecute his philosophical ideas in detail; though he continued to enrich the *Leipsic Acts* with a great variety of writings, partly on mathematics and partly on natural philosophy. In the year 1700, when the Elector of Brandenburg, afterwards King of Prussia, instituted an Academy of Sciences at Berlin, Leibnitz, whose reputation by this time had reached the highest point, was made President, and he was continued in the office, though he could only be occasionally present at the meetings of the Academy. In the Memoires of this Academy he published a number of papers both on mathematical and other subjects. The same year he was admitted a Member of the Academy of Sciences at Paris. His reputation by this time was so high that honours and pensions accumulated upon him from all quarters. He was Privy Counsellor of Justice to the Elector of Hanover,

Aulic Counsellor of the German Empire, and Privy Counsellor of Justice to the Czar of Russia, with a pension of a thousand ducats a year. In the year 1714, he came again to England to visit the Elector of Hanover, who was now raised to the throne of Great Britain. During this last visit he engaged in the well known controversy with Dr. Clarke, relating to the principles of natural philosophy and religion. He died on the 14th November 1716, in consequence of a violent fit of the gout, which produced, it is said, almost instant suffocation. A complete collection of his works was published at Geneva, in 1768, by M. Dutens, in seven quarto volumes. They are upon every subject. Metaphysics constituted a favourite subject of speculation with him: his metaphysical opinions are well known, and do no great honour either to his head or his heart.

He was fond of controversy, and during the course of his life was engaged in a great number of disputes. The most celebrated of all was his controversy about the invention of the differential calculus, which is absolutely the same with the Newtonian fluctionary calculus, the notation excepted. What Newton expressed by $\dot{x}$, Leibnitz expressed by dx. Had Leibnitz behaved with tolerable candour and moderation, such was the mildness and unassuming modesty of Sir Isaac Newton, that he might have passed off without question as, at least in some measure, the inventor of the differential method. But it is obvious from his letters that he had always a tendency to arrogate to himself the inventions of others. He even returned to Newton some of the Newtonian series that had been sent him, a little disguised, and claimed them as his own, till the remonstrances of his correspondents in England induced him to renounce his claim. He had previously had a kind of dispute with Dr. Wallis about the invention of the fluctionary calculus, and had admitted that Newton had the priority of invention by at least ten years. He did not begin to put in his claim for originality of invention till Dr. Wallis was dead, and all the old mathematicians who were acquainted with his correspondence with Oldenburg and Collins.

His differential calculus.

In the year 1704, the *Treatise of Quadratures*, by Newton, was published. This treatise had been written long before, many things being cited out of it in Newton's letters of October 24, and November 8, 1676. It related to the method of fluctions, and that it might not be taken for a new piece, Newton repeated what Dr. Wallis had published nine years before without being contradicted, namely, that this method was invented by degrees in the years 1665 and 1666. But the editors of the *Acta Lipsica* in their review of this book, and the author of the review was conceived to have been Leibnitz himself, represented Leibnitz as the first inventor of the method, and said that Newton had substituted fluctions for differences, just as Fabri, in his *Geometriæ Synopsis*, had substituted movement for the indivisibles of Cavalleri. This accusation gave a

His dispute with the British mathematicians.

beginning to the controversy. For Dr. Keill, in an epistle published in the Philosophical Transactions for September and October 1708, retorted the accusation, asserting that " all these things follow from the now so much celebrated method of fluctions, of which our Newton was doubtless the first inventor, as will be evident to any one who shall read his letters published by Dr. Wallis. Yet afterwards the same method was published by Mr. Leibnitz in the Acta Eruditorum; only changing the name and manner of notation."* Leibnitz, understanding this passage as a direct charge of plagiarism, complained of it as a calumny in a letter to Sir Hans Sloane, the Secretary, dated March 4, 1711, and moved that the Royal Society would cause Dr. Keill to make a public recantation. Dr. Keill chose rather to explain and defend what he had written; and though Newton was at first offended at his original paragraph, being apprehensive that it might occasion a controversy, yet when he was shown the accusation in the Acta Lipsica, he gave Keill liberty to maintain the opinions which he had advanced. Keill wrote a long letter to Sir Hans Sloane, in which he endeavours to demonstrate not only that Newton was the original discoverer, but that he had given Leibnitz so many hints of his method, that even a man of very ordinary abilities could hardly fail to make it out. This letter was sent to Leibnitz, who demanded that the Royal Society would put a stop to the accusations of a man too young to know what had passed between Newton and himself. Instead of making good his own accusation, as he ought to have done, that it might not be deemed a calumny, he insisted only on his own candour, and refused to tell how he came by his method. He said that the Acta Lipsica had given every man his due; that he had concealed the invention above nine years, that nobody might pretend to have been before him in it. He called Dr. Keill a novice who deserved to be silenced, and desired that Newton himself would give his opinion in the matter. Dr. Keill, in fact, had only repeated what Dr. Wallis had published 13 years before, and Newton had already given his opinion on the matter before the dispute began; and this opinion in all probability was the cause of the controversy, by giving origin to the severe and unjust treatment which the book had received in the Acta Lipsica.

The Royal Society being thus twice pressed by Mr. Leibnitz, and seeing no reason to condemn or censure Dr. Keill without inquiring into the matter; and that neither Newton nor Leibnitz (the only persons alive who knew and remembered any thing that had passed respecting these matters 40 years before) could be witnesses for or against Dr. Keill, appointed a committee, consisting of Dr. Arbuthnot, Mr. Hill, Dr. Halley, Mr. Jones, Mr. Machin, and Mr. Burnet, to search old letters and papers, and report their opinion on what they,

---

* Phil. Trans. 1708. Vol. XXVI. p. 174. The passage occurs in a Latin paper, by Dr. Keill, on *The Laws of Centripetal Force.*

might find, and ordered the letters and papers, with the report of their committee, to be published. This publication was the famous *Commercium Epistolicum*, containing the letters of Oldenburg, Collins, Leibnitz, Gregory, and Newton. The committee reported, that it appeared to them that Newton had the method in or before the year 1669, and it did not appear to them that Mr. Leibnitz had it before the year 1677, a year after the communication of a letter from Newton to Leibnitz, in which the method of fluctions was sufficiently described for every intelligent person. They add, that the differential method of Leibnitz was the same as the fluctionary calculus of Newton, the notation excepted; that they regard Newton as the first inventor of that method, and that Dr. Keill in saying so has done, in their opinion, no injury to Leibnitz.*

The dispute did not and could not well terminate here. Leibnitz complained bitterly of the *Commercium Epistolicum*, and threatened to publish a reply that would confound his antagonists. But he was wise enough not to attempt it. Indeed it would have been a hopeless and impossible task to have attempted to overturn the evidence, contained in the Commercium Epistolicum, of the priority of Newton as the inventor of fluctions. Several anonymous papers were published, either by Leibnitz or his friends, in which Newton was rather attacked than Leibnitz defended. Bernoulli even attempted to prove that Newton did not understand the differential method as far as the higher orders of fluctions are concerned. At last Leibnitz, associating himself with his friends the Bernoullis, had recourse to a method which, he conceived, would demonstrate his superiority over the British mathematicians, and thus give his claim of originality a greater chance of obtaining credit. This was to propose difficult problems to embarrass his adversaries. Newton had formerly solved two celebrated problems proposed by John Bernoulli in the year 1697, and an anonymous solution was published that year in the Philosophical Transactions. He received the first problem, proposed by the triumvirate, to confound the British mathematicians after he had undergone a good deal of fatigue in the Mint, and yet he solved it before he went to bed.

During the course of this dispute the extreme partiality which united Bernoulli to Leibnitz induced him to treat Newton with an unbecoming severity, and even injustice. He published, in the Acta Lipsica, under a disguised name, a most violent attack upon Keill, in which he endeavours to prove that Newton did not understand the rules of second differenciation. His opinion was founded upon a passage in the Treatise on the Quadrature of Curves, in which Newton from inadvertence had represented the different orders of fluctions by the terms

---

* The best edition of the Commercium Epistolicum is the octavo of 1722, which contains several articles not to be found in the original quarto edition.

of his binomial theorem.   On the other hand, Keill went too far when he attempted so show that Newton had committed no inadvertence whatever.   The dispute at last was silenced, if not terminated, by the death of Leibnitz in 1716.

Now that the violence which actuated both parties in this memorable dispute has subsided, and that all party feeling is, in some measure, at an end, we may be allowed to judge of the merits of both parties from the documents that remain.   Perhaps the safest way is to reject the arguments brought forward both by the British and German writers of the times, considering them as parties. French writers But there are several French writers who have given their opinions on the sub- on the dispute. ject, and who may be considered as more impartial.   The principal of these are Fontenelle, Buffon, Bossut, and Montucla.   Fontenelle is very unsatisfactory. He lived too near the time of the dispute, and had, in some measure, committed himself before hand in his *Eloge* on the Marquis De l'Hospital.   Buffon adopts the side of the English mathematicians, and seems to have taken their assertions for granted without ever having examined the documents.   Bossut adopts the side of Leibnitz; but it is equally evident that he has never examined the documents, but trusted to the assertions of the Bernoullis, and the other coadjutors of Leibnitz, with whose works he is obviously much better acquainted than with the Commercium Epistolicum, and the writings of Newton.   His History of Mathematics is written with much elegance, and does credit to his taste and talents; but from several particulars which it contains we may infer that it was written when he was a very young man, and that he gives many of the facts which his book contains only at second hand.   Montucla gives the most particular and the most impartial account of the dispute of any writer that we have seen; but even he does not appear to have perused all the documents, at least if we may be allowed to give that name to a very elaborate paper which appeared in the Philosophical Transactions, giving an account of the Commercium Epistolicum, and of the dispute between Leibnitz and Keill. Tradition ascribes this composition to Newton, and there is every reason from internal evidence to believe that he was the author.*   Now this paper brings forward several striking instances of Leibnitz having attempted to palm Newton's inventions, received in Newton's own writing, upon Newton himself as his own, and of his having desisted from his claim when the trick was pointed out to him.   It shows also that Leibnitz had formerly, during the life of Dr. Wallis, acknowledged that Newton was the original inventor of the calculus, and that he had retracted this admission after the death of Dr. Wallis. In short, this paper contains so many important facts, and such accurate reasoning, that

---

* Phil. Trans. 1714. Vol. XXIX. p. 173.

it ought to be perused by every person who wishes to form an accurate opinion on the dispute. We think that the following conclusions may be drawn with almost perfect certainty.

1. The method of fluctions, and the differential calculus, are absolutely the same in principle, and differ only in the notation. The reasoning of Newton is rigidly accurate; but that of Leibnitz loose and unsatisfactory. The Leibnitzian notation is conceived to have an advantage over the Newtonian: this advantage, as far as it is real, applies only to the mode of expressing the higher orders of fluctions; and this mode might be introduced, with facility, into the Newtonian notation. But it must be admitted, that the notation introduced by Leibnitz into the integral calculus has advantages, and, accordingly, it has been universally adopted by British writers; which is not the case with his mode of expressing the higher orders of fluctions.

2. Newton was in possession of his method of fluctions many years before Leibnitz thought of his differential calculus, or indeed before Leibnitz had made any great progress in mathematics. This is so obvious from the Commercium Epistolicum, that we believe no one will think of calling it in question.

3. Newton more than once announced to Leibnitz that he was in possession of the fluctionary calculus, and points out its advantages with peculiar emphasis; but, in none of his letters does he explain the nature of this calculus. There is an explanation of it, however, in Newton's *Analysis per Equationes Numero Terminorum Infinitas;* which was sent up to London by Dr. Barrow, in manuscript. There can be no doubt that Leibnitz saw this paper long before he wrote any thing respecting the differential calculus. The notice, we allow, is very brief; but it is much more than the information given to James Gregory, who, notwithstanding, succeeded in discovering Newton's method.

4. Newton himself admitted in his Principia, that Leibnitz had invented his differential calculus without receiving any information. This passage, which is of great importance to the subject in hand, is as follows:

" In literas quæ mihi cum geometra peritissimo *G. G. Leibnitio,* annis abhinc decem intercedebant, cum significarem me compotem esse methodi determinandi maximas et minimas, ducendi tangentes, et similia peragendi, quæ in terminis surdis æque ac in rationalibus procederent, et literis transpositis hanc sententiam involventibus *(data æquatione quotcunque fluentes quantitates involvente, fluxiones invenire, et vice versa)* eandem celarem: rescripsit vir clarissimus se quoque in ejusmodi methodum incidisse; et methodum suum communicavit a meo vix abludentem præterquam in verborum et notarum formulis, et idea generationis quantitatum utriusque fundamentum continetur in hoc lemmate."*

---

* Newton's Principia, Lib. II. Scholium, at the end of Lemma II. p. 226, of the edition of 1714.

This passage appeared in the first edition of the Principia, published in 1686; it was continued in the second edition, published in 1713, while the quarrel between Keill and Leibnitz was at the hottest,* but it is not to be found in the edition of 1722. From this passage, there is complete evidence that Newton admitted Leibnitz to have discovered the differential calculus; which he never would have done, if he had furnished him with a sufficient explanation of his own method to enable him to understand it.

5. Newton then was the original inventor; Leibnitz knew that he was in possession of some unknown method; this seems to have stimulated his invention. His differential calculus seems to be founded on the method of tangents of Barrow, from which it differs only in the notation. But it was the generalization of that method which constituted the chief merit of Leibnitz. Had Newton published his Treatise on Fluctions, when it was originally written, he would have had no competitor nor coadjutor, and Leibnitz's mathematical reputation would probably never have risen beyond mediocrity. All concealment in matters of science we consider as improper, and Nature usually punishes it by putting the same invention into the hands of some other person, who deprives the real discoverer of a part of the reputation which he would otherwise have acquired. Newton's hesitation, indeed, proceeded from an amiable and praise-worthy motive; yet it was a weakness, and, as such, was punished by raising up a competitor, who deprived him of his just share of reputation, and gave him more trouble and uneasiness than that which he sought to avoid by withholding his publication.

6. It has been attempted by some, to compare Leibnitz to Newton, and to hold him up as scarcely inferior to that illustrious man : but the comparison is very unequal. Leibnitz certainly was a very extraordinary man, and one of the greatest geniuses that ever appeared among mankind : the extent of his information was prodigious, his activity was indefatigable, and every region of knowledge which he traversed received marks of his original and inventive genius. But he was far inferior to Newton, both as a philosopher and as a man. None of his productions will bear a comparison with the *Principia,* or the *Optics,* of Sir Isaac Newton; nor do we think any of his mathematical writings equal to the Universal Arithmetic, or the Fluctions of Newton. We cannot

---

* Montucla assigns, as a reason why it was continued in that edition, that Cotes, the editor, published it without Newton's knowledge, and against his will. But there must be a mistake in this; for after the preface there is an advertisement, by Newton himself, mentioning a great number of additions and improvements, which he had made in that edition. A proof, not only that he knew of the edition, but that he had been at some pains in correcting and improving it. The paragraph then was retained, because Newton did not wish to erase it. It was doubtless scratched out of the third edition at the suggestion, of Keill, or of some other English mathematician engaged in the dispute. Newton's opinion of the dispute is·sufficiently evident from his account of the Commercium Epistolicum, in the Phil. Trans.

5

conclude this account better than by giving the comparison between the two men, drawn up by Newton himself, on occasion of this very controversy.

"It must be allowed that these two gentlemen differ very much in philoso- Newton and Leibnitz compared. phy. The one proceeds on the evidence arising from experiments and pheno- mena, and stops where such evidence is wanting; the other is taken up with hypotheses, and propounds them, not to be examined by experiments, but to be believed without examination. The one, for want of experiments to decide the question, does not affirm whether the cause of gravity be mechanical or not mechanical: the other, that it is a perpetual miracle, if it be not mechanical. The one, by way of inquiry, attributes it to the power of the Creator, that the least particles of matter are hard; the other, attributes the hardness of matter to conspiring motions, and calls it a perpetual miracle, if the cause of this hard- ness be other than mechanical. The one does not affirm that animal motion in man is purely mechanical: the other teaches that it is purely mechanical, the soul or mind (according to the hypothesis of a harmonia præstabilita) never acting on the body so as to alter or influence its motions. The one teaches that God (the God in whom we live, and move, and have our being,) is omnipre- sent; but not a soul of the world: the other, that he is not the soul of the world, but intelligentia supra mundana, an intelligence above the bounds of the world; whence it seems to follow, that he cannot do any thing within the bounds of the world, unless by an incredible miracle. The one teaches, that philosophers are to argue from phenomena and experiments to the causes thereof, and thence to the causes of those causes, and so on till we come to the first cause: the other, that all the actions of the first cause are miracles, and all the laws impressed on nature by the will of God are perpetual miracles and occult qualities, and, therefore, not to be considered in philo- sophy. But, must the constant and universal laws of nature, if derived from the power of God, or the action of a cause not yet known to us, be called mi- racles and occult qualities, that is to say, wonders and absurdities? Must all the arguments for a God, taken from the phenomena of nature, be exploded by new hard names? And must experimental philosophy be exploded as mira- culous and absurd, because it asserts nothing more than can be proved by expe- riments, and we cannot yet prove by experiments, that all the phenomena in nature can be solved by mere mechanical causes? Certainly these things deserve to be better considered."[*]

About five years after the death of Newton, Dr. Berkeley, the celebrated Berkeley's attack on fluctions. Bishop of Cloyne, induced, as it is said, by the scepticism of Dr. Halley, who

---

[*] Phil. Trans. 1714. Vol. XXIX. p. 173. To contribute as much as possible to the elucidation of this memorable dispute, we shall insert, in the Appendix No. III., the minutes of the Royal Society, as far as they relate to Sir Isaac Newton.

professed himself an unbeliever of the Christian religion, published a book, entitled, *The Minute Philosopher;* and afterwards another, entitled *The Analyst;* in which he undertook to prove, that mathematics was inconsistent with religion; that mathematicians were all unbelievers of the Christian religion; and yet, by a singular contradiction of the understanding, believed in the mysteries of the Newtonian calculus: that this calculus was false and obscure in its principles; that if it led to truth in geometrical investigations, it was because one error balanced another; and, finally, that Newton did not understand the principles of his own method. These singular assertions would only have served to entertain mathematicians, had not the metaphysical talents of the author, and his great celebrity, given them a certain degree of weight with the public, to which, in themselves, they were not entitled. On that account, various elaborate answers were speedily published. The first which appeared, in 1734, the same year in which *The Minute Philosopher* was laid before the public, was written by Drs. Middleton and Smyth, Professors at Cambridge; and was entitled, *Geometry, no Friend to Incredulity; or, a Defence of Sir Isaac Newton and the English Mathematicians,* &c. In this book, Bishop Berkeley is attacked, alternately, by the arms of ridicule, and of serious discussion. But this defence does not appear to have proved satisfactory to mathematicians. It was soon followed by another, written by Dr. Wilson, Professor of Mathematics in Dublin, entitled, *A Defence of the Principles of Fluctions.** But, one of the most important books which appeared on the subject was written by Mr. Benjamin Robins, and was entitled, *A Discourse concerning the Nature and Certainty of Sir Isaac Newton's Method of Fluctions,* &c.

Robins.

Benjamin Robins, the author of this treatise, has acquired too high a reputation by his mathematical labours to be passed over in silence. He was born at Bath, in 1707. His parents were Quakers, and did not, on that account, encourage their son in his pursuit of scientific knowledge. But he made, notwithstanding, surprising progress in different branches of learning; and some of his friends, who interested themselves in his progress, sent one of his mathematical papers to London, which was examined and highly approved of by Dr. Pemberton. That eminent cultivator of ancient geometry sent him down, as a further trial of his skill, some problems, of which he required elegant solutions, according to the ancient model, and not founded on algebraic calculations. The solutions of these problems by Mr. Robins gave a very advantageous idea of his taste as well as invention. After perusing most of the eminent ancient and modern mathematicians, he went up to London, where he supported himself for a number of years by teaching mathematics. He first distinguished

* This book I have never had an opportunity of seeing.

himself before the public by a refutation of the Leibnitzian doctrine of motion, composed by the celebrated John Bernoulli. This famous question, being in reality a dispute about a definition, was incapable of decision, and vanished entirely whenever mathematicians were at the trouble to explain the meaning of the terms which they employed. The Newtonian definition is obviously more simple than the Leibnitzian; and, on that account, better adapted to the purposes of philosophy. But neither are capable of demonstration. Both are true, and both are false, according to the meaning affixed to the terms employed.

In the year 1742, Mr. Robins published his most valuable performance, entitled *New Principles of Gunnery;* containing a great deal of new and valuable matter; and his experiments were conducted with the highest ingenuity and perspicuity. The reputation which he acquired by this performance induced the Prince of Orange to invite him over, in 1747, to assist in the defence of Bergen-op-zoom, at that time besieged by the French. He set out in consequence of this invitation; but the place was taken before his arrival. After the return of Lord Anson from his celebrated voyage round the world, the public were in eager expectation of an account of his adventures. Accordingly, Mr. Richard Walter, who had been Chaplain on board the Centurion during the greatest part of the expedition, was employed to draw up an account of the voyage. The manuscript was put into Mr. Robins's hands, who, after perusing it, gave it as his opinion that it was not proper for publication, and offered to write the whole over again himself. This offer was accepted; and the work was soon after published, entirely from the pen of Mr. Robins; employing Mr. Walter's manuscript merely as materials from which his information was obtained. This book is beyond comparison the most interesting account of any voyage ever published; and, as such, has been always a great favourite with the public, and has gone through a very great number of editions. It was probably the cause of the subsequent rise of Anson, who was First Lord of the Admiralty during one of the most successful wars in which Britain was ever engaged. This book also altered the circumstances of Mr. Robins himself, he was offered his choice of two very considerable places: the first was to go to Paris, as one of the Commissioners for adjusting the limits of *Acadia;* the other, was to be Engineer General to the East India Company; whose forts being in a most ruinous condition, required a man of abilities to put them into a proper posture of defence. This last situation he accepted, as best suited to his genius, and to his previous course of study. The Company settled on him a salary of 500*l.* a year for life, on condition that he continued in their service for five years. He set out for the East Indies at Christmas, 1749, having provided himself with a complete set of astronomical instruments; but the climate of the East Indies was too much for his delicate constitution. He was seized with a fever in September, 1750, and though he got the better of this malady, he gradually relapsed into a languishing condition, and died at Fort St. David, on the 29th of July, 1751, with his pen in his hand, as he was drawing up for the Company an account of the posture of their affairs.

Maclaurin.   But the most complete answer to Dr. Berkeley's objections was made by the celebrated Mr. Maclaurin, in his Treatise on Fluctions, published in 1742. Colin Maclaurin, one of the most eminent mathematicians that Britain has produced, was the son of a clergyman, and born at Kilmoddan, in Scotland, in 1698. He was sent to the University of Glasgow, in 1709, where he continued for five years, and applied himself to study with the most intense assiduity. He discovered his mathematical genius as early as the 12th year of his age. Having accidentally met with a Euclid, in a friend's chamber, he made himself master of the first six books in a few days, without any assistance. It is certain that, by the time he reached his 16th year, he had invented many of the propositions which were afterwards published under the title of *Geometrica Organica*. In his 15th year he took the degree of *Master of Arts;* on which occasion he composed, and publicly defended, a Thesis on the Power of Gravity, with great applause. After this, he quitted the University, and retired to a country seat of his uncle, who had the care of his education, for his parents had been dead for some time. Here he spent two or three years in pursuing his favourite studies; but, in 1717, he offered himself a candidate for the Professorship of Mathematics, in the Marischal College of Aberdeen; and obtained it, after a trial of ten days with a very able competitor. In 1719 he went to London, where he was elected a Fellow of the Royal Society, and where he acquired the friendship of Sir Isaac Newton, and of some others of the most eminent philosophers at that time in the British capital. In 1722, Lord Polworth, Plenipotentiary of the King of Great Britain, at the congress of Cambray, engaged Mr. Maclaurin to go as tutor and companion to his eldest son, who was just setting out on his travels. After a short stay at Paris, and some other towns in France, they fixed their residence at Lorrain, where Mr. Maclaurin wrote his Essay on the *Percussion of Forces;* which gained the prize of the Royal Academy of Sciences at Paris, for 1724.

His pupil dying soon after at Montpelier, Mr. Maclaurin returned immediately to his Prefessorship, at Aberdeen. But scarcely was he settled at that University, when he received an invitation to Edinburgh, the Curators of the University being desirous that he should supply the place of Mr. James Gregory, whose great age and infirmities had rendered him incapable of teaching. Mr. Maclaurin had some difficulties to encounter, arising from competitors, who had good interest with the Patrons of the University; and likewise from the want of an additional fund for the new Professor: but these were removed, chiefly by the interference of Sir Isaac Newton. He wrote a letter to Mr. Maclaurin, and another to the Lord Provost of Edinburgh on the subject, part of each of which we shall transcribe as curiosities. In his letter to Mr. Maclaurin he expresses himself thus: "I am very glad to hear that you have a prospect of being joined to Mr. James Gregory in the Professorship of the Mathematics at Edinburgh; not only because you are my friend, but principally because of your abilities; you being acquainted as well with the new improvements of mathematics as with the former state of those sciences. I heartily wish you good success, and shall be heartily glad to hear of your being elected." In his letter to the Lord Provost he writes thus: " I am glad to understand that Mr. Maclaurin is in good repute among you for his skill in mathematics, for, I think, he deserves it very well; and to satisfy you that I do not flatter you, and also to encourage him to accept the place of assisting Mr. Gregory, in order to succeed him, I am ready, if you please to give me leave, to contribute 20*l.* per annum towards a provision for him till Mr. Gregory's place becomes void, if I live so long, and I will pay it to his order in London."

In November 1725, he was introduced into the University of Edinburgh. The number of his pupils amounted to about 100 every year, and being of different standing and proficiency, he was obliged to divide them into four or five classes, in each of which he employed a full hour every day from the first of November to the first of June. In the first class he taught the first six Books of Euclid, plain trigonometry, practical geometry, the elements of fortification, and an introduction to algebra. In the second class he taught algebra, the 11th and 12th Books of Euclid, spherical trigonometry, conic sections, and the general principles of astronomy. The third class were instructed in the principles of astronomy and perspective, in Newton's Principia, and in the elements of fluctions. The fourth class were instructed further in fluctions, in the doctrine of chances, and read the remainder of Newton's Principia.

In 1734, when Dr. Berkeley published his *Analyst,* Mr. Maclaurin, thinking himself included in the charge of infidelity alleged against mathematicians in general, resolved to write an answer to the Bishop's book. But the work accumulating under his hands terminated in his celebrated Treatise on Fluctions, which was published at Edinburgh, in two quarto volumes, in 1742. In this book he demonstrates the whole doctrine of fluctions without having recourse to infinitely small quantities, or any other supposition capable of being contested; but, according to the rigid method of the ancients, following a mode of demonstration often employed by Archimedes in his works. But his demonstrations are often so long and complicated, and require such severe attention to follow them, that we believe they are seldom perused by the mathematicians of the present day, who, having turned almost the whole of their attention to the analytical method, are not so capable as their predecessors of following long synthetical demonstrations. But it will be acknowledged by every person who

6

peruses the book, that all the objections of Dr. Berkeley against the doctrine of fluctions are completely refuted, and whatever doubts the most captious metaphysicians may think proper hereafter to start about the nature of infinities, the mathematician has no more concern with them than with the famous sophisms about space and motion. In the year 1740, Mr. Maclaurin gained a prize of the Academy of Sciences, at Paris, for his Explanation of the *Motion of the Tides from the Theory of Gravitation.* He had only ten days to draw up this paper, and was obliged in consequence to send the first imperfect copy to the Academy. He afterwards revised the whole, and inserted it in his Treatise on Fluctions. His *Algebra* and his *Account of Sir Isaac Newton's Discoveries* were published after his death. In 1745, having been very active in fortifying the city of Edinburgh against the rebel army, he was obliged to take refuge in the North of England, where he was invited by Dr. Herring, then Archbishop of York, to reside with him during his stay in that country. During this expedition, being exposed to cold and hardships, he laid the foundation of an illness which soon after put an end to his life. This disease was an *ascites,* of which he died, on the 14th of June, in the forty-eighth year of his age. During his last moments he requested his friend, Dr. Monro, to account for the flashes of fire which seemed to dart from his eyes, while in the mean time his sight was failing so that he could scarcely distinguish one object from another. This singular request is a proof of the great calmness and serenity of mind which he enjoyed at his last moments. He left behind him two sons and two daughters. One of his sons was a lawyer and an eminent writer. He was appointed one of the Lords of Session, in Edinburgh, when he assumed the name of Lord Dreghorn.

Methodus Incrementorum.

Brook Taylor.

X. The *Methodus Incrementorum,* for the invention of which we are indebted to Dr. Brook Taylor, ought rather to have preceded than followed the invention of fluctions. Brook Taylor was born at Edmonton, in Middlesex, in the year 1685. In 1701, he entered St. John's College, Cambridge, and in 1708, he wrote his Tract on the Centre of Oscillation. This paper was published in the Philosophical Transactions for 1714, so that John Bernoulli's accusation of plagiarism, which he made with so much violence and indecorum, is totally groundless. In 1709, he took the degree of Bachelor of Laws. In 1712, he was elected a Fellow of the Royal Society, and the same year took his degree of LL.D. Dr. Taylor died in the year 1731, when only forty-six years of age. He published a variety of excellent papers in the Transactions, and was one of the chief writers in the memorable dispute between the Bernoullis and the British mathematicians respecting the discovery of fluctions. His great works were his *Principles of Linear Perspective,* in which he first established the true practice of the art on principles which have been followed ever since. This book was published in 1715. The same year he published his *Methodus*

Incrementorum Directa et Inversa. A book in which the calculus is explained so concisely, and with so little developement, that Bernoulli had reason on his side when he complained of it as excessively obscure It can scarcely be understood by any one, who is not as well acquainted with the method before he begins to read as the author himself. The subject was afterwards explained by Euler, in the first chapters of his *Institutiones Calculi Differentialis,* with that remarkable order and perspicuity which distinguishes all his writings. To that book therefore we refer those who wish to make themselves acquainted with the *Methodus Incrementorum* with the greatest facility. The subject was likewise treated by Emerson in his *Method of Increments,* published in 1763 This book, like all those written by Emerson, contains much valuable matter, but so very inelegantly put together, that it is scarcely possible to read it with patience. The *Methodus Incrementorum,* of Taylor, contains the famous theorem which The Taylorian retains his name. It consists in a particular series to express what any function theorem. of a variable quantity x becomes, when x acquires any increment whatever. Let Y be the function, and let x become $\Delta x,$* then, according to Taylor, Y becomes $Y + \frac{\Delta x d Y}{1.dx} + \frac{\Delta x^2 dd Y}{1.2.dx^2} + \frac{\Delta x^3 d^3 Y}{1.2.3.dx^3} + \frac{\Delta x^4 d^4 Y}{1.2.3\ 4dx^4},$ &c. If from this series, the law of which is sufficiently evident, we take the first term Y, the remainder will obviously express the simultaneous increment of Y. Maclaurin afterwards gave a demonstration of this theorem derived from the consideration of fluctions, and rigorously exact, but a demonstration upon pure analytical principles being considered as more elegant as well as more satisfactory, this has been lately given by M. L'Huillier of Geneva, in his *Principiorum Calculi Diff. et Int. Expositio Elementaris,* published in 1795.

XI. We have now brought our history of mathematics as low down as is requisite for the mathematical papers contained in the Philosophical Transactions. But several additional improvements have been made, chiefly by foreign mathematicians, which it will be sufficient merely to name here. The first is called Calculus of the *calculus of partial differences,* and is of great importance in many philoso-partial differences.phical discussions, such as the vibration of cords, the propagation of sound, &c. It was invented by Euler in the year 1734, and afterwards by D'Alembert, who was ignorant of what Euler had done, and who first had the merit of applying it to the solution of physico-mathematical questions. Euler has represented this method under a much simpler form than D'Alembert, and on that account mathematicians have followed the notation of Euler.

The *method of variations* is another improvement in the fluctionary calculus Method of of considerable importance. By means of it alone some of the most difficult variations.

---

* We use the foreign notation in this formula, the Newtonian not being fitted for expressing increments. Δx means the increment of x; dY, the fluctions of Y, &c.

problmos *de maximis et minimis* can be solved.   Suppose we have a function of
two or more variable quantities, the relation between which is expressed by a
particular law, this method teaches us how to find what the function becomes
when the law itself undergoes an infinitely small variation.   James Bernoulli,
in his celebrated isoperimetrical problem, which he proposed in consequence of
the quarrel between his brother and himself, had supposed an infinitely small
arc of the curve divided into three by two ordinates intermediate and equidis-
tant ; and making them vary, he had found the position which they ought to
have to fulfil the condition demanded, and by this means he resolved the problem.   This solution has been admired by all mathematicians as an astonishing
instance of patience and sagacity.   Euler afterwards generalized it in his work
entitled, *Methodus inveniendi Lineas Curvas Maximi Minimive Proprietati
gaudentes, seu Solutio Problematis Isoperimetrici latissimo Sensu accepti.*   But
he himself acknowledges that his method is destitute of that perfection which
it ought to possess.   La Grange, in 1755, invented a method purely analytical:
Euler acknowledged its great superiority, and explained it at some length in
two dissertations published in the Memoires of the Petersburgh Academy,
under the title of the *Method of Variations ;* for La Grange, satisfied with the
invention, had given it no name.   A detailed explanation of this method may
be seen in the work on the differential and integral calculus by Lacroix, and
in the book on the same subject by Bossut.   This last book is much better arranged, and may therefore be perused with much greater facility than the
former.   But no account of the method that we have seen is so good as that of
Euler in the 10th volume of the New Memoires of the Petersburgh Academy.[*]

Papers in the
Transactions.

XII.   It only remains to take a cursory view of the mathematical papers contained in the Philosophical Transactions.   We shall merely mention the most
important of them without entering into details which would be inconsistent
with the nature of this work.

1. The number of papers to which the title of Arithmetical is given in the
Transactions is 14.   There is an account of Dr. Wallis's edition of Archimedes'
dissertation, entitled, Ψαμμιτης, in which there is an account of the Greek
arithmetic lately explained at some length by Legendre.[†]   The next paper

---

[*] We have no history of mathematics in the English language.   Several have been published in
Germany and France.   The most detailed is that of Montucla, published in four quarto volumes.
The first two volumes of that work, which bring down the history to the invention of fluctions, are
excellent.   The history of mathematics during the 18th century, which occupies the first half of the
3d volume, is more defective.   Montucla died while it was printing, and the last part of the 3d
volume, and the whole of the 4th, were supplied by La Lande, and are greatly inferior to the rest of
the work.

[†] Phil. Trans. 1676. Vol. XI. p. 567.

consists of twelve problems on compound interest and annuities by Adam Martindale.[*]

The next is entitled, On Infinitely Finite Fractions, by Dr. Wood.[†] It consists of a number of infinite series of fractions with the corresponding numbers to which they are equal. The following table will give an idea of these:

$$1 = \tfrac{1}{2} + \tfrac{1}{4} + \tfrac{1}{8} + \tfrac{1}{16}, \ \&c.$$
$$\tfrac{1}{2} = \tfrac{1}{3} + \tfrac{1}{9} + \tfrac{1}{27} + \tfrac{1}{81}, \ \&c.$$
$$\tfrac{1}{3} = \tfrac{1}{4} + \tfrac{1}{16} + \tfrac{1}{64} + \tfrac{1}{256}, \ \&c.$$
$$\tfrac{1}{4} = \tfrac{1}{5} + \tfrac{1}{25} + \tfrac{1}{125} + \tfrac{1}{625}, \ \&c.$$
$$\tfrac{1}{5} = \tfrac{1}{6} + \tfrac{1}{36} + \tfrac{1}{216} + \tfrac{1}{1296}, \ \&c.$$

In these series it is evident that the numerator is always 1, and the denominators the powers of the first denominator.

The next paper is entitled, An Arithmetical Paradox concerning the Chances of Lotteries. By the Hon. Francis Roberts, Esq.[‡] The parodox is this. There are two lotteries, at either of which a gamester pays a shilling for a lot or throw. The first lottery, on a just computation of the odds, has 3 to 1 of the gamester; the second lottery but 2 to 1, nevertheless the gamester has the very same disadvantage, and no more, in playing at the first lottery as the second.

The next paper is entitled, The Doctrine of Combinations and Alternations, improved and completed. By Major Edward Thornycroft.[§] This is an elaborate and complete treatise, though the subject has been explained with more elegance by modern writers, especially by Euler.

The next paper is by De Moivre, and is entitled, De Mensura Sortis, seu de Probabilitate Eventuum in Ludis a casu fortuito pendentibus.[‖] This is the first sketch of a doctrine which was afterwards fully detailed in our author's celebrated treatise on the subject, published in 1718.

The next paper is entitled, A Short Account of Negativo-affirmative Arithmetic. By Mr. John Colson.[¶] This is a contrivance by means of which all numbers larger than five are excluded, and thus the common operations of arithmetic are performed with greater facility. But as the method has not come into practice, and never probably will, it is needless to attempt a particular description of it.

The next paper is The Description and Use of an Arithmetical Machine, invented by Christian Ludovicus Gersten, Professor of Mathematics, at Griefsen.[**] The first arithmetical machine was contrived by Pascal; another was in-

* Phil. Collections 1681. No. I. p. 34.　　† Phil. Col. 1681. No. III. p. 45.
‡ Phil. Trans. 1693. Vol. XVII. p. 677.　　§ Phil. Trans. 1705. Vol. XXIV. p. 1961.
‖ Phil. Trans. 1710. Vol. XXVII. p. 213.　　¶ Phil. Trans. 1726. Vol. XXXIV. p. 161.
** Phil. Trans. 1735. Vol. XXXIX. p. 79.

vented by Sir Samuel Moreland ; a third, by Leibnitz; a fourth, by the Mar-
chese Poleni ; and a fifth, by Mr. Leopold. This of Gersten is founded on
that of Leibnitz, but so complicated and operose as not to be entitled to any
attention.

The next paper is entitled, Of a New-invented Arithmetical Instrument, called
a Shwan-pan, or Chinese Account Table. By Gamaliel Smethurst.* This is
an instrument somewhat like the abacus of the ancients, adapted to our arithmetical notation, and conceived by Mr. Smethurst to be much superior to the
Shwan-pan of the Chinese. Such inventions are not worth much attention.
The common rules of arithmetic can be performed with sufficient facility without any such assistance.

The next paper is entitled, An Essay towards solving a Problem in the Doctrine of Chances. By the late Rev. Mr. Bayes.† The problem is as follows:
Having given the number of times an unknown event has happened and failed,
to find the chance that the probability of its happening should lie somewhere
between any two named degrees of probability. The solution is much too long
and intricate to be of much practical utility. Dr. Price published a supplement
to this paper in the succeeding volume of the Transactions.‡

The next paper is entitled, On the Theory of circulating Decimal Fractions.
By John Robertson, Lib. R. S.§ The subsequent publications of Dr. Hutton,
have deprived this paper of all its interest.

The next paper is by the same author, and is entitled, Investigations of Twenty
Cases of Compound Interest.|| These are investigations of cases of compound
interest previously engraven on a plate, by William Jones, Esq. ; and published
without demonstration by Gardiner and Dodson.

The last arithmetical paper we have to mention, is entitled, Some Properties
of the Sum of the Divisors of Numbers. By Edward Waring, M. D.¶
Waring was one of the profoundest mathematicians of the 18th century ; but
the inelegance and obscurity of his writings prevented him from obtaining that
reputation to which he was entitled. Except Emerson, there is scarcely any
writer whose works are so revolting as those of Waring. The great elegance
and admirable simplicity and order of all the writings of Euler contributed as
much to give him the great celebrity which he enjoyed, as his inventive genius.

2. The papers on Logarithms, in the Philosophical Transactions, amount to
11; and some of them are of very considerable importance. The first, is an
account of the Logarithmotechnia of Mercator ; a book which we noticed,
with due praise, while detailing the history of mathematics.

* Phil. Trans. 1749. Vol. XLVI. p. 22.   † Phil. Trans. 1763. Vol. LIII. p. 370.
‡ Phil. Trans. 1764. Vol. XIV. p. 296.   § Phil. Trans. 1768. Vol. LVIII. p. 207.
|| Phil. Trans. 1670. Vol. LX. p. 508.   ¶ Phil. Trans. 1788. Vol. LXXVIII. p. 388.

The next paper is an excellent one by Dr. Halley, on the Method of constructing Logarithms from the Nature of Numbers, without any regard to the Hyperbola.* There is another Method of constructing Logarithms, proposed some years after, by Mr. Craig, published in a subsequent volume;† and another by Mr. Long, in a subsequent volume of the Transactions.‡ But this method is, in some measure, mechanical; and is not to be compared to the methods of Halley and Craig.

Dr. Halley published an important paper, entitled, An Easy Demonstration of the Analogy of the Logarithmic Tangents to the Meridian Line, or Sum of the Secants; with various methods for computing the same to the utmost exactness.§ This paper, though we mention it here, is rather connected with navigation than logarithms.

The Logometria, of Cotes,‖ deserves to be mentioned on account of its importance. It was afterwards published in the collection of Mr. Cotes's works, in 1722, by Dr. Smith, under the title of *Harmonica Mensurarum*. Roger Cotes was one of the most eminent mathematicians of the age in which he lived. He was born in 1682, at Burbach, in Leicestershire. He was educated at St. Paul's School, London, and afterwards at Trinity College, Cambridge; where, in 1706, he was appointed the First Professor of Astronomy and Experimental Philosophy, on the foundation of Dr. Plume, Archdeacon of Rochester. In 1713 he published a new edition of Newton's Principia; his preface to which was much admired, and procured him great reputation. He unfortunately died in the year 1716, in the thirty-fourth year of his age, to the great regret of all lovers of the sciences. Newton had so high an opinion of his knowledge and genius, that he used to say, "had Cotes lived, we should have known something." The celebrated property of the circle, which he discovered, and which is of so much importance in the integral calculus, was not known till after his death, when it was found among his papers, and made out, with great difficulty, by his relation, Dr. Smith.

The next paper is An Account of the Method of protracting the Logarithmic Lines, on the common Gunter's Scale. By Mr. John Robertson.¶

The next paper is A Dissertation on Logarithms. By Mr. Jones.** It is remarkable for that great conciseness of expression, which distinguishes all the writings of that eminent mathematician. The view which he takes of the subject is very similar to one taken on logarithms by Euler, and which, as far as we can judge, seems to be the simplest and most perspicuous, as well as satisfactory way of viewing the subject. In a subsequent paper†† Mr. Hellins

Cotes

---

* Phil. Trans. 1695. Vol. XIX. p. 58.   † Phil. Trans. 1710. Vol. XXVII. p. 191.

‡ Phil. Trans. 1714. Vol. XXIX. p. 52.   § Phil. Trans. 1696. Vol. XIX. p. 202.

‖ Phil. Trans. 1713. Vol. XXIX. p. 5.   ¶ Phil. Trans. 1753. Vol. XLVIII. p. 96.

** Phil. Trans. 1771. Vol. LXI. p. 455.   †† Phil. Trans. 1780. Vol. LXX. p. 307.

gives Two Theorems, which he demonstrates, and which he shows to be of considerable utility in facilitating the computation of logarithms.

The last paper on this subject which we think requires mentioning, is one by Mr. Nicholson, in which he gives the principles and illustration of an advantageous method of arranging the differences of logarithms, on lines graduated for the purpose of computation.*

3. There are no fewer than 15 papers on the nature of equations. But as we have noticed the most important of these already, it will be sufficient merely to refer to them in a note.†

4. On the doctrine of series, which is one of the most important branches of mathematics, and which may be considered as in some measure the original of all the modern improvements in the science, there are 13 papers. The following are their titles, with the names of the authors:

Method of determining 2d, 3d, 4th, &c. Term of a Series taken in order. T. Simpson.

New Method of computing the Sums of certain Series. Landen.

On certain Infinite Series. Bayes.

Series for computing the Ratio of the Diameter of a Circle, and its Circumference. Hutton.

To find the Value of certain Infinite decreasing Series. Maseres.

To find a near Value of a slowly converging Series. Maseres.

New Methods of investigating the Sums of Infinite Series. Vince.

Summation of certain Series. Waring.

On Infinite Series. Waring.

On finding the Value of Algebraic Quantities, by converging Series. Waring.

On Infinite Series. Waring.

New Method of investigating the Sums of Infinite Series. Waring.

Method of computing the Values of slowly converging Series. Hellins.

5. There is only one paper on interpolation; the author of which is Dr. Waring. But Newton, in his Principia, has given an excellent method, probably the first ever contrived. The subject has been much discussed by the foreign mathematicians, but Newton's method still retains its utility.

6. On the doctrine of annuities, which was in a manner first broached in

---

* Phil. Trans. 1787. Vol. LXXVII. p. 246.

† Collins. Phil. Trans. 1669. Vol. IV. p. 929.—Halley. Ibid. 1687. Vol. XVI. p. 335 and 387; and Vol. XVIII. p. 136.—Colson. Ibid. 1707. Vol. XXV. p. 2353.—Demoivre. Ibid. Vol. XXV. p. 2368.—Maclaurin. Ibid. 1726. Vol. XXXIV. p. 104.—Milner. Ibid. 1778. Vol. LXVIII. p. 380.—Maseres. Ibid. Vol. LXVIII. p. 902.; and Vol. LXX. p. 85 and 221.—Waring. Ibid. 1779. Vol. LXIX. p. 86.—Lord Stanhope. Ibid. 1781. Vol. LXXI. p. 195.—Wales. Ibid. 1781. Vol. LXXI. p. 454.—Hellins. Ibid. 1782. Vol. LXXII. p. 417.—Wood. Ibid. 1798. Vol. LXXXVIII. p. 369.—Wilson. Ibid. 1799. Vol. LXXXIX. p. 265..

this country, and which has always been cultivated with much assiduity, there are eight papers. The following are their titles :

Value of an Annuity for Life, and Probability of Survivorship. Dodson.
Method of calculating Reversions depending on Survivorship. Price.
Theorems for Annuities. Price.
On the Probabilities of Survivorships, &c. Morgan.
Determination of Contingent Reversions. Morgan.
On determining the real Probabilities of Life, &c. Morgan.
Method of determining the Value of Contingent Reversions. Morgan.
Method of determining Contingent Reversions. Morgan.

7. There are three papers on the quadrature of the circle; not reckoning the account of James Gregory's book on the subject, and his paper in answer to the objections of Huygens. The first paper is by Leibnitz; the second is by Dr. Hutton, and has been already mentioned under the head of Series; the third paper is by Mr. Hellins; and is an improvement of Halley's method of computing the quadrature of the circle.

8. Upon the conic sections, and other curves of a higher order, we have a considerable number of papers. The following are the chief :

New Properties in Conic Sections. Waring.
Properties of the Conic Sections. Jones.
Length of the Arc of a Conic Hyperbola found. Landen.
Quadrature of the Hyperbola. Lord Brouncker.
New Quadrature of the Hyperbola. Demoivre.
Hyperbolical Cylindroid. Wren.
Testudo Veliformis Quadrabilis. David Gregory.
Synchronism of the Vibrations in a Cycloid.
Quadrable Cycloidal Spaces. Wallis
Quadrature of a Portion of an Epicycloid. Caswell.
Method of Measuring all Cycloids and Epicycloids. Halley.
History of the Cycloid. Wallis.
Properties of the Catenaria. Gregory.
Curve, called Cardioid. Castillion.
Curve of Swiftest Descent. Sault and Craig, and Machin, in three different papers. The problem, as first proposed by Bernoulli, is likewise resolved by Sir Isaac Newton.
Quadrature of the Logarithmic Curve. Craig.
Quadrature of the Lunula of Hippocrates. Pertus and Gregory.
Dimension of the Solids, produced by Hippocrates' Lunula. Demoivre.
Quadrature of Figures geometrically irrational. Craig.
Easy Method of measuring Curvelinear Figures. Wallis.
Of squaring some Kinds of Curves. Demoivre.
General Method of determining the Quadrature of Figures. Craig.

Quadrature of the Foliate, a Curve of the Third Order.   Demoivre.

Theorems of computing the Areas of certain Curves.   Landen.

General Method of describing Curves by the Intersection of Right Lines. Brakenridge.

Description of Curve Lines.   Maclaurin.

On finding Curve Lines from the Properties of the Variation of Curvature.

Two Species of Lines of the Third Order, not mentioned by Newton nor Stirling. Stone.

Length of Curve Lines.   Craig.

Measure of Curves, and their Construction.   Maclaurin.

Tangents to Curves.   Sluse.

Tangents of Curves, deduced from the Theory of Maxima & Minima. Ditton.

Collection of Secants.   Wallis.

9 On Plane and Spherical Trigonometry there are few papers in the Transactions; unless we were to include, under that title, some of those mentioned under the preceding heads, which might certainly be done without impropriety. The following are the only papers on that subject which I at present recollect :

Spherical Trigonometry reduced to Plane.   Blake.

Trigonometry abridged.   Murdoch.

Calculations in Spherical Trigonometry abridged.   Lyons.

10. We shall now give a list of a few papers, which could not, with propriety, be reduced under any of the preceding heads.

Equations for exhibiting the Resolution of Goniometrical Lines.   Jones.

Locus of three or four Lines, famous among the Ancient Geometricians. Pemberton.

Centre of Oscillation.   Brook Taylor.

Solid of least Resistance.   Craig.

On Centripetal Forces.   Waring.

On the Bases of Cells, where Bees lodge their Honey.   Maclaurin.

Properties of the Machine for turning Ovals demonstrated.   Ludlam.

The above lists may be considered as exhibiting the names of all the mathematical papers contained in the Philosophical Transactions, except a few papers on the fluctionary calculus, and subjects connected with it; the titles of which could not have been rendered intelligible, without entering into further details than is consistent with this work.   Here then we take our leave of this subject; upon which it would have been much easier to have extended our observations to a far greater length, than to have confined ourselves to the short space which our account of Mathematics occupies.   But we did not consider ourselves as at liberty to sacrifice the different branches of Mechanical Philosophy and Chemistry, which, to most of our readers will probably be more interesting than dry details respecting such abstract and difficult subjects.

BOOK III.

OF MECHANICAL PHILOSOPHY.

SUBSTANCES may either be examined in a state of rest, or as acting upon each other, and producing changes on each other. The knowledge derived from the first of these views is called *Natural History*; that which we obtain by the second, is distinguished by the name of *Science*. But bodies cannot act upon each other without producing motion, and the motions produced by such actions are of two kinds; either so great as to be visible to our senses, and capable of being measured by the space passed over; or so small as not to be distinguishable by our senses, except by the effects produced. The phenomena connected with the first of these kinds of motions constitute what is called *Natural Philosophy*, or *Mechanical Philosophy*, in this country, and on the Continent, *Physics*. The phenomena connected with the imperceptible motions belong to the science called *Chemistry*. *Nature of mechanical philosophy.*

Mechanical Philosophy comprehends under it a great many very important phenomena, and has been sub-divided into different subordinate heads. A variety of arrangements has been proposed, attended each with its peculiar advantages and defects. Our object is only to consider it as far as it is treated of in the Philosophical Transactions. Now all the topics connected with it, as far as they occupy a part of that work, may be arranged under the following heads: *Division.*

1. Astronomy.	4. Mechanics.	7. Navigation.
2. Optics.	5. Hydrodynamics.	8. Electricity.
3. Dynamics.	6. Acoustics.	9. Magnetism.

These will constitute the subject of the nine following chapters.

CHAP. I.

OF ASTRONOMY.

THE science of Astronomy furnishes the noblest proof of what the human intellect is capable. The only true mode of advancing, patient and assiduou observation, was early perceived by its cultivators; and for 2000 years past astronomers have devoted themselves to the observation of the heavenly phenomena, with a degree of persevering industry, which has only been equalled by the progress which they have made. No other science can boast of the same degree of perfection. It may be considered in some measure as complete. Astronomy constitutes one of the most prominent departments in the Philosophical Transactions; and owes much of its progress to the splendid exertions of some of the most conspicuous Members of the Royal Society. From the foundation of the Royal Observatory of Greenwich, in 1675, the place of Astronomer Royal has been filled by a succession of men of the first eminence, who have furnished a series of observations of infinite importance to navigation and astronomy. Whether that race be at length extinct by the death of Dr. Maskelyne, or whether succeeding Astronomers Royal will display any portion of the vigour and abilities which illustrated their predecessors, remains to be seen.

Origin of astronomy.

The History of Astronomy has been treated so amply and so well, in a great variety of books, which are in every body's hands, that we may be excused for passing it over in a very concise manner. The Chaldeans and Egyptians are conceived to have turned their attention to astronomy at a very early period; and attempts have been made to show that the Chinese, and the inhabitants of India, are in possession of tables indicating observations coeval with, or even anterior to, the commonly received era of the creation. But these observations, notwithstanding the zeal, the skill, and the acknowledged abilities of the ingenious supporters of them, have not met with that favourable reception from astronomers, which was expected by their propagators; and have been almost unanimously rejected as erroneous and impossible. The oldest astronomical observations on record were made at Babylon, 719 and 720 years before the commencement of the Christian era. Ptolemy, who has transmitted them to us, employed them for determining the period of the moon's mean motion; and therefore, in all probability, had none more ancient on which he could depend The Greeks borrowed their first notions of astronomy, as of all the other sciences, from the Egyptians. Thales, of Miletus, who lived in the seventh *Thales.* century before the Christian era, and who certainly acquired his knowledge from the Egyptian priests, predicted a famous eclipse of the sun, which put an end to a furious war that raged between Cyaxares, King of Media, and Alyattes, King of Lydia. A total eclipse of the sun happening during an engagement between the two armies, struck such a terror into each, that a peace was concluded.* This eclipse has given rise to much discussion among learned men; because, if the precise date at which it happened could be ascertained, it would be of great service in establishing or correcting *chronology.* No fewer than eight different dates have been proposed, varying from each other by a period of 43 years. We shall only mention the date recently assigned by Mr. Baily, as every thing concurs to render it most correct. According to him, it happened in the year 610 before Christ, on the 30th of September: and he has demonstrated, that this is the only total eclipse of the sun which took place in Asia Minor, where the battle was fought, between the years 650 and 580 before Christ.†

But the real foundation of astronomy was not laid by the Greeks till the establishment of the Alexandrian school by Ptolemy Soter, and his successors. Ptolemy Philadelphus built a magnificent observatory, and supplied it with the requisite books and instruments. The first astronomers, who were placed in this building, were Aristyllus and Timocharis, who lived about 300 years *Aristyllus and* before Christ, and observed, with accuracy, the position of the principal stars *Timocharis.* of the zodiac. The little that we know respecting their labours, we obtain from the writings of Ptolemy, who quotes them sometimes. About the same time flourished Aristarchus, of Samos, who is celebrated by the ancients for his genius and discoveries. He pointed out an ingenious way of determining the *Aristarchus.* comparative distances of the sun and moon from the earth; and demonstrated that the sun was at least 20 times further off than the moon. Though this determination was greatly below the truth, yet it was a considerable step at the time, and extended the bounds of the universe to the eyes of astronomers. He also supported the opinion, that the earth and the planets move round the sun; an opinion previously advanced by Pythagoras, and some of his immediate disciples; but which the imperfect state of astronomy and mathematics did not enable them to support by such arguments as were calculated to give it general credit.

Eratosthenes was another astronomer of the Alexandrian school, to whom *Eratosthenes.* the science was much indebted. He prevailed upon Ptolemy Euergetes to erect, in the portico of the observatory, different large instruments for observing the stars. It was by means of these instruments that most of the subsequent

---

* Herodotus. Lib. I. § 74.   † Phil. Trans. 1811. p. 220.

observations of the Grecian astronomers were made. Eratosthenes likewise observed the obliquity of the ecliptic, and found it 23° 51′ 13″; a determination which has been of some service to astronomers, in determining the change of the obliquity which has taken place since that time. But the most famous of his experiments was his measurement of a certain portion of the earth's surface. There was a deep pit at Syena, an island in the Nile, and almost at the southern extremity of Ptolemy's dominions. Eratosthenes had observed, that on the day of the summer solstice, the whole bottom of this pit was illuminated at mid-day; and, as objects for some distance round Syena cast no shadow on that day, he concluded that Syena was situated directly under the tropic of Cancer. He next concluded, that Alexandria was in the same meridian with Syena, and that the distance between them was 5000 stadia. He then determined, by observation, the difference between the latitude of the two places. From all this, he settled the length of a degree at 250,000 stadia : but as we do not know the length of the stadium which he used, this measurement is of no use at present. It cannot be very accurate, because several of the data, upon which it was founded, are erroneous. Alexandria and Syena are not under the same meridian, their longitude differs by several degrees. The distance between the two places was not measured, but only estimated, and of course could not be accurate. All the fragments of the works of Eratosthenes still remaining have been inserted in the magnificent edition of Eratus, published at Oxford in 1672.

But the true founder of astronomy, as a science, was Hipparchus, of Bithynia, who observed in Alexandria during 35 years, between 160 years before Christ, and 125 years B. C. We know very few particulars of his life; but his astronomical labours are well known, and constitute a memorable era in the science His first care was to determine the length of the year, which he did by a great many successive determinations of the exact times of the equinoxes and solstices. But, in order to be able to determine the point with greater exactness, he had recourse to the determination of the summer solstice, about 150 years before, by Aristarchus, of Samos. The result of his calculation was, that the year wants about five minutes of 365 days 6 hours. His next object was to determine the rate of the sun's motion ; and he showed, by numerous and accurate observations, that this luminary moves fastest in winter and slowest in summer ; so that the summer half year is about nine days longer than the winter half year. In order to account for this, Hipparchus supposed the sun to move uniformly in an eccentric circle, the distance of the centre of which, from that of the earth, was $\frac{1}{24}$th of radius, and he placed the apogee in the sixth degree of Gemini. In these determinations Ptolemy coincides with Hipparchus ; though it is now known that they made the eccentricity about $\frac{1}{4}$th too great.

His measurement of a degree.

Hipparchus.

Hipparchus also devoted much of his attention to the moon. He determined the length of a lunation, by the same means that he had ascertained the length of the year. He measured the inclination of the moon's orbit to the ecliptic, which he found amount to five degrees. He determined, likewise, the motion of the apsides and nodes with much greater exactness than had formerly been done. He thought of an ingenious mode of ascertaining the size and distance of the sun and moon; and his construction, though not sufficiently delicate for the purpose to which he applied it, is still employed in the calculation of eclipses. He found the distance of the sun 1200 semi-diameters of the earth, and his horizontal parallax 3'. The mean distance of the moon from the earth he found equal to 59 of these semi-diameters. From this he concluded that the diameter of the earth was $3\frac{1}{3}$ times as great as that of the moon; and the diameter of the sun $5\frac{1}{4}$ times as great as that of the earth. Though these numbers are far below the truth, they served, however, to extend considerably the bounds of the universe, and were a much nearer approximation than had been formerly obtained.

Hipparchus did not attempt to form any theory of the planetary motions; but he made many accurate observations respecting them, in order to put it in the power of his successors to contrive a theory to explain their motions. But perhaps the greatest of all the labours of Hipparchus was his catalogue of the stars, and the discovery of the precession of the equinoxes, to which he was led by that catalogue. It was the appearance of a new star in the heavens, during his time, that induced him to undertake that laborious task, in order to enable future astronomers to know whether any new stars had made their appearance after his time. He not only drew up a catalogue of the principal fixed stars, with their exact positions, but he appears to have constructed a solid sphere, and to have placed them upon it in their proper situations, and likewise to have projected this sphere upon paper, for the facility of transportation. On comparing his own observations with those of Arystillus and Timocharis, made 150 years before, he ascertained that all the stars had changed their place since that time two degrees in the order of the signs: a discovery which has been since amply confirmed, and is well known by the name of *The Precession of the Equinoxes*.

Makes a catalogue of the stars.

Hardly any additions were made to astronomy (for it is unnecessary to mention Posidonius, of whom we know little,) till the period of Ptolemy; whom antiquity, perhaps a little unjustly with regard to Hipparchus, have dignified with the name of the first of astronomers. Ptolemy was born in Ptolemais, in Egypt, and flourished during the reigns of Adrian and Antoninus, about the year 125 of our era. Ptolemy contrived the first system of astronomy, and consigned it in his great work, which still remains, and which is usually known by the name of *Almageste;* because it was first made known to Europeans by

Ptolemy.

translasions from the Arabic. He conceived the earth to be stationary in the middle of the universe, and the sun, and all the other planets, to move round her in the following order: the Moon, Mercury, Venus, the Sun, Mars, Jupiter, Saturn. He confirmed the discovery of Hipparchus of the precession of the equinoxes, and concluded it to be one degree in a century. It is now known to be somewhat greater than that, being estimated at one degree in 72 years. He has given us the longitudes and latitudes of 1022 fixed stars; partly from the labours of Hipparchus, and partly from his own observations. His theory of the moon must have cost him much labour, and is a favourable proof of his sagacity and patience. The lunar motions are so exceedingly complicated, and subjected to so many irregularities, that it is not more than a century since astronomers began to explain them in a satisfactory manner, setting out from the theory first sketched by Newton, in his Principia; and accurate lunar tables, notwithstanding their great importance to navigation, are of a much shorter date. Ptolemy conceived the moon to move round a small circle, while the centre of this circle moved in a circular orbit round the earth. The earth was placed out of the centre of the orbit, and the orbit itself had a particular motion. This complicated hypothesis satisfied the phenomena to a certain extent, and gave the moon's place in certain parts of the orbit with tolerable precision; though in the intermediate points it was faulty, as every false theory must naturally be. His explanation of the motions of the other planets was similar. They moved in epicycles, the centres of which described eccentric circles round the earth. He was obliged, in order to account for some parts of their motions, in some measure to give up a favourite idea of the ancient astronomers, that all the motions of the heavenly bodies were uniform. There is some difference in the details of his explanation of the movements of the superior and inferior planets. But it is not worth while to enter into further particulars concerning a system which can be considered in no other view than as a mathematical hypothesis, contrived to subject the motions of the planets to calculation.

Ptolemy was one of the most laborious of all the ancients. Besides his Almageste, which appears more than sufficient to employ the whole life of an ordinary man, he has left us a Geography, in eight books, the composition of which must have cost him immense labour. In it we find the first attempt to mark the situation of places by their latitudes and longitudes; an idea due to Hipparchus, and which is the only true method of pointing out the situation of places exactly. Most of Ptolemy's latitudes are founded on loose calculations, from the length of the days and other doubtful data, and therefore are often erroneous, as was unavoidable in those times, when the light of astronomy was almost confined to Egypt.

Ptolemy may be considered as the last of the Grecian astronomers; for his successors in the school of Alexandria, till the period of its destruction by the Mahometans, added nothing to his theory. The perpetual religious disputes which involved the Egyptians, after the introduction of the Christian religion, was injurious to the cultivation of the sciences. It is remarkable enough that Egypt has always been the focus of superstition, and that many of the absurdities which disfigure the Christian religion originated in that country. But the Arabians, after the rage of conquest was over, and their princes began to enjoy peace at home, began to turn their attention to the sciences which they had at first despised. The Caliph Abu Giafar, surnamed Almansor, who reigned about the middle of the eighth century, had considerable merit in promoting this happy revolution. His successors Haroun-al-Raschid, (the perpetual hero of the Arabian Tales) and Alamin, were influenced by the same spirit. But the Caliph Abdalla Almanon, second son of Haroun-al-Raschid, who began to reign at Bagdad in the year 814, was the prince to whom Arabian philosophy lies under the greatest obligations. Having defeated the Greek Emperor Michael III. he made it one of the conditions of peace, that the emperor should deliver to him copies of all the Greek books of eminence. He collected and encouraged by rewards a great number of translators, and thus put the Arabians in possession of all the scientific treasures accumulated by the Greeks. Among these books was the Almageste of Ptolemy, which was translated into Arabic by Alhazen-ben-Joseph, assisted by a Christian called Sergius. Not satisfied with this book, Almanon collected the most intelligent astronomers whom he could procure, and caused them to draw up a complete body of the science ; a work which still exists in manuscript. Almanon, likewise, employed a numerous band of astronomers to measure a degree of the meridian with greater exactness than had been done by the Greeks. A vast plain in Mesopotamia, called Sangior, was made choice of for the purpose. Here they divided themselves into two bands, one of which measured due north, the other due south, till each had completed a degree. The one measure was found to be 56 miles, the other 56 miles and two-thirds. Each of these miles contained 4,000 *yards*, containing each 27 inches ; and the inch was equal to six grains of barley placed side by side.

Encouraged by the munificence of the Califs, a vast number of Arabic astronomers made their appearance, who contributed considerably to the improvement of the science. But it would be useless to attempt a dry catalogue of names, as few of their works have been translated into the European languages; though many of them exist in manuscript in various libraries. Oxford is said to be particularly rich in Oriental manuscripts. Perhaps the most eminent of all the Arabian astronomers was Albatenius, who had been called the Ptolemy of the Arabians. He was born in Mesopotamia, and flourished about the year 880 of our era. He was not a Mahometan, but a Sabean, or worshipper of

The Arabians.

Albatenies.

1

the stars. He approached nearer the truth than Ptolemy respecting the precession of the equinoxes, making it one degree in 66 years. He made the eccentricity of the solar orbit 3,465 parts, supposing radius 100,000, which determination is very near the truth. He made the solar year too short by about two minutes, owing, as Dr. Halley informs us, to his trusting too implicitly to some erroneous observations of Ptolemy. He discovered the motion of the sun's apogee, and pointed out the insufficiency of Ptolemy's hypothesis of the lunar and planetary motions. Finally, he constructed a set of astronomical tables more correct than those of Ptolemy, which enjoyed a high and deserved reputation over all the East, and different editions of them have been published in Europe.

Without noticing the numerous astronomers who appeared among the Saracens and Persians, or in Spain while under the dominion of the Mahometans, let us turn our attention to Europe after the revival of letters. It is remarkable that for a period of nearly 300 years all the great astronomers were natives of Germany. We have already mentioned Purbach and Regiomontanus under the head of mathematics, and noticed the celebrity which they acquired by their astronomical observations. The next great man who appeared was Copernicus, who has given his name to the system of the motions of the heavenly bodies at present universally received as the true one. Nicolas Copernicus was born at Thorn, in Prussia, of a noble family, on the 19th of February, 1473. After having learned Greek and Latin in his father's house, he went to the University of Cracovia to continue his studies. Here he acquired the elements of astronomy; and, inflamed with ardour by the high reputation which Regiomontanus had acquired about 30 years before, he resolved to run the same career, and on that account travelled into Italy, the country to which every person who wished to acquire proficiency in the sciences at that time resorted. Here he studied at Bologna and at Rome; and such was his proficiency that, in this last city, he obtained a Professor's chair. He quitted Italy about the beginning of the 17th century, and his uncle, who was Bishop of Warmia, having made him a Canon in his Cathedral, he spent the rest of his life in that situation.

Here he turned his attention with ardour to the study of the heavens. The prodigious inconveniences of the Ptolemaic system struck him very forcibly. The great confusion and irregularity of the motions of the heavenly bodies, and the inconceivable and inconsistent velocity with which they must move, provided the celestial sphere makes a complete revolution round the earth every 24 hours, convinced him that astronomers had not yet succeeded in unravelling the true motions of the heavenly bodies, and set him to search whether some more satisfactory explanation was not to be found in the writings of the philosophers. Plutarch furnished him with the first idea of his system, by informing him that

Copernicus.

2

Philolaus, and other Pythagoreans, had placed the sun in the centre of the system, and made the earth turn round it ; and that they had accounted for the apparent diurnal motion of the heavens round the earth by making the earth turn round its axis. This last idea delighted him by its simplicity, and by its removing at once some of the harshest suppositions that accompanied the Ptolemaic system. He learned also from Martianus Capella that some philosophers had conceived that Mercury and Venus moved round the sun. This opinion quadrated so much better with the phenomena, that he could hardly hesitate to embrace it. Further, he observed himself that Saturn, Jupiter, and Mars, always appeared much larger when in opposition to the sun than when they were in any other part of their course. This he naturally ascribed to their being much nearer the earth at that period. This was a strong reason for suspecting that these planets did not move round the earth, otherwise it would be necessary to give them a prodigious excentricity. Hence it was much more probable that they moved round the sun. This opinion removed all difficulties, and explained the whole of their apparently intricate motions in a very satisfactory manner. These considerations induced Copernicus to conceive the sun at rest in the centre of the system, and the planets to revolve round him in the following order. Mercury, Venus, the Earth, Mars, Jupiter, Saturn. The moon was degraded from the place of a primary planet to that of a satellite moving round the earth, and accompanying that planet in its orbit.

Copernicus was not satisfied with the general agreement of his system with appearances. He thought it necessary to examine whether it answered equally well in every minute particular. To determine this important point, he undertook a long course of observations, which he continued without intermission for a period of 36 years before he ventured to publish his system. Nor did it appear at last without great reluctance on his part. But having communicated his ideas to some of his friends, Cardinal Schoenberg urged him not to conceal any longer such a treasure from the philosophic world. Rheticus, Professor at Wittemberg, attracted by the reputation of Copernicus, offered him his assistance to put the last hand to his work. Copernicus, thus assailed on all sides, could no longer hesitate. Accordingly his work appeared, in 1543, under the title of *De Revolutionibus Cælestibus,* in six books. In his preface, addressed to Paul III he presents his system only as a mathematical hypothesis, which tallies better with the phenomena than the Ptolemaic ; though from the arguments which he advances in the book itself, it is easy to see that he was a firm believer in the truth of his opinion. Copernicus died of a flux of blood on the 24th of May, 1543, a few days after the publication of his book. But his reputation will live as long as astronomy continues to be cultivated among mankind.

The Earth had enjoyed for so long a period the tranquil possession of the centre of the universe, that it could not be displaced without occasioning a violent contest. A very keen dispute accordingly took place. On the one side were arranged almost all the astronomers celebrated by their discoveries and their genius; on the other the schoolmen, and the Roman Catholic divines, who reasoned upon a subject which they did not understand, and advanced arguments not worth refuting. This dispute, for the credit of philosophy, had better be buried in oblivion, and on that account we shall not enter upon it here.

Tycho Brahé. We pass over the labours of William IV. Landgrave of Hesse Cassel, which he continued with unremitting attention for 20 years, and afterwards committed his observatory to the care of Rothman and Justus Byrge, and which were of considerable importance to the progress of astronomy, to speak of Tycho, one of the most celebrated names in the annals of the science. Tycho Brahé was born, in 1546, at Knadstrup, in Scania, of an illustrious family, still said to exist in Denmark. His passion for astronomy originated from an eclipse of the sun, which happened in the year 1560, when he was only 14 years of age. He was so much struck with the justness of the calculation of those who announced this phenomenon, that he resolved to make himself acquainted with the principles of so extraordinary a science. In 1562, he went to the University of Leipsic; and his tutor, by prohibiting him from turning his attention to astronomy, served only to increase his passion for the science. He sacrificed his pocket money to purchase the proper books, and devoted the night to the perusal of such as he could procure. By the year 1563, he had made such progress in the study as to be sensible of the imperfection of the tables which announced a conjunction of Jupiter and Saturn, and he conceived the project of perfecting the theory of the planets.

After remaining three years at Leipsic he returned to Copenhagen; but was speedily disgusted by the contempt which the nobility and his own relations expressed for his favourite study. This seems to have induced him to go abroad, and he spent the next four years in travelling through different parts of Germany. At Augsburg he contracted an acquaintance with two senators who were lovers of astronomy, and procured a variety of instruments to be made by the workmen of that city, who possessed considerable skill. On his return to his own country, his uncle, Steno Billé, furnished him with a commodious place for observing on one of his estates. Here he spent the year 1572, and divided his time between astronomy and chemistry. This last science possessed such attractions for him, that there was some risk of its diverting him from the contemplation of the heavens, when an unlooked for occurrence fixed him for ever to the study of astronomy. On the first of November, 1572, a new star appeared all of a sudden in the constellation Cassiopeia, and shone with peculiar lustre. Tycho was on his way to his laboratory, when happening to look up to the heavens, this new star attracted his attention, and struck him with astonishment. He ran to his observatory, measured the distance of this new star from several others, and continued to observe it, with the utmost attention, all the time that it continued visible, which was about a year and a half. Tycho had resolved to pay a new visit to Germany, and to travel into Italy, in 1573, but several circumstances induced him to put off his journey to 1574. His first journey was a visit to the Landgrave of Hesse Cassel, who honoured him ever after with his astronomical correspondence. When he came to Basil, he was so struck with the advantages of its situation on the borders of Germany, France, and Italy, that he formed the resolution of transporting to it his instruments, and of making it his abode during the rest of his life. But the munificence of Frederick, King of Denmark, at the instigation of the Landgrave of Hesse Cassel, induced him to alter his resolution. That prince sent one of his pages to him, requesting him to return to Copenhagen. Tycho having obeyed, Frederick communicated to him his project; offered him the little island of Huen at the entrance of the Baltic, as a convenient place for an observatory, and undertook to be answerable for the expence of his instruments, and to give him an annual pension amounting to a considerable sum. Tycho took possession of the isle of Huen in the summer of 1576, and laid the foundation of the observatory so famous under the name of Uraniburg. Here he was accompanied by his family, by a number of workmen who were employed in constructing his instruments, and by about 12 young men who assisted him in his observations. His reputation spread over Europe, and he was visited by many noble and learned men from all quarters; among others by James VI. King of Scotland, and afterwards of Great Britain, when he went to Copenhagen in order to marry the daughter of King Frederick.*

In this situation, so much to his taste, Tycho spent more than 20 years, employed in assiduous observations, till the death of Frederick, his patron, in 1597, exposed him to the attacks of that envy and malignity to which eminent men are always exposed. Christian IV. listening to the suggestions of his

---

* King James wrote a copy of verses in praise of Tycho Brahé, which have been published, and possess little merit. We shall insert them as a curiosity:

> Æthereis bis quinque globis, queis machina mundi
> Vertitur, ut celso est crustatus forice olympus
> Ignibus, et pictus fulgentibus undique lychnis:
> Pellucent vitreis domibus, vastisque planetæ
> Orbibus, ut geminant cursus, vi et sponte rotati,
> Ut miti aut torvo adspecta longe ante futura
> Præmonstrant, regnisque tonans quæ fata volutet:
> His tellure cupis, quæ vis, quis motus et ordo
> Cernere: sublimem deductumque æthera terræ
> Tychonis pandunt operæ? lege, disce, videbis
> Mira: domi mundum invenies, cælumque libello.

enemies, deprived him of his pension, and obliged him to leave Uraniburg. He transported his instruments to Copenhagen, and began to continue his observations in that capital; when Valkendorf, his inveterate adversary, procured an order for his discontinuing them. He was even threatened with being deprived of his instruments, the greatest loss which an astronomer can possibly incur. Tycho hired a vessel, and, having put on board his family, his books, and his instruments, set sail from his country with an intention never to return. He reached Rostock in 1597, and from that city went to the house of Count Rantzau, near Hamburg, where he finished his book, entitled *Astronomiæ Instauratæ Mechanica*; in which he gives a description of his different instruments; and, in order to procure to himself a new patron, he dedicated it to the Emperor Rodolphus II. His project succeeded. Rodolphus ordered his Vice-chancellor to write to him, and invite him to Prague; and making him offers with which he was perfectly satisfied, Tycho on the receipt of these offers set out for Prague, and reached that city in 1599. The Emperor received him in the most handsome manner, settled on him a pension of 3000 crowns per annum, and offered him his choice of three houses in the neighbourhood to establish an observatory. He made choice of Benatica; but some circumstances induced him to leave it, and return to Prague, where the Emperor procured for him the house of his old friend Curtius, in which he had resided formerly. Here, assisted by Kepler, and by his disciples and secretaries Joestelius and Longomontanus, he renewed his astronomical observations with ardour, when he was unfortunately cut off by a sudden death, on the 24th of October, 1601, in the 55th year of his age. Being at a ceremonious entertainment, in which, acording to the custom of the age and country he was obliged to drink with almost every person at table, he was seized with a strong desire to make water; which he resisted, out of modesty, till the repast was at an end. But it was now too late; he was afflicted with a retention of urine, which his medical attendants were unable to cure, and died in the course of a few days.

The improvements which he introduced into astronomical observations, and the precious collection of observations which he left behind him, have given him a reputation which will be as lasting as his favourite science. A short sketch of some of his most remarkable improvements will be sufficient for our purpose here. The astronomical instruments of the ancients were so imperfect, that little confidence could be put in their observations. Some improvements had been made in their instruments by Walther and Copernicus; but Tycho spent great sums of money, and devoted much time and ingenuity to this important part of astronomy; and perhaps went as far as he could well go without the assistance of telescopes and pendulum clocks, which were not introduced into observatories till after his time. Tycho was early sensible of the great defects of the Ptolemaic catalogue of the fixed stars. Indeed, the method employed by the antients to determine the longitude and latitude of the stars was so defective, that it was morally impossible for them to avoid errors. Tycho substituted the planet Venus for the moon in their method, and ascertained the places of 1000 stars, with greater accuracy than had been formerly done. It must be admitted that his determinations are not always correct; but they are perhaps as near the truth as it was possible to go, before the happy invention of pendulum clocks for the exact measurement of time. Before Tycho's time the comets had been considered by all astronomers as mere meteors, floating in the sublunary regions of the sky. This was the opinion of Aristotle, and of course was adopted by all the schoolmen. But the comet of 1577 furnished Tycho with an opportunity of correcting this erroneous notion. Examining that comet with attention, he found that it had no sensible parallax. Hence he drew two conclusions, both equally fatal to the Aristotelian doctrines; namely, that the comet was beyond the region of the moon; and that the heavens, considered at that time in the schools as solid, were permeable in every direction, and contained no matter so dense as air. Subsequent comets, which he had an opportunity of observing, confirmed him still farther in these opinions; and they were speedily acceded to by all astronomers as first principles. He even made some steps towards determining the orbits of the comets; though at that period, when the cycles and epicycles of Ptolemy were still considered as the courses of the planets, nothing could be expected but a very rude approximation. This opinion of Tycho, which shook the foundation of the Aristotelian doctrines, was not acceded to without a pretty keen debate. But it was a contest between learning and ignorance; between men of science and the votaries of the schools. The advocates for the Aristotelian opinion were, one Craig, a Scotchman; and Scipio Claramonti, a Professor of Pisa; who devoted a long life to stop the progress of the sciences, as far as his influence or exertions could extend. The advocates for the system of Tycho were, Kepler and Galileo; two of the greatest philosophers of the age in which they lived. They refuted triumphantly the miserable arguments of their opponents, and established the system of Tycho on a basis which could not be overturned. Tycho was likewise the first person who gave a theory of astronomical refraction. The existence of this refraction, at least when objects were near the horizon, had been observed by Ptolemy, and noticed by all succeeding observers; but they had ascribed it solely to the vapours which they conceived to exist in abundance near the surface of the earth. Tycho first showed that it was owing to the passage of the rays of light through strata of air increasing in density. He determined the quantity of refraction at the horizon pretty justly, and gave a table of the refractions up to 47°: but he fell into several errors upon the subject, which were afterwards pointed out. He conceived the refraction of the solar light to differ from that of the stars; and he supposed both to terminate at a certain height in the atmosphere. Both of these suppositions are contrary to truth. The refraction is the same with regard to all the celestial bodies; and it continues to the very zenith; though, before it reaches that point, it becomes insensible. Newton afterwards gave a much more accurate table of refractions, which was improved upon by Bradley; and the table of that accurate astronomer has been since improved upon by subsequent observers.

Tycho made the theory of the moon much more perfect, and discovered several irregularities which had escaped the ancients. He likewise made many observations on the planets, but, not being satisfied with their accuracy, did not publish them; though they were of much service to Kepler in his subsequent investigations. Tycho's precious collection of astronomical observations remained unpublished till 1666, when they were given to the world by Albert Curtius. The copy from which he printed them was incorrect in many respects. The original copy remained in Denmark, and was procured by Picard when he went to Uraniburg, in 1671. Picard put the manuscript into the hands of the French Academy of Sciences, who preserved it with proper care, and no doubt at present it is to be found in the collection of the French Institute. It remains only to say a few words on the system which Tycho attempted to substitute in place of the Ptolemaic and Copernican. Satisfied by the arguments of Copernicus that the Ptolemaic system was absurd; but being averse to the Copernican system, either from vanity or religious scruples, he placed the earth in the centre of the universe, and considered the sun to revolve round it, while the other planets revolved round the sun. This system will mathematically account for all the appearances as well as the Copernican; but it involved so many inconsistencies and improbabilities that it never had many supporters. There is another system called the Semitychonic, in which some of the absurdities are removed, by conceiving the earth to revolve round its axis every 24 hours. But the motion of the sun, carrying with it all the planets, round so small a body as the earth has too much improbability to be easily admitted. Accordingly, even the Semitychonic system had but few supporters.

Kepler.   We come now to the illustrious Kepler, who contributed more to the perfection of astronomy by his happy discoveries, than all his predecessors united together. John Kepler was born on the 12th of December, 1571, at Wiel, an Imperial city near the dutchy of Wirtemberg. His parents were noble, but ruined by the long-continued wars in which Germany had been involved. During the early years of his infancy he was left to the care of an infirm uncle, and experienced the greatest difficulty in acquiring the first rudiments of his education. At last, by the happy interference of the Duke of Wirtemberg, he was put to school, and he entered into the University of Tubingen in the year 1591. Kepler himself had turned his views towards the church; but Mæstlin, a Professor at Tubingen, exhorted him to devote himself to astronomy. Being some time after appointed Mathematical Professor at Gratz, and unwilling to disoblige his patron the Duke of Wirtemberg, he accepted of the place, and thus was under the necessity for some time of turning his attention to astronomy. But at last perceiving the vast extent, and the sublimity of the science upon which he had entered, he became sensible of its charms, and devoted himself to astronomical investigations with alacrity and enthusiasm. The first fruit of his studies was his *Prodromus Dissertationum Cosmographicarum*, &c.; published in 1596, in which, from some fanciful opinions respecting the mysteries of numbers, he endeavoured to explain why the number of the planets was only six. Tycho Brahé, who perused this work, perceived the abilities of its author, and exhorted him to lay aside speculating about causes, and turn his sole attention to observations.

Kepler married in 1597, and by that imprudent step (in a literary man destitute of fortune) involved the rest of his life in poverty and distress. In 1598, the religious disturbances which raged in Germany obliged him to emigrate into Hungary. He returned to Gratz in 1600; but found the country so desolated and disturbed by the calamities of war, that he was obliged to leave it again. He repaired to Prague to visit Tycho Brahé, who procured for him the title of Mathematician to the Emperor, together with a pension, which was always ill paid. Here he fell ill of an intermittent fever, and was confined for ten months; and quarrelled with Tycho because he refused to supply his wife with money during his illness. However, he speedily made up this quarrel, and was attached to Tycho by the Emperor, with a salary, to aid him in his calculations. His situation in consequence was becoming more comfortable, when the sudden death of Tycho threw him into his former destitute state. The next eleven years of his life were spent in the most abject poverty. At last he received the arrears of his pension, and was appointed Mathematical Professor at Lintz, in 1613. Here he passed about 16 years in tolerable tranquillity, during which he published a considerable number of his works. In 1629 he entered into the service of Albert, Duke of Friesland, and was appointed a Professor at Sagan. In 1630 he went to the Diet at Ratisbon, to solicit the payment of his pension; and here he died on the fifth of November, of that year, in the 59th year of his age. On his tomb-stone was engraved the following inscription:

Mensus eram cælos, nunc terræ metior umbras.
Mens cælestis erat, corporis umbra jacet.

In Christo piè obiit, anno salutis 1630, die quinto Novembris, ætatis suæ quinquagesimo nono.

His writings were numerous, and on various subjects, and all marked with the stamp of genius. But the most important of them all was the book pub- His discoveries.

3

lished in 1609, entitled *Astronomia Nova* αιτιολογητος, *sive Physica cælestis ; tradita Commentariis de Motibus Stellæ Martis.* This book contains his three grand discoveries, familiarly known under the name of Kepler's Three Laws. They are the following : 1. All the planets move in ellipses, having the sun in one of the foci ; 2. The radius vector describes equal areas in equal times ; 3. The squares of the periodic times of the planetary revolutions round the sun are as the cubes of their distances from that luminary. It was in consequence of a visit to Tycho, that Kepler was induced to make choice of Mars, in his researches to discover the orbits of the planets. He found Tycho engaged in correcting the orbit of Mars according to the Ptolemaic system of cycles and epicycles, and he immediately expressed the little confidence he had in that system. The great eccentricity of Mars rendered it easier for him to discover the elliptic orbit of that planet, than it would have been had he made choice of any of the others. Kepler's confidence in the mystical powers of numbers never forsook him ; and his discovery of his third law was in some measure owing to this whimsical notion. But we shall abstain from dwelling upon the foibles of one of the greatest men who has done honour to human nature.

Galileo.   The name of Galileo, the next astronomer that comes under our review, recalls to our mind every thing that is respectable and noble. He was the great founder of experimental science ; and, during the whole period of a long life, occupied the first place among the philosophers of Italy, at that time the centre and capital of the philosophy of Europe. Galileo was born at Pisa, on the 18th of February, 1564. His father, who was a noble Florentine, was versed in the mathematical sciences, and gave his son an education suitable to his views. Galileo was destined to be a physician ; but an irresistible impulse drew him towards mathematics and the mechanical sciences ; and in the year 1589, he obtained a Professor's Chair at Pisa. Here, however, he did not remain long: his experiments on the falling of heavy bodies raised the resentment of the Peripatetics, with whom that University was then filled, and obliged him to leave it. He was invited to Padua, where his celebrity raised him to a Professor's Chair. Here he remained till 1609, when the Grand Duke of Tuscany, unwilling that a foreign state should enjoy one of his subjects possessed of so much merit and celebrity, invited him back to Pisa, and made him the Chief Director of the University.

His splendid reputation, as far as foreign countries are concerned, may be dated from the year 1609. Being at Venice, and hearing that the telescope had been invented in Holland, though he received no account of the structure of the instrument but merely of its effects, he speedily divined it from his knowledge of optics, and soon made a telescope for himself which magnified about 33 times. He turned it towards the heavens, and began to examine the moon a few days after change. From the inequalities in the light and shade between the illuminated and dark part of that luminary, and from several other circumstances, he drew as a conclusion, that the moon was a body similar to our earth, and that it abounded with mountains and valleys. He even contrived to measure the height of one of the lunar mountains, and he showed that it was much higher than any mountain upon the earth. Examining the milky way and the nebulæ, he found that these parts of the heavens were covered with clusters of stars too small to be seen by the naked eye, and that they owed their luminous appearance to the multiplicity of these bodies. On the 8th of January, 1610, turning his telescope to the planet Jupiter, he observed three little stars, two on one side, and one on the other of the planet. These at first he considered as fixed stars; but observing the planet night after night, and finding that these stars continued always at the same distance from it notwithstanding its progressive motion, he no longer hesitated to consider them as moons, or satellites, moving round Jupiter, as our moon does round the earth. Soon after he discovered a fourth satellite. When he published an account of his discovery he called them the stars of the Medici, in honour of his patron the Grand Duke. Not satisfied with this discovery, he examined the movements of these satellites, and formed a theory of their revolutions so as to be able to predict the place of each two months beforehand. He even conceived the project of ascertaining the longitude by means of them; and, at the request of the Grand Duke, made some efforts to construct tables of their motions; but the project was never completed by him. On examining the planet Venus he found that it changed its phasis like the moon, as Copernicus had concluded that it must do. His telescope not possessing sufficient power to show him the figure of the ring which surrounds the body of Saturn, the appearance of that planet struck him with astonishment, but he was unable to explain it. His discovery of the spots on the sun's disc speedily followed those which we have mentioned, and contributed not a little to his celebrity.

Galileo was too much of a philosopher not to draw from these discoveries strong proofs of the truth of the Copernican system. He established the similarity of the other planets to the earth, and showed that satellites revolved round Jupiter as our moon does round the earth. The phases of Venus, and many other points which it is unnecessary to notice, were urged by him with irresistible force in favour of the motion of the earth, and the position of the sun in the centre of the system. These opinions were the occasion of the celebrated condemnation of Galileo by the Inquisition; a condemnation which contributed more than any thing else to throw a slur upon the ignorance and absurdity of the Roman Catholic religion. Foscarini, an Italian priest, and a convert to the system of Copernicus, was the innocent cause of this condemnation. In 1615, he wrote a letter to his general Fantoni, explaining the texts of scripture which appear contrary to the opinion of Copernicus, and showing, in a very ingenious

He is condemned by the Inquisition.

manner, how they ought to be understood. About the same time a Spanish clergyman, called Didace à Stunica, in a commentary on the book of Job, embraced the same opinion, and endeavoured to show that the texts of scripture, representing the motion of the sun, were merely written so in order to conform with the commonly received opinion of the world at the time when these books were published. These writings drew the attention of the Inquisition; they were condemned by the College of Cardinals; and those passages of the book of Copernicus, in which he asserts the reality of his system, were included in the condemnation. Galileo was too celebrated a man to escape notice upon this occasion. He was summoned before the Inquisition about the end of the year 1615; his opinions were condemned as heretical; and he was ordered to sub-scribe a formal recantation of them. Galileo did not choose to run the risk of perpetual imprisonment: he disavowed his opinions entirely, promised never again to maintain or propagate them, and was dismissed by the Inquisition in the beginning of the year 1616. Provoked at this unmerited treatment, Gali-leo meditated a retaliation, which he executed, in 1632, by the publication of his *Systema Cosmicum*. This book consists in a dialogue between three speakers; namely, Sagredo, a Venetian senator, his old friend; Galileo himself, under the name of Salviati; and a peripatetic, named Simplicio. The conversation turns upon the different systems respecting the revolutions of the heavenly bodies. Poor Simplicio defends of course the opinion of Aristotle, which places the earth in the centre of the universe, and brings in its defence all the argu-ments which the most consummate astronomer could advance in its favour; but he is sadly treated, and completely refuted, by his two antagonists, who support the Copernican system. Galileo probably expected that his patron, the Grand Duke, would defend him from the vengeance of the Inquisition. But he was mistaken. He was summoned the second time to appear at Rome on the 23d June, 1632, and on his arrival in that city he was arrested. He was confined in the house of Noailles, the French Ambassador, and obliged to subscribe a most humiliating recantation, which was published by the Inquisition. In con-sequence of his relapse he was condemned to perpetual imprisonment; but after the interval of a year he was allowed to return home. Some years before his death he lost the use of his eyes; but being accompanied by his two pupils, Torricelli and Viviani, he spent the latter part of his life pretty comfortably at his country house of Arcetri, which he called his prison. He died in 1642.

We come now to a period in which the number of astronomers become so great, and their discoveries so frequent, that we must confine ourselves chiefly to those who were Fellows of the Royal Society; and to those Fellows who contributed most essentially to the progress of the science; for almost all the celebrated astronomers in Europe enjoyed the honour of being associated with that illustrious body. The next person that comes to be mentioned is Huygens,

Huygens.

a man who occupied the very first rank among philosophers as a mathematician, an astronomer, and a mechanician, and who perhaps is entitled to the first place among the respectable list of literary men and philosophers that Holland has produced.   Christian Huygens, of Zelem and Zulichem, was born at the Hague on the 4th of April, 1629.   His father was not only a man of letters : but a philosopher and mathematician, and he was the first instructor of his son. Being intended for the bar, young Huygens was sent to the University of Leyden, where he found Schooten the commentator of Descartes ; and, profiting by the instructions of this skilful mathematician, and by the resources of his own genius, he soon made himself master of the most difficult parts of the Cartesian geometry.   Schooten, in his commentary, gives us some valuable remarks made at that time by Huygens.   The appearance of the famous quadrature of the circle by Gregory St. Vincent induced Huygens to commence author.   He refuted this pretended quadrature in a small pamphlet, to which the partizans of Gregory St. Vincent were never able to make any satisfactory reply.   The same year he published his *Theoremata de Circuli et Hyperbolæ Quadratura;* and, in the year 1659, he published his ingenious treatise *De Circuli Magnitudine Inventa Nova.*   By the year 1665, his reputation was so high, that when Louis XIV. founded the Royal Academy of Sciences at Paris, he considered it as an object to associate him as one of the Members, and offered him terms so honourable and advantageous, that Huygens accepted of them.   He came to Paris in 1666, and continued in that capital till 1681. During this period he enriched the Memoires of the French Academy with a great number of profound and ingenious discussions.   The revocation of the edict of Nantz was the occasion of his leaving France.   Though he was assured that he should enjoy the same liberty as before, and not be molested for his religious·opinions, he would not consent to live in a country where his religion was proscribed, and therefore returned to Holland.   After his return to his own country, he continued his labours, and published several important works.   He died on the 5th of June, 1695, in the sixty-sixth year of his age, having bequeathed his papers to the University of Leyden, and requested Burcher de Volder and Fullenius to select such of them as they thought of value, and give them to the public.   This task they executed in the year 1700, and some time after 's Gravesende published a complete edition of all the works of Huygens in four volumes in quarto.

The first, and not the least considerable of the astronomical discoveries of Huygens, was the real nature or rather figure of the luminous appearance which accompanies the planet Saturn, and which had puzzled all preceding astronomers.   Furnished with telescopes of his own construction, and much more perfect than those which had been employed by preceding observers, he perceived that what Galileo had taken for two detached globes were united to

His disco veries.

Saturn by two streaks of light. And continuing to observe the planet in different positions, he found that this luminous portion gradually assumed the appearance of an ellipse surrounding the body of the planet. This gave him the idea of a flat ring surrounding the body of Saturn at some distance. An opinion which has been confirmed and established by succeeding observations. Dr. Herschell has lately endeavoured to demonstrate that this ring, discovered by Huygens, consists in fact of two distinct rings, separated from each other by a dark interval. He has also made some interesting remarks upon the figure of the planet itself, which he has shown not to be round. Huygens likewise discovered one of the satellites of Saturn; and, misled by a false analogy, he concluded that this discovery completed the number of the planetary bodies. For he conceived that the number of the primary and of the secondary bodies must be just the same, and as there were six primary planets known, he took it for granted that the secondary planets must be just six also. Subsequent discoveries have shown us how erroneous this opinion was. The number of primary planets at present known amounts to eleven, and that of the secondary is considerably greater. Huygens himself was speedily undeceived. For Cassini discovered two others, one in 1671, and the other in 1672, lying on different sides of the Huygenian satellites. The excellent glasses of Campani enabled him afterwards to discover two others still nearer Saturn than any of the preceding. In the year 1784, Dr. Herschell, assisted by his excellent telescopes, discovered two other satellites circulating between Saturn and his ring; so that the whole number of satellites belonging to this planet amounts to seven; which, together with the ring, must supply the inhabitants of that remote planet with a very considerable quantity of light.

But perhaps the most precious presents which Huygens made to astronomers were his pendulum clock and his micrometer;* by the one furnishing them with an exact measure of time, the most precious element in all astronomical calculations; and by the other enabling them to measure small angular distances with much greater precision than would otherwise be possible. The micrometer was afterwards improved by Azout, and has been brought to a state of perfection by the improvements of modern times.

In the year 1671, the French Academy sent Richer to Cayenne, to settle, by more accurate observations than could be made in Europe, some of the elements of the theory of the sun. The most memorable and important of his observations was that a pendulum clock, which beat seconds at Paris, went too slow at Cayenne; and he found it necessary to shorten the pendulum. This observa-

---

* The micrometer was really first invented by Gascoigne, an astronomer, who was associated with Horrox, and who lost his life during the civil wars of Charles I. This is clearly proved by Townley and Flamsteed, in the Philosophical Transactions.

tion induced Huygens to conclude that the earth was an oblate spheroid flattened at the poles: an opinion which Newton demonstrated from the theory of gravitation, on the supposition that the earth had been originally in a fluid state.* This led to the famous controversy between Cassini and the Newtonians; which was decided in favour of Newton, by the celebrated measurements of a degree of the earth's meridian, in the torrid and frigid zones, by deputations from the French Academy of Sciences.

Cassini, one of the most eminent astronomers of the 17th century, to whom   Cassini. we are indebted for the discovery of four of the satellites of Saturn, bestowed much attention in observing the motions of Jupiter's satellites. He observed a curious circumstance respecting the eclipses of the first satellite. At some seasons of the year, they happen about 16 minutes too soon, according to the tables at that time used; while at the opposite seasons of the year, they happen about 16 minutes too late. They happen too soon when the earth is in that part of her orbit which is nearest Jupiter, and too late when in that part which is most remote. Cassini explained this anomaly at first, by ascribing it to the time taken up by the ray of light in passing from the satellite to our eye. If light moves from the satellite to our eye, it is obvious that it must take some time to traverse the space between them, and the greater the distance, the longer is the time which must elapse before it reaches the eye of the observer. But this opinion, which appeared at first sight so simple and satisfactory, Cassini afterwards gave up, because no such anomaly was perceptible in the eclipses of the other satellites, which he thought ought to have been the case if the retardation and acceleration were owing to the movement of light. Cassini's explanation was soon after taken up by Roemer, and illustrated with so much force and perspicuity, that it satisfied the astronomical world, who have unanimously concurred in the explanation of Roemer; an explanation which received much additional weight from Bradley's subsequent discovery of the aberration of light. The reason why the anomaly does not appear in the eclipses of the other satellites of Jupiter is, that these eclipses are not so regular as those of the first satellite, and the immersion is so slow that it is difficult to be sure of it within three or four minutes. From this theory, it follows, that light takes about eight minutes to travel the half of the earth's orbit, or to move from the sun to the earth, which is reckoned nearly 95,000,000 of English miles.

To this theory Roemer is indebted chiefly for his celebrity. Olaus Roemer   Roemer. was born at Copenhagen, the 25th of September, 1644. His parents were in poor circumstances; but this did not prevent their son from acquiring a liberal education, and from prosecuting the mathematical sciences, to which his genius

---

* Newton's *Principia*, in which this theory is developed, were published several years before Huygens gave his thoughts on the subject to the world.

chiefly led him. When Picard went to Uraniburg, he got acquainted with Roemer, and was so much pleased with his sagacity that he carried him with him to France. In 1672 he was admitted into the Academy of Sciences, and received a pension from the French King, Louis XIV., who was the most munificent patron of the sciences of the age in which he lived. In 1681, Roemer was recalled by the King of Denmark, and made Astronomer Royal. He spent many years in making observations, in order to discover the annual parallax of the fixed stars; observations which his successor Horrebow afterwards published, under the title of *Copernicus Triumphans;* and which he considered as demonstrating the annual parallax of the fixed stars. But they were afterwards shown to be inconsistent with the parallax which ought to arise from the annual motion of the earth; and Dr. Bradley was able to explain them upon different principles. Roemer died in 1710, on the 19th of September, in the 61st year of his age.

It is now time to turn our eyes to Great Britain, where, during the 17th and beginning of the 18th century, the number of astronomers was uncommonly great. We do not think it necessary to take any notice of Horrox or Gascoigne; though their merits, especially those of the former, were great; because they lived before the origin of the Royal Society. But Dr. Robert Hooke occupied too prominent a place, during the first thirty years of the Royal Society, and was too remarkable a man to be omitted. Though he paid considerable attention to astronomy, this was not the science to which he owed his great reputation. As a mechanical genius, he holds the very first rank of the age in which he lived, not even excepting Newton and Huygens; perhaps indeed he was the greatest mechanic that ever existed. Robert Hooke was born at Freshwater, on the west side of the Isle of Wight, on the 18th July, 1635. He was at first so very weak and infirm that he was not expected to live, and was therefore for several years educated at home by his father, who was a clergyman. He displayed from his earliest years a ready apprehension, a strong memory, and a surprising invention. After learning English, he began to apply himself to the Latin grammar; but falling ill, he was obliged to drop his studies, and his father had thoughts of binding him an apprentice to a watch-maker, or a painter, for which professions his mechanical turn seemed particularly to qualify him. But the death of his father, in 1648, prevented him from putting either of these schemes in execution. After this he was put to Westminster school, where he learned Greek and Latin, and made some proficiency in Hebrew. He went to the University of Oxford, in 1653, and took his degree of Master of Arts, in 1660. While at the University, he became conspicuous by his mechanical inventions; and the air pump which he contrived for Mr. Boyle gave him so much celebrity, that we find his name included in the first list of Members, chosen by the President and Council of the Royal Society, after they

Hooke.

had received their charter from Charles II. Soon after he was appointed Curator to the Society, and his business was to contrive and exhibit experiments at the meetings of that illustrious body. He was unremitting in his exertions, and exhibited a vast number of experiments and inventions, and read a great multitude of papers before the Royal Society, the list of which is so very long that we are under the necessity of omitting it.*

In the year 1664, Sir John Cutler gave a salary of 50*l.* per annum to Dr. Hooke, for reading a course of mechanical lectures, under the direction of the Royal Society. These lectures were afterwards published, under the name of *Lectiones Cutlerianæ.* Some time after he was appointed Professor of Astronomy, at Gresham College. We pass over his quarrel with Mr. Oldenburg, which originated in that jealousy for his inventions by which he was characterized. This quarrel did Dr. Hooke very little honour : Mr. Oldenburg was publicly supported by the Royal Society, and appears to have acted with perfect propriety. In the year 1677, when Mr. Oldenburg died, Dr. Hooke was chosen Secretary of the Society in his place ; and while he held that situation he published the seven numbers of *Philosophical Collections,* which have been always considered as a part of the Philosophical Transactions. Towards the end of his life his temper, which was always bad, became intolerable. He claimed as his own the inventions and discoveries of all the world ; being generally disappointed in his claims, he at last refused to communicate any of his discoveries to the Royal Society, or the public, and seems in fact to have been in a state bordering upon derangement. For the two or three last years of his life he is said to have sat night and day at a table, engrossed with his inventions and studies, and never to have gone to bed or even undressed. He died on the third of March, 1703, in the 67th year of his age. In his person he was small and deformed ; but he was exceedingly active, and spared no pains to execute all his mechanical contrivances.

His principal book is his *Micrographia.* His *Lampas* also possesses great merit, and deserves the perusal of the chemists of the present day. His posthumous works contain likewise much curious and valuable matter. The greatest of his mechanical inventions is the method of regulating watches by the balance spring, as a substitute for the pendulum ; a contrivance which has made the watch susceptible of much accuracy, as a measure of time. His object seems to have been to apply it to the discovery of the longitude at sea : and though he did not succeed in his attempts, the method has since been carried to great perfection, and time-keepers have been made which for more than two months

* Mr. Waller published a Life of Hooke, prefixed to his posthumous works ; there is also a Life of him in Ward's Lives of the Gresham Professors. Either of these gives a sufficient list of his papers and inventions. The greater number are also to be found in Birch's History of the Royal Society.

are capable of giving the longitude without any considerable error. Huygens has been considered as the author of this happy invention; but there is complete proof that Hooke anticipated him by about 14 years. As to his astronomical labours, they are not of so much importance as his mechanical inventions. He made some efforts to discover the annual parallax of the fixed stars. Bradley afterwards adopted his method, and was led by it to his celebrated discovery of the aberration of light. He gave directions respecting the best mode of measuring a degree of the earth's surface; and his method seems to have been afterwards adopted by the French astronomers. But the most ingenious of all his astronomical speculations was his notion respecting gravitation, which he conceived to be the power that kept the planets in their orbits. He even made some experiments to determine the law of gravitation; but his mathematical knowledge was not sufficiently great to enable him to succeed. When Newton's *Principia* appeared, Hooke, with that jealousy which was natural to him, claimed priority respecting the idea of gravitation. Newton, who was candour itself, admitted his claim, but showed at the same time that Hooke's notion of gravitation was different from his own, and that it did not coincide with the phenomena. In reality, the notion of gravitation is as ancient at least as the days of Lucretius, and is particularly noticed by Kepler. Newton's merit consisted, not in ascribing the planetary motions to gravitation, but in determining the law which gravitation follows, and in showing that it exactly accounts for all the planetary phenomena, which no other system does.

Flamsteed.
Astronomy lies under much greater obligations to Flamsteed, who devoted the whole course of his life to the most assiduous observations. John Flamsteed was born at Denby, in Derbyshire, on the 19th of August, 1646. He was educated at the Free School of Derby, and at the age of 14 fell into a severe illness, which impaired his constitution, and prevented him from going to the University, as had been intended. But this did not entirely obstruct his learning, as he applied himself to his studies with all the assiduity in his power; and in the year 1662, Sacro Bosco's book on the Sphere having fallen into his hands, he perused it with great attention, and contracted in consequence a strong taste for astronomical pursuits. For some time he had no other books on the subject but Sacro Bosco's Treatise, and Street's Caroline Tables; but Mr. Halton, a neighbouring mathematician, having contracted an acquaintance with him, in consequence of an eclipse of the sun, which he calculated in 1666, supplied him with Riccioli's Almagestum Novum, and Kepler's Rudolphine Tables. In the year 1669, he calculated a set of ephimerides for the succeeding year, and sent them to the Royal Society. They were read, and highly approved of by this illustrious body; and Flamsteed, in consequence, acquired the friendship and correspondence of all the eminent mathematicians and astronomers in London. In 1670 he paid a visit to London, in order to become

6

personally acquainted with these celebrated men. Here he got acquainted with Sir Jonas Moore, who continued ever after his zealous friend and patron; and here he procured a telescope, a micrometer, and several other instruments, with which, till that period, he had been entirely unprovided. On his return, he entered himself a student at Jesus College, Cambridge. In 1674 he wrote a theory, or rather a table, of the tides, at the request of Sir Jonas Moore, who presented it to King Charles II. When the Royal Observatory of Greenwich was founded, Flamsteed, chiefly by the influence of Sir Jonas Moore, was appointed Astronomer Royal, with a salary of 100*l.* a year. He had entered likewise into orders, and was presented with the Living of Burslow, in Surrey, which he retained as long as he lived. Flamsteed was of a morose, unsociable disposition; and, in the latter part of his life, was upon bad terms with most of his contemporaries. From some of his letters, it even appears that he complained of Sir Isaac Newton, as unreasonable in his demands of observations. Dr. Halley, in the preface of the *Historia Cœlestis Britannica*, which he published in 1712, draws rather an unfavourable picture of the disposition of Flamsteed; and I find, from one of Sir Hans Sloane's manuscripts, in the British Museum, that in the year 1710 he was expelled the Royal Society, because he refused to pay his annual contribution.

Flamsteed's observations were chiefly directed towards the moon, in order to complete the theory of that luminary, for the purposes of observation: and towards determining the place of the fixed stars, with more precision than had been done by preceding astronomers. Dr. Halley published an edition of his observations in 1712, in one volume folio, contrary to the wishes of Flamsteed, who never would acknowledge the work as his. He prepared an edition of the work himself, but died of a stranguary in 1719, before it was finished. It was afterwards printed and published by his widow, in 1725, in three volumes folio; and contains a prodigious number of observations, not only made by himself, but by Gascoigne, who was likewise a Derbyshire man, and had observed during the time of the civil wars. The third volume contains the position of 3000 stars, several of them telescopic: the situation of 76 stars in the zodiac is determined with particular care, because they may be eclipsed by the moon, or the planets. In 1729, Mr. Hodgeson published an Atlas Cœlestis from this catalogue; a work of very great service to astronomers.

Few men have led a more active life, or run a more successful and brilliant Halley. literary career, than Dr. Halley, who seems entitled to the second place among the British philosophers of the age in which he lived; and, after Newton, claims the first rank among the Members of the Royal Society. Edmond Halley was born at St. Leonard's, Shoreditch, on the 8th of November, 1656. His father was in affluent circumstances, and spared no pains to improve the promising genius of his son. He was educated at St. Paul's School, under Dr.

Gale, and made great proficiency in the languages and geometry. He went to Oxford in the 17th year of his age; and while in that University, he employed a great part of his time in the study of astronomy, to which he was particularly attached; his father having furnished him with a convenient apparatus. His progress was so great in that, his favourite study, that before he was 20 years of age he published several papers, on the subject, of real merit; which gave him considerable celebrity among astronomers. In the year 1677, he conceived the project of making a catalogue of the stars of the southern hemisphere, and addressed himself to Sir Joseph Williamson, Secretary of State, and Sir Jonas Moore, Surveyor of the Ordnance, who approved of his design, and prevailed upon King Charles II. to send him out for that object to St. Helena, which he had been told was a very convenient position for his purpose. This island, or rather rock, lying about the 17th degree of southern latitude, at an immense distance from every other land, he conceived would afford him every facility for viewing the heavens, and ascertaining the position of the stars. But the frequent fogs which hover over the island made the task much more difficult than he had expected, and it was only by embracing every opportunity which offered, during his abode on the island, that he was enabled to execute his purpose. He ascertained the position of 350 stars; and published his account of his labours in 1676, under the title of *Catalogus Stellarum Australium*. In honour of his royal patron, he formed a new southern constellation, to which he gave the name of *Robur Carolinum*, the Royal Oak. During his stay at St. Helena, he had an opportunity of observing the transit of Mercury over the sun's disk; an observation of some importance, because it could not be completely made in Europe, the sun not being risen in that country at the beginning of the transit.

Soon after his return to England, he was chosen a Fellow of the Royal Society. In 1679 he went to Dantzick, in order to satisfy himself and astronomers what degree of confidence ought to be put in the observations of Hevelius; and likewise to settle the dispute between Hevelius and Hooke, relative to the use of telescopic sights in astronomical observations. Halley made a number of observations together with Hevelius, and expressed himself highly satisfied with the accuracy of that astronomer. In 1680 he went to Paris, and from that capital to Rome, proposing to make the tour of Europe; but his private affairs obliged him to return home in 1681 He had not been long at home when he married a daughter of Mr. Tooke, Auditor of the Exchequer; established his residence at Islington, set up his instruments, and began to occupy himself assiduously in astronomical observations. But such was his assiduity, and such the versatility of his genius, that one department of science was not sufficient to occupy his attention. In 1683, he published his Theory of the Variation of the Compass; founded on the opinion, that the earth is a great magnet, having four magnetic poles, two near the north, and two near the south pole. In 1684, he turned his thoughts to the theory of the planetary motions; and gravity occurred to him, as it had done to Dr. Hooke, as the probable cause. But he could not satisfy himself as to the law according to which this power diminishes, and he applied to Dr. Hooke, and to Sir Christopher Wren, without obtaining any additional information. This induced him to make a journey to Cambridge, in order to consult Sir Isaac Newton. Sir Isaac communicated to him twelve theorems, which he had drawn up on the subject, comprehending the principles of what is now known by the name of the Newtonian theory of gravitation. Dr. Halley was delighted with them, and urged Newton so keenly not to withhold his theory from the world, that he prevailed upon him to draw up his *Principia;* which were published in 1686, and constitute such a memorable era in astronomy and science.

In 1685, he became Clerk to the Royal Society, and seems, for several years about that period, to have been the principal person employed in drawing up the Philosophical Transactions. In 1687, he published his curious paper on evaporation; in which he explained, in a satisfactory manner, why the Mediterranean Sea does not grow larger, notwithstanding the numerous rivers which flow into it. In 1692, he was candidate for the Savilian Professorship of Astronomy, at Oxford; but lost it because Bishop Stillingfleet refused to recommend him, on account of his opinions, which were considered as unfavourable to Christianity.

About the year 1698, he conceived the design of making a voyage, to determine the variations of the needle, in order to make a more correct theory of that important circumstance than he had hitherto been able to make. He had interest enough with King William to get himself appointed Captain of the Paramour Pink. He set sail from the River on the 20th of October, 1698. He advanced along the African coast, and then proceeded west to the coast of America, going some degrees to the south of the Line. But his officers becoming refractory, he proceeded to the West Indies, in order if possible to exchange them for others; and finding that could not be done, he returned again to England for that purpose about the beginning of July, 1699. He sailed again on the 16th of September, the same year, and went to about 53° of the southern latitude, where he was in a manner stopped by the ice. Having fully accomplished the object of his voyage, he returned to England, and anchored in Long Reach, on the 7th of September, 1700. Fom this voyage Dr. Halley acquired the title of Captain in the Navy, and Queen Caroline afterwards procured him half-pay; which he retained during the remainder of his life. The result of these voyages was his General Chart, published in 1701; showing, at one view, the variation of the compass, in all the seas which

His voyage to ascertain the variation.

2 X

British navigators frequented.* Very soon after, Dr. Halley went out again in the same ship, in order to observe the course of the tides in every part of the English channel, and to determine the latitudes and longitudes of the principal headlands.   This he accomplished with his usual expedition and precision, and, in 1702, published a large Chart of the British Channel.   In 1703, the Emperor of Germany sent for him, to fix upon a proper place for a harbour in the Gulf of Venice; but, upon the representations of the Dutch, the design was laid aside.

On the death of Dr. Wallis, in 1703, Dr. Halley succeeded him as Savilian Professor of Geometry, at Oxford; and, by the unanimous voice of the University, he had the degree of Doctor of Laws conferred upon him.   In 1713, he succeeded Sir Hans Sloane, as Secretary to the Royal Society; and in 1719, on the death of Mr. Flamsteed, he was appointed his successor, as Astronomer Royal at Greenwich.   This was a situation exactly to his taste, as it enabled him to devote himself to his favourite object, the theory of the moon.   If we consider that, when he was appointed to this laborious situation, he had reached his 64th year, and that he continued for more than 20 years without any assistance to observe so assiduously that he scarcely lost a meridian transit of the moon, we shall have some idea of the uncommon activity of mind and body which characterized this extraordinary man.   He died on the 14th of January, 1742, in the 85th year of his age.   While Astronomer Royal, he was frequently visited by Queen Caroline, who was a most distinguished patroness of literature and science.   The late Dr. Maskelyne used to relate a conversation that passed at one of these visits; upon what authority he told it we do not know.   Queen Caroline expressed her regret at the smallness of his salary as Astronomer Royal, and offered to use her influence with his Majesty to get it increased.   Dr. Halley requested the Queen not to make any such application: For, said he, the consequence of an increase of salary will be, that the place will come in time to be bestowed on some younger son of some nobleman, totally unfit for the office, but drawn to it merely by the emoluments.

In the preceding sketch, we have noticed only a very few of Dr. Halley's literary labours.   His mathematical and astronomical papers were very numerous and valuable; and published chiefly in the Philosophical Transactions.
His labours to complete the lunar theory. There is one branch of his astronomical labours, however, which he had so much at heart, and towards which he devoted so much of his time, that it would be improper to pass it over in total silence.   The perfection of the lunar

---

* In the year 1775, the original journals of Dr. Halley's two voyages were published, by Mr. Alexander Dalrymple, in a thin quarto volume. They are not of much value, and were obviously never intended for publication by Dr. Halley himself.

theory was always an important object with him.   Even when he published his
Catalogue of the Stars in the Southern Hemisphere, he mentioned several important discoveries which he made respecting the moon's motion. Far example,
that the moon moves more rapidly when the earth is in her aphelion, than in
her perihelion ; and, on that account, he introduced into the calculation of the
moon's place a new equation, depending on the distance of the earth from the
sun. He perceived that a great deal was yet wanting in order to render the
theory of the moon's motions complete; and that these additions could not be
expected from one man, nor from one age. He therefore thought of another
method of subjecting the inequalities of the motions of that luminary to calculation. The principal inequalities in the moon's motion depend upon her
position with respect to her apogee, to her nodes, and to the sun. Hence, if
a period of years could be found, at the end of which all these things were
brought to the same situation as at first, is is obvious that the inequalities of the
moon would renew themselves again in the same order as at first, and might
therefore be predicted exactly, provided they had been observed during the preceding period. But the most remote antiquity has furnished us with a period of
this kind. This consists of 223 lunations, and was known to the Chaldeans by
the name of *Saros*. This period consists of 18 Julian years 11 days 7 hours
42′ 45″, and the eclipses of the sun and moon succeed each other in every one
of these periods, nearly in the very same manner. Dr. Halley examined this
period, and found, that at the end of it the phenomena of the sun and moon
are renewed exactly in the same order, and within half an hour of the same
time. But this difference, which has some influence on the real place of the
moon, and on the time, has no sensible effect upon the equations. Hence, at
the end of the period, the difference between the calculated and real place of
the moon is not sensible. Halley conceived the idea of rectifying the theory of
the moon by means of this period, as early as the year 1680; and with that
view had made a continued series of observations upon the moon for 16 months
together, during the years 1682, 1683, and 1684 ; and he made the first essay
of his new method upon the eclipse of the sun, which happened in the month
of July, 1684, deducing all the circumstances of it from the eclipse which had
been observed in 1666 : and his calculation approached much nearer the truth
than any other deduced from the best tables. After being appointed Astronomer Royal, he renewed his observations at Greenwich, in the year 1722, and
resolved to make a continued set of observations on the moon during the whole
period of a *Saros*. After the expiration of one half of the period, he gave an
account of his plan, and of its progress, in a paper published in the Transactions of the Royal Society, for 1731. In this paper he announced, that what
had already been done would enable navigators to determine the longitude at
sea, by the place of the moon, within 20 leagues of the truth. This favourable

result induced several other astronomers to direct their attention to the same object. Delisle of Petersburgh, observed the moon for a course of 12 years without intermission. But M. Le Monnier is the astronomer who devoted the greatest share of attention to this subject: he finished the period of Halley, and began and finished a complete period himself.

His tables.

It would be improper not to mention the astronomical tables of Halley, which were partly drawn up by him in 1725; but labouring with much industry to render them more and more perfect, he did not publish them till he was prevented by death. They were published at last in 1749, and were for many years the best and most complete with which astronomers were furnished: though of late years other tables have been constructed still more perfect, and entitled to a greater degree of confidence.

But it is time to turn our attention to the various theories of the planetary motions which have been offered by philosophers. It is needless to specify the various speculations of the ancient philosophers on the subject. The first theory,

Theory of Descartes.

which attempted to account for the motions of the planets in detail, was that of Descartes, who conceived them to be placed in certain vortexes of matter revolving round the sun, and carrying them along while the secondary planets were carried round their primary by smaller vortexes of a similar kind. Newton demonstrated that these supposed vortexes were inconsistent with the phenomena; and notwithstanding the attempts of John Bernoulli to point out flaws in this demonstration, it has stood the test of the closest investigation, has been acquiesced in by the philosophic world, and has therefore overturned the Cartesian hypothesis. Newton himself adopted gravitation as the power which he

Newtonian theory.

conceived to retain the planets in their orbits; an opinion which seems to have been entertained by the Epicureans, since Lucretius assigns it as a reason for believing the world to be infinite, otherwise all the bodies of which it is composed would, according to him, have approached each other, and at last all united together. The same opinion was also embraced by Kepler; though in many parts of his works he seems to be better pleased with a kind of magnetic attraction. Dr. Hooke, likewise, had embraced the same opinion, and published a small paper explaining the planetary motions according to that doctrine. The opinion had occurred to Dr. Halley; and Newton, in his Principia, informs us with his usual candour, that the doctrine of gravitation had occurred to Hooke and Halley, about the same time that it did to himself.

It was in the year 1666, that Newton first began to suspect the existence of gravitation as the cause of the planetary motions. Having left Cambridge on account of the plague, and having retired to his own house in Lincolnshire, he began one day to consider that the cause which produces the fall of heavy bodies to the ground continues to act at the greatest possible distance to which we can go from the earth. It was probable therefore that it extended as far as the moon, and that it was the cause of the moon revolving continually round the earth. It was probable that this power continued to diminish according to the distance of the bodies on which it acted. The next point therefore was to determine the rate at which it diminishes. Now Newton considered that if gravity extended to the moon, it no doubt extended much further, and was the cause of the revolution of the planets round the sun. Now from comparing the periodic times of the planets with their distances, he found that the centrifugal force produced by the revolutions, and consequently the centripetal force which counterbalances it, is inversely as the square of the distance. Hence he concluded that the force which retains the moon in her orbit ought to be terrestrial gravitation, diminished as the square of the distance of the moon from the earth. But gravity at the earth's surface was represented by the space which a body falls through in a given instant of time, and the centripetal force which draws the moon to the earth is represented by the versed sine of the arc which that luminary describes in the same time. If the opinion be true, these two spaces ought to have to each other the ratio of inversely the square of the distance of the falling body and the moon from the earth's centre. But the distance of the moon from the earth was known to be equal to 60 semidiameters of the earth. Newton at that time being at a distance from books, and not being acquainted with the mensuration of a degree by Norwood, took 60 English miles as the length of a degree, as was commonly done at that time in books of geography. But the real length of a degree is about $69\frac{1}{10}$ miles. Hence when Newton came to compare the two forces he found them not to bear the same ratio to each other as he had expected. This induced him at the time to abandon his conjecture altogether.

He was induced to resume the subject again in 1676, perhaps in consequence of Dr. Hooke's paper, explaining the planetary motions by the principle of gravity. Picard, by this time, had measured a degree of the earth's meridian with tolerable exactness, and Newton employing this new measure, found that the space which the moon falls through in a minute towards the earth is just equal to what a heavy body at the earth's surface falls through in a second. Hence the first is just the 3,600th part of the last; which agrees exactly with the supposition. Newton having thus ascertained this truth, endeavoured in the next place to determine what curves bodies projected, and, obeying this law, ought to describe round the sun. He found in the first place that, whatever be the law according to which gravitation diminishes, bodies under its influence must move in such a manner that the radius vector between them and the sun must describe areas equal to the times. He found in the next place that if gravitation follow the law of the inverse of the square of the distance, the curve described must be a conic section. Hence it followed that when the curve returned into itself, as it does with regard to the orbits of the planets, it must be

6

either a circle or an ellipse.   Thus the two first of Kepler's laws followed at the outset from the theory of gravitation.

Newton had proceeded thus far when Dr. Halley, who had thought likewise of gravitation as a method of explaining the planetary motions, but could not determine the law according to which it diminished, paid him a visit at Cambridge.   Newton communicated to him his propositions on the subject, and Dr. Halley prevailed upon him to communicate them to the Royal Society, and to allow them to be published in the Philosophical Transactions.   But afterwards that illustrious body thought it better to endeavour to prevail upon Sir Isaac to draw up a full account of his discoveries, and lay them before the public, and Halley offered to take any pains whatever in superintending the publication.   These entreaties prevailed, and produced the *Principia Mathematica Philosophiæ Naturalis,* which Newton wrote in 18 months, and which were published, in 1687, in one volume in quarto.   Some time elapsed before the philosophical world were sensible of the merits of this extraordinary production, on account of some obscurity which runs through it.   But no sooner were its contents understood, than one universal cry of admiration burst forth from one extremity of Europe to the other; and Newton rose at once to the highest pinnacle of glory.   He has stood ever since in the front of the philosophic world, leaving the rest of mankind far behind him.   Not that the doctrine of gravitation was at once acceded to.   The Cartesian vortices, though annihilated in the second book of the Principia, found still some staunch supporters.   The dispute likewise about the invention of fluctions, which took place soon after, alienated the minds of Leibnitz and the Bernoullis, and disposed them to treat Newton himself, and of course his philosophical opinions, with harshness and dislike.   This conduct we find continued by their pupils for a considerable number of years.   Even Euler, notwithstanding his consummate merit and his candour in general, appears to have had a strong disposition to carp at the philosophical opinions of Newton.

His Principia.

Newton, having thus ascertained the existence of gravitation as a general principle, set himself to determine in what manner it acted.   He showed that if it followed the ratio of the distances of bodies, it would pass through the centre of gravity of the mass, and would be proportional to the distance from that centre.   The same holds when attraction follows the inverse of the square of the distances, provided the bodies attracting each other be spheres of the same density throughout, or at least composed of concentric layers of the same density.   But when the attracted body is within the sphere, he demonstrated that it is attracted directly as its distance from the centre.   When a particle of matter is placed within a hollow sphere, it is not acted upon at all, because the mutual and opposite attractions destroy each other.   When two spheres attract each other, they act precisely as if all the matter in each were accumulated in

3

their centres, a law which holds only when the attraction is inversely as the distance or inversely as the square of the distance.

Newton then showed that as attraction is mutual, not only the sun attracted the planets, but the planets the sun, and that they must all in fact revolve round their common centre of gravity. But the sun being so vast compared with all the planets put together, it happens that this common centre is not far removed from the body of the sun, and of course the motion of the sun itself is insensible. The common centre of gravity varies somewhat according to the position of the planets. Hence the motion of the apsides and of the nodes of the planets; motions which some persons without reason have urged as inconsistent with the doctrine of gravitation.

Newton deduced from his theory a very ingenious, and at the same time satisfactory, mode of determining the quantity of matter in the sun, and in all those planets which have satellites. The method does not apply to those that want satellites; but Newton deduced it with respect to them likewise, by observing that the density of the planets increased, the nearer they were to the sun, and by supposing that this density increases in proportion to the quantity of heat to which they are exposed. The following table exhibits the density and the quantity of matter in the sun and several of the planets, deduced according to Newton's method. The density of water is conceived to be 1.

	Density.	Quantity of matter.
Sun	$1\frac{9}{15}$	329630
Mercury	$9\frac{1}{2}$	0·135
Venus	$5\frac{1}{15}$	1·135
Earth	$4\frac{1}{9}$	1
Mars	$3\frac{2}{7}$	
Jupiter	$1\frac{1}{24}$	330·6
Saturn	$\frac{13}{32}$	103·95
Herschell	$1+$	
Moon		0 025

But it would occupy a much greater space than we can conveniently spare to attempt an abstract of all the important things pointed out in the Principia. Perhaps none of the discussions which it contains is more ingenious or important than the lunar theory, which Newton first sketched with infinite sagacity, and which has been, in some measure, perfected in our own days. He also first gave a theory of the comets, and pointed out the method of determing their parabolic orbits from three observations. Nor is his theory of the tides lees remarkable either for the sagacity displayed, or for its importance to navigation.

The astronomical papers in the Philosophical Transactions amount to 416. They are of an exceedingly miscellaneous nature; many of them being simple observations of eclipses or transits, while others involve the most important and abstruse topics in the theory of astronomy. On these accounts they are not susceptible of abridgment, and a particular detail of the contents of each paper would be both useless and exceedingly tiresome both to the writer and reader. On that account I shall select only a few of the most striking papers, and terminate the whole with a kind of tabular view of the planetary system, which will convey, in little compass, the sum and substance of a very considerable number of the astronomical papers contained in the Philosophical Transactions.

1. The first topic which will engage our attention is the aberration of the fixed stars, and the nutation of the earth's axis. Dr. James Bradley, to whom these discoveries gave celebrity, was one of the most ingenious and accurate astronomers of the last century. He was born at Sherborne, in Gloucestershire, in the year 1692; and, after receiving his grammar-school education, entered at Baliol College, Oxford, where he took a degree in arts, and then entered into orders. In 1719, he obtained the living of Bridstow, and afterwards that of Landewy Welfry. He became also curate to his uncle, Mr. James Pound, at Wansted, in Essex, and in this situation he gained his knowledge of mathematics and astronomy. His discovery of the aberration of the fixed stars, which he published in 1728, gave him celebrity; but he had been previously appointed the successor of Dr. John Keil, in 1721, as Savilian Professor of Astronomy at Oxford; in consequence of which, agreeably to the rules of the founder, he resigned all his church livings. In 1731, he succeeded Mr. Whiteside as Lecturer in Experimental Philosophy in the Museum at Oxford. On the death of Dr. Halley, he succeeded him as Astronomer Royal at Greenwich, in 1742. At the same time the University of Oxford presented him with the degree of D.D. In 1748, the King ordered 1,000l. to furnish the Greenwich observatory with better instruments. Dr. Bradley employed the celebrated artits Graham and Bird, who provided him with instruments entirely to his satisfaction; and of these he made a most assiduous use during all the remainder of his life. During his residence at the Royal Observatory he was offered the vacant living of Greenwich, which however he declined, alleging that the duty of a pastor was incompatible with his other studies and necessary engagements. On this his Majesty thought fit to grant him a pension of 250l. a year, over and above the Astronomer's original salary from the Board of Ordnance, a pension which has been continued to the Astronomer Royal ever since. Dr. Bradley died at Chalfont, in Gloucestershire, of a suppression of urine, in 1762, in the seventieth year of his age. He left behind him an immense number of astronomical observations in 13 folio volumes; which were

presented to the University of Oxford in 1776, on condition of their being printed and published, a condition however not yet complied with.

The aberration of the fixed stars was discovered in consequence of a set of observations made in order to verify Dr. Hooke's observations on the annual parallax of the fixed stars. The observations were begun by Mr. Samuel Molyneux, who got a very exact instrument constructed for the purpose by Mr. Graham, and erected it at Kew, in order to make a continued set of observations on γ Draconis, the star which Hooke had made choice of as passing near the zenith, and therefore being nearly free from refraction. Dr. Bradley assisted him in his observations. The apparatus was fitted up in November, 1725; and on the 3d of December γ Draconis was observed as it passed over the zenith, and its situation carefully taken with the instrument. Similar observations were made on the 5th, 11th, and 12th days of the same month; and there appearing no material difference in the place of the star, a further repetition of them at this season was deemed needless, it being a part of the year when no sensible alteration of parallax in this star could soon be expected. But on the 17th of December, Dr. Bradley, being at Kew, had the curiosity to repeat the observation, and perceived that the star passed a little more southerly than it had done when it had been observed before. This was at first ascribed to want of exactness in the observations. On this account it was resolved to repeat them again with as much precision as possible. This was done on the 20th of December, and it was found that the star passed still more to the south than on the 17th. This alteration being contrary to what it should have been, had it proceeded from the annual parallax of the star, puzzled them exceedingly, and they could think of no other means of accounting for it but some alteration in the instrument itself with which they observed. But a great variety of trials having convinced them of the accuracy of the instrument, and the star still continuing to advance towards the south, they at last concluded that there must be some regular cause producing this apparent motion. They resolved to examine the phenomena with the utmost care in order to discover this cause. About the beginning of March, 1726, the star was found to be 20″ more southerly than at the time of the first observation. It seemed now to have arrived at its utmost limits south, for on several trials no further alteration could be perceived. By the middle of April it appeared to be returning north, and about the beginning of June it passed at the same distance from the zenith as it had done in December when it was first observed. From the quick alteration of the star's declination about this time, it was conjectured that it would now proceed northward as it had formerly done southward, and this accordingly happened. It continued to move northward till September, being then nearly 20″ more northerly than in June. From September the star returned to the south, and in December it was precisely in the same place as it had been a twelvemonth before.

Aberration of the fixed stars.

2 Y

This was a sufficient proof that the instrument had not been the cause of this apparent motion of the star; but to find out the true cause was the difficulty. A nutation of the earth's axis was the first cause that presented itself; but, on examining the apparent motion of other stars, nearly in the same situation with γ Draconis, such a nutation was found inadequate to account for the phenomena. Dr. Bradley, that he might have it in his power to ascertain the phenomena with sufficient exactness, resolved to get an instrument of his own, and set it up at Wansted, where he lived, that he might, with the more ease and certainty, inquire into the laws of this new and unexpected motion. An instrument was made accordingly, by Mr. Graham, and fitted up on the 19th of August, 1727. Mr. Molyneux's original instrument was capable of very little change of position; but Dr. Bradley's was capable of extending about $6\frac{1}{4}°$ on each side the zenith, within which space there were about 200 stars that could be observed with precision. After continuing his observations for some time, he discovered what he considered as a general law with respect to these motions; namely, that each of the stars became stationary, or was farthest north or south when it passed over the zenith at six o'clock, either in the morning or evening. He perceived, likewise, that whatever situation the stars were in, with respect to the cardinal points of the ecliptic, the apparent motion of every one tended the same way when they passed the instrument about the same hour of the day or night; for they all moved southward while they passed in the day, and northward in the night; so that each was farthest north when it came about six o'clock in the evening, and farthest south when it came about six o'clock in the morning.

After completing a year's observations, and comparing them carefully with each other, so as to determine the general laws of the phenomena, he endeavoured to find out a cause for them. But his attempts, for some time, were very unsatisfactory. The nutation of the earth's axis had been already considered, and given up. The next thing that occurred was an alteration in the plumb-line, with which the instrument was constantly rectified; but this, upon trial, proved insufficient. He then considered how far refraction might account for this phenomena: but here nothing satisfactory occurred. At last, when he despaired of being able to account for the phenomena which he had observed, a satisfactory explanation of it occurred to him all at once, when he was not in search of it. He accompanied a pleasure party in a sail upon the river Thames. The boat in which they were was provided with a mast, which had a vane at the top of it. It blew a moderate wind, and the party sailed up and down the river for a considerable time. Dr Bradley remarked, that every time the boat put about, the vane at the top of the boat's mast shifted a little, as if there had been a slight change in the direction of the wind. He observed this three or four times, without speaking; at last he mentioned it to the sailors, and expressed his surprise that the wind should shift so regularly every time they put about. The sailors told him that the wind had not shifted, but, that the apparent change was owing to the change in the direction of the boat, and assured him that the same thing invariably happened in all cases. This accidental observation led him to conclude, that the phenomenon which had puzzled him so much, was owing to the combined motion of light and of the earth. Since light does not move instantaneously from one place to another, it is clear, that a spectator standing on the earth's surface, will not see a star by means of the same ray of light, if the earth be moving, that he would do if the earth were standing still. Suppose a tube to extend from the spectator's eye to the star, and that one extremity of this tube remains fixed at the star, and the other moves along with the eye, it is clear that the ray of light which proceeded in the original direction to the eye, will strike against the side of the tube, and be lost; and the ray, by which the star will be seen, will be that ray which issued in such a direction as to move always along the axis of the tube. Hence the star will not appear in its true place, but will be seen in the diagonal of the parallelogram, whose two sides are the velocity of light, and its direction from the star, and the velocity and direction of the earth in its orbit. From this explanation, it was evident that the stars at the pole of the ecliptic will annually describe a circle with a radius of $20\frac{1}{4}''$; that those between the pole of the ecliptic and the ecliptic will describe an ellipse, having the greater axis to the smaller, as radius to the sine of the latitude of the star, and of course at the ecliptic, where the stars have no latitude, they will describe annually, instead of a curve, a straight line. This explanation, which Bradley detailed in the Philosophical Transactions for 1728,* was immediately acceded to by all astronomers. Soon after, the subject was still farther developed by Mr. Thomas Simpson, and M. Clairaut, almost at the same time. The exact coincidence of the motions of all the stars with the hypothesis of the aberration of light, affords an unanswerable argument, both for the motion of light, and for the revolution of the earth round the sun. While the observations of Bradley show, that the annual parallax of the stars is insensible, and, therefore, that the orbit of the earth, when compared with the distance of the stars, is no more than a point. Now the diameter of the earth's orbit is nearly 190,000,000 of miles; hence we may form some conception of the immense distance of the fixed stars.

The discovery of the nutation of the earth's axis, which we likewise owe to the sagacity and assiduity of Dr. Bradley, cost him much more time than the discovery of the aberration, since he thought it necessary to make a continued series of observations for 20 years before he ventured to lay his discovery before the public. He had, indeed, suspected the truth long before the end of that

Nutation of the earth's axis.

* Vol. XXXV. p. 637.

period, and had even announced his opinion to several astronomers, among others to Lemonnier and Machin; but as the apparent motions of the stars which he had discovered seemed to depend upon the position of the moon's nodes, he thought it necessary to observe till the nodes returned to their original place, which they do in a period of little more than 18 years, in order to see whether the position of the stars would be restored to their original place when he began to observe, and this he had the satisfaction to verify. He published his account of his discovery at last in the Philosophical Transactions for 1747; together with tables indicating the change in the place of the stars, occasioned by this nutation.*

Soon after he had ascertained the aberration of the fixed stars, his attention was excited by another phenomenon, namely, an annual change of declination in some of the fixed stars, which appeared to be sensibly greater about that time, than a precession of the equinoxes of 50″ in a year would have occasioned. The quantity of difference, though small in itself, was rendered sensible by the exactness of his instrument, even in the first year of his observations; but being at a loss to guess from what cause that greater change of declination proceeded, he endeavoured to allow for it in his computations, leaving to time the discovery of the cause from which it proceeded. Though he left Wansted in 1732, yet his successor in the house that he occupied allowed his instrument to remain unaltered, and he went thither to make his observations at stated times annually, during a period of 20 years.

He had observed that the stars near the equinoctial colure showed a greater apparent declination than could arise from a precession of 50″ in a year; but, there appearing at the same time an effect of a quite contrary nature in some stars near the solstitial colure, which seemed to alter their declination less than a precession of 50″ required, it was obvious that the phenomenon could not be accounted for merely by supposing that he had assumed a wrong quantity for the precession of the equinoxes. At first he suspected that some alteration might have taken place in the position of the parts of his sector; and afterwards, that the appearances might be owing to the manner in which his plummet was suspended; but an accurate examination satisfied him that neither of these suppositions was well founded. During his residence at Wansted, from 1727 to 1732, he had made so many observations, and ascertained so many particulars, that he began to suspect the real cause of these appearances. He found that some of the stars near the solstitial colure had changed their declination 9″ or 10″ less than a precession of 50″ would have produced; and, at the same time, that others near the equinoctial colure had altered theirs about the same quantity, more than a like precession would have occasioned. The north pole of the equator seeming to have approached the stars which come to the meridian with the sun about the vernal equinox and winter solstice; and to have receded from those which come to the meridian with the sun about the autumnal equinox and the summer solstice. When he considered these circumstances, and the situation of the ascending node of the moon's orbit at the time when he began his observations, he suspected that the moon's action on the equatorial parts of the earth might produce these effects. For if the precession of the equinox be, according to Sir Isaac Newton's principles, caused by the action of the sun and moon on those parts, the plane of the moon's orbit being at one time above 10° more inclined to the plane of the equator than at another, it was reasonable to conclude, that the part of the whole annual precession which arises from her action would, in different years, be varied in its quantity. Whereas the plane of the ecliptic in which the sun appears keeping always the same inclination to the equator, that part of the precession which is owing to the sun's action, may be the same every year. Hence it would follow, that though the mean annual precession proceeding from the joint actions of the sun and moon were 50″, yet the apparent annual precession might sometimes exceed and sometimes fall short of that mean quantity, according to the various situations of the nodes of the moon's orbit.

This reasoning induced him to conclude, that the whole deviation was owing to the moon's action on the equatorial parts of the earth, which he conceived to produce a libratory motion of the earth's axis. But being unable to judge whether, after a complete revolution of the nodes, the axis would recover its original position, he thought it necessary to continue his observations during a complete revolution, and had the satisfaction to find that it did so. This leaving no doubt about the reality of his explanation, he no longer hesitated to publish it. He explains and shows how this nutation may be allowed for in calculating the true position of the stars, by a method which he ascribes to Mr. John Machin, at that time Secretary to the Royal Society. It consisted in supposing the true pole to describe a circle round the pole of the equator, the diameter of which is 18″. This circle is described in 18 years; and the pole is supposed to be always 90° distant from the nodes of the moon. D'Alembert afterwards demonstrated, that the real figure described by the pole was not a circle, but an ellipse, the larger axis of which was 18″, and the smaller 13″. In the year 1748, Lacaille published rules for calculating the effect of this nutation upon the place of the stars. Tables of the principal stars have even been published, with the places of each corrected according to the nutation, in order to save astronomers the trouble of making a troublesome calculation for each observation.

It may be proper to mention here, that Newton was the first person who deduced the precession of the equinoxes from the theory of gravitation. His method was exceedingly ingenious, but indirect; and the real figure of the earth not being so accurately known as it is at present, it was impossible for him to assign precisely the effect produced by the sun and by the moon. The problem first received a direct solution from D'Alembert. The effect of the sun is at present reckoned 17″, that of the moon 33″.

Transit of
Venus.

II. The next subject to which we shall turn our attention, is the transit of Venus over the sun's disc in 1761, and again in 1769. It constitutes one of the most memorable points in the history of astronomy, during the 18th century, and does infinite honour to the different learned societies, and to the liberality of the different monarchs, who reigned in Europe at the time.

In the year 1691, Dr. Halley published a paper in the Philosophical Transactions,* on the conjunction of the inferior planets with the sun. He gave tables of all the different transits of Mercury and Venus, for a considerable period of time; and mentioned the importance of an accurate observation of the transit of Venus, for determining the distance of the sun from the earth, and consequently the dimensions of the solar system. Mercury might be used for the same purpose, but his parallax is so small that observations, sufficiently accurate, could not be made. The transits of Mercury are frequent, but those of Venus happen very seldom. The first transit of this planet, ever observed, was that of November the 24th, 1639, which had been predicted by Horrox, and observed by him and Mr. Crabtree The next two transits did not happen till the years 1761 and 1769. In the year 1716, Dr. Halley published an elaborate paper on these transits, and described at length how they might be employed for determining the sun's parallax, conjuring the astronomers, who should be alive at that period, not to neglect so precious an opportunity to determine so very important an astronomical problem.† Dr. Halley committed some errors in his calculation; and the motion of the nodes of Venus not being known at the time when he wrote, he placed the transit of the planet too far from the sun's centre. Notwithstanding these trifling mistakes. the importance of the method continued undoubted, and, as the period of the first transit approached, astronomers set themselves earnestly to consider in what manner the intentions of Dr. Halley might best be accomplished. Dr. Halley had pointed out the northern parts of America, and the shores of the Ganges, as two places well adapted for the observation; but several circumstances, which he did not foresee at the time, rendered North America by no means well adapted for the purpose.

Transit of
1761.

The Royal Society sent out two different observers; namely, Dr. Maskelyne, to the island of St. Helena, and Messrs. Mason and Dixon to Bencoolen, in the island of Sumatra. But several circumstances prevented these latter gentlemen from arriving at their destined place soon enough. They therefore resolved to remain at the Cape of Good Hope, and observe the transit there, and it was fortunate that they did so; because Dr. Maskelyne was prevented, by bad weather, from making accurate observations, and Pingré, who had been sent by the French Academy to Rodrigues, committed an error of a minute in noting down the time of the internal contact. Joseph Delisle, who was in the habit of publishing details respecting every approaching astronomical phenomenon, printed a Chart, pointing out all the places where the transit of 1761 could be advantageously observed. This chart seems to have determined the measures of the Royal Academy of Sciences, at Paris. They sent the Abbé Chappe to Tobolsk, in Siberia; and Pingré to the island of Rodrigues, on the east side of Africa. Legentil had set out for the East Indies, as early as 1759, intending to observe the transit at Pondicherry. The Academy of Sciences, at Stockholm, sent observers into Lapland, and placed them in different parts of Sweden. The King of Denmark sent a party to Drontheim, in Norway; and the Petersburgh Academy dispatched astronomers as far as the borders of the empire of China. At Greenwich, Dr. Bradley was in too bad a state of health to be able to make the requisite observations; on that account he sent for Mr Bliss, who succeeded him soon after as Astronomer Royal; and that gentleman, together with Mr. Green, the Assistant at Greenwich, and Mr. Bird, observed the transit at the Royal Observatory. The transit was likewise observed by Mr. Short and Dr. Bevis, at Saville House; and by Mr. Hornsby and Mr. Phelps, at Shirburn Castle, the seat of Lord Macclesfield. The French astronomers at Paris were equally active. Owing to the unfavourable state of the weather, and the war which at that time existed between Britain and France, several of the intended observations were not made. Dr. Maskelyne failed from bad weather; Legentil was not enabled to observe at Pondicherry, on account of the war; and Mason and Dixon made their observations at the Cape of Good Hope instead of Bencoolen. After the observations had been all collected, and published in the Philosophical Transactions, Mr. Short and Mr. Hornsby drew up two excellent papers, deducing from the whole the solar parallax, which they found very nearly 8″·5. Pingré, relying upon his observations, at Rodrigues, published a paper in the Memoires of the Academy of Paris, in which he made the parallax rather more than 10″; but the coincidence of all the other observations convinced astronomers that Pingré had committed an error of a minute in marking down the time, and the transit of 1769 satisfied Pingré himself that this had been the case.

The transit of 1761 not having been so satisfactory as was expected, the approaching transit of 1769 was looked forward to with great expectation; and astronomers, warned by the errors into which they had fallen before, were now more able to take advantage of that transit, to obtain the wished for conclusions. Transit of 1769.

6

Mr. Hornsby published an excellent paper in the Philosophical Transactions, pointing out the places most proper for making the observations, and exhorting the Royal Society to make choice of some of the islands in the South Sea, as peculiarly fitted for the purpose. Accordingly, Messrs. Dymond and Wales were sent to Hudson's Bay; Mr. Green and Captain Cook, to Otaheite, being accompanied by Sir Joseph Banks and Dr. Solander; Mr. Call was sent to Madras; and Mr. Mason to the North Cape; and all the expences were munificently defrayed by his present Majesty. The French Academy sent Chappe to California, and Pingre to the island of St. Domingo. Veron also had been sent to India; but he died without having observed the transit. The Petersburgh Academy sent observers to different parts of Lapland. Indeed the number of observations made in different places, and the great interest taken in making them as complete as possible, by all the Sovereigns in Europe, one or two only excepted, make it next to impossible to give a complete account of them all. Dr. Bradley was dead; but Dr. Maskelyne, who afterwards succeeded Bliss, as Astronomer Royal, and whose accuracy as an observer is known to every one, observed at Greenwich. The result of all these observations, compared with each other, gave the solar parallax 8″·6; differing very little from what had been deduced by Short and Hornsby, from the Transit of 1761. Mr. Hornsby deduced it from the observations of 1769 to be 8″·78; which does not differ much from the preceding. He also gave us the following estimate of the dimensions of the solar system, which has been pretty generally acceded to He supposed the semi-diameter of the earth to be 3985 English miles.*

	Relative distance.	Absolute distances in English miles.
Mercury	387·10	36,281,700
Venus	723·33	67,795,500
Earth	1000·00	93,726,900
Mars	1523·69	142,818,000
Jupiter	5200·98	487,472,000
Saturn	9540·07	894,162,000

Density of the earth.

III. Sir Isaac Newton had calculated the density of the earth from the general theory of gravitation; but several circumstances concurred to render it probable that his result gave that density too small. It was therefore desirable to employ some direct method, by means of which that density could be deduced experimentally. Newton himself had suggested one, by mentioning the effect which a hill of a determinate size would have in drawing the plumb-line

* Phil. Trans. 1771. Vol. LXI. p. 574.

from the perpendicular. The French academicians, sent to measure a degree of the meridian in Peru, had endeavoured to determine the effect of Chimboraco upon their plumb-line, and had found it 8″; a quantity much smaller than was expected; but the imperfection of their instruments, and the exposed nature of the situation in which they made their observations, prevented the requisite degree of accuracy. It was therefore very desirable that the observations should be repeated upon some hill in Great Britain, and with instruments sufficiently delicate to determine the point with precision. In the year 1772, Dr. Maskelyne stated the importance of such an experiment in a paper given in to the Royal Society, and exhorted them to make it, mentioning at the same time some hills in Yorkshire and in Cumberland, which seemed well adapted for the purpose. The plan was embraced by the Council of the Royal Society, and Mr. Charles Mason was employed by them to make a journey through England and the Highlands of Scotland, in order to pitch upon a hill which would answer the purpose. He found the hills pointed out in Yorkshire and Cumberland not such as they had been described; but he recommended the mountain of Schehallien, in Perthshire, as the best for the purpose of any that he had met with. It was tolerably detached from other hills, very steep, and its direction, with respect to the greatest extent, was from east to west. Schehallien accordingly was chosen for the purpose, and Dr. Maskelyne was requested by the Royal Society to make the experiments, to which he readily consented. The Royal Society furnished him with the proper instruments, which they had procured for the transits of Venus of 1761 and 1769. His Majesty supplied them with the money requisite for defraying the expences, and was pleased to permit Dr. Maskelyne to absent himself from the Royal Observatory as long as would be necessary for the experiment. The summer of 1774 was chosen for the purpose. Dr. Maskelyne arrived at the mountain on the last day of June; and notwithstanding the badness of the season, which was uncommonly unfavourable, the astronomical observations, and the greatest part of the survey of the hill, were finished by the month of November. The result was that the sum of the attractions of the hill on the north and south sides amounted to 11″·6.*

(margin note) Experiments on Schehallien,

Dr. Hutton undertook, from the data thus obtained, to deduce the density of the earth; and after going through a most laborious set of calculations, he published a very curious and ingenious paper in the Philosophical Transactions,† giving an account of the methods which he followed, and deducing as a conclusion that the density of the earth was to that of the mountain Schehallien, as 17804 to 9933; or as 1434 to 800 nearly; or almost as 9 to 5. Dr. Hutton, proceeding on the supposition that the mean density of Schehallien was 2·5,

---

* Phil. Trans. 1775. Vol. LXV. p. 500.　　　　† Phil. Trans. 1778. Vol. LXVIII. p. 689.

deduced the density of the earth 4·481. But since that time Mr. Playfair has made an accurate survey of the mountain, in order to determine its specific gravity with as much accuracy as possible. He found that the rocks of which it is composed are of three different kinds; namely, mica slate, granular quartz (probably granite), and limestone; the specific gravity of all of which is considerably above 2·5. The mean specific gravity of the quartz rocks was found to be 2·639876, and that of the limestone and mica slate 2·81039. The granular quartz constitutes the upper part of the mountain, and the mica slate the lower. It being uncertain whether the granular quartz continues from the very summit to the base of the mountain, or whether the whole under part of the mountain consists of mica slate, and granular limestone, Mr. Playfair has calculated the density of the earth according to both suppositions. The first supposition gives us 4·55886 for the density of the earth; while the last (which appears the more probable supposition) gives us that density equal to 4·866997.* Mr. Cavendish, by a different method determined the density of the earth, at 5·48. Hence it is by no means unlikely that if other experiments similar to those on Schehallien were made upon a mountain composed of a homogeneous mass, the mean density of the earth might be found about 5.

Figure of the earth.

IV. The earth was supposed at first by astronomers to be an exact sphere; but after the oblate figure of Jupiter was observed, and after Richard had found that the second pendulum at Paris beat too slowly at Cayenne, then there were reasons for suspecting that the earth also might be an oblate spheroid. Huygens deduced from the variation of the pendulum that the earth was compressed at the poles. Newton, from the theory of gravitation, demonstrated that if the earth be of homogeneous density, the polar diameter must be to the equatorial as 329 to 330. These reasons were acquiesced in by astronomers till the finishing of the trigonometrical survey of France, which was begun in 1673, and concluded in 1716. From this survey, which had been made with great care, it appeared that the degrees of the meridian were shortest in the northern parts of the kingdom, and that they became gradually longer as they approached the southern extremity. The observation of this unexpected circumstance induced Cassini and Fontenelle to conclude that the earth, instead of being flat at the poles, had quite a different figure, and that the polar axis was longer than the equatorial. This occasioned a dispute between the British philosophers, who had unanimously adopted the opinion of Newton, and the French astronomers, who, relying upon the measurement made in France, adopted the opposite conclusion. To determine this dispute, the French Academy resolved upon measuring a degree in two different parts of the North of France, one near the west, and the other near the eastern extremity of that

* Phil. Trans. 1811. p. 347

kingdom, in order to compare the results obtained with what the length of a degree, in that latitude, ought to be according to each hypothesis. The result of this measurement was favourable to the hypothesis of Cassini and Fontenelle. But it made no change in the opinion of the Newtonians, who advanced reasons that appeared satisfactory for setting aside these new measurements as of no value.

At last it was proposed to determine the point by measuring a degree under the equator itself; and Count Maurepas procured the consent of the King of France to defray the expence of the whole with the utmost munificence. Bodin, Bouguer, and La Condamine, set sail for that purpose on the 16th of May, 1735, the King of Spain having given them liberty to make the proposed measurement in Peru, and having ordered Don George Juan, and Don Antonio de Ulloa, to accompany them in order to facilitate their operations. Another measurement was determined on at the polar circle, and Maupertuis, Clairaut, Camus, Lemonnier, and Outhier, were sent to accomplish it. The result of these measurements was entirely in favour of the Newtonian doctrine, which was in consequence acceded to by the whole philosophic world. But these measurements, owing to various circumstances, as unavoidable inaccuracies in the measurements, the attraction of mountains, and the effect of the various density of different strata near the earth's surface, &c. were insufficient to determine the true figure of the earth with the utmost precision. The following table exhibits the length of a degree in Paris toises according to the measurements of different astronomers:

Length of a degree.	Latitude.	Observers.	Years.
57422	66° 20 N.	Maupertuis, &c.	1736
57074	49 23	Maupertuis and Cassini	1739
57091	47 40	Liesganig..................	1768
57028	45 0	Cassini....................	1739
57069	44 44	Beccaria	1768
56979	43 0	Boscovich and Le Maire.....	1752
56888	39 12	Mason and Dixon...........	1764 and 1768
56750	0 0	Bougner and La Condamine..	1736
57037	33 18 S.	De La Caille..............	1752

But it ought to be observed that the first of these measurements has lately been repeated by the Swedish astronomers, and found considerably erroneous.

The measurement of degrees, from the many sources of uncertainty, not being sufficient to give us the true figure of the earth, philosophers have endeavoured to deduce that figure from physical principles. Mr. Murdoch demonstrated that the centrifugal force at the equator is equal to $\frac{1}{289}$th part of gravitation. The length of the pendulum that beats seconds in different latitudes has been found as follows. The lengths are given in French feet and lines:

	Feet.	Lines.
The equator	36	7·07
Lat. 9° 54′ N.	36	7·16
48 50	36	8·57
66 48	36	9·17

Captain Warren has lately determined the second pendulum at Pondicherry, in lat. 13° 4′ 12″ N. to be 39·026273 English inches. And he has deduced, from the experiments hitherto made, the following as the relative lengths of the pendulum beating seconds at the equator and at the pole in English inches.[*]

Equator 38·98726
Pole   39·20899

These facts show us that gravitation increases from the equator to the pole more than can be accounted for by the diminution of centrifugal force; and therefore demonstrate that the figure of the earth is an oblate spheroid. Sir Isaac Newton showed that if the earth be a homogeneous spheroid, the pendulum at the pole must be $\frac{1}{230}$th longer than at the equator, to vibrate seconds. But the difference by the preceding table is greater. Hence it follows that the supposition that the earth is a homogeous spheroid is not correct.

Clairaut showed that if the earth be a dense nucleus surrounded by a fluid, then the two fractions denoting the compression and the lengthening of the pendulum are always the same or $\frac{1}{230}$. The lengthening of the pendulum, according to the experiments of Lord Mulgrave, is $\frac{1}{115}$, which agrees very nearly with the preceding table. This subtracted from $\frac{1}{115}$ gives us the real shortening of the polar axis, and it lies between $\frac{1}{300}$ and $\frac{1}{310}$th of the axis. From a comparison of all the degrees measured, Mr. Playfair had deduced the number $\frac{1}{330}$ for the compression, which probably approaches very nearly to the truth.

We have not noticed the anomalies resulting from the present trigonometrical survey of Great Britain; because no satisfactory solution of them occurs. Mr. Playfair has suspected differences in the specific gravity of the strata under the surface of the earth in different parts of England; a conjecture very likely to be true, though it cannot well be verified by actual observation. Neither was it possible to notice the calculations of De la Place on the subject, because they are liable to some difficulties that have not yet been cleared up.

V. The discovery of five new planets, and of eight new satellites, belonging to the solar system, is not the least remarkable addition that has been made to astronomy during our own days. Dr. William Herschel, to whom the discovery

Herschel.

* Asiatic Researches, Vol. XI. p. 293.

of no fewer than nine of these bodies has given celebrity, was born in Hanover, in 1738, and came over to England as a musician in a Hanoverian regiment. He attracted some notice by his musical talents, and was appointed organist at Bath. Here he began to amuse himself with making telescopes; and being possessed of great industry as well as dexterity, he soon acquired so much skill as to make telescopes greatly superior to any that had been used before his time. Being in possession of these instruments, it was natural to turn them towards the heavens, and in a little time he acquired considerable skill as an observer, and a strong passion for the science of astronomy. On the 13th of March, 1781, he was struck by the appearance of a small star, which seemed to differ New Planet from those of the same size in its neighbourhood. On examining it with a more discovered. powerful magnifier, he found that it increased in diameter, which was not the case with the other stars near it, and on that account he suspected it to be a comet; and he was confirmed in this opinion by finding some days afterwards that it had changed its place. He communicated this piece of information to Dr. Maskelyne, who having examined it himself, and drawn the same conclusions that Dr. Herschel had done, sent information of the circumstance in the beginning of April to the astronomers at Paris. The supposed comet was assiduously observed for some time, and attempts were made to calculate its parabolic orbit. These attempts at first were very unsuccessful; but by degrees it was perceived that the distance of this body from the earth was very great, that its orbit was nearly circular, and of course that it ought not to be considered as a comet, but a planet. Its orbit was found to be about 83 years. It was perceived that the star in Taurus marked 34 in Flamsteed's catalogue, and the 964th star of Meyer's catalogue were no other than this planet. Hence astronomers were in possession of a whole century of observations on it at once, and were accordingly able to construct tables for it with surprising accuracy. Herschel, the discoverer, gave it the name of the *Georgium Sidus*; but this name has not been generally approved of. La Lande called it by the name of the discoverer Herschel, while the German Astronomers, preferring the ancient mode of assigning to the planets titles taken from the Heathen Mythology, gave it the name of Uranus, the father of the gods. This last name seems at present to stand the greatest chance of being generally adopted.

Dr. Herschell afterwards discovered six satellites moving round this planet; Its six satel- and what is very remarkable, their motion appears to be contrary to the order of lites. the signs. The following table exhibits the time of the revolution of each of these bodies as accurately as it has been ascertained.

1st	5 d.	21 h.	25 min.	4th	13 d.	12 h.	0 min.
2d	8	18	0	5th	38	1	49
3d	10	23	4	6th	107	16	40

New Planets between Mars and Jupiter.

The distance between Mars and Jupiter is so great when compared with the other planetary spaces, and the distance from the sun, that astronomers had been looking for some primary planet in that position. Of late years these expectations have been more than accomplished by the discovery of no fewer than four planetary bodies almost all in the same place; but so small that Dr. Herschel refuses to honour them with the name of planets, and chuses to call them asteroids, though for what reason it is not easy to determine, unless it be to deprive the discoverers of these bodies of any pretence for rating themselves as high in the list of astronomical discoverers as himself. These four bodies have received the names of Ceres, Pallas, Juno, and Vesta. They were discovered by Mr. Piazzi, Dr. Olbers, and Mr. Harding.

Some particulars respecting the solar system.

VI. We shall terminate these observations with a few tables representing some of the most remarkable particulars connected with the solar system.

1. Distances of the planets from the sun, that of the earth being 100.

Mercury 38	Vesta, ...237	Jupiter,............520
Venus,............72	Juno,............266	Saturn,............950
Earth,............100	Ceres, ...276	Herschel,.........1900
Mars,............150	Pallas, ...279	

2. Diameters of the sun and planets in English miles.

Sun,............883,246	Mars,............4,189	Pallas,............ 80
Mercury,.........3,224	Vesta,............ 238	Jupiter,........ 89,170
Venus,............7,687	Juno,............1,425	Saturn,........ 79,042
Earth,............7,911	Ceres,............ 163	Herschel, ... 35,112
Moon,............2,180		

3. Density of the planets and sun, that of water being one.

Sun,............$1\frac{2}{5}$	Earth,............ $4\frac{1}{2}$	Saturn,............ $\frac{1}{2}\frac{1}{2}$
Mercury,............$9\frac{1}{2}$	Mars,............ $3\frac{2}{7}$	Herschel, ... $1+$
Venus,............$5\frac{1}{2}$	Jupiter,............ $1\frac{1}{7}$	

But if we suppose the density of the earth to be five, this will alter a little all the numbers in the preceding table.

4. Quantity of matter in the planetary bodies, supposing that of the Earth to be one.

Sun,...... 329,630·	Earth,.......... 1·000	Saturn,........ 103·950
Mercury,........ 0·135	Moon,.......... 0·025	Herschel, ... 16·840
Venus, ... 1·135	Jupiter,.. .... 330·600	

5. Number of feet per second through which a heavy body would fall at the surface of each of these planets.

Sun, ... 420	Earth,.. .... 16+	Saturn, ... 15
Mercury,......... 12	Moon, ... 3	Herschel, ... 4·2
Venus,.......... 18	Jupiter,........... 42	

6. Rotation of the planets round their axis.

Sun, ... 25 d. 10 h.	Mars,.. 24 h. 39′ 22″	Saturn's ring, 10 h. 32′ 15″
Venus, ... 24 —	Jupiter, 9 55 33	Moon, ... A lunation.
Earth, ... 24	Saturn, 10 16 1	

7. Tropical revolution round the sun.

Mercury,.... 87 d.	23 h.	14 m.	33″	Jupiter, .. 4330 d.	14 h.	39 m.	2″
Venus,...... 224	16	41	24	Saturn, .. 10746	19	16	15
Earth,...... 365	5	48	48	Herschel, 30589	8	39	
Mars, ... 686	22	18	27	Moon,*.. 27	7	43	5

The orbits of all the planets formerly known are confined within the zodiac; but that of some of the newly discovered planets is so oblique as to pass without the zodiac.

CHAP. II.

OF OPTICS.

The science of Optics consists of two parts; the first of which treats of Vision, and the second of the Properties of Light. It is the most mathematical of all the sciences; its whole progress, after the establishment of certain fundamental points by experiment, depending upon the application of mathematical principles. It lies under the deepest obligations to the Royal Society, being one of the sciences most strenuously and successfully promoted by that illustrious body.

It would be in vain to look towards the ancients for much information respecting this important science. Ptolemy was acquainted with the existence of atmospherical refraction, and must therefore have had some knowledge of dioptrics; though he no where explains himself on the subject. The ancients were acquainted with plain mirrors, and with the use of glass spheres filled with

Optical knowledge of the ancients.

* Round the earth.

2

water, both as burning glasses, and for increasing the size of objects. They were used as magnifying glasses by the engravers on gems, as they still are even at present. The discovery of *spectacles*, one of the first and one of the most useful optical instruments ever brought into general use, seems to have been made about the end of the 13th century. It is obvious from the writings of Roger Bacon, published a little before that period, that spectacles were unknown in his time; and soon after the beginning of the 14th century they appear to have been in general use. The common and most probable opinion is that they were discovered in Italy, and Manni informs us that they were invented by a Florentine called Salvino degl'Armati. He even quotes an inscription on a tomb at Florence, which seems to leave no doubt of the fact. This inscription is as follows : *Qui diace Salvino d'Armato degl'Armati, Firenze, inventor di egl'Occhiali, anno* 1317. *Here lies Salvino d'Armato, a Florentine, the inventor of Spectacles, who died in* 1317.

Discovery of spectacles.

The nature of vision was first explained in a satisfactory manner by Kepler, in his *Astronomiæ Pars Optica*, published at Frankfort in 1604. He showed that the rays of light, coming from objects, were refracted in such a manner by the different humours of the eye, that a distinct inverted image of the object was delineated on the retina, and that vision was the consequence of this image In what manner this image occasions vision, it is beyond our faculties to explain : but the fact is certain. The retina is an expansion of the optic nerve; and no doubt the rays of light produce a certain effect upon it, and vision is the consequence of this effect. But all the metaphysical attempts to explain how, have terminated only in absurdities.

About the same time that Kepler's book on optics was published, another most important optical discovery was made in Holland, which has contributed perhaps more to the advancement of the mathematical sciences, than any other discovery whatever: we allude to the discovery of the telescope. Some modern writers have taken infinite pains to prove to us that this admirable instrument was known to the ancients; but their proofs are so weak, or so absurd, that they are not entitled to any notice. Indeed it is impossible, if that instrument had been known to the ancients, that not a single writer, either among the Greeks or Romans, should have made the least allusion to it. An enormous number of different stories have been told, respecting the history of this invention; but the account which seems supported by the best authority, is that of Mr. William Boreel, Envoy of the States of Holland to the British Court. He ascribes the invention to a maker of spectacles at Middleburg, called Zachary Jans, and places the date somewhere between 1590 and 1610. This man and his son having contrived a telescope, made a present of it to Prince Maurice, who requested them to conceal the invention, that he might be able to employ it exclusively in the war at that time raging between the

States of Holland and Spain. But the invention having transpired, some person or other went to Middleburg, to get exact information about it, and addressed himself, by mistake, to John Lapprey, likewise a maker of spectacles in the same town; that artist, in consequence of the questions that he put, divined the nature of the invention; and, on that account, came by some to be reckoned the inventor himself.

As to the way in which the discovery was made, we know nothing that can be implicitly depended on; though there can be little doubt that it was entirely owing to accident. It is said, that two of Jans's sons, while amusing themselves with different glasses in their father's shop, happened to put two lenses at the proper distance from each other, and looking through them to the weather-cock on the top of a neighbouring church, found that the size of the object was considerably increased. The father, that the observation might be easily repeated, fixed the glasses in their proper position; and afterwards, to exclude the side light, which was injurious to the effect, he fixed the glasses in a tube, which constituted the telescope. The original Dutch telescope consisted of two glasses: the object glass was convex, and the eye glass concave. It has the disadvantage of confining the field of vision within very narrow limits, especially if the telescope be of any considerable length; and, on that account, has been long laid aside for astronomical purposes.

Galileo was at Venice, as he informs us, when he heard the report of the discovery of the telescope, and probably that it consisted in the application of two lenses of glass to the extremities of a tube. A few trials enabled him to discover the lenses that answered the purpose, and after some labour and expence, he succeeded in constructing a telescope that magnified about 33 times; and with it he made his discoveries of the satellites of Jupiter, and the spots on the sun, which contributed so much to give him that celebrated name which he has left behind him. Kepler was the first person who pointed out the construction of what is usually called the astronomical telescope; which has two convex glasses, and which, as Kepler showed, has the property of reversing the object. But this illustrious philosopher did not attempt to construct such an instrument. Scheiner appears to have been the first person who ascertained the properties of such a telescope experimentally. In his *Rosa Ursina*, published in 1650, he describes its properties, and says, that as the stars are round, it may be used in examining them without any disadvantage, though it reverses objects. He points out, also, the method of restoring the original position of objects, by means of two convex eye glasses. These three kinds of telescopes were almost the only ones that were used by philosophers, for many years. At last, James Gregory proposed the reflecting telescope, and came to London on purpose to get it executed; but could not succeed. Sir Isaac Newton, after his discovery of the different refrangibility of the rays of light, concluded that it was impossible ever to bring refracting telescopes to any great degree of perfection, and therefore thought, likewise, of the reflecting telescope. Possessing a turn for mechanics, he did not satisfy himself with merely pointing out its advantages theoretically, but actually constructed one with his own hands, and sent it to the Royal Society, as a specimen of what telescopes of that construction could perform. Many years after, Mr. Dollond discovered the different dispersive power of different kinds of glass, and contrived in consequence a new kind of refracting telescope, much more perfect than the old ones, to which Dr. Bevis

Achromatic telescope. gave the name of *achromatic ;* because they represent objects free from those colours that arise from the difference of the refraction of the different rays. Dr. Herschel has of late years added greatly to the perfection of the Newtonian reflectors; nor is the convenience of the telescopes of his construction for use less remarkable, than their perfection in representing a very distinct image of the object viewed.

Microscope. The microscope was in some measure known to the ancients, since they made use of glass globes, filled with water, to magnify objects; and the simple microscope, as far as small sphericles of glass are employed, may be considered as not differing in principle from the glass globes of the ancients. Indeed all the simple microscopes may be referred either to such globules, or to lenses, the effect of which followed directly from the use of spectacles. But the compound microscope seems to have been invented by the same Zachary Jans, to whom we are indebted for the telescope. If we believe the testimony of Mr. Boreel, already mentioned, and it would not be easy to set it aside, the discovery of the microscope preceded that of the telescope. Boreel even informs us that he saw, in the possession of Cornelius Drebbel, the original microscope, made by Zachary Jans; which that artist had presented to the Archduke Albert, by whom it was given to Drebbel. This may perhaps be the reason why some have thought fit to ascribe the invention of the microscope to Drebbel himself.

Scarcely had the telescope been announced to the philosophic world, when Kepler undertook to give a theory of this important instrument. This was the

Dioptrics. object of his *Dioptrics,* published in 1611; a work of great merit, in which that illustrious philosopher laid the true foundation of Dioptrics, such as they exist at this day. He was not indeed acquainted with the law of refraction; but he assumed a law, deduced from experiment, which, though not perfectly accurate, yet, as far as the lenses of telescopes were concerned, he was able to

Law of refrac- make use of without falling into any error. The law of refraction was dis-
tion. covered by Snellius, a mathematician of the Low Countries, of some eminence. His book indeed, in which this discovery is contained, was never published; but Vossius informs us, that Hortensius had taught it publicly as the discovery of Snellius; and Huygens declares, in his Dioptrics, that he had read the manuscript of Snellius, which was in the hands of the heirs of Hortensius

Snellius announced the law in this way: " In the same medium, the secants of the complements of the angles of inclination and refraction have to each other the same ratio." Descartes afterwards announced the law in this manner: " In the same medium, the sines of the angles of inclination and refraction bear the same ratio to each other." This mode of expressing the law is more convenient, and has been generally adopted. It has been supposed, that Descartes acquired his knowledge of this law of refraction from Snellius's manuscript; and that he expressed it in a different manner, the better to disguise the theft. There can be no doubt that he actually perused the manuscript of Snellius, for Huygens expressly says so. Hence the probability is, that he derived his knowledge of the law in that manner. But at all events there can be no doubt that Snellius preceded him, and that the discovery belongs to that mathematician.

Descartes published his Optics in 1637, and it enjoyed, from its first appearance, a very high degree of celebrity. He conceived light to be produced by the action of luminous bodies on a subtile fluid which filled all space. As he believed that the transmission of light was instantaneous, it was necessary for him to conceive the parts of this fluid as absolutely inflexible. But his followers have since corrected this opinion, and have made the supposed fluid, by means of which light is propagated, a highly elastic fluid. This hypothesis, somewhat modified indeed, was afterwards adopted by Huygens, and supported with astonishing ingenuity. It has in it something very seducing, and is still embraced by several philosophers of the present day. But an objection has been urged against it which appears to be insurmountable; namely, that, as undulations propagate themselves not in straight lines, but in all directions, if the hypothesis be true, there ought to be no such thing as darkness at all. At midnight, and during a total eclipse of the sun, we ought to have just as much light as at noon day. Newton afterwards supported the hypothesis, that light moves from luminous bodies in straight lines. This opinion has been embraced by almost all succeeding philosophers; and even those who refuse to adopt it are under the necessity of employing its language when they explain the phenomena of optics.

Descartes's Optics.

Descartes attempted, likewise, to give a philosophical explanation of the cause of refraction, which occasioned a long protracted dispute between him and Fermat; a dispute never settled, and which Fermat continued after Descartes' death with some of his disciples. Fermat himself advanced an explanation equally false with the Cartesian; namely, that light always moves in the shortest course. Neither was the explanation of Leibnitz, that light always moves in the easiest direction, entitled to greater respect. We do not think it necessary to mention the explanations of Barrow, De la Hire, and various others. Newton afterwards gave the explanation which is considered as the true one. Descartes, in his Optics, added considerably to the explanation of the effect of lenses upon light. He showed that when the surface of these bodies is spherical, the light is not accurately collected into a point; but that this would be the case if the figure of these bodies were that of the ellipse, or hyperbola. He even made some attempts to grind glasses of a hyperbolic form, but they were not attended with success: nor were the subsequent attempts of other opticians more fortunate. But this need not occasion any regret, as we know, from the subsequent discoveries of Newton, that such glasses, even if they could be constructed, would not give us an image better defined than common lenses with spherical surfaces.

The rainbow.   We owe to Descartes, likewise, the first satisfactory explanation of the rainbow. De Dominis indeed, an Italian Clergyman, had about 40 years before shown that the inferior, or principal rainbow, was produced by one reflection of the ray of light against the bottom of the drop, and two refractions, one when entering, and the other when leaving the drop. But he had failed in explaining the upper rainbow, in which the colours are reversed. Descartes showed that in it the ray underwent two reflections and two refractions. He explained, also, why no more than these two rainbows can be seen; but he failed in his explanation of the colours of the rainbow; that part of the phenomenon being reserved for Newton.

After the publication of the Optics of Descartes, the science continued nearly stationary for a number of years; unless we are to reckon the discoveries of Kircher, such as the magic lanthorn, &c., as improvements in the science. But soon after the establishment of the Royal Society, the Members of that illustrious body directed their attention to optical subjects; and the result was a set of discoveries so important and unexpected, that they quite altered the aspect of the science. The first writer on the subject was James Gregory, whose observations upon optical instruments, particularly telescopes, are of the greatest value. We have already mentioned his reflecting telescope. He failed in his attempts to get it executed, because he wished to have his mirror polished with a parabolic surface; and no workman in London was able to give it this curvature. Barrow published his *Lectiones Opticæ* in 1674; and among other valuable things contained in that book, is a general theorem for the foci of lenses, of every possible shape.

Grimaldi's discoveries.   About this time a discovery of considerable importance was made in Italy, by Grimaldi, a Jesuit, who died in 1663, at the age of forty-four. His work in which his discovery is contained, was posthumous, and only published in 1665. It was entitled *Physicomathesis de Lumine, Coloribus, et Iride, aliisque annexis, Libri II.* Grimaldi let a small ray of light into a darkened room, by means of a hole in the window-shutter. Suspending a hair or some minute body in this ray, and measuring its shadow at a certain distance from the body, he found it larger than it ought to have been. In like manner, when he made a small hole in a metallic plate, and made the ray pass through the hole, he found that the image of the sun, formed by the light passing through this hole, was larger than it ought to have been. Hence he concluded, that the light had been bent out of its course by the bodies in its neighbourhood. This property is what has been since called the *deflection* of light.

we come now to the discoveries of Newton respecting light; which have The Newtonian discoveries. been justly reckoned to afford proofs of his sagacity equally striking with his discoveries respecting universal gravitation. Newton's optical discoveries were the first of his productions which saw the light. They were published in the Philosophical Transactions for the year 1672.* His experiments were made in the year 1666, when he was only 23 years of age. Being employed in grinding glasses for telescopes, he had the curiosity to purchase a glass prism, in order to observe the celebrated phenomena of the colours. That a ray of sunshine, passing through a glass prism, assumes beautiful colours like those of the rainbow, was known to the ancients, and had been observed by every optician. Newton having darkened his room, admitted a ray of light through his window-shutter, and applying his prism made the image fall upon the opposite wall, at the distance of 22 feet. For some time he was much amused with the brilliancy of the colours; but when he began to consider the spectrum more attentively, he was much struck with its great length. By the laws of refraction universally received, the spectrum ought to have been circular; but he found the length of it no less than five times greater than the breadth. The sides of it were terminated by straight lines; and the upper and lower extremities, as far as could be determined by observation, were semicircular. He first tried whether the thickness of the glass, or the size of the hole in the window-shutter, could have any effect in producing this enormous disproportion; but soon satisfied himself that it was owing to none of these causes. The unevenness of the glass, or some other contingent irregularity in the prism, suggested itself as a possible cause; but, upon applying another prism to the former, so that the light might be refracted contrary ways by each, he found that the light was now reduced to an orbicular form, with as much regularity as if it had not been refracted at all. Hence it followed, that the oblong image was not occasioned by any contingent irregularity in the glass. The effect of rays from different parts of the sun's disc was next examined, and found inadequate to produce the oblong image. He then began to suspect that the law of refraction, hitherto received, might not be quite accurate; but finding that a considerable change in the position of the prism made no sensible alteration in the length of the image, it was obvious that no small inaccuracy in the law of refraction could account for the phenomenon.

The gradual removal of these suspicions led him at last to an experimentum crucis, which furnished him with the true explanation of the phenomenon He took two boards, and placed one of them close behind the prism at the window, so that the light might pass through a small hole made in it for the purpose, and fall on the other board, which he placed at about 12 feet distance, having first made a small hole in it also for some of the incident light to pass through. Then he placed another prism behind this second board, that the light trajected through both the boards might pass through that also, and be again refracted before it arrived at the wall. This done, he took the first prism in his hand and turned it to and fro slowly about its axis, so much as to make the several parts of the image, cast on the second board, successively pass through the hole in it, that he might observe to what places on the wall the second prism would refract them. And he saw by the variation of these places, that the light tending to that end of the image towards which the refraction of the first prism was made, did in the second prism suffer a refraction considerably greater than the light tending to the other end. Thus he detected the true cause of the lengthened image to be no other than that light consists of rays differently refrangible, which without any respect to a difference in their incidence, were, according to their degrees of refrangibility, transmitted towards divers parts of the wall.

Thus Newton ascertained that light is composed of rays differing from each other in their refrangibility. The number of these rays he settled at seven, guided by the difference of their colour ; and he informs us in his Optics that the names of the colours were not imposed by himself, but by another person whose skill in distinguishing colours was much greater than his own. The least refrangible ray was the *red*, and the most refrangible the *violet*. The order of refrangibility, and the names of the rays are as follows : *Red, orange, yellow, green, blue, indigo, violet.* The proportional lengths of each of these rays in the spectrum, he found by measurement to be the following.

Red, ... $\frac{1}{8}$
Red and orange,............................. $\frac{1}{5}$
Red, orange, and yellow,................... $\frac{1}{3}$
Red, orange, yellow, and green,........... $\frac{1}{4}$
Red, orange, yellow, green, and blue,......... $\frac{2}{3}$
Red, orange, yellow, green, blue, indigo,.... $\frac{7}{9}$
Whole spectrum, ... 1

Colour.   These discoveries furnished him with an explanation of the different colours of bodies, a circumstance which had not hitherto been understood. Colours are not qualifications of light derived from refractions or reflections of natural bodies, but original and connate properties, which in different rays are different. To the same degree of refrangibility ever belongs the same colour, and to the same colour ever belongs the same refrangibility. Thus the red rays are always the least, and the violet rays the most, refrangible. And the same holds of all the others. The colours of any particular ray cannot be changed by refraction nor reflection from natural bodies, nor by any other cause that he could observe. Every possible method was tried to alter the colour of such a ray; it was refracted by prisms, and reflected from bodies possessing a different colour from itself; it was transmitted through variously coloured mediums; in consequence of this treatment it would, by contracting or dilating, become more brisk or faint, and, by the loss of many rays, in some cases very obscure and dark; but it never changed in the kind of colour which it exhibited. Yet seeming transmutations of colours may be made, where there is any mixture of different sorts of rays. For in such mixtures the common colours appear not; but by their mutual allaying one another, they constitute a middle colour, which by refraction may be resolved into all the constituents of which it is composed.

Thus there are two sorts of colours, the one original and simple, the other compounded of these. The original or primary colours are red, orange, yellow, green, blue, indigo, violet. The compounded ones are made by mixtures of the simple ones; thus blue and yellow make green; red and yellow, orange; red and indigo, violet; and so on. White is the most surprizing of these compound colours. It is formed by a mixture of all the simple colours in a due proportion. When all the rays after being refracted by the prism are made to converge together again, they constitute light perfectly white and free from every other colour. Hence it happens that whiteness is the usual colour of solar light. In it all the rays are mixed in their due proportion; but if any one ray predominate, the light will incline to its colour. Hence the reason why the light of a candle is yellow; that of burning sulphur, blue; that of Mars, red; and so on. These facts enabled him to explain the different colours which bodies possess. They are always of the colour of the ray or rays which they reflect to the eye of the spectator. The reason why different bodies appear of one colour by reflected, and of another by transmitted light, is equally evident; and why two transparent liquids of different colours become opaque when placed in contact. Nor is it more difficult to explain the colours produced by the prism or the colours of the rainbow. These things being so, it can no longer be disputed whether there be colours in the dark, nor whether they be the qualities of the objects we see, nor whether light be a body. For since colours are the qualities of light, having its rays for their entire and immediate subject, how can we think those rays qualities also, unless one quality may be the subject of and sustain another; which in effect is to call it a substance. We should not know bodies for substances, were it not for their sensible qualities, and the principle of those being now found due to something else, we have as good reason to believe that to be a substance also.

Such is a pretty full account of Newton's first discoveries respecting light, as given by himself in the Philosophical Transactions for 1672. No sooner had it made its appearance than it was attacked with considerable violence by different foreign mathematicians, chiefly disciples of Descartes. Father Ignatius Gaston Pardies, a Jesuit, and Professor of Mathematics in the Parisian College at Clermont, was the first who entered the lists. A set of animadversions by him on Newton's theory was inserted in the 84th number of the Philosophical Transactions published the same year with the original theory.* To this Newton wrote an answer, shewing the mistakes into which Pardies had fallen. Pardies wrote a second letter stating fresh objections, which was likewise answered by Newton. Upon which the French Philosopher declared that all his objections were removed, and that he was perfectly satisfied of the truth of the Newtonian theory of light and colours. He died the year following, in the 37th year of his age.

Newton's next antagonist was Dr. Hooke, who wrote a number of animadversions on the Newtonian theory, admitting the truth of most of the experiments, and explaining them by a hypothesis supposing light to be the undulations of a fluid. This paper was not published; but Newton's answer, which was very long, appeared in the same volume of the Philosophical Transactions with his preceding papers above mentioned.† This answer is a most masterly performance. He shows the absurdity of Dr. Hooke's hypothesis in a very clear manner, proves that his own doctrine is totally unconnected with any hypothesis, and merely expresses matter of fact. Finally he refutes Dr. Hooke's opinions about colours, in such a manner as to leave no room for reply. Dr. Hooke accordingly, notwithstanding the pertinacity with which he usually adhered to his opinions, did not venture to continue the contest.

Newton's next antagonist was an anonymous Frenchman, who called himself M. N. Some animadversions by him were published in the Philosophical Transactions for 1673.‡ To these animadversions Newton immediately replied. This reply was followed by another letter from M. N. upon the same subject, to which Newton having replied at considerable length, and pointed out the mistakes of his antagonist, the controversy dropped. The next antagonist was Franc. Linus, a peripatetic philosopher of some eminence in those days, but quite incapable of contending with Newton.§ To this only a very short answer was at first deemed necessary : Linus replied a second time, and after his death the controversy was still kept up by his pupil Mr. Gascoigne.|| To all of these

* Phil. Trans. Vol. VII. p. 4087.   † Page 5084.
‡ Vol. VIII. p. 6086.   § Phil. Trans. 1674. Vol. IX. p. 217.
|| Phil. Trans. 1676. Vol. X. p. 499.

Newton made appropriate answers, which were published in the Philosophical Transactions immediately after the animadversions of Linus himself. Mr. Gascoigne having employed Mr. Lucas, of Liege, to repeat the experiments to which Newton in his answers had appealed, that gentleman published an account of the result, together with a new set of animadversions upon the Newtonian doctrine.* The dispute between Linus and Newton turning chiefly upon the length of the spectrum, which Linus denied, we do not think it necessary to enter into minute details. To Mr. Lucas's animadversions Newton made an appropriate reply, pointing out the proper method of repeating his experiments, and shewing how he might satisfy himself of the truth of his doctrine of the different refrangibility of light, and of the nature of colours.

Thus terminated this long controversy, which had been carried on with considerable keenness for about four years. On all sides, however, the utmost good breeding was observed, unless some of the remarks of Dr. Hooke be considered as exceptions; and all his opponents uniformly expressed the greatest respect for the abilities of Newton. This controversy however had an unfortunate effect upon Newton's mind, and prevented him from laying his discoveries before the world as he had intended to do. An omission which, in a subsequent period of his life, occasioned other disputes of a very disagreeable nature; and which must have been the more unpleasant to him, as he must have been conscious that they proceeded entirely from his long hesitation in laying his mathematical discoveries before the world.

The first part of Newton's Optics, as he informs us in the preface, was drawn up in the year 1675, and read before the Royal Society. The last part was added about twelve years afterwards; it was not published till the year 1704.† A second edition, with some augmentations, was published in 1717. It was translated into Latin by Dr. Samuel Clarke; and no doubt has made its appearance in almost every language in Europe. It consists of three parts. In the first the different refrangibility of the rays of light is demonstrated by experiment. The second part treats of the colours of thin plates, of colour in general, and of some curious particulars respecting refraction and reflection. The third part treats of the inflection of light. The whole is terminated by a most important set of queries respecting the most abstruse parts of natural philosophy and chemistry. Any remarks upon this admirable book would be quite unnecessary, as it has long enjoyed that reputation to which it is so justly entitled. As an analytical investigation, it constitutes the finest model hitherto offered to the

Newton's Optics.

* Phil. Trans. 1676. Vol. XI. p. 692.

† Newton seems to have declined giving his work to the public till after the death of Dr. Hooke. probably in order to avoid reviving his disagreeable dispute with that caustic philosopher.

world. Various attempts have been made by philosophers to reduce the number of primary colours to three; namely, red, yellow, blue; and to prove that
all the other four are only mixtures of these three. Some of these attempts
have been exceedingly ingenious; and, in a metaphysical point of view, might
be admitted as satisfactory, if it were possible to get over the Newtonian experiments, which show the impossibility of resolving the orange, green, and violet, into more simple colours, and the objections which Newton started to the
admission of a nearly similar hypothesis advanced by Dr. Hooke.

Newton adopted the opinion that light consists of straight lines proceeding
in rays from luminous bodies, in contradistinction to the opinion of the undulations of a fluid advanced by Descartes, Huygens, and Hooke; but he does
not pertinaciously insist upon this opinion being adopted, and his Optics are
entirely independent of it.

Refraction and
reflection ex-
plained. It will be proper to notice, here, the way in which Newton explains refraction
and reflection in his Optics. It is exceedingly ingenious; but to a certain ex-
tent hypothetical, and by no means supported by the same evidence as the different refrangibility of light, and the theory of colours; both which are established beyond the possibility of being overturned. Newton begins by repeating some experiments first made by Grimaldi, and which it will be necessary to
mention. He made a very small hole in a plate of copper, and allowing a ray
of light to enter through this hole he suspended a hair in the ray. The shadow
of the hair received at a certain distance was much larger than it ought to have
been. Newton measured it, and found it 35 times broader than it ought to
have been by the simple divergence of the rays. Hence it was obvious that the
hair had acted upon the light, and pushed it out of its direction. Here then is
an example of bodies *repelling* light. That this increase of width in the shadow
might not be ascribed to an atmosphere surrounding the hair, he plunged it in
water contained between two glass plates, and found the effect not in the least
altered. Again, he took two sharp metallic edges, placed them parallel, and
brought them gradually within $\frac{1}{400}$th part of an inch of each other. The ray
of light that passed between them was now divided into two parts, which were
separated from each other, and thrown into the shadows of the respective
knives. Here it is obvious, that the light had been drawn out of its former
direction by the sharp metallic bodies, and that it had been drawn towards
them. This then is an example of light *attracted* by bodies. From an attentive consideration of the phenomena, Newton shows that these attractions and
repulsions, which are not sensible at moderate distances, become exceedingly
strong at very small distances, or in case of actual contact; and hence he infers,
that they do not follow the law of the inverse of the square of the distance,
but that they approach much more nearly to the inverse of the cube of the
distance.

From these data refraction is explained in the following manner. When a ray of light, moving in an oblique direction, approaches within the attracting distance of a denser body, it begins to be drawn towards, and of course describes a small curve concave towards, the body; and it enters the body in a direction nearer the perpendicular than it otherwise would have done. This curve direction continues, till it has advanced so far into the body as to be beyond the sphere of attraction of the medium which it had left when it entered the body; it then moves on in a straight line, but in a direction more approaching the perpendicular. Just the contrary happens when a ray of light passes from a denser to a rarer medium. And Newton demonstrates, that on these suppositions the sine of the angles of incidence and refraction, in the same medium, will always have to each other the same ratio.

There are two kinds of reflections; one by the posterior surface, another by the anterior of bodies. Newton demonstrates, that reflection is not produced by the elastic particles of light impinging against these surfaces. The reflection by the posterior surface is explained by the attraction of the body through which the ray passes, precisely as in the case of refraction. When the direction of the ray is so oblique that the posterior surface comes to be the apex of the curve, into which the motion of the ray is changed, then the smallest additional obliquity will oblige the ray to describe the other half of the curve, and so to be reflected in such a manner that the angle of reflection is equal to the angle of incidence. All these explanations are so happy, and explain the phenomena so completely, that there can be little hesitation in embracing them; but Newton does not appear equally fortunate in accounting for the reflection of light from the anterior surface of bodies. This he accounts for by supposing a repulsion between the body and the ray of light. In a mathematical point of view, these alternate attractions and repulsions of light by the same body do well enough; but when one begins to consider them metaphysically, it seems impossible to adopt them. Newton indeed, in one of his queries at the end of his Optics, suggests the action of a subtile elastic fluid as capable of accounting for all of these phenomena. But the question is too refined, and too incapable of solution, to be considered here.

Different persons, quite ignorant of mathematics, have, at different periods, started as adversaries to the Optics of Newton, and have published pretended demonstrations of the falsehood of the Newtonian principles. But in reality they are nothing else than demonstrations of the deplorable ignorance of these pretended reformers, who, in their knowledge of optics, were far behind what was even known in the time of Descartes. It is not worth while to notice their objections. They were chiefly Frenchmen; and the most remarkable of them were Gautier and Marat, who afterwards made so execrable a figure during the early period of the French revolution.

There is a curious branch of optics which has originated since the time of Newton, which we cannot with propriety avoid mentioning here. We mean the measurement of the intensity of the light of various bodies. Lambert, who has cultivated it with much sagacity, has given it the name of *photometry*, a name which has been adopted by those British writers who have treated of it Bouguer, a celebrated French mathematician and astronomer, may be considered. as the philosopher who laid the foundation of this branch of the science. He published an essay on the subject in 1729, entitled, *Essai d'Optique sur la Gradation de la Lumiere*. But he had prepared a much more complete work on the subject, which was published after his death by La Caille in 1760, under the title of *Traité d'Optique sur la Gradation de la Lumiere*. Bouguer showed, that if light passes through a transparent body, the rate of its diminution during its passage is measured by the ordinates of the logarithmic curve. Hence, if we know the diminution produced by a transparent body of a given thickness, we may easily calculate from a common table of logarithms the effect of every degree of thickness. Bouguer pointed out a very ingenious way of estimating the relative intensity of two different luminous bodies, by measuring their distances when the shadows which they cast were of equal intensities. We may mention some of the most remarkable results which he obtained. The light of the full moon he found to be to that of the sun, when at the same height above the horizon, as 1 to 300,000. But he deduces, in a very ingenious manner, that if the moon were all equally luminous with some parts of it, the degree of light given out by that luminary would be to that emitted by the sun as 1 to 90,000. Thus the moon only emits one-third of the light which it ought to emit. This is accounted for by the numerous spots upon her disc. He showed, that in winter, when the sun's height above the horizon was 19° 16', the intensity of his light was to that of the light emitted by him in summer, when at the height of 66° 11', as 1,681 to 2,500; or about ½ more in the one case than the other. This is not owing entirely to the obliquity of the rays, but to the greater atmospherical space which they pass through in the one case than the other. At the depth of 600 feet in the sea no sensible quantity of light penetrates. Air is 40,000 times more transparent than water. These are but a very small number of the curious particulars discovered by Bouguer; but we must refer to his treatise for further information on the subject: it is highly worthy the perusal of every person who takes any interest in photometry.

Euler, in the year 1750, published a curious paper on the light emitted from the sun, moon, planets, and stars. He made no experiments; but the method he employed to subject the light emitted by these bodies to calculation was highly ingenious. The following are the results which he obtained.

The light emitted by the full moon is one 374,000th part of that emitted by the sun.

The light of Saturn, in opposition, is 1,000,000 times more feeble than that of the moon.

The light of Jupiter, in opposition, is 46,000 times less than that of the moon, or 22 times greater than that of Saturn.

The light of Mars is 20,000 times less than that of the moon, or twice as great as that of Jupiter.

Venus, when most luminous, emits 3,000 times less light than the moon.

If the fixed stars be of the size of the sun, and emit the same quantity of light with Jupiter, they are 131,000 times further from the earth than the sun; and their parallax will be 3″.

Lambert published his treatise on this subject in 1760, under the title *Photometria; sive de Mensura et Gradibus Luminis, Colorum, et Umbræ*. It is a master-piece of sagacity, and contains a vast quantity of information of the most curious and important nature. He gives us the light emitted from the sun, moon, and planets; the relative light of different parts of the day and of twilight; shows the effect of different dilatations of the pupil of the eye; and gives a great number of the most curious and unexpected theorems; but they are all of such a nature that they could not be noticed without entering into much greater details than is consistent with the limits of this work. Of late years the subject of photometry has engaged the attention of Count Rumford, who has invented a new photometer, and ascertained a great number of curious facts which have been made known to the public in one of the volumes of the Philosophical Transactions, published since the commencement of the present century, and which therefore does not come under our consideration. Mr. Leslie has likewise occupied himself with the same inquiry, and has contrived a very simple and easily-managed photometer, depending upon the heat produced by the rays of light. This instrument, if its theory be accurate, will be of considerable service as a meteorological instrument, though it is not delicate enough to measure the light of the moon and planets, which produce no sensible degree of heat.

The optical papers in the Philosophical Transactions amount to 127; but of these there are 46 which may be fairly passed over as of little or no value, at least in the present state of the science. The remaining 91 are of a very miscellaneous nature, and of very various merit. We shall endeavour to give a view of the most important of them, omitting however those written by Newton and his opponents, because they have been already noticed in a preceding part of this chapter.

Optical papers in the Transactions.

1. The first paper we shall notice is an account of Villette's burning glass.* This glass was made at Lyons: it was about 30 inches in diameter, weighed a

Burning glasses.

---

* Phil. Trans. 1665. Vol. I. p. 95.

hundred weight, and its focus was at the distance of three feet, and was about the size of half a Louis d'or. Its effects as a burning glass were very powerful; setting fire to wood, melting all metals, and vitrifying those stony bodies that were exposed to its action. This glass was purchased from Villette by Louis XIV and is at present in the cabinet of the Jardin des Plantes at Paris. Villette afterwards made one of a much larger size which was purchased by the Landgrave of Hesse Cassel. But a much more powerful burning mirror was made by Tschirnhous in the year 1687. An account of it is also given in the Philosophical Transactions.* It was made of copper-plate about twice the thickness of the back of an ordinary knife, and though it was about three Leipsic ells in diameter it was very light. It was well polished, and exhibited all the effects of a common concave mirror. Its effect as a burning mirror was very powerful: no stony body tried being able to resist its action ; but being speedily melted into glass. Tschirnhous, not being satisfied with this mirror, got one made of glass, of three feet in diameter, which he contrived to polish by means of the requisite machinery. Its effects were very great as a burning glass, and have been described at length by different experimenters, in the Memoires of the French Academy. It was purchased by the Duke of Orleans, who made a present of it to the Academy of Sciences at Paris, in which city it is still to be found.

In the year 1747, Buffon contrived a kind of burning glass by means of 168 plane mirrors of six inches square each, and so placed, that the image of the sun, reflected by each, might coincide in the same place. By this instrument he burnt wood at the distance of 200 feet, and melted silver at the distance of 60 : thus showing the possibility of Archimedes burning the Roman fleet at the siege of Syracuse, which had been disbelieved as impossible by Descartes, and almost all preceding optical writers. There are no fewer than four accounts of this burning mirror of Buffon in the Philosophical Transactions ; the first by Tuberville Needham, the second by the Nicolini, the third by Maupertuis, and the fourth by Buffon himself.† The most powerful burning glass made in England was Parker's, by means of which Dr. Priestley was enabled to perform some of his most curious and important experiments. Unfortunately, it was sent by the British government as a present to the Emperor of China ; and of course cannot be expected to prove of any further service to the advancement of science.

Micrometers.   2. The next paper we shall notice is an account of Mr. Gascoigne's micrometer. Azout had announced, in a preceding number of the Transactions,† that he was in possession of a method of dividing a foot into 30,000 parts, and

---

* Phil. Trans. 1687. Vol. XVI. p. 352.

† Phil. Trans. 1747. Vol. XLIV. p. 493, 495, and Vol. XLV. p. 504.

‡ Phil. Trans. 1667. Vol. I. p. 373.

that he had applied this method to measuring the diameters of the sun and moon with greater exactness than had been formerly attained. This induced Mr. Townley to announce that Mr. Gascoigne had contrived a much more perfect instrument long before, and had used it for astronomical purposes, and that he himself had Gascoigne's original micrometer in his possession.[*] In a subsequent number of the Transactions, Dr. Hooke published a description and drawing of this micrometer :[†] it depended upon the divisions of a screw. Though micrometers have been greatly improved of late years, yet these papers still retain their value as curious historical documents.

3. The next paper we shall mention is a dissertation by Huygens, explaining Solar halos. the nature and origin of halos round the sun.[‡] This paper possesses great ingenuity; though the explanation given by Huygens, on account of his ignorance of the different refrangibility of the rays of light, is not quite accurate. He accounts for halos by supposing a number of small bodies floating about in the air, having an opaque nucleus, and a transparent outer ring, and he shews how such bodies must of necessity occasion a halo. What the nature of these bodies is, he does not attempt to explain; but he remarks that hail is often seen exactly similar to them.

4. The next paper we shall notice is the solution, or rather a number of Problem of solutions, of a problem of Alhazen, at that time very much celebrated among Alhazen. mathematicians. The problem is this; Having given a concave or convex speculum, as also the place of the eye and the visual point, to find the point of reflection. Huygens, at that time in Paris, sent a solution of it to the editor of the Philosophical Transactions, Mr. Oldenburg, in the year 1670. He sent it to Slusius, who informed him that the method was similar to one which he himself had fallen upon two or three years before, and he likewise sent a solution which he considered as easier. This produced a discussion between Huygens and Slusius; and various solutions were given by each. The whole of these interesting papers were afterwards published in the Transactions.[§]

5. The next paper deserving notice is an attempt, by Dr. Briggs, to explain vision. He endeavours to shew, that the upper part of the optic nerves consists of fibres which have a greater tension than those of the lower part, and he lays a great deal of stress on this circumstance. The only valuable observation in his essay is his reason why we see only single with two eyes. It is, he says, because the image in both eyes is formed on exactly the same part of the retina ; but wherever this does not happen, either from some disorder in the eyes, or from one of them being turned forcibly out of its natural direction, then we see double.[||]

---

[*] Phil. Trans. 1667. Vol. II. p. 457.   [†] Phil. Trans. 1667. Vol. II. p. 541.
[‡] Phil. Trans. 1670. Vol. V. p. 1065.   [§] Phil. Trans. 1673. Vol. VIII. p. 6119 and 6140.
[||] Phil. Coll. 1682. No. 6. p. 167, and Phil. Trans. 1683. Vol. XIII. p. 171.

6. The next paper deserving notice is a very curious one by Dr. Halley, *On the proportional Heat of the Sun in all Latitudes, with the Method of collecting the same.** It might have been placed under the head of *Meteorology*; but as the method employed by Dr. Halley depends upon optical principles, we thought it better to notice it here. He shews, that the heat communicated by the sun is always as the sine of his elevation, and as the length of the day; and he points out a method of determining it for a given day in every latitude. The following table exhibits the relative heat of the sun at the equinoxes and solstices, for every tenth degree of latitude, according to his calculation.

Lat.	Sun in ♈ ♎	Sun in ♋	Sun in ♑
0	20000	18341	18341
10	19696	20290	15834
20	18794	21737	13166
30	17321	22651	10124
40	15321	23048	6944
50	12855	22991	3798
60	10000	22773	1075
70	6840	23543	0000
80	3473	24673	0000
90	0000	25055	0000

It is obvious, that this table does not exactly coincide with the temperature of the different latitudes which it is meant to represent. If it did, the temperature at the pole, at the summer solstice, would be much greater than under the equator; which is quite inconsistent with experience. But there are two circumstances to be taken into consideration. The vast quantities of ice accumulated during winter, in the polar regions, prevent the sun from producing the effects which he otherwise would. Dr. Halley conceived cold to be a positive something, which acted in a way not yet understood; but there is no occasion for such a supposition; the accumulation is quite sufficient to account for the comparatively low temperature of the polar solstice. Another circumstance which Dr. Halley overlooked must have a considerable effect; namely, the obliquity of the sun's rays at the pole, and the greater quantity of atmospherical air and vapour through which his rays must in consequence pass. This must absorb a great deal of the heat which he would otherwise communicate.

7. The next paper is a very excellent one by Dr. Halley, giving formulas for the resolution of the problem of finding the foci of optic glasses universally.† This formula cannot well be understood without having recourse to a diagram, though it is now universally known to opticians; and has been considerably

---

* Phil. Trans. 1693. Vol. XVII. p. 878.   † Phil. Trans. 1693. Vol. XVII p. 960.

simplified. We may endeavour to render intelligible a single case of this problem. Suppose we have a concave mirror and a luminous object, it is required to find the focus of the mirror.

Let A B be the concave mirror, c its centre, and o the luminous object, it is required to find F, the focus. Let $oc = d$, $Gc = r$, $cF = x$, then by Halley's formula we have $x = \dfrac{dr}{2d + r}$ or $2d + r : d :: r : x$. When the rays are

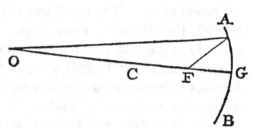

parallel, d becomes infinite, which reduces the formula to $x = \dfrac{dr}{2d} = \dfrac{r}{2}$. So that in the case of parallel rays, or when the mirror is presented to the sun, the focus is half way between the mirror and its centre. It may be demonstrated likewise that when the luminous object is placed in c, the focus also is in c. As o removes farther and farther from c, the focus advances nearer and nearer to the mirror, and when the distance of o is infinite, then F is always half way between c and G. When the mirror is convex, the same formula applies, only it becomes $x = \dfrac{dr}{r - 2d}$ and the focus in that case is manifestly virtual only.

In the case of lenses, if $r\, r'$ be the radii of the surfaces, and $\dfrac{m}{n}$ the ratio of the incidence and refraction of the luminous ray, then we have the formula changed into $x = \dfrac{drr'}{\left(\frac{m}{n} - 1\right)\left(dr + dr'\right) - r\,r'}$; and in the case of parallel rays

$x = \dfrac{rr'}{\left(\frac{m}{n} - 1\right)\left(r + r'\right)}$. In the case of glass we have $x = \dfrac{2r\,r'}{r + r'}$; and when the two radii of curvature are equal, as is commonly the case, then we have $x = \dfrac{2r^2}{2r} = r$, or the focus is the centre of curvature. In the plano convex, the formula becomes $x = \dfrac{dr}{\left(\frac{m}{n} - 1\right)d - r}$, and when the rays are parallel,

$x = \dfrac{r}{\frac{m}{n} - 1}$, and in the case of glass, $x = 2r$.

8. Mr. Stephen Gray described a very simple microscope, which deserves to be mentioned on account of the facility of its construction. It may be made in some measure at once, and therefore may be used in certain cases, where a more perfect one cannot be had. He took a thin brass plate, and made a hole in it the 30th of an inch in diameter. This hole was turned into two concave spherical surfaces on each side, and he filled it with a drop of clear water. The liquid assumed a spherical form, and answered the purpose of a micros-

Simple micros-cope.

cope.* His reflecting microscope cannot be considered as so happy a contrivance, and need not therefore be described.†

Refraction of
air.

9. Mr. Lowthorp made a very accurate experiment to determine the refracting power of air. The result was that if the sine of the angle of incidence be 100000, then the sine of the angle of refraction is 100036.‡ This coincides as nearly as possible with the result obtained by Newton, who found these sines to each other as 3200 : 3201.§ Lowthorp gives us also the proportion between the sines of incidence and refraction in water, as determined by Mr. Gascoigne, as the numbers 100000 : 134400. Newton's proportion is different : he found it to be 396 : 529, which gives us 100000 : 133585. This, no doubt, is the true ratio.

10. The next paper is a geometrical method of determining the colours and diameter of the rainbow from the given ratio of refraction, and the contrary, by Dr. Halley. It is an excellent paper, like all the rest furnished by that industrious and excellent philosopher, but not susceptible of abridgement.‖

Refraction of
liquids.

11. The next paper is a valuable set of experiments on the refraction of different liquids by Mr. Hauksbee. The following table exhibits the results which he obtained. The specific gravity of water is conceived to be 820.¶

Bodies.	Sp. Gr.	Ratio of refraction as 100,000 to	Bodies.	Sp. Gr.	Ratio of refraction as 100,000 to
Oil of sassafras	898	6475·8	Spirit of hartshorn	786	7468·3
——— turpentine	713·5	6441·8	——— vinegar	824·5	7468·3
——— bee's wax	662	6885·4	——— salammoniac	794·5	7475·2
——— caraways	752	6696·5	Succinic acid	825	7475·2
——— oranges	711	6741·2	Sulphuric acid	1510	7011·5
——— hyssop	769·5	6757·6	Nitrous acid	1166	7104
——— rosemary	747	6794·7	Aqua regia	987	7195
——— savine	789	6730·9	Aquafortis	1157	7120·5
——— oryganum	752	6770·2	Butter of antimony	1976	5941·3
——— pennyroyal	783	6730·9	Spirit of raw silk	916	7135
——— mint	780·5	6706·4	——— honey	716	as water.
——— spike	749	6807·3	Tincture of antimony	693	7294·3
——— fennel	798	6616·5	——— bark	720	7294·3
——— juniper	729	6757·6	——— balsam of Tolù	717	7219·3
——— cumin	766·5	6627·7	——— gum ammoniac	719	7257·3
——— tansey	757	6865·1	Vitreous humour of an ox's eye		as water.
——— dill	795·5	6582·7	Crystalline humour of ditto		6832·7
——— amber	783	6662·3	White of hen's egg		7401·3
——— cinnamon	828	6517·7	Jelly of hartshorn		7384·7
——— cloves	827	6606·8	Human saliva		as water.
——— nutmegs	759	6721·4	Human urine		7451·9
Spirit of wine	703·5	7287·9	French brandy		7338·6

* Phil. Trans. 1696. Vol. XIX. p. 353.   † Phil. Trans. 1696. Vol. XIX. p. 539.
‡ Phil. Trans. 1699. Vol. XXI. p. 339.   § Optics. p. 270.
‖ Phil. Trans. 1700. Vol. XXII. p. 714.   ¶ Phil. Trans. 1710. Vol. XXVII. p. 204.

12. The method of separating the different rays from each other by the prism completely not being understood, before the publication of Newton's Optics, in 1704, several persons who had attempted it could not succeed. This in particular was the case with M. Mariotte, who was induced in consequence to refuse his assent to the Newtonian theory. In a number of the Acta Eruditorum, Newton was requested to explain how the experiment should be made, and to remove the objections that had been started, in consequence of the want of success of other experimenters. This induced him to request Dr. Desaguliers to repeat the experiments as he had described them in his Optics, before the Royal Society. The request was complied with, and the accuracy of Newton's opinions fully confirmed. Dr. Desaguliers published three papers in the Transactions, describing at full length the method which he followed in repeating these experiments, and the success with which they were attended.* These papers deserve to be studied by every person who wishes to satisfy himself experimentally of the Newtonian theory of light and colours.

13. The next paper which we have to mention is a very important one, because it contains the first accurate table of the allowance to be made for atmospherical refraction in astronomical observations. This table was constructed by Newton, and the paper was written by Halley.† But, as astronomers are now in possession of tables of refraction, and of methods of calculating it still more precisely than this table, it is not necessary to transcribe it here. The following is the proportion which gives us an approximate value of this refraction for all heights. Let r = refraction, z = angle which a ray of light makes with the zenith line; then $33' : r :: \text{tang.} (90° - 33' + 3) : \text{tang.} \overline{z - 3r}$. This proportion only holds when the barometer stand at 30 inches, and the thermometer at 50°. When the barometer changes, let B = the actual height, then we may say $30 : B :: r : r'$ = the true refraction. When the thermometer changes, let t = actual height of it, then we may say $1 : 1 - (t - 50)\cdot00245 :: \frac{B}{30} r : r''$ = true refraction. So that the complete formula for refraction is $r'' = r + \frac{B}{30} + (1 - t - 50)\cdot00245$. The curve of a refracted ray nearly coincides with a circle whose diameter is seven times that of the earth.

14. We come now to the discoveries made by Mr. Dollond, which are the most important that have been made in optics since the time of Sir Isaac Newton. John Dollond, the author of these discoveries, was born in Spitalfields, London, in the year 1706. His father was a French refugee, who had left his country on the revocation of the edict of Nantz, and had established himself in Spitalfields as a silk weaver. His son John was brought up to the same

Atmospheric refraction.

Dollond.

---

* Phil. Trans. 1716. Vol. XXIX. p. 433 and 448; and Vol. XXXV. p. 596.
† Phil. Trans. 1721. Vol. XXXI. p. 169.

profession; but having a studious turn he contrived to get acquainted with mathematics, and afterwards to make considerable progress in astronomy and optics. Having married early, and being burthened with a family, he was obliged to bring up his eldest son Peter in his own profession. But the young man seeing the respect paid to his father for his knowledge by professional men, was induced to commence optician under his father's direction. Having succeeded in his undertaking, his father joined him in 1752; and, in consequence of his theoretical knowledge, soon became a proficient in the practical part of the art. His first attempt was to improve the eye glasses of refracting telescopes; and he contrived, by a combination of five eye glasses, to make telescopes greatly superior to those formerly in use. He gave an account of this improvement in the Philosophical Transactions.* He next made a very useful improvement in Savery's micrometers. Instead of employing two entire object glasses, as Savery and Bouguer had done, he used only one glass cut into two equal parts, one of them sliding laterally on the other. By which contrivance Mr. Short was enabled to apply it to the reflecting telescope with much advantage. Of this micrometer he likewise gave a description in the Philosophical Transactions.† These improvements gave him celebrity; and he was encouraged by the friendship and protection of the most eminent men of science in London.

He now engaged in a discussion, which at that time interested all the philosophers of Europe. Sir Isaac Newton had declared in his Optics, that all refracting substances diverged the prismatic colours in a constant proportion to their mean refraction, and drew as a conclusion that refraction could not be produced without colour, and consequently that no improvement could be expected in the refracting telescope. Every body acquiesced in the precision of Newton's experiments; but Euler was of opinion that his conclusion went too far, and maintained that in very small angles refraction might be obtained without colour. Mr. Dollond took up the other side of the question, and defended Newton's opinion with much ingenuity and skill; and drew as a conclusion from his arguments that, if the experiment be as Newton stated it, refraction could not be without colour. Mr. Dollond's letter, and Euler's answer, were both published in the Philosophical Transactions.‡

---

* 1753. Vol. XLVIII. p. 103.　　　　　† Vol. XLVIII. p. 178 and 551.

‡ Phil. Trans. 1753. Vol. XLVIII. p. 292. Euler informs us, in his letter, that he was led to his opinion by considering the structure of the eye, which he conceived refracted without any dispersion of the rays; and he thought that the eye was constructed of three different humours for that express purpose. But his hypothesis does not apply to the eye; and Dr. Maskelyne afterwards showed that the eye does not correct the dispersion of the rays; but, that the error is insensible, and that a more perfect structure would have been less adapted for the common purposes of life. Phil. Trans. 1789. Vol. LXXIX. p. 256.

But Mr. Dollond was not satisfied with bare reasoning on the subject. It occurred to him that it would be worth while to repeat the experiment of Sir Isaac Newton, upon which the presently received theory was founded. He took two thin glass plates, cemented them together at one of the edges, fitted brass plates to the ends, and filled this vessel with water, so as to make a water prism. Into this vessel he put a glass prism, with its reflecting angle uppermost. Thus he had two prisms, one of glass, and another of water, acting opposite ways; and by altering the angle of the water prism till an object seen through both prisms occupied its natural place, he made both to refract equally, so that they mutually destroyed each other's refractions. In this case, if Newton's principle were accurate, the object ought to be seen without colour; but, on the contrary, it was as much coloured as if seen through the glass prism alone. Hence it followed that Newton's principle was inaccurate; and that therefore refraction might be effected without inducing colour. To prove this, he put a glass prism, with a very small refracting angle, into his water prism; and altered the angle of the water prism, till objects seen through both were free from colour. In this case there was a considerable refraction. Having determined this important principle, his next object was to find whether there were different sorts of glass that possessed this difference in their dispersive power to a remarkable degree; and after trying various kinds, he found a remarkable difference between crown glass and flint glass in this respect. He found that two wedges of flint and crown glass, ground to such angles that the refraction of the flint was to that of the crown as two to three, when placed together, showed objects free from colour. Therefore a concave flint glass, and a convex crown glass lens, placed together, would act in this manner, supposing the focal distances of each inversely as the ratio of the refraction of the wedges. Upon trial, he was able, after overcoming various difficulties, to construct refracting telescopes in this manner greatly superior to any former ones.* To these new telescopes, Dr. Bevis gave the name of achromatic. They have been of great service to astronomy, by being applied to fixed instruments; and to navigation, by being applied to Hadley's sextant. For this important discovery, Mr. Dollond was presented with the annual gold medal by the Royal Society; though not yet a Member of that learned body. He was elected into the Royal Society in 1761, and appointed Optician to the King. But he did not long enjoy these honours; for on the 30th of November, of the same year, a fit of apoplexy terminated his life in a few hours, in the 55th year of his age.†

Achromatic glasses.

While upon this subject, we ought to mention a very excellent paper by Mr. Murdoch, in which he shows that Mr. Dollond's discovery is not inconsistent

---

* Phil. Trans. 1758. Vol. L. p. 733.   † Phil. Magazine. Vol. XVIII. p. 47.

with the doctrine of Sir Isaac Newton; but that a meaning had been affixed to his expressions which he never intended to convey.* Sufficient attention has not been paid by optical writers to this paper.

15. The next paper deserving notice, is a posthumous one, of Mr. Short, which he delivered sealed up to the Society, on the 30th of April, 1752. After his death it was opened by the council, and ordered to be printed. It is a minute description of the method which he employed for grinding the object glasses of refracting telescopes truly spherical; and must be of considerable importance to practical opticians.†

Objections to the opinion that light is a body answered.

16. The next paper is one by Dr. Horsley, to obviate certain objections started by Dr. Franklin, against the common opinion, that light is a body emitted from luminous substances. Dr. Franklin supposes, that if light were a body, its velocity is so great that it would have a momentum greater than that of a 24-pounder fired from a cannon; and that the sun, by the emission of such a vast quantity of matter, would shrink in its dimensions, and cease to act upon the revolving planets with the same energy. Dr. Horsley demonstrates that, on the supposition that the particles of light are sphericles, having a diameter equal to one millionth of one millionth of an inch, and a density three times as great as that of iron, the momentum of each will not be greater than that of a sphere of iron $\frac{1}{4}$th of an inch in diameter, and moving at the rate of less than an inch in 12,000 millions of millions of Egyptian years: that is, it will be absolutely insensible. He demonstrates likewise that, on the most unfavourable supposition possible, the sun could only lose $\frac{1}{13232}$ds of his matter in the space of 385,130,000 Egyptian years; and that at the end of that period the diminution of his bulk, and the alteration in the planetary motions, would be absolutely insensible.‡ Hence Dr. Franklin's opinions are obviously of no weight. But, even if the Newtonian doctrine of light be admitted, it is very conceivable that a great proportion of the light thus emitted is returned again into the sun. Newton himself was of opinion, that the light thus emitted gradually collected together, and formed masses of matter, which revolved round the sun. He conceived the comets to be masses of this kind, and that, after circulating for a given period, they at last fell into the sun's body, and thus restored all the loss of matter which that luminary had sustained. He even conceived that the very bright comet, which appeared in his time, and which is supposed to make a revolution in rather more than 500 years, would, from the extreme nearness of its approach to the sun, fall into it in the course of five or six revolutions; and the consequence, he supposed, would be such an increase of heat, as would destroy all the living beings on this earth.

* Phil. Trans. 1763. Vol. LIII. p. 173.   † Phil. Trans. 1769. Vol. LIX. p. 507.
‡ Phil. Trans. 1770. Vol. LX. p. 417.

There is a tendency among the British philosophers of the present day, to adopt the hypothesis of Huygens, or rather indeed of Dr. Hooke. This is chiefly owing to a paper by Dr. Wollaston, in which he shows, that the double refraction of Iceland crystal can be accounted for in a satisfactory manner on Huygens's principles, and on no other. This very ingenious paper is printed in a volume of the Transactions, which does not come under our review. Some curious discoveries of Malus, and some observations by La Place, in the two volumes of the Memoires d'Arcueil, lately published, have added still farther to the probability of this opinion. We must confess, however, that we think it impossible to get over the objections pointed out by Newton himself, in his answer to Dr. Hooke.

17. The next paper is a curious account of three brothers, who were born at Maryport, in Cumberland, who could distinguish the figure of objects perfectly well, and saw to as great a distance as other people, but could not distinguish colour. They knew the difference between light colours and dark colours, and never confounded white and black; but if two colours had the same degree of liveliness, they could not distinguish between them: thus green, red, and orange, they considered as the same.* This defect of sight is by no means uncommon. Mr. Dalton, of Manchester, who is himself affected with it, has given us a curious account of it in one of the volumes of the Memoirs of the Manchester Society. He mentions two other persons that he knew, who were affected with the same disorder. He conceives it to arise from the lens of the eye, or some of the humours, being tinged of a particular colour. *Persons incapable of distinguishing colours.*

18. The next paper is a very valuable one, by Mr. Mudge, giving directions for making the best composition for the metals of reflecting telescopes; with a description of the process for grinding, polishing, and giving the great speculum the true parabolic curve.† For this paper, Mr. Mudge was honoured with the annual gold medal of the Royal Society. The composition consists of two pounds of good copper, and $14\frac{1}{4}$ ounces of grain tin, melted together; and to prevent the speculum from being porous, it is always necessary to melt it twice over, giving it the second time just heat enough to melt it. Mr. Mudge's rules for grinding and polishing the speculum are excellent. But we must refer for an account of them to the paper itself. *Mirrors of telescopes.*

19. The next paper we shall notice is a very curious one, by Mr. Wilson, in which he proposes an experiment for determining, by the aberration of the fixed stars, whether the rays of light, in pervading different media, change their velocity according to the law which results from Sir Isaac Newton's ideas concerning the cause of refraction; and for ascertaining their velocity in every *Method of determining the velocity of the rays of light.*

---

* Phil. Trans. 1777. Vol. LXVII. p. 260.   † Phil. Trans. 1777. Vol. LXVII. p. 296.

medium whose refractive density is known.* This method is to employ a telescope, filled with water, and to observe with it the aberration of the stars in the
same way as Dr. Bradley did with his sector. Mr. Wilson demonstrates, that
if the light be accelerated in its passage through the water, its direction will
not be altered; and this sameness in the direction constitutes the expected
proof of the acceleration of light during its passage through the water, and
consequently of the truth of the Newtonian theory of refraction. Boscovich
had proposed a similar method of determining the point; but he had fallen into
an error respecting the alteration of the direction, which he conceived would
be produced by the water.

20. The last papers we shall notice are two very curious ones, by Mr.
Brougham, on the flection and reflection of light;† and one on the same subject by Prevost, of Geneva, opposing some of Mr. Brougham's conclusions,
and defending the Newtonian doctrine.‡ We regret that we cannot enter upon
the examination of these excellent papers. But to do them justice, or even to
render them intelligible to our readers, would require a length of discussion
totally inconsistent with the limits of this work. Here then we close our observations on optics. We acknowledge that various other papers would have been
entitled to notice; but we have still so much ground to travel over, and have
already exhausted so great a proportion of the limits allowed us, that any further details would be made at the expence of the sciences which we have yet to
notice.

CHAP. III.

OF DYNAMICS.

The term *Dynamics* is applied by natural philosophers to that preliminary
branch of their science which treats of the theory of moving bodies. Statics,
or the theory of equilibrium, is usually considered as a part of it, because it
depends upon similar principles. It is a very important branch of knowledge;
but it is so well understood, and treated so fully in all systems of natural philosophy, that we do not think it necessary to enter into details. We may mention, however, in a few sentences, the principal topics which enter into the

* Phil. Trans. 1782. Vol. LXXII. p. 58.
† Phil. Trans. 1796. Vol. LXXXVI. p. 227; and 1797. Vol. LXXXVII. p. 352.
‡ Phil. Trans. 1798. Vol. LXXXVIII. p. 311.

science called Dynamics. We shall then notice the few papers on the subject to be found in the Philosophical Transactions.

Motion is estimated by the *space* passed over in a given time, and the rate of Outline of dynamics. motion is called *velocity*. Let s denote the space, t the time, and v the velocity. Then, in uniform motions, we have 1. $s = tv$. 2. $v = \frac{s}{t}$. 3. $t = \frac{s}{v}$. These three simple theorems comprehend the whole doctrine of uniform motions. But motions are seldom uniform; they usually either increase or diminish. In such cases, philosophers have conceived the motion to be uniform during a very small time, and to receive a certain addition or diminution at the end of it. These additions or diminutions are called *increments*, or rather *fluctions*. And we have as before $\dot{s} = v\dot{t}$, $v = \frac{\dot{s}}{\dot{t}}$, and $\dot{t} = \frac{\dot{s}}{v}$. A body continues for ever either in a state of rest, or of uniform rectilineal motion, unless affected by some external cause. This cause is called a *force*, by writers on mechanics. Now forces are measured by the quantity of motion which they produce in a given time. When a body is acted on at once by two forces, if these forces be represented by lines meeting in an angle according to their direction, if we complete the parallelogram, of which the two lines constitute two sides, then the body will move in the diagonal of that parallelogram. Now these three lines constitute a triangle. Hence it follows, that if a body be acted on at once by three forces represented by the three sides of a triangle, it will remain in equilibrio.

The next branch of dynamics considers the collision of bodies: 1st, of such as are not elastic; 2d, of elastic bodies. In the first case, if A strike against B, whatever motion it communicates to B it loses itself, so that the quantity of motion continues unaltered. Suppose both bodies going the same way with different velocities, and let the velocity of A be a, and that of B, b; and let the common velocity after collision be x, then $x = \frac{Aa + Bb}{A + B}$. If the bodies move in opposite directions, then $x = \frac{Aa - Bb}{A + B}$. Supposing the bodies equal the formulas become $x = \frac{a + b}{2}$ and $x = \frac{a - b}{2}$. Or, in the first case half the sum, and in the second half the difference of the original motions.

If the bodies be elastic, they rebound after collision with a velocity equal to the stroke. In that case the velocity lost is double, and is represented by $\frac{2B(a - b)}{A + B}$. When the forces are equal, the bodies rebound back with interchanged velocities. In elastic bodies, the difference of velocities is the same after as before collision; and in them, as well as in non-elastic bodies, the sum of the motions is the same after as before collision. Hence the following proposition may be demonstrated: the square of the velocities, multiplied into the bodies, is the same before and after the collision. This is what is called the *vis viva* so much attended to by some of the continental writers.

Gravity.

The most important of all the accelerated motions is that from *gravity;* as the motion of a heavy body falling to the ground. During the first second of time it falls very nearly 16 feet and one inch, or 193 inches. Its velocity increases regularly with the time: hence the motion of gravity is called an *equably accelerated* motion. If the time be represented by the line AB, and the velocity at the end of AB by the line CB, then the space passed over is represented by the triangle ABC. Now as triangles are to each other as the squares of their homologous sides, it follows that the space passed over is as the square of AB, or as the square of the time, which is the great characteristic of equably accelerated motions. The space is also as the square of BC, or as the square of the acquired velocity. Hence it is obvious, that if it had set out at first with the acquired velocity BC, it

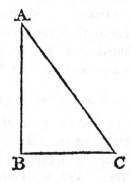

would have passed over double the space ABC. Hence the acquired velocity in a second is not measured by $16\frac{1}{12}$ feet, but by $32\frac{1}{6}$ feet. This velocity is usually represented by the letter g. The doctrine of equably accelerated motions is explained by the three following theorems, in which s, v, t, as before, represent the space, the velocity, and the time; and $g = 32\frac{1}{6}$ feet. 1. $v = gt$. 2. $s = \frac{g}{2}t^2$. 3. $2gs = v^2$. From these we may easily deduce every thing wanted respecting such motions.

When a body is thrown up from the surface of the earth it is uniformly retarded. Hence the same formulas apply to such motions as to uniformly accelerated motions. Let us suppose a body to be thrown up with a velocity c, and let t denote the time. If gravity did not act, in the time t, it would have moved ct: but during that time, by the action of gravity, it would have fallen $\frac{g}{2}t^2$. To obtain the real height we must subtract this quantity. It will be therefore $ct - \frac{g}{2}t^2$.

The short sketch which we have now given contains the elements of dynamics, stript indeed as much as possible of its mathematical dress; and we have been under the necessity of passing over those propositions which could not be rendered intelligible without entering into mathematical disquisitions. The first good system of dynamics that appeared was written by D'Alembert; and the best which we have ever seen in the English language is that which the late Professor Robison, of Edinburgh, inserted in the supplement of the third edition of the Encyclopædia Britannica, and which appeared in the first volume of his System of Natural Philosophy, a work unfortunately stopped almost at the outset by Dr. Robison's death. We shall now notice the most remarkable papers on dynamical subjects contained in the Philosophical Transactions. They amount in all to 40; but there are seven of these which can scarcely be considered as of any value, and some of the remaining 33 are rather analyses, or reviews of books, than original papers.

Papers in the Transactions.

1. The first three papers which occur in the Transactions contain the mathematical doctrine of the collision of bodies. This important part of dynamics was ascertained about the same time by Dr. Wallis, Sir Christopher Wren, and Mr. Huygens, who all communicated their theories of the subject to the Royal Society, within a very short space of time, and without any communication with each other. Dr. Wallis was first: his paper was read to the Society on the 29th of November, 1668. Sir Christopher Wren gave in his paper on the 17th of December, of the same year; and Mr. Huygens sent his on the 5th of January, 1669. So that the interval between them was very small.*

Collision of bodies.

2. The next paper deserving notice is a very valuable one by Dr. Halley on gravity and its properties. The law of the fall of heavy bodies was first determined by Galileo, and the subject was still further prosecuted by his pupil Torricelli. But Sir Isaac Newton was the person who brought it to perfection, and first demonstrated some of the propositions contained in our preceding sketch. Dr. Halley's paper was published before the appearance of the Principia, though it is plain that he was acquainted with many of the principles contained in that admirable book. Indeed he mentions it as almost ready for publication. Dr. Halley has fallen into some mistakes in this paper; though as a whole it possesses considerable value. It contains not only the general properties of gravity, but gives also the outlines of the mathematical theory of projectiles.†

3. The next paper is by Mr. Hauksbee, and is not strictly speaking dynamical, but consists of a set of experiments to prove the resistance of the air. I mention it here because the resistance of the air has considerable effect in certain dynamical experiments. Hauksbee found that a marble let fall on glass rebounded higher in vacuo than in common air, and higher in common air than in condensed air.‡

4. The next paper is by Mr. John Keill, entitled, *Of the Laws of Centripetal Force.*§ He begins by demonstrating the following theorem, which he informs us had been communicated to him by Dr. Halley, but discovered by Newton. If a body urged by a centripetal force move in any curve, then in every point of the curve that force will be in a ratio compounded of the direct ratio of the body's distance from the centre of force, and the reciprocal ratio

* Phil. Trans. Vol. III. p. 864, 867; and Vol. IV. p. 925.   † Phil. Trans. 1686. Vol. XVI. p. 3.
‡ Phil. Trans. 1705. Vol. XXIV. p. 1946.   § Phil. Trans. 1708. Vol. XXVI. p. 174.

of the cube of the perpendicular on the tangent to the same point of the curve, drawn into the radius of curvature of the same point. From this theorem, which he demonstrates, he deduces, in a very luminous manner, the laws of centripetal forces. But the paper is chiefly remarkable for containing the sentence about the discovery of fluctions, which gave so much offence to Leibnitz, and occasioned the famous controversy between the British and German mathematicians.

5. The next paper contains the repetition of the well known experiment of the guinea and feather falling from the top of an exhausted receiver, and reaching the bottom at the same time. Desaguliers, by joining together a number of glass receivers, contrived to make these bodies fall from the height of eight feet; and found, when the air was well exhausted, that both bodies fell to the bottom at the same instant of time.*

6. The Philosophical Transactions contain no less than 12 papers on the celebrated controversy, whether the force of moving bodies be proportional to their velocity, or to the square of their velocity. The authors of these papers are Dr. Pemberton,† Dr. Desaguliers,‡ Mr. Eames,§ Dr. Samuel Clarke,|| Dr. Jurin,¶ Dr. Reid,** and Mr. Milner.†† This, being a dispute about a definition, could not possibly be terminated, both parties being in the right, and both in the wrong. Newton's definition is simplest, and best adapted to mechanical philosophy, and therefore in every respect preferable to that of Leibnitz. Dr. Reid's paper above referred to explains the real nature of the controversy in a very luminous manner; and may be considered as sufficient to set the controversy for ever at rest.

7. In the next paper which deserves notice, Dr. Desaguliers demonstrates a property of the balance, which at first view appears paradoxical. If a man be counterpoised in one scale by a weight in the other, if he presses with a stick against the beam he will become heavier than the counterpoise. On the other hand, if the beam in which the man hangs be half way between the extremity of the balance, and the point of suspension of the balance, and if he presses against any part of the beam on the outside of the scale, he will become lighter than the counterpoise.‡‡ We likewise owe to Dr. Desaguliers some statical experiments described in a subsequent volume of the Transactions.§§

8. In the next paper Dr. Jurin gives a general theorem respecting the action of

* Phil. Trans. 1717. Vol. XXX. p. 717.   † Phil. Trans. 1722. Vol. XXXII. p. 57.
‡ Phil. Trans. 1723. Vol. XXXII. p. 269 and 285; and Vol. XXXVIII. p. 143.
§ Phil. Trans. 1726. Vol. XXXIV. p. 183; and Vol. XXXV. p. 843.
|| Phil. Trans. 1728. Vol. XXXV. p. 381.
¶ Phil. Trans. 1740. Vol. XLI. p. 607; 1745. Vol. XLIII. p. 423; and Vol. XLIV. p. 103.
** Phil. Trans. 1748. Vol. XLV. p. 505.   †† Phil. Trans. 1778. Vol. LXVIII. p. 344.
‡‡ Phil. Trans. 1729. Vol. XXXVI. p. 128.   §§ Phil. Trans. 1737. Vol. XL. p. 62.

springs, and deduces from it a great number of curious corollaries. The paper is not susceptible of abridgment.* The principle which regulates springs or elastic bodies was first pointed out by Dr. Hooke in this short sentence, *Ut tensio sic vis*, when he applied springs to regulate the motions of watches. The meaning of this sentence is, that the resistance of the spring is always proportional to the space by which it is driven from its natural situation. If a certain force remove it an inch from its natural situation, it will take twice the force to remove it two inches, and thrice the force to remove it three inches, and so on.

9. The next paper is a very elaborate one of Mr. Smeaton, entitled, *An experimental Examination of the Quantity and Proportion of Mechanical Power necessary to be employed in giving different degrees of velocity to heavy bodies from a state of rest.†* In this paper, Mr. Smeaton shows by experiment, that what he calls the *mechanical power* of a body increases as the square of the velocity. But we must carefully distinguish between this mechanical power and the Newtonian term *momentum* or quantity of motion. The one is measured by its instantaneous action, the other by its action for a certain time. The momentum by its definition is in the compound ratio of the mass of a body and its velocity, and therefore simply as the velocity in a given body; while the mechanical power by its definition is estimated by the mass compounded with the space it has described in acquiring its velocity. Now since the space fallen is as the square of the acquired velocity, it follows that this force must be as the square of the velocity in a given body.

10. The next paper is a very long and elaborate one by Mr. Landen, entitled, *A New Theory of the Rotatory Motion of Bodies affected by forces disturbing such Motion.‡* This is an excellent paper, like all the others communicated by that very eminent mathematician, and is of considerable importance in several astronomical discussions.

11. The next paper is by Mr. Bügge, Astronomer Royal at Copenhagen, and is entitled, *On the Theory of Pile Driving.§* This paper is very clearly written; but there are some mistakes in his assumptions, which prevents his ultimate conclusions from being quite accurate. His greatest error respects the friction, which he estimates too high, making it $(d+b)b$ instead of $(d+\frac{1}{4}b)b$.

12. The next paper is entitled, *An Investigation of the Principles of Progressive and Rotatory Motion.‖* Mr. Vince, the author of this excellent paper has carried the subject a good deal further than had been done by preceding mathematicians. There is also a set of experiments on the collision of bodies by Mr. Smeaton, in a subsequent volume of the Transactions,¶ in which that

* Phil. Trans. 1744. Vol. XLIII. p. 472. † Phil. Trans. 1776. Vol. LXVI. p. 450.
‡ Phil. Trans. 1777. Vol. LXVII. p. 266. § Phil. Trans. 1779. Vol. LXIX. p. 120.
‖ Phil. Trans. 1780. Vol. LXX. p. 546. ¶ Phil. Trans. 1782. Vol. LXXII. p. 337.

very ingenious practical engineer falls into some mistakes, from confounding his *mechanical force* with the Newtonian momentum.

13. We shall notice here a valuable paper by Sir George Shuckburgh Evelyn, on an invariable standard of weights and measures, because we do not know where else it can be better placed. Mr. Whitehurst had been employed, like Sir George, in endeavouring to determine an invariable standard of weights and measures, and the method, which occurred to him as the best, was to determine the difference between the length of two pendulums which vibrate with different degrees of velocity. He had constructed a machine for the purpose which Sir George Shuckburgh procured. The result was, that the difference between the length of two pendulums, which vibrate 42 and 84 times in a minute of mean time in the latitude of London at 113 feet above the level of the sea in the temperature of 68, and the barometer at 30 inches, is equal to 59·89358 inches of the parliamentary standard. From this all the measures of superficies and capacity are deducible. Agreeably to the same scale of inches, a cubic inch of pure distilled water, when the barometer is 29·74 inches, and the thermometer at 66°, weighs 252·422 parliamentary grains. From which all other weights may be easily deduced.[*]

CHAP. IV.

OF MECHANICS.

Explanation of the term.

THE term *Mechanics* is applied in different senses, or at least with very different degrees of latitude by different writers, sometimes comprehending almost the whole of natural philosophy, sometimes being confined to the constrained motions of bodies, and sometimes to the mechanical powers and machines. It will answer our purpose at present to take the word in the last sense; though we shall not scruple to class any paper in the Transactions, which happens to treat of the doctrine of constrained motion, if any such occur, under the present chapter.

The mechanical powers are ususally reduced under five heads; namely, the lever, the wheel and axle, the pulley, the wedge, and the screw. All these mechanical powers are mentioned by Pappus as known before his time, though

[*] Phil. Trans. 1798. Vol. LXXXVIII. p. 133.

as no preceding writer makes an enumeration of them, we do not know at what period or by whom they were discovered.

The *lever* is an inflexible rod moving about a fulcrum, and forces are ap- The lev r. plied to two or more points of it. The fundamental property by which it is distinguished is that, if it be supported on a fulcrum, and weights be suspended from its two extremities, these weights will be in equilibrio, if they be to each other inversely as their distances from the fulcrum. From this property the whole doctrine of the lever is easily deduced. Hence considerable pains have been taken to demonstrate it. It may be proper to mention that the lever is distinguished into two kinds, according to the position of the fulcrum; in the first kind the fulcrum lies between the two forces; in the second kind it is placed at one extremity of the lever. The common balance is nothing else than The balance. a lever with equal arms. A good balance should possess sensibility, correctness, and stability, properties which depend upon the following particulars. The arms must be of equal length : the friction of the beam about its axis should be the least possible : the beam should be exactly at right angles to the suspension : it should be so strong as not to bend under the weight, and it should be as long as possible : finally, the axis of suspension must be higher than the centre of gravity of the beam. Let a be one arm of the beam, b the distance between the centre of gravity and the point of suspension, l the load, w the weight of the beam, then the sensibility of the balance is represented by $\dfrac{a}{b(l+w)}$. Hence it is obvious that the sensibility increases with the length of the arm, and diminishes with the load, with the weight of the beam, and with the distance between the axis and the centre of gravity of the beam. But on the other hand, the stability diminishes as the length of the beam increases, and as the distance between the axis and centre of gravity diminishes; this prevents the sensibility from being carried beyond a certain point.

The *wheel* and *axle* is nothing else than a lever contrived to turn round an Wheel and axis, by which ingenious contrivance the greatest disadvantage of the lever is axle. got rid of; namely, the small height to which it can raise a weight. The wheel and axle consists of two wheels of very different diameters, attached to the same axis, the power is applied to the greater wheel, and the weight to be raised to the smaller. In order to produce equilibrio, the weight must be to the power inversely, as the diameter of the greater wheel is to that of the smaller.

The *pulley* is only a mechanical power when it is moveable. When there is The pulley. only a single moveable pulley, the equilibrium takes place when the power is to the weight as 1 to 2; and every additional pulley doubles the power. Hence if the number of pulleys be n, the advantage gained may be represented by 2^n.

The wedge.   The *wedge* is a triangular prism made of any tough materials. The fundamental property of the wedge, demonstrated by mechanical writers, is that, if
powers be applied perpendicularly to the faces, and to the back of a wedge,
they will balance each other, if they be proportional to the sides against which
they act. But it must be conceived that the directions of these forces are such,
that if the lines representing them be continued, they will meet in one point.
From this property it is obvious that the power of a wedge increases in proportion as it becomes sharper.

The screw.   The *screw* is nothing else than a wedge wrapped round a cylinder. We
generally use a lever to turn round the screw. Now, the power is to the resistance, as the circumference of the circle round which the lever moves to the
threads of the screw. The screw has this advantage over all the other mechanical powers, that it continues to act after the power is removed, in consequence
of the greatness of the friction.

There is one principle common to all the mechanical powers; namely, that
the power multiplied by its velocity is always equal to the weight multiplied by
its velocity. This is usually expressed by saying, that the *virtual* velocities are
always the same. Hence it is obvious that whatever is gained in power, is lost
in velocity.

These observations, which are quite elementary, will be sufficient for our purpose at present. For mechanics constitutes a very subordinate department in-
Papers in the   deed in the Philosophical Transactions. The papers on the subject amount to
Transactions.   51; but the greatest number of these are very insignificant. Scarcely more
than 13 or 14 can be considered as of any importance; and several even of these
are merely historical notices.

Watches.   1. The first papers that deserve notice are several by Huygens, Leibnitz, and
Hooke, on watches. Huygens was the first person who employed the pendulum
to equalize the motion of clocks, or watches, as he called them; and in order to
make all the vibrations of the same time, he made the pendulum vibrate between cycloidal cheeks. These watches were recommended by him for determining the longitude at sea; and several trials were made by British Naval
Officers, which turned out favourably. Though in process of time it was found
that the tossing of the ship, especially in stormy weather, was inconsistent with
the equal motion of these watches, and that they were constantly liable to stop
and be deranged. Hence it was speedily found that they could not be employed at sea with any advantage. This seems to have been the principal motive that induced Huygens to turn his attention to equalizing the motion of
watches by the application of a balance spring instead of a pendulum. This
happy thought occurred originally to Dr. Hooke, who communicated it to Lord
Brounker, and a plan was in agitation for some time to secure the profits of
the invention by a patent. But the scheme, (probably from the irritability of

Dr. Hooke's temper), was not put in practice, and Mr. Oldenburg affirms that none of the watches made upon Hooke's principle succeeded. Meanwhile Huygens thought of this method of regulating the motion of watches, communicated it to the Royal Society, published a description of his watch in the Journal des Scavans, and requested Mr. Oldenburg to insert this description in the Philosophical Transactions, which was complied with.* Upon this Dr Hooke published a pamphlet, entitled *A Description of Helioscopes, and some other Instruments, made by Robert Hooke, F. R. S.* In this publication he reflects upon Mr. Oldenburg for inserting Huygens' invention in the Transactions, without noticing that this invention was first found out by an Englishman, and long since published to the world; and complained on that account of *unhandsome proceedings.* Mr. Oldenburg inserted a vindication of himself in a subsequent number of the Transactions,† and took the insinuations of Dr. Hooke so much to heart, that he complained of them to the Royal Society, by whom his conduct was approved of, and Dr. Hooke was reprimanded for his insinuations of *unhandsome proceedings.* The proposal of Leibnitz‡ was also claimed by Dr. Hooke, though he acknowledged that he had never communicated it to any body, and therefore freed Leibnitz from all suspicion of having stolen it from him.

2. The next paper we shall mention contains a set of experiments to shew the advantage derived from making the fore wheels of a waggon as high as the hind ones.§ Dr. Wallis had previously pointed out the advantage of this practice, which is now well known; provided the axle of the wheels be not higher than the horse's breast, and provided the waggon is not to be drawn down a declivity.

3. The next paper which we shall mention is an account of some improvements in the crane, by Mr. Ralph Allen, Post Master of Bath, described by Dr. Desaguliers.‖ The crane is one of the most useful of all the simple machines. It is chiefly used for raising great weights to a certain height, and then lowering them down into a waggon or ship; or the contrary. The improvements described in the paper referred to are now so generally known, that we do not think it necessary to state them.

4. There are three papers in the Philosophical Transactions on friction, two Friction. by Dr. Desaguliers,¶ and one by Mr. Vince.** This last paper contains the best investigation of the subject which has yet appeared. The experiments

---

* Phil. Trans. 1675. Vol. IX. p. 272.   † Phil. Trans. 1675. Vol. X. p. 435.
‡ Phil. Trans. 1675. Vol. IX. p. 285.   § Phil. Trans. 1685. Vol. XV. p. 856
‖ Phil. Trans. 1729. Vol. XXXVI. p. 194.   ¶ Phil. Trans. 1732. Vol. XXXVII. p. 292 and 394.
** Phil. Trans. 1785. Vol. LXXV. p. 165.

3 E

seem to have been made with great care, and to be very accurate. He has ascertained by experiment, that friction is a uniformly retarding force; that it increases in a less ratio than the quantity of matter, or weight of the body; and that it increases with the surface of the rubbing bodies. Mr. Vince has shown also, by experience, that, when two bodies are allowed to remain at rest upon each other, they acquire a degree of cohesion which increases very much their apparent friction. In consequence of this, they require always a considerably greater force to put them in motion, than to keep them moving after motion has once begun.

When a body is placed upon a horizontal surface, it requires a considerable force to make it just ready to move upon the smallest impulse. This force is the measure of the friction. It varies from $\frac{1}{3}$d to $\frac{1}{4}$th of the weight to be put in motion. The smoother the surfaces the less the friction. It has been ascertained, that iron moves with less friction in brass than in any other metal. There are two sorts of frictions, that of bodies sliding on each other, and that of bodies rolling on each other. The first is much greater than the last. We may conceive friction to be occasioned by the hollows and protuberances of the bodies laid on each other, which enter into each other. Hence, before a body can move, its centre of gravity must rise a little: or, perhaps the protuberances yield a little; and a certain force is obviously necessary to produce this effect. When a body rolls upon another, the centre of gravity does not need to rise; but the one body is raised from the other: hence the reason of the less friction which takes place between rolling surfaces. We may here subjoin some of the most important practical remarks on friction that have been hitherto made.

Whenever motion is begun friction is very much diminished, and brought nearly to a half. The friction of wood upon wood is proportional to the weight; that of metal upon metal is proportional to the pressure. In woods the friction, when the bodies are in motion, is to that when at rest, as 2 to 9. But in metals the friction is not diminished by motion. The friction is less when one metal is made to rub against another, than when the same metals rub against each other. The greatest friction is that of a rope round a cylinder: it increases in a geometrical ratio for every turn of the rope. The great means of diminishing friction is by substituting the rolling for the sliding motion. Oils also are employed for the same purpose, and of all oils the best for the purpose is olive oil. This oil is apt to get thick in time, and this circumstance is the great cause which at present prevents time-pieces from going with the same accuracy almost as pendulum clocks. Black lead likewise has been found to diminish friction considerably.

5. There are two curious papers by Mr. Ellicott, on the effect which two

clocks had upon each other when placed very near. They were made as regulators, and on that account every part was constructed with very great care. The pendulums were heavy, and moved with much greater facility than usual. Mr. Ellicott found that when they were both set in motion together one of them speedily stopped. If the one that stopped was put in motion while the other was at rest, it soon began to act upon the pendulum at rest, and occasioned such a motion as soon set the clock going. This mutual influence was occasioned by the motion communicated from the one to the other by means of a rail against which both were leaning. If the clocks were removed from the rail, they ceased to act upon each other; but if a piece of wood was interposed between them, and wedged between both, the effect was greatly accelerated. The two clocks were found to go with unequal velocities; and the clock which went fastest was the one that was set in motion by the other.*

6 On the application of pendulums to clocks, it was conceived that the Pendulums: irregularities in their motions were owing to the unequal arcs described by the pendulum. This induced Huygens to suggest the application of cycloidal cheeks, which would completely put an end to every irregularity arising from unequal vibrations. But the irregularities still continuing, notwithstanding this improvement, it began to be suspected that they were owing to the effects produced upon the length of the pendulum by heat or cold. In the year 1726, Mr. George Graham gave a paper to the Royal Society on the subject, in which he mentions an idea that had struck him, that the irregularity from that cause might be prevented by making the pendulum rod of two metals, differing in their expansibility by heat; but, upon examining the expansion of metals by heat, he found them so nearly the same that he gave up all thoughts of a pendulum of that construction. He then describes a pendulum, which he contrived to answer the purpose by means of a column of mercury; which is usually known by the name of the quicksilver pendulum. In the year 1725, Mr. John Harrison, the celebrated inventor of the time-pieces, contrived a compensation pendulum, consisting of nine alternate wires of brass and steel, commonly known by the name of the gridiron pendulum. In the year 1752, Mr. John Ellicott gave a paper to the Royal Society, describing a compensation pendulum, which he had made by screwing a brass plate to an iron plate, and making the excess of the expansion of the brass above the iron lift up the ball of the pendulum, just as much as the iron had lengthened; by this contrivance keeping the pendulum always of the same length. This idea he informs us first struck him in the year 1732, and it was executed by him in the year 1738. Mr.

* Phil. Trans. 1739. Vol. XLI. p. 126.

Ellicott had found it necessary to make a set of experiments on the expansion of metals by heat. The following is the result which he obtained:*

Expansion of metals.

| Steel | 56 | Gold | 73 | Brass | 95 | Lead | 149 |
| Iron | 60 | Copper | 89 | Silver | 103 | | |

From this account, it appears that the persons to whom we are indebted for compensation pendulums, are Mr. Graham and Mr. Harrison; and that Mr. Ellicott contrived one on a principle a little different from that of Harrison.†

7. The next paper we shall notice, is an improvement on the common tackle of pulleys, by which the inequality from friction is prevented, and at the same time the pulleys so placed that the weight may be raised to a considerable height. The method is simple, but cannot easily be explained without a figure. Mr. Smeaton was the person to whom we are indebted for it.‡

8. There is a short investigation of the pressure of weights on moving machines, by Christian Hée, Professor of Mathematics and Experimental Philosophy, in the Marine Institution of Copenhagen.§ Such investigations are of little practical utility, on account of the effect of friction, which destroys all the mathematical deductions from the theory of machines. Mr. Hée demonstrates, that if we exclude the friction and the weight of the machine, the pressure in the case of the pulley is $= \frac{4\text{AB}}{\text{A}+\text{B}}$, A being the moving power, and B the weight to be raised.

9. Mr. Francis Blake has inserted in the Transactions a paper on the greatest effect of engines, with uniformly accelerated motion; and has shown that the effect is a maximum when the weight is to the power, as 0·618 to 1.‖

Lever.

10. Archimedes was the first person who attempted to demonstrate the fundamental property of the lever; but his demonstration is liable to certain difficulties, which have made writers on mechanics dissatisfied with it; and many mathematicians of the first eminence have attempted to substitute more satisfactory principles in its place. Thus Huygens, Newton, Maclaurin, Hamilton, and Landen, have all published demonstrations of their own. Mr. Landen's appears perfectly satisfactory, but is rather too tedious for so simple a subject. Mr. Vince has endeavoured, in a paper which he published in the Transactions, to remove the difficulties which have been started against the demonstration of

---

* Phil. Trans. 1751. Vol. XLVII. p. 479.
† Short. Phil. Trans. 1751. Vol. XLVII. p. 517. There is a curious paper on this subject also by Dr. Fordyce, in Phil. Trans. 1794. Vol. LXXXIV. p. 2.
‡ Phil. Trans. 1751. Vol. XLVII. p. 494.   § Phil. Trans. 1755. Vol. XLIX. p. 1.
‖ Phil. Trans. 1759. Vol. LI. p. 1.

Archimedes, and we think he has succeeded. He has rendered the demonstration perfectly satisfactory; and as it is the easiest of all the demonstrations of this important property hitherto proposed, perhaps it may, without impropriety, be considered as the best.*

11. The last paper we shall mention is a very useful one, by Mr. Atwood, for determining the times of vibration of watch balances, by investigations founded on the theory of motion.† The paper, from its nature, is not susceptible of abridgement; but deserves to be studied by those who wish to make improvements in time-pieces. Mr. Atwood has shown, that errors in the rate of going may arise from differences in the weights of the balance, too small to be determined experimentally.

CHAP. V.

OF HYDRODYNAMICS.

THE term *Hydrodynamics* has lately been applied, by writers on Natural Philosophy, to that branch of mechanical science which treats of fluids. Now fluids, as is well known, are of two very different kinds; namely, the *incompressible* and the *compressible*, or, *liquids* and *aerial fluids*. These possess properties so different, that it is necessary to consider them separately. We shall therefore divide this chapter into two sections. In the first we shall treat of *liquids*, as far as the papers in the Philosophical Transactions enable us to go. In the second we shall consider *elastic fluids*, or *airs*. {Division of the subject.}

SECTION I.—*Of Liquids.*

Liquids, in a mechanical point of view, may be considered in two states; a state of *rest* and a state of *motion*. This has induced mechanical writers to subdivide what relates to them into two parts, distinguished by the names of *hydrostatics*, which treats of liquids in a state of rest; and *hydraulics*, which treats of them in a state of motion. We ought to follow this example, were it not that the number of papers on the subject, in the Philosophical Transactions, is so small, that we may throw all the observations, which we can with propriety make on the subject, into one short section.

I. For the principles of the science of *hydrostatics*, we are chiefly indebted {Hydrostatics.}

---

* Phil. Trans. 1794. Vol. LXXXIV. p. 33.   † Phil. Trans. 1794. Vol. LXXXIV. p. 119

to Mr. Boyle, one of the original Members of the Royal Society, and one of the greatest ornaments of the age in which he lived. Robert Boyle was born in the province of Munster, in Ireland, on the 25th of January, 1627. He was the seventh son, and the 14th child of the Earl of Cork. After receiving the rudiments of education in Ireland, he was sent to Eaton, to be placed under the care of Sir Henry Wotton, who was the friend of his father. Here he continued three years, and afterwards his education was continued at home, by one of his father's chaplains. When very young he discovered, not only an excellent understanding, but a serious and religious turn of mind, and a strong desire to make himself acquainted with the original languages in which the Old and New Testament were written. This desire induced him to repair to the University of Leyden, in the year 1640, and he studied with such assiduity, that we are told by Bishop Burnet that he could quote every remarkable passage both in the Old and New Testament, in the original Hebrew and Greek. He left Leyden in order to make the tour of Europe; and during his travels through France, Switzerland, and Italy, he paid a good deal of attention to several of the Eastern languages. On his return from his travels, in 1655, he settled at Oxford, and devoted himself almost entirely to experimental researches. Chemistry, Pneumatics, and Hydrostatics, were the sciences which he chiefly cultivated, and which lie under the greatest obligations to him: but he made occasional excursions into the different departments of natural history, and every thing he touched felt the benefit of his experimental investigations. Few men were more respected during the period of their lives, or left behind them a more unblemished reputation. He was offered a Peerage by Charles II., but declined accepting it; and he refused to enter into orders, because he thought his religious writings would have a greater effect when they came from the hands of a layman, than if they had been written by a clergyman. His works were published in six enormous folio volumes; but they were afterwards abridged by Dr. Shaw, and published in three quartos, which, in consequence of their greater conciseness, while they contain all the facts discovered by Mr. Boyle, seem to be preferable to the original. He died on the 30th of December, 1691, in the sixty-fourth year of his age. We are chiefly indebted to Mr. Boyle for the air pump. The first rude machine of the kind indeed was contrived by Otto Guericke; but Mr. Boyle made it a much superior machine; and in his improvements he was assisted by the great mechanical genius of Dr. Hook

Boyle, not possessing any great extent of mathematical knowledge, could not be expected to add much to the hydrostatical investigations respecting stability, &c. Archimedes is the person to whom we are indebted for the establishment of hydrostatics on mathematical principles. His writings on that subject were more admired by his contemporaries than any other part of his profound investigations.

A FLUID is a body, the particles of which, when they are left to themselves, Fluid, what. are all in equilibrio, or are ready to move upon the smallest force being applied to any of them.   All the particles of fluids gravitate as well as those of solids. A perfect fluid ought to have no viscosity; but among liquid bodies we are acquainted with none that are entirely free from this property.   The following propositions comprehend the principles of hydrostatics.   The word *fluid* is used Outline of hydrostatics. in most of them, instead of *liquid*, because several of these propositions apply to aerial fluids as well as liquids; and stating such, generally, will save us some repetition hereafter.

1. The surface of every fluid when at rest is horizontal, or perpendicular to the direction of gravity.   It will at once be perceived, that when the extent of surface of the fluid is considerable, instead of being flat it will assume perceptibly the form of the segment of a sphere.   For example, if a pond extends two miles every way, it can be shewn that the centre is eight inches higher than the sides.   The quantity of curvature increases as the square of the arches described.   Hence in levelling, a correction is necessary for this curvature, and it is usually made in this way: If D be the distance in miles, two thirds of D^2 is equal to the correction in feet.

2. The fluid in a vessel being at rest, and subjected to the sole action of gravity, every particle of it is subjected to the same pressure every way: and the pressure is equal to the perpendicular column of water above the particle.

3. The pressure of water upon the sides of the vessel is equal to the greatest height of the water, without any regard to the extent of its upper surface. This is what is usuaslly called the hydrostatical paradox; because, in consequence of it, a very small quantity of water may be made to produce all the effects of a very great weight.   Mr. Bramah's very ingenious press is founded upon this property of fluids.   If the side and bottom of a vessel be equal, then the pressure on the bottom is twice as great as on the sides.   Hence, if we wish to inclose a pond of water by a wall, it it obvious that the thickness of the wall at the bottom must be greater than at the top, as it has a much greater pressure to withstand.   If the wall at the bottom by ⅔ths of the height of the water, it will just balance the pressure of the water.   A secure wall would require to be thicker.

4. When a body floats in water, it loses a portion of its own weight, just equal to that of the water which it displaces.   The same proposition holds if the body be plunged entirely under the surface; in which case the bulk of water displaced is just equal to the bulk of the body immersed.   This principle was first observed by Archimedes, and he founded on it the method of ascertaining the specific gravity of bodies, as at present practised.   Let w be the weight of any body in air, w' its weight when immersed in water, then w — w' is the weight of the water equal to the bulk of the body.   Let 1 be the specific gra-

vity of water, and s the specific gravity of w, then we have $1 : s :: w - w' : w$; which gives us $s = \dfrac{w}{w - w'}$, the specific gravity required. As the weight of water varies with the temperature, it is usual to take specific gravities at the temperature of 60°.

5. When a body floats in water it affects a particular position, and this position is such, that the line which joins the centre of gravity of the body and the centre of gravity of the immersed part is always vertical. Hence a body floating in water may have such a form as to have *no stability*, but to float indifferently in any position whatever. The form must be such, that the centre of gravity of the immersed part always retains the very same position, whatever part of the body is under water. This is obviously the case with a sphere: or a body floating in water may have a great deal of stability or tendency to return to a particular position: or it may have a particular position which it prefers, but from which it is very easily driven. Stability is obviously an important requisite in ship-building. Now the lower the centre of gravity of a ship is below the deck, the greater its stability; but if it be too low, the ship returns to the vertical position with such celerity as to endanger the masts.*

Papers in the Transactions. The papers on hydrostatics, in the Philosophical Transactions, amount only to 17; and of these there are no fewer than seven which may be passed over as trifling. We shall notice the most remarkable of the remaining ten: but nothing like the general principles of the science is to be expected in them; they contain only a few detached remarks.

1. The first paper contains a set of experiments, by Mr. Boyle, to show that water, when weighed in water, has exactly the same weight as when weighed in air. He blew a glass ball, with a long stem, rarified the air in it, and then sealed it hermetically. By attaching a piece of lead to this bubble, it was counterpoised under the surface of water, and the point of the stem being then broken off, water rushed in and filled it half full, and it became so much heavier than before that four drachms thirty-eight grains were requisite to restore the equilibrium. The water being expelled from the bubble, it was found to weigh four drachms and thirty grains; the remaining eight were accounted for by the pieces of glass broken off and lost, and by the water that could not be expelled from the bubble.†

Hydrometers. 2. Various instruments have been contrived for determining the specific gravity of liquids, and likewise of solids, immersed in liquids. These instruments have received the name of *hydrometers* and *areometers*. The first of these instruments is said to have been contrived by the celebrated Hypatia, the Egyptian

---

* There are two excellent papers in the Transactions on the stability of ships, by Mr. Atwood. Phil. Trans. 1796. Vol. LXXXVI. p. 46; and 1798. Vol. LXXXVIII. p. 201.

† Phil. Trans. 1669. Vol. IV. p. 1001.

mathematician, who afterwards lost her life in an insurrection of that superstitious people. There are various of these instruments described in the Philosophical Transactions: one by Mr. Boyle, with the mode which he employed in adjusting it.* He used it chiefly for determining the purity of metals, and did not attempt to obtain the real specific gravity, but merely a comparative value A much superior contrivance of the same kind is described by Fahrenheit, calculated only to determine the specific gravity of liquids.† The principles laid down in this paper have been followed in almost all the instruments of the same kind proposed since this paper was published. An instrument of the same kind made of copper, contrived by Mr. Clark, and intended for determining the specific gravity of spirits, is described in a subsequent number of the Transactions.‡ This instrument is very complicated and inconvenient; but it was till very lately, if it be not still, employed by the excise officers of this country for regulating the amount of the duties charged on spirits.

The methods of determining the specific gravity either of liquids or solids by such instruments as these, even when in the greatest state of perfection to which they have been brought, are greatly inferior in accuracy to the common hydrostatic balance for solids, and the weighing bottle for liquids. There is an adhesion between the particles of water and all other liquids, which prevents the instrument from acquiring much sensibility; and as this adhesion is apt to be overcome by a sort of jerk, it requires much time to reach even the degree of precision which such instruments are capable of attaining.

3. There are a variety of papers on the specific gravity of bodies in the Philosophical Transactions, which we do not consider as necessary to be transcribed, on account of the inaccuracies apparent in them. We shall therefore only mention them, and refer to the volumes where they may be found. The first paper is anonymous, and contains a naked list of the specific gravity of 77 bodies without any observations.§ The second is by Hauksbee, and gives the specific gravity of six metals, gold, silver, copper, brass, lead, iron. But the metals were obviously impure.‖ The third paper is by Fahrenheit, and contains the specific gravity of 29 bodies. The experiments appear to have been accurately made; but we have no proofs that the substances employed were pure.¶ The fourth paper is of very considerable value. It gives a detailed history of all that had been previously done by philosophers to determine the specific gravity of bodies, and then gives a copious table of specific gravities extracted from all preceding authors. This paper is drawn up by Dr. Davies, and deserves to be consulted even by modern writers. Many of the specific

Specific gravity of bodies.

* Phil. Trans. 1675. Vol. X. p. 329. † Phil. Trans. 1724. Vol. XXXII. p. 140.
‡ Phil. Trans. 1730. Vol. XXXVI. p. 277. § Phil. Trans. 1693. Vol. XVII. p. 694.
‖ Phil. Trans. 1712. Vol. XXVII. p. 511. ¶ Phil Trans. 1724. Vol. XXXIII. p. 114

gravities indeed are inaccurate; but the gradual approximation to accuracy which may be seen in them is curious; while others have anticipated some of the most accurate modern experiments.* There is another paper on a subject connected with specific gravity which deserves to be mentioned; though the experiments which it contains are not quite so satisfactory as they might have easily been made. The object of Dr. Wilkinson, the author of the paper, was to determine how much cork was necessary to buoy up a man in water. For this purpose he attached pieces of cork to a certain weight of lead, and ascertained the buoyancy of that substance both in river and sea water. There were considerable differences, as might be expected, from the diversity in the compactness of different pieces of cork. The next object was to determine how much cork would buoy up a man in river water. A thin man was taken five feet two inches high, and weighing 104 pounds. It was found that 12 ounces, five drachms, and two scruples, or 6100 grains of cork, were just sufficient to buoy him up in the river Thames. Such a quantity of cork was found capable of buoying up 60 ounces, three drachms, and 21 grains.† If these data be accurate, it follows that the specific gravity of the man subjected to this trial was 1·0508; which was unusually great. Many persons, indeed the greater number, are specifically lighter than water, and would have required no cork whatever to buoy them up. From this experiment then we may safely conclude, that a cork jacket weighing 15 ounces will be sufficient to give abundance of buoyancy to any man whatever.

Water compressible.

4. The last paper which we have to mention, connected with this subject, is a very ingenious one by Mr. Canton, in which he shows by a simple, but conclusive set of experiments, that water, and indeed all liquids, possess elasticity, and are compressible. His method was to fill a glass ball, having a slender tube attached to it, with the liquids to be tried, and to put them under the receiver of an air pump. When the receiver was exhausted, he observed how much they increased in bulk. This increase was owing to the expansion produced by the removal of the pressure of the atmosphere. It disappeared when the atmospherical air was let into the receiver. The following table exhibits the increase of bulk produced by the removal of the aerial pressure in all the liquids examined by Mr. Canton:

	Millionth parts.	Sp. gravity.
Alcohol	66	0·846
Olive oil	48	0·918
Rain water	46	1·000
Sea water	40	1·028
Mercury	3	13·595

* Phil. Trans. 1748. Vol. XLV. p. 416.   † Phil. Trans. 1765. Vol. LV. p. 95

Mr. Canton observed, what it is difficult to account for, that water is more compressible when near the freezing point than when near the temperature of 60°. In all the other liquids the compressibility increased with the temperature. From this curious discovery it follows, that the sea, where it is two miles deep, is compressed by its own weight 69 feet two inches.*

II. There is a curious subject which has been usually arranged by natural philosophers under the head of hydrostatics, we mean *the suspension of water in capillary tubes.* As several papers upon this subject occur in the Philosophical Transactions, they come naturally to be noticed here. We shall begin by giving a brief statement of the phenomena; observing that the whole doctrine of capillary attraction depends upon experiment, and that almost all the experiments hitherto made on the subject are to be found in the Transactions. From them chiefly, of course, our information is derived.

Suspension of water in capillary tubes.

Water contained in vessels is observed to stand higher at the sides than in the middle of the vessel. If you diminish the size of the vessel, the concave surfaces gradually approach each other; and when the diameter of the vessel is reduced to about ⅟₄th of an inch, there is no level surface at all remaining. If the bore be still diminished, the water rises in the tube, and its height is always inversely as the diameter of the tube. These tubes, in consequence of the smallness of their bore, are called capillary tubes.

If two glass plates be placed in contact at one edge, and separated from each other at the opposite edge by a small wedge, and in this state be immersed in water, the water rises between them as it does in a capillary tube, and it rises the higher the nearer the plates approach each other. Hence the upper surface of this capillary water forms a hyperbola, having the bottom and the close edge of the glass plates as assymptotes.

When water has risen in a capillary tube plunged in a vessel of water, we may take the tube out of the vessel, and the water will still continue in it. If we incline the tube (supposing it wet within) the capillary water will move freely from one end of the tube to the other, but when it comes very near either extremity it shows some reluctance to advance any further.

However small the bore of the tube the water never rises to the top, unless we immerse the tube altogether under water. If you take a wide tube, terminating in a very fine capillary bore, the water will stand as high in it as if the whole tube were of the same capillary bore.

Such are the phenomena; let us now state the explanation of them that is considered by philosophers as most satisfactory. It is obvious that the force which sustains the water in the capillary tubes is proportional to the diameter of the tubes. Let r, r' denote the semidiameters of the tubes, and a, a' the re-

* Phil. Trans. 1762. Vol. LII. p. 640; and 1764. Vol. LIV. p. 261.

spective height of water, we have $r : r' :: a' : a$, aad therefore $ra = r'a'$. Now the areas of the tubes are $\Pi a r^2$ and $\Pi a' r'^2$. Let the weights of the water in each be w, w', then we have $w : w' :: \Pi a r^2 : \Pi a' r'^2$. Hence $w : w' :: r : r'$. This result naturally leads to the conclusion that the water is sustained in the tube by a certain attraction between the inner surface of the glass tube and the water; a supposition which explains the phenomena in a satisfactory manner. The water in the capillary tube will move easily along the tube, because it is equally attracted both ways, till it comes nearer the end of the tube than the distance at which glass can attract water; at this distance being drawn more strongly in one direction than in the other, it will not easily be induced to move nearer the extremity. When a tube is immersed in water, this capillary attraction diminishes the gravitation of the column of water immediately under the tube: hence it rises in the tube.

Papers in the
Transactions. The number of papers on capillary attraction in the Philosophical Transactions amount to ten; and, one only excepted, they all possess a certain degree of value. Mr. Hauksbee shewed that the phenomena are quite the same under an exhausted receiver as in the open air.* In another paper he described the rising of water between two glass plates, a phenomenon of which, he was the original discoverer. He likewise found that oil and alcohol rise between glass plates as well as water, and that plates of marble and of brass exhibit the same phenomena as glass plates. Finally, he ascertained that water would rise to the height of 32 inches, (the whole length of his tube), in a glass tube filled with ashes and well crammed.† In another paper he shewed that if a drop of oil of oranges be placed between two glass plates touching at one side, and raised a little from each other by a wedge at the other side, and laid horizontally on a table, the drop of oil will immediately begin to move towards the side where the glasses touch,‡ and the nearer it approaches to that side, the higher is it necessary to elevate that edge of the glass to make the drop stationary.§ Dr. Brook Taylor appears to have been the person who observed the hyperbolic figures of the upper surface of water between two glass plates touching at one side, and separated a little at the other.|| The experiment was repeated by Mr. Hauksbee,¶ who tried also the same experiment with alcohol instead of water;|||| and he varied the common experiment by trying it with success under an exhausted receiver.‡‡

Dr. Jurin, in consequence of a proposal made to him about a perpetual motion, was induced to examine the subject of capillary attraction. He endeavoured to demonstrate that the water was suspended by the attraction of the

---

* Phil. Trans. 1706. Vol. XXV. p. 2223.   † Phil. Trans. 1709. Vol. XXVI. p. 258.
‡ Phil. Trans. 1711. Vol. XXVII. p. 395.   § Phil. Trans. 1712. Vol. XXVII. p. 473.
|| Phil. Trans. 1712. Vol. XXVII p. 538.   ¶ Phil. Trans. 1712. Vol. XXVII. p. 539.
|||| Phil. Trans. 1713. Vol. XXVIII. p. 151.   ‡‡ Phil. Trans. 1713. Vol. XXVIII. p. 153.

upper film of the glass tube to which the water reaches.* In consequence of some objections made to his explanation, he was induced to draw up another paper upon the subject, in which he has given a detailed theory of capillary attraction.† He informs us that Sir Isaac Newton entertained the same opinion respecting capillary attraction that he himself did. His explanation is very ingenious, and some of his experiments have an imposing aspect. But his theory is now generally considered as erroneous, and the attraction which suspends the water is considered as placed in the lower surface of the tube, instead of the upper.

III. Hydraulics is a much more difficult science than Hydrostatics, and has by no means made the same progress. To attempt an intelligible outline of this subject would necessarily subject us to details incompatible with the nature of this work. A very few observations therefore will suffice. *Hydraulics.*

The object of Hydraulics is to consider water in a state of motion. Now water may move various ways; it may issue from an orifice in the bottom or the sides of a vessel; it may move along a pipe, or an open channel like a river; and it may be employed as a moving force to drive Hydraulic machines. Hydraulics is employed in considering each of these motions. *Its object.*

Water issues out of a hole at the bottom of a vessel with a velocity equal to that which it would have acquired by falling from the surface. When water issues out at the side of a vessel, it can be demonstrated that the quantity issuing out in a given time, is as the square root of the depth, and that the velocity, per second for example, is always equal to $8\sqrt{D}$, D being the depth. If the area of the orifice be represented by a^2, then the quantity of water issuing out in a second of time will be $8a^2\sqrt{D}$ in cubic feet: hence we have $a^2 = \frac{Q}{8\sqrt{D}}$, Q being the quantity of water issuing per second. Such are the theoretical results. But several circumstances prevent the issues of water in these cases from agreeing with theory. The principle of these is the friction of the particles of water against each other, and against the sides of the hole, which reduces the velocity from $8\sqrt{D}$, to $5\sqrt{D}$. The addition of a short tube to the hole at which the water issues, is found to increase the discharge considerably; and this tube acts with most advantage when it is of a conical form. The water, after it has gone a small distance from the hole, contracts its diameter very sensibly; an observation first made by Sir Isaac Newton, and quite obvious to any person who will make the experiment.

When water passes through pipes for any considerable distance, the quantity discharged is still farther reduced, owing to the great friction against the sides of the tube. Desaguliers found that when water passes for a mile or two through leaden pipes on uneven ground, the quantity of water discharged was

---

* Phil. Trans. 1718. Vol. XXX. p. 739.   † Phil. Trans. 1719. Vol. XXX. p. 1097.

reduced to $\frac{1}{10}$th part, so that the preceding formula in that case must be changed into $Q = \frac{8a^3 \sqrt{D}}{10} = \frac{4}{5} a^2 \sqrt{D}$. When the water flows in iron pipes, the deduction to be made is much greater. It is obvious that the quantity of friction must increase in all cases with the length of the tubes. Suppose the diameter of the pipe to be x, then a^2 (or the area of the pipe), $= \Pi . x^2$. Hence we have $\Pi . x^2 = \frac{Q}{8\sqrt{D}}$; and of consequence $x = \sqrt{\frac{Q}{8\,\Pi\,\sqrt{D}}}$. Such is the theoretical expression for the diameter of the tube; but experiment gives us a very different expression, from it we have $x = \sqrt{\frac{5Q}{4\,\Pi\,\sqrt{D}}}$. When the Magistrates of Edinburgh first brought water into that city by leaden pipes, they applied to Dr. Desaguliers, and to Mr. Maclaurin, to inform them what quantity of water they could obtain by means of a pipe of a given diameter. After the pipes were laid, the quantity of water discharged was measured and found to be only $\frac{1}{5}$th of Desagulier's estimate, and $\frac{1}{11}$th of the estimate of Maclaurin.

Motion of water in rivers. The motion of water in rivers was first accurately investigated by Du Buat. It exhibits the greatness of the friction in a very striking point of view; for the motion of rivers soon become equable: Hence the whole force of gravity must be nearly employed in overcoming the force of friction. The mean velocity of most rivers may be reckoned about four feet per second, or not quite three miles in the hour. The water next the bottom moves slowest, and the rapidity gradually and equably increases from the bottom to the surface, which moves with the greatest rapidity. The same thing holds from the centre of the stream to the edges of the river. When a river winds, its velocity is diminished, and the diminution is the greater the more abruptly the direction is changed. The greatest velocity of a winding river is usually at its concave edge, which is contrary to what one would naturally expect.

Hydraulic machines. Hydraulic machines are usually divided into three kinds; those that are put in motion by the *impulse*, by the *weight*, and by the *reaction* of water.

Unless the wheel go slower than the water, it will be able to do no work. If the velocity be too small, it will also do very little work. Hence there is some intermediate velocity which will give the greatest advantage. Theory gives us the greatest advantage when the velocity of the wheel is one third that of the stream. But the experiments of Mr. Smeaton, to whom we are indebted for the first accurate experimental investigation of the theory of the different kinds of water wheels, have shewn that the velocity of the wheel may be a little greater than $\frac{1}{3}$d of that of the stream.

Theoretical reasoning gives us the work done by such a wheel equal to $\frac{4}{17}$ths of the water expended. But practice gives us the double of this, or $\frac{8}{17}$ths, nearly $\frac{1}{2}$d of the water expended.

There are two kinds of water wheels; namely, the *undershot*, in which the Water wheels. water escapes below the wheel, and acts by pressure; and the *overshot*, which is moved by the weight of the water filling the buckets on one side, and of course turning the wheel round.   An overshot wheel must go slowly, otherwise part of the water is thrown off and lost by the centrifugal force.   This wheel is much more powerful than the undershot, the effect produced being equal to 2 thirds of the water expended.   The greatest effect is produced when the circumference of the wheel moves with a velocity of three feet per second.

The Hydraulic machines set in motion by the reaction of water, have the advantage of requiring very little apparatus; but a considerable fall is necessary to enable them to act with efficacy.   I do not know any example of such a hydraulic engine in Great Britain: though there are certainly many spots where they might be erected with advantage.   Suppose a cylinder A B terminating in a box, and moveable about an axis, D E.   Suppose a hole to be opened at c, if the whole be filled with water, the fluid issuing out at c, from the reaction of the column of water on the side opposite to c will cause the whole to move round.   At F there is another hole on the opposite side of the box.   There may be another box with similar holes crossing the box c F at right

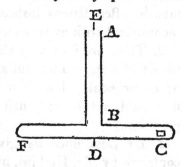

angles.   It is obvious that if a stream of water constantly enter at E sufficient to keep the cylinder and boxes full, the whole will turn round, and may be made to put in motion any apparatus proportional to the force with which it moves.

When a body moves through water, the resistance is rather greater than the square of the velocity; and it varies exceedingly according to the shape of the body in motion.   Thus the resistance to a sphere is much greater than to a sphere cut in two, with a cylinder interposed between them.   A long body with the hinder end pointed, is much less resisted than when that end is flat. The reason is, because the water displaced does not close in upon the body instantaneously, and of course the pressure from behind must be diminished as the flatness of that surface increases.*

The papers on Hydraulics in the Philosophical Transactions amount to 34. Papers in the Transactions. But of these there are no fewer than 15 of so little value, that we may without impropriety leave them out of consideration.   The remaining 19 are of various degrees of importance.   Some of the statements given in the preceding pages

* A very curious and valuable set of experiments on this subject have been made by Col. Beaufoy, F. R. S. of the Tower Hamlet Militia.   But unfortunately he has not thought proper to communicate his results to the public.

have been derived from them. We shall take a view of the contents of these papers in the same manner as we have done of those concerning the preceding branches of this subject.

1. We may mention two books on the subject of Hydraulics, of which an account is given in the Philosophical Transactions; the first is the *Traité du Mouvement des Eaux et des autres Corps fluides*, by Mariotte. This book contains some good experiments on the quantity of water discharged from a given orifice in the side of a vessel.* The other is entitled *De Motu Aquæ mixto*, and is written by Poleni, a Professor at Padua. It contains a very good historical account of the progress of discoveries relative to Hydraulics before the publication of this book. He deduces a theorem from a set of experiments which he made, and concludes from it that diminishing the breadth of rivers at the mouth, often lessens instead of increases their velocity. A proposition quite inconsistent with experience, and therefore inadmissible.†

2. There are four very elaborate papers by Dr. Jurin, in which he gives the theory of water issuing out at a hole from a vessel. The theory is not confirmed by experiments; but there is no doubt that it is very nearly just, and that it would not differ much from the result of experiments, if made with precision.‡

3. Dr. Desaguliers has given us a description of a pump for raising water, contrived by Mr. Hoskins, in which, instead of the common piston with leather, the vacuum is accomplished by means of mercury. In the common pumps the loss of water is about ⅓th at an average, and often more. In this new pump the whole water given by the capacity of the engine is discharged.§ But the great expence of mercury, and the probability that it would be speedily oxydized in such a situation, will ever prevent contrivances of this nature from coming into common use.

4. The discharge of water through long pipes is much smaller than through short. This difference is partly occasioned by the friction. But Dr. Desaguliers turning his thoughts to the subject, found that there was another cause; namely, the air confined in the upper part of the pipes, which diminished their bore, and of course might lessen the quantity of water discharged almost to any amount. With the assistance of his friends, he contrived a very ingenious apparatus for letting out this air, whenever it accumulated, to which he gave the very fanciful appellation of *Jack in a box*.‖

5. Mr. Beighton, a celebrated civil engineer of the time, has given us an accurate description of the water-works at London Bridge, which consist of a

---

* Phil. Trans. 1686. Vol. XVI p. 119.   † Phil. Trans. 1717. Vol. XXX. p. 723.

‡ Phil. Trans 1718. Vol. XXX. p. 748; and 1722. Vol. XXXII. p. 179; and 1739. Vol. XLI. p. 5 and 65.

§ Phil. Trans. 1722. Vol. XXXII. p. 5.   ‖ Phil. Trans. 1726. Vol. XXXIV. p. 77.

set of forcing pumps, driven by the tide, for raising water out of the river Thames, for the use of that part of the city which is in the neighbourhood. The quantity of water, which these pumps raise to the height of 120 feet, amounts to 1563 hogsheads in the hour; supposing a loss of $\frac{1}{4}$th of the quantity calculated from the capacity of the cylinders.*

6. Mr. Robertson has given us the formula for calculating the fall of water under bridges, in consequence of the diminution of the breadth of the current by the piers; and has applied his formula to London and Westminster bridges, the only two existing at London at the time when he wrote. The breadth of the Thames, at London bridge, is 926 feet. The sum of the water ways, at the time of the greatest fall, is 236 feet. The fall at London bridge is four feet nine inches. The breadth of the Thames, at Westminster bridge, is 1220 feet; but at the time of the greatest fall there is water only through the 13 large arches, which amount to 820 feet; to which, adding the breadth of the 12 intermediate piers, equal to 174 feet, we obtain 994 feet for the breadth of the river at that time. The fall amounts only to an inch.†

7. The next paper which we have to mention is a very valuable one, by Mr. Smeaton, entitled, An Experimental Inquiry concerning the natural Powers of Water and Wind to turn Mills and other Machines depending on a circular Motion.‡ This paper is so very long, that we cannot attempt giving a general view of its contents. It was from it that the first accurate views respecting overshot and undershot wheels were obtained. The principles laid down at the commencement of this article were derived from this paper.

8. The next paper we shall mention is a dissertation by M. Mallet, of Geneva, on the most advantageous construction of water wheels.§

9. Mr. Whitehurst has given a curious account of a contrivance for raising water, employed at Oulton, Cheshire, the seat of Philip Egerton, Esq. The water was contained in a reservoir, from the bottom of which there passed a pipe to the kitchen, sixteen feet below the reservoir. This pipe had two extremities; one of them furnished with a stop-cock, was for the use of the kitchen; the other, furnished with a valve, terminated near the bottom of a stout vessel containing air. From the bottom of this air vessel, there passed a tube to another reservoir, higher than the original reservoir, and destined for the brew-house. When water was drawn for the kitchen, the water in the pipe acquired, by running, a considerable velocity. Hence, when the stop-cock was shut, it acted on the valve, forced it open, and rushing into the air vessel, compressed the air which it contained This happening every time that water was drawn for the use of the kitchen, which was very frequently, the

---

* Phil. Trans. 1731. Vol. XXXVII. p. 5.   † Phil. Trans. 1758. Vol. L. p. 492.
‡ Phil. Trans. 1759. Vol. LI. p. 100.   § Phil. Trans. 1767. Vol. LVII. p. 372.

water made its way into the brew-house reservoir, and supplied it suffi-cientl .*

10. Major Rennel, so well known by his important geographical labours, has given us a very curious and entertaining account of the Ganges and Bur-rampooter; two mighty rivers, which, rising from opposite sides of the moun-tains of Thibet, run, the one west; and the other east, and afterwards meeting in Bengal, flow united into the sea, discharging a most enormous mass of water. The average breadth of the Ganges is about a mile; that of the Burrampooter, between three and five miles. When the Ganges swells, in the rainy season, the average rise is about 31 feet. The fall of the Ganges is about four inches per mile; and the river flows at the rate of about three miles in the hour; but in the rainy season the rate is increased to six miles in the hour. The average quantity of water discharged by the Ganges into the sea, is 80,000 cubic feet per second; but, during the rainy season, the quantity discharged amounts to 405,000 cubic feet. The Ganges varies its channel very much during its course through Bengal, wearing away the banks on one side, while land is formed on the other side. It empties itself into the sea by eight great channels. Major Rennel was the original discoverer that the Sanpoo, of Thibet, is the same with the Burrampooter. Before that time, the Sanpoo had been supposed to discharge itself into the sea by the Gulph of Ava. To him, then, we are in-debted for our knowledge of the Burrampooter, as one of the largest rivers of Asia.†

11. The last paper we shall mention, consists of a valuable set of experi-ments, on the resistance of bodies moving in fluids; made by Mr. Vince. He shows, that the resistance is much greater than that given by theory; and hence deduces, very justly, that every theory of such resistances can only be deduced from experiment.‡

Section II.—*Of Pneumatics.*

Under the word *Pneumatics*, in our language, two different branches of Hydrodynamics are usually included; namely, the consideration of elastic fluids in a state of *rest*, and in a state of *motion*. Some persons, indeed, have lately distinguished the first branch by the name of *Aerostatics;* and the second, by that of *Pneumatics.* As they occupy only a very subordinate place in the Philosophical Transactions, it will not be necessary for our purpose to divide Pneumatics into two separate sections.

Aerostatics.

I. Galileo may be considered as the founder of this science; for he was the first person that supposed the air to have weight. This truth was established by

* Phil. Trans. 1775. Vol. LXV. p. 277.   † Phil. Trans. 1781. Vol. LXXI. p. 87.
‡ Phil. Trans. 1798. Vol. LXXXVIII. p. 1.

his pupil Torricelli; and Pascal showed how the barometer might be applied to the mensuration of heights. Mr. Boyle seems to have been the first person who determined the law of the elasticity of air, by direct experiment. Its weight was nearly determined both by Boyle and Hauksbee; though perhaps Sir George Shuckburgh's determination is the only one that can be securely relied on.

The number of elastic fluids, at present known, amounts to about 24. But, as in their mechanical properties they all resemble atmospheric air very closely, that fluid has been chosen, by mechanical experimenters, as the one upon which almost all their trials have been made; because it is most easily procured, and always exactly in the same state of purity.

The atmosphere is an invisible fluid, which surrounds the earth to a height *The atmos-* considerably exceeding that of the highest mountain. It presses the same in all *phere.* directions, as is the case with water. A stratum of air is pressed upon by all the air above it. When a body is plunged into air, it loses a portion of its weight, equal to that of the air displaced.

According to the experiments of Sir George Shuckburgh, at the temperature *Weight of air.* of 60°, and when the barometer stands at 30 inches, a cubic inch of atmospherical air weighs 0.305 grains. Hence 100 cubic inches weigh 30.5 grains. It has been observed, that at the temperature of 32°, when a barometer is elevated 1.448 fathoms, the mercury sinks $\frac{1}{100}$th of an inch. Hence we have the weight of air at 32 to that of mercury, as $\frac{1}{100}$ inch to 1.448 fathom, or as $\frac{1}{100}$: 104.256, or as 1 : 10425.6. But the density of mercury at 32° has been considered as 14.019. Hence it is easy to compare the weight of air and of water with each other at 32°. It follows, from the preceding numbers, that air at that temperature is $\frac{1}{741.73}$ of the weight of water. But Sir George Shuckburgh's method appears susceptible of greater accuracy than this, and therefore more to be depended on.

The elasticity of air is such, that its bulk is always inversely as the pressure; as was first determined by Mr. Boyle. Let b, $b + \text{P}$ be the compressing forces of air; D and D$'$ its density under these pressures; E and E$'$ its elasticities under the same circumstances; s and s$'$ the spaces occupied: then this law gives us the following analogies:

1. $b : b + \text{P} :: s' : s$
2. $\text{D} : \text{D}' \qquad :: s' : s$
3. $b : b + \text{P} :: \text{D} : \text{D}'$
4. $\text{E} : \text{E}' \qquad :: b : b + \text{P}$
5. $\text{E} : \text{E}' \qquad :: \text{D} : \text{D}'$

This property of air has furnished the best means hitherto thought of, as well *Method of* as in most cases the most convenient, of ascertaining the height of mountains. *measuring heights by the barometer.*

It was first shown by Pascal, that a barometer, when carried to the top of a mountain, descends. The quantity of descent was soon perceived to depend upon the height, and hence it naturally occurred as a measure of that height. The first person that attempted to put this method in practice in Scotland, was Professor Sinclair, of Glasgow, so well known in the 17th century for some absurd speculations which he published. He carried a barometer to the top of Tinto, and estimated the height of that mountain simply by the Rule of Three: a method incapable of leading to any accurate result. Dr. Halley soon after showed the application of logarithms to the mensuration of heights by the barometer, and gave formulas; which, from ignorance of the effect of heat on the bulk of air and mercury, were not capable of giving accurate results, except at particular temperatures. These rules were put to the test of experiment in Switzerland, and shown to be inaccurate; and no less than three methods were proposed, by three different foreign mathematicians :- namely, Mariotte, whose method was antecedent to that of Dr. Halley; Cassini, whose method was published in the Memoires of the French Academy for 1705; and Scheutzer, whose experiments were made in 1709, in Switzerland. Notwithstanding all these attempts, the method still remained imperfect till Mr. Deluc, with that assiduity which has characterized his labours, during the long period which he has devoted to philosophy, with so much credit to himself and advantage to the world, made a set of very laborious experiments, to determine the correction for the temperature. These experiments brought the method much nearer to perfection. Deluc's formulas were adapted to the English measures, by Dr. Maskelyne; and Dr. Horsley published very simple formulas, founded on them; by means of which, the height of mountains might be determined by the barometer. Sir George Shuckburgh and General Roy set themselves, about the same time, to ascertain the accuracy of Deluc's corrections; and to determine the effect of heat upon air and upon mercury, with the utmost possible exactness. In consequence of the labours of these distinguished experimenters, the method has been brought to a degree of precision which answers all the purposes required. When the air is still, the error, if the observations be carefully made, will not exceed a foot or two on a height of 4, or 5000 feet; which is as near the truth as we can well be expected to reach. Mr. Playfair published an excellent paper on this subject, in the Transactions of the Royal Society of Edinburgh.

The formulas followed in this country are nearly the same with those given by Dr. Horsley, only corrected by the subsequent improvements of Shuckburgh and Roy. De Laplace has lately given a much more complicated formula, taking in the diminution of gravity, and some other corrections which we omit in this country. It has the appearance of being more correct than the British method; but in fact the only advantage it has over that method is in being much more complicated, and of course more difficult. Mr. Playfair, in order to compare them, took one of the Pyrenean mountains, which is given as the foundation of the French calculation, and calculating its height by the British formulas, the difference was only about half a foot; which, in a height of 8 or 9,000 feet, may be fairly estimated as of no importance whatever. The French seem to have adopted this new method, though inferior to ours, merely that they might not be beholden to England for the foundation of so important a thing as the mensuration of heights by the barometer.

I shall now give the method as at present practised in this country as briefly as possible. It has been demonstrated that the densities of the air at different heights follow a geometrical progression, supposing the heights to follow an arithmetical. Hence to find the height of a mountain we have only to determine the height of the barometer at the bottom and at the top, these heights representing the densities. Take the logarithms of these heights, and subtract the one from the other, and multiply the remainder by m, the product is the height in fathoms. Let the height of the barometer at the bottom of the mountain be b, and the height of it at the top β, then the height of the mountain is equal to m (L.b — L.β.) Now it has been ascertained by experiment that when the thermometer stands at 32° m is equal to 10,000. Therefore when the thermometer stands at 32° the rule is this. From the logarithm of the height of the barometer at the bottom of the hill subtract the logarithm of its height at the top, reckoning the first four figures of both of these logarithms integers, the remainder gives the height of the mountain in fathoms, and of course if we multiply by 6 we have the height in feet.

But this method is accurate only when the thermometer stands at 32°. At all other temperatures two corrections are necessary, one for the temperature of the mercury, and another for the temperature of the air. To observe the temperature of the mercury there is always a thermometer fixed to the barometer used as near the mercury as possible. The correction for the mercury is thus made. It has been observed that for every degree of cold (measured by Fahrenheit's scale) the column of mercury is shorter than it ought to be by 0 000105. Multiply 0 000105 by the number of degrees of cold, and let the product be multiplied by the height of the barometer. This product gives us the correction which much be added to the height of the barometer. Let the accurate height thus obtained be denoted by β'.

The correction for the temperature of the air is thus made. Observe the temperature of the air by means of a thermometer at the bottom of the hill, and then at the top; let these be t and t'; add them together, and divide the sum by 2. If the quotient be 32° no correction is necessary. If not, subtract 32, and multiply the remainder by 0·00244 (the mean of the change in the bulk of air occasioned by one degree of Fahrenheit according to the experiments of Shuckburgh and Roy.) This expressed symbolically is $\left(\frac{t+t'}{2} - 32\right) 0\cdot00244$. Thus it appears that at all other temperatures besides 32° the height of the mountain is obtained by this formula $\left(1 + \left(\frac{t+t'}{2} - 32\right) 0\cdot00245\right) \mathrm{L}\frac{b}{\beta}$. This very simple formula gives us the height in fathoms by reckoning the first four figures of the logarithms integers.

Height of the atmosphere. If the atmosphere were homogeneous its height would be 26062·5 feet, which is nearly five miles. But we are certain that it greatly exceeds that height. It begins to refract light at the height of about 50 miles. At the height of three miles the barometer stands at 15 inches, while its mean height at the sea shore is 30 inches. So that at that height we have passed through one half of the atmosphere in point of quantity.

Sir Isaac Newton first demonstrated, that the particles of air repel each other with forces inversely as their distance from each other. This property obviously gives us the law of the elasticity and density of the atmosphere under different pressures, and at different heights.

Steam. There is another elastic fluid besides air, which requires to be mentioned here, in consequence of its great importance as a moving power; we mean *steam*, or the vapour of boiling water. This elastic fluid possesses all the mechanical properties of air, but it differs in one essential particular: the application of cold makes it lose its elasticity, and converts it into water. Mr. Watt found that a cubic inch of water is converted into a cubic foot of steam, or its bulk by this change is increased 1728 times. Hence steam is to air nearly as 5 to 11. or more accurately, if the specific gravity of air be reckoned 1, then that of steam will be 0·467.

The elasticity of steam increases with the temperature; at 212° it is equal to that of the atmosphere, or capable of balancing a column of mercury 30 inches high. According to the common opinion the elasticity of steam doubles for every 30° of additional heat; so that at 242° it is equal to that of two atmospheres. But Mr. Dalton's experiments give the increase of elasticity somewhat less than this. According to him, at the temperature of 242°, the elasticity of steam is capable only of balancing a column of mercury 51¼ inches high. According to the ordinary opinion, if the elasticity of steam at 212° be 1, and its elasticity at any other temperature (t) be y; then $y = (1\cdot0234)^{t-212}$ or Log. $y = t - 212 \times$ Log. $1\cdot0234$.

Steam engine. Steam is employed as a moving power in the important mechanical engine known by the name of the steam engine. The first hint of this engine was given by the Marquis of Worcester in his Century of Inventions. But as far as we know, the first steam engine was constructed by Captain Savery. Newcomen greatly improved it by introducing the method of condensing the steam in the cylinder. But the great improver of it, the person to whom we are indebted for the general introduction of it into our mines and manufactories, is the celebrated Mr. Watt of Birmingham. He was employed, at Glasgow, to put in order a model of Newcomen's engine belonging to the university. He perceived the defects of this model, and, in endeavouring to correct them, gradually determined the properties of steam, and contrived his own engine, which may be considered as the finest example of mechanical invention upon a large scale that has ever been exhibited. So admirably indeed are the motions of this machine contrived, that they bear a closer resemblance to the motions of living bodies than to that of ordinary machines.

In Newcomen's engine the pressure did not exceed 7lbs. upon every square inch; in Watt's it amounts to 12lbs. Let the radius of the piston be s, then its surface is Πs^2, and of course the whole pressure upon the piston in lbs. is $12\Pi s^2$. Let L be the length of a stroke of the engine, as the best engines work both in going down and in coming up, to obtain the work done in lbs. we must multiply $12\Pi s^2$ by 2L. The product, or $24\Pi s^2$L, is the work done per stroke. If we multiply this by the number of strokes in a minute, we get the work done per minute in lbs. Mr. Watt estimates the amount of the work done by his engines by calling them machines of so many horse power, supposing that a horse is capable of raising 33,000lbs. to the height of one foot in a minute.

An engine of 40 horse power burns 11,000lbs. of Staffordshire coals in 24 hours, which amounts to 275lbs. for every horse power. For every bushel (or 84lbs.) of Newcastle coals consumed by such an engine it will raise 30 millions of lbs. one foot high, or it will drive 1,000 cotton spindles with all the requisite machinery.

II. The second part of Pneumatics, considers elastic fluids as a power to put bodies in motion. Its theory is still so imperfect that a few general observations are all that we can offer.

Air is the power which moves ships at sea. It acts by striking against the sails of the vessel; and every thing depends upon the direction of the sails, and upon the shape of the vessel. The shape and position of the vessel are of importance, chiefly because the resistance of the water to the ship's motion depends on them. Maclaurin first demonstrated, that when a ship's sails are properly trimmed it may be made to move faster than the wind. *Motion of ships at sea.*

The resistance of the air to the motion of the sails of a wind-mill, has been found rather greater than the square of the velocity.

But the most important moving force, among the elastic fluids, is the air generated by the inflammation of gunpowder. We do not know who the inventor of gunpowder was. It is mentioned by Friar Bacon, and its composition described in such a way, as to lead to the supposition that he had dis- *Gunpowder.*

covered it himself   It is said to have been first openly manufactured for sale by Swartz, a German, to whom the credit of the invention, on that account, has been sometimes given.   When gunpowder was first used, in the art of war, cannot be determined.   Roger Bacon, who died in 1292, does not appear to have been acquainted with this application of it.   Barbour, in his Life of Robert Bruce, informs us that guns were first used by the English, in the year 1320, in the beginning of the reign of Edward the Third.   There is reason to believe that cannons were used at the battle of Cressy, which was fought in 1346.   It is said, that there is a piece of ordnance at Amberg, in Germany, on which is inscribed the year 1303.   All these circumstances combined, render it almost certain that gunpowder began to be applied to the purposes of war at the very commencement of the 14th century.

Count Saluces first showed, that air is extricated from gunpowder when it is fired.   But it is to the experiments of Robins that we owe the first accurate information on the subject.   He found that the air, extricated from gunpowder by combustion, occupies 250 times the space of the powder.   He supposed the heat of burning gunpowder to be 1000°.   By this heat, air is expanded four times.   Hence it follows, that the elasticity of air, just extricated from gunpowder, is 1000 times greater than that of common air; or capable of sustaining a column of mercury 2500 feet high.   But, from Count Rumford's experiments, there is reason to conclude that the elasticity of air, from gunpowder, is ten times greater than this, or capable of supporting a column of mercury 25,000 feet high.

But the air extracted from gunpowder exerts this prodigious effect only for an instant.   It loses heat at a great rate ; and as its bulk increases by expelling the ball, its capacity for heat increases at the same time, which must contribute still more to lower its temperature.   Hence the effect which it exerts upon the ball, while moving along the barrel, is diminishing constantly at a great rate. The best experiments made upon this subject are those of Dr. Hutton, of Woolwich.   They were made upon a sufficiently large scale.   We shall therefore give a short summary of them in this place.

Dr. Hutton's experiments.   Dr. Hutton, in his experiments, employed a one-pounder.   The velocity was a maximum when the powder employed amounted to half the weight of the ball, and was then 2000 feet per second.   But the velocity depends, in some measure, upon the length of the piece ; for the shorter the piece the less the velocity.   If the weight of the powder be P, and that of the ball B, then the accelerating force is $\frac{P}{B}$.   As we do not know the elasticity, let us suppose this quantity, multiplied by an indeterminate quantity m ; and let L be the length of the gun, not occupied by the charge.   By the principles of Dynamics, formerly explained, we have $2gs = v^2$. Substituting the preceding quantities in this equation, we get $\frac{2m \text{ P}}{\text{B}} \text{ L} = v^2$. Instead of v^2 let us put c^2, c being 1700 feet per second, and we have $\left(\text{since } \frac{\text{P}}{\text{B}} = \frac{1}{7}\right) 2m \times \frac{1}{7} \times \text{L} = c^2$, or which is the same thing $m\text{L} = c^2$. Hence $\text{L} = \frac{c^2}{m}$ and $m = \frac{1700^2}{\text{L}}$. In Dr. Hutton's experiments L was 2·5 feet, hence $m = \frac{1700^2}{2·5}$. Substituting this value of m in the original equation, we obtain $\frac{2(1700)^2}{2·5} \times \frac{\text{PL}}{\text{B}} = v^2$, which gives us $v = 1700$ $\sqrt{\frac{4\text{PL}}{5\text{B}}} = 3400 \sqrt{\frac{\text{PL}}{\text{B}}}$. Hence, v being the initial velocity, we have always $v : v' :: \sqrt{\frac{\text{P}}{\text{B}}} : \sqrt{\frac{\text{P}'}{\text{B}'}}.$

We must not fill the gun with powder beyond a certain point, otherwise the effect is diminished. The initial velocity is the greater the longer the gun is: but the advantage is small after the piece has reached a certain length. The range, by theory, ought to increase with the powder, and to be proportional to the square of the velocity. But, in reality, it is only as the square root of the velocity. No advantage is gained by increasing the weight of the gun, or by altering the wadding, or by ramming the powder hard, or leaving it loose, or by firing it in different places; but the windage, or the difference between the diameter of the ball and that of the piece, occasions a very great disadvantage. When the difference amounts to $\frac{1}{20}$th, the loss amounts to about $\frac{1}{3}$d. Besides, when the windage is considerable, the direction is uncertain.

The loss from windage may be estimated as the square of the empty space between the ball and the piece. The change of direction in the ball may be owing to various causes, as the windage, the resistance of the air, the weight of one side of the ball being greater than that of the other, &c. The deviation in Dr. Hutton's experiments, owing no doubt to all these united causes, amounted to 15 degrees.

The penetration of balls into wood was found to be as the logarithms of their velocity. A very odd and unaccountable coincidence.

The resistance of the air was found to increase as the square of the velocity. A ball of the diameter 1·905 inches, when moving with the velocity of 1000 feet per second, was resisted by a force equal to 350 ounces. Its own weight was about 16 ounces, so that the resistance was 20 times the weight of the ball. Had it moved with double the velocity, the resistance would have been 80 times the weight of the ball. Let d be the diameter of the ball, v the velocity, r the resistance of the air. We have $2^2 \times 1000 : d^2 v^2 :: 350 : r$. Hence $r = \frac{350 d^2 v^2}{4000}.$

Papers on
pneumatics in
the Transac-
tions.

III. The papers on pneumatics, in the Philosophical Transactions, amount to 57.   But of these there are no fewer than 34 that may be considered as of little or no value.   Even among the remaining 23, the importance of several may be fairly questioned.   It is needless, therefore, to observe, that pneumatics occupies a very subordinate place in the Philosophical Transactions.   A few remarks on some of the more important papers is all that will be necessary here.

Barometer.

1. The barometer was an instrument that proceeded directly from the Torricellian experiment, and is now so familiarly known to every body, that it is viewed without surprise.   But, at the time when the Royal Society was founded, this was far from being the case.   Accordingly, we find many papers on it in the early volumes of the Transactions; some endeavouring to account for its rising and falling, and some describing improvements in it.   Thus we have observations on the barometer, by Dr. Beal;* the wheel barometer, by Dr. Hook;† on the capillary attraction, observable in narrow tubes, by Huygens‡ and Wallis;§ a contrivance for measuring the height of the mercury in the barometer, by a circle on one of the weather plates, by Dr Derham,‖ and by Mr. Gray;¶ the marine barometer, by Hooke;** a new barometer, by Mr. Caswell, of Oxford;†† and some others, which it is needless to mention, because it has been found that the common barometer, well purged of air, and accurately divided into inches and decimals, yields as good a method of measuring the variation as any other, while it is free from several inconveniences to which they are liable.

2. Several of the Members of the Royal Society ascertained, by experiment, that air is compressed into half its bulk, by the pressure of a perpendicular column of Thames water, 33 feet high ‡‡

Mensuration
of heights.

3. In the Philosophical Transactions, for 1686,§§ we have Dr. Halley's paper on the method of measuring heights by the barometer.   He shows, that the heights are as the logarithms of the heights of the mercury; and gives two tables, showing the altitude of the place according to the height of the mercury, and the heights of the mercury at given altitudes.   These are remarkable, as being the first tables of the kind that ever were published; and though Halley's method, from his not being aware of the corrections necessary for the temperature, was necessarily inaccurate, yet his paper deserves to be remembered, as the first proper commencement of the method of measuring heights by the barometer.   He ascribes the variations in the height of the barometer in this

* Phil. Trans. 1666. Vol. I. p. 153.   † Phil. Trans. 1666. Vol. I. p. 218.
‡ Phil. Trans. 1672. Vol. VII. p. 5027.   § Phil. Trans. 1672. Vol. VII. p. 5160.
‖ Phil. Trans. 1698. Vol. XX. p. 45.   ¶ Phil. Trans. 1698. Vol. XX. p. 176.
** Phil. Trans. 1701. Vol. XXII. p. 791.   †† Phil. Trans. 1704. Vol. XXIV. p. 1597.
‡‡ Phil. Trans. 1671. Vol. VI. p. 2192 and 2239.   §§ Vol. XVI. p. 104.

paper to variable winds.   Dr. Halley carried a barometer to the top of Snowdon, and observed its height, and afterwards at the bottom of that mountain,
and he proposes the difference which he found, namely, 3·8 inches, to serve as
a standard in other cases, the height of Snowdon having been determined geometrically to be 1240 yards.*

4. Dr. Papin relates an experiment, from which he deduces the velocity with
which air rushes into a vacuum, or rather the method of calculating that velocity. He makes it 1305 feet in a second.†   But his numbers are not quite
accurate.   Besides, we know that this velocity must vary with the temperature,
an element which at that time could not be introduced into the calculus.

5. Dr. Wallis published an elaborate paper on the resistance of air to bodies
moving in it.‡   But he founds the whole of his reasoning upon the principle
that the resistance is proportional to the velocity; a principle which is erroneous, Newton having demonstrated that the resistance is as the square of the
velocity.   This error vitiates his whole reasoning, and destroys the value of his
paper.

6. Dr. Derham, by experiments made at the monument, found the mercury in
the barometer to sink $\frac{1}{10}$th of an inch by an ascent of 82 feet.§   The experiment
is not susceptible of accuracy, because the height corresponding to $\frac{1}{10}$th inch
varies with the temperature of the air.   As Derham's experiment was made in
September, we may conceive the thermometer to have stood at 65°; in that case
the height ought to have been 93 feet.   But Derham's barometer being a simple tube with slips of paper pasted on, the change in the height of the mercury
could not be estimated with much precision.

7. Mr. Hauksbee showed by an experiment made before the Royal Society,
in the month of May, that the weight of air is to that of water as 1 to 885.‖
This result makes air considerably lighter than it is in reality.

8. Dr. Desaguliers relates several experiments to estimate the resistance of the
air to falling bodies, by measuring the time taken by bodies of very different
densities to fall from the same height.¶

9. In the Transactions for 1728, Dr. Scheuchzer of Switzerland gives a long
historical account of the attempts made before that time to determine proper
formulas for measuring heights by the barometer.   The paper is only valuable
as an historical document.**

10. Mr. Boyle's air pump continued in general use till about the year 1751,
when Mr. Smeaton contrived one upon a much better principle, and capable of
carrying exhaustion much farther.   He has published a detailed account of this

* Phil. Trans. 1697. Vol. XIX. p. 582.   † Phil. Trans. 1686. Vol. XVI. p. 193.
‡ Phil. Trans. 1687. Vol. XVI. p. 269.   § Phil. Trans. 1687. Vol. XX. p. 2.
‖ Phil. Trans. 1706. Vol. XXV. p. 2221.   ¶ Phil. Trans. 1719. Vol. XXX. p. 1071, and 1075.
** Vol. XXXV. p. 537.

invention,* which has been superseded by the still superior air pump of Cuthberton.

11. Various methods have been thought of for measuring the velocity of the wind; but none hitherto proposed answers the purpose completely. Mr. Brice's method, when it can be practised, undoubtedly gives the true velocity. But unfortunately it is only at times that it can be employed. His method is to measure the space passed over by the shadow of a cloud in a given time.†

12. In this paper we have Deluc's rules for measuring heights by the barometer reduced to the English standard by Dr. Maskelyne.‡ From the publication of this paper the mensuration of heights in this country with tolerable accuracy may be dated.

13. Mr. Deluc gives a set of barometrical measurements of the mines in the Hartz, and shows that they coincide with his rules.

14. The next paper we have to mention is entitled *Observations made in Savoy to ascertain the Heights of Mountains by means of the Barometer ; being an Examination of Mr. Deluc's Rules delivered in his Recherches sur la Modifications de l'Atmosphere.*‖ This paper by Sir George Shuckburgh, is exceedingly valuable, and was one of the chief steps in bringing the method of measuring heights by the barometer to its present precision. The specific gravity of air, and the rate of its expansion by heat, are determined with great delicacy. It may be fairly considered as the most valuable of all Sir George's publications. It is rivalled, however, by the no less important paper of General Roy, published in the same volume of the Transactions, and entitled *Experiments and Observations made in Britain, in order to obtain a Rule for measuring Heights with the Barometer.*¶ In this paper the expansion of mercury and air by heat are determined with the most rigid accuracy, and the heights measured by the barometer were determined by the laborious, but precise mode of levelling. It was General Roy chiefly that contrived the formulas for measuring heights that have been already given in a previous part of this chapter. The remarkable coincidence between the results obtained by Shuckburgh and Roy could not fail to strike the most careless observer. Sir George published a short comparison between his own experiments and those of General Roy, in a subsequent volume, and pointed out their very close resemblance to each other.**

15. The last paper we have to mention on this subject contains a set of experiments on the resistance of air, by Mr Lovell Edgeworth, in which he

---

* Phil. Trans. 1751. Vol. XLVII. p. 415.   † Phil. Trans. 1766. Vol. LVI. p. 224.
‡ Phil. Trans. 1774. Vol. LXIV. p. 158.   § Phil. Trans. 1777. Vol. LXVII. p. 401.
‖ Phil. Trans. 1777. Vol. LXVII. p. 513.   ¶ Phil. Trans. 1777. Vol. LXVII. p. 658.
** Phil. Trans. 1778. Vol. LXVIII. p. 681.

shews that it depends upon the position of the bodies, and that it increases at a greater rate than the surface of the bodies resisted.*

We may here add twelve papers, partly upon the force of gunpowder, partly upon gunnery, and partly upon projectiles.

Experiments for improving the art of gunnery. By Sir Robert Moray. Phil. Trans. 1667. Vol. II. p. 473.

Experiments at Woolwich, March 18, 1651, for trying the force of great guns. By Mr. Greaves. Ibid. 1685. Vol. XV. p. 1090.

Solution of a problem of great use in gunnery. By E. Halley. Ibid. 1685. Vol. XVI. p. 3.

A proposition of general use in the art of gunnery, showing the rule of laying a mortar to pass in order to strike any object above or below the horizon. By E. Halley. Ibid. 1695. Vol. XIX. p. 68.

Some propositions respecting the parabolic motion of projectiles, written in 1710. By Brook Taylor, LL.D. Ibid. 1721. Vol. XXXI. p. 151.

Report of the Committee of the Royal Society appointed to examine some questions in gunnery. Ibid. 1742. Vol. XLII. p. 172.

An account of a book, entitled, New Principles of Gunnery, containing the determination of the force of gunpowder, and an investigation of the resisting power of the air to swift and slow motions. By Benjamin Robins. Ibid. 1743. Vol. XLII. p. 437.

The motion of projectiles near the earth's surface considered, independent of the properties of the conic sections. By Mr. Thomas Simpson. Ibid. 1748. Vol. XLV. p. 137.

Observations on the height to which rockets ascend. By Mr. Benj. Robins. Ibid. 1749. Vol. XLVI. p. 131 and 578.

The force of fired gunpowder, and the initial velocities of cannon balls, determined by experiments; from which is also deduced the relation of the initial velocity to the weight of the shot, and the quantity of powder. By Mr. Charles Hutton of the Military Academy at Woolwich. Ibid. 1778. Vol. LXVIII. p. 50.

New experiments on gunpowder. By Benj. Thompson, Esq. Ibid. 1781. Vol. LXXI. p. 229.

Experiments to determine the force of fired gunpowder. By Benjamin Count of Rumford. Ibid. 1797. Vol. LXXXVII. p. 222.

CHAP. VI.

OF ACOUSTICS.

By *Acoustics* is meant that branch of science which treats of *sound:* a very curious and important part of mechanical philosophy. Though it occupies but a very subordinate place in the Philosophical Transactions.

Sound is produced by the vibrations of hard and elastic bodies; and, in com-   Production of sound.

---

* Phil. Trans. 1783. Vol. LXXIII. p. 136.

mon cases, is conveyed to the ear by the vibrations of the air. But water, metals, wood, and all similar bodies, are equally capable of being the vehicles of sound. From the experiments made by Dr. Derham, compared with those of the French academicians in 1738, it has been concluded that sound is propagated through air at the rate of 1,130 feet in a second of time, and it is sensibly affected by the direction and violence of the winds that happen to blow. Theory gives the rate of the propagation of sound through air only 946 feet in a second. There was therefore a discordancy between theory and observation which could not be accounted for. La Place lately suggested that this discordancy might be occasioned by the heat evolved during the undulations into which the air is thrown. This heat must increase its elasticity, and of course the velocity of its motion. This is probably the real cause of the acceleration. It is certain that heat must be evolved though we have no means of estimating the quantity experimentally.

The theory of sound in different bodies was first given by Newton. Its velocity in air was settled by Derham and the French academicians. To Chladni we are indebted for some curious experiments on its propagation through other bodies. The vibrations of a string were first explained by Dr. Brook Taylor, and much more completely by D'Alembert. To Euler, D. Bernoulli, La Grange, and La Place, we are likewise indebted for much curious information on the same subject.

A string stretched by the same weight performs all its vibrations in the same time. Let t be the time of one vibration, w the weight of an inch of string in grains, f the force of tension, L the length of the string, g the force of gravity equal to 386 inches; then $t = 2L\sqrt{\dfrac{w}{gf}}$, or $t = 2L\sqrt{\dfrac{w}{386f}}$. This is an accurate formula for determining the number of vibrations in a musical string. If the number of vibrations in a second be n, then $n = \dfrac{\sqrt{386f}}{2L\sqrt{w}}$. Hence it appears that the number of vibrations is inversely as the length. If the length remains the same, the number of vibrations increases as the square root of the force of tension.

If v denote the velocity of air rushing into a vacuum, e its elasticity, d its density, then $v = \sqrt{\dfrac{e}{d}}$. Hence it is obvious that the velocity of hydrogen gas must be much greater than that of common air, and that sound must be propagated much more rapidly through hydrogen gas than common air.

Papers on acoustics in the Transactions.

The number of papers on acoustics, in the Philosophical Transactions, amounts to 25, of which there are seven that we may pass over as of little or no value. A very few observations on the remainder is all that we can venture to make.

1. The first paper we shall mention gives an account of the speaking trumpet, invented by Sir Samuel Moreland.*   This trumpet was soon after improved by Mr. John Conyers.†

2. In the next paper Dr. Wallis gives us an account of a discovery in acoustics, made by Mr. William Noble, of Oxford, in the year 1674. The discovery was this.   If the string A be an upper octave to B. and therefore in unison with its two halves, when A is struck, each of the halves of B vibrates, but the middle point remains at rest.   The same thing happens in similar cases.‡

3. In the next paper we have a dissertation on the trumpet, by Mr. Roberts, and a reason assigned for the defective notes in it.§

4. We have two curious papers by Dr. Wallis.   In the first he gives an account of the divisions of the monochord on mathematical principles.‖   In the next he gives an explanation of the defects of the organ;¶ defects which have been remedied by a late improvement of that instrument by Mr. Liston.

5. Hauksbee found by experiment that the sound of a bell was nearly doubled by doubling the density of the air,** and that it was very much diminished by rarifying the air.††   He found likewise that sound is not propagated through a vacuum, but that it passes readily through water.‡‡

6. In the next paper Mr. Salmon gives the exact comparative length of the strings which sound the different notes of music.§§

7. The next paper which we have to mention is the most important on the subject of acoustics in the Philosophical Transactions.   It is by Dr. Derham, and gives an account of the experiments that had been made before his time, in order to determine the velocity of sound through air.   He then relates his own experiments on the subject, which appear to have been made with great care. The result was, that sound moves at the rate of 1142 feet in a second; that the rate is not altered by the state of the air, nor the situation of the place; but that a favourable wind accelerates, while an unfavourable i nd retards the motion of sound.‖‖

8. The next paper on the subject of acoustics is by Guido Grandi, and contains an elaborate discussion respecting the propagation of sounds, and hypothetic demonstrations of their motion in hyperbolic curves.¶¶

9. The next paper contains Dr. Brook Taylor's discoveries respecting tense strings, and his theory founded on them.   He demonstrates that the curvature of

---

* Phil. Trans. 1672. Vol. VI. p. 3056.
† Phil. Trans. 1678. Vol. XII. p. 1027.
‡ Phil. Trans. 1677. Vol. XII. p. 839.
§ Phil. Trans. 1692. Vol. XVII. p. 559.
‖ Phil. Trans. 1698. Vol. XX. p. 80.
¶ Phil. Trans. 1698. Vol. XX. p. 249.
** Phil. Trans. 1698. Vol. XX. p. 433.
†† Phil. Trans. 1705. Vol. XXIV. p. 1904.
‡‡ Phil. Trans. 1709. Vol. XXVI. p. 367.
§§ Phil. Trans. 1705. Vol. XXIV. p. 2072.
‖‖ Phil. Trans. 1708. Vol. XXVI. p. 2.
¶¶ Phil. Trans. 1709. Vol. XXVI. p. 270.

vibrating strings is as the distance of the straight ; that the velocity is the same ; that every part performs its motion with the same velocity ; and finds the velocity, the force, &c. being given.*

10. The next paper is a very curious and entertaining one, by the Hon. Daines Barrington, entitled Experiments and Observations on the singing of Birds.† It contains the most complete collection of facts upon the subject that has ever appeared, and does great credit both to the assiduity and the musical ear of the author. We shall content ourselves with transcribing his table, exhibiting the comparative musical powers of the different singing birds, in which the number 20 denotes the point of absolute perfection.

	Mellowness of tone.	Sprightly notes.	Plaintive notes.	Compass.	Execution.
Nightingale,	19	14	19	19	19
Skylark,	4	19	4	18	18
Woodlark,	18	4	17	12	8
Titlark,	12	12	12	12	12
Linnet,	12	16	12	16	18
Goldfinch,	4	19	4	12	12
Chaffinch,	4	12	4	8	8
Greenfinch,	4	4	4	4	6
Hedge Sparrow,	6	0	6	4	4
Aberdavine, (or Siskin),	2	4	0	4	4
Redpoll,	0	4	0	4	4
Thrush,	4	4	4	4	4
Blackbird,	4	4	0	2	2
Robin,	6	16	12	12	12
Wren,	0	12	0	4	4
Reed Sparrow,	0	4	0	2	2
Black Cap, or Norfolk Mock Nightingale,	14	12	12	14	14

11. The next paper which occurs on the subject of acoustics is by Mr. Caallo, and is entitled *On the Temperature of those Musical Instruments in which the Tones, Keys, or Frets, are fixed; as in the Harpsichord, Organ, Guitar, &c.‡* This paper contains a curious account of the music of the ancients, and comparison of it with that of the moderns.

12. The last paper on acoustics which we have to mention is by Dr. Young. It is entitled Outlines of Experiments and Inquiries respecting Sound and Light.§ This is a very important paper, and contains a pretty full outline of the whole of acoustics. But it is so long that we are under the necessity for want of room of satisfying ourselves with merely mentioning it, and referring our readers for full information on the subject to the paper itself.

* Phil. Trans. 1713. Vol. XXVIII. p. 26.   † Phil. Trans. 1773. Vol. LXIII. p. 249.
† Phil. Trans. 1788. Vol. LXXVIII. p. 238.   § Phil. Trans. 1800. Vol. LC. p. 106.

CHAP. VII.

OF NAVIGATION.

Under the term *Navigation,* we mean to arrange the papers in the Philosophical Transactions respecting the tides.   This subject is usually considered as a branch of astronomy, because the theory of the tides depends upon astronomical principles.   We would have followed the same arrangement, had there not been a few papers in the Transactions professedly on the subject of navigation, which though they would not have afforded matter for a separate chapter, could not with propriety have been omitted.   On that account we have joined them with the papers relative to the tides, and shall notice them of course in this chapter.

The relation between the tides and the moon has been observed from the remotest ages ; but Descartes seems to have been the first who attempted to account for this relation.   According to him the moon gravitates upon the waters over which she is vertical.   But waters pressed down at one place necessarily rise in another.   Hence the tides.   This explanation is not only inconsistent with the philosophy of Descartes, but also contrary to the phenomena of the tides, and of course incorrect.   Galileo had recourse to the rotatory motion of the earth round its axis, and its annual motion round the sun, for an explanation.   When these two motions coincided, the waters rose ; when they opposed each other, the waters fell.   This explanation leaves the action of the moon entirely out of view, and it would occasion high water always at noon and midnight, and low water at the rising and setting of the sun, which consequences being contrary to fact, destroy the theory.   Wallis approached much nearer to the true explanation, though he did not reach it.   He observed that it was the centre of gravity of the earth and moon that in fact moved round the earth, and he shows how the sun and moon acting together increase the tides, and how, when they act in opposition to each other, they diminish them.   Newton in fine gave the real theory of the tides in his *Principia,* a theory which was almost immediately adopted by all philosophers.   Three very elaborate dissertations on the tides by Maclaurin, Daniel Bernoulli, and Euler, jointly gained the prize of the French Academy in 1738.   Since that time the subject has been treated of much in detail, and with the profoundest skill by La Place, in the Memoires of the French Academy for the years 1775, 1789, and 1790 ; and still more lately in his *Mechanique Cæleste.*   But notwithstanding the profound sagacity of this consummate mathematician, he has not been able to apply his discussions to the

Theory of the tides.

3 I

sea, as it exists in nature; but to a hypothetical sea of a determinate depth, and covering the whole earth.

Phenomena of the tides. Before proceeding to the papers contained in the Philosophical Transactions, it will be proper to give a short sketch of the phenomena of the tides, and of the Newtonian explanation of them, founded on the united action of the sun and moon.

There is a tide every 6 hours and 12 minutes, or the interval between high water and high water is 12 hours and 24 minutes. The time from the tide of one day to the corresponding tide of the next day is 24 hours and 48 minutes. Now this is exactly equal to the time of the moon's diurnal revolution round the earth. The high water then makes a revolution round the globe in 24 hours and 48 minutes. Hence we are certain that it is not a current, but an undulation. A current, moving with such rapidity, would sweep every thing before it.* Small seas are not affected by the tides. In the Caspian and Baltic they are insensible. In the Mediterranean they do not exceed a foot.

The tides rise highest at the new and full moons, not exactly at the full and change, but 37 hours after these; or it is the third tide after the new and full moon, which is the greatest. Such tides in this country are called *spring tides.* The spring tides are not always equal in magnitude: they are always greatest when the diameter of the moon is greatest. The smallest tides happen at the quarters, and are called *neap tides.* At the equinoxes the tides are *cæteris paribus* greatest; but the difference is not considerable. It is always high water at every particular place when the moon is in the meridian of that place, or directly under the meridian.

Newton explained the tides by the attraction of the sun and moon elevating somewhat the waters of the ocean. At the new and full moon, the sun and moon act in conjunction, and of course produce the greatest effect; while at the quarters they act in opposition, and of course diminish the effect which would otherwise be produced. The effect of the sun is much smaller than that of the moon, because he is much farther off. If gravity at the surface of the earth be 1, then the sun's action on the waters may be demontrated to be $\frac{1}{12868200}$ of gravity. Now the centrifugal force, at the surface of the earth, is $\frac{1}{289}$th part of that of gravity, and raises the water 54692 feet. We have therefore this proportion, which gives us the effect of the sun in raising the waters of the ocean in feet $12868200 : 289 :: 54692 : 1\cdot2$. So that the tide produced by the sun is $1\cdot2$ feet. The spring tide is to the neap tide as 9 to 5. Now let s be the solar tide, and l the lunar tide, we have $l = \frac{7s}{2} = 4\cdot2$ feet.

---

* Had Bernardin St. Pierre attended to this simple consideration, he never would have proposed his explanation of the tides by the melting of the ice at the North Pole.

So that the tide produced by the moon amounts to 4·2 feet. Hence the spring tides ought to be 5 4 feet, and the neap tides 3 feet. This agrees pretty well with what happens in the open sea; but the effect is greatly increased by the shores, and by a variety of local circumstances; so that in many harbours the tides are greatly beyond these numbers. At London, and at Leith, the spring tides are about 16 feet; at Brest they are 21 feet; and at Bristol no less than 48 feet.

The tides round Great Britain are derivative, and not occasioned by the direct action of the sun and moon. They start at Ushant from the tide of the ocean, and move in two separate undulations. The first, which passes through St. George's Channel, takes 24 hours to go round by the North of Scotland, and come round to Dover; the second, passes up the English Channel, and reaches Dover in 12 hours.

The papers on the tides, in the Philosophical Transactions, including a few upon other subjects connected with navigation, amount to 48. But of these there are 33 which are either of no importance, or cannot with propriety be noticed in a general view. We shall make a few observations on the remaining 15, or at least the chief of them. Papers in the Transactions on the tides.

1. The first paper we shall mention is by Sir Robert Moray, and gives an account of an irregularity in the tide at Berneray, in Long Island, on the west coast of Scotland. For four days, at the quarters, the tide runs east for 12 hours, both flow and ebb, and west 12 hours. At other times the flow is east, the ebb west.*

2. The next paper deserving notice, is Dr. Wallis's ingenious hypothesis, to account for the tides.† As this hypothesis is now universally abandoned, it is not necessary to give any further account of it than what has been stated at the beginning of this chapter. In a subsequent volume of the Transactions, some animadversions are made on it by Mr. Childrey.‡

3. The next paper we shall mention, is a pretty full account of Witsen's Naval Architecture and Conduct;§ a work written in Dutch, and giving a most elaborate and learned account of ships, ancient and modern; of the improvements in naval architecture; and the modes practised at the time by the most considerable civilized states of Europe. In the same volume, occurs an account of Meibomius's Treatise on the Vessels of the Ancients;‖ a subject not yet made out in a very satisfactory manner, notwithstanding the pains that have been bestowed upon it.

4. Mr. Flamsteed published a correct table, showing the time of high water at London bridge, for every day of the year; which, though now so common

---

* Phil. Trans. 1665. Vol. I. p. 53.   † Phil. Trans. 1665. Vol. I. p. 263.
‡ Phil. Trans. 1670. Vol. V. p. 2061.   § Phil. Trans. 1671. Vol. VI. p. 3006
‖ Phil. Trans. 1671. Vol. VI. p. 3064.

as not to be worth transcribing, yet at that period was of considerable importance, because the errors in the almanacs amounted usually to about two hours.[*] Mr. Flamsteed published a similar table, for the year 1684; and gave at the same time a table, by means of which it could be accommodated to most parts in Britain, and to some of those in France and Holland.[†]

5. The next paper we shall mention, is a very good popular explanation of the tides, according to the Newtonian principles, by Dr. Halley. It is remarkable, because it was written and presented to King James II., along with a copy of the Principia, when that book was first published.[‡]

6. The next paper that we shall notice, is by Mr. Parks, and gives an account of an easy practical method of dividing the nautical meridian in Mercator s projection. This paper is remarkable likewise for containing demonstrations of some of the most striking properties of the catenaria.[§]

7. In the next paper we have a set of experiments, made to measure the velocity with which the tide moves in Lambeth Reach, in the Thames, during every time of tide. The experiments are by Mr. Henry de Saumarez. They are curious, but too long to transcribe. It appears, that when the tide is flowing with the greatest velocity, during spring tides, the rate is not quite two miles in the hour; and during the whole of the ebb, the velocity is still less.[||]

8. The next paper gives an account of some very singular circumstances which attend the tides in the river Forth. There exist in that river certain irregular motions, called *leakies* by the common people. When the river is flowing, before high water it intermits and ebbs for a considerable time, after which it resumes its former course, and flows till high water: and in ebbing, before low water the river flows again for some time, and then ebbs till low water. The leakey begins at Queensferry, nine miles above Leith, at neap tide and low water, and goes to the house of Maner, 25 miles above Queensferry by water, but only four by land. At neap tide and high water, as also at spring tide and low water, the leakey extends as far as Craigforth, three miles above Stirling; which is nearly as far as the tide ascends up the river. At Queensferry there are no leakies at neaps and springs at high water, nor springs at low water; they begin between Barrowstoness and the mouth of the river Carron, about 10 miles above Queensferry.[¶]

9. Captain Cook gave a table of the height to which the tides rise in different harbours of New Zealand and New Holland. The least was 4½ feet, the greatest 16 feet.[**] He observed, that at the spring tides, the evening tide upon the coast of New Holland was considerably greater than the morning tide. In the

---

[*] Phil. Trans. 1683. Vol. XIII. p. 10.
[†] Phil. Trans. 1683. Vol. XIV. p. 458.
[‡] Phil. Trans. 1697. Vol. XIX. p. 445.
[§] Phil. Trans. 1715. Vol. XXIX. p. 331.
[||] Phil. Trans. 1726. Vol. XXXIV. p. 68.
[¶] Phil. Trans. 1750. Vol. XLVI. p. 412.
[**] Phil. Trans. 1772. Vol. LXII. p. 357.

Endeavour river the evening spring tide was nine feet, the morning scarcely seven feet.*

10. In the Adriatic Sea the spring tides amount to between three and four feet; the neap tides scarcely exceed three inches, according to the observations of Toaldo. The tides in that sea are highest in winter.†

11. The last paper we shall mention is by Major Rennel, giving an account of a current which sets in towards the Scilly Islands, from the south; and which he conceives occasioned the loss of Sir Cloudesley Shovel, and several ships of his fleet. This current sets eastward along the north coast of Spain, passes along the west coast of France, and thence crosses the mouth of the British Channel towards the Scilly Islands. It is only dangerous after continued westerly winds, and ought to be known to seamen in order to be guarded against.‡

CHAP. VIII.

OF ELECTRICITY.

THIS science, as far as the Royal Society is concerned, differs from every other. It originated after the establishment of the Royal Society; and almost every electrical discovery of importance was made by Members of the Royal Society, and is to be found registered in the Transactions. We would not be understood to assert, that all who have made discoveries in electricity were natives of Britain, though the greatest number of them certainly were; but merely that they were Members of the Royal Society. The whole science of electricity then, as far as we are acquainted with it, being to be found in the Philosophical Transactions, we may follow a different plan with respect to it from what we have done with the preceding sciences, and trace the progress of discovery pretty accurately by following the order of the papers. *This science chiefly cultivated by Fellows of the Society.*

The papers on electricity, in the Philosophical Transactions, amount to 211. But of these there are 25 which I consider as of no value; there are seven merely giving an account of death, produced by lightning; and 21 giving descriptions of houses and ships struck with lightning. These papers, though they have their utility in some points of view, cannot well be considered, when we mean merely to trace the progress of the science. Though there is one example in *Papers on it in the Transactions.*

---

* Phil. Trans. 1776. Vol. LXVI. p. 447.   † Phil. Trans. 1777. Vol. LXVII. p. 141.
‡ Phil. Trans. 1793. Vol. LXXXIII. p. 182.

the Transactions of death by lightning being brought forward to prove a very important electrical theory.

The ancients knew that amber, *(electrum,)* when rubbed, acquired the property of attracting light bodies to it; and this property they distinguished by the name of electricity. Dr. Gilbert, of Colchester, in his book on Magnetism, published about the beginning of the 17th century, gives us a catalogue of bodies possessing the same property. Some facts also were observed by Boyle. Such was the state of the subject at the beginning of the 18th century. It could not be called a science, but merely the collection of a few insulated facts.

Hauksbee. The first experimenter on electricity was Mr. Hauksbee, and he made some discoveries of importance, though he did not suspect at first that electricity was the agent which he employed. In the year 1705, he found that when mercury is shaken in glass vessels it produced light, (a fact known before,) and that when the air is rarified to at least one half, the light is very vivid. He ascertained, likewise, that it was the friction of the mercury against the glass, or the air, (for he could not determine which) that occasioned the light.* Soon after, he ascertained that light was produced when the following bodies were rubbed against each other in vacuo; namely, amber against flannel, glass against flannel, glass against oyster shells, woollen against woollen, and glass against glass. Glass, he found, when rubbed against glass in the open air, or under water, likewise produced light; but that was not the case with any of the other bodies tried.†

Thus he determined that many bodies, under certain circumstances, have the property of emitting light; but he did not suspect that this light was connected with electricity. Still he continued his experiments, and found that when an exhausted glass globe was rubbed with the hand, a vivid light was produced · that mercury agitated in a varnished vessel, under an exhausted receiver, produced light; that loaf sugar, when broke, emitted light; that a lump of *calomel*, broken in the same way, also occasioned light. At first he suspected that this light was owing to a particular quality in the glass. When air was let into the glass globe the light diminished; but if the friction was continued, and a person brought his finger near the globe, light appeared on the point of the finger.‡

At last Mr. Hauksbee recognised the electricity of glass by friction. He found that a glass tube, when rubbed, attracted brass leaf, &c., and gave sparks. When the tube was exhausted of air, the light emitted was increased; but it lost the attracting power. On admitting the air, the attracting power returned. He employed a pretty large glass cylinder, turned by a winch, and rubbed by the

* Phil. Trans. Vol. XXIV. p. 2129.   † Phil. Trans. 1705. Vol. XXIV. p. 2165.
‡ Phil Trans. 1706. Vol. XXV. p. 2277.

hand, and found the effects the same. In this may be recognised the first rudiments of the electrical machine. Strings of linen, suspended all round, were attracted towards the axis of the cylinder.* Soon afterwards Mr. Hauksbee found, that after the threads were attracted by the glass, they were repelled on making a finger approach them. The same thing happened if the strings were suspended within the glass, and the finger without. When an exhausted glass was brought near a rubbed tube, and turned round, light appeared in it.† He next found that the approach of paper, &c., to an excited tube, or filling it with sand, prevented it from attracting light bodies.‡ When an exhausted glass is brought near an excited tube, it is not necessary to move the exhausted vessel in order to produce light, but the light is stronger if it be moved.§

About this time an electrical paper was published, by Dr. Wall, which deserves to be mentioned, though it interrupts the detail of Hauksbee's discoveries. Dr. Wall found that amber, diamond, gum lac, and sealing-wax become luminous when rubbed; and he affirms that all electric bodies do the same. Thus a second general property of electric bodies came to be generally known; namely, that they become luminous as well as attract light bodies. Dr. Wall compares the crackling of these excited bodies, and the flashes emitted by them, to thunder and lightning.‖

Mr. Hauksbee next found that sealing-wax, rosin, and sulphur, coating a cylinder of wood, could be excited by friction. The electricity was increased by heat. Bodies attracted by such excited cylinders he found to be repelled by sealing-wax and amber, without rubbing.¶ Mr. Hauksbee next lined a globe with sealing-wax, and found, that when exhausted and rubbed it became luminous within. This globe was found to attract even when exhausted.** This experiment was repeated with some alterations: one half of the inside of the globe was lined with pitch, when the globe was exhausted, and the pitched part rubbed, the hand was seen luminous through the pitch †† If sulphur, or wax, be substituted for pitch, the experiment succeeds equally well.‡‡ Brass cannot be excited by friction.§§

Such are the electrical discoveries made by Mr. Hauksbee. They are of no great importance indeed in themselves; but they constituted the beginning of the science, and, by drawing the attention of philosophers to that particular subject, were doubtless of considerable service in promoting electrical investigations. The next electrical experimenter, who appears upon the stage, was Mr. Stephen Gray, and his discoveries were much more important than those of Gray.

---

* Phil. Trans. 1706. Vol. XXV. p. 2327. † Phil. Trans. 1707. Vol. XXV. p. 2372.
‡ Phil. Trans. 1708. Vol. XXVI. p. 82. § Phil. Trans. 1707. Vol. XXV. p. 2313.
‖ Phil. Trans. 1708. Vol. XXVI. p. 69. ¶ Phil. Trans. 1708. Vol. XXVI. p. 87.
** Phil. Trans. 1708. Vol. XXVI. p. 219. †† Phil. Trans. 1709. Vol. XXVI. p. 391.
‡‡ Phil. Trans. 1709. Vol. XXVI. p. 439. §§ Phil. Trans. 1711. Vol. XXVII. p. 328.

Hauksbee. So important indeed were they that he may lay claim, in a great measure, to have established the science of electricity upon a sure foundation, and to have constituted it, in some measure, what it is at this day. It is remarkable that no biographical memoirs remain of a man to whom electricity lies under such obligations. From some observations made by Desaguliers, it appears that his character was very particular, and by no means amiable.

In his first paper, Mr. Gray describes a set of experiments, made in order to determine what bodies may be excited by friction. He found that hair, feathers, silk, linen, woollen, paper, leather, firwood shavings, parchment, and gold beater's leaf, became electrical when rubbed with the hand.* Soon after, he found that metals cannot be excited by friction. Thus he divided all the bodies in nature into two classes; namely, *electrics*, or those which can be excited by friction, and *non electrics*, or those which cannot be excited by friction.

When a tube is excited, it communicates its electricity to cork, fir, ivory, packthread, iron, brass, gold, silver, copper, tin, stones, bricks, different kinds of wood, animals, and water. If one end of any of these bodies be inserted into the excited glass tube, or held near it, the other end (though the body be 1,000 feet long) attracts and repels precisely like the excited tube itself, provided the body thus placed be suspended by silk or hair, but not if by linen or metal.† Water also may be rendered electrical by insulating a dish of it, and bringing an excited tube near it. Water is attracted by an excited tube when brought sufficiently near.‡ An excited conductor will excite another insulated conductor though at some distance.§ Thus Mr. Gray discovered three fundamental and very important laws in the science of electricity; namely, that *non electrics* may be excited by being brought into the neighbourhood of excited *electrics*; that *non electrics* are *conductors* of electricity; that *electrics* are *non conductors*, and may be employed to insulate conductors so as to enable them to carry the electric energy to a very considerable distance. These discoveries do much credit both to the sagacity and industry of Mr. Gray. His next discovery was that sulphur, wax, and several other resins, when melted, become electric in cooling, and retain their electricity for months. Excited electrics attract equally in vacuo, and in the open air.‖

We must here interrupt the detail of Mr. Gray's discoveries to mention an important paper by M. Dufay, containing two great discoveries in electricity. 1. That electrical bodies when excited attract those that are not excited, communicate electricity to them, and then repel them. 2. That there are two kinds of electricity, the *resinous* and the *vitreous*; the first belonging to amber, copal, lac. silk, thread, paper, &c.; the second to glass, rock crystal, precious stones,

Marginal notes:
Divides bodies into electrics and non electrics.

Conductors and non conductors.

Dufay.

Two kinds of electricity.

---

* Phil. Trans. 1720. Vol. XXXI. p. 104. † Phil. Trans. 1731. Vol. XXXVII. p. 81.
‡ Phil. Trans. 1731. Vol. XXXVII. p. 227. § Phil. Trans. 1732. Vol. XXXVII. p. 397.
‖ Phil. Trans. 1732. Vol. XXXVII. p. 285.

hair, wool, &c. Bodies having the *same* electricity, when excited, repel; having different electricities, they attract. Bodies communicate their own electricity to conductors. When the finger is brought near an excited conductor a spark and snap are perceived. *

Dufay's important experiments were repeated by Mr. Gray. He found that knobs gave stronger sparks than points. He draws as a conclusion from his observations that electrical fire is the same thing as thunder and -lightning.† Wood cannot be so fully excited by communication as metal. An excited conductor, when brought in contact with an insulated conductor, divides its electricity with it.‡ The last paper communicated by Mr. Gray, on the subject of electricity, is a fanciful one, describing the revolution of small bodies round great electrics like the planets round the sun.§ This paper was dictated to Dr Cromwell Mortimer, by Mr. Gray, while lying on his death bed. He died in February, 1736, before he had time to finish his experiments. Had he lived to examine the subject completely, he would no doubt have abandoned his opinions respecting these fanciful revolutions, which do not exist in nature.

The next experimenter on electricity, who enters upon the field, is Mr. Wheler. Wheler, who had been in the habit of making experiments on the subject in company with Mr. Gray. His paper was published in 1738, but he informs us that the experiments which it contains were made in 1732. It contains two facts which, if they had been published in 1732, would have been new; namely, 1. That bodies excited by the same electric repel the electric and each other. 2. That bodies excited by friction repel each other.‖ These observations had been already published by Dufay, who is entitled to the credit of them. For posthumous claims of this kind never can be admitted unless they are supported by much better evidence than the mere assertions of the posthumous publisher. In a subsequent paper, Mr. Wheler gives an account of his observations on the supposed circular revolutions of balls round electrics described by Mr. Gray, and shows that they were the effect of the volution of the experimenter, who held the strings that supported these revolving balls.¶

When electricity was laid down by Gray and Wheler, it was taken up by Desaguliers. Dr. Desaguliers, who published no fewer than 13 papers upon the subject. His discoveries were not of much importance; but his papers possessed value, because he arranged all the facts discovered by his predecessors in a luminous order, and thus facilitated the means of becoming acquainted with them. To his papers, likewise, we owe the introduction of various terms into electricity, which have since become technical. He divides all substances into two classes; namely, *electrics*, which can be excited by friction, and *conductors*, which can-

* Phil. Trans. 1734. Vol. XXXVIII. p. 258.   † Phil. Trans. 1735. Vol. XXXIX. p. 16.
‡ Phil. Trans. 1735. Vol. XXXIX. p. 166.   § Phil. Trans. 1736. Vol. XXXIX. p. 400.
‖ Phil. Trans. 1739. Vol. XLI. p. 98.   ¶ Phil. Trans. 1739. Vol. XLI. p. 112 and 118.

not be excited by friction. Electrics insulate, conductors do not. These facts had been discovered by Gray; but had not been placed by him in so clear a point of view. Desaguliers gives a list of *electrics* and of *conductors*. When an electric touches or approaches very near a conductor, it communicates part of its electricity. Hence the reason of the spark which is perceived. Desaguliers confirmed Dufay's discovery of the *resinous* and *vitreous* electricities.[*] Electrics, when in contact with conductors, as water, cannot be excited. From this fact, Desaguliers explains why glass tubes cannot easily be excited in moist weather.[†] The most important discovery of Desaguliers was that air is an electric. From this fact, he endeavours to account for the rise of vapours on electrical principles.[‡]

Electricity used to kindle combustibles.

Electricity began about this time to be employed as the means of setting fire to combustibles, and several amusing and popular experiments were contrived, which contributed not a little to give the science celebrity, and to draw to it the general attention. Mr. Hollman gives us an account of a set of experiments of this kind made at Leyden. Electricity was employed to set fire to gunpowder, to pitch, to sulphur, and to sealing-wax, previously well heated. Alcohol also was kindled in the same manner.[§] Dr. Miles kindled phosphorus by an electric spark.[||] These experiments were repeated with some variations by Bozes.[¶]

Watson.

Dr. Watson, afterwards Sir William Watson, first appeared upon the field as an electrician in 1745. He prosecuted the subject with much ardour and success for a long period of years, and is indebted to it chiefly for the high reputation which he acquired. For though his botanical papers do him credit, they would not of themselves have been sufficient to give him celebrity. In his first paper, relative to this science, he gives an account of the method of firing bodies by electricity, and explains it very neatly and precisely.[**] In his second paper he shows that ice is electric, draws a parallel between electricity and light, and remarks that cold does not diminish the spark.[††] In a subsequent paper, Dr. Watson describes an electrical machine, proves that the flame of alcohol, and the vapour of oil of turpentine, are conductors of electricity. He

Leyden phial.

likewise gives an account of the Leyden phial, which had been accidentally discovered a little before by Muschenbroeck, by means of which electricity could be accumulated in such quantities as to produce effects that filled all Europe with astonishment. Dr. Watson endeavours, though without success, to give a theory of this phial. He affirms, that all bodies, in their natural state,

---

* Phil. Trans. 1739. Vol XLI. p. 186, 193, 196, 198, 200, 209, 634, 637, 639, 661.
† Phil. Trans. 1742. Vol. XLII. p. 14.   ‡ Phil. Trans. 1742. Vol. XLII. p. 140.
§ Phil. Trans. 1744. Vol. XLIII. p. 239. || Phil. Trans. 1745. Vol. XLIII. p. 290.
¶ Phil. Trans. 1745. Vol. XLIII. p. 419.   ** Phil. Trans. 1745. Vol. XLIII. p. 481.
†† Phil. Trans. 1746. Vol. XLIV. p. 41; and 1747. Vol. XLIV. p. 695.

3

contain a certain quantity of electricity. When the quantity of it is accumulated, and increases beyond the due proportion, it has a tendency to fly off;
and is carried off by conductors in proportion to their size, supposing them insulated. Hence the attractions and repulsions observed in excited bodies are
explained in a satisfactory manner.*

Dr. Watson's next paper, which is a very long one, contains a great deal of
very curious information. He describes experiments made to settle the relative
goodness of conductors. He describes a set of ι ials made by himself, and several other Members of the Royal Society, to determine how far electrical shocks
can be conveyed by means of conductors, and whether any sensible time elapses
during the motion of the electricity through very long conductors. They made
the shock pass across the Thames at Westminster, the circuit being completed
by means of wire carried along the side of the bridge held in the hand of the
experimenters on each side of the river, and an iron rod was dipped by each into
the water which served to unite the two experimenters.

They contrived by means of wires to make the shock pass through a circuit
of two miles, and in one case of nearly six miles. No interval of time was perceived to elapse during the passage of the electricity through this long circuit.
Hence, as far as experiment goes, we must conclude the shock to be instantaneous. In this paper there occurs also the first mention of Dr. Franklin, and of
his theory of *positive* and *negative* electricity. Dr. Watson informs us, that a
similar theory had occurred to himself, and that he had communicated a sketch
of it to the Royal Society before any information of Dr. Franklin's peculiar
opinions on the subject had reached the country.†

It was about this period that Dr. Franklin became known to the world as a   Franklin.
philosopher, and he stands indebted for his celebrity to his ingenious experiments and speculations on electricity. Dr. Franklin is one of the most eminent
men that the 18th century, so fruitful in men of genius, has produced. He was
born in Boston, in New England, afterwards settled in Philadelphia as a printer,
and spent a great many years in London. But he has left a biographical account of himself, written with his own hand, one of the most interesting documents of the kind ever laid before the public. This book being in the hands of
every reader, renders it unnecessary for us to take up a portion of room which
we can but ill spare with an abstract of his life.

The first of Dr. Franklin's letters is dated 1747, and is to Mr. Peter Collinson. They were speedily translated into all the languages of Europe; and the
particular view of the subject, which he took, very soon gained currency in the
science, and came to be distinguished by the name of the *Franklinian theory of
electricity*. His discoveries were very numerous, but the most important of

* Phil. Trans. 1746. Vol. XLIV. p. 704.   † Phil. Trans. 1748. Vol. XLV. p. 49 and 491.

them were the three following: 1. What Dufay considered as two distinct species of electricities, the *vitreous* and *resinous*, Dr. Franklin conceived to be two different states of the same electricity. All substances, according to him, in their natural state, contain a certain quantity of electricity. They may be excited two ways, either by *increasing* the natural quantity, or by *diminishing* it. If a man standing upon wax rub a glass tube, both the tube and the man are excited, and give a spark to any other person presenting his hand within a certain distance of them. If the man approach the tube, a still stronger spark will pass between them, and the electricity of both will in consequence become insensible. Hence it appears that in this case the electricity had divided itself unequally between the tube and the man, that the one contained more than the usual quantity, and the other less. A variety of circumstances, which it would be tedious to enumerate, led him to conclude, that the tube contained more than its usual quantity, and the man less. Hence he said that the tube was electrified *positively*, or *plus*; and the man *negatively*, or *minus*. What Dufay called *vitreous* electricity, Dr. Franklin called *positive*; and what Dufay called *resinous*, he termed *negative*. This, which constitutes the foundation of the present theory of electricity, is usually called the Franklinian theory. It belongs, however, equally to Dr. Watson, as he had communicated it to the Royal Society before Dr. Franklin's peculiar opinions on the subject reached this country.

Identity of lightning and electricity.

2. Dr. Franklin's second discovery is by far his greatest, and the one for which he will be indebted for his celebrity among posterity. He demonstrated the identity of common electricity and lightning. This identity had been previously suspected by different electricians, but no means had been thought of to determine the truth or the falsehood of the opinion. Dr. Franklin drew up a catalogue of the points of resemblance between electricity and lightning, and pointed out a method of putting the hypothesis of their identity to the test of experiment. The experiment was first tried, and its truth ascertained, in France; but he himself thought of a method of trying it in Philadelphia, long before any accounts of the success of the experiment in France had time to reach him. Seeing a thunder storm approach, he went out into the fields with his son, and raised a boy's kite, which he made by means of a silk handkerchief stretched over two cross sticks; to the end of the string was attached a key, and from the key there passed a silk string, by which he kept the kite in his hand. He took it for granted that as soon as the string was wetted by the rain, it would conduct the electricity to the key, from which, by taking sparks, he could easily verify his doctrine, and establish the identity of lightning and electricity. At first, no signs of electricity appeared, and the anxiety of the experimenter may be easily conceived. A very promising cloud passed over the kite without producing any effect. At last, perceiving the down stand

upright from the string, he began to conceive hopes, and applying his hand to the key was gratified with a spark. Abundance of sparks were soon procured, and he easily repeated the usual experiments, and verified the identity of electricity and lightning. He soon after erected an apparatus in his own house, at Philadelphia, for the purpose of collecting electricity from the clouds; and that he might lose no opportunity of making experiments with it, he contrived a set of bells, which, by their ringing, always gave him notice whenever his apparatus was sufficiently loaded with electricity for his purpose. He conceived, likewise, a method of protecting houses from the effects of thunder storms, founded on this discovery. This was nothing else than a pointed metallic conductor, elevated a little higher than the building, and terminating in the moist earth, or rather in the nearest water. He examined the state of the electricity during a thunder storm, and found it sometimes negative and sometimes positive.

3. His third discovery was the analysis of the Leyden phial, which at that time excited the astonishment of electricians, and contributed most materially to the popularity which the science of electricity acquired. If a glass jar be coated within and without with any conducting substance, to within a certain distance of the top; if it be half filled for instance with water, or mercury, or iron filings, while the outside is coated with tinfoil, in that case the phial is capable of receiving a charge. If you connect the inside coating with the conductor of an electrical machine, while the outside coating is in contact with the hand, or with some conducting substance, and drive the machine, a considerable quantity of electricity is accumulated in the jar. If you now bring the inside coating in contact with the outside by means of a conductor, a snap is heard, and if the conductor is an animal, a shock is received. The charge is now gone, and the glass jar is in the state in which it was before the experiment was made. By uniting together a number of such jars, an electrical battery is formed, and by charging all these at once a shock is produced great enough to kill small animals, and to melt metallic wires. Such is the Leyden phial, concerning which nothing accurate was known till Dr. Franklin explained it. He demonstrated, that the electricity is neither accumulated in the inside coating, nor in the outside coating, but in the glass itself. He coated a jar in the inside, by filling it to a certain height with water, then charged it, and lifting up the jar poured out the water. This water was found to possess no marks of electricity. The jar still retained its charge, and fresh water being poured in, gave the same shock that it would have done if the original water had not been displaced. In the same way he removed the outside coating of tinfoil from a jar after charging it, and coated it again without affecting the charge in the smallest degree. He next showed that a jar cannot be charged unless its outside coating be in contact with a conductor, or very near one; and that, for every spark of electricity which enters into the inside of the glass, an equal quantity of electricity leaves

Analysis of the Leyden phial.

the outside. The jar therefore contains just the same quantity of electricity which it did at first. But this electricity is distributed in a different manner. More than the usual quantity is accumulated in the inside, and less than the usual quantity on the outside. The inside is charged *plus,* and the outside *minus ;* and as glass is a non-conductor of electricity, the surplus cannot pass from the inside to the outside through the glass. But it immediately discharges itself through any conducting substance. Hence the snap, and hence the reason that the electricity disappears. Such was the luminous and satisfactory explanation of the Leyden phial ; which served essentially to confirm the Franklinian theory of *plus* and *minus* electricity, though that theory was founded on other arguments.

Le Monnier.   We may here notice M. Le Monnier, whose experiments, though of no great consequence in themselves, yet attracted considerable attention at first, and are detailed at considerable length in the Philosophical Transactions. In the first paper his electrical machine is described ; which appears to have been more powerful than any machine at that time employed in England. We have also a long list of his experiments ; none of which is new, except his killing a sparrow by means of a charged Leyden phial.* In a subsequent paper, Le Monnier relates his experiments on the passage of electricity through water and through wires, in order to determine the velocity of its motion. But, like the British electricians who preceded him in these experiments, he was unable to perceive the elapse of any time during the passage of electricity through the longest circuit which he employed. He determines, from his experiments, that conductors receive electricity in proportion to their surface, but that length has greater effect than breadth.† But these opinions were soon after examined and refuted by Dr. Watson.‡

Mr. Cook, in a subsequent paper, informs us that new flannel acquires the property of sparkling by wearing, but loses it on being washed. He ascribes the circumstance to the sulphur which new flannel contains.§

Nollet.   We now come to the Abbé Nollet, a most active and ingenious electrician, who prosecuted the study with unremitting attention for more than 30 years, and was even at the expence of a journey through Italy, to see with his own eyes a set of electrical experiments, which he had unsuccessfully attempted to repeat. He persisted in his theory with the most persevering obstinacy, against all the arguments brought forward by his antagonists, in which number may be reckoned almost all the eminent electrical philosophers of Europe. It is no easy matter to form a very adequate idea of this theory, which has been long since abandoned by every person. When an electric is excited, electricity flows

* Phil. Trans. 1746. Vol. XLIV. p. 247.   † Phil. Trans. 1746. Vol. XLIV. p. 290.
‡ Phil. Trans. 1747. Vol. XLIV. p. 388.
§ Phil. Trans. 1747. Vol. XLIV. p. 457 ; and Vol. XLV. p. 394.

to it from all quarters, and when thus *affluent* (as he termed it) it drives light bodies before it. Hence the reason why excited bodies attract. When the electricity is *effluent*, the light bodies are of course driven from the electric, which in that state appears to repel. He conceived every electric to be possessed of two different kinds of pores, one for the emission of the electric matter, and the other for its reception.

We have in the Transactions the result of a great number of experiments, made by Nollet, on the effect produced by electricity on the flowing of water through capillary tubes; on the evaporation of liquids; the transpiration of vegetables; and the respiration of animals. He found that electricity makes water flow from capillary tubes with an acceleration inversely as the bore; that it promotes the evaporation of liquids; the transpiration of vegetables; and the respiration of animals.* These last experiments have been often repeated since, and the results drawn by the Abbé are not considered as established.

It may be proper to mention here a theory of electricity, proposed about this time by Mr. Ellicott; though as this theory was not adopted, it attracted but little attention. It was founded on the three following data. 1. Electrical phenomena are produced by effluvia. 2. These effluvia repel each other. 3. They are attracted by all other matter.† If we substitute the word *fluid* for effluvia, these data are absolutely the same with those afterwards adopted by Epinus and Cavendish, and made the basis of the only satisfactory theory of electricity which has hitherto been proposed.

Notwithstanding the great number of experiments made upon electricity in England, and the vast accumulation of new facts, little attention was paid by British Philosophers to medical electricity, till a set of surprising experiments made in Venice, and repeated with success at Leipsic, drew their attention to this branch of the subject in a very particular manner. Signior Joannes Francisco Pivati, a person of eminence at Venice, published a pamphlet on the subject in 1747. He concealed a quantity of balsam of Peru in a glass cylinder, electrified with it a man who had a pain in his side, and who had previously applied hyssop to it by the advice of his physician. This man when he went home sweated and fell asleep, and the power of the balsam was so dispersed that even his clothes, the bed, and the chamber, all smelled of it. When he combed his head he found that the balsam had penetrated his hair, so that the very comb was perfumed. Next day Signior Pivati electrified a man in health in the same manner, who knew nothing of what had been done before. On going into company half an hour after, he found a gradual warmth diffused through his body, and he became more lively and cheerful than before. The company were surprised at an odour, and could not conceive whence it proceeded; but

Marginal note: Ellicott.

Marginal note: Medical Electricity.

---

* Phil. Trans. 1748. Vol. XLV. p. 187.   † Phil. Trans. 1748. Vol. XLV. p. 195.

he was conscious that it issued out of his body, and could not account for it, not being aware that it could be owing to the previous electrization performed on him by Signior Pivati. These and several other similar cases described by Pivati, induced Mr. Winkler, Professor of Philosophy at Leipsic, to repeat them, and his experiments were attended with considerable success. When sulphur was confined in a glass globe, and the glass excited, fumes of sulphur issued out, and were perceived at the distance of ten feet. The same thing succeeded when cinnamon was substituted for sulphur.*

The Abbe Nollet having attempted to repeat these experiments in vain, made a journey through Italy, on purpose to see them repeated by the Italians themselves, and returned home convinced that the whole was a deception. Dr. Watson, and the other British Electricians, endeavoured with equal want of success to repeat the experiment of Mr. Winkler. They went so far as to request him to send them some of his own glass globes and tubes, by means of which he had made his experiments, together with specific directions how to proceed. But though they obtained all this, and even tried methods which they thought better calculated for diffusing odours than those described by Winkler, still their trials were all unsuccessful. Hence we are entitled to consider the whole account of this diffusion of odours by electricity as fabulous.† As to the medical virtue of electricity, a great number of trials were made both in Britain and on the continent. The diseases chiefly experimented on were paralysis, deafness, blindness, St. Vitus's dance, and other similar affections. In some cases the patients received considerable benefit. But in general no advantage whatever was obtained. What happens in all cases of new medicines was equally observable here. Those physicians who first recommended electricity as a medicine, and who had the strongest belief in its good effects, were sure to perform the greatest cures; while those who had no great opinion of its efficacy, tried it in vain in cases exactly similar to those which were cured by their contemporaries. This shows clearly that the good effects which electricity produces are owing entirely to the faith of the patient. For in medicine as well as in religion, faith is essential for receiving a cure.

Mr. Bose of Wittemberg, in a paper published about this time, informs us that glass vessels which have been used for distillation are more readily and powerfully excited than any others.‡ Mr. Bose's success must have been owing to some accidental circumstances, and not to the use to which the vessels had been put. For I have often tried the experiment with glass retorts which had been repeatedly used for chemical purposes, and which had in some cases been

* Phil. Trans. 1748. Vol. XLV. p. 262, and 270.   † Phil. Trans. 1750. Vol. XLVI. p. 348;
  1751. Vol. XLVII. p. 231 ; and 1750. Vol. XLVI. p. 368.
  Phil. Trans. 1749. Vol. XLVI. p. 189.

exposed to a red heat, but never found any of them superior to retorts which had never been used at all.

It was at this period that Dr. Franklin made his first appearance as a writer in the Philosophical Transactions. Dr. Watson gave an account of his treatise on electricity, and praised it highly.* In this treatise Dr. Franklin first explains the efficacy of points in drawing off electricity. In a subsequent paper Dr. Franklin gives an explanation of the comazants which appear burning upon a ship's mast before she is struck with lightning. He describes the melting of wire, and of tin foil by electricity. It is plain, from what he says in this paper, that he employed a cushion as a rubber to his electrical machine, and that he coated his Leyden phials with tin foil.†

In the next paper Dr. Watson describes a set of experiments which he had made to determine the passage of electricity through a vacuum. He shows that it passes through a vacuum of any length, and therefore infers, that were it not for the air, electricity could not be accumulated in electrics.‡ It has been since demonstrated by the experiments of Mr. Morgan, that a perfect Torricellian vacuum is a non-conductor of electricity, but that the admission of a very small quantity of air renders it a conductor. The quantity of air present may in some measure be guessed at from the colour of the electric light. When the quantity of air is very small, the light is green; as the quantity of air increases the light changes to blue, to indigo, and at last to purple.§ Thus it appears that air in its usual state of density is a non conductor, rarified air is a conductor, but when in a state of extreme rarification it becomes a non-conductor again. When electricity moves through rarified air light is always visible. Hence, no doubt, originate many of those luminous appearances so frequent at a certain height above the surface of the earth.

The next paper which appears in the Transactions deserves notice, merely Eeles. because it is the production of Mr. Eeles. He explained thunder upon the same principles as Franklin had done before him, and he promises to explain evaporation, the reflection and refraction of light and gravitation upon the principles of electricity.|| Mr. Eeles afterwards published a separate work on this subject, in which he accuses the Royal Society of allowing the members to steal discoveries from his papers. These accusations are very apt to be made by speculative men, and the more common-place their speculations are, the more readily do they indulge themselves in such accusations; because there is a greater probability of their being anticipated by other persons.

Mr. Canton was one of the most successful experimenters of that golden age Canton.

---

* Phil. Trans. 1751. Vol. XLVII. p. 202.   † Phil. Trans. 1751. Vol. XLVII. p. 289.
‡ Phil. Trans. 1751. Vol. XLVII. p. 362.   § Phil. Trans. 1785. Vol. LXXV. p. 272.
|| Phil. Trans. 1751. Vol. XLVII. p. 524.

of electricity in England. He was the first successful repeater of Dr. Franklin's method of drawing lightning from thunder clouds by means of a kite. But he did not stop here, but continued his experiments upon the electricity of clouds for a considerable period. He observed them sometimes negative, and sometimes positive. He supposed that rarified air gave out electricity to the clouds, while condensed air abstracted it from them. He found that glass when rubbed becomes positive or negative, according to the smoothness or roughness of its surface, and according to the kind of rubber employed. He was capable of electrifying air positively or negatively at pleasure.* Mazeas made a set of continued experiments on the electricity of the air. In general it is sensible, but weak; it is not affected by winds, but much increased by stormy clouds.† Dr. Lining, of Charlestown, in South Carolina, in imitation of Dr. Franklin, sent up a kite into a thunder cloud, and by that means dissipated the lightning.‡

Death of Professor Rich-mau.

The event connected with electricity, which drew the greatest attention of mankind in general, was the death of Professor Richman, at Petersburg, on the 6th of August, 1753, by a thunder clap, which was supposed to have been attracted towards his apartment by an apparatus which he had constructed to observe the degree of electricity of the clouds. Richman was a Swede, born at Pernau, on the 11th of July, 1711. He studied in the universities of Halle and Jena, and made mathematics and philosophy his principal objects. He was made a member of the Imperial Academy in 1735, extraordinary professor in 1741, and at last in 1745, ordinary professor of experimental philosophy. He was occupied with a work on electricity, and was in consequence very assiduous in observing the degree of electricity in the air during thunder storms. For this purpose he had constructed in his chamber what he called an electrical gnomon. It consisted of a metallic rod fixed in a glass vessel. To this rod was attached a linen thread, the elevation of which, marked on an accompanying quadrant, the degree of electricity which the rod received. This rod was connected by means of a chain totally surrounded by electrics, with a metal rod passing through a perforated bottle, and fixed upon the roof of the house. During a thunder storm, the electricity collected by this elevated rod was expected to excite the gnomon, and the effect was measured by the elevation of the linen thread upon the quadrant.

The day of his death, Professor Richman had attended an ordinary meeting of the Academy, and hearing some thunder claps at a considerable distance, he hastened home in order to make some observations by means of his gnomon, and he carried with him the engraver Sokolow, who had the charge of the figures for

---

* Phil. Trans. 1753. Vol. XLVIII. p. 350 ; and 1754. Vol. XLVIII. p. 780.
† Phil. Trans. 1753. Vol. XLVIII. p. 377.   ‡ Phil. Trans. 1754. Vol. XLVIII. p. 757.

his projected work, that in consequence of seeing the effects of his gnomon, he might be the better able to represent them upon paper. The linen thread marked 4° upon the quadrant, and the Professor was describing the dangerous effects which might result from its rising so high as 45°, when a terrible clap of thunder alarmed all the inhabitants of Petersburg. His gnomon was broken in pieces, and a ball of fire leaped from the rod to the Professor's head, who was standing at about a foot distance, and stooping towards the apparatus. He immediately fell backwards upon a chest, and was instantly dead. Sokolow was stupified by the shock, and the fragments of the red hot wires were thrown upon his coat, but he was not otherwise injured. The door and part of the ceiling near the conducting chain were rent. There can be little doubt that the light=ning must have been conducted to that particular room in preference by the Professor's apparatus. But in all probability the thunder would have struck his house if no apparatus whatever had been present; and either he himself or some one else of the family might have fallen victims. There are two very particular accounts of Professor Richman's death in the Philosophical Transactions, the first and best by Dr. Watson,* the other translated from the German.†

The next experimenter who appears on the stage is Mr. Delaval. To him Delaval. we are indebted for several notable discoveries. He found that the oxides of metals are non-conductors, a curious circumstance which has since been generalized by Mr. Davy. Powdered dry Portland stone (granular limestone) is a non-conductor : the residue after the complete combustion of animal or vegetable substances is a non-conductor. Portland stone, and tobacco pipes, become non-conductors, when just so hot as to be handled; when hotter or colder they conduct.‡ These last facts were afterwards explained in a satisfactory manner by Mr. Canton. Tobacco-pipe conducts only in consequence of its moisture; hence when heated so as to be rendered dry, it becomes a non-conductor ; when heated red hot, the hot air conducts the electricity ; when cooled, the pipe absorbs moisture and becomes a conductor again.§ Mr. Delaval found that Iceland crystal in some cases becomes a non-conductor, and capable of being excited by friction. When moderately heated it always became a conductor. But these experiments did not succeed with Bergman, who repeated them with great care; he found Iceland crystal a conductor in all cases, to what degree soever of cold it was exposed.‖

It is now time to mention Mr. Wilson, who occupied so conspicuous a figure Wilson. on the stage of the Royal Society for a considerable number of years, chiefly on account of the obstinacy with which he opposed the use of pointed thunder

---

* Phil. Trans. 1754. Vol. XLVIII. p. 765.   † Phil. Trans. 1755. Vol. XLIX. p. 61.
‡ Phil. Trans. 1759. Vol. LI. p. 83 ; and 1761. Vol. LII. p. 353.
§ Phil. Trans. 1762. Vol. LII. p. 457.   ‖ Phil. Trans. 1763. Vol. LIII. p. 97.

rods for the protection of buildings, and maintained the superiority of balls over points. It was by his obstinacy and improper conduct that he introduced those unhappy divisions which had so unfortunate an effect upon the Royal Society, and were so disgraceful to the cause of science and philosophy. Mr. Wilson's best paper appears to be the one which he wrote on the tourmalin. This is a crystalized mineral, very common in primitive mountains in every part of the earth, but at that time brought chiefly, if not entirely, from the East Indies. Its electrical properties are mentioned by Theophrastus, but doubted by Pliny. When the tourmalin is heated to a certain temperature, it becomes always strongly electric; but loses that property either by being too much heated, or by being allowed to cool. Mr. Wilson described the phenomena with considerable accuracy. The subject was afterwards further investigated by Æpinus, and an excellent paper on the tourmalin was written by Bergman. Mr. Wilson likewise observed that when two electrics are rubbed together, the one always becomes electrified plus, and the other minus. The harder of the two, according to him, becomes usually plus, the softer minus.* When insulated silver and glass are rubbed, the silver becomes minus, the glass plus.† Mr. Wilson found some other crystalized minerals to possess the electrical properties of the tourmalin. He describes them as having the hardness of the topaz, and as being red, and some of them inclining to orange.‡ They were no doubt varieties of tourmalin, which the imperfect state of mineralogy did not enable Mr. Wilson to recognise.

Miscellaneous experiments.

I shall here mention some miscellaneous electrical observations which were made about this time, and which do not well admit of being arranged under distinct heads. Symmers observed that stockings of different colours, black and white silk, black and white worsted, black silk and white worsted, &c. when worn together on the leg become strongly electrical, the black *minus*, and the white *plus*. They cohere together with considerable force, and when forcibly separated emit sparks.§ Bergman, from a set of experiments, concluded that water is a very imperfect conductor, when compared with metal. He shows too that ice is a conductor, though it conducts worse than water.‖ It was afterwards ascertained that ice when cooled down to 14° becomes a non-conductor. Mr. Canton observed that when glass is dipped into mercury and taken out again, it is electrified plus, and the mercury minus.¶ This curious observation may be considered as the commencement of a set of investigations afterwards carried to a very considerable length by Volta, and which terminated in the discovery of the galvanic battery. Mr. Kinnersley, of Philadelphia, in

---

* Phil. Trans. 1759. Vol. LI. p. 308.   † Phil. Trans. 1760 Vol. LI. p. 896
‡ Phil. Trans. 1762. Vol. LII. p. 443.   § Phil. Trans. 1759. Vol. LI. p. 340; and 390.
‖ Phil. Trans. 1760. Vol. LI. p. 907.   ¶ Phil. Trans. 1762. Vol. LII. p. 457.

2

North America, who had been the original associate of Dr. Franklin, in electrical experiments, continued afterwards to prosecute the subject by himself, and published a curious paper in the Philosophical Transactions. The following are the most important of his experiments. A coated flask containing boiling water cannot be charged, the electricity passing off with the steam; but when the water gets cold, the flask may be charged as usual. A person in a negative state of electricity standing upon an electric, and holding up a long sharp needle out of doors in the dark, observes light upon the point of it. No heat is produced by electrizing a thermometer, nor by passing shocks through large wire. But small wire is heated red hot, expanded, and melted.*

Dr. Priestley did not make his appearance in public till the year 1767, when he   Priestley. published his *History of Electricity :* a work which is rather carelessly and hastily executed, but which nevertheless must have been of advantage to the science. Almost the whole of his historical facts are taken from the Philosophical Transactions, which may be considered as the great storehouse of electrical information. For the English Electricians of that time, not satisfied with their own original papers, though numerous and important, introduced into the Transactions detailed accounts of all the important books on electricity, which made their appearance abroad. Dr. Priestley at the end of his history gives us a number of original experiments of his own. The most important of all his electrical discoveries was that charcoal is a conductor of electricity, and so good a conductor that it vies even with the metals themselves. This is a curious and unexpected analogy between the metals and charcoal, and chemistry furnishes us with several others no less striking and important. When the conducting power of charcoal was tried by succeeding electricians it was found to vary in the most unaccountable manner, sometimes scarcely conducting at all, sometimes imperfectly, and sometimes remarkably well. This diversity indicates a difference in the nature of different specimens of charcoal. Hence we may be assured that charcoal is a compound, and that it often varies in the proportion of some one or other of its ingredients.

Dr. Priestley's electrical papers in the Philosophical Transactions are not of much importance. The following are the facts which he establishes. He describes coloured circles produced by receiving shocks from 21 square feet of glass on metal plates.† Mr. Canton had previously produced the same colours by melting wires upon glass plates, and the subject was long after carried a great deal further by Van Marum, assisted by his great Teylerian machine When an electrical battery is discharged, if light bodies be placed near the electrical circuit, they are moved. This motion Dr. Priestley ascribes to what he calls the force of the lateral explosion, and he conceives it to depend upon the

---

* Phil. Trans. 1763. Vol. LIII. p. 84.   † Phil. Trans. 1768. Vol. LVIII. p. 68.

sudden elasticity given to the air.*   Finally, Dr. Priestley found that a long circuit conducts much worse than a short circuit, even when the conductors are the same.†   And that when the circuit contains an imperfect conductor a spark passes to bodies near, yet no electricity is communicated.‡

The remaining papers on electricity in the Philosophical Transactions are of so miscellaneous a nature that it is impossible to connect them into any regular detail.   We shall therefore divide the remainder of this chapter into three parts.   In the first we shall give a short outline of the principal facts contained in the various papers; in the second we shall give a sketch of that part of electricity which is called *galvanism,* and which makes its appearance for the first time in the last of the volumes of the Transactions which comes under our review.   In the last part we shall give a view of the theory of electricity as it has been established by a most important paper published by Mr. Cavendish in the Philosophical Transactions, and by Æpinus in a work which made its appearance rather before the publication of Mr. Cavendish's paper, but which Mr. Cavendish assures us he had not seen till after his paper was drawn up, and ready to be presented to the Society.

Remaining papers in the Transactions.

I.   There are about 38 papers on electricity still remaining, which it would be improper to pass over in silence, as they contain each at least something that deserves to be remembered.   The following is the shortest account of them which it is in our power to give.

1. Fogs are usually electric.   During winter the air is often in an excited state.   Ronayne.   Phil. Trans. 1772. Vol. LXII. p. 137.

2. Dr. Priestley gives a description of Henley's electrometer, then newly contrived.   It is a semicircle of ivory divided into degrees, and attached to a piece of brass which is fixed into the principal conductor of the machine.   From the centre of the semicircle a piece of pith ball is suspended by a very light piece of wood so contrived as to be very moveable about the centre.   When the machine is driven the pith ball is repelled, and marks upon the semicircle the degree of repulsion which it sustains by the height to which it rises.   Ibid. p. 359.

3. Charcoal, being examined by Mr. Kennersley, was found to vary in its conducting power.   Oak, beech, and maple charcoal, he found to conduct well, but the charcoal from the pine would not conduct at all.   A line drawn upon paper by a black lead pencil conducted pretty well.   Phil. Trans. 1773. Vol. LXIII. p. 38.

4. The Board of Ordnance having consulted the Royal Society about the best mode of securing the powder magazine, at Purfleet, from the effects of

---

* Phil. Trans. 1769. Vol. LIX. p. 57.   † Phil. Trans. 1769. Vol. LIX. p. 63.
‡ Phil. Trans. 1770. Vol. LX. p. 192.

lightning, the Society appointed Mr. Cavendish, Dr. Watson, Dr. Franklin, Mr. Robertson, and Mr. Wilson, a committee to examine the building, and report upon it; these gentlemen went accordingly, and the first four recommended the erecting of pointed conductors in particular parts of the building, as a means which they thought would afford complete security.* Mr. Wilson dissented from the other gentlemen, being of opinion that the conductors ought not to be pointed but blunt, because pointed conductors solicit and draw down the lightning which might otherwise pass by. He published a long paper on the subject, assigning a great variety of reasons for his preference.† It was this dissent of Mr. Wilson which produced the controversy between the electricians of the Royal Society respecting the comparative merits of pointed and blunt conductors; a controversy which continued for a number of years, and a variety of papers in support of both sides of the question made their successive appearance in the Philosophical Transactions. Indeed it occupies almost the exclusive attention of the writers on electricity for several successive volumes of that work.

5. The next paper deserving notice is by Mr. Walsh on the electric properties of the torpedo. This gentleman went to Rochelle on purpose to examine the extraordinary shocks which it was known the torpedo was capable of giving. The result of his examination was that these shocks were purely electrical. It was found upon examination that the fish possessed a peculiar set of organs constructed for the express purpose of communicating these shocks. These organs were dissected and described by Mr. John Hunter. Phil. Trans. 1773. Vol. LXIII. p. 461.

6. Mr. Nairne gives a set of experiments showing that the shock is diminished by lengthening the circuit, and demonstrating the superiority of points over balls as conductors. Phil. Trans. 1774. Vol. LXIV. p. 79.

7. Mr. Henley proceeds on the same side, and gives a set of experiments showing that pointed conductors ought to be preferred to blunt ones for securing buildings. Ibid. p. 133.

8. The next paper by Mr. Henley is a very long and curious one. He shows that the vapour of water is a conductor of electricity, that when the flame of a candle is introduced into the circuit, and a charged Leyden phial discharged through it, the flame always inclines towards the negative side. He gives a great many curious experiments showing the superiority of points to balls as conductors. He made likewise a set of experiments to determine the relative conducting power of the different metals by the quantity of a wire of each of a given size melted by equal electrical shocks passed through them. He found

---

* Phil. Trans. 1773. Vol. LXIII. p. 42.   † Phil. Trans. 1773. Vol. LXIII. p. 49

the metals as conductors to hold the following order. Gold, brass, copper silvered, silver, iron. Phil. Trans. 1774. Vol. LXIV. p. 389.

9. Mr. Walsh's observations and experiments on the torpedo, seemed to leave no doubt that the shock which it gave was the effect of electricity. But there were several circumstances that required explanation. The shock was given though the fish was under water, yet it could not be made to pass through a chain, no light was ever observed, and it never was sufficient to produce any divergence in the lightest cork balls. These difficulties induced Mr. Cavendish to make a kind of artificial torpedo, by means of which, and a large glass battery, he was enabled to imitate the natural torpedo very closely. He demonstrates, that the length through which the electricity is able to make its way in air depends upon the degree to which the battery is charged. Four jars charged only half as high as two jars will give the same intensity of shock, but the electricity will only pass through half the interval of air. This paper, like all the other papers of Mr. Cavendish, contains a great deal of very curious matter, and all the experiments are made with the utmost exactness. We may mention a few of the most striking facts. Metals conduct 400,000,000 times better than water; sea water conducts 100 times better than fresh water; and water saturated with salt 720 times better than fresh water. The human body is a better conductor than fresh water. Phil. Trans. 1776. Vol. LXVI. p. 196.

10. A pyed bullock was struck with lightning; the white hair came off, the red remained unchanged. Ibid. p. 493.

11. We have in this paper a description of Volta's electrophorus, with an explanation of it by Mr. Henley. Ibid p. 513.

12. In this paper we have a set of experiments, by Mr. Swift, showing the superiority of pointed conductors in securing buildings against lightning. Phil. Trans. 1778. Vol. LXVIII. p. 155.

13. And in this we have a vindication of Mr. Wilson's opinion respecting blunt conductors, by Dr. Musgrave. Ibid. p. 801.

14. Dr. Higgins proposes amalgam of zinc as a substitute for amalgam of tin, for electrical excitation. The amalgam at present usually employed, is amalgam of zinc with a certain quantity of amalgam of tin mixed with it. Ibid. p. 861.

15. Mr. Wilson, who still retained his original opinions, gives us here a long set of experiments, to demonstrate that points are struck in preference to balls, and therefore, that pointed conductors are not so good as blunt ones. Ibid. p 999.

16. Dr. Ingenhousz, who was the contriver of some very amusing experiments, both electrical and chemical, gives us, in this paper, a method of lighting a

candle by means of electricity. Surround a wire with cotton, roll it in powdered resin, discharge a small vial through it; the resin takes fire and will light
a candle. Phil. Trans. 1778. Vol. LXVIII. p. 1022.

17. Dr. Ingenhousz, in the Bakerian lecture, gives an ingenious theory of
the electrophorus of Volta, upon the Franklinian theory of positive and negative electricity. Ibid. p. 1027.

18. This theory is confirmed by Henley; who shows, likewise, that electricity cannot pass through glass. Ibid. p. 1049.

19. The plate-glass machine, which has since been carried to such perfection
by Mr. Cuthbertson, was proposed by Dr. Ingenhousz. He even constructed
one of varnished pasteboard discs. Phil. Trans. 1779. Vol. LXIX. p. 659.

20. Mr. Nairne shows that electricity shortens wire; and that copper conducts
better than iron. Phil. Trans. 1780. Vol. LXX. p. 334.

21. We are more indebted to Mr. Volta than to any other philosopher of the
present age for the introduction of new and important electrical apparatus.
In this paper he gives us an account of his condenser; a very useful instrument
for detecting the presence of small quantities of electricity. By means of it,
Lavoisier, La Place, and himself, succeeded in ascertaining the existence of
negative electricity in the vapour of water, the smoke of burning coals, the air
produced by the solution of iron in weak sulphuric acid, &c. Phil. Trans.
1782. Vol. LXXII. p. 237.

22. Mr. Nairne gives an example of a wire shortened by lightning. Phil.
Trans. 1783. Vol. LXXIII. p. 223.

23. The next paper that comes to be noticed is a description of that very
useful and sensible electrical instrument, the gold-leaf electrometer, of Mr.
Bennet. The contriver gives also some very curious experiments in proof of its
sensibility. Phil. Trans. 1787. Vol. LXXVII. p. 26.

24. In the year 1787, as James Lauder was driving a cart from the bridge Lord Stan-
at Coldstream, in Berwickshire, both he and his horses were killed by light- hope's return-
ning. Yet it appeared from the accurate account of the whole, drawn up by emplified.
Mr. Brydone, and sent to the Royal Society, that there was no discharge of
thunder from a thunder cloud that was not at some miles distance. Lord
Stanhope explained the accident in a very satisfactory manner, by the theory
of the returning stroke, which he had first given in the year 1779. He conceives, that when a large cloud is charged with electricity, it drives out a considerable portion of the usual electricity from the stratum of air in its neighbourhood. When this cloud is discharged, the electricity returns into the
portion of air from which it had been driven, by the influence of the electricity
in the cloud; and it may return with such violence, and in such quantity, as
to destroy the life of any person who may be exposed to its action. Lord
Stanhope gives satisfactory proofs that this must have been the way in which

James Lauder and his horses were killed. Phil. Trans. 1787. Vol. LXXVII p. 130.

25. In this paper Mr. Cavallo points out several defects in Bennet's doubler; and gives various other ways of detecting the presence of small quantities of electricity. He attempts, likewise, to explain in what manner electricity is accumulated by friction. Phil. Trans. 1788. Vol. LXXVIII. p. 1.

26. In this paper Dr. Gray endeavours to explain the way in which glass is charged by the electrical fluid, and discharged. Phil. Trans. 1788. Vol. LXXVIII. p. 121.

27. In this paper Mr. Cavallo gives a description of an instrument capable of collecting together a diffused or little condensed quantity of electricity. He considers it as an improvement upon Volta's condenser. Ibid. p. 255.

28. In this paper we have some curious experiments, by Mr. Nicholson, on the excitation of electricity; on the luminous appearances of electricity; on the action of points; and on what he calls compensated electricity. Some of the remarks in this paper are of considerable value to the practical electrician Phil. Trans. 1789. Vol. LXXIX. p. 265.

29. Dr. Withering, in this paper, gives an account of some extraordinary effects produced by lightning. Phil. Trans. 1790. Vol. LXXX. p. 293.

30. In this paper, and another published afterwards, we have a journal of the state of atmospherical electricity, kept for two whole years, at Knightsbridge, near London, by Mr. John Read. The journal is very curious, but not susceptible of abridgement. It deserves to be studied by meteorologists. Phil. Trans. 1791. Vol. LXXXI. p. 185; and 1792. Vol. LXXXII. p. 225.

31. The last paper we have to mention is likewise by Mr John Read; and consists of a set of experiments made upon Mr. Bennet's doubler, in order to determine its real utility in the investigation of the electricity of atmospheric air, in different states of purity. Mr. Read found air rendered disagreeable by the breathing of a great number of people strongly negative. Phil. Trans. 1794. Vol. LXXXIV. p. 266.

Account of
galvanism.

II. The introduction of what is called the Galvanic Apparatus has constituted an era in electricity, and has considerably altered the views of philosophers respecting the nature and properties of the supposed electrical fluid. What is called *Galvanism* took its name from Galvani, Professor of Anatomy, at Bononia, in Italy, who first observed the phenomena, and published a book on the subject in 1791. The discovery is said to have originated from an accident. Some frogs newly skinned were lying in the dissecting room, in which also there happened to be an electrifying machine. While the machine was at work, some person accidentally touched the thigh of a frog with a probe, the consequence was that the whole leg was thrown into convulsions. Galvani afterwards discovered that the same convulsions may be produced by forming a chain of

Galvani's
discovery.

conducting substances between the outside of the muscles of the leg of a frog and the crural nerve of the same leg. He explained the phenomenon by conceiving the muscles to resemble a charged Leyden phial, having electricity accumulated in the inside, while the outside was minus. The nerve he conceived to be connected with the inside: when it was united with the outside by conductors, the surplus electricity was discharged; and hence the motions of the limb.

Volta, to whose discoveries in this department of science we lie under such obligations, was the first person who added any thing of importance to the original discoveries of Galvani. His letters to Cavallo on the subject were published in the Philosophical Transactions for the year 1793.* He found that the convulsions were produced by touching two parts of the same nerve or muscle, provided two metals be employed to form the electric circuit. These observations destroyed the hypothesis of Galvani, and induced him to conclude that the electricity was produced by the action of the two metals on each other. An opinion which he investigated with much attention for seven years, and was at last led, by means of it, to the fortunate discovery of the Galvanic pile. Volta found, that when two metals in contact were made to touch different parts of the tongue, a pretty strong acid taste was perceived. The metals which were found to act most powerfully after repeated trials, were zinc and silver.

In the same year, 1793, appeared Dr Fowler's Essay on Galvanism. He found that the contractions may be excited by making the metals touch under water, even at a distance from the animal. He succeeded in exciting the heart to contract; but could not produce the same effect upon the stomach and intestines. He found, that when the two metals were applied to the eye, or, what was better, when they were thrust up between the teeth and the lips, and then made to touch, a flash of light is perceived.

Dr. Wells published a valuable paper on the subject in 1795. He found that one metal and a piece of charcoal might be substituted for two metals; and that even one metal would answer if it had been previously rubbed with another metal. He objected to Volta's hypothesis; and conceived the phenomena to depend upon electricity accumulated in the muscles.

Fabroni soon after wrote a curious paper on the subject of galvanism; the object of which was to show that the whole depended upon the chemical action of the metals on each other. He affirmed that when two metals were placed in contact with liquids, certain chemical changes were produced, owing entirely to the presence and the action of the metals.

It would be tedious and uninteresting to give an account of the numerous

Marginal note: Volta's discoveries.

* Vol. LXXXIII. p. 10.

papers on the subject of galvanism which made their appearance in Germany, during the ten years which elapsed between the original discovery of Galvani, and that of the Galvanic pile, by Volta. They are so numerous and intricate as scarcely to admit of generalization ; and often so inconsistent with one another, that no great degree of confidence can be put in their accuracy. The most voluminous writers on the subject were Humboldt, Pfaff, and Ritter.

In the year 1800, Mr. Volta sent an account of his new-discovered Galvanic pile to the Royal Society. Owing to the war, which at that time existed between Britain and France, one portion of the paper came to the Royal Society several months before any opportunity occurred of sending the remainder This delayed the publication of the discovery longer than would otherwise have taken place, and, in consequence of that circumstance, the apparatus was constructed, and various curious experiments made by different persons in London, before the original paper was laid before the public. The

Voltaic pile. Galvanic pile, discovered by Volta, consisted of three conductors of electricity, two perfect, and one imperfect, piled above each other in the same order, in repeated sets. The perfect conductors were silver and zinc, or copper and zinc ; the imperfect, a piece of card or leather, soked in salt water, and a little smaller than the metal plates. They were piled above each other in the following order : silver, zinc, card ; silver, zinc, card, and so on, employing as many sets as can conveniently stand above each other without risk of falling. The greater the number, the more powerful the effects produced. If the finger of one hand be made to touch the uppermost disc of metal, and the finger of the other hand the undermost disc, a shock is perceived, similar to the shock of a Leyden phial, or still more similar to an electrical battery weakly charged ; and this shock is repeated for a long time whenever the circuit is completed, by placing the fingers as above mentioned.*

Discoveries of Nicholson and Carlisle. The first British experimenters on this interesting subject were Messrs. Nicholson and Carlisle. They discovered that the zinc end of the pile was positive ; and the silver, or copper end, negative. But the most important discovery which they made, was the chemical action of the pile. When the circuit is made by gold, or platinum wires, if a vessel of water be placed in this circuit, and the wires be plunged into it, the water is decomposed with considerable rapidity ; oxygen gas being disengaged from the wire connected with the positive extremity of the pile, and hydrogen gas from the wire connected with the negative extremity. If the wires be of any other metal except gold, or platinum, in that case the negative wire gives out hydrogen gas as usual ; but no gas is evolved from the positive, which undergoes a rapid oxydation.

of Cruickshanks. The next experimenter was Mr. Cruickshanks, of Woolwich He observed

---

* Phil. Trans. 1800. Vol. XC. p. 403.

2

that the positive wire reddened the infusion of litmus, and the negative wire the infusion of Brazil wood.   Hence he concluded, that an acid was evolved at the positive wire; and an alkali at the negative.   The acid, from several circumstances, he conceived to be the nitric; and the alkali, ammonia; and he explained their formation from the common air, or rather the azotic gas, which was contained in a state of solution in the water that he employed.   It was known already, that when nascent oxygen gas comes in contact with azote, it forms nitric acid; that nascent hydrogen, when it comes in contact with the same azote, forms ammonia.   Hence the evolution of these substances, in his experiments, become evident.   Mr. Cruickshanks found, likewise, that the negative wire has the property of reviving metals held in solution by acid.   He found all liquids tried, which contained no oxygen in their composition, non-conductors. But we lie under the greatest obligation to Mr. Cruickshanks for the change which he made in the galvanic apparatus; substituting a wooden trough into which the pairs of plates (previously soldered together) were cemented at regular intervals, and the intervening cells were filled with the imperfect conductor, salt water for example, or water accumulated by means of an acid.   This has been gradually improved into the very convenient apparatus at present used in this country; which, being familiarly known, we do not consider it as necessary to describe.

The next experimenter was Dr. Henry, of Manchester.   He decomposed of Henry ammonia by the galvanic energy, and found that more oxygen was given out when the wires were plunged into sulphuric, nitric, or muriatic acids, than when plunged into pure water.   The gases, as might have been expected, were found by him to be non-conductors.   We may here mention some curious facts established by Colonel Haldane.   He found that the pile does not act under water, nor in vacuo, nor when surrounded with azotic gas; and that it is more powerful than usual when surrounded with oxygen gas.   But the two last circumstances depend upon the nature of the perfect conductor employed.

We now come to Mr. Davy, whose first discoveries were of some importance. of Davy He found that the oxygen and hydrogen gases, obtained from the decomposition of water, might be evolved in different vessels, at a considerable distance from each other.   He found, that if the water interposed between the plates be pure, the pile does not act; that the action is proportional to the oxydation of the zinc; that it is increased by acids; and when they are employed it acts even in vacuo. A pile of zinc and charcoal possesses considerable energy.   He constructed a galvanic trough, consisting of two imperfect conductors; namely, nitrous acid, and a solution of sulphuret of potash in water, and one metal.   This trough, indeed, did not retain its energy any great length of time, because the imperfect conductors decomposed one another; but its action was quite sensible at first. He even found that a trough might be constructed possessed of sensible energy without the presence of any metal whatever, substituting a piece of charcoal for the metal.

of Fourcroy and Vauquelin. The next experimenters that deserve to be mentioned are Fourcroy and Vauquelin. They thought of substituting large square plates of zinc and copper for the small discs that had been hitherto employed. The consequence of this change was the discovery of several remarkable phenomena: the burning of gold leaf, dutch leaf, tinfoil, iron wire, &c., and above all the combustion of charcoal, with a degree of brilliancy that rivals the light of the sun. These chemical effects were found to increase at a great rate with the size of the plates; whereas the violence of the shocks depended more upon the number of plates than upon their size. The different effects of a great number of small plates, and of a small number of very large plates, has been lately placed in a striking point of view, by a set of experiments published in a late volume of the Philosophical Transactions, by Mr. Children. He found, that when the circuit was composed of very good conductors, the great plates produced a very powerful effect; but the intensity was so small, that the electricity was unable to penetrate through imperfect conductors. He showed, that the quantity of electricity increases with the size of the plates, while its intensity increases with the number of pairs.

of Wollaston. The next experimenter, who comes to be mentioned, is Dr. Wollaston. He showed, that oxydation has the property of disengaging electricity; and hence explained the way in which amalgam of zinc acts in increasing the energy of an electric machine. He conceived that it became speedily oxydated by the action of the air, and that during the change it evolved abundance of electricity, which made its way to the prime conductor. He found, that a machine could not be excited when surrounded with carbonic acid gas; a fact which he considered as corroborating his hypothesis respecting the use of the amalgam. Finally, Dr. Wollaston succeeded in producing, by the common electrifying machine, the decomposition of water, and other chemical decompositions, which, till that time, were considered as peculiar to galvanic electricity. Thus he removed the grounds which had induced many persons to consider electricity and galvanism as two distinct species of energies.

General law discovered by Davy. But the great discovery respecting the agency of galvanism was made by Mr. Davy, and published in the Philosophical Transactions, in a paper which gained the prize proposed on galvanism by the French Emperor This discovery may be expressed in the following sentence: the galvanic energy has the property of decomposing all compound substances, (supposing the battery sufficiently powerful,) and the constituents arrange themselves round the wires, passing from the two extremities of the battery, according to the following law: oxygen and acids arrange themselves round the positive wire; hydrogen, alkalies, earths, and metals, round the negative wire. From this very important discovery, Mr. Davy drew some very plausible inferences. Oxygen and acids, since they are attracted towards the positive wire, are naturally negative; while, on the other hand, hydrogen, alkalies, and metals, being attracted to the negative wire, are naturally positive. When two substances are chemically combined, they are in different states of electricity; and the more completely opposite these states are, the more intimately they are united. To separate the two constituents of bodies from each other, we have only to bring them to the same electrical state; and this is the effect which galvanism produces. Chemical affinity is nothing else than the attraction which exists between bodies in different states of electricity. The decomposition of the fixed alkalies, the alkaline earths, and boracic acid, soon after accomplished by Mr. Davy, was the natural consequence of his original discovery. These, though very striking and important, are not to be compared, in point of value, to his original discovery of the decomposing power of galvanism; which has made us acquainted with a new power in nature, and put in our possession a much more efficient chemical agent than any with which we were before acquainted. This is the discovery which does Mr. Davy so much honour, and puts him on a level with the small number of individuals, who have been fortunate enough to lay open to the world a new law of nature. Mr. Davy's discoveries have been invidiously compared with those of Newton; and have been affirmed to consist entirely in insulated facts, unconnected with any general principle. The allegation is unjust. The decomposition of the alkalies and boracic acid are insulated facts; but Mr. Davy's great discovery is this general law, that galvanism decomposes all compound bodies, and that the decomposition takes place in a certain determinate manner.

III. The most rigid and satisfactory explanation of the phenomena of elec- *Theory of* tricity, which has hitherto appeared in any language, is contained in a very *electricity.* long, but most masterly paper of Mr. Cavendish, published in the Philosophical Transactions for 1771.* It is a very remarkable, and to me an unaccountable circumstance, that notwithstanding the great number of treatises on electricity which have appeared since the publication of this paper, which is, beyond dispute, the most important treatise on the subject that has ever been published, no one, so far as I recollect, has ever taken the least notice of Mr. Cavendish's labours, far less given a detailed account of his theory. Whether this be owing to the mathematical dress in which Mr. Cavendish was obliged to clothe his theory, or to the popular and elementary nature of the treatises which have been published, I shall not pretend to determine; but at all events it is a thing very much to be regretted. Mr. Cavendish's paper is so long, so concise, and so mathematical, that it is quite impossible to give a full account of it

* Vol. LXI. p. 584.

in this place. I shall attempt merely to give a general outline of his theory. This will be most perspicuously done in detached propositions. It may be proper to mention, that the same theory is also given by Æpinus, though not in so complete a form as by Mr. Cavendish. It appears also that Æpinus's treatise on the subject was published first.

1. A fluid exists in all bodies, distinguished by the name of the electric fluid.

2. Its particles repel each other with a force varying inversely as the square of the distance. Mr. Cavendish was not acquainted with the rate of variation when he wrote; but I consider it as established by the experiments of Coulomb.

3. They attract some other ingredient in all other bodies with a force varying inversely as the square of the distance.

4. The electric fluid is dispersed through the pores of bodies; and, owing to some unknown cause, it moves through them with various degrees of facility.

5. Through one set of bodies called *conductors,* and sometimes *non electrics,* it moves with great facility; while through another set called *electrics* or *non conductors,* it moves with great difficulty, or not at all.

6. The phenomena of electricity arise from the attractions and repulsions of the electric fluid, and from its passage from one body to another.

7. Bodies are in their natural state when the repulsion of the electricity is balanced by the attraction of the matter. In that case they are said to be *saturated* with fluid. Two bodies in their natural state neither attract nor repel.

8. Redundant electricity is called *positive* or *plus;* deficient electricity is called *negative* or *minus.*

9. When a body contains more than its natural proportion of electricity the surplus is repelled till the body is reduced to its natural state.

10. When a body is less than saturated, the *matter* attracts *fluid* till it is brought to its natural state.

11. This efflux or influx may be prevented either by surrounding the body with substances through the pores of which the fluid cannot pass, or by the body itself being of that nature.

12. When a body is over-saturated at one end, and under-saturated at the other end, the fluid flows towards the under-saturated end till the equal distribution is restored; because it is repelled by the over-charged, and attracted by the under-charged end.

13. The same attractions and repulsions exist when two bodies, the one over-charged the other under-charged, are separated by a non-conducting substance; but in that case the flow cannot take place.

14. These bodies may contain such proportions of fluid and matter, that the fluid shall have no tendency to enter or escape from them. The bodies in that case are called *neutral.* Both ends cannot be neutral at once. Either may attract or repel.

15. Let us suppose two bodies A and B, in their natural state, containing the fluid F, f; the matter M, m; and the matter attracting the fluid with a force z. A tends towards B with a force $= Fmz + Mf\dot{z} - Ff\dot{z}$. Now Fmz is manifestly equal to $Ff\dot{z}$. Consequently they ought to move towards each other with a force $= Mf\dot{z}$. But as no such attraction is perceptible, it is obvious that it must be opposed by some equal repulsion. Hence it follows that M must repel m; or at least some part of M must repel m with a force equal to $Mf\dot{z}$.

16. When A and B contain the redundant matter M$'$, m', they repel with a force equal to M$'m'z$.

17. When A contains redundant fluid F$'$, and B redundant matter m', they attract with a force equal to F$'m'z$.

18. If either A or B be in its natural state, the bodies neither attract nor repel.

19. When A is over-charged, and B partly over-charged, partly under-charged, A is attracted, repelled, or neutral to B according to the proportion and distance. But when the fluid is unequally distributed in both bodies the phenomena become more complicated.

20. If A when over-charged be brought near B, the fluid in B is driven towards the side most remote from A, so that the side of B nearest A becomes undercharged and is attracted. The fluid in B increases in density from the side nearest A to the other extremity, and there is a point in it where this density is natural. This is called the *neutral point*.

21. If an electrified body A be brought near another body B, then B becomes electrified by position, and its electrical state is always opposite to that of A. In this case it is obvious that nothing is communicated. Therefore it follows that there is only one electrical fluid and not two, as is supposed by some philosophers.

22. When an insulated body B is brought near an electrified body A, a spark is observed to pass between them; B becomes electrified permanently, and the electricity of A is diminished. This passage of the electric fluid from the one body to the other is instantaneous.

23. When a body has imparted electricity to another, it constantly repels it, because both are in the same state.

24. In the preceding case it is obvious that something is transferred from one body to the other. The vivacity of the phenomena is proportional to the quantity of fluid transferred. It is greatest when conductors are employed, because the discharge is then most complete.

25. The action of electricity is not interrupted by the interposition of non conductors, but only the passage of it from one body to another.

26. Strong electricity may exist in bodies without being sensible.

Such is a brief outline of the principal points in the theory of electricity. For the demonstration of these propositions, and for a comparison of them with the phenomena of nature, we refer to Mr. Cavendish's paper.

CHAP. IX.

OF MAGNETISM.

ALTHOUGH magnetism is of more practical utility to mankind than electricity, it cannot be said, as a science, to have made so much progress. Æpinus indeed has given us a mathematical explanation of it, founded on an hypothesis, similar to his electrical. But the best treatise on magnetism, which we have seen, is contained in the supplement to the third edition of the Encyclopædia Britannica, and was drawn up by the late Professor Robison, of Edinburgh. His hypothesis is the same as that of Æpinus, but he has given us many original and valuable experiments of his own. We shall begin this chapter with a short statement of the principal facts that have been discovered relative to magnetism, and then give a short historical sketch of the discoveries belonging to the science as far as we are acquainted with them.

Phenomena.
By *Magnetism* is meant the property which certain metallic bodies have of attracting and repelling bodies of the same nature with themselves, and of assuming a particular direction when suspended freely.

We are at present acquainted with four metals possessed of these properties; namely, *iron, nickel, cobalt,* and *chromium.* But the three last of these metals being very scarce, and it being difficult to procure them in a state of purity, almost all the experiments on magnetism have been made with iron.

There is an ore of iron frequently met with which possesses the magnetic virtues in a considerable degree. This ore is called *magnetic iron ore,* or sometimes *loadstone;* and it was by means of it that the magnetic virtue of iron came first to be known. This ore, according to the analysis of Bucholz, is a combination or mixture of the black and red oxides of iron, and usually containing a portion of silica.

Pure iron is not susceptible of acquiring the permanent magnetic virtue. It must be combined with a portion of carbon, phosphorus, or sulphur. When saturated with any of these bodies, it becomes incapable of exhibiting the magnetic energy. For these curious facts we are indebted to Mr. Hatchett.

A bar of steel may be converted into a magnet, either by hammering it while placed in a direction parallel to the axis of the earth, or simply by stroking it with a magnet in the same direction.

A magnet has two poles, the north and the south. When the magnet is suspended freely, the first turns towards the north, and the second towards the south. Dissimilar poles attract, similar poles repel: and from the experiments of Lambert, Robison, and Coulomb, it appears that these attractions and repulsions vary inversely as the square of the distance.

Every piece of iron, when brought near a magnet, becomes a magnet. Soft iron retains the magnetic virtue only while in that situation. But hard steel when once converted into a magnet continues so, though removed to a distance from the magnet, to which it was originally indebted for the communication of that quality.

Magnets arrange themselves with respect to each other in a particular direction. Let a magnet, a, and a needle, b, fixed on a pivot, be placed near each other, the magnet will cause the needle to assume a particular direction. The force which the magnet thus exerts is called the *directive force;* the tendency of the needle to assume a particular direction, is called its *polarity.*

The ratio of the directive to the attractive force is increased by diminishing the length of b. Hence the polarity of a small magnet may continue when its attraction is insensible.

The directive force causes a needle, when carried round a magnet, to arrange itself as the tangent of a *magnetic curve.* When a needle is carried round two magnets, it becomes a tangent to a compound magnetic curve.

When a needle is drawn out of its direction, the force which restores it is proportional to the sine of the angle of inclination.

The power of magnets is increased by communicating magnetism to other bars. Therefore they lose nothing by this communication; or nothing is transferred from one bar to the other.

The earth is a great magnet. Hence the reason why iron in the proper position becomes magnetical.

The needle properly poised points to the north; but its direction in the same place is not fixed, but varies in an irregular manner. At one time at London, for example, it points due north; from this it gradually deviates westwards, till in the course of a series of years it points perhaps 25° west; here it becomes stationary for some time, then begins to move eastward, and does not stop till it has got as far to the east of the meridian, as it was west before. This variation is called the *declination* of the needle. No satisfactory explanation has yet been given of it; the hypothesis of Halley not having been received by the philosophic world. The needle has likewise a diurnal variation, west in the morning and east in the afternoon. This diurnal variation has been sufficiently accounted for by the effect of the solar rays.

There is a certain great circle of the earth cutting the equator in two nodes, and deviating about 10° from the equator. This great circle is called the *magnetic equator.* Its poles are at about latitude 80°. The north magnetic pole seems to be in north latitude 80°, and west longitude from Greenwich 35° or 40°

A needle placed any where upon the magnetic equator remains horizontal; but if it be north or south from this equator, it deviates from the horizontal direction, *dipping* towards the nearest magnetic pole. This dip is called the *inclination* of the needle. It is more regular than the *declination*, and has been subjected to calculation. The dip increases at first rapidly, and afterwards more slowly, as the latitude increases.

The magnetic intensity increases as we advance from the equator towards the poles.

Such are the principal magnetical facts hitherto ascertained. Let us now take a rapid view of the history of magnetism.

The ancients were acquainted with the attractive property of magnets, but they were ignorant of their directive property. Nor do we know when this property was discovered, or who first thought of applying it to the purposes of Mariners' compass navigation. But we are certain that the compass was known and employed in the year 1260. Its history is unknown, though it is supposed to have been borrowed from the Arabians. Flavio Gioio, an inhabitant of Melphi, in Italy, is said to have first suspended the needle on a pivot, and Dr. Wallis claims for the English the first application of the convenient method, at present practised, of having the needle attached to a card. But the evidence which he adduces in favour of this opinion amounts to nothing at all. Various improvements on the mariners' compass have been proposed by different persons in almost all the nations of Europe; as by Middleton, in 1738; by Knight and Smeaton, in 1750; by Lahire, in 1716; by Mean, in 1731; by Buache, in 1732; by Lemaire, in 1747; by Gaule, in 1777; by Coulomb, in 1780; not to mention the improvements proposed by Cassini, Zecher, Van Swinden, &c. The first dipping needle was made by Norman: Mitchell's dipping needle was improved by Nairne, in 1772: Lorimer's needle for the dip and variation was constructed in 1775; and Macculloch's sea compasses, in 1788.

Variation.   The variation is usually said to have been first discovered by Columbus; but this can only mean that he first observed that the magnetic meridians, in different longitudes, do not make the same angles with the meridians of longitude. The discovery, that the needle varies in its declination in the same place, was made in England; though the name of the discoverer is not accurately known. According to Wallis, it was made by Gellibrand, in 1645; according to Bond, it was made by Mr. John Mair. The first person who collected observations on the declination in different places, and made a chart representing them by curves, was Dr. Halley, in 1700. Since that time another very complete one was published by Mountain and Dodson, first in 1744, and afterwards in 1756. Wilke made one in 1772; Lambert another, in 1779; and Churchman a very complete one, in 1794. Various attempts have been made to reduce the declination to calculation; but none of them have been successful. There is a very extraordinary paper in a late volume of the Transactions, which appears inconsistent with all our previous knowledge of the declination. It is a demonstration

1

that there has been no alteration in the variation of the compass in Jamaica for these 140 years. We know that both on the west and east side of Jamaica, the variation has changed prodigiously within that time: and one can conceive no reason why this should not have happened also in Jamaica.

The diurnal variation of the needle was discovered by Mr. George Graham, in 1724, and it was explained soon after in a satisfactory manner, by Canton. Van Swinden and Coulomb have also given us explanations of it, but not so satisfactory as that of Canton. *Diurnal Variation.*

The dip of the needle was first observed by Norman, and little addition has since been made to his hypothesis and collection of facts. From the united observations which have been made in different times and places, it appears that the dip is subject to much less variation than the declination. Some persons, indeed, have supposed that the dip does not vary at all; but the very careful observations of Gilpin, lately published, demonstrate, that at London it is at present diminishing. The magnetic equator, as laid down by Wilke and Lemonnier, is nearly the very same which has been lately deduced from the observations of Humboldt, by Biot. To Biot we are likewise indebted for a mathematical hypothesis, (though not quite correct), and for a very good table of observations. *Dip.*

It was Graham that first thought of measuring the magnetic intensity by the vibrations of the needle; a method afterwards used by Coulomb, and considered by many as invented by him. From the observations made by Humboldt, and Gay Lussac, in this manner, Biot has deduced the variation of intensity in different latitudes.

The magnetic laws were first generalized and explained by Dr. Gilbert, whose book on magnetism, published in 1600, is one of the finest examples of inductive philosophy that has ever been presented to the world. It is the more remarkable, because it preceded the *Novum Organum* of Bacon, in which the inductive method of pholosophizing was first explained. Dr. Brook Taylor, and afterwards Muschenbroeck, attempted without success to determine by experiment the rate at which the magnetic attractions and repulsions vary. This rate was successfully investigated by the subsequent experiments of Lambert, Robison, and Coulomb. The nature of magnetic curves was first satisfactorily explained by Lambert, Robison, and Playfair.

The first attempt to make artificial magnets, as far as I know, was by Sellers, in 1667.* There is said to be a paper on this subject by Reaumur in the Memoires of the French Academy for the year 1723. In the year 1730, Servington Savery succeeded in making them, and in showing them to be preferable to loadstones. Dr. Gowan Knight soon after made very powerful artifi- *Artificial magnets.*

* Phil Trans. Vol. II. p. 478.

cial magnets. He concealed his method; but after his death it was published, and said to consist in making up a paste of elutriated iron filings and linseed oil. His experiments were published in 1744. Mr. Mitchell, of Cambridge, published an improved method in 1750, and Mr. Canton another in 1751. Antheaume's method was published, in the year 1760, in a separate treatise, but it had appeared before in the Memoires of the French Academy for 1753. Æpinus and Coulomb proposed good methods of making artificial magnets, but later than any of those which have been already enumerated.

From the preceding statements it appears that almost all the discoveries relative to magnetism have been made in England. The most distinguished names as discoverers are those of Gilbert, Norman, Gellibrand, Halley, Savery, Graham, Knight, Mitchell, and Canton. Lambert, Æpinus, and Coulomb, hold the most distinguished place among foreigners.

Papers in the Transactions.

The papers on magnetism, in the Philosophical Transactions, amount to seventy-eight; but of these twenty-one may be passed over as trifling. The following is a sketch of the topics treated of in the remaining papers, omitting those that have been already noticed.

1. Colepress found that heat weakened the magnetic energy. Phil. Trans. 1667. Vol. II. p. 502.

2. Mr. Bond published a table of the variation of the needles, at London, for the next 50 years. But his hypothesis turned out inaccurate. Ibid. p. 789.

3. Dr. Halley attempted to explain the declination of the needle by the hypothesis of the earth being a great magnet with four poles, two in the north and two in the south. The change of declination he accounted for, by supposing the earth a hollow sphere enclosing another sphere at its centre. Each sphere has two poles, and the internal has a slow rotatory motion west. This ingenious hypothesis never gained much ground, and at present, we believe, has no supporter. Phil. Trans. 1683. Vol. XIII. p. 208; and 1692. Vol. XVII. p. 563.

4. An iron bar held in the proper direction becomes a magnet. Phil. Trans. 1685. Vol. XV. p. 1213.

5. Twisting a wire destroys its magnetism. Splitting a wire often inverts the poles or destroys the polarity. The result is affected by the side of the wire laid uppermost. Derham. Phil. Trans. 1705. Vol. XXIV. p. 2136.

6. Dr. Brook Taylor made experiments to ascertain the law of magnetic attraction. At small distances it was nearly as $\frac{1}{d^2}$; but at greater distances as $\frac{1}{d^3}$ or more. He gave four poles to a wire by touching it at one end, or at various parts. Phil. Trans. 1715. Vol. XXIX. p 294; and 1721. Vol. XXXI. p. 204.

7. Mr. Graham found the dip at London, in 1723, to be 74° 40'. In the year 1747, he found it 73° 30'. In the year 1723, the needle made 100 vibra-

tions in 5′ 35″ at London. Phil. Trans. 1725. Vol. XXXIII. p. 332; and 1748. Vol. XLV. p. 279.

8. Mr. Savery describes his method of making artificial magnets. Phil. Trans. 1730. Vol. XXXVI. p. 295.

9. Captain Middleton observed that needles would not traverse in the neighbourhood of ice, and that cold had the same effect; but they recovered their qualities when heated. Phil. Trans. 1738. Vol. XL. p. 310.

10. The magnetic experiments of Dr. Gowan Knight are circumstantially related. The only novelties which they exhibit are the great magnetic power which the magnets employed possessed, and the facility with which a powerful magnet reverses the poles of a smaller one. Phil. Trans. 1744. Vol. XLIII. p. 161, 361; and Vol. XLIV. p. 656.

11. The diurnal variation of the needle was first observed by Graham, then by Celsius, then by Wargentine, Canton, &c. Phil. Trans. 1751. Vol. XLVII. p. 126.

12. The diurnal variation is least in December, and greatest in June at London. It is westward from eight in the morning till one or two o'clock, then returns east again till the evening. Mr. Canton accounts for it in a satisfactory manner by showing that heat diminishes the magnetic power of a needle. The diurnal variation becomes irregular when an aurora borealis appears. The following table gives us the mean diurnal variation for each month of the year 1759, as observed by Mr. Canton.

January	7′	8″	July	13′	14″
February	8	58	August	12	19
March	11	17	September	11	43
April	12	26	October	10	36
May	13	0	November	8	9
June	13	21	December	6	58

Thus we see that it is greatest in June, and least in December. Phil. Trans. 1759. Vol. LI. p. 398.

13. Various tables of the variation occur in the Philosophical Transactions. The following, not reckoning Halley's which has been already quoted, are the most remarkable of these. Montaine. Phil. Trans. 1766. Vol. LVI. p. 216 Ross. Ibid. p. 218. Wales. Ibid. 1769. Vol. LIX. p. 467. Cook. Ibid. 1771. Vol. LXI. p. 422.

14. In the year 1772, the dip, at London, appears from the experiments of Mr. Nairne to have been about 72° 18′ Phil. Trans. 1772. Vol. LXII. p. 476

15. Mr. Hutchins gives us the dip in the year 1774, in different places. The following are the principal :

At Stromness   Lat. 58° 59'   Long. 3° 20' W   Dip 75° 51'
        Hudson's Straits Lat. 62   3   Long. 69 W.   Dip 82   42
        Moose Fort   Lat. 51   20   Long. 82   30 W.   Dip 80   13
Phil. Trans. 1775. Vol. LXV. p. 129.

16. A particular description of Dr. Gowan Knight's great magnetical maga-zine is given, and likewise an account of the method which he employed in making his artificial magnets. Phil. Trans. 1776. Vol. LXVI p. 591; and 1779. Vol. LXIX. p. 51.

17. Cavallo found that brass becomes magnetic by hammering, though it contains no sensible quantity of iron. The magnetism is destroyed by heating the brass red hot, or by partially melting it. Phil. Trans. 1786. Vol. LXXVI p. 62.

18. Cavallo shows that the magnetism of bodies may be best tried by putting them on the surface of clean mercury, and then bringing a magnet near them Phil. Trans. 1787. Vol. LXXVII. p. 6.

19. Mr. Bennet proposes the suspension of a fine sewing needle by a spider's thread in the glass of a gold leaf electrometer, as an excellent means of disco-vering very minute quantities of magnetic attraction. Phil. Trans. 1792. Vol. LXXXII. p. 81.

20. Mr. Macdonald gives us an account of the diurnal variation of the needle in Sumatra. It was less than Mr. Canton observed it in London, not exceeding 2' or 3'; but the whole variation of the needle at Fort Marlbough, in 1794, was only 1° 8' east. Mr. Macdonald considers the diurnal variation as a con-firmation of Halley's hypothesis of the four magnetic poles of the earth. Phil. Trans. 1796. Vol. LXXXVI. p. 340.

21. The same gentleman has given us the diurnal variation at St. Helena. He found it to be 3' 55'' in November when the whole variation was 15° 48' 34¼'' west. Phil. Trans. 1798. Vol. LXXXVIII. p. 397.

BOOK IV.

OF CHEMISTRY.

WE shall divide this Book into three Chapters. In the first we shall treat of Chemistry, strictly so called; in the second, of Meteorology; and in the third, of those Chemical Arts and Manufactures which are described in the Philosophical Transactions.

CHAP. I.

OF CHEMISTRY.

THE origin and primitive meaning of the word *Chemistry* are not accurately known. The science itself is of modern date, and scarcely preceded the origin of the Royal Society. It sprung from a sect of fanatics, who occupied themselves assiduously in endeavouring to convert the baser metals into gold and silver. These men, in the course of the numberless experiments which they made upon the metals, and upon other bodies capable of bearing the heat of their fires, or of altering the appearance of their metals, discovered a variety of powerful agents hitherto unknown; and ascertained many curious and important properties of the substances on which they experimented. These they registered in their writings, in a language at once pedantic and obscure, and thickly scattered with the most childish and ridiculous absurdities. Some of these *alchymists*, as they were called, being medical men, began to think of applying some of the preparations which had been thus discovered in medicine. The effects of these new medicines, particularly the preparations of antimony, the first chemical medicines brought into general use in Europe, were so powerful, as not only to draw the attention of the medical faculty, but even to occupy the deliberations of whole kingdoms. Paracelsus, who carried these chemical medicines to an extravagant length, and who boasted that he was in possession of one, capable of curing all diseases, and of prolonging life to an indefinite

Origin of the science.

The alchymists.

Paracelsus.

3 O

period; and who was peculiarly fitted by his pomposity and chicanery, for making a noise in the world, and laying the foundation of a new sect, first gave public lectures on chemistry; and impressed mankind with an idea of the importance of attending to the science for medical purposes, independent of the grand alchymistical pursuit of making gold.

The sect established by Paracelsus did not long outlive the founder; but medical chemistry, notwithstanding, continued to be valued, and chemical medicines to be employed, by the most judicious and skilful physicians in Europe. Various persons appeared in different countries, who distinguished themselves, either by the extent of their views, by chemical discoveries useful in the arts, or by contriving preparations of importance in medicine; such as Agricola, Libavius, Van Helmont, &c. Small treatises on chemistry were even drawn up and published; containing an account of the metals, of sulphur, of alkalies and acids, of precious stones, of a few salts, and some rude attempts towards explaining the constituents of the animal and vegetable kingdom. Still chemistry could hardly be considered as constituting a distinct science; but, as depending upon pharmacy and mining. It was chiefly cultivated in Germany, where pharmacy and mining had been most attended to, and where they had been brought to the greatest perfection.

Boyle.

Perhaps Mr. Boyle may be considered as the first person, neither connected with pharmacy nor mining, who devoted a considerable degree of attention to chemical pursuits. Mr. Boyle, though, in common with the literary men of his age, he may be accused of credulity, was both very laborious and intelligent; and his chemical pursuits, which were various and extensive, and intended solely to develope the truth, without any regard to previously conceived opinions, contributed essentially to set chemistry free from the trammels of absurdity and superstition, in which it had been hitherto enveloped; and to recommend it to philosophers as a science deserving to be studied, on account of the important information which it was qualified to convey. His refutation of the alchymistical opinions respecting the constituents of bodies, his observations on cold, on the air, on phosphorus, and on ether, deserve particularly to be mentioned, as doing him much honour. We have no regular account of any one substance, or of any class of bodies, in Mr. Boyle, similar to those which at present are considered as belonging exclusively to the province of chemistry. Neither did he attempt to systematize the phenomena, or to subject them to any hypothetical explanation.

Hooke.

But his contemporary, Dr. Hooke, who had a particular predilection for hypothesis, sketched, in his *Microgaphia*, a very beautiful theoretical explanation of combustion; and promised to develope his doctrine more fully in a subsequent book: a promise which he never fulfilled; though in his *Lampas*, published

3

about 20 years afterwards, he has given a very beautiful explanation of the way in which a candle burns.   Mayow, in his Essays, published at Oxford about ten years after the *Micrographia,* embraced the hypothesis of Dr. Hooke, without acknowledgement ; but clogged it with so many absurd additions of his own, as greatly to obscure its lustre and diminish its beauty.   Mayow's first and principal Essay contains some happy experiments on respiration and air, and some fortunate conjectures respecting the combustion of the metals ; but the most valuable part of the whole, is the chapter on affinities ; in which he appears to have gone much farther than any other chemists of his day, and to have anticipated some of the best established doctrines of his successors.   Sir Isaac Newton, to whom all the sciences lie under such great obligations, made **Newton.** two most important contributions to chemistry, which constitute, as it were, the foundation stones of its two great divisions.   The first, was pointing out a method of graduating thermometers, so as to be comparable with each other in whatever part of the world observations with them are made.   The second, was by pointing out the nature of *chemical affinity,* and showing that it consisted in an *attraction,* by which, the constituents of bodies were drawn towards each other, and kept united.   Thus destroying the previous hypothesis of the hooks, and points, and rings, and wedges, by means of which the different constituents of bodies were conceived to be kept together.

Such were the chemical views and improvements of the British philosophers during the latter part of the 17th century.   The German philosophers, with views less enlightened and correct, but much better acquainted with mining and pharmacy, or what may be called the rudiments of chemical analysis, took a different and less philosophical road ; but a road so pleasing and spacious, that it captivated the fancy of all the chemists in Europe, and contributed very much to the popularity and the advancement of the science.   John Joachim Beccher, the contriver of the first chemical theory, which gained celebrity, and **Beccher.** constituted it a separate and independent science, was born at Spires, in the year 1625, and seems to have been of Jewish extraction.   He was first a Professor of Medicine, then Physician to the Elector of Mentz ; and afterwards to the Elector of Bavaria.   Towards the end of his life he went to England, and died in London, in great poverty, in the year 1682.   The greatest part of his time was spent, as he informs us in the preface to one of his books, in going down into mines, and examining the structure of the earth, and the changes which the minerals gradually underwent.   A slight inspection of his writings is sufficient to show that he had paid very little attention to the mathematical sciences, for he falls into mistakes which a very slight acquaintance with the principles of mechanical philosophy would have enabled him to avoid.   His great work, in which he developes his chemical theory, was his *Physica Sub-*

terranea, published at Frankfort, in 1669. The book is curious in a mineralogical point of view, and would deserve the perusal of some of the keen theoretic geologists of the present day. Beccher's theory, stripped of every thing
but the naked statement, may be expressed in the following sentence : besides
water and *air,** there are three other substances, called *earths,* which enter
into the composition of bodies ; namely, the *fusible,* or *vitrifiable* earth ; the
inflammable, or *sulphureous ;* and the *mercurial.* By the intimate combination
of *earths* with *water* is formed a universal acid, from which proceed all other
acid bodies : *stones* are produced by the combination of *certain earths: metals,*
by the combination of all the *three earths,* in proportions, which vary according
to the metal.

Stahl.   Ernest Stahl adopted the chemical theory of Beccher, and improved it so
much that the reputation of the master was lost in the celebrity of the disciple ;
and the theory was ever after distinguished by the name of the Stahlian Theory.
Stahl was a German, first Physician to the King of Prussia, and Professor of
Medicine and of Chemistry. He was born in 1660, and seems to have been
actuated, from his very infancy, by a violent passion for chemistry. At the age
of 15, he could repeat by heart the whole of the *Chemica Philosophica* of Ber-
nerus, at that time the most popular book on the subject. Stahl discarded the
mercurial earth of Beccher, and to his *inflammable earth* he gave the name of
phlogiston, and made it the foundation of his doctrine of *combustion,* which
constitutes the fundamental part of his chemical theory. Phlogiston, according
to him, constitutes a constituent of all combustible bodies : during combustion
the phlogiston makes its escape, and is the cause of the heat and the light which
accompany combustion. What remains after the combustion is at an end is the
other constituent of the combustible body. For example, *sulphur,* when burnt,
leaves behind it a quantity of *sulphuric acid.* Hence sulphur is composed of
sulphuric acid and phlogiston. The Stahlian theory hung so well together,
and was supported by such plausible and satisfactory experiments, that it was
almost immediately embraced by the chemical world, and for a period of more
than 50 years, from its original propagation, the sole object of chemists seems
to have been to confirm and extend it.

The chemistry of Stahl, as it was cultivated in Germany, and France, and
other countries of Europe, scarcely aspired beyond the bounds within which it
had been circumscribed by its original founder. A few important facts indeed
were added, but they were either connected with medical preparations, or
attracted attention solely as objects of curiosity. The great and tempting field

---

* I am not quite clear about Beccher's opinions relative to water and air. They do not appear
consistent.

of Philosophical Chemistry lay unexplored, when it was entered upon with ardour by Dr. Cullen, who first perceived its value, and whose genius and in- Cullen. dustry, had they not been turned into another channel, would in all probability have been crowned with the richest discoveries. But though Dr. Cullen soon abandoned his chemical pursuits for those of medicine, he was fortunate enough to have initiated into the science a man, whose discoveries formed an era in Chemistry, and who first struck out a new and brilliant path, which was afterwards fully laid open and traversed with so much eclat by the British Philosophers who followed his career.

This fortunate pupil of Dr. Cullen was Dr. Joseph Black, who was born at Black. Bourdeaux, in France, in the year 1728. His father was a native of Belfast, but of a Scottish family, and he resided at the time of his son's birth at Bourdeaux, where he was engaged in the wine trade. As soon as he had reached the age of 12, young Black was sent to Belfast, that he might have the advantage of a grammar school education; and four years after, or in the year 1746, he was sent to the University of Glasgow. Here he got acquainted with Dr. Cullen, who was at that time Lecturer on Chemistry in the University of Glasgow, became enamoured with the science, and the zealous assistant and friend of his master. Young Black, having made choice of medicine as the profession best suited to his views, went to Edinburgh in 1750, to finish his education. At that time the professors entertained different opinions about the action of lithonthriptic medicines, particularly lime-water; and the students as usual entered eagerly into the controversy. It seems to have been this circumstance which first led Mr. Black to investigate the cause of the difference between lime-stone and quick-lime, and to examine the nature and properties of magnesia. These investigations appear to have been completed about the end of the year 1752 or 1753. The result he first sketched in his inaugural dissertation, and afterwards detailed at full length in his Essay on *Magnesia,* published a few years afterwards. Lime-stone he demonstrated to be a compound of *lime,* and an aerial substance, to which he gave the name of *fixed air.* This fixed air has the property also of combining with the alkalies, and with magnesia. When these bodies contain it, they are in what was called at that time their *mild state,* and they effervesce, when placed in contact with acids; because all acids displace the fixed air, and drive it away. The alkalies, lime, and magnesia, when deprived of this fixed air, are in their *caustic* state. This important discovery constituted the commencement of the investigation of gaseous bodies. Its accuracy was at first disputed in Germany, where Meyer had advanced another and more complicated explanation. But after some controversy the hypothesis of Meyer was laid aside, and the explanation of Dr. Black finally acceded to.

When Dr. Cullen was removed to Edinburgh, in 1756, Dr. Black was appointed his successor as Professor of Anatomy, and Lecturer on Chemistry, in the University of Glasgow. Not considering himself as fully qualified to fill the anatomical chair, he was induced to change tasks with the Professor of Medicine. While in Glasgow, therefore, his chief business was delivering lectures on the institutes of medicine. His reputation as a Professor increased every year, and he became a favourite practitioner in that rich and active city. It was between the years 1759 and 1762, that he brought to maturity his important speculations concerning heat, which had occupied his attention at intervals from the very first dawn of his philosophical investigations Besides constituting the foundation of every thing philosophical at present known respecting heat, we are indebted to these speculations for the steam engine, the most brilliant and valuable present ever made by the sciences to the arts. Dr. Black's discovery may be summed up in the following proposition. When a solid body is converted into a fluid, there enters into it, and unites with it, a quantity of heat, the presence of which is not indicated by the thermometer: on the other hand, when a fluid body is converted into a solid, a quantity of heat separates from it, the presence of which was not before indicated by the thermometer The heat which thus combines, or disengages itself, Dr. Black called *latent heat*, because its presence is not indicated by the thermometer.

In the year 1766, when Dr. Cullen was made Professor of Medicine, in the University of Edinburgh, Dr. Black was appointed his successor in the chemical chair. This situation he filled for more than 30 years with uncommon success. His lectures were remarkably perspicuous, and delivered with such ease, and in so pleasing a manner, totally free from affectation, as to produce universal delight. At last, on the 26th of November, 1799, he expired, in the seventy-first year of his age, precisely in the way that he had often wished might happen. Being at table, with his usual fare, some bread, a few prunes, and a measured quantity of milk diluted with water, he set the cup down on his knees which were joined together, and kept it steady with his hand in the manner of a person perfectly at ease; and in this attitude he expired without spilling a drop, and without a writhe on his countenance, as if an experiment had been required to shew to his friends the facility with which he departed. His servant opened the door to tell him that some one had left his name, but getting no answer he stepped about half way towards him, and seeing him sitting in that easy posture, supporting his bason of milk with one hand, he thought that he had dropped asleep, which sometimes happened after his meals. He went back and shut the door, but before he went down stairs some anxiety made him return, and look at his master Even then he was satisfied, and turned to go away, but returning again, and coming close up to him, he found him without life.

Dr. Black's discoveries were all made before he reached the age of 34. Partly from indolence, and partly from want of health, he never attempted to prosecute the vast and tempting career which he had laid open. But the subject was soon after taken up by Mr. Henry Cavendish, and prosecuted with a degree Cavendish. of precision and success, which has scarcely ever been equalled, and never surpassed. His first dissertation was upon the *fixed air* of Dr. Black, or *carbonic acid gas,* as it was afterwards called ; and upon *hydrogen gas.* He ascertained the properties of both these aerial bodies with remarkable precision ; and though this paper, published in the Transactions for 1765, may be considered as the first dissertation on gaseous bodies ever published, every property was determided with the utmost exactness. Of hydrogen gas, Mr. Cavendish may be considered as the discoverer. It had indeed been collected by former chemists, and its inflammability exhibited, but all its other properties remained unknown till they were discovered by Mr. Cavendish. Some of the properties of carbonic acid had been previously determined by Dr. Black, but Mr. Cavendish's additions were of the utmost consequence. To this important dissertation likewise we are indebted, in some measure, for the apparatus by means of which Chemists have been enabled to investigate the nature of gaseous bodies, though this apparatus was very much improved by the subsequent labours of Dr. Priestley.

Mr. Cavendish's next discovery was a very important one; no less than the constituents of which water is composed. He ascertained that when two measures of hydrogen gas, and one of oxygen gas, are set on fire, they burn with an explosion, and in place of them is found a quantity of water, just equal to them in weight. Hence it was concluded that water is a compound of oxygen and hydrogen, in the proportion of one measure of the first to two of the second, or that it is composed of about $7\frac{1}{4}$ parts by weight of oxygen, and one part of hydrogen, Mr. Watt had previously drawn the same conclusion from the experiments of Dr. Priestley and Mr. Warltire; but his opinions were unknown to Mr. Cavendish, when he laid his paper on the subject before the Royal Society. Mr. Cavendish's next discovery was the composition of atmospherical air, which he demonstrated to be a mixture of nearly 21 measures of oxygen gas, and 79 measures of azotic gas. He found also that these proportions were not liable to any variation, and that the supposed differences observed by Dr. Ingenhouz, and others, were owing to errors committed in making the experiments.

Mr. Cavendish's last, and perhaps his greatest, chemical discovery, was the composition of nitrous acid, which he showed to be a compound of one part by weight of azote, and about $2\frac{1}{2}$ parts of oxygen. During the last part of his life, he appears to have in a great measure relinquished his chemical pursuits. He was a decided enemy to the French new chemical nomenclature, and wrote a dissertation to expose its imperfections, which he was persuaded not to publish, on the supposition that the new nomenclature could not be permanent. A supposition very likely to be verified. Mr. Cavendish died in the year 1810 in a very advanced age, leaving behind him an immense fortune. He was never married, and in his domestic habits was very particular.

Priestley. Dr. Priestley, who began his chemical career a few years later than Mr. Cavendish, was as rapid and precipitate, as the latter was cautious and slow. But from his splendid talents and happy turn of mind, he contributed no less essentially to the progress of the science, and certainly more than any other British Philosopher of his time to its popularity. No man ever entered upon the study of chemistry with more disadvantages than Dr. Priestley, and yet few men have occupied a more dignified station in it, or contributed a greater number of new and important facts. He was an obscure dissenting clergyman, struggling with a very limited income, unsuccessful as a preacher from a defect in his voice, unacceptable to his hearers on account of his theological opinions, which involved him in almost perpetual controversy, loaded with a family, ignorant of chemistry, unacquainted with the necessary apparatus, and so situated that he could not conveniently procure information. These circumstances in his situation, which seemed to present an insurmountable bar to his entering upon experimental investigations, were probably in a great measure the causes of his splendid success. The career which he selected was new, and he entered upon it free from those prejudices which warped the mind, and limited the views of those who had been regularly bred to the science. His invention was set to work to contrive new and appropriate and cheap instruments of investigation. His profession ensured his temperance and sobriety, his family and limited income stimulated his activity, while his controversial discussions kept up his spirits, and inflamed his vanity and love of celebrity. Add to all this a facility of expressing himself, and of writing, which he had acquired by the practice of teaching, and of early composition; an activity of mind, and an ardour of curiosity which it was impossible to damp; a sagacity capable of overcoming every obstacle; and a turn for observation which enabled him to profit by all the phenomena that presented themselves to his view. His habits of regularity were such that every thing was registered as soon as observed. He was perfectly sincere and unaffected, and the discovery of truth seems in every case to have been his real and undisguised object.

The events of Dr. Priestley's life are recent and well known. Indeed a particular account of them has been lately laid before the public, partly drawn up by himself, and partly by his son. The errors into which he fell, and which drew upon him so much obloquy, ought rather, I think, to be ascribed to the simplicity of his character, and to his almost total ignorance of human nature, than to any thing radically wrong either in his disposition or his understanding.

Joseph Priestley was born in 1733, at Fieldhead, about six miles from Leeds, in Yorkshire. His father was a dresser of woollen cloth. Our author, who

5

was the eldest son, lost his mother in 1739, when he was only six years of age, and after that period he was taken home, and supported by a sister of his father. She was a respectable dissenter, and her house was visited by the dissenting clergy in the neighbourhood. Joseph was educated at a public school, and at 16 had made considerable progress in Latin, Greek, and Hebrew. Having shown a passion for books at a very early age, his aunt had hopes of his becoming a dissenting clergyman, which she considered as the most respectable of all professions; and he himself entered very eagerly into her views. But his health declining about this period, his aunt became apprehensive that public speaking would not suit him. In consequence of these apprehensions he turned his thoughts to trade, and learned French, Italian, and German, without a master. But getting stronger soon afterwards, his former plan of becoming a clergyman was resumed, and in 1752 he was sent to the Academy of Daventry, to study under Mr. Ashworth; after having, as he informs us, made some progress in mechanical philosophy and metaphysics, and learned Chaldee, Syriac, and Arabic. At Daventry he spent three years, engaged keenly in theological studies, and there he wrote some of his earliest tracts on religion. After leaving Daventry he settled at Needham, in Suffolk, as assistant in a small, obscure, dissenting meeting-house, where his income did not exceed thirty pounds a year. His hearers fell off in consequence of his religious opinions, which speedily began to deviate from the rigid system of Calvinism, in which he had been educated; and he was partly supported by money, procured from charitable funds, by means of Dr. Benson and Dr. Kippis. Though several vacancies occurred in his neighbourhood, he was considered as too contemptible to be permitted to supply them. Even the dissenting clergy, who lived near him, considered it as a degradation to associate with him. In 1758 he went to Sheffield, as a candidate for a vacant meeting-house; but was rejected. Mr. Haines, however, procured him a meeting-house at Nantwich, in Cheshire. Here he kept a school, and taught privately besides, and by that means improved his income. He taught the school from seven in the morning till four in the afternoon, and he taught privately from four to seven; so that his time it would appear was completely occupied: yet he contrived to find leisure to write his Grammar, and some other tracts. In 1761 he succeeded Dr. Aiken, as Teacher of Languages in the dissenting academy, at Warrington. In this situation he continued six years, and wrote several of his works, especially his History of Electricity, which first brought him into notice as an experimental philosopher, and procured him celebrity. In 1767, he removed from Warrington to a dissenting meeting-house, at Leeds. Here he continued six years, wrote many books, and commenced his great chemical career, which soon drew universal attention, and raised him to the first rank among British philosophers. He was led accidentally to think of pneumatic chemistry, by living in the immediate vicinity of a brewery.

In consequence probably of the celebrity which he had acquired, the Earl of Shelburn (afterwards Marquis of Lansdown) engaged him to live with him, at a salary of 250 pounds a year. But, in consequence of circumstances not very well understood, they soon parted, and Dr. Priestley settled in a dissenting meeting-house, at Birmingham. The writings which he published on theology and politics, in consequence of the unfortunate turn which the minds of men took after the commencement of the French revolution, gradually drew upon him the obloquy of the well-wishers to the British government. He himself affirmed that the minds of the people of Birmingham were artificially excited. Be that as it may, a riot took place in 1791, his house was burnt down, and he narrowly escaped with his life. He went to London, and settled at Hackney; but the obloquy still continuing, and some of his oldest friends refusing to associate with him, he left England in 1795, and settled in Northumberland, in America. Here he continued actively engaged in his literary pursuits, and amply supported by the liberality of his patrons in England; and here he died in 1804, a few hours after he had arranged all his literary concerns, inspected some proof sheets of his last theological work, and given instructions to his son how it should be printed.

Dr. Priestley's chemical career was rendered illustrious by a great number of important discoveries, chiefly of gaseous bodies and their properties. He discovered *oxygen gas, nitrous gas, nitrous oxyde gas, nitrous vapour, carbonic oxyde gas, sulphurous acid gas, fluoric acid gas, muriatic acid gas, ammoniacal gas.* He ascertained various properties of *carbureted hydrogen gas, azotic gas, and hydrogen gas.* His experiments may in general be sufficiently relied on for accuracy; but we look in vain in them for that precise chemical knowledge, which distinguished the experiments of some of his contemporaries. He never attempted to determine the constituents of his gases, nor their specific gravity, nor any other numerical results. He was not even in possession of the means of determining the purity of those gaseous bodies which he examined. This is the more extraordinary, as he was himself the discoverer of some of the best means which are at present in our power. The revolution in chemical theory, produced by Lavoisier, and the greater parade of accuracy which appeared in the writings of the French chemists, and which gave them for a time a preponderancy in the chemical world to which they were not entitled, contributed materially to injure Dr. Priestley's reputation towards the latter part of his life. At one time, perhaps his reputation as a chemist stood higher than his intrinsic merit entitled him to rank; but certainly afterwards he sunk as much too low.

Black, Cavendish, and Priestley, were the three men who pointed out the importance of chemistry, and showed the great part which it acts in the phenomena of nature; who drew to it the attention of the philosophic world, and gave it that popularity which it has ever since enjoyed. But while they were running their great chemical career in Britain, two men appeared in Sweden not less eminent, and who contributed as essentially to the progress and improvement of the science. These men were Bergman and Scheele.

Torbern Bergman was born in 1735, at Catherinberg, in West Gothland. Bergman. His father was receiver of the revenues in that province, a post neither very lucrative nor respectable; which induced him to urge his son to assume the profession of law or divinity. Young Bergman accordingly went to the University of Upsal with that intention, and commenced his studies under the inspection of a friend. But he very soon testified a dislike for both the professions recommended by his father, while he manifested a violent passion for mathematics and philosophy. His friend remonstrated, pointed out the absurdity of his choice, told him that law and divinity were the roads to profit and preferment, while mathematics and philosophy had nothing but celebrity to bestow upon their votaries. Our young philosopher listened to these remonstrances in silence; but still persevered in his favourite pursuits. His friend deprived him of his books, restricted his studies, and left him only to chuse between law and divinity. This restraint almost proved fatal. Bergman's health declined: it was found necessary for him to leave the University and return home. His relations, finding it in vain to struggle with his inclination, at last indulged it, and left him at liberty to pursue those studies of which he was so enamoured.

At that time Linnæus, after having surmounted obstacles sufficient to have crushed an ordinary man, was in the height of his glory, and was revered every where as the patriarch of natural history. He had infused the enthusiasm which actuated his own breast into the minds of his pupils; and, at Upsal, every student was a natural historian. Bergman in particular attached himself to Linnæus, and bestowed much pains on botany and entomology. On his return to the University, he imparted to Linnæus his discoveries respecting leeches, who sent them, with a very flattering panegyric, to the Academy of Stockholm. *Vidi et obstupui.*

In the year 1761, Bergman was appointed Teacher of Mathematics and Natural Philosophy, at Upsal. He was in this situation in 1767, when Wallerius, the Professor of Chemistry, resigned his chair. Bergman, though it does not appear that before that time he had paid any particular attention to chemistry, offered himself as a candidate for the vacant professorship. But Wallerius had other plans. There was a relation of his own whom he wished to succeed him, and the Professors at Upsal had so much influence, that few doubted but he would accomplish his plan.

Bergman, to show himself qualified for the place to which he aspired, pub lished two Dissertations on Alum, which were attacked by Wallerius in a style of acrimony unworthy of his reputation. Gustavus IV., afterwards King of Sweden, was at that time Prince Royal, and Chancellor of the University. The character and abilities of this extraordinary man, and the zeal with which he sometimes supported men of science, are well known. Fortunately for chemistry, and for the reputation of Sweden, Gustavus took the part of Bergman, on the recommendation, it is said, of Von Swab, and he was so keen on the subject, that he supported his cause in person before the Senate. Wallerius and his party were baffled by the influence of this prince, and Bergman got the chair. He filled it with the greatest glory for 17 years; and, during that period, his numerous publications entirely altered the state of chemistry. He introduced an order, a perspicuity, an exactness, which were unknown before, and which contributed more than any thing else to the subsequent rapid progress of the science. To Bergman we are indebted for the invention of the method of analysing mineral waters, and mineral bodies in general, such as is practised at present. Some feeble attempts indeed had been made by Boyle, Duclos, and some succeeding chemists, to determine the constituents of mineral waters. Macbride had been more successful, and had explained the cause of the briskness of certain acidulous waters, by showing that they contained carbonic acid gas in solution. But these efforts of preceding chemists vanish into nothing, when contrasted with the exact and truly admirable labours of Bergman, who not only determined the constituents of mineral waters, but pointed out the method of imitating them with exactness. Margraaff had made some essays towards the analysis of minerals by solution; but they are insignificant when compared with the labours of Bergman, who gave detailed analyses of the precious stones, and described, at large, the method of determining the constituents of all the metallic ores. It is true that the methods of Bergman have been greatly improved and altered; and the science is chiefly indebted to Klaproth for these ameliorations. He has devoted a long life to this important object; and has given us the analysis of a greater number of minerals than all the other chemists of Europe united together. But this does not detract from the merit of Bergman, to whom the place of founder of this most important department of chemistry unquestionably belongs.

Bergman's first care, after obtaining the chemical chair, was to collect all the different chemical substances, and form them into a cabinet. Another cabinet contained the minerals of Sweden, arranged according to the places from which they were obtained; while a third cabinet was filled with the models of the different instruments employed in chemistry, and in the manufactures depending on it. These were designed for the instruction of his pupils, whom he encouraged and inspired with that enthusiastic ardour which is necessary for those who devote themselves to the improvement of the sciences. One day he was told of a young man who resided in the house of an apothecary, at Upsal, and who was reproached for neglecting the duties of his profession, while he devoted the whole of his time to the study of chemistry. Bergman's curiosity was excited. He paid the young man a visit, and was astonished at the knowledge which he displayed, and at the profound researches in which he was engaged, notwithstanding the poverty under which he laboured, and the restraint under which his situation placed him. He encouraged his ardour, and made him his favourite pupil and friend. This young man was Scheele, one of the most extraordinary chemists that ever existed.

Charles William Scheele was born in 1742, in Stralsund, the capital of Swedish Pomerania, where his father was a tradesman. Young Scheele, at a very early period of his life, showed a strong inclination to learn pharmacy, and was accordingly bound an apprentice to Mr. Bauch, an apothecary at Gottenburg, with whom he lived about eight years. Here he first felt the impulse of that genius which afterwards made him so conspicuous. He durst not indeed devote himself openly to chemical experiments; but he contrived to make himself master of the science by dedicating those hours to study which were assigned him for sleep. According to the testimony of Mr. Grunberg, his fellow apprentice, he was of a very reserved and serious disposition, but uncommonly diligent. He attended very minutely to all the processes, reflected on them while alone, especially during the night time, and read every thing relating to the subject in the writings of Neuman, Lemery, Kunkel, and Stahl. Kunkel's Laboratory was his favourite book. He used secretly, during the night time, to repeat experiments out of it. On one of these occasions, as he was employed in making pyrophorus, another apprentice, without his knowledge, put some fulminating powder into the mixture. The consequence was a violent explosion; the whole family was thrown into confusion, and our young chemist was severely chastised.*

From Gottenburg, Scheele went to Malmo, and two years after to Stockholm. In 1773 he went to Upsal, and resided for some time in the house of Mr. Loock. Here Bergman first found him, saw his merit and encouraged it, adopted his opinions, defended him with zeal, and took upon himself the charge of publishing his treatises. He gave him free access to his laboratory, and procured him a salary from the Swedish Academy. Encouraged by this magnanimous conduct, the genius of Scheele, though unassisted by education, or by wealth, burst forth with the most brilliant lustre, and produced a suite of discoveries which at once astonished and enlightened the world. He ascertained the nature of manganese; discovered the existence and the singular properties

Scheele.

---

* The late Mr. Driander, who lived at Gottenburg at the time, used to say that the house in which Scheele served his apprenticeship was a common dram shop.

of *oxymuriatic acid*; and gave a theory of the composition of muriatic acid, which promises fair to be the true one.   He discovered a new earth, which was afterwards called *barytes*; and he determined the constituents of the volatile alkali.   All these brilliant discoveries are related in one paper, published about the year 1772.   He discovered and ascertained the properties of arsenic acid, oxalic acid, tartaric acid, malic acid, citric acid, fluoric acid, gallic acid, prussic acid, tungstic acid, molybdic acid, lactic acid, and uric acid.   He ascertained the nature of plumbago, and molybdena; analysed fluor spar, which had eluded the researches of Margraaff, and all preceding chemists; and determined the constituents of *tungstate of lime*.   His two essays on the prussic acid are particularly interesting, and display the resources of his mind, and his patient industry, in a very remarkable point of view.   His different papers on animal and vegetable substances are particularly interesting, and replete with valuable and accurate information.

Scheele and Bergman cannot well be compared together.   Their zeal indeed for the science, their candour, their love of truth, and their industry, were equal: but Bergman was distinguished by the extent of his views, and by the plans which he formed for the general improvement of the science; Scheele, by the skill with which he conducted particular analyses, and the acuteness with which he distinguished substances by their properties.   Bergman's views were general; Scheele's were particular.   Bergman was the man for drawing the outline; but Scheele understood best how to fill it up.   On one occasion indeed, I mean in his treatise *on fire*, Scheele attempted the very difficult and general subject of combustion; but his attempt was not crowned with success. However, the acuteness with which he treated it deserves our admiration; and the vast number of new and important facts, which he brought forward in support of his hypothesis, is truly astonishing, and perhaps could not have been brought together by any other man than Scheele.   He discovered oxygen gas, and ascertained the composition of the atmosphere, without any knowledge of what had been previously done by Priestley.   His views respecting the nature of atmospheric air were much more correct than those of Priestley; and his experiments on vegetation and respiration, founded on these views, were possessed of considerable value.

Chemistry was deprived of these two illustrious philosophers, Bergman and Scheele, almost at the same time.   Bergman died in 1784, at the baths of Medwi, in the forty-ninth year of his age, crowned with glory, after a life spent in the most laborious industry.   His works have been published in six octavo volumes; the first four of which were given to the world by himself, and the last two were collected and published after his death.

In 1775, on the death of Mr. Pohlen, apothecary at Koping, Scheele was appointed by the Medical College to succeed him; and, in 1777, the widow sold him the shop and business; but they still continued house-keeping at their joint expence. In this situation he continued till his death, which happened in 1786, when he was only 44 years of age. For about six months previous to his death, he had been severely afflicted with the gout. When he began to perceive that his end was approaching, he was induced, from a principle of gratitude, to marry the widow of his predecessor. The ceremony was performed on the 19th of May, while Scheele lay on his death bed. On the 21st, he left her by his will the whole of his property, and on the same day on which he had so tenderly provided for her he died.

The exertions of the British and Swedish chemists had developed a prodigious number of new facts, for which Stahl's theory had not provided, and which therefore it was unable to explain in a satisfactory manner. Its defects and imperfections in consequence came to be universally felt, and various attempts were made in different quarters to amend and improve it. Mr. Lavoisier followed a different plan, and after about 18 years of most assiduous application to the science, succeeded in overturning the Stahlian theory altogether, and established at the same time a new school, which, aided by the fury of the French revolution, domineered for more than 15 years over the science with the most absolute and insolent sway.

Lavoisier was born at Paris in 1743, and his father, who was opulent, spared no expence in his education. His taste for the sciences was decided, and the progress which he made in them extremely rapid. When only 21 years of age, he obtained the prize offered by government for the best essay on lighting the streets of Paris. In the year 1768, he was admitted a Member of the Academy of Sciences. At that time he was become conscious of his own strength, but hesitated to which of the sciences he should devote his attention. Mathematics at first caught his fancy; but the brilliant discoveries of Dr. Black and the British chemists, as he informs us himself, decided him to make choice of chemistry. In 1772, he published a volume of essays, in which he gives an historical view of the progress of the science, and finishes with an elaborate defence of Dr. Black's discoveries respecting fixed air. His great wealth, his situation, his excellent education, his mathematical precision, his general views, and his persevering industry, all contributed to ensure his success. A comparison of the Stahlian theory with the discoveries of the British philosophers soon satisfied him that it was insufficient to explain the appearances of chemistry. And he devoted the whole of his attention to the invention of another theory which might be more satisfactory. On this account his efforts were not directed towards the discovery of new substances, but towards a more accurate knowledge of those that had been already discovered. He repeated the experiments of the British and Swedish chemists with greater attention than they had paid to the numerical results; and he frequently varied the methods of his predeces-

5

sors, or contrived new ones better calculated to attain his objects. He ascertained the constituents of sulphuric acid, phosphoric acid, and carbonic acid; of the metallic oxides; of alcohol, oils, and sugar; determined the ultimate ingredients in animal and vegetable substances; and finally established this general law of nature, that in all cases of *combustion oxygen unites with the burning body*. There is a clearness and precision in his reasoning, and an air of accuracy in his experiments, which contribute very much to give us a confidence in them. He was fond of drawing general conclusions, and sometimes was very happy in his deductions, though at other times he has evidently failed. His speculations respecting heat are neither very satisfactory nor intelligible, and bear striking marks of having originated from imperfect accounts respecting the opinions of Dr. Black on the subject delivered in his lectures. His theory respecting the *acidifying principle* was very plausible when he first proposed it; but it has not been verified by subsequent discoveries, and indeed cannot be admitted unless we alter and restrict the meaning of the word *acid* more than has hitherto been done. His experiments on respiration and perspiration not having been completed, we cannot judge of them with propriety. They certainly possess considerable merit, though the methods employed were neither the best nor the simplest that might have been thought of.

All the papers of Lavoisier, written on chemical subjects, were published in the Memoires of the French Academy, except his first volumes of essays abovementioned. In the year 1789, he published a kind of general view of all that he had done, under the title of *Elements of Chemistry*, a work which can scarcely be praised too highly, whether we consider the arrangement or the matter. The third part indeed contains the description of various expensive and useless pieces of machinery; but they were well calculated to impose upon the ignorant, and thereby to augment the effect which he wanted to produce.

During the horrors of the dictatorship of Robespierre, Lavoisier forsaw that he should be stripped of his fortune, and informed La Lande (to whom we are indebted for the only account of him hitherto published) that he intended to support himself by commencing apothecary. But he was accused of defrauding the public revenue, thrown into prison, and on the 8th of May, 1794, suffered on the scaffold at the age of 51. As a philosopher, Lavoisier must always rank in the first class; and those who knew him best bestow no less praise on him as a man. He was mild, sociable, obliging, and extremely active: his manners were plain and unaffected: he was the patron of men of science, and equally attentive to the wants of the poor.

Lavoiserian theory.

Mr. Lavoisier's particular opinions began to attract attention as early as the year 1780, and some of the older French chemists took umbrage at his threatening to overturn the doctrine of phlogiston. Macquer in particular, who had acquired considerable celebrity, chiefly by the elegance of his writings, declared that if Mr. Lavoisier were to succeed in his plans he would rather renounce the science altogether than adopt the new opinions. Indeed the objections to the Lavoiserian theory were so striking and important, that no person, even in France, far less in any other country of Europe, was disposed to embrace it. But Mr. Cavendish's two great discoveries, of the constituents of water and nitrous acid, enabled Mr. Lavoisier to answer all the objections brought against him in the most triumphant manner. In these answers he was assisted by De la Place, and nothing that Lavoisier ever did is more admirable than the acuteness and ingenuity which he displayed on this occasion. At last, at a meeting of the Academy in 1785, Berthollet, who, after Lavoisier, was the most celebrated Berthollet. chemist in France, induced partly by the force of Mr. Lavoisier's arguments, and partly by his own experiments on *oxymuriatic acid*, renounced the doctrine of phlogiston, and declared himself a convert to the Lavoiserian theory. Four- Fourcroy. croy, at that time Professor of Chemistry at the King's Garden in Paris, and who, by his activity and the elegance of his manner as a lecturer, had also obtained a high reputation, followed the example of Berthollet, and united himself with Lavoisier. Guyton Morveau, who at that time resided in Dijon, and Morveau. was the editor of the chemical part of the Encyclopedie Methodique, and possessed likewise of considerable celebrity, was prevailed upon, during a visit which he made to Paris in 1787, to follow the example of these celebrated men, and declare himself a convert to the Lavoiserian doctrines.

Lavoisier, who was not less of a politician than a philosopher, and who aspired to the honour of being the founder of a new theory which should bear his name, and alter the whole face of the science, resolved to spare no labour nor expence to accomplish his objects. Three Herculean labours were undertaken at the same time, all of them well calculated to attain the object in view, and all of them accomplished with consummate skill and address. 1. The New nomen‑ language of chemistry was far from precise; it had been formed at different clature. times, by different persons, and from different motives. But as most of the terms had been contrived by persons under the influence of the Stahlian theory, they were ill adapted to express the Lavoiserian opinions, and opposed nearly an insurmountable bar to their ever being generally received by the chemical world. Fortunately the defects of the nomenclature were universally felt, and attempts had been already made by various chemists, as for example by Bergman and Morveau, to contrive a new one better adapted to the state of the science. Lavoisier, pointing out the deplorable state of the nomenclature, proposed to fabricate a new one better adapted for his purpose, and for his opinions.

To construct this nomenclature, he associated with himself Berthollet, Fourcroy, and Morveau, the new converts to his opinions. They accordingly contrived a new nomenclature, founded entirely on the Lavoiserian theory, which was published in 1787, under the title of the *New French Chemical Nomen-*

clature ; a title which, by pleasing the vanity of the French Nation, and giving them importance in their own eyes, contributed more than any thing to spread the Lavoisierian opinions, or, as they delighted to call them, the *new French theory*, through the whole kingdom of France. This new chemical nomenclature was undoubtedly possessed of great merit, and adapted with the utmost sagacity to the state of the science in the year 1787; and if the science had stopped short in that state, it might have been considered as unexceptionable. But as this nomenclature was of necessity adapted only to the opinions entertained in 1787, as the science at that period was only in an infant state, as many of the opinions entertained by Lavoisier and his associates must have been imperfect, or even erroneous, and as new facts must be perpetually occurring, for which no systematic nomenclature can provide, it is obvious that every contrivance of the sort must be imperfect, and that its imperfections must become more and more conspicuous as the science advances. This has been the fate of the French nomenclature. It is already exceedingly ill adapted for expressing the present state of the science, and its imperfections must of necessity become more glaring every day. So that in no great number of years it will be absolutely necessary either to discard it entirely, or to alter it altogether. Indeed, the establishment of a new nomenclature in any science ought to be considered as high treason against our ancestors, as it is nothing less than an attempt to render their writings unintelligible, to annihilate their discoveries, and to claim the whole as our own property.

The Annales de Chimie.

2. Though the new nomenclature was thus contrived and published, a much more difficult task remained ; namely, to prevail upon the chemical world to embrace it, and discard the old and familiar terms. Even in France it was opposed by different individuals, all of them keen, and some of them possessed of celebrity, while in the other nations of Europe the prejudice against it might be considered as strong, if not invincible. The method fallen upon by Lavoisier and his associates, to overcome this difficulty, was exceedingly ingenious, and well calculated to answer the purpose. A journal was established, published once a month, and edited by the united body of the French Antiphlogistic Chemists, as they termed themselves. This journal contained all the chemical papers which they published, and notices respecting almost all the writings of importance which appeared on the subject in every country in Europe, and the *new nomenclature* only was employed. This obliged all the French chemists at least to make themselves acquainted with the new nomenclature, as it was by means of this journal, the *Annales de Chimie*, that they got acquainted with the improvements in the science. The celebrity of the editors, and the valuable matter which the journal contained, secured it the attention of all chemists in every country in Europe. So that in a short space of time the new nomenclature became generally known and understood over the

whole civilized world.   If to this we add the fervour of the French Revolution, and the violent passion for innovation which at that time seized not only France, but every country of Europe, so that even England herself, notwithstanding the excellence of her government, and the steady attachment to old customs and opinions which characterize her people, was in danger of being involved in the vortex of revolution, we shall not be surprized that the new nomenclature made its way so speedily through every country in Europe, which at any other period of history would have been difficult, if not impossible.

3. The third means which Lavoisier had recourse to in order to establish his theory, was not less efficacious than the preceding, while it was highly necessary for the speedy propagation of his opinions.   Mr. Kirwan had acquired a _{Kirwan.} high and merited reputation by his essays on *affinity*, on *sulphureted hydrogen*, on the *constituents of the salts*, and the *strength of acids*, and by his *system of mineralogy*.   He published an elaborate defence of the doctrine of *phlogiston*, _{His defence of phlogiston.} founded almost entirely on the experiments of the German chemists, and on those of Dr. Priestley.   His object was to prove that all combustible substances contain a common inflammable principle, and this inflammable principle he conceived to be *hydrogen*.   According to this doctrine, hydrogen is a constituent of charcoal, sulphur, phosphorus, and of all the metals, and during combustion it separates from these bodies, and enters into new combinations.   This modification of the phlogistic theory was favourably received by the most eminent chemists in Europe, and there seems to have been a general tendency in Britain, Germany, and Sweden, to adopt it.   It occurred to Mr. Lavoisier and his associates, that a satisfactory refutation of this theory, associated as it was with the greatest names in Europe, would be a death blow to the doctrine of phlogiston, and would give an eclat to the *French theory*, as they were fond of calling it, which would render it invincible.

Accordingly, Mr. Kirwan's *Essay on Phlogiston* was translated into French, _{Refuted.} and the different sections parcelled out among the apostles of the new system, that they might write a refutation of each.   Lavoisier took charge of four sections, Berthollet of three, Fourcroy of three, Morveau of two, and Monge of one.   The book was then published with a refutation at the end of each section. The effect was precisely what might have been expected.   The experiments of the French chemists appeared so accurate, their reasoning so precise, and their refutation so complete, that Mr. Kirwan himself, with that candour which distinguishes superior minds, acknowledged himself overcome, and embraced the doctrine of his adversaries.

This victory drove all the British chemists out of the field.   Dr. Black had _{British Chemists relinquish the study.} been long inactive as an experimenter; but he adopted the theory of Lavoisier, and taught it in his lectures.   Dr. Priestley still continued to adhere to his old opinions, and to write in support of them; but he was obliged to leave the country with diminished reputation and diminished activity. Mr. Cavendish had published a vindication of the phlogistic theory, which it was impossible to refute, and which the French chemists were wise enough not to notice. He had even written an essay against the French Nomenclature, which he was prevailed upon not to publish. But finding the current setting in strongly in favour of the Lavoiserian theory, he relinquished the study of chemistry altogether, and betook himself to his old pursuits, connected with mathematics and mechanical philosophy. For some years Britain contained scarcely an active chemist, except Dr. Crawford, whose celebrity depended upon another branch of the science; and Dr. Austin, who trode nearly in the footsteps of the French chemists. All our information was drawn from France. The very books, by means of which the elements of the science were acquired, were all translations from that language.

New British chemists.

By degrees, however, a new race of chemists arose, founded at first, as one may say, on the French School, and moulded after their model. Dr. Pearson led the way by his translation of the French Nomenclature, and by the various valuable papers which he published in the Philosophical Transactions. By degrees the reverence, almost approaching to absurdity, which was paid to the French School, wore off; and the natural genius and invention of British Philosophers began to appear. It would be scarcely consistent with propriety to particularize the discoveries which have been the result of this happy revolution, and which have restored Great Britain to the first rank among the improvers of chemistry which she formerly held, as the celebrated men to whom we are indebted for them are still alive, and actively engaged in extending the bounds of the science. But we may mention that three distinct schools, (if we may use the expression) have been established by three gentlemen, certainly not the least celebrated among British chemists. These are, Dr. Wollaston who possesses an uncommon neatness of hand, and who has invented a very ingenious method of determining the properties and constituents of very minute quantities of matter. This is attended with several great advantages; it requires but little apparatus, and therefore the experiments may be performed in almost any situation; it saves a great deal of time and a great deal of expence; while the numerous discoveries made by Dr. Wollaston demonstrate the precision of which his method is susceptible. The second of these schools has been established by Mr. Davy, whose galvanic discoveries and analyses of the alkalies have constituted a new era in the science, and furnished us with a totally new method of experimenting. Mr. Davy is the most active as well as celebrated chemist in Great Britain, and many more discoveries may be expected from him. The third school has been established by Mr. Dalton, who has not yet obtained that degree of celebrity in this country to which he is entitled, and which he will

5

infallibly acquire. His atomic theory is one of the greatest steps that the science has hitherto made; and it appears to be established by the most indisputable evidence. We could with much pleasure to ourselves, and with advantage to the science, descant upon the discoveries and the ingenuity of Mr. Tennant, Dr. Henry, Mr. Hatchett, Mr. Smithson, Mr. Gregor, Mr. Brande, Mr. Howard, Dr Marcet, &c. all of them Members of the Royal Society, and active and successful chemists, were we not withheld by a regard to our limits, and by the impropriety of discussing the merits of living philosophers.

The German chemists did not yield with the same facility as the British. *German chemists.* They considered the credit of their country as in some measure connected with the Stahlian theory, and therefore defended it with much obstinacy. And when they were beat out of their strong holds, they invented new theories, or new modifications of the old, which they defended with as much pertinacity as ever. But the champions of phlogiston in Germany were not equal in abilities to their French opponents. Wiglieb was a laborious, but seems to have been a weak man. His experiments cannot be implicitly relied on. Gren possessed greater abilities, but he was always occupied with so many things at the same time, that he never had leisure to prosecute his ideas. Klaproth, the most eminent of the German chemists, never engaged in the controversy. About the year 1791, he repeated the fundamental experiments before the Berlin Academy, and adopted in consequence the theory of Lavoisier Since the ardour of the controversy has been at an' end in Germany, a new race of chemists has arisen in the North of Germany, who have been of the greatest service to the science, and have contributed a multitude of laborious and difficult experiments. The chief of these, after Klaproth, who still holds the first place, are Bucholz, whose analyses are remarkably correct; Richter and Rose, both lately dead; Gehlen, Trommsdorf, and John. But the French invasion of Prussia, and the introduction of the French system of police, has had a most deleterious effect upon the state of science in the North of Germany.

In Sweden, the death of Bergman and Scheele, which happened about the *Swedish chemists.* time of the introduction of the Lavoiserian theory into the science, produced a kind of interregnum in chemistry; and the chemists, who afterwards appeared, seem to have early embraced the Antiphlogistic Creed. The chief of them were Hielm, Gadolin, Ekeberg, and, above all, Berzelius, who has distinguished himself eminently in both the departments which have been embraced by Davy and Dalton.

The horrors of the dictatorship of Robespierre at first scattered the French *French chemists.* chemists, and filled them with terror and dismay. But after his death, they again collected and resumed their labours, and, except Lavoisier and Vandermonde, who was rather a mathematician than a chemist, their number was still complete. Since that period several new chemists of first rate abilities, and who have acquired high and deserved reputations, have been added to their number. Since Lavoisier's death they have not paid much attention to theory, but endeavoured, by new experiments, to enlarge the bounds of the science, and render those parts already known more complete: except Berthollet, whose acute genius has always given him a tendency to theory. One of his most valuable papers, in a theoretic point of view, is his essay on sulphureted hydrogen, and his account of the hydrosulphurets. His work on chemical affinity possesses great merit; though he has perhaps carried some of his doctrines rather too far, especially his opinions respecting carbonic oxide, and the indefinite proportions in which bodies may combine. Even his fundamental doctrine respecting the non-existence of elective attraction has been called in question, and the experiments, by means of which he overturned it, have been repeated by Mr. Davy, as he informed me himself, without success. Thenard also, and Gay Lussac, have dipped pretty deep into theory, and have followed the routes laid open by Davy and by Dalton, with much ingenuity and success. Their late work would have done them more credit if they had been less anxious to show their priority to Davy in minute particulars of no great consequence, and to point out trifling mistakes into which first discoverers are almost always sure to fall. One of the most active and illustrious of all the French chemists is Vauquelin. He has never speculated upon any theoretical subject whatever, but devoted himself to mineral, vegetable, and animal analyses; the number of of which, published by him, is prodigious, and his accuracy as an experimenter has never been surpassed. It would be easy to dwell upon the merits of a good many other French chemists; but the same reasons that induced us to be silent respecting living British chemists, equally forbid us to touch upon those in France.

Papers in the Transactions.   The chemical papers in the Philosophical Transactions, down to the end of the year 1800, amount to 406. But of these there are no fewer than 118, occurring chiefly in the earlier volumes, that may be passed over as useless or trifling. Of the 288 papers that remain, it will be impossible to give either a particular, or a full account. The following observations are all that we can with propriety venture upon. The papers are of so miscellaneous a nature as to be hardly susceptible of arrangement. However, as every arrangement, even though imperfect, is better than none at all, we shall venture to throw the papers under the following heads.

I. GASSES. The papers on this subject in the Philosophical Transactions are curious in an historical point of view. We see the first rude notions on these bodies gradually improving, till this branch of the science acquired the perfection and the beauty which it has now reached. The papers on the subject in the Transactions amount to about 30. We shall merely state the object of the

Damps.   chief of them. What are called *damps*, in coal-pits and other mines, are now known to be various species of gases, chiefly carbureted hydrogen, and carbonic acid We have various papers on them in the early volumes of the Transactions, by Mr. Jessop and Mr. Beaumont. There are four kinds of damps in mines, Mr. Jessop says, namely, 1. The common (obviously carbonic acid gas.) 2 The peas blossom, so called from its smell, and concerning the nature of which nothing accurate is known. 3 The bag; this is described as a visible round ball, hanging at the top of the mine, which if it be broken, disperses itself and suffocates all the miners. We know nothing about its nature. 4. The *fire damp*, which has been examined repeatedly from coal-pits, and found to be carbureted hydrogen.* We have some curious† and rather important experiments on air, by Mr. Boyle, and by Huygens and Papin. Among other facts, Huygens and Papin found that air, produced by the combustion of gunpowder, amounted to $\frac{1}{7}$th of the weight of the powder.‡ Hauksbee afterwards made some experiments on this subject, the result of which was, that one grain weight of gunpowder yielded one cubic inch of air, $\frac{12}{48}$ths of which was slowly imbibed by water.§

Mr. Hauksbee made a curious set of experiments on the change produced upon air, by passing it through red hot iron and brass. He found that it was incapable of supporting the lives of animals plunged in it. The reason was, that he had deprived it of its oxygen, and converted it into azote. Brass heated to 212° produced no change upon air.‖

Mr. Greenwood, Professor of Mathematics, at Cambridge, in New England, gives some striking proofs of the fatal effects of damp (carbonic acid gas) upon the lives of persons exposed to it.¶ Sir James Lowther, in the year 1733, collected the fire damp (carbureted hydrogen) from a coal-mine, in bladders, and set it on fire for the amusement of his friends.** He even brought bladders filled with it to London, and exhibited the combustion before the Royal Society. This induced Mr. John Maud to collect the hydrogen gas, which Hydrogen gas. escapes during the solution of iron filings in diluted sulphuric acid, and exhibit its combustion in like manner before the Society. This experiment deserves attention, as being probably the first attempt ever made in England, to make experiments upon pure hydrogen gas.†† The same year Mr. Clayton, in imitation of Sir John Lowther, collected the gas which is produced during the distillation of pit-coal, and showed its combustibility. This is considered at

* Phil. Trans. 1675. Vol. X. p. 391 and 450.   † Phil. Collect. No. I. p. 6.
‡ Phil. Trans. 1675. Vol. X. p. 467, 533, 544.
§ Phil. Trans. 1705. Vol. XXIV. p. 1806, 1807; and 1707. Vol. XXV. p. 2409.
‖ Phil. Trans. 1710. Vol. XXVII. p. 199.   ¶ Phil. Trans. 1729. Vol. XXXVI. p. 184.
** Phil. Trans. 1733. Vol. XXXVIII. p. 109.   †† Phil. Trans. 1736. Vol. XXXIX. p. 282.

present as chiefly consisting of carbureted hydrogen, and is employed for lighting manufactories under the name of gas ligh s.*

The experiments of Dr. Hales had by this time laid open a vast field to experimental philosophers, respecting the air contained in different bodies.   Dr.
Brownrig seems to have been one of the first who availed himself of these new
views to explain several phenomena hitherto unaccounted for.   He demonstrated
that Spa water contains half its bulk of an air in which animals cannot live:
the same, he thinks, as choak damp.†   His experiments and observations were of
considerable importance, though they vanish into nothing when compared with
the experiments which Mr. Cavendish soon after communicated, on hydrogen
Carbonic acid gas and carbonic acid.   He ascertained the specific gravity and combustibility
gas.   of the first, and determined the quantity yielded during the solution of different
metals in acid.   He likewise ascertained the specific gravity of the second, and
its absorbability by water.   He determined the gas, produced during fermentation, to be carbonic acid; while putrefaction he found to evolve an inflammable
gas.‡   The investigation of the properties of carbonic acid, thus begun, could
not stop here.   Mr. Lane showed, that when in solution in water, it has the
property of dissolving iron, and concluded that chalybeats commonly consist of
this solution; an opinion fully verified by succeeding observations.§

The unhealthy effects produced by marshes, especially in warm climates,
have been long known, and many attempts were made by physicians to account
for the fact.   Dr. Priestley threw some light upon the subject, by showing that
putrid water rapidly absorbs the oxygen of the atmosphere.‖

Fontana gives us a valuable set of trials upon the breathing of hydrogen gas,
and shows, that when diluted with common air, it occasions no inconvenience;
but when perfectly pure it cannot be breathed without occasioning death.¶

Dr. Ingenhousz, who resided for some years in London, was the contriver of
a variety of curious experiments, both electrical and chemical, rather calculated
to amuse, than to improve the science.   He shows that ether speedily is converted into an elastic fluid; that when a drop of it is let into a tube filled with
common air, or with oxygen gas, the mixture explodes with great violence
when brought in contact with a burning body.   He even attempts, on this experiment, to found a theory of gunpowder; but his effort was not attended with
success.**

Fontana examined airs extracted from different kinds of water.   They con-

---

* Phil. Trans. 1739. Vol. XLI. p. 59.   † Phil. Trans. 1765. Vol. LV. p. 218.
‡ Phil. Trans. 1766. Vol. LVI. p. 141.   § Phil. Trans. 1769. Vol. LIX. p. 216.
‖ Phil. Trans. 1774. Vol. LXIV. p. 90.   ¶ Phil. Trans. 1779. Vol. LXIX. p. 337.
** Phil. Trans. 1779. Vol. LXIX. p. 376.

sisted partly of carbonic acid, and partly of air having a greater proportion of oxygen gas than common air.   He shows that the atmospherical air is but little variable in the proportion of its constituents.*   Dr. Ingenhousz, on the other hand, endeavoured to prove that the air over the sea contained more oxygen than the air over the land; but his conclusions were deduced from faulty experiments.†   To this philosopher we are indebted for another curious and important remark.   He found that the conferva rivularis, when put into boiled water and exposed to the action of the solar rays, gave out oxygen gas ‡

When nitrous gas and common air are mixed, they become reddish yellow. and diminish in bulk if the experiment be made over water.   The diminution is much greater if nitrous gas and oxygen gas be mixed together.   It bears some relation to the proportion of oxygen in the air examined, always being greater as it increases.   This was first observed by Dr. Priestley, and employed by him under the name of eudiometer, as a means of determining the purity of Eudiometer. air.   Priestley's method was adopted by Fontana, Ingenhousz, and by chemists in general.   But there was a striking want of correspondence in the results obtained, indicating a defect in the methods employed.   This induced Mr. Cavendish to make a very exact set of experiments on the subject.   He pointed out the method of making the observations in such a way as to obtain uniform results, and determined atmospherical air to be a compound of very nearly 21 measures of oxygen gas, and 79 of azotic gas.§

Dr. Darwin demonstrated, by a set of experiments, that whenever air expands, its temperature sinks.‖   This subject is now very well understood.   The diminution of temperature is owing to the increased capacity of air for heat, which keeps pace with the expansion.

Mr. Milner made a curious set of experiments on the effect of passing nitric acid and nitrous gas through red hot iron.   It was decomposed, and when the process was conducted slowly, nothing was obtained but azotic gas.   When nitrous oxyde gas was passed in this manner, he observed symptoms of the formation of ammonia.   This led him to attempt to reverse the experiment, and to try to produce nitrous gas by passing ammonia through red hot black oxide of manganese.   The experiment was attended with the wished for success.   He obtained nitrous gas likewise by passing ammonia through red hot calcined green sulphate of iron.   But he could not succeed with red lead, or with alum.¶ It is said that advantage of this experiment was taken by the French chemists, during the fervour of the revolution, to procure nitric acid for the manufacture of gunpowder.   But the story does not seem entitled to much credit.

Mr. Smithson Tennant was the first person that succeeded in decomposing

* Phil. Trans. 1779. Vol. LXIX. p. 433.   † Phil. Trans. 1780. Vol. LXX. p. 354.
‡ Phil. Trans. 1782. Vol. LXXII. p. 426.   § Phil. Trans. 1783. Vol. LXXIII. p. 106.
‖ Phil. Trans. 1788. Vol. LXXVIII. p. 43.   ¶ Phil. Trans. 1789. Vol. LXXIX. p. 300.

carbonic acid, and separating charcoal from it. His method was to pass phosphorus through red hot marble. A black powder was obtained consisting chiefly of a mixture of phosphate of lime and charcoal.* These experiments were confirmed by Dr. Pearson, who, by similar processes, decomposed carbonate of potash, of soda, of lime, of barytes, and of magnesia, and obtained charcoal from them all. Dr. Pearson's paper is remarkable for containing the first account of phosphuret of lime, and of its singular properties. He was the discoverer of this substance.†

To Dr. Pearson likewise we are indebted for a set of accurate experiments on the nature of the gases obtained by passing electric shocks through water. The experiments were made with the assistance of Mr. Cuthbertson; the gases were found to be pure oxygen and hydrogen, in the proportion of two measures of the latter to one of the former. These experiments confirmed Mr. Cavendish's discovery of the composition of water.‡

The last paper we have to mention on this subject is a very valuable one by Dr. Henry, of Manchester, accounting for the expansion of carbureted hydrogen gas by electricity, observed by Dr. Austin. This expansion is owing to the decomposition of the water held in solution by the gas, the oxygen of which unites to the carbon of the inflammable gas, and forms carbonic acid, while the hydrogen assumes the elastic form, and occasions the great expansion.§

New substances.
II. NEW SUBSTANCES. The number of chemical substances has been constantly increasing ever since the origin of the science, owing to the increased analytical and synthetical skill of practical chemists. A great many of these substances were originally described in papers published in the Philosophical Transactions. We shall give a catalogue of the most important of them.

1. Formic acid: first observed by Mr. Hulse and Dr. Fisher, in 1671. Phil. Trans. Vol. V p. 2063.

2. Baldwin's phosphorus (nitrate of lime heated) presented by him to King Charles II. in 1677. Phil. Trans. Vol. XI. p. 788.

3. A catalogue of shining substances is given. They are Baldwin's phosphorus, Bolonian stone,‖ (sulphate of barytes) phosphorus smaragdinus, (fluor spar) and common phosphorus. Phil. Trans. 1677. Vol. XII. p. 867.

4. Dr. Slare makes a set of experiments on common phosphorus, observes its combustibility, the acid formed, and the orange residue left behind. Phil. Collect. No. III. p. 48; No. IV. p. 84; and Phil. Trans. 1683. Vol. XIII. p. 289.

5. Mr. Boyle left his method of preparing phosphorus sealed up with the

---

* Phil. Trans. 1791. Vol. LXXXI. p. 182.   † Phil. Trans. 1792. Vol. LXXXII. p. 289.
‡ Phil. Trans. 1797. Vol. LXXXVII. p. 142.   § Phil. Trans. 1797. Vol. LXXXVII. p. 401.
‖ See Phil. Collect. No. III. p. 77.

Royal Society in 1680. It was opened after his death, and published in 1693. It is as follows. Evaporate human urine to dryness, mix the residue with thrice its weight of sand, and distill with an intense heat. Phil. Trans. Vol. XVII. p. 583.

6. Grew describes the properties of *sulphate of magnesia*, proves its nature to be peculiar, and that its base is neither lime nor an alkali. Phil. Trans. 1695. Vol. XIX. p. 76.

7. Brown describes the method of obtaining sulphate of magnesia from sea water, and mentions *muriate of magnesia*, describing several of its properties. Phil. Trans. 1723. Vol. XXXII. p. 348.

8. The method of making *Prussian blue*, described by Dr. Woodward. Mr. Brown shows that iron is a necessary ingredient, and describes the colour given by prussiate of potash to other metallic solutions. Phil. Trans. 1724. Vol. XXXIII. pp. 15, 17.

9. Newman first ascertained the properties of camphor. He obtained a camphor from oil of thyme, which Mr. Brown showed to differ in its properties from oriental camphor. Phil. Trans. 1725. Vol. XXXIII. pp. 321, 361 ; and 1733. Vol. XXXVIII. p. 202.

10. Frobenius, in an absurd, pompous paper, describes the properties of sulphuric *ether*; the combustion of phosphorus, and the properties of its acid. Phil. Trans. 1730. Vol. XXXVI. p. 283; and 1733. Vol. XXXVIII. p. 55.

11. Boerhaave converted mercury into *black oxide* by exposing it for some months to the temperature of 180°. He likewise formed it by agitating the mercury. In a stronger heat he obtained red oxide. Phil. Trans. 1733. Vol. XXXVIII. p. 145; and 1736. Vol. XXXIX. p 343.

12. Geoffroy describes the method of obtaining Seignette's salt and boracic acid, by mixing borax and sulphuric acid. Of this last process he was the inventor. Phil. Trans. 1735. Vol. XXXIX. p. 37.

13. Frobenius's process for making *ether* described. It is nearly the common process. Phil. Trans. 1741. Vol. XLI. p. 864.

14. Seehl gives two processes for procuring *sulphurous acid*, by distilling a mixture of sulphuret of potash and sulphuric acid. The product should contain sulphureted hydrogen. Phil. Trans. 1744. Vol. XLIII. p. 1.

15. The first account of platina by Ch. Wood, Brownrig, and Watson. Phil. Trans. 1750. Vol. XLVI. p. 584.

16. Dr. Lewis published two admirable papers on platina, in which he demonstrates it to be a peculiar metal, and determines its properties. Phil. Trans. 1754. Vol. XLVIII. p. 638; and 1757. Vol. L. p. 148.

17. Dr. Heberden found carbonate of soda on the Peake of Teneriffe. In this paper soda and potash are distinguished from each other, and a set of ex-

periments by Mr. Cavendish is given showing that potash has a stronger affinity than soda for the three mineral acids.   Phil. Trans. 1765. Vol. LV. p. 57.

18. Dr. Monro, in an elaborate paper on the subject, demonstrated, that there are different species of *vegetable acids;* and that the acids of amber and benzoin differ from all the rest, and possess peculiar properties.   Phil. Trans. 1767. Vol. LVII. p. 479.

19. Lord Glenbervie describes the properties of the blue substance found in peat mosses ('phosphate of iron.)   Phil. Trans. 1768. Vol. LVIII. p. 181.

20. Canton's pyrophorus is described.   It consists of calcined oyster-shells and flowers of sulphur mixed, and heated to redness in a crucible.   Ibid. p. 337.

21. Fontana gives an account of the *poison* of the *viper,* and the vegetable poison called *tacunas.*   They are similar in their action, kill when mixed with the blood, but not when applied to the nerves.   They prevent the blood from coagulating.   They kill also when swallowed in considerable quantity.   Lauro-cerasus water kills instantly when swallowed; but not when mixed with the blood, or applied to the nerves.   Phil. Trans. 1780. Vol. LXX. p. 163.

22. Swediauer proves that *ambergris* is the hardened excrement of a species of whale.   Phil. Trans. 1783. Vol. LXXIII. p. 226.

23. Wedgewood shows that *blackwad,* a mineral that takes fire when mixed with linseed oil, is a mixture of oxides of iron and manganese.   Ibid. p. 284.

24. Dr. Pearson gives us a curious set of experiments on *white lac,* and on an acid which it contains, and to which he has given the name of *laccic* acid. Phil. Trans. 1794. Vol. LXXXIV. p. 383.

25. The properties of *strontian,* then newly discovered, are given by Schmeisser.   Phil. Trans. 1794. Vol. LXXXIV. p. 418.

26. Mr. Tennant ascertained the existence of a kind of limestone, composed of carbonates of lime and of magnesia united.   Dolomite is a limestone of this kind.   He showed that this lime when used in considerable quantities is injurious to vegetation, because the magnesia does not absorb carbonic acid, and continues to act upon the plants.   The walls and cathedral of York are built with this limestone.   Phil. Trans. 1799. Vol. LXXXIX. p. 305.

27. Mr. Howard describes the properties of a new *fulminating mercury,* which he discovered by dissolving a little mercury in a considerable portion of nitric acid, and pouring the solution into alcohol. It possesses some very singular properties.   Phil. Trans. 1800. Vol. XC. p. 204.

Apparatus.   III. APPARATUS.   Under this head we do not mean to notice slight improvements in chemical apparatus; but only the introduction of new and important instruments, or new methods.   This will reduce the number of papers to a very few.   The following are the chief:

1. The common mercurial thermometer is incapable of measuring heats

higher than the boiling point of mercury; which renders it useless for a great variety of chemical purposes. This has turned the attention of philosophers to the contrivance of instruments capable of measuring the fusing points of the different metals, and other high temperatures. The expansion which rods of metal undergo when exposed to heat is one of the simplest methods, and therefore was first had recourse to. Dr. Mortimer describes a thermometer of this kind made in the year 1734 by Mr. Jackson, and gives us an engraving of it, and the fusing point of some of the metals, as antimony, tin, and lead.* But this instrument, which was made of a bar of iron, never came into general use. Indeed the ease with which iron is altered by heat would soon make it quite useless. Morveau has lately made a similar instrument, but substituted platinum for iron.

2. It has always been an object of some consequence to get thermometers so contrived as to indicate the greatest heat or cold which has taken place during the observer's absence. Lord Charles Cavendish contrived several such instruments;† but as more convenient ones have been invented since, it is unnecessary to give an account of his Lordship's inventions here.

3. One of the most valuable improvements in chemical apparatus, which has taken place in modern times, is the union of a number of receivers communicating with each other, known by the name of Woulfe's apparatus. An account of this apparatus is given by the inventor himself in the Philosophical Transactions.‡ It has since been improved in various ways, especially by connecting it with a pneumatic trough, by which means the gaseous products are observed and collected.

4. Dr. Nooth's apparatus for impregnating water with carbonic acid deserves to be mentioned on account of its ingenuity,§ though it has been entirely superseded by other contrivances, which answer the purpose much better.

5. No precise method being followed by workmen for graduating thermometers, it was found that they often differed from each other several degrees. To prevent this Mr. Cavendish suggested the propriety of appointing a committee of the Royal Society, to consider the best means of graduating these important instruments correctly. Accordingly Mr. Cavendish, Dr. Heberden, Mr. Aubert, Mr. Deluc, Dr. Maskelyne, Dr Horsley, and Mr. Planta, were appointed a committee for that purpose. After considering the subject thoroughly, they published a most excellent report, giving specific directions for the proper graduation of thermometers. This report is not susceptible of abridgement. But deserves to be studied carefully by all thermometer makers, and by all those who use these instruments.||

* Phil Trans. 1747. Vol. XLIV. p. 484.   † Phil. Trans. 1757. Vol. L. p. 300.
‡ Phil. Trans. 1767. Vol. LVII. p. 517.   § Phil. Trans. 1775. Vol. LXV. p. 59.
|| Phil. Trans. 1777. Vol. LXVII. p. 816.

6. Mr. Wedgewood, by a happy application of clay pieces baked with peculiar care, simply in a red heat, contrived a pyrometer which measured high degrees of heat in such a manner that the result of different trials could be compared together. The clay pieces contracted according to the heat, and the degree of contraction was measured by dipping the pieces between two brass rules graduated, and inclining towards each other. We have no less than three papers on this important subject by Mr. Wedgewood in the Transactions.* Since his death the instrument unfortunately has been laid aside. The pieces manufactured by his sons were found not to agree with each other. This induced them to give over manufacturing them. And though various attempts have been made by other persons to imitate the process of Wedgewood, no one has hitherto succeeded. Sir James Hall has been occupied with the subject for several years, and from his perseverance and sagacity we have every reason to look for the happiest results.

7. The last instrument we shall mention is a bottle contrived by Schmeisser, for the purpose of taking the specific gravity of liquids.† It does not appear to be so convenient as a simple glass bottle with a glass stopper ground to it, and perforated with a small hole.

IV. ANALYSES. These are so numerous that we cannot specify them all. Some of the oldest attempts of that kind, being the most curious in a historical point of view, deserve to be mentioned.

Maydew.

1. What in this country is called *maydew*, is a sweet excrementitious substance from the *Aphides*, insects which in such great numbers haunt the leaves of trees. Mr. Henshaw from a quantity of this substance extracted a salt, which he says had the shape of *rock petre*, by which probably he means common salt.‡

2. Beccher extracted iron from clay, by heating it with linseed oil. This was considered at the time as a demonstration of the transmutation of one substance into another.§

3. Mr. Boyle first pointed out the method of detecting the presence of common salt in water by means of nitrate of silver.||

Juices of plants.

4. Dr. Lister published a long paper in an early volume of the Transactions, containing some curious and important observations on the constituents of the juices of plants. Among other things it contains a minute description of the method of making bird-lime from the bark of the holly.¶

5. Redi gives a valuable table of the ashes and salts obtained from various vegetables by combustion.**

* Phil. Trans. 1782. Vol. LXXII. p. 305; and 1784. Vol. LXXIV. p. 358; and 1786. Vol. LXXVI. p. 390.

† Phil. Trans. 1793. Vol. LXXXIII. p. 164. ‡ Phil. Trans. 1665. Vol. I. p. 36.
§ Phil. Trans. 1671. Vol. VI. p. 2231. || Phil. Trans. 1693. Vol. XVII. p. 627.
¶ Phil. Trans. 1697. Vol. XIX. p. 365. ** Phil. Trans. 1698. Vol. XX. p. 281.

6. Dr. Slare endeavours to prove that the waters of Spa and Pyrmont contain no acid. It is needless to say that his arguments are wholly inadequate to prove his position.*

7. We have a curious paper by Dr. Hales on the constituents of various purging waters near London, obtained by evaporation. Though it is far from being an exact analysis of these waters, yet the paper possesses value, and is greatly superior to any of the efforts that had preceded it.†

8. Dr. Rutty gives an account of the waters of Amlwch, in the Isle of Anglesey, and of Hartfell Spa, at Moffat, and of several other waters, containing sulphate of iron.‡

9. But the first accurate analysis of a mineral water, which occurs in the Philosophical Transactions, is that of Rathbone Place, London, by Mr. Cavendish. He found it to contain abundance of carbonates of lime and magnesia, sulphate of magnesia and common salt. He demonstrated that the carbonates were kept in solution by an excess of carbonic acid. He found in it also some ammonia.§

10. Dr. Monro shows that the waters of Castleloed are sulphureous; and that those of Pitkeathly contain sea salt, muriate of lime, and muriate of magnesia.‖

11. Dr. Fordyce shows, that, in what is called auriferous pyrites, the gold is in a native state ; and that *vitreous silver ore* is a *sulphuret of silver*.¶

12. Dr. Fordyce proposes to examine copper ores by dissolving them in an acid, and precipitating the copper by a metal.**

13. Dr. Withering gives us the analysis of two minerals, distinguished in the North of England by the names of *Rowley ragstone*, and *toadstone*, and which from his description appear to be varieties of *greenstone*. The Rowley ragstone was composed of

Silica	47¼
Alumina	32¼
Oxide of iron	20
	100

The toadstone was composed of

Silica	63·5
Oxide of iron	16·0
Lime	7·5
Alumina	14·8
	101·8

* Phil. Trans. 1713. Vol. XXVIII. p. 247 ; and 1717. Vol. XXX. p. 564.
† Phil. Trans. 1750. Vol. XLVI. p. 446.   ‡ Phil. Trans. 1760. Vol. LI. p. 470.
§ Phil. Trans. 1767. Vol. LVII. p. 92.   ‖ Phil Trans. 1772. Vol. LXII. p. 15.
¶ Phil. Trans. 1779. Vol. LXIX. p. 527.   ** Phil. Trans. 1780. Vol. LXX. p. 30.

As no analysis of this rock has been made by any modern chemist, we do not know how far the preceding proportions can be relied upon.*

14. By the same chemist we have the first accurate analysis of carbonate of barytes, of which he was the discoverer, and of sulphate of barytes. The first he found a compound of

Barytes............78·6
Carbonic acid..........20·8
Sulphate of barytes..... 0·6

100·0

The second of

Barytes..............67·2
Sulphuric acid.........32·8

100·0

These results agree very nearly with the best and most recent analysis of these salts.†

15. Elliot shows that common salt decomposes soap of lead in alcohol, but not in water.‡

16. Hassenfratz shows that sulphureted hydrogen is formed by passing hydrogen gas over melted sulphur.§

17. Dr. Austin gives some curious examples of the artificial formation of ammonia, and gives the constituents of that substance, nearly in the same proportion as they had been before determined by Berthollet.||

18. In a subsequent paper Dr. Austin endeavoured to demonstrate that heavy inflammable air (carbureted hydrogen) is a compound of hydrogen and azote.¶ But his opinion was afterwards overturned by the more accurate experiments of Dr. Henry of Manchester.

19. Mr. Wedgewood, in an elaborate analysis of a mineral from Sydney Cove, in New Holland, announced the discovery of a new earth, to which Mr. Kirwan afterwards gave the name of *Sydneia*. Its peculiar character was solubility in muriatic acid, but in no other acid.** This supposed discovery was afterwards refuted by the more accurate experiments of Klaproth and Hatchett. The experiments of Mr. Hatchett could leave no doubt, as they were made upon a part of the very same specimen which Mr. Wedgewood had received from Sir Joseph Banks.

* Phil. Trans. 1782. Vol. LXXII. p. 327.   † Phil. Trans. 1784. Vol. LXXIV. p. 293.
‡ Phil. Trans. 1786. Vol. LXXVI. p. 155.   § Phil. Trans. 1787. Vol. LXXVII. p. 305.
|| Phil. Trans. 1788. Vol. LXXVIII. p. 379.   ¶ Phil. Trans. 1790. Vol. LXXX p. 51.
** Phil. Trans. 1790. Vol. LXXX. p. 306.

20. Dr. Pearson made a set of experiments on James's powder, and showed that it was a mixture of phosphate of lime and oxide of antimony ; but that the proportions of these constituents varied considerably in different specimens.* The subject was afterwards resumed and investigated with much precision, by Mr. Chenevix.

21. Mr. Smithson made an elaborate set of experiments on the tabasheer, a concretion obtained from the bamboo. The result of his experiments demonstrated it to consist chiefly of silica.† But there were some differences between its properties and those of pure silica, which were accounted for by the subsequent experiments of Fourcroy and Vauquelin, who found it a compound of silica and potash.

22. Dr. Pearson gives an elaborate set of experiments on wootz, a metal from Bombay, which he shows to be a species of steel made directly from the ore, without ever having been in the state of soft iron.‡

23. The next paper that requires to be mentioned is a most excellent one by Mr. Hatchett, on the molybdate of lead ; containing not only the analysis of this curious ore, but a very valuable set of experiments on the molybdic acid itself. The molybdate of lead he found composed of

Oxide of lead . . . . 58·40
Molybdic acid . . · . . . . 38·00
Oxide of iron . . . . 2·08
Silica. . . . . . · . . . . 0·28
                                        ————
                                        98·76

24. Mr. Tennant made a set of curious experiments on the diamond, and showed it to be composed of pure carbon. These experiments have been confirmed by all the subsequent labours of chemists on this difficult but curious subject.||

25. The next paper which we shall notice is an examination of urinary and gouty calculi, by Dr. Wollaston. It is by far the most valuable paper which has hitherto appeared on the subject; not even excepting the very long and elaborate papers of Fourcroy and Vauquelin. Dr Wollaston discovered a variety of new ingredients in these bodies. He discovered the presence of phosphate of magnesia and ammonia, of oxalate of lime, and calculi composed of phosphate of lime alone. Gouty concretions he found to consist of urate of soda.¶ It is quite unaccountable that this important paper was never noticed by Fourcroy and Vauquelin, though they went over the very same ground.

* Phil. Trans. 1791. Vol. LXXXI. p. 317.   † Phil. Trans. 1791. Vol. LXXXI. p. 368.
‡ Phil. Trans. 1795. Vol. LXXXV. p. 322.   § Phil. Trans. 1796. Vol. LXXXVI. p. 285.
|| Phil. Trans. 1797. Vol. LXXXVII. p. 123.   ¶ Phil. Trans. 1797. Vol. LXXXVII. p. 386.

26. There is a curious paper by Crell, on the analysis of boracic acid. His method appears very preposterous, and not calculated to attain his object. Yet the substance which he extracted bears so many analogies to the *boracium,* which has been recently ascertained to be the basis of this acid, that it is difficult to refuse him the honour of having first recognised it. The subject deserves to be examined anew. It would even be worth while to repeat his experiments, in order to discover the way in which the oxymuriatic acid acted on the boracic acid.*

27. The last papers we shall mention are two excellent ones, by Mr. Hatchett, on the constituents of shells, bones, and zoophites. Shells he found composed of carbonate of lime and animal matter; crusts, of carbonate of lime, some phosphate of lime, and an animal matter; bones, of phosphate of lime, a little carbonate, and animal matter. Some zoophites resembled shells; some crusts, in their composition; and some, like horns and hoofs, were composed almost entirely of animal matter, without any interposed salt.†

Processes.

V. Processes. Under this name we may arrange certain branches of chemistry, which are treated of in the Transactions considerably in detail. We shall only give a small number of examples; for it would extend this chapter to too great a length to mention the whole.

Refining gold by antimony.

1. Dr. Goddard, who was one of the most eminent chemists in the Royal Society, at its first institution, made a set of experiments to determine the goodness of the method of refining gold by melting it with antimony. The result was a constant loss of about $\frac{1}{10}$th of the gold.‡ This method is tedious, and less profitable than the common process of cupellation, which is always employed by the moderns to free gold from the baser metals. To free it from silver, the only method is the process of quartation, or alloying with a considerable proportion of silver, and subsequent solution in nitric acid. The pure gold remains in the state of a fine powder.

White vitriol.

2 There is a paper by Dr. Lister, in the Transactions for 1699, in which he complains of the ignorance of chemists respecting *white vitriol.* Nothing more was known, he says, than that it was obtained from a species of lead ore.§ Neuman first showed that this salt contained zinc; and Brandt, some time after, proved that it was a compound of sulphuric acid and oxide of zinc.‖ It is obtained, not from an ore of lead, but an ore of zinc. Though the ore in question, *blende,* is almost always in company with lead ore, and hence doubtless arose the mistake.

Phil. Trans. 1799. Vol. LXXXIX. p. 56.

† Phil. Trans. 1799. Vol. LXXXIX. p. 554; and 1800. Vol. XC. p. 327.

‡ Phil. Trans. 1678. Vol. XII. p. 953.   § Phil. Trans. Vol. XXI. p. 331.

‖ Pott published a dissertation on the subject in 1743, in which he points out the constituents of the salt, and claims the facts stated as a discovery of his own.

3. Dr. Krieg gives us an account of the process for making *smalt*; a sort of Smalt. blue pigment, employed for painting stone ware, for giving a blue colour to glass, and other well-known purposes It is prepared at Schneeberg, in Germany, from an ore of cobalt, which is found in considerable quantity in those parts. The ore is powdered, and then roasted in a furnace, to drive off the arsenic and sulphur with which it is mixed. The roasted ore is mixed with potash and pounded flints, and melted into a blue glass, which is afterwards reduced to a very fine powder, and sold under the name of *smalt*.*

4. In the Transactions for 1748, there is a long paper by Dr. Mitchell, on Potash. the preparation and uses of various kinds of potash. The methods employed in different countries to obtain this alkali, from the ashes of vegetables, are distinctly detailed. So that the paper is of some value on that account. It deserves to be remarked, that in this paper potash and soda are confounded together. Of course the difference between them was unknown in London in the year 1748.†

5. The Hudson's Bay Indians dye a bright *yellow* with the roots of *helleborus trifolius*; and *red*, with *gallium tinctorium*. The leaves of the cascalote tree dye the finest and most lasting black.‡

6. Mr. Jackson, in a paper of considerable length, gives us the method of Isinglass preparing isinglass employed by the Russians. It is now well known to be only the dried sounds of the sturgeon.§ Mr. Jackson's name ought not to be passed over without some notice. He was originally an apothecary, on Tower-hill, and amassed a large fortune by teaching the London brewers the method of substituting other and cheaper ingredients for malt and hops in the manufacture of porter. By this unfortunate and wicked practice, he destroyed the goodness of our national liquor for ever. He afterwards set up as a petty-fogging justice, in Woolwich; and when his infamous practices became too notorious he retired to Tottenham, where he persisted in the same trade till the time of his death in 1801.

7. The method of making ice in India, about the latitude of the tropic of Cancer, Ice. or a little farther north, is very ingenious. In that country it hardly ever freezes naturally. They dig pits in the ground about two feet deep, which they line with dried sugar canes, or Indian corn. On this they place very shallow dishes, made of unglazed and very porous earthen-ware, and filled with soft water, that has been boiled. They are deposited in the evening, and, in consequence of the evaporation from the outside of the dishes, a considerable part of the water is found frozen next morning. The ice is collected before sun-rise, and rammed

* Phil. Trans. 1704. Vol. XXIV. p. 1753.　　† Phil. Trans. Vol. XLV. p. 541.
‡ Forster. Phil. Trans. 1772. Vol. LXII. p. 54.　§ Phil Trans. 1773. Vol. LXIII. p. 1

into a cellar under ground, and lined with straw, where, in consequence of its own accumulated cold, it freezes into a solid mass.*

Red cabbage.　　8. It is well known to chemists that one of the most delicate tests of the presence of acid or alkaline substances, is the change of colour produced by them upon vegetable blue colours: acids changing them into red, and alkalies to green. Various colours have been chosen for the purpose, as violets, mallows, litmus, &c. But these (except litmus, which is not affected by alkalies,) can only be procured at certain seasons of the year. Mr Watt, therefore, conferred a benefit on chemists by proposing, as a substitute, the infusion of red cabbage leaves, which can be procured with great facility in winter, and which is affected as easily both by acids and alkalies as any vegetable infusion whatever.†

Ink.　　9. From a set of experiments, by Sir Charles Blagden, it appears that the inks employed in writing, from the 9th century to the present time, were composed of the same ingredients as our present writing ink. Sir Charles suggested a very ingenious way of restoring colour to those ancient manuscripts which had become so pale as scarcely to be legible. Touch the writing with a diluted acid, and then with a solution of prussiate of potash. If any iron remain, the letters will become of an intense blue colour.‡

Iron.　　10. Dr. Beddoes has given us an interesting account of the process of converting cast into malleable iron, by means of the reverberating furnace, now commonly practised in this country. In about half an hour the metal melted. It was then stirred by a workman, the flame being let off and on at intervals, in order to regulate the heat properly. By this stirring, it was gradually brought into a state like sand. In about an hour the mass began to heave, and a deep blue flame to cover its surface. And though the flame had been kept off for some time, the iron manifestly became hotter than it had been before. By degrees it collected in lumps, the heaving and blue flame ceased, it no longer stuck to the iron tool employed to stir it. The workman, therefore, considering the change as accomplished, stopped the process. The iron was taken and hammered, or rather rolled out into bars.§ The theory of the process of converting cast iron into malleable iron is still imperfect, notwithstanding the labours of chemists to throw light on the subject. The experiments of Bergman and of the French chemists prove that cast iron contains charcoal and oxygen. Now, if these be the only foreign bodies, and if they exist in the proper proportion, the preceding operation admits of an easy explanation. We have only to conceive that the oxygen and carbon unite, and form carbonic oxide. This accounts for the heaving of the iron, and for the blue flame upon the surface,

---

* Barker. Phil. Trans. 1775. Vol. LXV. p. 252.　　† Phil. Trans. 1784. Vol. LXXIV. p. 419.
‡ Phil. Trans. 1787. Vol. LXXVII. p. 451.　　§ Phil. Trans. 1791. Vol. LXXXI. p. 173.

and for the change from cast iron into malleable iron. But the manufacture of iron still stands in need of a set of chemical experiments to elucidate it. The present would be a favourable time to make them. For being in some measure deprived of Swedish iron, and British iron not being proper for making steel, some experiments are necessary in order to ascertain the reason of this curious distinction. I think Dr. Beddoes's experiments leave no doubt that the gas, extricated from cast iron, is carbonic oxide; and it appears to be extricated at a very low temperature.*

11. What is called cast steel is nothing else than steel which has undergone the process of steelification twice over, and has become in consequence more fusible than common steel. It is employed for making a great variety of cutting instruments; and when heated to the temperature at which iron welds, it runs under the hammer like sand. Hence it was considered as incapable of being welded to soft iron; but Sir Thomas Frankland, upon making the experiment, found that when iron is heated to a welding heat, and cast steel to a white heat, they may be readily welded together.†

VI. New Experiments. Notwithstanding the many chemical papers already noticed, there remain a great number not susceptible of being reduced under any one general head, and many of them of great value. It will be impossible to notice them all, or even the greater number of them. I shall therefore satisfy myself with pointing out a few of those which seem entitled to the greatest attention.

1. The methods of producing artificial cold, by the solution of salts in water, Artificial cold. were first put in practice by Mr. Boyle. He used pounded sal ammoniac, and usually dissolved about a pound at a time, in four times its weight of water.‡ The subject was carried much farther by Fahrenheit; and a number of curious experiments, made by him and by Boerhaave, may be seen in the first volume of Boerhaave's Chemistry. But it is to Mr. Walker, of Oxford, that we are indebted for the most complete and instructive set of experiments on the subject hitherto published. He was able, by a judicious combination of freezing mixtures, to congeal mercury even in summer.§ To Mr. Cavendish, likewise, we are indebted for some most important facts respecting the congelation of mercury and the mineral acids.‖ To Sir Charles Blagden we owe some curious facts respecting the effect of salts to lower the freezing point of water :¶ and a very complete history of the discovery of the congelation of mercury.**

---

* Phil. Trans. 1792. Vol. LXXXII. p. 257.   † Phil. Trans. 1795. Vol. LXXXV. p. 296.
‡ Phil. Trans. 1666. Vol. I. p. 255.
§ Phil. Trans. 1787. Vol LXXVII. p. 282; 1788. Vol. LXXVIII. p. 395; 1789. Vol. LXXIX. p. 199; 1795. Vol. LXXXV. p. 270.
‖ Phil. Trans. 1786. Vol. LXXVI. p. 241; 1788. Vol. LXXVIII. p. 166.
¶ Phil. Trans. 1788. Vol. LXXVIII. p. 277.   ** Phil. Trans. 1783. Vol. LXXIII. p. 329.

Light.

2. The emission of light from luminous bodies has long puzzled philosophers The two means of accounting for it have been mentioned in a former part of this work, under the head of Optics. Some modern authors, who have revived the Huygenian system, which explains light by the undulations of an elastic fluid, have affirmed, that the quantity of light emitted depends in all cases upon the temperature of the luminous body Not recollecting that many bodies shine without any sensible elevation of temperature, and that when the temperature is artificially raised, the bodies soon cease to shine. This is the case with those luminous creatures that inhabit the sea. Fish and wood, likewise, often become luminous without any increase of temperature. Mr. Boyle made some experiments which help to explain the shining of those bodies. He found that they lost their light in vacuo, and recovered it again on the introduction of air.[*] The subject was afterwards carried much farther by the experiments of Canton[†] and Hulme [‡]

3. Dr. Halley found that water, of about the temperature of 40°, expands $\frac{1}{16}$th part when heated to the boiling point: mercury, by the same heat, expands only $\frac{1}{71}$d part of its bulk: and alcohol, when raised from that temperature to its own boiling point, expands $\frac{1}{12}$th of its bulk.[§]

Oils inflamed by nitric acid.

4. Dr. Slare made a very extensive set of experiments on the effect produced by nitric acid when poured upon oils. The following were set on fire by it:

Oil of carui seeds.	Oil of box.	Oil of hartshorn.
cloves.	camphor.	man's scull.
sassafras.	Jamaica pepper.	hoofs.
guaiacum.	cinnamon.	human blood.

The following occasioned a violent effervescence, but without any visible combustion.

Oil of cummin.	Oil of wormwood.	Oil of rue.
fennel.	angelica.	sage.
dill.	hyssop.	savin.
bay.	lavender.	lemons.
juniper.	rosemary.	oranges.
thyme.	penny royal.	nutmegs.

The following were not affected by nitric acid.

Oils of amber, petroleum, Barbadoes tar, bees' wax.

Spirit of wine also burns, and so does balsam of sulphur, (solution of sulphur in oil of turpentine.)[‖]

---

[*] Phil. Trans. 1667. Vol. II. p. 581; 1672. Vol. VII. p. 5108.
[†] Phil. Trans. 1769. Vol. LIX. p. 446.   [‡] Phil. Trans. 1800. Vol. XC. p. 161.
[§] Phil. Trans. 1693. Vol. XVII. p. 650.   [‖] Phil. Trans. 1694. Vol. XVIII. p. 200.

5. The phosphorescence of fluor spar, when exposed to heat, is noticed by Sir Robert Southwell, in 1698. When the fact was discovered we cannot say; but this is the first time we have met with it.*

6. Hauksbee observed that the shining of phosphorus is increased by rarifying the air.† This deserves to be noticed, because it has been announced of late years as a modern discovery.

7. Hanckewitz found that phosphorus is soluble in ether, and that when a little of the solution is poured upon water, the surface becomes luminous in the dark. This also has been announced as a modern discovery. Hanckewitz likewise ascertained that phosphoric acid is heavier than the phosphorus from which it was obtained; that it is very fixed; and that, when strongly heated, it assumes the appearance of glass.‡

8. The curious fact, that blackening the bulb of a thermometer makes it, when exposed to the rays of the sun, rise higher than when its bulb is left clean, was discovered in the year 1772, by Dr. Watson, the present Bishop of Landaff. He exposed a thermometer to the sun: it rose to 108°. He then blackened the bulb with China ink; it rose in consequence to 118°.§

9. Dr. Higgins discovered that when moist pounded nitrate of copper is wrapped up in tinfoil, the tin acts with such violence upon the salt that actual combustion is produced.‖ The experiment is a very striking one.

10. Mr. Keir accidentally discovered that when glass is allowed to cool slowly in the pots, it crystallizes. He describes the crystals with great accuracy, and infers, from his observations, that basalt is probably a species of lava:¶ an opinion zealously maintained by many geologists, and under a modified form by the Huttonians, who extend it to greenstone; and Sir James Hall's experiments are considered as decisive proofs of the truth of their opinion.

11. Sir George Shuckburgh, with his usual accuracy, made a set of experiments to determine the boiling points of water, at different heights of the barometer The following were the first which he made:

Boiling point of water.

Height of barometer.	Boiling point of water.	Height of barometer.	Boiling point of water.
30·21 inches........	213·5°	25·75 inches........	205·1°
28·60.......... ...	210·4	24·03.............	201·2
26·61............	207·3	23·91.............	201·1

Afterwards, with a very delicate thermometer, made by Ramsden, capable

* Phil. Trans. Vol. XX. p. 363.
† Phil. Trans. 1733. Vol. XXXVIII. p. 58.
‖ Phil Trans. 1773. Vol. LXIII p. 137.
† Phil. Trans. 1705. Vol. XXIV. p. 1865.
§ Phil. Trans. 1772. Vol. LXII. p. 40.
¶ Phil. Trans. 1776. Vol. LXVI. p. 530.

1

of measuring to the fiftieth part of a degree, he constructed the following table:

Height of barometer.	Correction of the boiling point of water.	Height of barometer.	Correction of the boiling point of water.
26 inches..	— 7°·09	29 inches..	— 1°·72
26·5	— 6 ·18	29·5	— 0 ·85
27	— 5 ·27	30	0 ·00
27·5	— 4 ·37	30·5	+ 0 ·85
28	— 3 ·48	31	+ 1 ·69
28·5..	— 2 ·59		

This table must be of considerable service to thermometer makers.*

12. Cavallo points out an easy method of freezing water in a tube, by dropping ether on it. The great effect of ether in producing cold, by its rapid evaporation, was originally observed by Dr. Cullen. Ether may be purified by agitation in water. By this agitation it seems to lose a portion of alcohol, with which it is usually united. Mercury expands $\frac{1}{100000}$th parts per degree of Fahrenheit's thermometrical barometer.†

Sulphureted hydrogen gas.

13. It would be improper to omit mentioning Mr. Kirwan's curious experiments on sulphureted hydrogen gas. He pointed out its acid properties, its combination with alkalies, its effect on metals, and many other important particulars. In the same paper, likewise, he gave an account of phosphoreted hydrogen gas; of which he was one of the original discoverers. This paper is perhaps the most valuable that Mr. Kirwan has produced during the whole course of his distinguished chemical career.‡

14. Neither can we omit mentioning the important labours of Sir Charles Blagden, and Mr. Gilpin, in determining the specific gravities of various mixtures of alcohol and water, in order to regulate the method of levying the duty on spirits. These experiments possess an uncommon degree of accuracy, and are the most valuable that have hitherto been presented to the world. They are so long that it would be impossible even to give the slightest account of them in this place.§

15. The most complete set of experiments hitherto made on the production of light from bodies by heat, attrition, &c., is by Mr. Thomas Wedgwood.‖ They are too numerous to be detailed here; but we would advise those philo-

---

* Phil. Trans. 1779. Vol. LXIX. p. 362.   † Phil. Trans. 1781. Vol. LXXI. p. 501.

‡ Phil. Trans. 1786. Vol. LXXVI. p. 118.

§ Phil. Trans. 1790. Vol. LXXX. p. 321; 1792. Vol. LXXXII. pp. 425, 439; 1794. Vol. LXXXIV p. 275.

‖ Phil. Trans. 1792. Vol. LXXXII. p. 28 and 270.

sophers, who are endeavouring to draw conclusions respecting light from the theory of its being the undulations of an elastic fluid, to study the experiments of Mr. Wedgewood and some others, that have been already mentioned in this chapter.

16. There are several papers by Count Rumford, which would deserve particular notice, if we had any more room to spare. For example, the experiments by which he demonstrated that heat does not alter the weight of bodies; the experiments made to show that the chemical effects, usually ascribed to the action of light, are owing to the temperature produced by the light, and take place whenever that temperature is produced by any other means. But we have already extended this chapter much beyond our original intention. We must, therefore, reluctantly omit the particular mention of these, and many other papers, of first rate merit. Indeed there is perhaps no other work that contains so rich a collection of chemical papers as the Philosophical Transactions.

CHAP. II.

OF METEOROLOGY.

THE object of Meteorology is to ascertain the changes which take place in Object. the atmosphere, upon which the *weather* of every country, and of course the comforts of the inhabitants, entirely depend. These changes are owing, partly to electrical, but chiefly to chemical causes; and cannot, therefore, be fully explained till we be thoroughly acquainted with the constituents of the atmosphere, and with the chemical action of these constituents upon each other.

Since the discovery of the different meteorological instruments, which took place in the 17th century, though they were only brought to a state of tolerable perfection in the 18th century, and some still continue extremely defective, a vast number of meteorological observations have been made, and registers of the weather kept in many places, for a long series of years, with a view to discover some empyrical rules to determine the changes of the weather, on the supposition that these changes obey some regular, though unknown, laws, and that the same changes, after a certain number of years, return again in exactly the same order. But, though several meteorologists, Toaldo for example, have been sanguine in their belief that they had detected such a period, and have generally pitched upon one of about 18 years, yet, observation has not confirmed their hypothesis; and we are still as far as ever from being able to predict what sort of weather is likely to happen during the ensuing year. Indeed, if we consider that the weather must be influenced, to a certain degree, by the face of the country, and that this face is perpetually changing by the draining of marshes, the cutting down or growing up of woods, and by the improvements which are taking place in the cultivation of most civilized countries, we cannot expect that the weather, in such countries, should observe any regular order. It must alter, in some degree, with the country. Accordingly it has been always observed that the climate is ameliorated with the progress of cultivation.

Papers in the Transactions.

The number of papers on meteorological subjects, in the Philosophical Transactions, amounts to 281; but of these there is a considerable proportion consisting of bare diaries of the weather in particular spots, or of insulated meteorological observations, which, though they may all have their utility, and be of service towards constructing a system of meteorology, cannot with propriety be noticed in a work of this kind. We must confine ourselves to those papers alone, which either attempt to explain some important branch of meteorology, or which make us acquainted with new and important meteorological facts. All the meteorological papers in the Transactions, which can with propriety be mentioned, will be included under the following general heads.

Weight of the atmosphere.

I. WEIGHT OF THE ATMOSPHERE. The weight of the atmosphere remained unknown till the time of Galileo, and was demonstrated by the barometer first constructed by Torricelli. The height of the mercury, in the barometer tube, measures the weight of the atmosphere, because this mercurial column just balances that weight. Now it was soon observed, that though a barometer be kept always in the same place, its height did not continue stationary. Sometimes it fell, sometimes it rose. These changes were sometimes slow, sometimes rapid. It was soon observed, that they depended in some measure upon the latitude of the place. Under the equator they are barely perceptible; but they increase as we advance to either pole, till at the latitude of 60° the range of the barometer exceeds three inches. At the level of the sea, in every latitude, the mean height of the barometer is very near 30 inches. The height diminishes of course as the height of the situation increases; and, as most inhabited places are higher than the level of the sea, the mean height of the barometer in most places is under 30 inches. The range of the barometer in Jamaica, which lies within the torrid zone, though considerably north of the equator, is only 0 3 inches, according to the observations of Sir William Beeston, Governor of that island.*

Rise and fall of the barometer.

accounted *r.*

Dr. Halley, who embraced a greater variety of studies than almost any other man of his time, published a valuable paper, in which he endeavoured to account for the range of the barometer by the variable winds. His explanation

* Phil. Trans. 1696. Vol. XIX. p. 225.

is perhaps the most satisfactory that has hitherto been offered to the public. In the torrid zone the wind blows regularly all the year round in the same direction: hence the weight of the atmosphere cannot vary much, and accordingly the range of the barometer is small. Sometimes indeed, in the case of hurricanes, it sinks extremely low, but soon rises again, and this only happens about once in two years. In any country from which the wind blows in different directions at the same time, the barometer must necessarily fall. Thus if it blows a south wind in Scotland, and a north wind in France, the atmosphere over England will be attenuated, and the barometer will fall. On the contrary, if the winds blow from different directions towards any country, the barometer must rise. Hence the reason why it generally rises in this country when it blows a north-east wind. The south-west wind blowing almost perpetually in the Atlantic Ocean, the north-east is sure to be checked by it, and this will occasion an accumulation of air over Great Britain. The greater range in northern than in more southern climates seems to depend upon the greater violence with which the wind blows in these northern regions.*

Such is a sketch of Dr. Halley's ingenious explanation. Many phenomena might be mentioned which confirm this view of the subject. For example, the well known fact, that whenever the barometer falls suddenly we are always certain of a storm of wind; not always indeed in the place where the observation is made, but within the distance of a few hundred miles at most. The only other paper on this subject, which requires to be mentioned, is a mathematical dissertation by Gersten, to explain the rising and falling of the barometer, and particularly why the barometer sinks when the wind blows from the south-west. He endeavours to demonstrate, that air is dilated by tremulous motion, and that this dilatation is greatest when a wind blows contrary to the natural current of the air, which is towards the west. This dilatation, he conceives, diminishes the weight of the air, and therefore accounts for the sinking of the barometer when a south-west wind blows.† This explanation is neither very intelligible nor satisfactory. Indeed the most provoking of all books are mathematical ones founded upon erroneous data. The reasoning may be perfectly legitimate and demonstrative, yet the whole leads only to error.

II. Temperature. The temperature of the air is liable to as great or rather Temperature. to greater changes than the weight of the atmosphere, and it obviously depends upon the influence of the sun. Hence the reason why it is greatest in summer, why the temperature is highest at the equator, and why it diminishes as we advance towards the poles. It depends not upon the sun's action on the air, which being transparent receives no increase of temperature from the sun's rays, but upon his action upon the earth, which communicates the temperature it has

* Phil. Trans. 1686. Vol. XVI. p. 104.   † Phil. Trans. 1733. Vol. XXXVIII. p. 43.

received to the air. Hence the reason that the heat diminishes as we ascend in
the atmosphere, the thermometer sinking about one degree for every 280 feet of
ascent. Hence also the reason why the temperature of islands is much more
equable than that of continents in the same latitude, the surrounding ocean
preventing them from being either heated or cooled beyond a certain point.

In Great Britain the mean temperature diminishes nearly a degree for every
degree of latitude. But when places are situated on the sea coast, the rule does
not hold. The winter becomes milder and the summer colder. Thus for example, the difference between the latitudes of London and Edinburgh is nearly
$4\frac{1}{8}°$, but the difference between the mean temperature of the two cities is only
3°. The winter in Edinburgh is warmer than the winter in London; the mean
difference in favour of Edinburgh is 3°; but on the other hand, the mean summer heat in London is 6° greater than in Edinburgh.* It is probable that in
those parts of Scotland, which are further from the sea than Edinburgh, the
winters will be colder, and the summers warmer.

There is only one paper on temperature in the Philosophical Transactions
which requires to be noticed, and that is an attempt by Dr. Thomas Heberden
to determine the rate at which the temperature diminishes as we ascend in the
atmosphere. His observations were made in Teneriffe, and the result was that
the thermometer sinks a degree for every 190 feet of elevation.† This is considerably more than the result obtained by General Roy, from experiments made
chiefly at Edinburgh, and with great care. He found that one degree of diminution of heat was equivalent to an elevation of 280 feet.

Evaporation. III. Evaporation. That vapour is constantly rising from the surface of
water, and the moist surface of the earth, and that vegetables exhale a great
portion of moisture has been long known. Dr. Halley was the first person who
made accurate experiments on the subject. He showed, by an experiment exhibited before the Royal Society, that water kept at the highest summer temperature of London (which we may suppose about 75°) loses from its surface $\frac{1}{10}$th
of an inch in thickness in 12 hours.‡ Were the evaporation to continue at this
rate all the year round for 12 hours every day, it would amount in a year to
36·5 inches. Now it is a curious, though no doubt an accidental circumstance,
that this estimate coincides almost exactly with the experiments made by Dr.
Dobson, at Liverpool, for an interval of four years, to determine the quantity
evaporated from the surface of water. His mean result was 36·78 inches.§
Mr. Hunt, the Operator of the Royal Society, made a set of experiments, the
result of which was that water when left in a still place, and excluded from the
action of the sun and wind, evaporates only a quantity amounting to eight

* Roebuck. Phil. Trans. 1775. Vol. LXV. p. 459.   † Phil. Trans. 1765. Vol. LV. p. 126.
‡ Phil. Trans. 1687. Vol. XVI. p. 366.   § Phil. Trans. 1777. Vol. LXVII. p. 244.

cubic inches in a year.* This striking difference shows us the great effect of the wind in increasing the quantity of vapour raised. It acts, as Dr. Halley pointed out, by preventing the vapour from stagnating on the surface of the water. For a stratum of vapour effectually prevents all additional evaporation till it is itself removed.

From the quantity of vapour which thus rises from the surface of the sea and the land, Dr. Halley accounts, in a satisfactory manner, for the origin of springs and rivers, and shows the reason why, notwithstanding the great quantity of water which flows into the Mediterranean Sea, it never overflows its banks; the evaporation being more than sufficient to compensate for all the water which it receives.

From the most exact experiments hitherto made it appears that the annual evaporation from the surface of the earth, in England, amounts to 30 inches, and from the surface of the sea which borders on it 36¾ inches. We shall not err much if we take 35 inches as the mean quantity evaporated from the whole surface of the earth during the course of a year.

There is a curious experiment by Count Rumford, somewhat connected with this subject, which may be mentioned here, as a proof of the different disposition of various bodies to absorb moisture from the atmosphere. He left in a damp place for a given time the same weights of wool and of cotton. The wool absorbed 16 per cent. of its weight of moisture, the cotton only eight per cent.†

IV. Rain. There are a great many papers in the Philosophical Transactions giving the mean quantity of rain which falls in different places during a year It would be tedious to transcribe all these accounts; but we shall give one or two by way of specimen.

At Townley, in Lancashire, from a mean of 15 years observations by Mr. Townley, the annual fall of rain amounts to 41·62 inches.‡ At Upminster, in Essex, from a mean of 18 years observations by Dr. Derham, the annual fall of rain amounts to only 19·79 inches.§ Dr. Derham has also given us the mean quantity of rain which fell at Paris for 15 years from the observations of M. De la Hire. It amounted to 20·19 inches, which differs but little from the fall at Upminster.

A very curious discovery, respecting the quantity of rain which falls at different heights, was made by the late Dr. Heberden. Two rain gages fixed up in London about the distance of a mile from each other, made by the same artist with the same care, being compared together, it was found that the quantity of rain collected by the one was constantly greater than what was collected

---

* Phil. Trans. 1694. Vol. XVIII. p. 183. † Phil. Trans. 1787. Vol. LXXVII. p. 240.
‡ Phil. Trans. 1694. Vol. XVIII. p. 51.   § Phil. Trans. 1714. Vol. XXIX. p. 130

in the other.   No other cause being perceptible, it was conjectured that this difference might depend upon the difference in height of each, the one being placed above the tops of the chimneys, and the other lower down.   To determine this, two gages were placed in a garden nearly in the same spot, but the one considerably higher than the other.   The same difference was found between them as in the former case.   A rain gage was placed on the top of Westminster Abbey.   The following were the quantities of rain collected in each of these situations during the course of a year.

Low down.	House top.	Westminster Abbey.
Inches.	Inches.	Inches.
22·608	18·139.	12·099.

Thus it appears that nearly twice as much rain was collected in the gage placed near the ground as in the one upon the top of Westminster Abbey.* This difference seems to be connected rather with the proximity to the ground than with the quantity of atmosphere through which the rain falls.   For Mr. Barrington having placed a rain gage on the top of Rennig, a mountain in Wales, and another at its bottom, nearly the same quantity of rain was collected in each.†

V. Dew.   Dew is nothing else than a portion of the vapour formerly suspended in the atmosphere, condensed by means of the cold of the evening.   It has been observed with surprise that when a number of bodies are exposed together to the dew, some are quite wetted with it while others remain dry.   This circumstance probably depends upon the goodness of the body as a conductor of heat.   Good conductors will part with their heat more readily, and will therefore evaporate the dew again, whereas it will remain upon bad conductors, which will not so easily part with their heat   If this explanation be the true one, it follows that bodies exposed to the dew and dry, must have a lower temperature than those which remain moist.   The following table by Dr. Stocke, exhibits the relative quantity of dew remaining upon a variety of bodies:‡

Bodies.	Effect.	Bodies.	Effect.
Glass.	Much dew.	Polished lead.	A little.
Polished brass.	Very little.	Silver	None.
Rough brass.	A little more.	Silver gilded	None.
Lattin or iron tinned.	A little.	Blue porcelain	None.
Ditto rough.	Very much.	A stone slab.	Much.
Ditto smooth.	Scarce any.	Basket of Indian cane	A little.
Ditto rusty	None.	Smooth white oaken plank.	Very much.
Quicksilver	None.	Ditto black	Much less.
Smooth tin	None.	Smooth fir plank	Little.
Rough lead	Much.		

* Phil. Trans. 1769. Vol. LIX. p. 359.
† Phil. Trans. 1771. Vol. LXI. p. 294.
‡ Phil. Trans. 1742. Vol. XLII. p. 112.

If these experiments be accurate, it is obvious that the goodness of the bodies as conductors of heat will not account for the phenomenon. Perhaps we must look for it in the goodness of the bodies as conductors of electricity.

VI. WINDS. The winds in those parts of the ocean which lie within the Wind. tropics, and which are open both to the north and south, blow all the year nearly in the same direction, and are known by the name of the *trade* winds. But in the Indian Ocean, which is partly shut up on the south by New Holland, and wholly on the north by the continent of Asia, the winds shift every six months, blowing one half of the year in the usual direction of the trade winds, and the other half of the year in the opposite direction. These winds are known by the name of *Monsoons.* We have two elaborate explanations of the cause of these Trade winds accounted for regular winds in the Philosophical Transactions. The first by Dr. Halley, who gives a very minute and exact description both of the trade winds and monsoons; and accounts for them by the action of the sun's rays in rarifying the atmosphere, which occasions a motion of the whole westward, and a current from the north and south to supply the air thus moving.* The second is by Mr. Hadley, who explains the trade winds by the diurnal motion of the earth. The air which comes from the north and south to supply the place of the rarified air, which has ascended in the torrid zone, not having acquired the same velocity of rotation with that part of the earth which constitutes the torrid zone, the high parts of the earth strike against it, and of course occasion an east wind.† Both causes, no doubt, contribute to produce the trade wind, but if we consider the slowness of the motion of the air southwards, we cannot suppose that the cause assigned by Hadley can produce any great effect.

On the west coast of Africa, between the equator and north latitude 15°, Harmattan there is an extraordinary wind which blows several times a year from the north east, sometimes for four or five days together, and sometimes for fourteen or fifteen. It is always accompanied by a considerable fog, so that the sun is hardly visible. The matter of this fog deposites itself upon substances, and gives them a white appearance. This wind is extremely dry. No dew falls during its continuance, and evaporation goes on with more than twice its usual rapidity. Wood shrinks, and ships become leaky. The scarf skin falls off. This wind is in other respects remarkably healthy. But by its drying qualities it injures, or even destroys, the vegetation of plants. It appears to proceed from that part of Africa which lies in 15° north latitude, and 25° east longitude, where according to Ptolemy, the mountains of Caphas are situated.‡

What is called the Gulf stream, is a current obviously occasioned by the Gulf stream. trade winds; it sets in through the Gulf of Florida, and moves along the coast

---

* Phil. Trans. 1686. Vol. XVI. p. 153.   † Phil. Trans. 1735. Vol. XXXIX. p. 58.
‡ Phil. Trans. 1781. Vol. LXXI. p. 46.

of North America with considerable velocity. Sir Charles Blagden crossed it in 1776, about north latitude 33°. In that place it was about 20 leagues broad, and at least six degrees hotter than the surrounding ocean.[*]

Aurora Borea-
lis.

VII. Aurora Borealis. The name *aurora borealis* was given by Gassendi to an appearance familiar, we presume, to most of our readers. Streaks of light which appear during the night in the serene sky, and which, altering their position and form with surprizing rapidity, afford a beautiful and amusing spectacle. They are much more brilliant in countries situated nearer the pole than Britain; and their brilliancy seems to depend on that situation, and not on the coldness of the winter. For in Hudson's Bay, though the winter is incomparably more severe than in England, the Aurora Borealis is not so brilliant. Similar appearances were seen by Captain Cook in the southern polar regions, and called by him Aurora Australis. Mention is made of these appearances by some of the ancients, and they were fancifully compared to fiery spears and swords, and to armies of men combating in the sky. Thus at the death of Cæsar they made their appearance, and were considered as portending that tragical event. They were still more brilliant before the siege of Jerusalem, by Titus Vespasian, and are minutely described by the Jewish historians, and conceived by them to have been warnings sent from heaven of the approaching fate of Jerusalem, and of the destruction of the Jewish nation. They are minutely described by Cambden and Stow, as having made their appearance in the year 1574, during the reign of Queen Elizabeth, and several subsequent appearances of them are noticed by the writers in the Low Countries, and in France; especially a very brilliant Aurora Borealis seen all over France in the year 1621, on the 2d of February, is described by Gassendi. A minute account of a very brilliant Aurora, which appeared on the 6th of March 1716, is described by Dr. Halley with much more accuracy than any description that had been given of preceding Auroras. He shows that they must be at a great height above the earth's surface; and endeavours to account for them by conceiving them to be connected with magnetic effluvia, which he supposed issued out at the magnetic poles of the earth, and circulated in the magnetic curves.[†] That they may have some connection with magnetism may be true; for Canton observed that they always affected the diurnal variation of the needle. But the magnetic curves are no proofs of the circulation of a fluid through the poles of magnets. They are merely the consequence of the mutual attractions of the two poles. The small filings of iron are converted each into a temporary magnet, and in consequence of the mutual attractions, each assumes a particular position which constitute together the magnetic curve.

Since Dr. Halley's description, the Aurora Borealis has been exceedingly com-

---

[*] Phil. Trans. 1781. Vol. LXXI. p. 334.   [†] Phil. Trans. 1716. Vol. XXIX. p. 406.

mon, and a great many subsequent descriptions of it occur in the Philosophical Transactions. But it is not worth while to give a particular account of these papers, as they throw no additional light whatever on the nature of the phenomenon. It seems to be connected with electricity, and may in some measure depend upon the facility with which electrical light passes through highly rarified air. There is one observation, however, occurring in the Philosophical Transactions, which deserves to be stated. Mr. Winn says, that the Aurora Borealis is constantly succeeded by a south-west wind, with rain. The gale, he says, begins about three hours after the Aurora.*

There occur in the Philosophical Transactions a great many accounts of fire Fire balls. balls seen in the atmosphere, and it appears demonstrated by the observations made, that they must have been often seen at the height of more than forty miles above the surface of the earth. These fire balls are considered at present as connected with the numerous stony bodies which so frequently fall from the sky. Various attempts have been made to account for them; but hitherto no explanation in the least degree satisfactory has been offered. On this account it does not seem necessary to take up room, which is precious to us for other purposes, with a dry detail of a number of similar phenomena that do not admit of a rational explanation.

CHAP. III.

OF CHEMICAL ARTS AND MANUFACTURES.

The papers on these subjects which occur in the Philosophical Transactions Papers in the amount to 87. But of these there are 48 that may be passed over as of very Transactions. little value. We shall, in this chapter, give some account of the more important among the remaining 39 papers. For the whole of even these are not of such consequence as to be entitled to particular notice.

1. The first paper we shall mention contains a receipt by Mr. A. Kircher, for staining white marble so as to represent different pictures on it that shall sink to a considerable depth. The ingredients are nitrate of silver, and nitromuriate of gold, and these liquids are to be laid on day after day till they sink to the requisite depth.†

2. Mr. Ball describes the method employed in his time in Italy for preserving

---

Phil. Trans. 1774. Vol. LXIV. p. 128.   † Phil. Trans. 1665. Vol. I. p. 125.

ice during summer, and now well understood in Britain. It consists in contriving the ice house so that no water shall lodge in it, and lining it well with chaff or straw; which, being bad conductors, prevent the heat from penetrating and melting the ice.*

Salt making.

3. We have a particular description of the method followed in France of making *bay salt* by the heat of the sun. The sea water is let into square compartments, having a great surface and very shallow. Here it gradually evaporates away by the heat of the summer, and as it becomes more concentrated, it is advanced to compartments at a greater distance from the mouth. Here the salt gradually crystallizes, and is taken out from the bottom. It is in large, very hard, but somewhat dark-coloured crystals; because it is not quite free from the mud in the bottom of the square where the crystals are deposited.†

The method of making salt in Cheshire by boiling down their brine, obtained either from salt springs, or from rock salt dissolved in water, is also described. It is unnecessary to enter into details, as the process is nothing more than simply boiling the brine to dryness in an open pan.‡ The same may be said of the salt made from the salt springs in Worcestershire.§

It may be proper to notice here Dr. Brownrig's Treatise on Salt Making, of which a particular account is given in the Philosophical Transactions.‖ It was written in order to improve the process of salt making in Great Britain, and, though not free from mistakes, contains a great deal of valuable information, and certainly is well entitled to the attention of the manufacturers of salt.

Vinegar making.

4. The method followed in France for making vinegar is detailed in an early volume of the Philosophical Transactions. From the System of Chemistry, published by the Dijon Academy, we learn that this method is still practised in various parts of France. The method is as follows. Two large casks are taken, open at top; into the bottom of each is put a trivet, about a foot high, on which vine twigs are laid; then the casks are filled with *rape*, or the dried foot stalks of grapes. Wine is poured into one cask half way up, into the other nearly to the top; and the wine is alternately poured from the one cask to the other every day. In about twelve days the wine is converted into vinegar.¶

Smelting iron.

5. One of the oldest manufactories of cast iron in England is in the forest of Dean, in Glocestershire. We have an account of the method of smelting the ore in that work in an early volume of the Transactions. We may presume, though Mr. Powle, to whom we owe the account, takes no notice of the circumstance, that limestone was employed in the process as it is at present; for it is not likely that without its assistance the iron would have separated from the

---

* Phil. Trans. 1666. Vol. I. p. 139.   † Phil. Trans. 1669. Vol. IV. p. 1025.
‡ Phil. Trans. 1669. Vol. IV. p. 1060.   § Phil. Trans. 1678. Vol. XII. p. 1059.
‖ Phil. Trans. 1748. Vol. XLV. p. 351.   ¶ Phil. Trans. 1670. Vol. V. p. 2002.

clay with which it mixed. Unless, indeed, the ore had contained the requisite proportion of limestone, which may possibly have been the case. The ore is first roasted, by which it loses about a third of its weight. It is then mixed with charcoal and limestone, and put into the blast furnace, where it is gradually reduced to the state of cast iron by the united action of the heat and charcoal; and the clay being reduced to liquid glass by means of the lime, the iron tumbles down to the bottom in a liquid state, and is let out and cast into pigs.*

5. We have a description of the method of making *white lead*, by Sir Philiberto Vernati, which differs in some particulars from the methods employed in our best manufactories at present, though the nature of the process is still the same. The lead, in thin slips, or rolled up in such a way that the rolls do not touch, is placed above small pots, containing vinegar, and standing in horsedung or wet oak bark; the heat volatilizes the vinegar, which slowly corrodes the lead, and converts it into carbonate on the surface. The carbonate is rubbed off, ground fine, and by a great deal of washing with water, is freed from the part which still continues in the state of lead, and then it is fit for use.† Unless the lead used be very pure, or at least free from copper, it will not make white lead. This is the reason why many manufacturers fail in the process. White lead.

6. Alum began to be manufactured in Great Britain during the reign of Queen Elizabeth. There are two kinds of mineral from which it is prepared in this country. 1. The *slate clay*, or rather *shale* of old coal pits. Two very extensive alum works are carried on near Glasgow from this material, which seems to be richer in the constituents of alum than the common *ore* of alum, as it is called. 2. Alum slate, which is the common ingredient employed in most countries for the manufacture of this salt. It occurs in great abundance on the east coast of Yorkshire, and, as far as we know, the only alum works at present in existence in England are situated in that quarter. We do not know the constituents of alum slate exactly, no regular analysis having been made; but there can be no doubt that it contains alumina, iron, and sulphur; besides other ingredients. There is a description of the mode followed in the making of alum in the Yorkshire alum-works, in the year 1678, published in the Philosophical Transactions by Mr. Colwall;‡ and the same method nearly is still followed. The alum slate is first burnt in a smothered fire, and, according to Mr. Colwall, much depends upon the way in which the burning is conducted. If the slate, after being burnt, has a black or a red colour, it is a proof that the process has been improperly conducted. The colour ought to be white. The burnt slate is put into pits, and water let on it. This water is pumped from one pit to another till it is conceived to be sufficiently impregnated with the soluble parts of the slate. It is then boiled down considerably, and mixed with kelp Alum.

---

* Phil. Trans. 1678. Vol. XII. p. 931.   † Phil. Trans. 1678. Vol. XII. p. 935.
‡ Phil. Trans. 1678. Vol. XXII. p. 1052.

dissolved in water, or stale urine. An effervescence takes place because the liquor contains an excess of sulphuric acid. The whole is then allowed to remain till the sediment (consisting of suboxysulphate of iron) subsides. The clear liquid is then drawn off, boiled down sufficiently, and poured into the vessel in which the alum is to crystallize. The crystals of alum, at the time when Mr. Colwall wrote, were again dissolved and crystallized afresh. Such is the process as described by Mr. Colwall. Alum is a triple salt composed of sulphuric acid, alumina and potash, or ammonia. The alum-slate supplies the first two of these constituents; the kelp or the urine the other. At present muriate of potash, and sulphate of potash, are usually employed by the alum-makers.

Copperas.

7. Copperas, or sulphate of iron, is a salt very much employed for a variety of purposes. It is one of the chief ingredients in the mixture which constitutes writing ink, and it is an essential article in dyeing black. It is usually prepared from pyrites, or sulphuret of iron, which being placed in pits, and exposed for years to the action of the sun and rain, gradually combines with oxygen, and becomes, in some measure, soluble in the water. This water contains a great excess of acid. It is put into leaden boilers, and mixed with abundance of iron. Heat is applied, and the evaporation continued till it has dissolved enough of iron, and is sufficiently concentrated. It is then let off, and allowed to crystallize.* Manufacturers prefer the darkest and most opaque specimens. It is supposed that they contain less acid, though this does not appear to be established by any satisfactory evidence.

Malt.

8. In the same volume of the Transactions with the preceding details we have an account of the process followed in Scotland for making malt described by Sir Robert Moray. The very same method continued nearly unaltered till within these 30 years; but it has been improved since. Sir Robert says that there are two kinds of barley sown in Scotland, one kind with four rows on the ear, and another with two rows. In reality there are three varieties known in Scotland. 1. The two rowed, called *barley;* 2. The four rowed, called *bear;* 3. The six rowed, called *big.* The first, which has the largest *grains,* answers much better for malt than either of the other two varieties. Sir Robert's process is as follows: The barley is steeped in water for three days, then thrown out upon the floor, and the water allowed to drain off. Next it is made up into a heap called the *coming* heap, and upon the management of it a great deal depends. Here it heats and the roots spring out. As soon as they appear the grain is to be spread upon the floor about five inches deep, and turned over every four or five hours for several days, till it has been converted into malt till the acrospire extends nearly to the end of the barley corn.) It is then to

* Colwall. Phil. Trans. 1678. Vol. XII. p. 1056.

be collected into one heap, and left till it becomes so hot that the hand can scarcely bear the temperature. After this it is to be cooled by spreading it on the floor, and finally dried upon the kiln.*

9. In a subsequent volume of the Transactions we have an imperfect account Brass. of the method of making brass in England by Mr. Povey.† This method of making brass was known both to the ancient Greeks and Romans. Dr. Watson has shown that it was distinguished among the Latin writers by the name of *orichalcum.* For many years the process was confined to Germany, where thousands of workmen were continually employed in making this useful alloy. Like all the other metallurgic processes it was gradually imported from Germany to Britain, and at present our manufacturers make it of a superior quality to the brass of any other nation. Brass is a compound of copper and zinc; nearly in the proportion of three parts of copper, and one part of zinc; but it is liable to a good deal of variation in the proportion. The ore of zinc called *calamine,* (which is an oxide) previously roasted, and as pure as possible, is taken and mixed with the requisite proportions of granular copper and charcoal powder. This mixture is put into large crucibles, and let down with tongs into a wind furnace, eight feet deep. Here it remains for eight or ten hours. The zinc is gradually reduced by the action of the charcoal to the metallic state, and combines readily with the grains of copper. At last the furnace is raised to a heat sufficiently high to melt the whole, and then the brass is poured out into moulds and cast into sheets.

10. The next process we shall mention is the conversion of iron into steel, Steel. which occupied a good deal of the attention of the early Members of the Royal Society. Dr. Lister gives us the process described by Agricola, in his treatise *De Re Metallica;* which probably was still followed when Dr. Lister wrote. It consisted in filling a crucible with powder of charcoal, heating it red hot in a smith's forge, putting into it a mixture of pieces of iron and some stone that readily melts, urging the fire till this mixture is brought into fusion, and then putting into it the iron bars which are to be converted into steel, and boiling them for some time in the melted mass. They are afterwards taken out, hammered into bars, and quenched while red hot in cold water.‡ The method at present followed is much simpler and better. Steel, as has been ascertained by experiment, is a compound of iron and carbon. Bars of Swedish iron (which alone make good steel) are inclosed in oblong stone vessels, filled with charcoal, and kept about eight days in a red heat. They imbibe the requisite quantity of carbon, and are thus converted into steel. When the process is twice repeated

* Phil. Trans. 1678. Vol. XII. p. 1069.   † Phil. Trans. 1693. Vol. XVII. p. 735.
‡ Phil. Trans. 1693. Vol. XVII. p. 865.

2

the steel becomes more fusible, and is known by the name of cast steel, because it is melted and cast into moulds.

Coal tar.

11. Most of our readers are probably acquainted with the tar, extracted by Lord Dundonald from pit-coal, with the application of it to the bottoms of ships, and with the expensive works which he erected for the manufacture of it. But it does not appear to be known that the same thing had been practised as early as the year 1697. This, however, appears from a paper published in the Philosophical Transactions, by Mr. Martin Ele, the inventor of the process. Instead of coal, he employed *bituminous shale*, which was ground to powder and boiled in large caldrons of water. The stony matter fell to the bottom, and the melted tar or pitch swam on the surface. He even mentions an oil which may be obtained and used as a substitute for oil of turpentine. Such an oil is at present got from coals, and used by japanners as a substitute for that essential oil.*

Diving bell.

12. The diving bell is a very old invention; but it was greatly improved by a very simple addition made to it by D Halley. He contrived a large diving bell of wood, in the form of a truncated cone, three feet diameter above, and five below, with a glass window in the top to let in the light. It was so balanced with lead that it would sink empty, with its mouth perpendicularly downwards, so as to retain all the air which it contained. There was a cock at the top of the machine to let out the hot air, that had been breathed, when requisite. A couple of barrels were fixed to a rope, and so loaded with lead that they could be let down alternately empty, in a perpendicular direction. There was an open hole in the bottom to let in the water as the barrel descended, and a long tube in the top, which hung down below the bottom of the barrel, and through which of course no air could escape till it was lifted up. When a barrel was let down, a man in the bell lifted up this tube, and let the air of the barrel into the bell; it was then drawn up, and the other barrel let down and emptied of the air in the same manner. By this constant supply it was possible to remain below as long as the divers pleased, without inconvenience. By means of a long tube, fitted to a leaden cap, a diver could go to a distance from the bell and yet be supplied with air sufficient for the purposes of respiration.† Some improvements were afterwards made upon Dr. Halley's ingenious invention by Mr. Triewald.‡

13. Nothing is more striking than a comparison of the present state of most of the chemical manufactures of Great Britain, with what it was 70 or 80 years

---

* Phil. Trans. 1697. Vol. XIX. p. 544.
† Phil. Trans. 1716. Vol. XXIX. p. 492; and 1721. Vol. XXXI. p. 177.
‡ Phil. Trans. 1736. Vol. XXXIX. p. 377.

ago. The present superiority is surprising. The article we are to metion here affords a good example of it. In the year 1728, no tin-plate was manufactured Tin-plate. in Great Britain : notwithstanding the greatness of the consumption of that useful article, the whole of it was imported from Germany. The manufacture had been recently introduced into France by the Germans, and Reaumur considered it as of sufficient consequence to bestow a great deal of pains on it, and to publish an account of the process in the Memoires of the French Academy. This account Dr. Rutty, at that time Secretary to the Royal Society, translated into English, and published it in the Philosophical Transactions.* The process was very simple. Iron was hammered out into plates of the requisite thickness, steeped in a sour infusion of rye for about two days, to remove the vitriform oxide which covered the outside of the iron, and prevented the tin from adhering. The plates were then scoured and put into clean water till the moment they were to be tinned. The tin was melted in a crucible, larger and deeper than the iron plates, so that the plates could be plunged wholly under the tin. It was requisite that the tin should possess a certain degree of heat only; otherwise the process did not succeed. The surface of the tin was covered with a coat of suet, that had been previously heated, which gave it a black appearance. When the manufacture was introduced into England, about the middle of the last century, an improvement was made in it which at once gave the English tin-plate a decided superiority over the German. Instead of hammering out the iron plates, they were rolled out, to the requisite size and thinness. This made them much smoother, and consequently the tin-plate more beautiful, and it prevented the inequality between the smoothness of the two sides, which occasioned a striking blemish in the German tin-plate.

14. The making of sal ammoniac is another chemical manufacture which has Sal ammoniac been introduced into Great Britain within these sixty years. It was first begun in Edinburgh by the late Dr. Hutton and Mr. Davy, according to a process which they contrived while attending the University in that city. Dr. Hutton from the University of Edinburgh went to Paris, while in the meantime Mr. Davy set the process on foot, and admitted Dr. Hutton a partner on his return. Before that period all the sal ammoniac used for the various purposes to which it is applied in the arts was brought from Egypt. In the year 1760, the Egyptian method of making sal ammoniac was published in the Philosophical Transactions, by Linnæus, from an account sent him by Hasselquist at that time in Egypt.† In Egypt fuel is so scarce that they burn the dung of their cattle as a substitute, and it is from the soot of this dung that the salt is made. They make it only during the spring when the cattle feed upon fresh grass. At no other season of the year will the dung answer their purpose. This soot is

---

* Phil. Trans. 1728. Vol. XXXV. p. 630.   † Phil. Trans. 1760. Vol. LI. p. 504.

1

put into globular glasses having a neck open at top, and arranged by 50 on the top of a furnace. The fire is kindled below them, and is at first gentle, but on the third day they make it very intense. The sal ammoniac sublimes and coats the upper part of the glasses. The cakes are obtained by breaking these glasses. Such is the process as described by Hasselquist. But there is reason to suspect that it is imperfect. As no source presents itself whence the muriatic acid, one of the constituents of sal ammoniac, could be obtained; unless indeed we were to conceive the salt to exist ready formed in the soot, which is highly improbable. The method followed in Britain, and we presume in every manufactory of sal ammoniac in Europe, is to mix sulphate of ammonia and common salt in glass vessels, and apply heat. A double decomposition takes place, sal ammoniac sublimes, and glauber's salt remains behind in the bottom of the glass. The difficulty of the process consists in the procuring of sulphate of ammonia. It exists ready formed in the soot of our chimneys, proceeding from the pit coal employed as an article of fuel; and this is the source from which it is usually obtained.

Salop.

15. Among the nourishing and agreeable articles of food prepared from the vegetable kingdom, *salop* has always held a distinguished rank. It is usually imported from Persia, and is known to consist merely of the bulbs of different species of orchis dried in a particular way. Mr. Moult has given us a method of preparing these bulbs, which he says from experience answers perfectly well. The plant is to be dug up when the seed is ripe, and the stalk going to fall. The fresh bulb is to be cleaned, and freed from its skin, by dipping it in warm water, and rubbing it with a coarse cloth. The bulbs thus prepared are arranged on a tin plate, and introduced into an oven of the heat requisite for baking bread, left in it for ten minutes, and then taken out. They have lost their milky appearance, and acquired a semi-transparency. They are then to be dried slowly, either by simple exposure to the air, or by a moderate artificial heat.* The use of the oven seems to be to destroy the life of the bulb, and thus prevent it from growing instead of drying.

Tokay mine.

16. The last paper which we shall mention relating to chemical manufactures, is an account of Tokay wine by Lord Glenbervie. Tokay is a village which lies in Hungary at the confluence of the rivers Bodrog and Theis. The hills round it occupy the space of about 12 miles, and all of them produce the grapes from which Tokay wine is made. They are all white grapes. The vintage is as late as possible, because the frosts of September are considered as of service to the flavour. The consequence is that many of the grapes are shrivelled at the time of pulling them, and look like half dried resins. These shrivelled grapes are carefully picked out, and thrown into a perforated cask. The

* Phil. Trans. 1769. Vol. LIX. p. 1.

juice that flows out in consequence of the pressure of their own weight is fermented by itself, and constitutes that kind of wine which at Tokay is known by the name of *essence*. It is thick and clammy, and is chiefly used to mix with the other sorts. Upon the shrivelled grapes thus deprived of a part of their juice, the juice of the unshrivelled grapes is poured, and the whole is subjected to pressure. The juice thus expressed constitutes, when fermented, the wine called *auspruch*, which is the kind of Tokay usually exported, and in fact the best More juice of fresh grapes is poured upon the shrivelled grapes, and they are strongly squeezed by the hands. The juice thus obtained, when fermented constitutes the wine called *masslasch*; it is of rather inferior quality, but when mixed with a portion of essence it becomes as good as the auspruch. The common people mix all the grapes together at once, and make a species of wine which is usually drunk in the country, and constitutes the vin du pays. The following, according to Lord Glenbervie, are the best rules for judging of the qualities of Tokay wine. 1. The colour should neither be reddish nor very pale, but a light silver. 2. In trying it, you should not swallow it immediately, but only wet your palate, and the tip of the tongue. If it discovers any acrimony to the tongue it is not good. The taste ought to be soft and mild. 3. It should, when poured out, form globules in the glass, and have an oily appearance. 4. When genuine, the strongest is always of the best quality. 5. When swallowed it should have an astringent taste in the mouth, which they call the taste of the root. Tokay wine has a strong aromatic flavour, quite peculiar, by which it is easily distinguished from all other wine. It keeps a very long period. Lord Glenbervic drank of some that had been above 80 years in a cellar in Vienna.*

* Phil. Trans. 1773. Vol. LXIII. p. 292.

BOOK V.

MISCELLANEOUS ARTICLES.

NOTWITHSTANDING the numerous topics comprehended under the divisions in the preceding books, the subjects treated of in the Philosophical Transactions are of so various a nature, that it was not possible, without using violence, to include the whole of them under any of the great branches into which natural science has been divided. On that account we have thought it better to throw several detached subjects together under the comprehensive title of miscellaneous. To some of these, names may be given; others are so unconnected, that they will not admit of any common title, while they are not of sufficient importance to constitute separate chapters. We shall divide this book into four chapters, in which we shall treat respectively of *Weights and Measures*, of *Political Arithmetic*, of *Antiquities*, and of *Miscellaneous Articles*, so far as these subjects occupy a place in the Philosophical Transactions.

CHAP. I.

OF WEIGHTS AND MEASURES.

THE diversity of weights and measures in different countries occasion not a little inconvenience to scientific men; as it is necessary, before it be possible to reap any advantage from experiments in which weights or measures are concerned, to reduce them to some common standard. The uncertainty of the weights and measures employed by the ancient nations have rendered many of their experiments and observations useless to the moderns. To remedy these defects as much as possible, two things have occupied a good deal of the attention of men of science; namely, to determine some unalterable standard of weights and measures, which may be determined with exactness, which may be had recourse to at all times, and by means of which the weights and the measures at present in use, supposing them hereafter to be lost, may be recovered and

3 x 2

ascertained with perfect accuracy. And, secondly, to compare the weights and measures used by different nations together, and determine how much they differ from each other.

Standard
weights and
Measures. With respect to the first of these particulars, the French, when they established their revolutionary weights and measures, had recourse to the ten millionth part of the distance between the north pole and the equator as their standard fundamental measure, to which they gave the name of *metre*, and from which all their other measures and weights are deduced. Now the length of a metre has been ascertained to be equal to 39·37100 English inches. The method adopted by Sir George Shuckburgh Evelyn is easier, and seems susceptible of much greater accuracy. It has this great advantage, that it furnishes a method of recovering the English weights and measures with perfect accuracy if they should hereafter be lost. He determined that the difference between two pendulums vibrating 42 and 84 times in a minute of mean time in the latitude of London, at 113 feet above the level of the sea, in the temperature of 60°, and when the barometer is at 30 inches, is = 59·89358 inches of parliamentary standard. A cubic inch of distilled water, when the barometer is at 29·5 inches, and the thermometer at 60°, weighs according to a mean of Sir George Shuckburgh's experiments 252·506 grains troy; and in vacuo 252·806 grains.*

We shall now state the points respecting the weights and measures of different nations which are to be found in the Philosophical Transactions.

Old English
weights. Before the conquest the standard English weight was the sterling penny, which was declared equal to 32 grains of dry wheat. Twenty of these pennies were equal to an ounce, 12 ounces to a pound, 8 pounds were a gallon of wine, and 8 of these gallons made a London bushel, which is the 8th part of a quarter. These weights and measures were confirmed by William the Conqueror, and by succeeding kings, and they continued till the reign of Henry VII From the experiments of Mr. Norris, it appears that 32 grains of dry wheat of ordinary size weigh very nearly 22¼ troy grains. Hence it follows that the Saxon pound of 240 penny weights continued down to the time of Henry VII. weighed 5400 of our present troy grains. From this it is easy to deduce the weight of the gallon, bushel, and quarter.

Troy weight. Henry VII, for what reason does not appear, altered the old pound, and substituted in place of it the present troy pound, and the old pound was finally abolished by Henry VIII. who states, in the act of parliament by which he establishes the troy pound, that it is ¾ ounce heavier than the old pound. The present troy pound consists of 5760 grains, and is divided in the same way as the old pound.†

Avoirdupois
weight. When the avoirdupois weight, at present in use for almost all articles in this

---

* Phil. Trans. 1798. Vol. LXXXVIII. p. 133.   † Phil. Trans. 1775. Vol. LXV. p. 48.

country, was introduced, does not appear.   Nor is it easy to conceive how the term avoirdupois (literally, *to have weight*) was introduced.   It is first employed in some acts of parliament, in the time of Edward III., relative to weights and measures.   Neither is it known when the present difference between the wine and ale gallon originated; but they were recognised by act of parliament soon after the establishment of the excise on ale and beer.   By the report of the Committee of the House of Commons, in 1759, it appears that the avoirdupois pound weighs 7002 troy grains.   It would appear, though there are different opinions on the subject, that avoirdupois weight was first made legal by Henry VIII.   Every body knows that the avoirdupois pound is divided into 16 ounces; and the ounce into 16 drachms.

Mr. Barlow has endeavoured to show that a cubic foot of water is the foundation of the English weights.   It weighs 62¼lbs, and constitutes the bushel.   Eight bushels make a quarter, and four quarters make a ton.   So that a ton is $= 62\frac{1}{4} \times 32 = 2000$lbs; and the word quarter was used to denote the fourth part of a ton.*

Mr. Reynardson, in a long and learned paper published in the Philosophical Transactions, in the year 1749, endeavours to prove that the avoirdupois was the ancient English standard.   But he does not appear to have established that the weights used before the time of Henry VIII. were the avoirdupois; but only that the word avoirdupois occurs in our laws, relating to weights and measures, as early as the reign of Edward III.†   And the subsequent observations of Mr. Norris seem to have established the difference of the Saxon, or original British pound, both from the troy and the avoirdupois.

Standard weights and measures are kept at the Exchequer, at the Tower, and at the Mint.   A set was made for the House of Commons, and another for the Royal Society.   These have been very carefully and accurately compared with each other, first by Mr. Graham,‡ and afterwards by Sir George Shuckburgh Evelyn,§ and the results recorded: but it does not seem necessary to specify them here; it will be sufficient to refer the reader to the papers themselves, for minute information upon this curious subject.

The yard, which is the English standard measure, is said to have been taken from the arm of Henry I.   It is divided into three feet, and every foot into twelve inches.

The French weights before the revolution, or rather the Paris weights, which French weights. were always used by philosophers, were the following:

Phil. Trans. 1740. Vol. XLI. p. 457.   † Phil. Trans. 1749. Vol. XLVI. p. 54.
Phil. Trans. 1743. Vol. XLII. p. 541.   § Phil. Trans. 1798. Vol. LXXXVIII. p. 133.

24 grains = 1 denier, or scruple.
3 deniers = 1 gros.
8 gros   = 1 ounce.
8 ounces   = 1 marc.
2 marcs   = 1 pound.

Their measures of length were lines, inches, feet, toises. We have two papers in the Philosophical Transactions comparing the French with the English weights and measures.

It appears from a very exact measurement, made by Dr. Maskelyne, that at the temperature of 61°, the Paris toise is equal to 76·7344 English inches.*

From the experiments of the Royal Society, it appears that the Paris pound, of two marcs, or sixteen ounces, weighs 7560 grains troy.† From these data it is easy to determine the proportion between all the other French weights and measures.

Roman measures. From a paper, by Mr. Martin Folkes, it appears that the ancient Roman foot, as determined by measures still remaining in the city of Rome, was equal to 0·967 of the English foot; while the Grecian foot was equal to 1·006 of the English foot.‡ These measures were afterwards determined with greater precision by Sir George Shuckburgh Evelyn. He found the ancient Roman foot = 11·6063 English inches; the Grecian foot = 12·09 English inches.§

Greek and Roman money. In the year 1771, Mr. Raper published, in the Philosophical Transactions, a very curious dissertation on the ancient Greek and Roman money, which comes to be mentioned here, because these coins were used as weights as well as coins; and from them, therefore, we derive the only means which we possess of determining the weights of these ancient nations. The Attic coins were all silver during the existence of the republic. They coined copper, indeed, during the Pelopennesian war, but it was soon cried down again. The first, or standard coin, was the drachm. The mina contained 100 drachms; the talent 60 minas: and these were not only coins but weights. The drachm was divided into six oboles. The drachm was equal to half a Persian daric. The philippic, a gold coin of Philip, of Macedon, is conceived to have been equal to the daric in weight. Mr. Raper, from a number of philippics which he weighed, shows that the weight of one was at least 133 grains troy; and hence concludes that the Attic drachm weighed 66¼ grains.

Silver was first coined at Rome, in the 485th year of the city, five years before the first Punic war; and the denarius was made to pass for ten pounds

* Phil. Trans. 1768. Vol. LVIII. p. 274.   † Phil. Trans. 1742. Vol. XLII. p. 185.
‡ Phil. Trans. 1738. Vol. XXXIX. p. 262.   § Phil. Trans. 1798. Vol. LXXXVIII. p. 133.

of copper, the quinarius for five, and the sesterce for two and a half; but the weight of the as was reduced in the first Punic war, when the republic, being unable to defray its expences, resolved to coin six asses out of the pound; by which they gained five parts, and paid their debts.   The stamp of the as was a double-faced Janus on one side, and the prow of a ship on the other: on the triens and quadrans, a boat.· After this, when they were pressed by Hannibal, the as was reduced to one ounce; and the silver denarius made to pass for six-teen asses; the quinarius for eight; and the sesterce for four: and the republic gained one half.   But, in the pay of the army, the soldier received a silver de-narius for ten asses.   The stamp of the silver money was a chariot and a pair, or a chariot and four horses: whence they were called *bigati* and *quadrigati*. The as was soon after reduced to half an ounce, by the Papirian law.   What is now called the *victoriat* was coined under the Clodian law; before which it was imported from Illyricum, as merchandize.   Its stamp is a Victory, whence it takes its name.   The gold money was coined sixty years after the silver, and the scruple passed for 20 sesterces; which, as the sesterce was reckoned at that time ($2\frac{1}{2}$ asses), made the pound of gold worth 900 silver denarii (of 16 asses each).   It was afterwards thought proper to coin 40 pieces out of the pound of gold; and the weight was by degrees diminished to 45 in the pound.   Such is the account given by Pliny of the Roman coinage.   It was afterwards entirely altered by Constantine.   The Roman coins were not, like the Greek, used as weights: on the contrary, they depended upon the Roman weights.   From a number of gold coins, weighed by Mr. Raper, and from many corroborating circumstances, he concludes that the Roman pound was equal to 5040 troy grains.*   If this be the true weight of the Roman pound, it follows, that the ancient Saxon pound, which weighed 5400 troy grains, was not the Roman pound as has been conjectured; but considerably heavier.

The last paper in the Transactions, on the subject of weights and measures, Jewish weights and measures. is an account of an essay by Dr. Cumberland, upon the Jewish weights and measures.   He endeavours to prove, and his proofs seem conclusive, that the ancient Egyptian cubit is the same as the modern.   There is every reason to believe that the Jews borrowed their weights and measures from the Egyptians. Now the present Egyptian cubit is = 21·9 English inches.   Hence it is easy to deduce all the other Jewish measures, whose relation to the cubit is known.   He endeavours to prove that the *epha* was equal to $\frac{1}{4}$th of the *ardub*, or cube of the Egyptian cubit; so that it held $7\frac{1}{2}$ gallons and half a pint, or was nearly equal to the English cubic foot, which (of water) weighs nearly 1000 avoirdu-pois ounces.   Hence it is easy to deduce the measures of capacity, whose rela-

* Phil. Trans. 1771. Vol. LXI. p. 462.

tion to the epha is known.   The shekel, he determines as equal to half an ounce avoirdupois.   The value of the coin called the shekel was, he says, two shillings and four pence halfpenny.   Hence we may deduce all the weights and coins.*

CHAP. II.

OF POLITICAL ARITHMETIC.

This subject might, without impropriety, have been discussed under the head of Mathematics ; but we left it out, partly because mathematics was too much crowded with important articles, and partly because several of the most curious papers in the Philosophical Transactions do not take a mathematical view of the subject.   We mean to restrict the term *Political Arithmetic* here within much narrower limits than it usually bears.   The papers in the Philosophical Transactions relate simply to the population of different places ; and some of them touch upon the causes of the population.   As to the doctrine of annuities, founded on the probability of human life, that is a doctrine purely mathematical, depending upon the theory of compound interest, and cannot with propriety be touched upon here.

Healthiness of London.

One of the very first things that strike one is the surprizing difference in the healthiness of London at present, notwithstanding its great increase of population, and what it was during the 17th century.   At present the number of inhabitants in London, and the contiguous villages, which in fact make a part of it, is 1,090,000.   We do not know exactly what it was in the 17th century ; but, as it has been increasing ever since, and, as in the year 1753, the number of inhabitants did non exceed 750,000,† we shall not probably err very much if we reckon the inhabitants of London at the revolution, in 1688, about half a million.   Yet, in the years 1685, 1686, and 1687, the births and deaths were to each other as follows :‡

	Births.	Deaths.
1685	14730	23222
1686	14694	22609
1687	14951	21460

So that at this period the deaths exceeded the births by no less a quantity

* Phil. Trans. 1686. Vol. XVI. p. 33.   † Phil. Trans. 1754. Vol. XLVIII. p. 788.
‡ Phil. Trans. 1685. Vol. XV. p. 1245; and Vol. XVII. p. 445.

than 7639, or more than one half of the whole births. At present, the number of births exceed that of deaths. This must be owing to the different mode of living, and the improvements in the width of the streets, and in cleanliness. The same improvements having taken place in every part of Great Britain, it is probable that the value of human life is rather increased in this island.

We had no very accurate means of estimating the population of Great Bri- Population of Great Britain. tain, previous to the late returns made to Parliament, in the years 1802 and 1811, so that the estimates made are probably incorrect. But there can be no doubt that the population of the island has been constantly on the increase, at least since the revolution, and probably for a long period before it. In the year 1755, Brakenridge reckoned the whole inhabitants of England at 6 millions;* while Forster reckoned them at 7¼ millions.† The last, notwithstanding the plausible arguments of Brakenridge to the contrary, was probably nearest the truth. At present, the population is above 10 millions. It has increased (at least the population of Great Britain) 1,600,000 within the last 10 years. And supposing the whole population of Great Britain to be 12 millions, which is not very far from the truth, then it follows that, at the present rate of increase, the number of inhabitants doubles in 70 years. This is a degree of rapidity which no body was aware of before the late parliamentary returns. It ought to check that spirit of despondency which some of our political writers are apt to indulge. Great Britain is at present in such a state that, with frugality and wisdom, we could continue a contest with all Europe for 50 years to come, without any material injury to the nation. Indeed, in some points of view, the present state of exclusion from the continent may be considered as advantageous. To those politicians indeed, who consider money as the only article of value in a nation, nothing can be said: but, to those wiser men, who regard the moral principles and spirit of a nation as of more consequence than wealth, it may afford consolation to consider that, by this exclusion, our young men of fortune are prevented from imbibing the profligate principles which are so common on the continent, and which they were imbibing with such eagerness, that a few years of peace would have brought Great Britain to the state of profligacy so glaring in France; or to the state of imbecility and meanness so deplorable among the higher ranks in Spain.

Between the years 1740 and 1750, the number of inhabitants in Bristol, as ascertained with a good deal of accuracy by Mr. Browning, was 43,692.‡ Since that period the population has increased considerably, but not in the same proportion as that of several other towns in Great Britain. Various causes have checked the rapid increase of trade in Bristol; the chief probably is, that the

---

* Phil. Trans. 1755. Vol. XLIX. p. 268.   † Phil. Trans. 1757. Vol. L. p. 457.
‡ Phil. Trans. 1753. Vol. XLVIII. p. 217.

trade of the town has got into the hands of a few rich merchants, who do not encourage the adventurous spirit of young men destitute of fortune, upon whom the increase of most places in trade and wealth, and consequently in population, depends. Bristol was formerly the second town in Great Britain, in point of population; now it is exceeded by nine or ten. Two of the most remarkable examples of increase of population are Liverpool and Sheffield. Liverpool, in Queen Elizabeth's time, was hardly inhabited at all: at present its population is not much short of 100,000 Sheffield, in Queen Elizabeth's time, did not contain more than 2000 inhabitants: its present population exceeds 53,000. In the year 1773, the number of inhabitants in Manchester, from an actual survey, was 27,246. By the returns made to Parliament, in 1811, they amount to about 100,000; and if we include Salford, which in reality constitutes a part of the town, as much as Southwark does of London, the present population of Manchester is about 128,000.

Chester is one of the oldest towns in Great Britain. From the name, and from many other circumstances, there can be no doubt that it was a Roman station. It has long been resorted to as an agreeable retirement for persons of small incomes, both on account of the comparative cheapness of the place, and the agreeableness of the country. There is another inducement of no less importance, of which perhaps people in general are not aware. From a comparison of the bills of mortality at Chester with those in other places, Dr. Haygarth has shown that it is one of the healthiest spots in Great Britain.*

Dr. White, by comparing the births in York, between 1728 and 1735, with those between 1770 and 1776, has shown that the city has increased prodigiously in healthiness. At the former period the burials exceeded the births by 98 annually. The average number of burials being 498. At the latter period the births exceeded the burials by $21\frac{1}{2}$ annually. The decrease of burials had amounted to $44\frac{1}{2}$ annually, and the increase of births to $74\frac{1}{2}$ annually. He shows that at the period when he wrote, York rather exceeded the healthiness of any great town of which registers had been published, the deaths being only 1 in $28\frac{1}{2}$, while in Vienna they were 1 in $19\frac{1}{4}$, and in London 1 in $20\frac{3}{4}$. Liverpool and Manchester approached nearest to York in point of healthiness; the deaths in the former being 1 in $27\frac{7}{10}$, and the latter 1 in 28.†

It appears, from a paper by Mr. Panton, that the population in the Isle of Anglesey had increased in the year 1773. But his statements are not sufficient to give us any exact data on the subject. He conceives that the increase may be owing to the substitution of potatoes for herrings as an article of food, a notion too vague to merit any serious attention.‡

* Phil. Trans. 1774. Vol. LXIV. p. 67; Vol. LXV. p. 85; and Vol. LXVIII. p. 131.
† Phil. Trans. 1782. Vol. LXXII. p. 35.   ‡ Phil. Trans. 1773. Vol. LXIII p. 180.

3

We have a very exact account of the population of Madeira in the year 1767, by Dr. Thomas Heberden, from an actual survey which he caused to be made, on purpose to ascertain the point. The result was as follows :—In the year 1743, the number of inhabitants amounted to 53,057; in the year 1767 to 64,614. Hence it follows that the increase is 1.0082 per cent. per annum, and the numbers in the island double every 84 years.*

In the year 1738, a curious book was published by Mr. Kersseboom on the number of people in Holland and West Friesland, as also in Haarlem, Gauda, and the Hague; a particular account of which is given in several papers in the Philosophical Transactions. We shall give the number of inhabitants in Holland and West Friesland, and in different cities, according to this author.†

	Inhabitants.		Inhabitants.
Holland and West Friesland	980,000	Gauda	20,000
Amsterdam	241,000	Hague	41,500
Haarlem	50,500	London	653,600

He is at much pains to prove that the number of inhabitants in Paris is greater than in London. But his proofs are unsatisfactory, and were triumphantly refuted by Mr. Maitland.‡ At present the number of inhabitants in London approaches to double the number of the inhabitants in Paris. In travelling through Britain and France, the state of the towns is no less strikingly different than that of the country. In Britain every town abounds with new houses, and buildings are every where rising in great abundance: in France the houses in every town are old, and hardly such a thing as a new house is any where to be seen. More new houses have been built in London alone within these 12 years than in all France for 50 years back.

Dr. Price, in a curious paper which he published in the Philosophical Transactions for the year 1775, gives strong reasons for believing that there is a prodigious preponderancy in favour of the country above the most healthy cities. In Manchester, for example, the number of deaths is one in 28; but in the neighbouring country only one in 56 dies. He has brought evidence that nearly the same disproportion holds in other cases. From a pretty long table given by Dr. Price, in this paper, it appears that the number of male births to that of females is as 20 to 19 very nearly. But, in some measure to counteract this, the chance of life in females is greater than in males.§ It may be supposed perhaps that the risk which females run from parturition counterbalances this greater chance of life. But from the observations of Dr. Bland

* Phil. Trans. 1767. Vol. LVII. p. 461.   † Phil. Trans. 1738. Vol. XI. p. 401
‡ Phil. Trans. 1738. Vol. XL. p. 407.   § Phil. Trans. 1775. Vol. LXV. p. 424.

made at the Westminster Dispensatory, it appears that the deaths from parturition amount only to one in 270.*

We shall finish this part of our subject with noticing a curious paper by Dr. Arbuthnot, the celebrated satirical writer, and associate of Swift and Pope, in which he demonstrates, that the constant equality kept up between the sexes is an unanswerable argument in favour of the existence of a Divine Providence. For the chances are so much against this equality, that it could not be supposed to take place unless it were so ordered by the will of the Divine Being, for the express purpose of preventing the possibility of the extinction of the human species.†

CHAP. III.

OF ANTIQUITIES.

There are no fewer than 120 papers in the Philosophical Transactions on the subject of antiquities. But the greater number of them consist of descriptions of coins, urns, baths, &c. which do not appear of sufficient importance to merit notice here. The papers which we shall notice are of a different kind, and deserve attention, either in a historical point of view, or as connected with the progress of science, and the improvement of human knowledge.

Cæsar's expedition into Britain.

1. The first paper we shall mention is a curious one by Dr. Halley to determine the time and the place of Cæsar's first landing in Britain. The only authors who treat of the subject are, Cæsar himself in the 4th Book of his Commentaries, and Dion Cassius in his 39th Book. From the circumstances mentioned by Cæsar, Dr. Halley shows that the day of the landing was the 26th of August in the afternoon, in the year 55, before the beginning of the Christian æra. The Portus Icius from which he sailed agrees best with Calais. From the distances between different places and the British coast, mentioned by Ptolemy and other ancient writers, Dr. Halley thinks that no other place but Calais will suit. But no great stress can be put on such determinations. Davila, who is a modern historian, and who having spent the greater part of his life in France, ought to have been better informed, says, that the distance between Britain and Calais is 30 leagues, whereas it does not exceed 26 English miles. Now if Davila, who lived almost on the spot, could fall into so great an error, what confidence can be put in Ptolemy, who lived at so great a distance as

Egypt. Cæsar indeed says, that the distance was 30 miles; and his determination, as he possessed a military eye, and as he sailed over the space, may be relied on as nearly exact. Now the distance between Calais and Dover is just 28½ Roman miles, which agrees sufficiently well with Cæsar's estimate, and seems to settle the point pretty correctly. From the wind with which Cæsar sailed, from the time of the day, and the direction of the tide, together with the description which he gives, it seems quite certain that he first made land at the cliffs of Dover; but that the Britons preventing him from coming on shore in that place, he sailed round the North Foreland, and landed about eight miles from Dover in the Downs. The only difficulty attending this opinion is Dion Cassius's account of the place where the Britons skirmished with Cæsar, ες τα τεναγη, which is commonly translated *among the marshes*. Dr. Halley removes this difficulty by translating the phrase *at the water's edge*, which he shows satisfactorily that it will bear.*

2. We have three papers in an early volume of the Philosophical Transactions, Ruins of Palmyra. giving an account of the ruins of Tadmor or Palmyra, a city built in a kind of Oasis, in the desert of Arabia, about 150 miles from Aleppo, and 60 from the river Euphrates. There is every reason to believe that it was built by Solomon, who is said (1 Kings IX. 18, and 2 Chron. VIII. 6,) to have founded a city of that name (תדמר) in the desert. It did not probably remain long in possession of the Jews, and doubtless fell successively under the Babylonian, the Persian, and the Grecian monarchies; till at last when the Roman conquests were stopped by the Parthians, being a kind of frontier town, it was allowed to retain its liberty; and during this period it seems to have acquired considerable wealth and grandeur. But when Trajan took possession of Ctesiphon, the capital of the Parthian empire, the inhabitants of Palmyra, being surrounded on all sides by the Roman empire, submitted themselves to the emperor Adrian about the year 130, when he made his progress through Syria into Egypt. Adrian being pleased with the situation and the place, thought proper to rebuild and adorn it, and probably presented it with some of those marble and porphyry columns, which have been so much admired by all who have visited its ruins.

From the time of Adrian to that of Aurelian, Palmyra continued to flourish and increase in wealth and power. When the Emperor Valerian was taken prisoner by Sapor, king of Persia, Odænathus, one of the lords of Palmyra, was able, while Gallienus neglected his duty both to his father and country, to bring a powerful army into the field, to recover Mesopotamia from the Persians, and to penetrate as far as Ctesiphon. Gallienus reckoned the service thus performed so important, that he considered himself as obliged to give Odænathus

* Phil. Trans. 1691. Vol. XVII. p. 495.

a share in the empire.   Being murdered soon after, together with his son Hero-
cles, and dying with the title of Augustus, his widow Zenobia took upon herself the government of the East, in right of her son Waballathus, then a minor,
and managed it to admiration.   Gallienus being soon after murdered, she seized
upon the government of Egypt, and held it during the short reign of the Emperor Claudius Gothicus.   But when Aurelian was advanced to the imperial
dignity, he refused to allow Waballathus to bear the title of Augustus, though
he permitted him to continue his authority under the name of Vice Cæsaris.
But Zenobia refusing to accept of any thing less than a participation of the
empire, Aurelian marched against her, defeated her forces, and shut her up and
besieged her in Palmyra.   The city speedily surrendered, Zenobia and her son
flying were taken, and the celebrated Longinus, her minister, was put to death.
Aurelian spared the city, and leaving a small garrison in it, marched back to
Rome.   But the inhabitants, believing he would not again return, murdered
the garrison, and set up for themselves.   Aurelian, though by this time he had
got into Europe, marched directly back, collected an army on his way, besieged
the city, took it by storm, and gave it up to be plundered by his troops.

The city never recovered this disaster.   For though Aurelian does not appear
to have burnt it, but merely to have demolished its walls, yet it never afterwards
made any figure in history.   When it was finally destroyed we do not know;
though it might probably be during the wars of the Saracens.   The destruction
seems to have been sudden and complete.   At present it is inhabited by about
40 families, who occupy sorry huts within the precincts of the ancient palaces
no better than pig-sties.   The ruins have remained without much dilapidation,
except what proceeded from the superstition of the Mahometans, because it is
too far off from every other city to make it worth while to carry off the stones
for building.   As to the description of the ruins, which appear to be very fine,
we refer the reader to the papers in the Transactions,* and to various books of
travels in which they are particularly described and figured.

Age of Homer.   3. The oldest profane poets in existence are Hesiod and Homer, who are generally considered as nearly contemporaries, though Hesiod is allowed to have
the precedence in point of antiquity by a few years.   Homer's poems were unknown in Greece till they were brought to Athens by Solon, who seems to have
met with them in some of the Grecian cities of Asia.   We are informed by Cicero, that it was Pisistratus who first arranged the writings of Homer in the
order in which we find them at present.   Now Pisistratus, according to Scaliger, seized the chief government of Athens in the 50th Olympiad, or 577 years
before the commencement of the Christian æra.   The Oxford marble places the
commencement of his reign at 557 years before Christ, a date which agrees well

* Phil. Trans. 1695. Vol. XIX. p. 83, 129, and 160

with the account given by Plutarch. In what year he digested the poems of Homer we do not know; but it certainly was not before that period, as Solon did not return to Athens till Pisistratus was preparing to assume the government of the city. Thus it appears that the period when Homer became known as a poet in Greece was not earlier than 557 years before the commencement of the Christian æra. Homer himself certainly lived before that period, and a circumstance related by Laertius would lead us to believe that he preceded it a considerable time. According to that writer, Solon made use of the following lines of Homer, to prove that the Athenians had a right to the island of Salamis.

> Ἀιας δ' ἐκ Σαλαμῖνος ἀγεν δυοκαιδεκα νῆας,
> Στῆσε δ' ἀγων, ἱν' Αθηναιων ἱσταντο φαλαγγες.

Now it is not consistent with common sense, to suppose that he would bring such a proof from the writings of a contemporary poet: or that it would have been listened to with patience, if he had.

As to the exact period when Homer lived, nothing very satisfactory is known. Velleius Paterculus places him 920 years before the commencement of the Christian æra. Herodotus, a much older authority, places him four hundred years before his own time, or about 831 years before the Christian æra. Petavius places Homer 1000 years before the commencement of the Christian æra, and Sir Isaac Newton places Hesiod 879 years before the same æra, while Mr. Costard brings down Homer to within 557 years of the Christian æra, which seems too low, if we attend to the circumstances above mentioned. Sir Isaac Newton, and astronomical writers in general, have been led to fix upon the above period for the age of Hesiod, because in one of his poems he mentions that the star Arcturus rose achronically 60 days after the winter solstice. But several circumstances prevent this from marking a very precise period. The winter solstice could not be very exactly known in Hesiod's time, and the difference of a day or two would make a very material alteration in the period to which Hesiod refers. By rising achronically must mean not strictly rising the instant the sun sets; because the stars are never visible at that period: it must mean rising at such an interval after sunset as will admit the stars to be seen. Now this gives such a latitude to the poet that we have no means of even guessing how long it might be. Upon the whole, the period assigned by Newton seems as likely to be correct as any other. At any rate we are certain that Homer lived before the year 557 before Christ, and that his poems have been known and admired ever since, or for a period of 2369 years.*

4. The famous pillar composed of a single mass of stone, usually called Pom- Pompey's Pillar.

---

* Phil. Trans. 1754. Vol. XLVIII. p. 441.

pey's Pillar situated about a mile from the walls of Alexandria, in Egypt, has been commonly considered as having been erected to the memory of Pompey. The remarks of Mr. Wortley Montague, proving this opinion to be inaccurate, therefore deserve to be noticed. The following are the measurements of the different parts of this immense pillar, as ascertained by Mr. Montague

Capital of the pillar 9 feet, 7 inches.	The pedestal .... 10 feet, 5¾ inches.
The shaft........ 66 1¾	Height from the ground 92 0
The base........ 5 9¾	Its diameter 9 1

Mr. Montague was surprised to find the pedestal so inferior to the pillar itself. It was composed of small and great stones of different sorts, and was of bad and weak masonry. Taking out a stone he found that this pedestal was hollow, and after the labour of many days he made a hole in it large enough to let him in. He found that the pillar stood upon a reversed obelisk, 5 feet square, and 4 feet and 1 inch thick. This obelisk was covered with hieroglyphics reversed, a proof that it had been used for the purpose of supporting the pillar when these were no longer the sacred characters of Egypt.

Observing that the cement or mortar, which closes the small separation of the shaft from the base, was quite destroyed in one part, Mr. Montague was curious to see if any thing was made use of to fasten or tie the shaft to the base. He soon saw that there was something, and being desirous to know if it was lead, he introduced a pretty large hanger, in order if possible to cut off a part of the grapple. He then discovered a dark spot at the distance of more than a foot, within the circumference of the pillar, and by striking it with the hanger, he discovered that it was something stuck fast to the base. After striking it several times he detached it from its place, and found it to be a medal of Vespasian in fine order. On one side was this inscription ΑΛΤ. ΚΑΙΣ. ΣΕΒΑ. ΟΛΕΣΠ. The reverse is Victoria gradiens; dextra spicas, sinis. palmam. From this medal, which he conceived must have been placed where he found it at the time the pillar was erected, Mr. Montague concludes that it was erected during the time of Vespasian, and no doubt in honour of that emperor. This lateness of the erection accounts for the silence of Strabo, which would have been singular had the pillar been erected in memory of Pompey.*

Visit to the written mountains.

5. Mr. Wortley Montague, whose singular character is well known, spent a great part of his time in eastern countries, and was as well acquainted with Hebrew, Arabic, and Persian, as with his own language. On that account his journey from Cairo to the written mountains, undertaken on purpose to trace the rout of the Israelites out of Egypt, deserves considerable attention. Few

* Phil. Trans. 1767. Vol. LVII. p. 488.

people were, by knowledge of the language, so well qualified for the task, and he seems to have been at the requisite pains. His account of his journey is somewhat confused; but such as it is we shall mention the most striking particulars which it contains.

He set out from Cairo by the road known by the name of *Tauriche Beni Israel, Road of the Children of Israel.* After about 20 hours travelling at the rate of three miles an hour, he passed by an opening in the mountains on the right hand, the mountains Maxatree. There are two roads, one to the northward of this, which the Mecca pilgrims go; and one to the south between the mountains. But this last road is seldom travelled, because it does not lead to Suez. The children of Israel are said to have taken the most northerly road. From Suez, Mr. Montague went to Tor, by sea. At Suez the tide flows six feet, at spring tides nine feet, and sometimes, when the south wind blows strong, twelve feet. It is high water when the moon is in the meridian. The Egyptian shore of the red sea, from Badeah to opposite Tor, is all mountainous and steep, and at Elim, the northernmost point of the bay of Tor, ends the ridge of mountains which begin on the eastern shore of this western branch at Karondel. From this place Mr. Montague crossed the plain in about eight hours, and entered the mountains of Sinai. They are, he says, of granite of different colours. He observed some writings upon the mountains which he did not examine particularly, conceiving them to be, comparatively speaking, modern. He gives us an account of the print of the foot of Mahomet's camel observed on the mountain, so famous among the Mahometans. It is, he says, a very curious lusus naturæ, and bears evident marks of having never been touched by man. For the coat of granite is entire and unbroken in every part. What he means by the *coat of granite* it is difficult to say. But surely it is not inconceivable that a piece of granite cut above 1200 years ago, and exposed ever since to the action of the weather, may have acquired the very same appearance as any other part of the mountain.

Mr. Montague next describes Meribah, the place where Moses smote the rock, and procured water for the Israelites. He says it is surprisingly striking. He examined the lips of its mouths, and found that no chissel had ever worked there. The channel is plainly worn by the course of the water, and the bare inspection of it is sufficient to convince any one that it is not the work of man. He never, he says, met with any thing like this, except that at Jerusalem, and the two cracks in the rock which Moses struck twice. Mr. Montague, from his enquiries and observations, considers it as certain that the present *Sharme* is the *Midian* of the Old Testament, and *Meenah El Dzahab, Eziongeber.*

There are two roads from Mount Sinai to Jerusalem; the one through Pharan; the other by the way of Dzahab. The first is eleven days' journey; but the second is longer, because it is more mountainous. He set out from Mount

3 Z

Sinai, by the way of Scheich Salem, and after passing Mahomet's stone, came to a beautiful valley. Here, he says, he discovered the manna upon which the children of Israel fed. Soon after he came to the rock which Moses struck twice, and from which proceeds a river. This river runs by a number of ruins which Mr. Montague considers as the ancient Kadesh Barnea. This is the river which Eratosthenes supposed to be formed by the Arabian lakes, not being acquainted with its miraculous source.

From this place Mr. Montague went down a large valley to the west towards the sea, and passed the head of a valley, a part of the Desert of Sin, which separates the mountains of Pharan from those which run along the coast, and the same plain which they had passed from Tor. Here they came to the written mountains. On examining the characters which covered these mountains, he was greatly disappointed at finding them every where interspersed with figures of men and beasts, which convinced him that they were not written by the Israelites. Neither could they have been written by the Mahometans. For the religious opinions of both prevent them from making figures of men or animals. He thinks it probable that they were engraven by the first Jewish converts to Christianity, when probably pilgrimages from Jerusalem to Mount Sinai were fashionable and frequent. The characters he conceives to be those that were used by the Jews at the time of our Saviour. There are a few Greek, Arabic, and Saracen, inscriptions on the same mountains, which merely say such a one was here at such a time. Probably the others, which cannot be made out, allude to the same thing. So that they are in reality not worth the trouble of decyphering.

From the appearance of the country, and from the names of the places, which all allude to the wonderful event; as, Tauriche Beni Israel, road of the children of Israel; Attacah, deliverance; Badeah, new thing or miracle; Bachorel Polsum, sea of destruction; Mr. Montague is convinced that the Israelites entered the sea at Badeah, and no where else. The current too sets from this place, where he encamped, towards the opposite shore into the pool Birque Pharaone, Pool of Pharaoh, where the tradition is that his host was drowned; a current formed, he supposes, by the falling and rushing of one watery wall on the other, and driving it down.*

atacombs of Rome.

6. The catacombs at Rome and Naples have long been the admiration of antiquarians. They are long galleries under ground, extending an immense way, from which others proceed, some of them merely niches, others in the form of small rooms or chapels. We have an account of the catacombs at Rome in an early volume of the Transactions, by Mr. John Monro. From his description it would appear that they extended a great many miles, and that the

* Phil. Trans. 1766. Vol. LVI. p. 40.

whole of Rome, and a considerable space round it, was hollow underneath. In his time the Roman Catholic priests were continually employed in digging into these repositories, alleging that the bodies of a great number of Christian martyrs had been deposited there. Aphial, together with the letters Xp, which they translated pro Christo, was the mark of a martyr, and whenever they found them they carried the accompanying relics to the palace. There were two opinions entertained respecting the origin of these catacombs, both of which are too extravagant to be credited. The first is, that they were dug by the primitive Christians as places of refuge in times of persecution. That they held their meetings in them when it was unlawful for them to meet any where else; and that in them they were in the habit of depositing the bodies of their dead. But the vastness of the catacombs is an insuperable objection to the admission of this account, as it was impossible for any person to have dug them, without the knowledge of government, who would not have permitted the Christians to undertake such a work. Besides had the catacombs been dug so recently as after the introduction of the Christian religion, the work would have been too notorious to have escaped the notice of historians. The other opinion is, that they were the puteoli into which the Romans were accustomed to throw the dead bodies of their slaves, to save the expence of burning them. But the immence size of the catacombs is inconsistent with such an opinion. The digging of them would be more expensive than burning the dead bodies of their slaves for many years. Besides, the care with which the bodies were deposited, and the small rooms or chapels so common in them, show a much greater degree of care and attention than would have been bestowed upon places intended for depositing the bodies of slaves. The puteoli were nothing else than deep pits into which the bodies of the slaves were carelessly thrown. The catacombs must have been dug at an early period, since no account of them is to be traced in history. They must have been dug at a time when it was customary to bury and not to burn the dead, and therefore before the period of the Roman power.*

7. In the county of Cornwall there are a great many conical tumuli known **Barrows in Cornwall.** by the name of *barrows*. They differ in height from 4 feet to 30, and in breath from 15 to 130. Some of them have a circle of stones at the top, others at the bottom. Some are surrounded with a ditch, while others are without that accompaniment. They are all composed of artificial earth, and have a small pit at their centre filled with black fat earth. The general opinion entertained was, that they were ancient tombs, and this opinion was verified by Dr. Williams, who digging into a large one found under three flat stones near one of the borders of the barrow, a large urn of baked earth, filled with half burnt

* Phil. Trans. 1700. Vol. XXII. p. 643.

bones and ashes. At what period these barrows were raised we have no means of knowing. There is some reason to suspect that Cornwall was never in the possession of the Romans. It long maintained its independency against the Saxons; but at last became tributary to them, though without intermixing much. Hence it would appear, that this county continued to be inhabited by the original inhabitants. But on account of their connection with the Phœnicians and the Greeks, who traded to Cornwall for tin, they might easily adopt the custom of burning the dead which prevailed in these nations.*

Herculaneum.　8. About the year 1713, the city of Herculaneum, which is supposed to have been buried under the lava of mount Vesuvius during the reign of Titus Vespasian, was discovered almost under the town of Portici, where the king of Naples had a summer palace, and about half a mile from the sea side. The original entrance was by the town well, a perpendicular pit of considerable depth, at the bottom of which there was a passage that led into the city. The king of Naples employed workmen to dig their way into this subterranean city, which they did with great labour and difficulty. As the whole was filled with lava which required to be removed before they could make any observations on the nature of the buildings. There are no fewer than 14 papers in the Philosophical Transactions, giving an account of the discoveries made in Herculaneum, during the first fifty years of digging; and as the labour was continued long after, and as the Prince of Wales took some charge of unrolling and deciphering the manuscripts, a book has been lately published in this country, giving an account of what has been done since the publication of the papers in the Transactions. A very short notice of some of the discoveries is all that will be expected here.

It would appear that the inhabitants of the city had time to save themselves, though they were obliged to leave their furniture behind them, as very few skeletons were discovered. Some few however were met with. Particularly one in a running posture at the door of a house, with a purse of silver coin in his hand, completely inclosed in lava. If we consider the slowness with which the lava from a volcano moves, the existence of a skeleton in such an attitude must appear extraordinary. The man must have been instantaneously surrounded with the lava, while in the act of making his escape: and the lava must have been too cold to melt silver; for the coins were found unaltered in a hollow under the left hand of the skeleton.

A great many pictures were found (about 1,500 in all) all painted on stucco in water colours in fresco. The colours were remarkably beautiful, and appeared as fresh as when newly painted. The greater number of these pictures were poorly executed, being compared, by some of those who describe them,

* Phil. Trans. 1740. Vol. XLI. p. 465.

to the modern pictures on Chinese screens. A few were tolerably well done, and two or three are praised as excellent. One of the best is a picture of Theseus, after having slain the Minotaur, with that monster dead at his feet, and the children, whose life he had saved, returning him thanks for his exertions in their favour. There is another, representing a groupe of characters, and describing some solemn and melancholy scene, as the accusation of Virginia, by the Decemvir Appius Claudius, which is also described as very fine.

A considerable number of statues were also found in most of the ruins, very much mutilated, though some entire, and of exquisite workmanship. A great number of ancient utensils were also dug out and described. Among other articles ink-stands, with pens in them. The pens were of wood, and had no slit in them like our modern pens, because the ancients did not join their letters as we do, but wrote them all separate. Their mode of writing must have been very tedious.

But the discovery, which excited the greatest attention and expectation, was made in an apartment of a palace, into which the workmen made their way. It was about 150 rolls of papyrus, as black as charcoal, and all sticking together, which proved on examination to be so many ancient books. For a long time all attempts to unroll these manuscripts, and of course to form any conception of their contents, failed. At last, the King of Naples was advised to send for Father Antonio, a writer in the Vatican, as the only man capable of executing so difficult a task. This man contrived a kind of machine for the purpose, and with infinite pains and patience, succeeded in unrolling one roll. It was a Greek treatise on musick, by an author named Philodemus, abusing it as enervating the mind, and injuring the moral dispositions. Another treatise on rhetoric, by the same author, was also unrolled. Father Antonio copied fac similes of these two treatises, with all the defects arising from the injury which the papyrus had sustained. These defects being single words were easily supplied, and the books were published. Unfortunately they are of no great value; and though several other rolls have since been decyphered, the public has hitherto been disappointed in their expectation of recovering any of the lost works of some of the celebrated authors of antiquity.*

9. Much has been done by the admirers of the fine arts, in this country, to make us acquainted with the antiquities of Athens; which, next to those of Rome, seem to be the most deserving of study of any that exist. Indeed during the latter part of the last century, a great many Grecian antiquities have been brought to this country; and a considerable change has no doubt been pro-

Antiquities of Greece.

* Phil. Trans. 1740. Vol. XLI. p. 484, 489, 493; 1749. Vol. XLVI. p. 14; 1751. Vol. XLVII. p. 131, 150; 1754. Vol. XLVIII. p. 634, 821; 1755. Vol. XLIX. p. 112, 490; 1757. Vol. L. p. 49, 88, 619.

duced by the dilapidations of the Turks. But the antiquities of Athens are composed of so bulky and so solid materials, that a considerable part of them must still remain. We have a short account of the most remarkable particulars to be seen in Greece, in an early volume of the Philosophical Transactions, by Mr. Vernon, who travelled over-land from Venice to Smyrna. It may be worth while to notice some of the most remarkable particulars that he describes, that we may have an opportunity of knowing the changes which have taken place in that country, since about the year 1675, when that journey was probably made.

At Spalatro, in Dalmatia, Dioclesian's palace, to which he retired after resigning the empire, was still remaining. It is represented as an immense building, as large as the whole town of Spalatro, which was built out of its ruins, and took its name from that circumstance. Within it was an entire temple of Jupiter, of eight sides, with noble porphyry pillars, and admirable cornices. In Athens, the temple of Minerva is represented as very entire; but being situated within the castle of Athens, in which was a Turkish garrison, it was difficult to examine it with exactness. The length of the cella, or body of the temple without, is 168 English feet; the breadth 71 feet. The portico, of the Doric order, which runs round it, has 8 pillars in front, and 17 on the sides; its length is 230 feet. The shaft of the pillars is $19\frac{1}{4}$ feet in circumference: the intercolumnium $1\frac{1}{4}$ of the diameter of the pillars. The temple of Theseus was likewise entire; though much less, yet built after the same model. The length of its cella is 73 feet; its breadth 26. The whole length of the portico which goes round it 123 feet. It is a Doric building, as is that of Minerva, and both are of white marble. About the cornice, on the outside of the temple of Minerva, is a basso relievo of men on horseback, others in chariots, and a whole procession of people going to a sacrifice, of very curious sculpture. On the front is the history of the birth of Minerva. In the temple of Theseus, at the west end, is the battle of the Centaurs; and at the east end, there seems to be a continuation of the same story. There is a temple of Hercules, a round building, only six feet diameter; but of neat architecture. The pillars are of the Corinthian order, supporting an architrave and piaze, wherein are done the labours of Hercules, in relievo. The top is one single stone, wrought like a shield, with a flower on the outside which rises like a plume of feathers. The tower of Andronicus Cirrhestes was still standing. It is an octagon, with the figures of the eight Winds, which are large, and of good workmanship; and the names of the Winds remain legible in fair Greek characters, each pointing to its own quarter of the heavens. Several other ancient buildings are described, but these are the principal.

Thebes is still a large town, about 50 miles from Athens; but contains few antiquities except a portion of the old wall, and one gate, which, it is said, were spared by Alexander the Great when he destroyed the town. Sparta is quite forsaken : Mestra, four miles distant from it, being the town which is inhabited.   Large ruins are seen there about almost all the walls; several towers and foundations of temples, with pillars and chapiters, being demolished. A theatre remains pretty entire.   The city might have been about five miles in compass, and about a quarter of a mile from the river Eurotas.   The plain of Sparta and Laconia is about 80 miles long, well watered, and very fruitful. But the plain of Calamatta, anciently Messene, is rather richer.   Alpheus is by far the largest river in Greece.   Arcadia is a pleasant champaign, and full of cattle ; but encompassed with rugged hills.*

10. The custom which prevailed among the ancient Egyptians of embalming their dead, and the mummies which, in such great numbers have been brought from that country, are well known. A great number of these mummies have been examined at various times, and, from the whole, it appears clear that the methods of embalming were not always the same.   In the case of a mummy belonging to the Royal Society, and examined at London, in 1763, by Dr. Hadley, it appears that the body had, in the first place, been reduced to the state of a skeleton, all except the feet.   It seems then to have been kept for some time in boiling pitch.   Finally, it was wrapped round with fillets of linen, dipped in melted pitch, and in some gummy substance, and all the cavities were filled with melted pitch which had been poured in.   The bones had been penetrated with the pitch, and many of their cavities filled with it.   This pitch burned with an aromatic odour, and was evidently a vegetable production, and probably a mixture of various ingredients.   When burned, it left behind it a considerable quantity of fixed alkali.   The nails of the toes were entire, and below one of the feet, there was a bulbous root, quite fresh looking.   Dr. Hadley does not say what this root was.† *(margin: Mummics.)*

11. Mr. Swinton, who has published in the Philosophical Transactions a great many descriptions of ancient coins, has mentioned one which deserves particular notice.   It belonged to Mr. Godwyn, Fellow of Balliol College, Oxford. On one side, is the veiled head of a woman, with the Greek word ΒΑΣΙΛΙΣΣΑΣ; and on the reverse, the name ΦΙΛΙΣΤΙΔΟΣ, in the exergue.   He shows, that Philistidis was Queen of Malta and Gozo, and that the coin must have been struck before the Carthaginians came into the possession of these islands, and while they were possessed by the Greeks and Phœnicians.‡ *(margin: Ancient coin.)*

12. There is another very curious paper by Mr. Swinton, in which he gives an account of the numeral characters used by the ancient inhabitants of Sidon ; *(margin: Numerals of Sidon.)*

* Phil. Trans. 1676. Vol. XI. p. 575.   † Phil. Trans. 1764. Vol. LIV. p. 1.
‡ Phil. Trans. 1770. Vol. LX. p. 80.

which he made out from the Sidonian coins, in consequence of their great similarity to the Palmyrene numerals, which were known. We are under the necessity of referring to the paper itself, as it would be impossible to make the numerals intelligible without engravings.*

Chronology of Newton.

13. We shall conclude this chapter with a few words on the chronology of Sir Isaac Newton; concerning which there are three papers in the Philosophical Transactions, one written by Sir Isaac himself, and two by Dr. Halley. Sir Isaac informs us, that at the request of a very great personage, (probably Queen Caroline,) he drew up a compend of chronology, with which subject he had been accustomed to amuse himself at Cambridge, when tired with other studies. At the request of this person a copy of the Newtonian Chronology was given to the Abbe Conti, a noble Venetian, at that time in London, on the express stipulation that the Abbé should keep it secret, as it was not meant for publication. But when he went to Paris, he communicated copies of it to different persons. P. Souciet translated it into French, and published it with a refutation at the end, though Newton had expressly refused to give his consent to any such publication. In this refutation, Newton himself is treated with very little ceremony, and even turned into ridicule. When a copy of this book was sent to Newton, he was hurt at the treatment, and at the improper conduct of the French translator and refuter, who had published his paper which never had been intended for publication, not only without leave, but in spite of an express prohibition. On that account, Newton's paper, in which he points out the mistakes of the French translator, and vindicates his own system, is written with considerable severity. Nor does he spare the character of the Abbé Conti himself, to whose improper conduct the publication was owing. He had assumed the character, he said, of Newton's friend, and had pretended to mediate in the dispute between him and Leibnitz; but, in reality, had been a partizan of Leibnitz: and that singular man endeavoured, in consequence, to involve Newton in new disputes about occult qualities, universal gravity, the sensorium of God, space, time, vacuum, atoms, the perfection of the world, supramundane intelligence, and mathematical problems. And what he has been doing in Italy, concludes Newton, may be understood by the disputes raised there by one of his friends, who denies many of my optical experiments, though they have been all tried in France with success. But I hope that these things, and the perpetual motion, will be the last efforts of this kind.†

Dr. Halley took up the subject after Newton's death, and displayed upon it his usual learning and acuteness, pointing out abundance of mistakes, into which

Phil. Trans. 1758. Vol. L. p. 791.   † Phil. Trans. 1725. Vol. XXXIII. p. 315.

P. Souciet had fallen, and showing that the Newtonian conclusions had not been drawn from loose or inadequate data.*

By the Newtonian calculation, the expedition of the Argonauts, upon which the whole of his chronology is founded, happened 955 years before Christ: but from other data, he places that expedition 937 years before Christ. It is at present the general opinion that this date crowds the subsequent events rather too closely upon each other; and therefore the ordinary chronology, which places the Argonautic expedition 1250 years before the commencement of the Christian era, is at present followed. It is about half between the Newtonian date and that of Souciet.

CHAP. IV.

OF MISCELLANEOUS ARTICLES.

Besides the subjects treated of already in the preceding chapters, there are a few others of so miscellaneous a nature, that they cannot be included under any single head. We shall notice the chief of them in this chapter.

I. *Printing.*

We have three papers on the discovery of printing in the Philosophical Transactions. As the subject is curious, and the authors appear to have been at considerable pains in collecting information, we shall state here the substance of their opinions.

Printing was discovered at Haarlem, in Holland, by Coster, and the first book was printed in the year 1430. It was a Dutch piece of theology, printed only on one side of the page, and in imitation of manuscript. The first attempts at printing were upon loose leaves, and the printed part was accompanied with cuts, somewhat in the manner of our present ballads. Coster's method was to cut out the letters upon a wooden block. He took for an apprentice John Fust, and bound him to secrecy. But Fust ran away with his master's materials, and set up for himself at Mentz. He had a servant called Peter Schoeffer, who first invented separate metal types. Fust, upon seeing them was so delighted, that he gave Schoeffer his daughter in marriage, and made him his partner. The first book they printed is said to have been Cicero de Officiis, which bears the date of 1465. But other books are mentioned with

* Phil. Trans. 1727. Vol. XXXIV. p. 205; and Vol. XXXV. p. 296.

earlier dates, 1457, 1442. They printed a number of bibles, in imitation of manuscript, and Fust carried them to Paris for sale. The Parisians, upon comparing together the different copies, were confounded at the exact similarity they bore to each other in every part; a similarity so great, that the most exact copyist could not have attained it. They accused Fust of being possessed of some diabolical art. This at once obliged him to discover the secret; and gave origin to the story of Dr. Faustus.

After the discovery of the art of printing, thus brought about at Paris, it quickly made its way over the whole of Europe. The first book printed in England, is said to have been Rufinus on the Creed; printed at Oxford, in 1468.

At first the impression was taken off with a list, coiled up, as the card makers use at this day. But when they came to use single types, they employed stronger paper, with vellum and parchment. At last the press was introduced, and brought gradually to its present state. The same observation applies to the ink: at first, the common writing ink was employed, and the printing ink, of lamp-black and oil, at present used, was introduced by degrees. Rolling press printing was not used in England till the time of King James the First; and then it was brought from Antwerp by the industrious John Speed.*

II. *The Stylus and Paper of the Ancients.*

A number of the instruments used by the ancients for writing, called *styli,* having been discovered near the wall of Antoninus Pius, in Scotland, Sir John Clerk was induced, in consequence, to write a learned dissertation on the subject.

These styli were pointed instruments of some length, made of gold, silver, brass, iron, or bone. They were always flat at the other end, for the purpose of rubbing out what had been written, in order to correct it. These styli were sometimes used by way of daggers. Thus, Julius Cæsar is said to have wounded Cassius in the arm *graphio;* and Caligula contrived to get his enemies murdered, when they came into the Senate House, *graphiis.* With these styli the ancients wrote upon tablets of wax, or plates of lead.

The papers first used for writing on by the ancients were made of bark of trees, or skins, or were such as were called *pugillares.* The oldest were the inner barks of trees, called *liber* in Latin, hence the name for a book. The papyrus† also, or the thin pellicles of the *cyperus papyrus,* may be considered as nearly similar. The skins were dressed either like glove leather or parchment. The first sort was commonly used by the Jews, for writing the law of Moses on

---

* Phil. Trans. 1703. Vol. XXIII. p. 1416, 1507; and 1707. Vol. XXV. p. 2397.

† It was called βίβλος, by the Greeks. Hence their books got the name of βίβλοι.

it; and, from the rolling up of these skins, comes the word *volumen*. Parchment is said to have been invented by Eumenes, King of Pergamus, when Ptolemy prohibited the exportation of papyrus from Egypt. The pugillares are mentioned by Homer; they consisted of pieces of wood, ivory, or skin, covered over with wax. The Greeks, during the time of the empire, invented the method of making paper from cotton. Paper, from linen rags, is an European invention of the middle ages; but the exact time is unknown. Charters granted by Edward III. written upon paper, still exist.

The ink of the ancients was of various kinds, but usually black, and this was made sometimes from soot, sometimes from the liquor of the cuttle fish. They used red ink for the titles of their chapters; hence the titles of the Roman laws are called rubricæ. Their purpura, obtained from a species of murex, was much in vogue with the Byzantine writers, and called by them κιυυαϐαρις. The term purple was given by the Romans to all very splendid colours.*

III. *Hatching of Chickens at Cairo.*

At Cairo, in Egypt, chickens are hatched by means of artificial heat, in great numbers at a time. They have a double row of ovens, about 28 in number, in which the hatching is conducted. Each oven consists of two compartments, one above another. The lower compartment is on a level with the ground. It is covered with mats, and the eggs are laid upon these in two layers, one above another, except just under the hearths, where three eggs are laid above one another. The fire is lighted on hearths, in the upper ovens, and is so proportioned that the temperature is kept as near as possible at 105°. The eggs are kept 14 days in the lower ovens, and are carefully turned and shifted every day, in order to give to all the same degree of heat. The fires are then allowed to go out, and the eggs are shifted into the upper ovens, the holes of which are stopped up with flax, and the eggs are turned four times a day; about the 21st day the chickens are hatched. They are put into the under oven on the mats, which have a quantity of bran under them to dry the chickens. The first day they eat not. Next day they are carried away by women, and afterwards fed with corn. The proprietor of the ovens has one third of the chickens for his trouble, and two thirds go to the owners of the eggs.†

IV. *Encaustic Painting.*

There is a species of painting mentioned by Pliny, and some other ancient writers, under the name of encaustic painting, which was conceived to have been lost by the moderns. Count Caylus, a celebrated antiquary, and a great patron of the arts, made experiments in order to revive it, and succeeded. His

---

* Phil. Trans. 1731. Vol. XXXVII. p. 157.   † Phil. Trans. 1677. Vol. XII. p. 923.

method, as described by the Abbé Mazeas, was as follows. The cloth or wood
to be painted is waxed, by only rubbing it simply with a piece of bee's wax.
It is then rubbed over with Spanish white, after which the picture is painted in
the usual way with water colours. When the picture is dry it is brought near
the fire, by which the wax melts, and absorbs all the colours.* This description is so concise that it is not quite sufficient to satisfy those who wish to make
the requisite trials. On that account Mr. Colebrooke made a set of experiments
on the subject, some of which were attended with considerable success. But
for the particulars we refer to his paper on the subject.†

V. *Teaching the Deaf and Dumb to speak.*

This very important art was first contrived by Dr. Wallis, whose versatile
talents and industry displayed themselves in so many ways, and gave birth to so
many admirable inventions. He has himself given us an account of the method
which he practised. The person, on whom he began the trial, was Mr. Daniel
Whalley, who had once been able to speak, but having lost his hearing when
only five years of age, he had gradually in consequence left off speaking, and
forgotten all language. The task of teaching a deaf person to speak consists of
two parts, quite independent of each other, and rendered on that account more
difficult. The first consists in teaching the scholar the position of the tongue,
and other organs of speech, in pronouncing the different sounds; the second in
making him acquainted with the meaning of words, and with the use of language. The commencement of both of these parts is most difficult. The position of the organs is explained as precisely as possible by signs, and when a particular sound is once attained, the scholar by frequent repetition is prevented
from forgetting it. In explaining the meaning of words Dr. Wallis arranged
the most material words in classes, and pointed out their meaning, either by
showing the objects of which they were the names, or by some other equivalent
means. In about two months Mr. Whalley was able to pronounce most words,
and to understand an English book on common subjects. He was on the 21st
of May, 1662, present at a meeting of the Royal Society, and in their presence
and to their great satisfaction, pronounced, distinctly enough, such words as by
the company were proposed to him; and though not altogether with the usual
tone and accent, yet so as to be easily understood. He did the same thing before King Charles II, Prince Rupert, and several of the nobility. In the space
of one year, which was the whole time of his stay with Dr. Wallis, he had
read over a great part of the English Bible, and had attained so much skill,
as to be able to express himself intelligibly in ordinary affairs, to understand

letters written to him, and to write answers to them, though not elegantly, yet so as to be understood, and in the presence of foreigners he pronounced the most difficult words of their language that could be proposed to him.*

VI. *The Romansh Language.*

We have a curious paper by Mr. Planta on the Romansh language, which, he informs us, is still used with little or no alteration by the inhabitants of the Grisons.   About 400 years before the commencement of the Christian era, the Gauls overran the north of Italy, and treated the nobility of the country with such harshness, that they thought proper to emigrate with the best part of their effects, and their dependants; and making their way into the mountains of the Alps, settled in the vallies of the Grisons, where they were well received by the aborigines of the country.   At subsequent periods other inhabitants of Italy emigrated to the same country, during the civil wars and other convulsions into which Italy was repeatedly thrown.   This occasioned the two dialects, which still continue in the country, namely the *Cialover*, and the *Ladin*.

Mr. Planta gives strong reasons for concluding that the inhabitants of the Grisons, called Rhætia by the Romans, from Rhætus the leader of the original emigrants, were never conquered, or intermingled with any foreign power. Though the Princes of the Merovingian race, and the Emperors of Germany, held them under a nominal subjection, they were in fact governed by native bishops and dukes, who were merely nominated by these foreign princes, and who at last treated their subjects with so much cruelty, that a league was formed by a number of old men, clad in the grey cloth of their country (hence the name Grisons) to resist and throw off their tyranny.   The league was successful, and was soon after followed by two others of a similar nature and effect.

The language used all over France was of the same nature with that of the Grisons, and known by the same name.   It made its way into England, and in consequence of the conquest of William I, became the prevailing language.   It would seem also that it made its way equally into Italy and Spain, and that it was the real parent of the French, Italian, and Spanish languages, which were gradually formed out of it by the refinements and the writings of the various countries.   While in the country of the Grisons, no such refinements having taken place, the language still retains the very same state that it was in nine centuries ago.   Mr. Planta thinks that the Romansh was likewise the origin of the Lingua Franca, or language used in the intercourse between the Europeans and Eastern nations.†

* Phil. Trans. 1670. Vol. V. p. 1087; and 1698. Vol. XX. p. 353.
† Phil. Trans. 1776. Vol. LXVI. p. 129.

VII. *Of Mozart, a remarkable Infant Musician.*

Joannes Chrysostomus Wolfgangus Theophilus Mozart, was born at Saltz-burg, in Bavaria, on the 17th of January, 1756. At the age of four he was not only capable of executing lessons on his favourite instrument, the harpsi-chord, but composed some in an easy style and taste, which were very much ad-mired. His extraordinary musical talents soon reached the ears of the Empress Dowager, who used to place him on her knees, while he played on the harpsi-chord. At seven years of age his father carried him to Paris, where he was so much admired, that a drawing was made of him. From Paris he came to Lon-don, and continued in England about a year. During this time Mr. Barring-ton got acquainted with him, and made several trials of his musical powers. He carried to him a manuscript duet, which was composed by an English gen-tleman to some favourite words in Metastasio's opera of Demopoonte. The whole score was in five parts; namely, accompaniments for a first and second violin, the two vocal parts, and a base. The score was no sooner put upon his desk than he began to play the symphony in a most masterly manner, as well as in the time and style which corresponded with the intention of the composer. The symphony ended, he took the upper part, leaving the under one to his fa-ther. His voice, in the tone of it, was thin and infantine, but nothing could exceed the masterly manner in which he sang. His father was once or twice out, on which occasions the son looked back upon him with some anger, point-ing out to him his mistakes, and setting him right. He not only did complete justice to the duet by singing his own part in the truest taste, and with the greatest precision; but he also threw in the accompaniments of the two violins wherever they were most necessary. It is well known that none but the most capital musicians are capable of accompanying in this superior style.

Mr. Barrington hearing that he was often visited with musical ideas, to which even in the middle of the night he would give utterance on his harpsichord, re quested the boy to give him an extempore love song. Mozart looked back with much archness, continuing to sit at his harpsichord, and immediately began five or six lines of a jargon recitative, proper to introduce a love song. He then played a symphony which might correspond with an air composed to the single word *Assetto*. It had a first and a second part, and together with the sympho-nies was of the length that opera songs generally last. If this extempore com-position was not amazingly capital, yet it was really above mediocrity, and showed most extraordinary readiness of invention. Mr. Barrington next re-quested an extempore song of rage, which was also complied with, and in the middle of it the boy had worked himself up to such a pitch, that he beat the harpsichord like a person possessed, rising sometimes in his chair. The word he pitched upon for this second extempore composition, was *perfido*.

In the year 1769, Mozart was at Saltzburg. The Prince of Saltzburg not crediting that such masterly compositions, as several oratorios that went by his name, were really those of a child, shut him up for a week, during which he was not permitted to see any one, and was left only with music paper, and the words of an oratorio. During this short time he composed a very capital oratorio, which was most highly approved on being performed.*

VIII. *Of Crotch, another Infant Musician.*

William Crotch was born at Norwich, in 1775. His father, by trade a carpenter, having a passion for music, of which however he had no knowledge, undertook to build an organ, on which he learned to play two or three common tunes, with which and such chords as were pleasing to his ear, he used to try the perfection of his instrument. About Christmas, 1776, when the child was only a year and a half old, he discovered a great inclination for music, by leaving even his food to attend to it, when the organ was playing; and about midsummer, 1777, he would touch the key note of his particular favourite tunes in order to persuade his father to play them. Soon after this, as he was unable to name the tunes, he would play the two or three first notes of them when he thought the key note did not sufficiently explain which he wished to have played. It seems to have been owing to his having heard the superior performance of Mrs. Lulman, a musical lady who came to try his father's organ, and who not only played on it, but sang to her own accompaniment, that he first attempted to play a tune himself. That evening in passing through the dining room he screamed and struggled violently to go to the organ, in which when he was indulged, he eagerly beat down the keys with his little fists. Next day being left with his brother, a youth of 14, in the dining room, he would not let him rest till he blew the bellows of the organ, while he sat on his knees and beat down the keys at first promiscuously; but presently with one hand he played enough of *God Save the King* to awaken the curiosity of his father; who being in a garret, which was his workshop, hastened down stairs to inform himself who was playing this tune on the organ. When he found it was the child, he could hardly believe what he heard and saw. At this time he was exactly two years and three weeks old. Next day he made himself master of the treble of the second part, and the day after he attempted the base. In the beginning of November, 1777, he played both the treble and base of " Hope, thou nurse of young desire."

On the parents relating this extraordinary circumstance to some of their neighbours, they were laughed at, and advised by no means to repeat such

* Phil. Trans. 1770. Vol. LX. p. 54.

marvellous stories, as they would only expose them to ridicule.   However, a few days after, Mr. Crotch being ill and unable to go out to work, Mr. Paul, a master weaver, by whom he was employed, passing accidentally by the door, and hearing the organ, fancied that he had been deceived, and that Mr. Crotch had staid at home in order to divert himself on his favourite instrument.   Fully prepossessed with this idea, he entered the house, and suddenly opening the dining room door, saw the child playing on the organ while his brother was blowing the bellows.   Mr. Paul thought the performance so extraordinary, that he immediately brought two or three of the neighbours to hear it, who propagating the news, a crowd of near 100 people came the next day to hear the young performer, and on the following days a still greater number flocked to the house from all quarters of the city.   Till at length the child's parents were forced to limit his exhibition to certain days and hours, in order to lessen his fatigue, and exempt themselves from the inconvenience of constant attendance on the curious multitude.

Dr. Burney, to whom we are indebted for the preceding account, enters into particulars relative to the performance of this infant musician, and accounts for the defects which he displayed by the bad models from which he had copied.*

CONCLUSION.

We have now taken a view of the contents of the Philosophical Transactions down to the end of the year 1800.   The following short recapitulation will give the reader some idea of the relative space occupied by the departments of science in this voluminous and important work.

The papers in the Philosophical Transactions amount to 4166.   They are distributed as follows.

	No. of Papers.		No. of Papers.		No. of Papers.
1. Botany	107	11. Mining	29	20. Navigation	48
2. Vegetable Physiology	82	12. Geography and Topography	67	21. Electricity	211
3. Agriculture	44	13. Mathematics	208	22. Magnetism	71
4. Zoology	290	14. Astronomy	416	23. Chemistry	406
5. Anatomy	131	15. Optics	137	24. Meteorology	281
6. Comparative Anatomy	90	16. Dynamics	40	25. Chemical arts	87
7. Physiology	220	17. Mechanics	51	26. Weights and Measures	12
8. Medicine and Surgery	478	18. Hydrodynamics	120	27. Political Arithmetic	39
9. Mineralogy	38	19. Acoustics	26	28. Antiquities	120
10. Geognosy	251			29. Miscellaneous	66

From this table it appears, that Medicine, Astronomy, and Chemistry, are the sciences which furnish the greatest number of papers ; though Electricity is the science which is most completely developed.

* Phil. Trans. 1779. Vol. LXIX. p. 189.

APPENDIX.

APPENDIX.

No. I.

Charter of the Royal Society.

" Carolus secundus, Dei gratia Angliæ, Scotiæ, Franciæ, & Hiberniæ, rex, fidei defensor, &c. om-
" nibus, ad quos hæ literæ nostræ patentes pervenerint, salutem. Diu multumque apud nos statuimus,
" ut imperii fines, sic etiam artes atque scientias ipsas promovere. Favemus itaque omnibus disciplinis,
" particulari autem gratia indulgemus philosophicis studiis, præsertim iis, quæ solidis experimentis co-
" nantur aut novam extundere philosophiam, aut expolire veterem. Ut igitur inclarescant apud nostros hujus-
" modi studia, quæ nusquam terrarum adhuc satis emicuerunt ; utque nos tandem universus literarum orbis
" non solum fidei defensorem, sed etiam veritatis omnimodæ & cultorem ubique & patronum semper ag-
" noscat : Sciatis, quod nos de gratia nostra speciali ac ex certa scientia & mero motu nostris ordinavi-
" mus constituimus & concessimus, ac per præsentes pro nobis heredibus & successoribus nostris ordina-
" vimus constituimus & concedimus, quod de cætero in perpetuum erit societatis de præside concilio
" & sodalibus consistens, qui vocabuntur & nuncupabuntur *Præses Concilium & Sodales Regalis Socie-*
" *tatis Londini pro scientia naturali promovenda* (cujus quidem societatis nos ipsos fundatorem & patro-
" num per præsentes declaramus) & eandem societatem, per nomen Præsidis Concilii & Sodalium Rega-
" lis Societatis Londini pro scientia naturali promovenda, unum corpus corporatum & politicum in re
" facto & nomine realiter & ad plenum pro nobis heredibus & successoribus nostris facimus ordinamus
" creamus & constituimus per præsentes, & quod per idem nomen habeant successionem perpetuam : Et
" quod ipsi & eorum successores (quorum studia ad rerum naturalium artiumque utilium scientias expe-
" rimentorum fide ulterius promovendas in Dei Creatoris gloriam, & generis humani commodum appli-
" canda sunt) per idem nomen *Præsidis Concilii & Sodalium Regalis Societatis Londini pro scientia*
" *naturali promovenda*, sint & erunt perpetuis futuris temporibus personæ habiles & in lege capaces ad
" habendum perquirendum percipiendum & possidendum terras tenementa prata pascua pasturas libertates
" privilegia franchesias jurisdictiones & hereditamenta quæcunque sibi & successoribus suis in feodo &
" perpetuitate, vel pro termino vitæ vitarum vel annorum, seu aliter quocunque modo, ac etiam bona &
" catalla ac omnes alias res cujuscunque fuerint generis naturæ speciei sive qualitatis (statuto *De aliena-*
" *tione in manum mortuam* non obstante) necnon ad dandum concedendum & assignandum eadem terras
" tenementa & hereditamenta bona & catalla, & omnia facta et res necessarias faciendum & exequendum
" de & concernentia eadem, per nomen prædictum : Et quod per nomen *Præsidis Concilii & Sodalium*
" *Regalis Societatis Londini pro scientia naturali promovenda* prædictum placitare & implacitari, respon-
" dere & responderi, defendere & defendi, de cætero in perpetuum valeant & possint, in quibuscunque curiis
" placeis & locis, & coram quibuscunque judicibus & justiciariis & aliis personis & officiariis nostris heredum
" & successorum nostrorum, in omnibus & singulis actionibus, tum realibus tum personalibus, placitis
" sectis querelis causis materiis rebus & demandis quibuscunque, cujuscunque sint aut erunt generis na-

" turæ vel speciei, eisdem modo & forma, prout aliqui ligei nostri intra hoc regnum nostrum Angliæ,
" personæ habiles & in lege capaces, aut ut aliquod corpus corporatum vel politicum intra hoc regnum
" nostrum Angliæ, habere perquirere recipere possidere dare & concedere, placitare & implacitari, res-
" pondere & responderi, defendere & defendi, valeant & possint, valeat & possit : Et quod iidem præses
" consilium & sodales Regalis Societatis prædictæ & successores sui habeant in perpetuum commune
" sigillum, pro causis & negotiis suis & successorum suorum quibuscunque agendis deserviturum ; &
" quod bene liceat & licebit eisdem præsidi concilio & sodalibus Regalis Societatis prædictæ, & succes-
" soribus suis pro tempore existentibus, sigillum illud de tempore in tempus frangere mutare & de novo
" facere, prout eis melius fore videbitur experiri. Damus insuper & concedimus per præsentes præsidi
" concilio & sodalibus Regalis Societatis prædictæ eorumque in perpetuum successoribus, in favoris nostri
" regii erga ipsos nostræque de ipsis peculiaris existimationis præsenti & futuris ætatibus testimonium,
" hæc honoris insignia sequentia, viz.   In parmæ argenteæ angulo dextro tres leones nostros Anglicos ;
" & pro crista galeam corona flosculis interstincta adornatam, cui supereminet aquila nativi coloris, altero
" pede scutum leonibus nostris insignitum tenens ; telamones scutarios, duos canes sagaces albos, colla
" coronis cinctos (prout in margine luculentius videre est) a prædictis præside concilio & sodalibus ipso-
" rumque successoribus, prout feret occasio, in perpetuum gestanda producenda possidenda.   Et, quod
" intentio nostra regia meliorem sortiatur effectum, ac pro bono regimine & gubernatione prædictæ Re-
" galis Societatis de tempore in tempus, volumus, ac per præsentes pro nobis heredibus & successoribus
" nostris concedimus eisdem præsidi concilio concilio & sodalibus Societatis Regalis prædictæ & succes-
" soribus suis, quod de cætero in perpetuum concilium prædictum erit & consistet ex viginti & una per-
" sonis (quarum præsidem pro tempore existentem, vel ejus deputatum, semper unum esse volumus) &
" quod omnes & singulæ aliæ personæ, quæ intra duos menses proxime sequentes post datum præsentium
" per præsidem & concilium, vel per aliquos undecem vel plures eorum (quorum præsidem pro tempore
" existentem, vel ejus deputatum, semper unum esse volumus) vel per duas tertias partes vel plures præ-
" dictorum undecim vel plurium ; & in omni tempore sequenti per præsidem concilium & sodales, sive
" per aliquos viginti & unum vel plures eorum (quorum præsidem pro tempore existentem, vel ejus de-
" putatum, semper unum esse volumus) sive per duas tertias partes vel plures prædictorum viginti &
" unius vel plurium ; in eandem societatem accipientur & admittentur ut membra Regalis Societatis
" prædictæ, & in registro per ipsos conservando annotatæ fuerint erunt vocabuntur & nuncupabuntur so-
" dales Regalis Societatis prædictæ, quamdiu vixerint, nisi ob causam aliquam rationabilem, secundum
" statuta Regalis Societatis prædictæ condenda, quemvis eorum amoveri contigerit : quos quanto eminen-
" tius omnis generis doctrinæ bonarumque literarum studio clarescant, quanto ardentius hujusce societatis
" honorem studia & emolumentum promoveri cupiant, quanto vitæ integritatæ morumque probitate ac
" pietate emineant, & fidelitate animique erga nos coronam & dignitatem nostram sincero affectu pol-
" leant ; eo magis idoneos & dignos, qui in sodalium ejusdem societatis numerum adsciscantur, omnino
" censeri volumus.   Et, pro meliori executione voluntatis & concessionis nostræ in hac parte, assignavi-
" mus nominavimus constituimus & fecimus, ac per præsentes pro nobis heredibus & successoribus nostris
" assignamus nominamus constituimus & facimus, prædilectum & fidelem nobis WILLIELMUM viceco-
" mitem BROUNCKER, cancellarium præcharissimæ consortis nostræ reginæ CATHARINÆ, esse primum
" & modernum præsidem Regalis Societatis prædictæ ; volentes quod prædictus WILLIELMUS vicecomes
" BROUNCKER in officio præsidis Regalis Societatis prædictæ a datu præsentium usque ad festum
" Sancti ANDREÆ proximum sequentem [1] post datum præsentium continuabit, & quousque unus alius
" de concilio Regalis Societatis prædictæ pro tempore existente ad officium illud debito modo electus
" præfectus & juratus fuerit, juxta ordinationem & provisionem in his præsentibus inferius expressam &
" declaratam (si prædictus WILLIELMUS vicecomes BROUNCKER tam diu vixerit) sacramento corporali
" in omnibus & per omnia officium illud tangentia bene & fideliter exequendum, secundum veram inten-
" tionem harum præsentium, coram prædilecto & perquam fideli consanguineo & consiliario nostro

[1] Sic in Authent. & aliquoties infra.

" EDWARDO comite CLARENDON cancellario nostro prædicto sacramentum prædictum administrare
" plenam potestatem & authoritatem damus & concedimus, in hæc verba sequentia, viz. *I WILLIAM*
" *viscount BROUNCKER do promise to deal faithfully and honestly in all things belonging to the trust*
" *committed to me, as president of the Royal Society of London for improving natural knowledge,*
" *during my employment in that capacity. So help me God.* Assignavimus etiam constituimus & feci-
" mus, ac per præsentes pro nobis heredibus & successoribus nostris facimus dilectos nobis & fideles
" ROBERTUM MORAY militem unum a secretoribus nostris conciliis in regno nostro Scotiæ ROBERTUM
" BOYLE armigerum, WILLIELMUM BRERETON armigerum filium primogenitum baronis de BRERETON,
" KENELMUM DIGBY militem præcharissimæ matri nostræ MARIÆ reginæ cancellarium, GILBERTUM
" TALBOT militem jocalium nostrorum thesaurarium, PAULUM NEILE militem unum ostiariorum ca-
" meræ privatæ nostræ, HENRICUM SLINGESBY armigerum unum generosorum prædictæ privatæ ca-
" meræ nostræ, WILLIELMUM PETTY militem, TIMOTHEUM CLARKE in medicinis doctorem, &
" unum medicorum nostrorum, JOHANNEM WILKINS in theologia doctorem, GEORGIUM ENT in medi-
" cinis doctorem, WILLIELMUM AERSKINE unum a poculis nostris, JONATHAN GODDARD in medicinis
" doctorem & professorem Collegii de Gresham, WILLIELMUM BALLE armigerum, MATTHÆUM
" WREN armigerum, JOHANNEM EVELYN armigerum, THOMAM HENSHAW armigerum, DUDLY
" PALMER de Grey's-Inn in comitatu nostro Middlesexiæ armigerum, ABRAHAMUM HILL de London
" armigerum, & HENRICUM OLDENBURG armigerum, una cum præside prædicto, fore & esse primos
" & modernos viginti & unum de concilio & sodalibus Regalis Societatis prædictæ, continuandos in
" officiis concilii prædicti a datu præsentium usque ad prædictum festum sancti ANDREÆ apostoli proxi-
" mum sequentem, & deinde quousque aliæ idoneæ personæ & habiles & sufficientes in officia prædicta
" electæ præfectæ et juratæ fuerint (si tam diu vixerint, aut pro aliqua justa & rationabili causa non
" amotæ ¹ fuerint) sacramentis corporalibus coram præside pro tempore existente prædictæ Regalis So-
" cietatis ad officia sua bene & fideliter in omnibus & per omnia officia illa tangentia exequendum prius
" præstandis, secundum formam & effectum prædicti sacramenti, mutatis mutandis, præsidi Regalis So-
" cietatis prædictæ per cancellarium nostrum Angliæ administrandi (cui quidem præsidi pro tempore
" existenti sacramenta prædicta administrare personis prædictis, & aliis quibuscunque in posterum de
" tempore in tempus in concilium prædictum eligendis, plenam potestatem & authoritatem pro nobis he-
" redibus & successoribus nostris damus & concedimus per præsentes) & quod eædem personæ sic, ut
" præfertur, ad concilium prædictæ Regalis Societatis electæ præfectæ & juratæ, & in posterum eligendæ
" præficiendæ & jurandæ de tempore in tempus, erunt & existent auxiliantes consulentes & assistentes
" in omnibus materiis rebus & negotiis, meliores ² regulationem gubernationem & directionem prædictæ
" Regalis Societatis, & cujuslibet membri ejusdem, tangentibus seu concernentibus. Concedimus etiam
" præsidi concilio & sodalibus Societatis prædictæ & eorum in perpetuum successoribus, quod ipsi &
" successores eorum, seu aliqui novem vel plures eorum (quorum præsidem pro tempore existentem, vel
" ejus deputatum, semper unum esse volumus) conventus seu congregationes de seipsis pro experimento-
" rum & rerum naturalium cognitione & indagine aliisque negotiis ad Societatem prædictam spectantibus,
" quoties & quando opus fuerit, licite facere & habere possint in collegio sive aula sive alio loco com-
" modo intra civitatem nostram Londini, vel in aliquo alio loco commodo intra decem milliaria ab eadem
" civitate nostra. Et ulterius volumus, ac per præsentes pro nobis heredibus & successoribus nostris
" concedimus præfatis præsidi concilio & sodalibus Regalis Societatis prædictæ & successoribus suis, quod
" præses concilium & sodales Regalis Societatis prædictæ pro tempore existentes, sive aliqui triginta &
" unus vel plures eorum (quorum præsidem pro tempore existentem, vel ejus deputatum, unum esse vo-
" lumus) seu major pars prædictorum triginta & unius vel plurium, de tempore in tempus perpetuis fu-
" turis temporibus potestatem & authoritatem habeant & habebunt nominandi & eligendi, & quod eli-
" gere & nominare possint & valeant, quolibet anno in prædicto festo sancti Andreæ, unum de concilio

" prædictæ Regalis Societatis pro tempore existente, qui sit & erit præses Regalis Societatis prædictæ
" usque ad festum sancti Andreæ apostoli exinde proximum sequentem (si tam diu vixerit, aut interim
" pro aliqua justa & rationabili causa non amotus fuerit) & exinde, quousque unus alius in officium præ-
" sidis Regalis Societatis prædictæ electus præfectus & nominatus fuerit; quodque ille postquam sic, ut
" præfertur, electus & nominatus fuerit in officium præsidis Regalis Societatis prædictæ, antequam ad
" officium illud admittatur, sacramentum corporale coram concilio ejusdem Regalis Societatis, aut ali-
" quibus septem vel pluribus eorum, ad officium illud recte bene & fideliter in omnibus officium illud
" tangentibus exequendum præstabit, secundum formam & effectum prædicti sacramenti, mutatis mu-
" tandis (cui quidem concilio, aut aliquibus septem vel pluribus eorum, sacramentum prædictum adminis-
" trare pro nobis heredibus & successoribus nostris plenam potestatem & authoritatem de tempore in tem-
" pus, quotiescunque præsidem eligere opus fuerit, damus & concedimus per præsentes) & quod post hujus-
" modi sacramentum sic, ut præfertur, præstitum officium præsidis Regalis Societatis prædictæ usque ad
" festum sancti Andreæ apostoli exinde proximum sequentem exequi valeat & possit : & si contigerit præsi-
" dem Regalis Societatis prædictæ pro tempore existentem aliquo tempore, quamdiu fuerit in officio præsidis
" ejusdem Regalis Societatis, obire, decedere, vel ab officio suo amoveri; quod tunc & toties bene liceat &
" licebit concilio Regalis Societatis prædictæ eorumque in perpetuum successoribus, sive aliquibus undecim
" vel pluribus eorum, convenire vel congregari ad eligendum unum de prædicto numero concilii prædicti
" in præsidem Regalis Societatis prædictæ; & quod ille, qui per concilium prædictum, vel per prædictos
" undecim vel plures, vel per majorem partem prædictorum undecim & plurium, electus fuerit &
" juratus, ut præfertur, officium illud habeat & exerceat durante residuo ejusdem anni, quousque alius
" ad officium illud debito modo electus & juratus fuerit, sacramento corporali in forma supra specificata
" prius præstando ; & sic toties quoties casus sic acciderit. Et ulterius volumus, quod quandocunque
" contigerit aliquem vel aliquos de concilio Regalis Societatis prædictæ pro tempore existente mori, vel
" ab officio illo amoveri, vel decedere ; quos quidem de concilio Regalis Societatis prædictæ & eorum
" quemlibet pro male se gerendis aut aliqua alia rationabili causa amobiles esse volumus, ad beneplacitum
" præsidis & cæterorum de concilio prædicto (quorum præsidem pro tempore existentem, vel ejus depu-
" tatum, unum esse volumus) vel majoris partis eorundem : quod tunc & toties bene liceat & licebit
" præfatis præsidi concilio & sodalibus Regalis Societatis prædictæ eorumque in perpetuum successoribus,
" vel aliquibus viginti uni vel pluribus eorundem (quorum præsidem Regalis Societatis prædictæ pro
" tempore existentem, vel ejus deputatum, unum esse volumus) vel majori parti prædictorum viginti &
" unius vel plurium, unum alium vel plures alios de sodalibus Regalis Societatis prædictæ, loco sive locis
" ipsius vel ipsorum sic mortuorum decedentium vel amotorum, ad supplendum prædictum numerum vi-
" ginti & unius personarum de concilio Regalis Societatis prædictæ nominare eligere & præficere : &
" quod illi sive ille sic in officio illo electi & præfecti idem officium habeat & habeant usque ad festum
" sancti Andreæ apostoli tunc proximum sequentem, & exinde, quousque unus alius vel plures alii electus
" præfectus & nominatus fuerit electi præfecti & nominati fuerint; sacramento corporali ad officium
" illud in omnibus & per omnia officium illud tangentia coram præside & concilio Regalis Societatis præ-
" dictæ, vel aliquibus septem vel pluribus eorum (quorum præsidem pro tempore existentem, vel ejus
" deputatum, semper unum esse volumus) bene & fideliter exequendum, secundum veram intentionem
" præsentium, prius præstando. Et ulterius volumus, ac per præsentes pro nobis heredibus & successoribus
" nostris concedimus præfatis præsidi concilio & sodalibus prædictæ Regalis Societatis & successoribus
" suis, quod ipsi & successores sui, sive aliqui triginta & unus vel plures eorum (quorum præsidem pro
" tempore existentem, vel ejus deputatum, semper unum esse volumus) sive major pars prædictorum tri-
" ginta & unius vel plurium quolibet anno, in prædicto festo sancti Andreæ apostoli, plenam potestatem
" & authoritatem habeant & habebunt eligendi nominandi præficiendi & mutandi decem de sodalibus
" Regalis Societatis prædictæ, ad supplendum loca & officia decem prædicti numeri viginti & unius de
" concilio Regalis Societatis prædictæ; quoniam regiam voluntatem nostram esse declaramus, ac per
" præsentes pro nobis heredibus & successoribus nostris concedimus, quod decem de concilio prædicto,

4

" & non amplius, per præsidem concilium & sodales Regalis Societatis prædictæ annuatim mutati &
" amoti fuerint. Volumus etiam, & pro nobis heredibus & successoribus nostris concedimu præfatis
" præsidi concilio & sodalibus prædictæ Regalis Societatis, & successoribus suis in perpetuum, quod si
" contigerit præsidem ejusdem Regalis Societatis pro tempore existentem ægritudine vel infirmitate de-
" tineri, vel in servitio nostro heredum vel successorum nostrorum versari, vel aliter esse occupatum, ita
" quod necessariis negotiis ejusdem Regalis Societatis officium præsidis tangentibus attendere non po-
" terit; quod tunc & toties bene liceat & licebit eidem præsidi sic detento versato vel occupato unum de
" concilio prædictæ Regalis Societatis pro tempore existente, fore & esse deputatum ejusdem præsidis,
" nominare & appunctuare: qui quidem deputatus, in officio deputati præsidis prædicti sic faciendus &
" constituendus, sit & erit deputatus ejusdem præsidis de tempore in tempus, toties quoties prædictus
" præses sic abesse contigerit, durante toto tempore, quo prædictus præses in officio præsidis continua-
" verit; nisi interim prædictus præses Regalis Societatis prædictæ pro tempore existens unum alium de
" prædicto concilio ejus deputatum fecerit & constituerit: & quod quilibet hujusmodi deputatus præ-
" dicti præsidis sic, ut præfertur, faciendus & constituendus omnia & singula, quæ ad officium præsidis
" prædictæ Regalis Societatis pertinent seu pertinere debent, vel per prædictum præsidem virtute harum
" literarum nostrarum patentium limitata & appunctuata fore [1] facienda & exquenda de tempore in
" tempus, toties quoties prædictus præses sic abesse contigerit, durante tali tempore, quo deputatus præ-
" dicti præsidis continuaverit, facere & exequi valeat & possit, vigore harum literarum nostrarum paten-
" tium, adeo plene libere & integre, ac in tam amplis modo & forma, prout præses prædictus, si præ-
" sens esset, illa facere & exequi valeret & posset; sacramento corporali supra sancta Dei evangelia in
" forma & effectu supra specificatis per hujusmodi deputatum ad omnia & singula, quæ ad officium præ-
" sidis pertinent, bene & fideliter exequendum, coram præfato concilio prædictæ Regalis Societatis, vel
" aliquibus septem vel pluribus eorum, prius præstando; & sic toties quoties casus sic acciderit: cui
" quidem concilio, vel aliquibus septem vel pluribus eorum pro tempore existente,[2] sacramentum præ-
" dictum administrare potestatem & authoritatem, quoties casus sic acciderit, damus & concedimus per
" præsentes, absque brevi commissione sive ulteriori warranto in ea parte a nobis heredibus vel successo-
" ribus nostris procurando seu obtinendo. Et ulterius volumus, ac per præsentes pro nobis heredibus &
" successoribus nostris concedimus præfatis præsidi concilio & sodalibus Regalis Societatis prædictæ &
" successoribus suis, quod ipsi & successores sui de cætero in perpetuum habeant & habebunt unum the-
" saurarium, duos secretarios, curatores experimentorum duos vel plures, clericum unum vel plures, & præ-
" terea duos servientes ad clavas qui de tempore in tempus super præsidem attendant; quodque prædicti the-
" saurarius secretarii curatores clericus vel clerici & servientes ad clavas per præsidem concilium & sodales
" Regalis Societatis prædictæ, sive per aliquos triginta & unum vel plures eorum (quorum præsidem pro
" tempore existentem, vel ejus deputatum, unum esse volumus) vel per majorem partem prædictorum tri-
" ginta & unius vel plurium, eligendi & nominandi, antequam ad officia sua speciali & respectiva exequen-
" dum admittantur, sacramenta sua corporalia in forma & effectu supra specificatis, coram præside vel ejus
" deputato, & concilio ejusdem Regalis Societatis, aut aliquibus septem vel pluribus eorum, officia sua se-
" paralia & respectiva in omnibus illa tangentibus recte bene & fideliter exequendum præstabunt; &
" quod post hujusmodi sacramenta sic, ut præfertur, præstita officia sua respectiva exerceant & utantur:
" quibus quidem præsidi & concilio, aut aliquibus septem vel pluribus eorum, sacramenta prædicta de
" tempore in tempus administrare prædictis separalibus & respectivis officiariis & successoribus suis ple-
" nam potestatem & authoritatem damus & concedimus per præsentes: & assignavimus nominavimus
" elegimus creavimus constituimus & fecimus, ac per præsentes pro nobis heredibus & successoribus
" nostris assignamus nominamus eligimus creamus constituimus & facimus, dilectos subditos nostros
" prædictum WILLIELMUM BALLE armigerum fore & esse primum & modernum thesaurarium, & esse
" prædictum [3] JOHANNEM WILKINS & HENRICUM OLDENBURG fore & esse primos & modernos secre-

" tarios prædictæ Regalis Societatis, continuandos in eisdem officiis usque ad prædictum festum sancti An-
" dreæ apostoli proximum sequentem post datum præsentium : quodque de tempore in tempus & ad omnia
" tempora in prædicto festo sancti Andreæ apostoli (si non fuerit dies dominicus, & si fuerit dies dominicus,
" tunc die proximo sequente) præses concilium & sodales prædictæ Regalis Societatis pro tempore
" existentes, sive aliqui triginta & unus vel plures eorum (quorum præsidem pro tempore exis-
" tentem, vel ejus deputatum, unum esse volumus) sive major pars prædictorum triginta & unius
" vel plurium, probos & discretos viros de tempore in tempus in thesaurarium & secretarios, qui sunt &
" erunt de numero concilii Regalis Societatis prædictæ, eligere nominare & præficere valeant & possint ;
" quodque illi, qui in separalia & respectiva officia prædicta sic, ut præfertur, electi præfecti & jurati fuerint,
" officia illa respectiva exercere & gaudere possint & valeant usque ad prædictum festum sancti Andreæ
" extunc proximum sequentem, sacramentis suis prædictis sic, ut præfertur, prius præstandis ; & sic to-
" ties quoties casus sic acciderit : Et si contigerit electiones prædictas præsidis concilii thesaurarii secre-
" tariorum, vel alicujus vel aliquorum eorum, in festo sancti Andreæ prædicto commode fieri vel perfici
" non posse, damus & concedimus prædictis præsidi concilio & sodalibus, & successoribus eorum in per-
" petuum, quod ipsi, vel aliqui triginta & unus vel plures eorum (quorum præsidem pro tempore exis-
" tentem, vel ejus deputatum, unum esse volumus) vel major pars prædictorum triginta & unius vel plu-
" rium, licite possint nominare & assignare unum alium diem, quam proxime ad festum sancti Andreæ
" prædictum commodé fieri poterit, pro electionibus prædictis faciendis vel perficiendis ; & sic de die in
" diem, donec prædictæ electiones perficiantur : Et si contigerit aliquem vel aliquos officiariorum præ-
" dictorum ejusdem Regalis Societatis obire, decedere, vel ab officiis suis respectivis amoveri ; quod tunc
" & toties bene liceat & licebit præsidi concilio & sodalibus prædictæ Regalis Societatis, & eorum suc-
" cessoribus in perpetuum, sive aliquibus viginti & uni vel pluribus eorum (quorum præsidem pro tem-
" pore existentem, vel ejus deputatum, unum esse volumus) seu majori parti prædictorum viginti &
" unius vel plurium, alium vel alios in officium sive officia illarum personarum sic defunctarum deceden-
" tium sive amotarum eligere & præficere ; & quod ille sive illi sic electus & præfectus electi & præfecti
" officia prædicta respectiva habeant & exerceant durante residuo ejusdem anni, & quousque alius sive
" alii ad officia illa respectiva debito modo electus & juratus fuerit electi & jurati fuerint ; & sic toties
" quoties casus sic acciderit. Et insuper volumus, ac de gratia nostra speciali ac ex certa scientia &
" mero motu nostris concedimus præfatis præsidi concilio & sodalibus Regalis Societatis prædictæ, & suc-
" cessoribus suis in perpetuum, quod præses & concilium prædictæ Regalis Societatis pro tempore ex-
" istentes (præmissa semper in conventibus extraordinariis omnium membrorum concilii prædicti debita
" seu legitima summonitione vel citatione) sive aliqui novem vel plures eorum (quorum præsidem pro
" tempore existentem, vel ejus deputatum, unum esse volumus) pariter congregare & assemblare possint
" & valeant in collegio sive aula sive alio loco conveniente intra civitatem nostram Londini, vel in aliquo
" alio loco conveniente intra decem milliaria ab eadem civitate nostra ; & quod ipsi sic congregati & as-
" semblati, sive major pars eorum, habebunt & habeant plenam authoritatem potestatem & facultatem
" de tempore in tempus condendi constituendi ordinandi faciendi & stabiliendi hujusmodi leges statuta
" jura ordinationes & constitutiones, quæ eis, aut eorum majori parti, bona salubria utilia honesta & ne-
" cessaria juxta eorum sanas discretiones fore videbuntur, pro meliori gubernatione regulatione & direc-
" tione Regalis Societatis prædictæ & cujuslibet membri ejusdem, omniaque ad gubernationem res
" bona facultates redditus terras tenementa hereditamenta & negotia Regalis Societatis prædictæ spec-
" tantia agendi & faciendi ; quæ omnia & singula leges statuta jura ordinationes & constitutiones sic, ut
" præfertur, facienda volumus, & per præsentes pro nobis heredibus & successoribus nostris firmiter in-
" jungendo præcipimus & mandamus, quod de tempore in tempus inviolabiliter observata fuerint, secun-
" dum tenorem & effectum eorundem : ita tamen, quod prædictæ leges statuta jura ordinationes & con-
" stitutiones sic, ut præfertur, facienda, & eorum quælibet, sint rationabilia, & non sint repugnantia nec
" contraria legibus consuetudinibus juribus sive statutis hujus regni nostri Angliæ. Et ulterius de ampli-
" ori gratia nostra speciali ac ex certa scientia & mero motu nostris dedimus & concessimus, ac per præ-

3

" sentes pro nobis heredibus & snccessoribus nostris damus & concedimus, præfatis præsidi concilio & so-
" dalibus prædictæ Regalis Societatis, & successoribus suis in perpetuum, sive aliquibus viginti & uni vel
" pluribus eorum (quorum præsidem pro tempore existentem, vel ejus deputatum, semper unum esse vo-
" lumus) seu majori parti prædictorum viginti & unius vel plurium, plenam potestatem & authoritatem
" de tempore in tempus eligendi nominandi & constituendi unum vel plures typographos sive impressores,
" & chalcographos seu sculptores, & ipsi vel ipsis per scriptum communi sigillo prædictæ Regalis Socie-
" tatis sigillatum, & manu præsidis pro tempore existentis signatum, facultatem concedendi, ut impri-
" mant tales res materias & negotia prædictam Regalem Societatem tangentia vel concernentia, quales
" prædicto typographo vel impressori, chalcographo seu sculptori, vel typographis vel impressoribus,
" chalcographis vel sculptoribus, de tempore in tempus per præsidem & concilium prædictæ Regalis So-
" cietatis, vel aliquos septem vel plures eorum (quorum præsidem pro tempore existentem, vel ejus de-
" putatum, unum esse volumus) vel per majorem partem prædictorum septem vel plurium, commissæ
" fuerint; sacramentis suis corporalibus, antequam ad officia sua exercenda admittantur, coram præside
" & concilio pro tempore existente, vel aliquibus septem vel pluribus eorum, in forma & effectu ultimis [1]
" specificatis, prius præstandis; quibus quidem præsidi & concilio, vel aliquibus septem vel pluribus
" eorum, sacramenta prædicta administrare plenam potestatem & authoritatem damus & concedimus
" per præsentes. Et ulterius, quod prædicti præses concilium & sodales prædictæ Regalis Societatis in
" philosophicis suis studiis meliorem sortiantur effectum, de ampliori gratia nostra speciali ac ex certa
" scientia & mero motu nostris dedimus & concessimus, ac per præsentes pro nobis heredibus & succes-
" soribus nostris damus & concedimus, prædictis præsidi concilio &. sodalibus prædictæ Regalis So-
" cietatis, & successoribus suis in perpetuum, quod ipsi & successores sui, sive aliqui novem vel plures
" eorum (quorum præsidem pro tempore existentem, vel ejus deputatum, unum esse volumus) sive major
" pars prædictorum novem vel plurium, de tempore in tempus, habeant & habebunt plenam potestatem
" & authoritatem de tempore in tempus, & ad talia tempestiva tempora, secundum eorum discretionem,
" per assignatum vel assignatos suos requirere capere & recipere cadavera talium personarum, quæ mor-
" tem manu carnificis passæ fuerunt, & ea anatomizare, in tam amplis modo & forma, & ad omnes inten-
" tiones & proposita, prout præsidens Collegii Medicorum & Societatis Chirurgorum civitatis nostræ
" London (quibuscunque nominibus duæ prædictæ corporationes insignitæ fuerint) eisdem cadaveribus
" usi vel gavisi fuerunt, aut uti vel gaudere valeant & possint. Et ulterius, pro melioratione experimen-
" torum artium & scientiarum prædictæ Regalis Societatis, de abundantiori gratia nostra speciali ac ex
" certa scientia & mero motu nostris dedimus & concessimus, ac per præsentes pro nobis heredibus
" & successoribus nostris damus & concedimus, præfatis præsidi concilio & sodalibus prædictæ
" Regalis Societatis, & successoribus suis in perpetuum, quod ipsi & successores sui, sive aliqui
" novem vel plures eorum (quorum præsidem pro tempore existentem, vel ejus deputatum, unum esse
" volumus) sive major pars prædictorum novem vel plurium, de tempore in tempus habeant & habebunt
" plenam potestatem & authoritatem per literas vel epistolas, sub manu prædicti præsidis vel ejus depu-
" tati, in præsentia concilii vel aliquorum septem vel plurium eorum, & in nomine Regalis Societatis,
" mutuis intelligentiis fruentur [2] & negotiis cum omnibus & omnimodis peregrinis & alienis, utrum pri-
" vatis vel collegiatis, corporatis vel politicis, absque aliqua molestatione interruptione vel inquietate [3]
" quacunque: proviso tamen, quod hæc indulgentia nostra sic, ut præfertur, concessa ad ulteriorem non
" extendatur usum, quam particulare beneficium & interesse prædictæ Regalis Societatis in materiis seu
" rebus philosophicis mathematicis aut mechanicis. Et ulterius dedimus & concessimus, ac per præ-
" sentes pro nobis heredibus & successoribus nostris damus & concedimus, præfatis præsidi concilio &
" sodalibus Regalis Societatis prædictæ, & successoribus suis in perpetuum, sive præsidi & concilio Re-
" galis Societatis prædictæ, vel majori parti eorum, plenam potestatem & authoritatem erigendi edifi-
" candi & extruendi, aut erigi ædificari & extrui faciendi vel causandi, intra civitatem nostram Londini,

---

[1] *Sic in Authent.*   [2] *Sic in Authent.*   [3] *Sic in Authent.*

" vel decem milliaria ab eadem, unum vel plura collegium vel collegia cujuscunque modi & qualitatis,
" pro habitatione assemblatione & congregatione prædictorum præsidis concilii & sodalium prædictæ
" Regalis Societatis & successorum suorum, ad negotia sua & alias res eandem Regalem Societatem con-
" cernentes ordinandum & disponendum. Et ulterius volumus, ac per præsentes pro nobis heredibus &
" successoribus nostris ordinamus constituimus & appunctuamus, quod si aliqui abusus vel discrepantiæ
" in posterum orientur & accident de gubernatione aut aliis rebus vel negotiis prædictæ Regalis So-
" cietatis, unde ejusdem constitutioni stabilimini & studiorum progressui vel rebus & negotiis aliqua in-
" feratur injuria vel impedimentum ; quod tunc & toties per præsentes pro nobis heredibus & successo-
" ribus nostris authorizamus nominamus assignamus & constituimus præfatum prædilectum & perquam
" fidelem consanguineum & consiliarium nostrum EDWARDUM comitem de CLARENDON cancellarium
" nostrum regni nostri Angliæ, per seipsum durante vita sua, & post ejus mortem tunc archiepiscopum
" Cantuariensem, cancellarium vel custodem privati sigilli, episcopum Londinensem, & duos principales
" secretarios pro tempore existentes, aut aliquos quatuor vel plures eorum, easdem discrepantias &
" abusus reconciliare componere & reducere. Et ulterius volumus, ac per præsentes pro nobis here-
" dibus & successoribus nostris firmiter injungendo præcipimus & mandamus omnibus & singulis justi-
" tiariis, majoribus, aldermannis, vicecomitibus, ballivis, constabulariis, & aliis officiariis ministris &
" subditis nostris heredum & successorum nostrorum quibuscunque, quod de tempore in tempus sint
" auxiliantes & assistentes prædictis præsidi concilio & sodalibus Regalis Societatis prædictæ, eorumque
" in perpetuum successoribus, in omnibus & per omnia, secundum veram intentionem harum literarum
" nostrarum patentium. Eo, quod expressa mentio de vero valore annuo, vel de certitudine præmisso-
" rum sive eorum alicujus, aut de aliis donis sive concessionibus per nos seu per aliquem progenitorum
" sive prædecessorum nostrorum præfatis præsidi concilio & sodalibus Regalis Societatis prædictæ ante
" hæc tempora factis, in præsentibus minime facta existit, aut aliquo statuto actu ordinatione provisione
" proclamatione sive restrictione in contrarium inde antehac habitis factis editis ordinatis sive provisis, aut
" aliqua alia re causa vel materia quacunque, in aliquo non obstante. In cujus rei testimonium has literas
" nostras fieri fecimus patentes. Teste me ipso apud Westmonasterium, vicesimo secundo die Aprilis,
" anno regni nostri decimo quinto.

" Per breve de privato sigillo.

" HOWARD."

No. II.

Patent granting Chelsea to the Royal Society, together with some additional privileges and powers.

" CAROLUS SECUNDUS Dei gratia Angliæ Scotiæ Franciæ et Hiberniæ Rex, fidei Defensor, &c. omni-
" bus, ad quos hæ Literæ nostræ patentes pervenerint, salutem. SCIATIS, quod nos de gratia nostra
" speciali ac ex certa scientia et mero motu nostris dedimus et concessimus, ac per præsentes pro nobis
" heredibus et successoribus nostris damus et concedimus, dilectis et fidelibus nostris *Præsidi Concilio*
" *et Sodalibus Regalis Societatis Londini pro scientia naturali promovenda*, et successoribus suis in per-
" petuum, totam illam peciam terræ arabilis vocatam Teamshott, continentem per æstimationem viginti
" acras, jacentem inter viam nostram ducentem a Westmonasterio versus Chelsey ex parte boreali et
" occidentali ; et peciam prati continentem per æstimationem quatuor acras, parcellam octodecim
" acrarum prati nuper in tenura Comitis Nottinghamiensis, vel assignatorum suorum, ex parte australi ;
" ac clausum prati vocatum Stonebridge Close ex parte orientali ; et peciam terræ arabilis nuper in oc-

" cupatione Thomæ Evans, vel assignatorum suorum, ex parte occidentali ; per particularia inde men-
" tionata esse annualis redditus, sive valoris, viginti trium solidorum et quatuor denariorum : Nec non
" totum illum prædictum clausum prati vocatum Stony Bridge Close, continentem per æstimationem
" quatuor acras, nuper in occupatione Johannis Deakes, vel assignatorum suorum, jacentem inter
" rivum vocatum Le Common Sewer ex parte orientali ; et prædictam peciam terræ vocatam Teamshott
" ex parte occidentali ; et pontem vocatum Stony Bridge ex parte boreali ; per particularia inde mentio-
" nata esse annualis redditus, sive valoris, viginti solidorum : Nec non totam illam unam peciam terræ ara-
" bilis in communi campo vocato East Field, continentem per æstimationem tres acras, nuper in occu-
" patione Thomæ Frances, vel assignatorum suorum, jacentem inter prædictam peciam terræ vocatam
" Teamshott ex parte orientali ; peciam terræ arabilis nuper in tenura Comitis Lincolniensis, vel assigna-
" torum suorum, ex parte occidentali ; parcellam prati de Earles Court Land ex parte australi ; et viam
" nostram ducentem a Westmonasterio versus Chelsey prædictam ex parte boreali et occidentali ; per
" particularia inde mentionata esse annualis redditus, sive valoris, quatuor solidorum (quæ quidem præ-
" missa sunt aut olim fuerunt parcella terræ nostræ in Chelsey, existentis parcellæ terræ Dominicalis
" Manerii de Chelsey prædicta, ac nuper fuerunt parcella possessionum Johannis nuper Ducis Nor-
" thumbriæ ; et quæ nuper per præcharissimum avum nostrum beatæ memoriæ Jacobum Regem per
" Literas suas patentes, gerentes datum apud Westmonasterium octavo die Maii, anno regni sui Angliæ
" octavo et Scotiæ quadragesimo tertio, concessa fuerunt, aut mentionata esse concessa, Præposito et
" Sociis Collegii Regis Jacobi in Chelsey prope London ex fundatione ejusdem Jacobi Regis Angliæ,
" et successoribus suis in perpetuum ; tenendum de præfato Jacobo Rege, ut de manerio suo de East
" Greenwich in comitatu Cantiæ, per fidelitatem tantum, in libero et communi soccagio, et non in capite,
" nec per servitium militare) : Ac etiam omnia et singula domus ædificia structuras boscos subboscos ar-
" bores, ac totam terram fundam et solum eorundem boscorum subboscorum et arborum, ac omnia alia
" jura jurisdictiones franchesias privilegia libertates proficua commoditates advantagia emolumenta et
" hereditamenta nostra quæcunque, cum eorum pertinentiis universis, cujuscunque sint generis naturæ
" seu speciei, seu quibuscunque nominibus sciantur censeantur nuncupentur seu cognoscantur, scituata
" jacentia et existentia provenientia crescentia renovantia sive emergentia infra comitatum villas campos
" loca sive hamlettas prædicta, vel alibi ubicunque, prædictis terris et cæteris præmissis vel alicui inde
" parcellæ quoquo modo spectantia [1] : Necnon reversionem et reversiones omnium et singulorum præ-
" missorum superius per præsentes præconcessorum, et cujuslibet inde parcellæ, dependentes vel expec-
" tantes de in vel super aliquam dimissionem vel concessionem pro termino vel terminis vitæ vel vitarum
" vel annorum, aut aliter de præmissis superius per præsentes præconcessis seu de aliqua inde parcella
" quoquo modo factam, existentem de recordo vel non de recordo : Necnon omnia et singula redditus et
" annualia proficua quæcunque reservata super quibuscunque dimissionibus vel concessionibus de et su-
" per præmissa per præsentes præconcessa, vel de et super aliquam inde parcellam. Dedimus etiam et
" concessimus, ac per præsentes pro nobis heredibus et successoribus nostris damus et concedimus, præ-
" fatis Præsidi Concilio et Sodalibus Regalis Societatis Londini pro scientia naturali promovenda, et suc-
" cessoribus suis in perpetuum, quod ipsi et eorum successores de cætero in perpetuum habeant teneant
" et gaudeant, ac habere tenere et gaudere valeant et possint, infra præmissa superius per præsentes
" præconcessa, ac infra quamlibet inde parcellam, tot tanta talia eadem hujusmodi et consimilia jura ju-
" risdictiones libertates franchesias consuetudines privilegia proficua commoditates advantagia emolumenta
" et hereditamenta quæcunque, quot quanta qualia et quæ, ac adeo plene libere et integre, ac in tam
" amplis modo et forma, prout prædictus Johannes nuper Dux Northumbriæ, aut prædictus Præpositus
" et Socii Collegii Regis Jacobi in Chelsey prope London ex fundatione ejusdem Jacobi Regis Angliæ,
" aut aliquis alius sive aliqui alii, prædicta terras tenementa et cætera præmissa cum suis pertinentiis, aut
" aliquam inde parcellam, unquam antehac habentes possidentes aut seisiti inde existentes, habens possi-

---

[1] *Sic in* Authent.

" dens aut seisitus inde exstens, unquam habuerunt tenuerunt usi vel gavisi fuerunt, habuit te-
" nuit usus vel gavisus fuit, seu habere tenere uti vel gaudere debuerunt aut debuit, in præmissis
" superius per præsentes præconcessis, aut aliqua inde parcella, ratione vel prætextu alicujus chartæ
" doni concessionis vel confirmationis per nos seu aliquem progenitorum vel antecessorum nostro-
" rum nuper Regum vel Reginarum Angliæ antehac habitorum factorum vel concessorum seu confirma-
" torum, aut ratione vel prætextu alicujus actus Parliamenti vel aliquorum actuum Parliamentorum, aut
" ratione vel prætextu alicujus legitimæ præscriptionis usus seu consuetudinis antehac habitorum seu
" usitatorum, aut aliter quocunque legali modo jure seu titulo, ac adeo plene libere et integre, ac in tam
" amplis modo et forma, prout nos aut aliquis progenitorum vel antecessorum nostrorum nuper Regum
" vel Reginarum Angliæ prædicta terras tenementa et cætera præmissa, aut aliquam inde parcellam, ha-
" buimus et gavisi fuimus aut habuerunt et gavisi fuerunt, seu habere et gaudere debuimus aut habere et
" gaudere debuerunt aut debuit. DAMUS ulterius, ac per præsentes pro nobis heredibus et successoribus
" nostris concedimus, præfatis Præsidi Concilio et Sodalibus Regalis Societatis Londini pro scientia natu-
" rali promovenda, et eorum successoribus, omnia et singula præmissa superius per præsentes præcon-
" cessa, cum eorum pertinentiis universis, adeo plene libere et integre, ac in tam amplis modo et forma,
" prout ea omnia et singula præmissa, aut aliqui inde parcella, ad manus nostras, seu ad manus aliquorum
" progenitorum vel antecessorum nostrorum nuper Regum vel Reginarum Angliæ, ratione vel prætextu
" dissolutionis vel sursum redditionis alicujus nuper monasterii prioratus sive hospitalis, aut ratione vel
" prætextu alicujus actus Parliamenti vel aliquorum actuum Parliamentorum, aut ratione alicujus at-
" tincturæ sive forisfacturæ, aut ratione alicujus excambii vel perquisiti, aut alicujus doni vel concessionis,
" aut ratione eschaetæ, aut quocunque alio legali modo jure seu titulo devenerunt seu devenire debuerunt,
" ac in manibus nostris jam existunt seu existere debent vel debuerunt : habendum tenendum et gauden-
" dum prædicta terras tenementa et hereditamenta, ac cætera omnia et singula præmissa superius per
" præsentes præconcessa, cum eorum pertinentiis universis, præfato Præsidi Concilio et Sodalibus Rega-
" lis Societatis Londini pro scientia naturali promovenda, et successoribus suis in perpetuum ; tenendum
" de nobis heredibus et successoribus nostris, ut de manerio nostro de East Greenwich in comitatu nostro
" Cantiæ, per fidelitatem tantum, in libero et communi soccagio, et non in capite, nec per servitium mi-
" litare ; ac reddendum annuatim nobis heredibus et successoribus nostris de et pro prædicta terra arabili
" vocata Teamshott viginti tres solidos et quatuor denarios, ac de et pro prædicto clauso prati vocato
" Stony Bridge Close viginti solidos, ac de et pro prædicta pecia terræ arabilis in communi campo vo-
" cato East Field quatuor solidos legalis monetæ Angliæ, ad festa Sancti MICHAELIS Archangeli et An-
" nunciationis beatæ MARIÆ Virginis, ad receptam Scaccarii nostri Westmonasterii heredum et succes-
" sorum nostrorum, seu ad manus Ballivorum, seu Receptorum præmissorum pro tempore existente,[i] per
" æquales portiones annuatim solvendas in perpetuum. ET ULTERIUS de uberiori gratia nostra speciali
" ac ex certa scientia et mero motu nostris volumus, ac per præsentes pro nobis heredibus et successoribus
" nostris concedimus præfato Præsidi Concilio et Sodalibus Regalis Societatis prædictæ et successoribus
" suis, quod nos heredes et successores nostri de cætero in perpetuum annuatim, et de tempore in tem-
" pus, exonerabimus acquietabimus et indempnes conservabimus, tam præfatos Præsidem Concilium et
" Sodales Regalis Societatis prædictæ et successores suos, quam prædicta terras tenementa et cætera
" omnia et singula præmissa superius expressa et specificata ac per præsentes præconcessa, et quamlibet
" inde parcellam, cum eorum pertinentiis universis, de et ab omnibus et omnimodis corrodiis redditibus
" feodis servitiis annuitatibus pensionibus portionibus ac denariorum summis ac oneribus quibuscunque de
" præmissis, seu aliqua inde parcella, nobis heredibus vel successoribus nostris exeuntibus vel solvendis,
" vel superinde versus nos heredes vel successores nostros oneratis vel onerandis ; præterquam de redditi-
" bus servitiis et tenuris superius in his præsentibus nobis heredibus et successoribus nostris reservatis, ac
" præterquam de dimissionibus et concessionibus de præmissis seu de aliqua inde parcella antehac factis,
" ac conventionibus et conditionibus in iisdem existentibus, ac conventionibus et oneribus, quæ aliquis

" firmarius seu aliqui firmarii præmissorum ratione indenturarum et dimissionum suarum facere et ex-
" onerare tenetur seu tenentur. VOLUMUS etiam, ac per præsentes pro nobis heredibus et successoribus
" nostris firmiter injungendo præcipimus, tam Commissionariis pro thesauro nostro, Thesaurario, Came-
" rario, Subthesaurario, et Baronibus Scaccarii nostri heredum et successorum nostrorum pro tempore ex-
" istente, quam omnibus et singulis Auditoribus, et aliis officiariis, et ministris nostris heredum et succes-
" sorum nostrorum quibuscunque pro tempore existente; quod ipsi et eorum quilibet super solam demon-
" strationem harum Literarum nostrarum patentium, vel irrotulamenti earundem, absque aliquo alio
" brevi seu warranto a nobis heredibus vel successoribus nostris quoquo modo impetrando seu prosequendo,
" plenam integram debitamque allocationem et exonerationem manifestam de et ab omnibus et omnimodis
" hujusmodi corrodiis redditibus feodis pensionibus portionibus et denariorum summis ac oneribus quibus-
" cunque (præterquam de servitiis redditibus tenuris ac arreragiis redditus ac cæteris præmissis in his
" præsentibus, ut præfertur, reservatis, et per præfatos Præsidem Concilium et Sodales Regalis Societatis
" prædictæ et successores suos solubilibus fiendis [1] seu performandis) de præmissis per præsentes præcon-
" cessis, seu de aliqua inde parte vel parcella, nobis heredibus vel successoribus nostris exeuntibus seu
" solvendis, vel superinde versus nos heredes vel successores nostros oneratis seu onerandis, præfatis Præ-
" sidi Concilio et Sodalibus Regalis Societatis prædictæ et successoribus suis facient, et de tempore in
" tempus fieri causabunt: Et hæ Literæ nostræ patentes, vel irrotulamentum earundem, erunt de tem-
" pore in tempus, tam dictis Commissionariis pro Thesauro nostro, Thesaurario, Cancellario, et Baronibus
" Scaccarii nostri heredum et successorum nostrorum pro tempore existente, quam omnibus et singulis
" Auditoribus, et aliis officiariis, et ministris nostris heredum et successorum nostrorum quibuscunque
" pro tempore existente, sufficiens warrantum et exoneratio in hac parte. ET CUM NOS per Literas nos-
" tras patentes, gerentes datum apud Westmonasterium vicesimo secundo die Aprilis, anno regni nostri
" decimo quinto, Præsidi Concilio et Sodalibus Regalis Societatis prædictæ factas, inter alia concessimus
" præfatis Præsidi Concilio et Sodalibus prædictæ Regalis Societatis, et successoribus suis in perpetuum,
" quod si contigerit Præsidem ejusdem Regalis Societatis pro tempore existente ægritudine vel infirmi-
" tate detineri, vel in servitio nostro heredum vel successorum nostrorum versari, vel aliter esse occupa-
" tum, ita quod necessariis negotiis ejusdem Regalis Societatis officium Præsidis tangentibus attendere
" non poterit; quod tunc et toties bene liceat et licebit eidem Præsidi sic detento versato vel occupato
" unum de Concilio prædictæ Regalis Societatis pro tempore existente, fore et esse Deputatum ejusdem
" Præsidis, nominare et appunctuare: qui quidem Deputatus, in officio Deputati Præsidis prædicti sic
" faciendus et constituendus, sit et esset Deputatus ejusdem Præsidis de tempore in tempus, toties quo-
" ties prædictus Præses sic abesse contigerit, durante toto tempore, quo prædictus Præses in officio præ-
" sidis continuaverit; nisi interim prædictus Præses Regalis Societatis prædictæ pro tempore existente
" unum alium de prædicto Concilio ejus Deputatum fecerit et constituerit: Et quod quilibet, hujusmodi
" Deputatus prædicti Præsidis sic, ut præfertur, faciendus et constituendus omnia et singula, quæ ad offi-
" cium Præsidis prædictæ Regalis Societatis pertinent seu pertinere debent, vel per prædictum Præsidem
" virtute istarum Literarum nostrarum patentium limitata et appunctuata fore [2] facienda et exequenda
" de tempore in tempus, toties quoties prædictus Præses sic abesse contigerit, durante tali tempore, quo
" Deputatus prædicti Præsidis continuaverit, facere et exequi valeat et possit, vigore istarum Literarum
" nostrarum patentium, adeo plene libere et integre, ac in tam amplis modo et forma, prout Præses præ-
" dictus, si præsens esset, illa facere et exequi valeat et possit; sacramento corporali super sancta Dei
" Evangelia, in forma et effectu in eisdem Literis nostris patentibus specificatis, per hujusmodi Deputa-
" tum ad omnia et singula, quæ ad officium Præsidis pertinent, bene et fideliter exequendum, coram
" præfato Concilio prædictæ Regalis Societatis, vel aliquibus septem vel pluribus eorum, prius præstando;
" et sic toties quoties casus sic acciderit: cui quidem Concilio, vel aliquibus septem vel pluribus eorum
" pro tempore existente, sacramentum prædictum administrare potestatem, et authoritatem, quoties casus

" sic acciderit, dedimus et concessimus per easdem Literas nostras patentes, absque brevi commissione
" sive ulteriori warranto in ea parte a nobis heredibus et successoribus nostris procurandis seu obtinendis :
" Ac quod ipsi et successores eorum, seu aliqui novem vel plures eorum (quorum Præsidem pro tempore
" existente, vel ejus deputatum, semper unum esse volumus) conventus seu congregationes de seipsis pro
" experimentorum et rerum naturalium cognitione et indagine aliisque negotiis ad Societatem prædictam
" spectantibus, quoties et quando opus fuerit, licite facere et habere possint in collegio sive aula sive alio
" loco commodo intra civitatem nostram London, vel in aliquo alio loco commodo intra decem milliaria
" ab eadem civitate nostra : Et cum diversa et varia res potestates libertates et privilegia in iisdem Literis
" nostris patentibus præfato Præsidi Concilio ac Sodalibus Regalis Societatis prædictæ concessa, virtute
" istarum Literarum nostrarum patentium, non sunt exercenda facienda performanda seu exequenda,
" nisi per prædictum Præsidem et Concilium, aut aliquos septem vel plures eorum : Et cum ulterius per
" prædictas Literas nostras patentes pro nobis heredibus et successoribus nostris dedimus et concessimus
" Præfatis Præsidi Concilio et Sodalibus prædictæ Regalis Societatis, et successoribus suis in perpetuum,
" sive aliquibus viginti et uni vel pluribus eorum (quorum Præsidem pro tempore existente, vel ejus De-
" putatum, semper unum esse volumus) seu majori parti prædictorum viginti et unius vel plurium, ple-
" nam potestatem et authoritatem de tempore in tempus eligendi nominandi et constituendi unum vel
" plures Typographos sive Impressores, et Chalcographos seu Sculptores, et ipsi vel ipsis per scriptum
" communi Sigillo prædictæ Regalis Societatis sigillatum, et manu Præsidis pro tempore existente sig-
" natum, facultatem concedendi, ut imprimant tales res materias et negotia prædictam Regalem So-
" cietatem tangentia vel concernentia, quales prædictis Typographo vel Impressori, Chalcographo vel
" Sculptori, vel Typographis vel Impressoribus, Chalcographis vel Sculptoribus, de tempore in tempus
" per Præsidem et Concilium prædictæ Regalis Societatis, vel aliquos septem vel plures eorum (quorum
" Præsidem pro tempore existente, vel ejus Deputatum, unum esse volumus) vel per majorem partem
" prædictorum septem vel plurium commissa[1] fuerint; sacramentis suis corporalibus, antequam ad
" officia sua exercenda admittantur, coram Præside et Concilio pro tempore existente, vel aliquibus
" septem vel pluribus eorum, prius præstandis ; cui quidem Præsidi et Concilio, vel aliquibus septem vel
" pluribus eorum pro tempore existente, sacramenta prædicta administrare plenam potestatem et autho-
" ritatem dedimus et concessimus per prædictas Literas nostras patentes; prout in eisdem Literis nostris
" patentibus, relatione inde habita, plenius liquet et apparet : Nos, de abundantiori gratia nostra speciali
" ac ex certa scientia et mero motu nostris, DEDIMUS et concessimus, ac per præsentes pro nobis here-
" dibus et successoribus nostris damus et concedimus, præfatis Præsidi Concilio et Sodalibus prædictæ
" Regalis Societatis, et successoribus suis in perpetuum, quod de cætero in perpetuum, si contigerit Præsi-
" dem ejusdem Regalis Societatis pro tempore existente ægritudine vel infirmitate detineri, vel in servitio
" nostro heredum vel successorum nostrorum versari, vel aliter esse occupatum, ita quod necessariis nogo-
" tiis ejusdem Regalis Societatis officium Præsidis tangentibus attendere non poterit ; quod tunc et toties
" bene liceat et licebit eidem Præsidi sic detento versato vel occupato unum de Concilio prædictæ Rega-
" lis Societatis pro tempore existente, fore et esse Deputatum ejusdem Præsidis, nominare et appunc-
" tuare ; qui quidem Deputatus, in officio Deputati Præsidis prædicti sic faciendus et constituendus, sit et
" erit Deputatus ejusdem Præsidis de tempore in tempus, toties quoties prædictus Præses sic abesse conti-
" gerit, durante toto tempore, quo prædictus Præses in officio Præsidis continuaverit, etiamsi interim
" Præses Regalis Societatis prædictæ pro tempore existente unum alium vel plures alios de prædicto
" Concilio ejus Deputatum et Deputatos fecerit et constituerit; cui quidem Præsidi pro tempore ex-
" istente duos vel plures de prædicto Concilio ejus Deputatos ipso et eodem tempore facere et constituere
" potestatem et authoritatem, quoties ei placuerit, damus et concedimus per præsentes pro nobis here-
" dibus et successoribus nostris : Et quod quilibet hujusmodi Deputatus et Deputati prædicti Præsidis sic,
" ut præfertur, faciendi et constituendi omnia et singula, quæ ad officium Præsidis prædictæ Regalis So-

[1] Sic in Authent.

" cietatis pertinent seu pertinere debent, vel per prædictum Præsidem virtute prædictarum Literarum nos-
" trarum patentium, vel præsentium, limitata et appunctuata fore ¹ facienda et exequenda de tempore in
" tempus, toties quoties prædictus Præses sic abesse contigerit, durante tali tempore, quo Deputatus et
" Deputati prædicti Præsidis continuaverit et continuaverint, facere et exequi valeat et possit valeant et
" possint, vigore harum Literarum nostrarum patentium, adeo plene libere et integre, ac in tam amplis
" modo et forma, prout Præses prædictus, si præsens esset, illa facere et exequi valeret et posset; sacra-
" mento corporali super sancta Dei Evangelia, in forma et effectu in eisdem Literis nostris patentibus
" specificatis, per hujusmodi Deputatum et Deputatos ad omnia et singula, quæ ad officium Præsidis per-
" tinent, bene et fideliter exequendum, coram præfato Concilio prædictæ Regalis Societatis, vel aliquibus
" quinque vel pluribus eorum, prius præstando; et sic toties quoties casus sic acciderit: cui quidem Con-
" cilio, vel aliquibus quinque vel pluribus eorum pro tempore existente, sacramentum, prædictum admi-
" nistrare potestatem et authoritatem quoties casus sic acciderit, damus et concedimus per præsentes,
" absque brevi commissione sive ulteriori warranto in ea parte a nobis heredibus et successoribus nostris
" procurandis seu obtinendis: Ac ulterius, quod de cætero in perpetuum ipsi et successores eorum, seu
" aliqui novem vel plures eorum (quorum Præsidem pro tempore existente, vel ejus Deputatum, semper
" unum esse volumus) conventus seu congregationes de seipsis pro experimentorum et rerum naturalium
" cognitione et indagine aliisque negotiis ad Societatem prædictam spectantibus quoties et quando opus
" fuerit, licite facere et habere possint in collegio sive aula sive alio loco commodo infra regnum nostrum
" Angliæ: Ac ulterius, quod omnia et singula res potestates libertates et privilegia in prædictis Literis nos-
" tris patentibus præfatis Præsidi Concilio et Sodalibus Regalis Societatis prædicta concessa, virtute
" istarum Literarum nostrarum patentium, quæ non sunt exercenda facienda performanda seu exe-
" quenda, nisi per prædictum Præsidem et Consilium, aut aliquos septem vel plures eorum; de cætero
" in perpetuum exerceri fieri performari seu exequi possint et valeant per prædictum Præsidem et Conci-
" lium, aut aliquos quinque vel plures eorum. Ac ULTERIUS de uberiori gratia nostra dedimus et con-
" cessimus, ac per præsentes pro nobis heredibus et successoribus nostris damus et concedimus, præfatis
" Præsidi Concilio et Sodalibus prædictæ Regalis Societatis, et successoribus suis in perpetuum, quod de
" cætero in perpetuum bene liceat et licebit Præsidi Regalis Societatis prædictæ pro tempore existente,
" de tempore in tempus eligere nominare et constituere aliquem vel aliquos Typographum sive Impres-
" sorem, Typographos sive Impressores, et Chalcographum seu Sculptorem, Chalcographos seu Sculptores,
" et ipsi vel ipsis facultatem concedere, ut imprimant tales res materias et negotia prædictam Regalem
" Societatem tangentia vel concernentia, quales prædictis Typographo vel Impressori, Chalcographo seu
" Sculptori, vel Typographis vel Impressoribus, Chalcographis vel Sculptoribus, de tempore in tempus per
" Præsidem et Concilium prædictæ Regalis Societatis, vel aliquos quinque vel plures eorum (quorum
" Præsidem pro tempore existente, vel ejus Deputatum, unum esse volumus) vel per majorem partem
" prædictorum quinque vel plurium commissæ fuerint; sacramentis suis corporalibus, antequam ad officia
" sua exercenda admittantur, coram Præside et Concilio pro tempore existente, vel' aliquibus quinque vel
" pluribus eorum, prius præstandis; et sic toties quoties casus sic acciderit: cui quidem Præsidi et Con-
" cilio pro tempore existente, vel aliquibus quinque vel pluribus eorum, sacramenta prædicta administrare
" plenam potestatem et authoritatem damus et concedimus per præsentes. ET ULTERIUS volumus, ac
" per præsentes pro nobis heredibus et successoribus nostris concedimus præfatis Præsidi Concilio et Soda-
" libus Regalis Societatis prædictæ et successoribus suis, quod hæ Literæ nostræ patentes, vel irrotula-
" mentum earundem, stabunt et erunt in omnibus et per omnia bonæ firmæ validæ sufficientes et effec-
" tuales in lege ad omnes respectus proposita constructiones et intentiones erga et contra nos heredes et
" successores nostros, tam in omnibus curiis nostris, quam alibi infra regnum nostrum Angliæ, absque ali-
" quibus confirmationibus licentiis vel tolerationibus de nobis heredibus vel successoribus nostris quoquo-
" modo in posterum procurandis aut obtinendis. NON OBSTANTE male nominando vel male recitando,

¹ Sic in Authent.

" aut non recitando prædicta terras tenementa et cætara præmissa, vel aliquam inde parcellam. Et non
" obstante non inveniendo officium aut inquisitionem præmissorum, aut alicujus inde parcellæ, per quæ
" titulus noster inveniri debuit, ante confectionem harum Literarum nostrarum patentium. Et non obstante
" male recitando vel non recitando aliquam dimissionem vel concessionem de præmissis vel de aliqua inde
" parcella factam, existentem de recordo vel non de recordo. Et non obstante male nominando vel non
" nominando aliquam villam hamletam parochiam locum vel comitatum, in quibus præmissa vel aliqua
" inde parcella existunt vel existit. Et non obstante, quod de nobis tenentium firmariorum sive occupa-
" torum præmissorum, vel alicujus inde parcellæ, plena vera et certa non fit mentio. Et non obstante
" aliquibus defectibus [1] de certitudine vel computatione aut declaratione veri annui valoris præmissorum,
" aut alicujus inde parcellæ; aut annualis redditus reservati de et super præmissis, vel de et super aliqua
" inde parcella, in his Literis nostris patentibus expressis et contentis. Et non obstante statuto in Par-
" liamento Domini HENRICI nuper Regis Angliæ sexti progenitoris nostri, anno regni sui decimo oc-
" tavo, facto et edito. Et non obstante aliquibus aliis defectis,[2] in non certe nominando naturam genus
" speciei [3] quantitatem aut qualitatem præmissorum, aut alicujus inde parcellæ. Er non obstante statuto
" de terris et tenementis ad manum mortuam non ponendis; aut aliquo alio statuto actu ordinatione
" proclamatione provisione sive restrictione in contrarium inde antehac habitis factis editis ordinatis seu
" provisis, in aliquo non obstante. SALVO tamen ANDREÆ COLE Armigero et omnibus aliis personis
" quibuscunque, præterquam nos heredes et successores nostros, talia jus clameum interesse et demanda
" quæcunque; qualia ipse vel ipsi seu eorum aliquis habet seu habeant, aut de jure habere debent,[4] de
" et in præmissis, seu aliqua parte vel parcella inde. ET ULTERIUS volumus, et per præsentes pro
" nobis heredibus et successoribus nostris ordinamus et firmiter injungendo præcipimus, quod Præses
" Societatis prædictæ pro tempore existente, et Deputati ejus, antequam ipsi aut eorum aliqui ad execu-
" tionem officii illius admittantur, tam sacramentum corporale communiter vocatum *The oath of obe-*
" *dience,* quam sacramentum corporale communiter vocatum *The oath of supremacy,* super sacrosanctis
" Dei Evangeliis præstabunt, et eorum quilibet præstabit, coram Concilio ejusdem Societatis, aut aliqui-
" bus septem vel pluribus eorum; cui quidem Concilio, aut aliquibus septem vel pluribus eorum, sacra-
" menta prædicta administrare pro nobis heredibus et successoribus nostris plenam potestatem et autho-
" ritatem de tempore in tempus, quotiescunque opus fuerit, damus et concedimus per præsentes. PRO-
" VISO semper, et voluntas et intentio nostra regia est, quod terræ et præmissa prædicta per præsentes,
" ut præfertur, concessa, seu eorum aliqua, non alienabuntur vel vendentur alicui personæ sive aliquibus
" personis quibuscunque, aliquo in præsentibus contento in contrarium inde non obstante. Eo, quod
" expressa mentio de vero valore annuo vel de certitudine præmissorum sive eorum alicujus, aut de aliis
" donis sive concessionibus per nos seu per aliquem progenitorum sive prædecessorum nostrorum præfa-
" tis Præsidi Concilio et Sodalibus Regalis Societatis de London, et successoribus, ante hæc tempora
" factis, in præsentibus minime facta existit; aut aliquo statuto actu ordinatione provisione proclamatione
" sive restrictione in contrarium inde antehac habitis factis editis ordinatis sive provisis, aut aliqua alia re
" causa vel materia quacunque, in aliquo non obstante. IN CUJUS rei testimonium has Literas nostras
" fieri fecimus patentes. TESTE Me ipso apud Westmonasterium, octavo die Aprilis, anno regni nostri
" vicesimo primo.

" Per breve de privato sigillo.

" PIGOTT."

---

[1] *Sic in* Authent. [2] *Sic* ibid. [3] *Sic* ibid. [4] *Sic* ibid.

No. III.

Minutes of the Royal Society respecting Sir Isaac Newton.

1671. Dec. 23. The Lord Bishop of Sarum proposed for candidate Mr. Isaac Newton, Professor of Mathematics at Cambridge.

January 11. Mr. Isaac Newton was elected. At that meeting mention was made of his improvement of telescopes, by contracting them, and that that, which himself had sent thither to be examined, had been seen by the King, and considered, also by the President, Sir Robert Murray, Sir Paul Neile, Dr. Christopher Wren, and Mr. Hook, at Whitehall; and that they had so good an opinion of it, as that they concluded a description, and scheme of it, should be sent by the Secretary, in a letter on purpose, to Mr. Huygins at Paris, thereby to secure this contrivance to the author, who had also written a letter to Mr. Oldenburgh from Cambridge, (Jan. 6, 1671,) altering and enlarging the description of his instrument, which had been sent from hence for his review, before it should go abroad. This description was read, and ordered to be entered in the register-book, together with the scheme. (Vide Phil. Trans. No. 80 —83.)

The Curator said he did endeavour to make such a telescope himself, and to find out a metal not obnoxious to tarnishing. It was ordered that a letter should be wrote by the Secretary to Mr. Newton, to signify to him his election, and also to thank him for the communication of his telescope, and to assure him, that the Society would take care that all right should be done him in the matter of this invention.

January 18. Mr. Newton's new telescope was examined and applauded.

——— 25. Mr. Oldenburgh read a letter of Mr. Newton's, written to him from Cambridge January 18, concerning an intimation, 1st, Of a way of preparing a fit metalline matter for reflecting concaves; 2dly, Of a considerable philosophical discovery he intends to send to this Society, to be considered and examined.

February 1. There was again produced the four-foot telescope of Mr. Newton's, which was now better than the last time.

February 8. The third letter from Mr. Newton from Cambridge, of Feb. 6, about his discovery of the nature of light, refractions, and colours; importing that light (and colours) was not a similar, but heterogeneous thing, consisting of difform rays, which had essentially different refractions, abstracted from bodies they pass through: and that colours are produced from such and such rays, whereof some in their own nature are disposed to be red, others blue, others purple, &c. and that whiteness is nothing else but a mixture of all sorts of colours, or that it is produced by all sorts of colours blended together. (Vide Phil. Trans. No. LXXX.)

Ordered, that the author be solemnly thanked, in the name of the Society, for this very ingenious discourse, and be made acquainted that the Society think very fit, if he consents, to have it forthwith printed, as well for the greater conveniency of having it well considered by philosophers, as for securing the considerable notices thereof to the author against the arrogations of others. Ordered also, that the discourse be entered in the register-book, and that the Bishop of Salisbury, Mr. Boyle, and Mr. Hook, be desired to pursue and consider it, and bring in a report of it to the Society.

1672. March 28. There was read a letter of Mr. Newton's, written to Mr. Oldenburgh from Cambridge, (March 26,) containing some more particulars relating to his new telescope, especially the proportions of the apertures, and changes for several lengths of that sort of telescope. (Phil. Trans. No. LXXXII.)

3

April 4. The Secretary read a letter of Mr. Newton's, written to him from Cambridge, March 30, concerning his answers to the difficulties objected by Mons. ———, about his reflecting telescope, as also to the queries of Mons. Denys concerning the same; together with his proposal of a way of using, instead of a little oval metal in that telescope, a crystal figured like a triangular prism. (Phil. Trans. No. LXXXII.) Ordered, that the Curator take care to make such a crystalline prism for the design mentioned, and to try the same.

May 15. Mr. Hook made some experiments relating to Mr. Newton's theory of light and colours, which he was desired to bring in writing to be registered.

May 22. Mr. Hook made some more experiments with two prisms, confirming what Mr. Newton hath written in his discourse about light and colours, viz. that the rays of light being separated by one prism into distinct colours, the refractions made by another prism doth not alter these colours. (Phil. Trans. No. LXXXIII.)

1675. Nov. 18. Mr. Newton offering to send to the Society, in a letter dated November 13, a discourse of his about colours, when it shall be thought convenient, the Society ordered the Secretary to thank him for his offer, and to desire him to send that discourse as soon as he pleased.

December 9. There was produced a manuscript of Mr. Newton's, touching his theory of light and colours, containing partly an hypothesis to explain the properties of light, by him discoursed of in his former papers, and partly the principal phœnomena of the various colours exhibited by thin plates or bubbles, esteemed by him to be of a more difficult consideration, yet to depend also on the said properties of light. Of the hypothesis there was read only the first part, giving an account of the refraction, reflexion, transparency and opacity: the second part, explaining colours, was referred to the next meeting.

December 16. The sequel of his hypothesis, which was began to be read the last day, was read to the end. To which Mr. Hook said, that the main of it was contained in his Micrography, which Mr. Newton, in some particulars, had only carried further.

January 20. Read a letter of Mr. Newton's, written to Mr. Oldenburgh, December 21, 1675, stating the difference betwixt his hypothesis and that of Mr. Hook, in his Micrography; the result of which is, that he (Mr. Newton) had nothing in common with Mr. Hook: but a supposition that Ether is a medium susceptible of vibrations; of which supposition Mr. Newton saith he makes quite a different use; Mr. Hook supposing it light itself, which Mr. Newton does not. Besides, that he explains very differently from Mr. Hook the manner of refraction and reflexion, and the nature and production of colours in all cases, and even in the colours of transparent substances. Mr. Newton says he explains every thing in a way so differing from Mr. Hook, that the experiments he grounds his discourse upon destroy 1 Mr. Hook saith about them. And that the two main experiments, without which the manner of production of those colours is not to be found out, were not only unknown to Mr. Hook, when he wrote his Micrography; but even last spring, as he understood by mentioning them to the said Mr. Hook. Read the beginning of Mr. Newton's discourse, containing such observations as conduce to further discoveries, for compleating his theory of light and colours, especially as to the constitution of natural bodies, on which their colour and transparency depend; in which discourse he first describes the principal of his observations, and then considers, and makes use of them.

January 20. At this time were read the first 15 observations, which did so well please the company, that they ordered the Secretary to desire the author would permit them to be published; together with the rest, which they presumed did correspond to those that had now been read to them.

January 27. Mr. Newton's letter of January 5 was read, wherein he acknowledges the favour of the Society, in the kind acceptance of his late papers, and declares that he knows not how to deny any thing which they desire should be done; only he desires, that the printing his observations upon colours may be suspended a while, because he has some thoughts of writing another set of observations, for determining the manner of production of colours by the prism; which observations, he says, ought to precede those now in our hands, and will do best to be joined with them.

February 3. The reading Mr. Newton's discourse was continued, viz. that part wherein he explains the simplest of colours by the more compounded.

February 10. There was read the last part of Mr. Newton's discourse, wherein is considered, in nine propositions, how the phenomena of thin transparent plates stand related to those of all natural bodies, in which he inquires after their constitutions, whereby they reflect some rays more copiously than others.

1684. *December 10.* Mr. Halley gave an account that he had lately seen Mr. Newton at Cambridge, and that he had shewed him a curious treatise *de Motu Corporum*; which, upon his desire, he said, was promised to be sent to the Society to be entered upon their register. Mr. Halley was desired to put Mr. Newton in mind of his promise, for securing his invention to himself, till such time as he can be at leisure to publish it. Mr. Paget was desired to join with Mr. Halley.

February 25. A letter was read from Mr. Newton concerning his willingness to promote a philosophical meeting at Cambridge, the entering in our register his notions about motion, and his intentions to fit them suddenly for the press.

1686. *April 28.* Dr. Vincent presented the Society with a MS. treatise, entitled *Philosophiæ naturalis Principia mathematica*, and dedicated to the Society by Mr. Is. Newton; wherein he gives a mathematical demonstration of the Copernican hypothesis, as proposed by Kepler, and makes out all the phenomena of the celestial motions, by the only supposition of a gravitation towards the centre of the sun, decreasing as the squares of the distances therefrom reciprocally.

May 19. Ordered, that Mr. Newton's book be printed forthwith, in a 4to. of a fair letter; and that a letter be written to him to signify the Society's resolution, and to desire his opinion as to the print, volume, cutts, &c.

June 2. Ordered that Mr. Newton's book be printed, and that Mr. Edm. Halley shall undertake the business of looking after it, and printing it at his own charge; which he engaged to do. (Vide Journal of Council.)

January 26. Ordered, that Mr. Newton be consulted, whether he designs to treat of the opposition of the medium to bodies moving in it, in his treatise *de Motu Corporum* now in the press.

1692. *February 1.* There was produced Mr. Newton's and Dr. Gregory's quadrature of curve lines; both which will be printed in Dr. Wallis's Latin edition of his Algebra. It was chiefly a proposition sent to Dr. Gregory from Mr. Newton, much about the time, Dr. Gregory says, he discovered it himself; being a method of squaring all curve lines, that are expressible in any binomial. Mr. Newton has subjoined a like rule, when it cannot be expressed under a trinomial; and mentioned that his process will go on *ad infinitum*; and square the curve, when the ordinate cannot be expressed without an infinite series.

1694. *July 4.* Ordered, that a letter be written to Mr. Newton, praying that he will please to communicate to the Society, in order to be printed, his Treatise of Light and Colours; and what other mathematical or physical treatises he has ready by him.

October 31. Dr. Halley said that Mr. Newton had lately told him, that there was reason to conclude, that the bulk of the earth did grow and increase in magnitude, by the perpetual accession of new particles, attracted out of the ether, by its gravitating power: and he supposed, and proposed to the Society, that this increase of the moles of the earth would occasion an acceleration of the moon's motion, she being, at this time, attracted by a stronger *vis centripetæ* than in remote ages.

A letter from Mr. Leibnitz to Mr. Bridges was read, wherein he recommends to the Society to use their endeavours to induce Mr. Newton to publish his farther thoughts and improvements on the subject of his late book, *Principia Philosophiæ mathematicæ*, and his other physical and mathematical discoveries; lest by his death they should happen to be lost.

1703. *November 4.* Sir Isaac Newton chosen of the Council and President the same day.

February 16. The President presented his book of Opticks to the Society; Mr. Halley was desired to peruse it; and to give an abstract of it; and the Society gave the President thanks for the book, and for being pleased to publish it.

1711. April 5. Mr. Keill observed that, in the Leipswick *Acta Eruditorum* for the year 1705, there is an unfair account given of Sir Isaac Newton's Discourse of Quadratures, asserting the method of demonstration by him there made use of, to Mr. Leibnitz, &c. Upon which the President gave a short account of that matter, with the particular time of his first mentioning or discovering his invention, referring to some letters published by Dr. Wallis; upon which Mr. Keil was desired to draw up an account of the matter in dispute, and set it in a just light.

May 24. A letter from Mr. John Keil to Dr. Sloane was produced and read, relating to the dispute concerning the priority of invention of the arithmetick of fluxions, between Sir Isaac Newton and Mr. Leibnitz, wherein Mr. Keil asserts the President's claim, &c. A copy of this letter was ordered to be sent to Mr. Leibnitz, and Dr. Sloane was desired to draw up a letter to accompany it, before it was made public in the Transactions, which should not be till after the receipt of Mr. Leibnitz's answer.

1711. January 31. A letter from Mons. Leibnitz to Dr. Sloane was read, in which he complains of Mr. Keil's unfair dealing with him in his last letter, relating to the dispute between him and Sir Isaac Newton; the letter was delivered to the President to consider the contents thereof.

March 11. Upon account of Mons. Leibnitz's letter to Dr. Sloane, concerning the disputes formerly mentioned, a committee was appointed by the Society to inspect the letters and papers relating thereto; viz. Dr. Arbuthnot, Mr. Hill, Dr. Halley, Mr. Jones, Mr. Machin, and Mr. Burnet, who were to make their report to the Society.

1712. April 24. The Committee, appointed to inspect the papers, letters, and books of the Society, on account of the dispute between Mr. Leibnitz and Mr. Keil, delivered in their report, which was read as follows:

" We have consulted the letters and letter-books in the custody of the Royal Society, and those found amongst the papers of Mr. John Collins, dated between the years 1669 and 1677 inclusive, and shewed them to such as knew and avowed the hands of Mr. Barrow, Mr. Collins, Mr. Oldenburgh, and Mr. Leibnitz, and compared those of Mr. Gregory with one another, and with copies of some of them taken in the hand of Mr. Collins, and have extracted from them what relates to the matter referred to us, all which extracts herewith delivered to you, we believe to be genuine and authentic; and by these letters and papers we find, 1st, That Mr. Leibnitz was in London in the beginning of the year 1673, and went thence in or about March, to Paris, where he kept a correspondence with Mr. Collins, by means of Mr. Oldenburgh, till about September, 1676, and then returned by London and Amsterdam to Hanover, and that Mr. Collins was very free in communicating to able mathematicians what he had received from Mr. Newton and Mr. Gregory. 2dly, That when Mr. Leibnitz was the first time in London, he contended for the invention of another differential method, properly so called; and notwithstanding that he was shewn by Dr. Pell, that it was Newton's method, he persisted in maintaining it to be his own invention, by reason that he found it by himself, without knowing what Newton had done before, and had much improved it; and we find no mention of his having any other differential method than Newton's, before his letter of June 21, 1677, which was a year after a copy of Mr. Newton's letter of December 10, 1672 had been sent to Paris, to be communicated to him, and above four years after Mr. Collins began to communicate that letter to his correspondents; in which letter the method of fluxions was sufficiently described to any intelligent person. 3dly, That by Mr. Newton's letter of June 13, 1676, it appears, that he had the method of fluxions above five years before the writing of that letter; and by his *Analysis per Operationes numero terminorum infinitas*, communicated by Dr. Barrow to Mr. Collins in July, 1669, we find that he had invented the method before that time. 4thly, That the differential method is one and the same with the method of fluxions, excepting the name and mode of notation, Mr. Leibnitz calling these quantities differences, which Mr. Newton calls moments, or fluxions, and marking them with the letter d, a mark not used by Mr. Newton. We therefore take the proper question to be, not who invented this or that method, but who was the first inventor of the method; and we believe that those who had reputed Mr. Leibnitz the first inventor, know little or nothing of his correspondence with Mr. Collins and Mr. Oldenburgh, long before, nor of Mr. Newton's having that method above fifteen years before Mr. Leibnitz began to publish it in the *Acta Eruditorum* of Leipswick."

" For which reasons we reckon Mr. Newton the first inventor, and are of opinion that Mr. Keil, in asserting the same, has been no way injurious to Mr. Leibnitz; and we submit to the judgment of the Society, whether the extract of the letters and papers, now presented, together with what is extant to the same purpose in Dr. Wallis's third volume, may not deserve to be made public."

To which report the Society agreed *nem. con.* and ordered that the whole matter from the beginning, with the extracts of all the letters relating thereto, and Mr. Keil and Mr. Leibnitz's letters, be published with all convenient speed that may be, together with the report of the said Committee.

Ordered, that Dr. Halley, Mr. Jones, and Mr. Machin, be desired to take care of the said impression, (which they promised,) and Mr. Jones, to make an estimate of the charges, against the next meeting.

1713. January 8. Some copies of the book intituled *Commercium Epistolicum, &c.* printed by the Society's order being brought, the President ordered one to be delivered to each person of the Committee, appointed for that purpose, to examine it before its publication."

No. IV.

List of the Fellows of the Royal Society from its first Institution, to the year 1812, in the order of their Election.

THE FIRST PRESIDENT AND COUNCIL OF THE ROYAL SOCIETY, NAMED BY HIS MAJESTY'S CHARTER, DATED 22d APRIL, 1663.

BORN.	DIED.	
1620	April 5, 1684	William Lord Viscount Brouncker, President
	July 4, 1673	Sir Robert Moray, Kt.
Feb. 25, 1627	Dec. 31, 1691	Robert Boyle, Esq.
	1697	William Brereton, Esq.—afterwards Lord Brereton.
July 11, 1603	June 11, 1665	Sir Kenelme Digby, Kt.
		Sir Gilbert Talbot, Kt.
		Sir Paule Neile, Kt.
		Henry Slingesby, Esq.—expelled June 24, 1675.
May 16, 1623	Dec. 16, 1687	Sir William Petty, Kt.
		Timothy Clarke, M. D.
1614	Nov. 19, 1672	John Wilkins, D. D.—afterwards Bishop of Chester.
1604	Oct. 13, 1689	George Ent, M. D.—afterwards Sir George Ent, Kt.
		William Erskine, Esq.
1617	March 24, 1675	Jonathan Goddart, M. D. Prof. Med. Gresh.
		William Balle, Esq. Treasurer.
Aug. 20, 1629	June 11, 1672	Matthew Wren, Esq.
Oct. 31, 1620	Feb. 27, 1706	John Evelyn, Esq.
1617	1699	Thomas Henshaw, Esq.
1602	1666	Dudley Palmer, Esq.
		Abraham Hill, Esq.
	Sept. 1677	Henry Oldenburg, Esq. Secretary.

Which Council did, at a Meeting held May 20, 1663, by virtue of the power given them by the charter for two months, declare the following Persons Members of the Society.

	1690	James Lord Annesley—afterwards Earl of Anglesey.
		John Alleyn, Esq.
May 23, 1617	May 18, 1692	Elias Ashmole, Esq.
		John Austen, Esq.
Nov. 3, 1625	1697	John Aubrey, Esq.
1620	1687	George Duke of Buckingham.
	Oct. 10, 1698	George Lord Berkeley—afterwards Earl of Berkeley.
	Oct. 20, 1685	Robert Lord Bruce—afterwards Earl of Aylesbury.
		Richard Boyle, Esq.
1622	Sept. 5, 1681	Thomas Bayne, M. D.—afterwards Kt.
Oct. 1630	May 4, 1677	Isaac Barrow, B. D.—afterwards D. D.
		Peter Balle, M. D.
		John Brook, Esq.—afterwards Sir John Brook, Bart.
		David Bruce, M. D.—expelled Nov. 18, 1675.

3

BORN.	DIED.	
1608	April 19, 1669	George Bate, M. D.
Jan. 25, 1640	Aug. 18, 1707	William Lord Cavendish—afterwards Duke of Devonshire.
1619	1707	Walter Charleton, M. D.
	Nov. 11, 1671	Edward Cotton, D. D. Archdeacon of Cornwall.
	1690	Daniel Colwall, Esq.
		John Clayton, Esq.—afterwards Sir John Clayton, Kt.
		Thomas Coxe, M. D.
	Oct. 12, 1684	William Croone, M. D.
	1676	John Earl of Crawford and Lindsay.
	1680	Henry Marquis of Dorchester.
1617	Nov. 23, 1684	William Earl of Devonshire.
1615	March 28, 1669	Sir John Denham, Knight of the Bath.
1631	1701	Mr. John Dryden, the poet.
1618	Jan. 6, 1689	Seth, Lord Bishop of Exeter—afterwards of Salisbury.
	1673	Andrew Ellis, Esq.
		Sir Francis Feane, Knight of the Bath.
1621	Sept. 5, 1680	Sir John Finch, Kt.
		Mons. Le Febure.
	Oct. 14, 1677	Francis Glisson, M.D.
1620	April 18, 1674	John Graunt, Esq.
	1670	Christopher Lord Hatton.
		Charles Howard, Esq.
		William Hoare, M. D.
	1673	Sir Robert Harley, Kt.
		Nathaniel Henshaw, M.D.
		James Hayes, Esq.—afterwards Kt.
1615	Jan. 24, 1697	William Holder, D. D.
	1690	Theodore Haake, Esq.
		William Hammond, Esq.
		John Hoskyns, Esq.—afterwards Bart
July 18, 1635	March 3, 1702	Robert Hooke, M. A.—afterwards LL. D.
		Richard Jones, Esq.—afterwards Earl of Ranelaugh.
	1680	Alexander, Earl of Kincardin.
	1678	Sir Andrew King, Kt.
		John Lord Lucas.
		James Long, Esq.—afterwards Bart.
		Anthony Lowther, Esq.
	1695	John, Lord Viscount Massarene.
		Sir Anthony Morgan, Kt.
1614	1695	Christopher Merret, M. D.
	Dec. 15, 1681	James, Earl of Northampton.
		Sir Thomas Nott, Kt.—expelled Nov. 18, 1675.
Dec. 7, 1637	Aug. 25, 1670	William Neile, Esq.
1622	Oct. 31, 1679	Jaspar Needham, M. D.
		Sir William Persall, Kt.
		Sir Richard Powle, Knight of the Bath.
	1682	Sir Robert Paston, Knight of the Bath—afterwards Earl of Yarmouth.
		Sir Peter Pett, Kt.—expelled Nov. 18, 1675
	June 1714	Walter Pope, M. D.
March 1, 1610	Dec. 12, 1685	John Pell, D. D.
		Peter Pett, Esq.
		Henry Powle, Esq.
		Thomas Povey, Esq.
		Henry Proby, Esq.

BORN.	DIED.	
	Dec. 24, 1686	Philip Packer, Esq.
		William Quatremaine, M.D.
July 27, [1625	May 28, 1672	Edward, Earl of Sandwich.
		Sir James Shaen, Kt.
1618	July 26, 1693	Charles Scarburgh, M.D.—afterwards Sir C. Scarburgh, Kt.
	April 12, 1678	Thomas Stanley, Esq.
		George Smyth, M.D.
		Alex. Stanhope, Esq.—withdrawn March 8, 1681.
1636	Sept. 11, 1702	Robert Southwell, Esq.—afterwards Sir Robert Southwell, Kt.
		William Schroter, Esq.
1636	May 20, 1718	Thomas Sprat, M.A.—afterwards D.D. and Bishop of Rochester.
		Christopher Terne, M.D.
		Samuel Tuke, Esq.—afterwards Kt.
		Cornelius Vermuyden, Esq.—afterwards Kt.
	Dec. 29, 1707	Sir Cyril Wyche, Kt.
		Sir Peter Wyche, Kt.
1616	1703	John Wallis, D.D.
1606	Oct. 31, 1687	Edmund Waller, Esq.
1631	Sept. 3, 1701	Joseph Williamson, Esq.—afterwards Kt.
1636	July 3, 1672	Francis Willughby, Esq.
		William Winde, Esq.
		John Winthrop, Esq.
		Thomas Wren, M.D.
Oct. 20, 1632	Feb. 25, 1723	Christopher Wren, LL.D.—afterwards Kt.
		Edmund Wylde, Esq.
	May 11, 1684	Daniel Whistler, M.D.
	Dec. 15, 1679	Sir Edward Bysshe, Kt.
1615	Dec. 4, 1679	Sir John Birkenhead, Kt.
April 14, 1629	1695	Mons. Christian Huygens, of Zulichem.
Sept. 17, 1615	April 9, 1670	Mons. Samuel Sorbiere.

The two months during which the Council of the Royal Society were empowered to nominate Members being expired, the Society proceeded from time to time to elect by virtue of their Charter.

BORN.	DIED.		ELECTED.	ADMITTED.
Jan. 20, 1610	March 2, 1674	Sir Justinian Isham, Bart.	July 1, 1663	1663
	1673	Henry Power, M.D.	.. ..	July 1
	May 3, 1681	Alexander Fraizer, M.D.; afterwards Kt. ..	July 8	
		Mons. Victor Beaufort Yabres de Fresars	15	15
	1700	Sir Edward Horley, Knight of the Bath	22	22
		Henry Ford, Esq.; afterwards Kt.	22	22
1630	May 14, 1714	Sir John Talbot, Kt.	29	Jan. 6, 1664
	May 30, 1670	Edward Waterhouse, Esq.	.. ..	Aug. 5, 1663
	1683	{ Antony Ashley Cooper, Ld. Ashley; afterwards } { Earl of Shaftesbury }	Aug. 5	Dec. 30
	Oct. 25, 1698	Sir John Pettus, Bart.	19	
1620	June 14, 1704	{ Ralph Bathurst, M.D.; afterwards D.D. and } { Dean of Wells. }	.. ..	Aug. 19

BORN.	DIED.		ELECTED.	ADMITTED.
		Mr. John Beal; afterwards D. D.		
	1685	Archibald, Earl of Argyle	Oct. 28, 1663	Nov. 4, 1663
		Thomas Coxe, Esq	Nov. 4	11
1594	April, 1678	Francis Potter, B. D.	.. ..	11
		William Goneldon, Senior.	11	18
	June 5, 1675	Henry Earl of Peterborough	18	18
1621	Nov. 11, 1675	Thomas Willis, M. D.	.. ..	Oct. 24, 1667
		Sir Ellis Leighton, Kt.	Dec. 9	Dec. 16, 1663
		John Creed, Esq.	16	16
	1695	Charles Viscount Dungarvan	Jan. 6	Jan. 13
Feb. 20, 1605	Nov. 30, 1675	Sir John Lowther, Bart.	27	27
		Edward Smyth, Esq.	.. ..	June 8, 1664
	1700	John Earl of Tweedale	Feb.. 3	Feb. 3, 1663
		Mr. Roger Williams	.. ..	17
Sept. 18, 1643	March 17, 1714	{ Mr. Gilbert Burnet; afterwards D. D. and Bishop of Sarum	March 23	
	Expelled Nov. 18, 1675 }	Mr. James Carkes		
1621	1687	Monsieur Joannes Hevelius	March 30, 1664	
	Feb. 11, 1688	Monsieur Isaacus Vessius, D D.	April 20	April 20, 1664
1614	1687	Henry More, D. D.	May 25	June 1
		Thomas Neale, Esq.	June 1	15
		James Hoare, Esq.	Nov. 2	Nov. 30
		{ William Godolphin, Esq.; afterwards Sir Wm. Godolphin, Kt.	.. ..	23
		John Newburgh, Esq.	.. ..	2
		Samuel Woodford, Esq.	.. ..	16
1621	1709	Sir Robert Atkyns, Kt.	Nov. 9	16
		Sir John Cutler, Kt.		
	March 7, 1664	Nicholas Bagenall, Esq.	23	23
1630	July 28, 1714	{ Thomas Thynne, Esq.; afterwards Bart. and afterwards Viscount Weymouth	.. ..	30
	1679	John Harvey, Esq.	Dec. 7	
1636	Oct. 4, 1680	Jos. Glanvill, A. M.; afterwards B. D. ..	14	Dec. 14
		Thomas Rolt, Esq.	.. ..	14
		Sir Nicholas Stanning, Kt.	21	March 1
		{ Sir Wm. Portman, Kt. Bart. and Kt. of the Bath	28	Jan. 4
	March 26, 1688	Sir Winston Churchill, Kt.	.. ..	4
	1685	King Charles the II. (Patron)	Jan. 9	
	1701	{ James, Duke of York; afterwards King James II. (Patron)	.. ..	
		George, Duke of Albermarle	18	
May 22, 1636	April 23, 1687	Ferdinand Albert, Duke of Brunswick	25	25
	Dec. 19, 1674	{ Edward, Earl of Clarendon, Lord High Chancellor	Feb. 8	
	Dec. 29, 1680	William, Lord Viscount Stafford	Jan. 18	Jan 25
		Col. Thomas Blunt	Feb. 8	Feb. 8
	May 16, 1703	Samuel Pepys, Esq.	15	15
		{ Philip Carteret, Esq.; afterwards Sir Philip Carteret, Kt.	22	March 1
1597	Nov. 9, 1677	Gilbert Sheldon, Lord Archbishop of Canterbury	March 29, 1665	
	Jan. 19, 1730	Daniel Coxe, M. D.		
1596	June 18, 1683	Richard Sterne, Lord Archbishop of York		
	October, 1675	Humphry Henchman, Lord Bishop of London		
1624	April 11, 1686	{ John Dolben, D. D. Dean of Westminster; afterwards Bishop of Rochester, and at last Archbishop of York		
Sept. 16, 1622	August 1677	Richard, Earl of Dorset	May 3	May 3
		Sir Richard Corbett, Kt.		
		Sir Theodore de Vaux, Kt.	10	24
1602	May 5, 1671	Edward, Earl of Manchester		
		Sir William Hayward	17	
	Withdrawn March 8, 1681 }	Malachy Thruston, M. D.	24	24
1630	Feb. 26, 1686	Charles, Earl of Carlisle	June 14	
1641	1671	{ Mons. Hughes Louis de Lionne, son of Mons. de Lionne, Sec. of State in France	28	June 28
		{ Mons. Vital Dumas, agent of the French merchants	28	28

Here the Meetings of the Society were interrupted by the calamity of the Plague, but began again on the 14th March, 1665.

BORN.	DIED.		ELECTED.	ADMITTED.
		Captain George Cock	March 21, 166⅝	April 4, 1666
		William Harrington, Esq.		March 28
		John Copplestone, Esq.	May 9, 1666	
	May 20, 1713	{ John Lord Yester; afterwards Marquis of Tweedale	May 23	
	1691	Mons. Adrian Auzout		
	April 27, 1680	Thomas Crisp, Esq.; afterwards Kt.	.. ..	June 27
1641	March 20, 1688	{ Samuel Parker, A. M.; afterwards D. D. afterwards Bishop of Oxford	June 13	
		Edward Nelthorpe, Esq.	27	
1629	May 29, 1709	{ Mr. Edmond King; afterwards M. D. and Sir E. King, Kt.		
		David, Lord Viscount Stormont	Aug. 29	Aug. 29
	July 17, 1685	{ John, Lord Robartes; afterwards Lord Privy Seal, afterwards Earl of Radnor	Nov. 14	
	Jan. 24, 1674	{ Benjamin Laney, Lord Bishop of Lincoln; afterwards of Ely		
		Mr. Nicholas Mercator	.. ..	Nov. 25, 1669
	May 8, 1701	{ Robert, Earl of Lindsey, Lord Great Chamberlain of England	21	
July 12, 1628	Jan. 11, 1683	{ Henry Howard, of Norfolk; afterwards Earl of Norwich and Earl Marshal, at last Duke of Norfolk	28	Nov. 28, 1666
1628	Nov. 16, 1700	{ Paul Rycaut, Esq.; afterwards Sir P. Rycaut, Kt.	Dec. 12	
		Thomas Lake, Esq.	Feb. 14, 1667	March 7, 1667
Feb. 12, 161⅔	July 16, 1686	{ John Pearson, D. D.; afterwards Lord Bishop of Chester	March 14	April 25
		Sir Clifford Clifton, Kt.	March 28	
1616	1687	Mons. Ismael Bullialdus	April 4	
		Mons. Pierre Pettit		
		Francis Smethwick, Esq.	.. ..	May 16
		{ Philip Skippon, Esq.; afterwards Sir P. Skippon, Kt.	May 16	16
		Thomas Harley, Esq.	30	
		Sir Barnard Gascon, Kt.	June 20	June 20
	1691	Walter Needham, M. D.		April 6, 1671
		Sir William Curtius, Kt. and Bart.	Oct. 3	
		Sir Maurice Berkley, Kt.	17	
		Col. Bullen Reymer	17	Nov. 7, 1667
	Expelled June 24, 1675, died 1691	} Richard Lower, M. D.	.. ..	Oct. 17
March 5, 162⅘	Nov. 10, 1683	Mr. John Collins	.. ..	24
	Withdrawn March 8, 1681	} Sir Nicholas Stewart, Bart.	24	24
		Mons. Theodore de Beringhen		24
		Henry Clerke, M. D.	.. ..	Nov. 7
Nov. 29, 1628	Jan. 17, 170⅘	John Wray, M. D.—Johannes Raius	Nov. 7	7
		William Aglionby, M. D.		14
		Mons. Leyonbergh, Swedish President ..	21	21
		William Soame, Esq.		21
		Count Charles of Ubaldini, of Montefeltri ..	.. ..	28
	Sept 24, 1710	{ Sir Charles Berkley, K. B.; afterwards Ld. Dursley, at last Earl of Berkley	.. ..	28
	Dec. 25, 1681	Nicholas Oudart, Esq.		30
	Expelled June 24, 1675	} Jaques du Moulin, M. D.	Dec. 5	Dec. 5
		John Downs, M. D.	12	12
1642	Aug. 27, 1708	{ Edward Brown, M. D.; afterwards President of the College of Physicians	Jan. 2	Jan. 2
	1683	Edward, Lord Conway		
		Sir Maurice Eustace	9	
		Charles Hotham, Esq.		
June 8, 1626	Oct. 16, 1695	William, Earl of Strafford	Feb. 6.	Feb. 6
	1684	Thomas Allen, M. D.	.. ..	13
		William Le Hunt, Esq.	13	13
		Esay Ward, A. M.	20	20
		——— Flower, Esq.		
		Don Antonio Alvares da Cunha	April 9, 1668	
July 10, 1614	April 6, 1686	Arthur, Earl of Anglesey	23	
		Sir Erasmus Hardy, Bart.	.. ..	May 7, 1668
1635	Dec. 8, 1688	Thomas Flatman, Esq.	30	
	1711	Benjamin Woodroffe, M. A.; afterwards D. D.	7	7
		John Colwall, Esq.	14	21

d

BORN.	DIED.		ELECTED.	ADMITTED.
	Expelled Nov. 18, 1675	Colonel Thomas Collepepyr, Esq.	May 28, 1668	June 25, 1668
	1675	Mr. James Gregory .. .. .. , ...	June 11	11
	1691	{ Mr. James Arderne, A. M.; afterwards D.D. and Dean of Chester	18	18
		Peter Courthope, Esq.	18	Nov. 26
		Edward Howard, Esq. of Norfolk	Aug. 6	26
		Maurice, Lord Viscount Fitzharding	Nov. 5	5
		Sir Kingsmill Lucy, Bart.	26	30
1632	1704	John Lock, Esq.	.. ..	26
		Daniel Finch, Esq.		
	1703	{ Edward Chamberlayne, Esq.; afterwards LL. D.	Dec. 3	Dec. 3
		Sir John Banks, Bart.	10	31
	Dec. 1704	Colonel Silas Titus	Jan. 14	
		James Hoare, jun. Esq.	21	Feb. 18, 1674
1641	Jan. 1697	Anthony Horneck, M. A.; afterwards D. D.	28	March 4, 1668
1635	1673	George Morley, Lord Bishop of Winchester	Unknown	
	Oct. 12, 1673	George Castle, M. D.	Feb. 4	Feb. 4
1627	Nov. 29, 1694	Sig. Marcellus Malpighi	March 4	
		Edward Jeffreys, Esq.	Oct. 28, 1669	Nov. 4, 1669
		Thomas Barrington, Esq.	Nov. 18	25
		{ Don Gaspar Merez de Souza, Professor of Mathematics, in the University of Coimbra	18	
1641	1724	{ Dr. Hirbanus Hiaerne; afterwards Archbishop to the King of Sweden	.. ..	Dec. 2
1598	April 22, 1672	{ Mons. George Stiernhielm, Præses Collegii Antiquitatis, Stockholm	Dec. 9	
		Mons. Gustavus Helmfeld	April 21, 1670	April 21, 1670
		Mons. Andreas Monceaux	Dec. 15	
	June 17, 1723	Sir John Williams, Kt. and Bart.	March 23	Nov. 30, 1671
		Sir Philip Mathews		
		{ Robert Redding, Esq.; afterwards Sir R. Redding, Bart.	Nov. 2, 1671	30
	1711	Mr. Martin Lyster; afterwards M. D.		
	1711	Nehemiah Grew, M. D.	.. ..	30
Dec. 25, 1642	March 20, 1727	{ Mr. Isaac Newton; afterwards Sir I. Newton, Kt. and P. R. S.	Jan. 11	Feb. 18, 1674
	1672	Sir Frechville Hollis	18	
1629	Nov. 22, 1694	{ John Tillotson, D. D. Dean of Canterbury; afterwards Archbishop of Canterbury	25	March 14, 1671
June 8, 1625	Sept. 14, 1712	Joannes Dominicus Cassini	May, 22, 1672	
1637	1676	Francis Vernon, Esq.	.. ..	June 12, 1672
	April 2, 1701	{ Henry, Lord Howard, of Castle Riding; afterwards Earl of Arundel, at last Duke of Norfolk	Oct. 30	Oct. 30
	Nov. 1689	{ Thomas Howard, of Norfolk, second son of the Earl Marshal; afterwards called Lord Howard	Nov. 6	Nov. 6
	1694	{ Edward Bernard, B. D. Professor of Astronomy, Oxon	April 9, 1673	
June 23, 1646	Nov. 14, 1716	Mons. Gothofredus Gulielmus Leibnitz		
1660	July 13, 1698	{ Charles, Lord Herbert, eldest son to the Marquis of Worcester; afterwards stiled Marquis of Worcester	June 4	
		{ John Stafford Howard, Esq. son of the Lord Viscount Stafford	Nov. 6	
		Sir Justinian Isham, Bart.		
	July 23, 1718	Sir John Lawrence, Kt. Alderman of London	27	Dec. 1, 1673
		Sir Richard Ford, Kt. Alderman of London	.. ..	1
		{ Sir Thomas Player, Kt. Chamberlain of London	Nov. 27	1
		Rowland Winn, Esq.	.. ..	1
		Mr. Andrew Birch; afterwards M. D.	.. ..	1
		Francis Robartes, Esq.	Dec. 11	Feb. 5
	Feb. 3, 1711½ 1675	Colonel Giles Strangeways		
		Doctor John Le Gassick		
		{ Renatus Franciscus Slusius, Canonicus Leodiensis	Unknown	
	1697	{ Henry Jenkes, M. A. Profes. of Rhet. at Gresh. College; afterwards D. D.	Nov. 30, 1674	
1615	1681	Sir Jonas Moore, Kt.	Dec. 3	Dec. 3, 1674
	Dec. 23, 1721	Sir Paul Whichcote, Bart.	Jan. 14	Feb. 11
		Daniel Milles, B. D.	.. ..	11
	1685	Edmond Castell, D. D.	Unknown	Jan. 28

1

BORN.	DIED.		ELECTED.	ADMITTED.
1656	1680	Sir Philip Percivale, Bart.	Feb. 18, 1674	April 29, 1675
	April 5, 1695	{ George, Lord Viscount Halifax; afterwards } Earl, and afterwards Marquis	Nov. 30, 1675	Dec. 2
June 15, 1631	Nov. 10, 1721	John Mapletoft, M.D.; afterwards D.D. ..	Feb. 10	Feb. 10
	April 21, 1710	Captain Henry Sheers		
		Henry Hall, Esq.	.. ..	May 18, 1676
		Sig. Franciscus Fravagino		
1639	1688	Sir Richard Edgecumbe, Bart.	Nov. 30, 1676	Nov. 30
		Sir Thomas Clutterbuck, Kt.		
	Oct. 28, 1681	{ Mr. John King, Rhetorick Professor, at Gre- } sham College, M.B.		
	1680	Sir George Croke, Kt.	Feb. 8	Feb. 8
June 21, 1632	1680 or 82	Dr. Christ. Adolphus Baldwin		
1645	Dec. 31, 1719	Mr. John Flamsteed, Astronomer Royal ..	.. ..	Feb. 13, 1678
	1619	George Ent, Esq.		

(The minutes of four Meetings are wanting here)

BORN.	DIED.		ELECTED.	ADMITTED.
		Mr. Oliver Hill.		
		Mr. ——— Wyndham	Dec. 6, 1677	
1635	April 8, 1702	Thomas Gale, D.D.; afterwards Dean of York	.. ..	Dec. 6, 1677
1641	April 30, 1696	Robert Plott, LL.D.	.. ..	6
	1702	Thomas Smith, B.D.; afterwards D.D. ..	.. ..	6
		Sir Peter Colleton, Kt.	13	
		John Herbert, Esq.		
		{ George Wheeler, Esq; afterwards Sir Geo. } Wheeler, D.D. Bart.	.. ..	Jan. 3
	1699	Sir James Langham, Bart.	Jan. 17	
		Mons. Theodorus Kerckringius		
	1693	Walter Chetwynd, Esq.	31	
1625	July 3, 1707	Edmond Dickenson, D.M.		
		Joseph Lane, Esq.		
		Francis Aston, Esq.	Nov. 30, 1678	Nov. 30, 1678
	1679	John Mayow, M.D.		
		Mr. David Hanuisius, Bibliothecarius Bruns.		
		John Bemde, Esq.	.. ..	30
	Sept. 1696	William Perry, A.M...	.. ..	30
		Mr. Dethlevus Cluverus		
Nov. 8, 1656	Jan. 25, 1742	{ Mr. Edmond Halley; afterwards LL.D. } Savill. Prof. and Astro. Royal	Nov. 30	
		Mr. Joseph Moxon	.. ..	30
1629	Nov. 14, 1710	{ Mons. Ezekiel de Spanheim, Envoy of Bran- } denburgh	Feb. 6, 1679	March 13
		Thomas Sheridan, Esq.	.. ..	Feb. 13
		Sir William Waller	20	
	Aug. 1, 1708	Edward Tyson, M.D.	Dec. 1	
1625	1695	Henry Paman, M.D.		
		William Naper, Esq.	.. ..	Dec. 18, 1679
		Sig. Jo. Ambrosius Sarotti, Nob. Ven. ..	.. ..	4
		William Bridgman, Esq.	Dec. 18	March 18
		Mr. Thomas Pigot, Fellow of Wadham College	.. ..	July 13, 1681
		{ Mons. John Christian Heusch, M.D. Physi- } cian to the Elector Palatine	Jan. 29, 1680	Feb. 5, 1679
		Mr. Thomas Firmin	.. ..	26
		Mr. John Houghton	.. ..	5
1632	1723	Mons. Antonine Van Leuwenhoeck		
		Mr. Peter Perkins	Feb. 5	5
		Jonas Moore, Esq.	March 11	March 25, 1680
1656	1715	Robert Nelson, Esq.	April 1	Nov. 30, 1695
		Andrew Clenche, M.D.	.. ..	April 22, 1680
		John Wood, Esq.	.. ..	29
1647	1683	{ Dr. Jacobus Pighius Veronensis, Professor } of Anatomy, at Padua	29	
	1708	Dionisius Papin, M.D.	Unknown	
		Frederick Slare, M.D.	Unknown	Dec. 16
	April 9, 1685	{ Robert Wood, LL.D. Master of Christ's } Hospital	April 6, 1681	
		Sir Anthony Deane, Kt.		
	1728	Hugh Chamberlen, M.D.	.. ..	Nov. 2, 1691
1660	1686	Sir John Percivale, Bart.	.. ..	30
1637	1700	{ Roger Meredith, Esq. Prof. of Civil Law, } Gresh.; afterwards Master in Chancery	.. ..	May 4
		John Rogers, M.A.		
		Mr. Oliver Salusbury	.. ..	July 13
		Dr. Russell		

BORN.	DIED.		ELECTED.	ADMITTED.
		Richard Waller, Esq.	April 27, 1681	May 4, 1681
	April 27, 1705	Jeremy Sambrooke, Esq.; afterwards Kt. ..	.. ..	June 1
		——— Braddon		
		Mr. Goodwyn	27	
		Joannes Philippus Jordis, M. D. Francfort.		
		Sir Patience Ward, Kt. Alderman of London	Unknown	
		Henry Eve, Esq.	Nov. 9	Nov. 23
		Mr. William Payne, A. M.	.. ..	23
1644	July 30, 1718	William Penn, Esq.		
	Jan. 25, 1726	Sir Rowland Guynne, Kt.	23	23
		Mr. Jodocus Crull		30
	Jan. 30, 1733	Richard Robinson, M. D.	.. ..	30
		Mr. Francis Lodwick	30	
1630	1701	Sig. Gregorio Leti	.. ..	30
		Mr. Isaac Dorislaus		
	Sept. 25, 1693	Mons. Henry Justell	Dec. 7	Dec. 7
		Mr. Samuel Blackburne	14	
		{ Sidi Hamet Ben Hamet Ben Haddu Otter, } { Ambassador from Morocco	April 26, 1682	
	Jan. 7, 1726	Walter Mills, M. D.	July 12	Nov. 1, 1682
		John Turner, Esq.	Nov. 8	15
		Edward Paget, M. A.		
May 20, 1660	May 22, 1729	Marc. Antonio Principe Borghese	29	
		{ Don Joseppe de Faria, Envoy from the } { King of Portugal	30	30
1643	Dec. 25, 1712	Sir John Chardin, Kt.		30
		{ Mons. Conrad van Beuninghen, Envoy from } { the States General	Dec. 13	Dec. 13
	Withdrawn Nov. 12, 1707	{ Robert Pitt, M. D. Prof. of Anatomy, at } { Oxford	20	
	Oct. 1686	Mr. William Gould	May 2, 1683	May 9, 1683
		Edward Haynes, Esq.		
1635	Nov. 12, 1713	{ Edward Wetenhall, Lord Bishop of Cork } { and Rosse, and afterwards of Kilmore	July 4	
		Allen Moulin, M. D.	18	July 25
		Charles Willughby, M. D.	25	25
		Nathaniel Vincent, D. D.	Nov. 30	Dec. 5
		Mr. Arthur Bailey	.. ..	19
	Dec. 23, 1721	William Musgrave, LL. D.	March 19, 1684	Dec. 1, 1684
		Mons. Fremont d'Ablancourt	Nov. 12	
		James Monson, Esq.		
		Richard Beaumont, Esq.	19	1
	1690	Mr. Thomas Baker		
	Oct. 19, 1728	Alexander Pitfield, Esq.	.. ..	1
	March 29, 1748	Tancred Robinson, M. D.		
	Oct. 31, 1709	Henry, Earl of Clarendon	Dec. 1	1
	1713	{ John, Lord Vaughan; afterwards Earl of } { Carberry	Jan. 21, 1685	
April 16, 1660	Jan. 11, 1753	{ Hans Sloane, M. D.; afterwards Sir H. } { Sloane, Bart. and P. R. S.		
		Benjamin Von Münchausen		
	Jan. 22, 173¾	Thomas, Earl of Pembroke and Montgomery	May 13	
		Mons. Esprit Cabart de Villermont		
		Mr. John Beaumont		
		Mr. Charles Leigh, M. A.; afterwards M. D.		
		Sir Richard Bulkley, Kt.	Nov. 25	Nov. 25, 1685
		Sir Robert Gordon, Kt.	Feb. 3, 1686	Feb. 3
1656	1698	William Molyneux, Esq.		
	1718	{ St. George Ash, A. M.; afterwards D. D. } { and Lord Bishop of Clogher, at last Lord } { Bishop of Derry	.. ..	June 2, 1686
		Thomas Molyneux, Esq.; afterwards M. D.	Nov. 3	
		John Harwood, A. M.; afterwards LL. D.	3	March 2
		Clopton Havers, M. D.	17	Dec. 15
		——— Mears, Esq.	Dec. 8	
		——— Sylvius, M. D. of Dublin	March 9, 1687	
		{ Mons. Jean Weichard Valvasor, Liber. Baro, } { of Carniola	Dec. 14	
		William Wotton, M. A.; afterwards D. D.	.. ..	Feb. 1, 1687
		Benjamin Middleton, Esq.	.. ..	Jan. 18
	May 10, 1753	Mr. Nicholaus Facio de Duillier	.. ..	May 2, 1688
1646	April 11, 1706	Mons. Joannes Nicholaus Pechlin, M. D.	Nov. 30, 1688	
1641	1715	Mons. Raymond Vieussens, M. D.		
	1742 or 3	Mons. Js. Adamus Stampfer, of Stiria		
		——— Adair, Esq.		
		——— Clayton, Esq.		

BORN.	DIED.		ELECTED.	ADMITTED.
1660	1708	Mr. Edward Stillingfleet		
	April 4, 1718	Charles Gresham, M. A. Rhet. Prof. Gresh. Col.		
		{ Mynheer Nicholaus Witsen, Burgomaster of } Amsterdam	Nov. 30, 1689	
		{ William Stanley, D. D.; afterwards Dean } of St. Asaph	.. ..	Dec. 18, 1689
		Joseph Raphson, A. M.	.. ..	4
		Mr. George Moult	.. ..	Nov. 30
	1691	Sig. Jacomo Grandi, of Venice	Dec. 1, 1690	
	April 8, 1709	Sir Godfrey Copley, Kt. and Bart.	Nov. 30, 1691	
		Thomas Day, Esq.		
July 10, 1658	Nov. 1, 1730	Mons. Lewis Ferdinand Count Marsigli		
Oct. 2, 1667	Feb. 3, 1716	{ Alex. Torriano, M. A.; afterwards LL. D. } Prof. Ast. Gresh.	.. ..	Jan. 27, 1691
1638	1716	Sir William Trumball, Kt.	Nov. 30, 1692	
		Charles Isaac, Esq.; afterwards Kt.	.. ..	April 4, 1694
		Mons. John Theodore Heinson, of Hanover	.. ..	Nov. 30, 1692
1667	Aug. 9, 1728	{ Edward Lany, M. A. Prof. of Divinity, } Gresh. College; afterwards D. D.	.. ..	30
		Edward Southwell, Esq.	.. ..	30
	Oct. 12, 1708	David Gregory, M. D.	.. ..	April 5, 1693
		Jonathan Blackwell, Esq.		
		Lord George Douglas	.. ..	Dec. 14, 1692
Sept. 8, 1650	1707	Joannes Delceus, M. D.		
		Mr. Ralph Lane	.. ..	Jan. 25
1630	1714	George Viscount Tarbat	Unknown	
		Louis Paule, M. D.	Do.	
1665	April 25, 1728	{ John Woodward, M. D. Prof. Med. Gresh. } College	Nov. 30, 1693	Nov. 30, 1693
1660	Dec. 22, 1718	Robert Briggs, M. A. Prof. Ast. Gresh. Coll.	.. ..	30
		Thomas Kirke, Esq.	.. ..	30
		John Henley, Esq.	.. ..	30
	Aug. 3, 1723	Charles Bodvill, Earl of Radnor		
	April 2, 1729	{ Sir Thomas Willoughby, Bart.; afterwards } Lord Middleton		
May 3, 1644	June 14, 1704	{ Georgius Franck de Franckenaw, Prof. of } Med. Wittenburg		
		{ Christopher Wren, Jun. Esq. eldest son of Sir } Christopher	.. ..	Nov. 30, 1698
	Aug. 9, 1744	{ James Brydges, Esq.; afterwards Duke of } Chandos	Nov. 30, 1694	
		John Jackson, Esq.		
		Patrick Gordon, A. M.		
	1742 or 3	Sig. Dominico Bottini, M. D.	Oct. 23, 1695	
	1706	Mr. Samuel Doody	Nov. 27	Nov. 27, 1695
	April 4, 1718	{ Mr. James Petiver, Apothecary to the Char- } ter-house	27	27
1665	1698	Bernard Connor, M. D.	.. ..	27
		Sig. Tommaso Del Bene	30	30
April 16, 1661	May 19, 1715	{ Rt. Hon. Charles Montague, Esq.; after- } wards Lord Halifax, at last Earl of Halifax	.. ..	30
Jan. 27, 166½	July 14, 1742	Mr. Richard Bentley; afterwards D. D ..	.. ..	30
		Mr. Moyses Pujolas	.. ..	Dec. 18
	Nov. 4, 1720	{ Edward Smyth, D. D. Dean of St. Patrick's; } afterwards Lord Bishop of Down and } Connor	.. ..	April 29, 1696
		Mons. Pomponius Baron de Scarlotti	April 29, 1696	
April 5, 1622	Sept. 22, 1703	Sig. Vincentio Viviani, of Florence		
		Sig. Fornassari, of Bononia		
	Aug. 14, 1721	{ Orlando Bridgeman, Esq.; afterwards Sir } O. Bridgeman, Kt.	.. ..	29
		William Byrd, Esq.	.. ..	29
	Sept. 7, 1719	John Harris, M. A.; afterwards D. D.		
		Sig. Bon Figlilio		
1649	1713	{ Godofridus Bidloo, M. D. Prof. of Anatomy } in the University of Leyden	.. ..	Jan. 21, 1701
	Jan. 22, 1733	{ Thomas Foley, Jun. Esq. of Whitley; after- } wards Lord Foley	July 15	July 22, 1696
		Ralph Lowardes, Esq.		
		Mr. ——— Chadwick		
		Mr. Philip Ryley		
		Sig. Francisco Spoleto, M. Prof. Pa.		
	April 17, 1751	{ Henry Petty, Esq.; afterwards Viscount } Shellburne, at last Earl of Shellburne	Nov. 30	Nov. 30
		{ Mr. Charles Bernard; afterwards Sergeant } Surgeon	.. ..	Dec. 16

BORN.	DIED.		ELECTED.	ADMITTED.
Dec. 4, 1664	Sept. 13, 1735	John Newey, A. M.	.. ..	Nov. 30, 1696
	Nov. 16, 1739	William Cockburn, M. D.	.. ..	Dec. 2
	March 17, 1738	Hugh Howard, Esq.	Nov. 30, 1696	Nov. 30
		John Hutton, M. D.	30, 1697	30, 1697
May 26, 1667	Nov. 27, 1754	Mr. Abram de Moivre		
1673	Sept. 15, 1707	George Stepney, Esq.	.. ..	30
	1723	Mons. Jacques Basnage de Beauval		
1658	Sept. 13, 1725	Ralph Thoresby, Esq.		
Feb. 10, 1677	April 15, 1755	Mons. Jacques Cassini	March 23	March 23
1665	1721	Matthew Prior, Esq.		
	Nov. 19, 1759	Sir Berkeley Lucy, Bart.	.. ..	March 30, 1698
Sept. 27, 1655	July 12, 1710	Sig. Dominice Guglielmini		
		Orlando Bridgeman, Esq.	.. ..	Nov. 16
		Maurice Emmett, Esq.	.. ..	March 23, 1697
		John Fryer, M. D		
1656	May 30, 1725	{ Robert Molesworth, Esq. ; afterwards Lord { M., of Swords	.. ..	April 6, 1698
		Mons. Balthasar Becker, M. D.	July 6, 1698	
1668	1708	Sign. Georgio Baglivi		
Feb. 13, 1672	Jan. 6, 1731	Mons Estienne Francois Geoffrey		July 6
	March 21, 1748	George Lord Reay	Nov. 9	Nov. 9
		{ James, Lord Viscount Seafield ; afterwards { Earl of Finlater	.. ..	16
		Edward Norris, M. D.	.. ..	9
		{ John Stanley, Esq. ; afterwards Sir John { Stanley, Kt. and Bart.	.. ..	30
	Oct. 9, 1731	Thomas Isted, Esq.	.. ..	16
	Jan. 16, 1744	Edward Haistwell, Esq.		
1652	April 26, 1716	John, Lord Somers, Lord Chancellor	30	
	April 19, 1722	{ Charles, Lord Spencer ; afterwards Earl of { Sunderland	30	June 6, 1705
		Anthony Hammond, Esq.	.. ..	Oct. 30, 1700
Jan. 24, 1637	June 29, 1705⅚	Charles, Earl of Dorset	Jan 11	
		Mr. R. Shirley	.. ..	Jan. 11, 1698
	March 7, 1733	George Worth, Esq.	.. ..	11
		Mr. Agricola		
		David Krieg., M. D.	.. ..	1L
		Mr. Williams		
	1710	Mr. William Cooper, Surgeon	.. .. ..	Feb. 1
	1710	Thomas Browne, M. D.	Nov. 30, 1699	
		Martin Bowes, Esq.	.. ..	Dec. 6, 1699
		Mr. Paul Bussiere	.. ..	May 22, 1700
		Mons. Pierre Silvester		
1575	1732	Mons. Jo. Burchard Menkenius		
	July 6, 1700	John, Viscount Lonsdale, Lord Privy Seal ..	.. ..	Dec. 20, 1699
	1724	Mr. James Pound, M. B.	.. ..	July 30, 1713
		Mr. James Cunningham		
Jan. 3, 1634	March 18, 1715	Mr. Otto Sperling, of Copenhagen	Nov. 30, 1700	
	1725	Abraham Cyprianus, M. D.	.. ..	Dec. 11, 1700
	Withdrawn May 21, 1707 }	Sir Philip Sydenham, Bart.	.. ..	Feb. 26
		Charles du Bois, Esq.	.. ..	May 27, 1714
1671	Aug. 29, 1721	{ John Keill, M. A. ; afterwards M. D. Sav. { Prof. of Astron.	.. ..	Feb. 25, 1701
	April 8, 1710	Mons. Christ. Leyon Crona	Nov. 30, 1701	
		{ Sir John Percivale, Bart. ; afterwards Lord { Percivale, at last Earl of Egmont		
		Ciril Arthington, Esq.	.. ..	Dec. 31
		Mr. Michael Le Oassor	.. ..	Jan. 14
	March 2, 1707	James Drake, M. D.	.. ..	21
	Jan. 4, 1747	John Shadwell, M. D. since Sir J. Shadwell, Kt.	.. ..	Dec. 3
	April 12, 1743	Mr. George Cheyne ; afterwards M. D. ..	March 18	May 6, 1702
		M. De la Pryme		
		Robert Thompson, Esq.	May 6, 1702	May 13
		Mr. Vernon		
	1723	John Chamberlayne, Esq. .. .. ~ ..	.. ..	20
	Sept. 2, 1724	John Lowthorp, M. A.	Nov. 30	Nov. 30
		Mr. James Young		
		Mr. Ludlow		
		Mons. Jean Chardelon	.. ..	30
1657	April 5, 1735	Mr. Wm. Derham		Feb. 3, 1703
Aug. 5, 1580	Dec. 12, 1764	Johannes Philippus Breynius, M. D. of Danzig	May 2, 1703	
		John Hickes, Esq.	Nov. 30	Jan. 26
		Joseph Morland, M. D.	.. ..	Dec. 8
	April 13, 1721	Sir Matthew Dudley, Bart.	.. ..	Jan. 26

BORN.	DIED.		ELECTED.	ADMITTED.
		John Morton, M. A.		
Aug. 11, 1673	Feb. 16, 1754	Richard Mead, M. D.	.. ..	Nov. 30, 1703
		Robert Areskyne, M. D.	.. ..	30
	March 27, 1716	William Oliver, M. D.	.. ..	Jan. 5
		Joseph Shaw, Esq.	.. ..	20
1672	June 25, 1755	Mr. James Hodgson	.. ..	Dec. 15
	1738	Philip Stubs, M. A.; afterwards B. D. ..	.. ..	Feb. 23
May 3, 1661	Jan. 28, 1730	Sign. Antonio Valisnieri		
Dec. 9, 1652	Dec. 30, 1723	{ Augustus Quirinus Rivinius, M. D. Prof. of Med. in the University of Leipzig }		
Aug. 2, 1672	June 23, 1733	Joannes Jacobus Scheuchzerus, M. D.		
		Sign. Emanuel Timone, M. D.		
		Mr. P. Boothe	.. ..	May 31, 1704
		Russell Robartes, Esq.	.. ..	June 28
1653	Oct. 28, 1708	Prince George of Denmark	Nov. 30, 1704	
		Mr. Samuel Morland	.. ..	Dec. 6
		John Fuller, Esq.	.. ..	June 8, 1709
	Withdrawn Mar. 12, 1706 {	William Fellowes, Esq.	.. ..	Nov. 30, 1704
1673	Jan. 20, 1731	{ Andrew Tooke, M. A. Prof. of Geom. G Coll. resh. }	.. ..	30
		——— Annesley		
		Walter Clavell, Esq.	.. ..	Dec. 30
1681	Feb 27, 1734½	John Arbuthnot, M. D.	.. ..	13
1651	Jan. 25, 1733	Sir Gilbert Heathcote, Kt.	Nov. 30, 1705	
		John Mortimer, Esq.	.. ..	3, 1705
		D'Acre Barrett Lennard, Esq.		
		Mr. Francis Hawksbee	.. ..	Nov. 30
		John Thorpe, M. B.; afterwards M. D. ..	.. ..	30
		Henry Worsley, Esq.	.. ..	Dec. 5
1650	May 7, 1729	Wm. King, Lord Archbishop of Dublin		
	Feb. 14, 1727	{ Wm. Nicolson, Lord Bishop of Carlisle; afterwards Lord Bishop of Derry, at last Archbishop of Cassill }	.. ..	5
		Mr. Buys, Esq. of Holland	Feb. 13, 1706	
	1729	Wm. Burnett, Esq.	.. ..	Feb. 13
	1721	{ Mr. Philip Bisse; afterwards D. D. and Lord Bishop of St. David's, at last Lord Bishop of Hereford }	.. ..	Jan. 21, 1707
		Captain Thomas Savery	.. ..	April 10, 1706
	Oct. 10, 1723	{ Wm. Cowper, Esq. Lord Keeper; afterwards Earl Cowper }	April 3	
1674	June 21, 1738	Charles, Lord Viscount Townshend		
	May 23, 1743	John, Lord Powlet		
Aug, 1676	Aug. 28, 1731	Charles, Earl of Orrery	.. ..	June 26
		Mons. Jean Christoph. Facio de Duillier ..	3	
	Oct. 25, 1709	{ Mons. Le Compte de Briançon, Envoy from the Duke of Savoy }	June 5	12
	March 11, 1711	Mons. Antoine Marquis de Guiscard	.. ..	12
	July 3, 1711	{ Mons. Van Vrijberge, Envoy from the States General }	.. ..	19
	Jan. 18, 1719	Samuel Garth, M. D.; afterwards Kt.		
		Sign. Gallucci	Nov. 20	
Oct. 26, 1654	Feb. 20, 1720	Sign. Giovanni Maria Lancisi, M. D.		
	April 2, 1742	James Douglas, M. D.	.. ..	Dec. 4
1671	July 6, 1726	Mr. Humphrey Wanley	.. ..	4
	Nov. 24, 1745	Robert Shippen, M. A. since D. D.	.. ..	May 14, 1707
		Col. Francis Nicholson	Dec. 4	Dec. 4, 1706
	Feb. 12, 1738	Mr. James Sherrard	.. ..	18
		Thomas Forster, Esq.	.. ..	18
		John Knight, Esq.	.. ..	18
		Mr. Thomas Frankland	March 19, 1707	
	Nov. 20, 1714	William Frankland, Esq.	.. ..	June 11, 1707
	Feb. 11, 1719	Rowland Holt, Esq.	.. ..	April 2
		Mr. Benjamin Morland	.. ..	March 26
	1742 or 3	John, Duke of Roxburgh	May 28	June 11
	Jan. 7, 1742	James, Duke of Montrose	.. ..	11
1658	June 19, 1726	{ Sir Thomas Trevor, Kt. Lord Chief Justice of Common Pleas: afterwards Ld. Trevor }	Dec. 1	Feb. 25
		Mr. Thomas Ayres	.. ..	Dec. 1
		James Venables, Esq.	.. ..	Jan. 28
		Thomas Hoy, M. D. Med. Prof. Reg. Oxon		
	Nov. 26, 1746	Henry Plumptre, M. D.	.. ..	Dec. 1
	1715	Benjamin Pratt, D. D.	April 7, 1708	April 7, 1708
	May 26, 1759	Mr. Thomas Woodford	.. ..	Nov. 10

BORN.	DIED.		ELECTED.	ADMITTED.
		Dr. Mills, Greek Prof. of Oxford		
	June 9, 1736	Sir George Markham, Bart.	.. ..	April 14, 1708
	Aug. 28, 1721	Sir David Hamilton, Kt. M. D.	.. ..	May 5
		—— Williams		
	1724	{John Bridges, Esq. Commissioner of the Customs	.. ..	June 15, 1709
		Mr. Thomas Whalley		
	Jan. 31, 1719-20	Thomas, Earl of Stamford	May 12, 1708	Nov. 10, 1708
	Jan 28, 171¾	Philip, Earl of Chesterfield	Nov. 30	
		{Sign. Francisco Comaro, Venetian Ambassadbr	.. ..	Dec. 15
		Mons. Jean Rodrigue Lavater, M. D.	.. ..	Nov. 30
	Jan. 9, 1724	William Fellows, Esq.		
		Robert Balle, Esq.	.. ..	May 4, 1709
		Mr. Richard Tighe	.. ..	Nov. 22, 1710
		Sir Edward Lawrence, Kt.		
	March 27, 1732	Richard Foley, Esq.	.. ..	June 8, 1709
April 10, 1655	March 13, 1740	{Sign. Michael Angelo Tilli, M. D. Bot. Prof. Pisan.		
		{Mr. Owen Lloyd, Keeper of the Ashmol. Mus. at Oxford, signed his name so in the Charter book		
		Mr. Charles Nicholas Ayres		
	Aug. 12, 1740	Archibald Hutcheson, Esq.	.. ..	May 18
		{Count Lorenzo Magalotti, Counsellor of State to the Grand Duke of Tuscany	May 4, 1709	
		Henry Cressener, M. A.	.. ..	May 11
	July 29, 1715	{Henry Newton, LL. D.; afterwards Sir H. Newton, Kt.	.. ..	Nov. 8, 1711
		—— Leopold, M. D. of Lubeck	4	
	1742 or 3	Sign. Guido Graudi, of Florence		
		Robert Hunter, Esq.	.. ..	May 18, 1709
	Dec. 2, 1758	Samuel Tufnell, Esq.		11
	1742 or 3	{Sign. Vendramin Bianchi, President from Venice	Nov. 8, 1710	
Jan 24, 1679	April 9, 1754	Mons. Christian Wolfius		
Jan. 10, 1654	Aug. 3, 1712	Joshua Barnes, Greek Professor, at Cambridge	.. ..	Nov. 15, 1710
		Owen Brigstock, Esq.	30	30
	June 9, 1751	Mr. John Machin, Prof. Astronomy, Gresham	.. ..	30
	Nov. 19, 1724	Mr. Joseph Tanner	.. ..	30
Aug. 23, 1683	Nov. 15, 1761	Sign. Joannes Poleni, Prof. of Astron. Padua		
	July 13, 1727	Mr. Alexander Geekie, Surgeon	.. ..	30
	Feb. 25, 1758	Mr. Samuel Hill	Nov. 30, 1711	
	1718	Mr. Philbert Collet		
1679	1746	{Mons. Charles Count Gyllenberg, Envoy from Sweden	.. ..	Dec. 13, 1711
		Alexander Sandilands, M. D.	.. ..	Nov. 30, 1711
	Nov. 2, 1722	Sir Marmaduke Wyvil, Bart.		
		Mr. Linda		
		Walter Douglas, Esq.	.. ..	Feb. 25, 1712
		{Mons. Louis Frid. Bonet, the King of Prussia's Minister	.. ..	
	April 12, 1752, Withdrawn	Mr. Fettiplace Bellers	.. ..	April 17, 1712
1688	1745	Mr. William Cheselden, Surgeon	.. ..	Nov. 30
		Mr. John Craig		
	July 3, 1749	William Jones, Esq.	.. ..	Nov. 30
		Roger Cotes, M. A. Plum. Pr. Ast. Cambrig.	30	May 2, 1714
		Thomas Greene, Esq.	.. ..	Dec. 13, 1711
	April 28, 1732	{Sir Thomas Parker, Lord Chief Justice; afterwards Earl of Macclesfield	March 20, 1712	Oct. 23, 1712
	July 27, 1758	{Lord Viscount Duplin; afterwards Earl of Kinoul		
Dec. 5, 1661	May 21, 1724	Robt. Lord Harley; afterwards Earl of Oxford		
	Nov. 30, 1731	Brook Taylor, LL.B.; afterwards LL. D.	.. ..	March 27
1676	1728	John Friend, M. D.	.. ..	April 3
	May 10, 1720	{Thomas Spratt, M. A. Archdeacon of Rochester; afterwards Bishop of Roffons	.. ..	March 27
1670	Dec. 19, 1746	{John Fortescue Aland, Esq.; afterwards Sir J. Fortescue Aland, Kt.	.. ..	May 1
	Sept. 24, 1729	Peter Le Neve, Esq. Norroy King of Arms	.. ..	March 27
		Mr. John Kempe		27
	July 4, 1744	Thomas Pellet, M. D.	.. ..	April 10
	March 1752	{Sign. Pietro Grimani, Venetian Ambassador; afterwards Doge of Venice	Oct. 23	Dec. 1

3

BORN.	DIED.		ELECTED.	ADMITTED.
	1742 or 3	Sign. Rinaldo de Duliolo, Med. Prof. Bonon.		
		Richard Richardson, D. M.		
	March 27, 1733	Richard Myddleton Massey, M.D.	.. ..	Feb. 19, 1712
		Thomas Rawlinson, Esq.	.. ..	19
		Thomas Bower, M. D.	.. ..	Nov. 13
	1742 or 3	Sign. Josephus Averinus, Prof. Jur. of Pisa		
	April 13, 1728	Samuel Molyneux, Esq.	Dec. 1, 1712	Dec. 1
Aug. 7, 1667	Jan. 1, 1748	Mons. Jean Bernoulli, Math. Prof. Basil.		
	Aug. 15, 1761	William Tempest, Esq.	.. ..	1
		Mr. Patrick Blair; afterwards M. D.		
	Nov. 4, 1732	{ Mr. Richard Bradley; afterwards Bot. Prof. Cant.	Dec. 1	1
	May 8, 1740	John Inglis, M. D.	.. ..	Jan. 22
		Sign. Joannes Antonius Count Baldini	Jan. 29	29
Dec. 13, 1662	March 2, 1729	{ Sign. Francesco Bianchini, signed Franc. Blanchinus	.. ..	29
1667	1723	Louis d'Aumont de Rochebaron Duc d'Aumont	May 21, 1713	May 21, 1713
		{ Mons. Pierre de Mellarede, Minister of the Duke of Savoy	June 11	
	1742 or 3	{ Mons. Kreienberg, Minister of the Elector of Hanover		
		{ Mons. Iver Baron Rosenkrantz, Minister of the King of Denmark		
		Richard Barrett, Esq.	.. ..	Jan. 14
	Dec. 9, 1719	Charles Oliphant, M. D.		
		Mr. George Tollett	.. ..	June 25
	1741	Mons. Daniel Ernest Jablouski		
	Jan. 20, 1760	Mr. John Colson; afterwards Luc. Prof. Cant.		
		Mr. William Brattle	March 11	
		Col. John Leveret		
		Edmond Turner, Esq.		
		Thomas Bromfield, M. D.	.. ..	March 25, 1714
	Sept. 10, 1719	Robert Keck, Esq.	.. ..	25
	1759	{ Mons. Nicholas Bernoulli, JU. M. D. Math. Professor Patav. dein Juris Prof. Basil		
		Prince Alexander de Mensicoff	July 29, 1714	
Oct. 29, 1690	June 28, 1754	Martin Folkes, Esq.; afterwards P. R. S.	.. ..	Nov. 11
1654	Dec. 22, 1722	Mons. Pierre Varignon		
	April 5, 1755	Richard Rawlinson, M. A.; afterwards LL. D.		
	March 1744	{ John Theophilus Desaguliers, M. A.; afterwards LL. D.	.. ..	Oct. 28
		{ John George Steigertahl, M. D. his Majesty's Physician	Nov. 11	Nov. 18
	July 1, 1730	Thomas Jett, Esq.	30	Dec. 2
	1742 or 3	Alexander Stuart, Esq.	.. ..	Nov. 30
		Mr. Thomas Watkins		30
		Thomas Hodges, Esq.	June 9, 1715	June 16, 1715
	1742	Mr. Wilhel. Jacobus Gravesende		16
March 23, 1638	Feb. 22, 1731	Mons. Frederick Ruysh, M. D.		
		Mons. Levinus Vincent, of Haarlem		
		Mons. Marpurgher		
	Dec. 7, 1733	{ George, Lord Newborough; afterwards Earl of Cholmondelley		30
July 14, 1671	Sept. 10, 1732	{ Mons. Jacques Eugene d'Allonville, Chevalier de Louville		
Aug. 8, 1685	March 9, 1752	Mons. Claude Joseph Geoffroy		
Oct. 27, 1678	Oct. 7, 1719	Mons. Pierre Remonde de Montmort		
		John Sherlock, Esq.		
		Mr. Pemberton		
		{ James, Lord Paisley; afterwards Earl of Abercorne	Nov. 10	Nov 17
		Sign. Nicolo Trou, Venetian Ambassador	.. ..	17
1657	1729	{ Michael Bernardus Valentini, Prof. of Medicine in the University of Giessen		
		Sign. Antoni, Comes de Comitibus of Venice	.. ..	17
		Mr. John Godfrey		
		John Moore, Esq.	.. ..	Nov. 24
		Mons. Justus van Effen	Nov. 30	Dec. 8
		{ Mons. Francois Wicardel, Chev. de Fleury, son to the Sicilian Ambassador		
1656	1713	Bruno Tozzi, Abbas Vallumbrosanus	Unknown	
		Henry Nicholson, M. D.	April 5, 1716	
	May 1774	John Churchill Wicksted, Esq.	.. ..	April 12, 1716
		Mr. ——— Cartwright	5	

e

BORN.	DIED.		ELECTED.	ADMITTED.
		Otto Christop. Count Volkra, Imperial Envoy	. .. .	April 19, 1716
		Mons. Le Baron d'Iscaut		
		Mons. le Marquis de Monte Leon		
	April 1, 1722	Sir Joseph Hodges, Bart.		
		Sign. Antonio Maria Salvini, of Florence		
	July 6, 1740	Claudius Amyand, Esq.	.. ..	April 19, 1717
	June 12, 1765	Col. John Guise; afterwards Lieut.-General	Nov. 30, 1716	Feb. 27
		Sign. Joseph Marquis d'Oroi, of Bologna		
		Mr. William Simon	.. ..	Jan. 10, 1716
	June 19, 1762	Mr. Robert Paul .. .. _	.. ..	Dec. 6
	Feb. 15, 1744	John Hadley, Esq.	March 21	March 28, 1717
		Samuel Scheurer, D. D. Professor of Bern ..	March 28, 1717	April 4
	June 25, 1744	Roger Gale, Esq.		11
		Dr. Joh. Aug. Hugo, his Majesty's Physican	Nov. 14	
	1750	{ Sign. Luigi Antonio Muratori, Library } keeper to the Grand Duke of Tuscany		
	March 22, 1750	{ James Jurin, M. D.; afterwards President } of the College of Physicians	.. ..	Nov. 21
		Mr. Henry Barham	.. ..	21
		Walter Jeffreys, Esq.	Nov. 30	30
		Edmond Littlehales, M. D.	.. ..	Dec. 12
		Orlando Gee, Esq.	.. ..	Nov. 30
	July 8, 1721	Elihu Yale, Esq.		
	Feb. 1, 1735	Robert Walsted, M. D.	March 3	March 20, 1718
	May 5, 1725	William Wagstaffe, M. D.	.. ..	March 27
1687	1765	William Stukeley, M. D.		
Sept. 7, 1677	Jan. 4, 1761	Stephen Hales, B. D.	.. ..	Nov. 20
	July 16, 1749	John, Duke of Montague		Feb. 15, 1721
		James Mickleton, Esq.	July 3, 1718	
	Oct. 22, 1729	John Whiteside, M. A.	.. ..	Oct. 30, 1718
		Sign. Ludovicus a Ripa		
		Caleb Colesworth, M. D.		
1692	July 13, 1762	{ James Bradley, M. A. since Professor of } Astronomy Oxford, and Ast. Royal	Nov. 6	
1682	April 21, 1739	{ Nicholas Sanderson, M. A. Professor Math. } Cambridge	.. ..	May 21, 1719
	Oct. 31, 1738	Mr. Robert Gay		
	1722	{ Sir Thomas Bury, Lord Chief Baron of the } Exchequer	Dec. 1	April 9
		Charles Cadogan, Esq. since Lord Cadogan ..	.. ..	Dec. 4, 1718
1688	May 23, 1737	John Conduitt, Esq.	.. ..	11
	Jan. 16, 1733	James Campbell, M. D.	.. ..	1
		William Stephens, M. D.	.. ..	11
		Samuel Crawys, Esq.	.. ..	11
	April 12, 1733	Mons. Thomas Fantet de Languy		
		Mr. Thomas Bates	Dec. 11	Jan. 8
		Mr. John Bamber, since M. D.	.. ..	22
		John Hollier, M. D.	.. ..	May 14, 1719
	July 13, 1737	Sir Wilfrid Lawson, Bart.	.. ..	Jan 15, 1718
		Mr. William Beckett	Dec. 11	Dec. 18
July 8, 1686	April 22, 1758	Mons Antoine de Jussieu, M. D.		
	1742 or 3	Sign. Pietro Antonio Michaellotti, of Venice		
		Mons. Lenck, of Leipsic		
		Mr. John Bellers	Feb. 5	Feb. 12, 1719
	Feb. 2, 1768	{ Robert Smith, M. A. Plum. Prof. Astron. } Camb. since LL. D.	.. ..	May 21, 1718
1688	1743	Mons. John George Keyssler, of Hanover ..	.. ..	Feb. 12, 1719
	Feb. 22, 1736	Sir John Meres, Bart.	March 12	May 14
		Mr. Dalrymple		
	Dec. 14, 1751	Lord James Cavendish	April 9, 1719	April 16
		Mr. John Bushby	.. ..	16
		James Hill, Esq.	.. ..	30
	1743	Mr. Isaac Rand	Nov. 5	Dec. 3
	April 9, 1742	Moses Williams, M. A.	.. ..	May 11, 1721
		John Strackey, Esq.	.. ..	Dec. 10, 1719
		{ Mons. Albert Henri de Salengre, Auditor } Surveyor of the Bank of Holland		
	1746	{ Mr. Colin M'Laurin, Prof. Math. Aber- } deen; afterwards Prof. Math. Edinburgh	.. ..	Nov. 5
		John Georges, Esq.	Nov. 30	30
	Sept. 17, 1730	Charles Bale, M. D.		
		Oliver St. John, Esq.	Feb. 25	March 3
		George Stanley, Esq.	.. ..	17
		Charles Stuart, M. D.	.. ..	3
	Aug. 12, 1728	William Sherrard, LL. D.	.. ..	April 7, 1720

BORN.	DIED.		ELECTED.	ADMITTED.
	Feb. 25, 1720	Jeffrey Palmer, Esq.		March 10, 1719
	Dec. 8, 1744	Abel Ketelbey, Esq.	Feb. 25, 1719	10
	March 1759	{ Richard Manningham, M. D. since Sir R. Manningham, Kt. }	March 10	24
	Ejected June 9, 1757	John Warburton, Esq.	.. ..	24
		William Mathew, Esq.	.. ..	March 10, 1720
		Sign. J. Baptista Recanati	June 30, 1720	
	Oct. 31, 1734	William, Lord North and Grey	March 31	April 7
	Oct. 27, 1727	{ Henry Heathcote, Esq. son of Sir Gilbert Heathcote }	June 30	July 7
1690	June 29. 1774	{ Zachary Pearce, M. A. since D. D. and Bishop of Rochester }	.. ..	7
	Ejected June 9, 1757	{ Alexander Cuming, Esq. since Sir Alexander Comyns, Bart. }	.. ..	7
	June 10, 1730	William Rutty, M. A.	.. ..	Oct. 27
	Feb. 26, 1762	David Papillon, Esq.	.. ..	Nov. 30
		Samuel Sanders, Esq.	Nov. 3	Jan. 25, 1721
		Mr. Henry Beighton	.. ..	March 2
	1739	Sir Thomas Dereham, Bart.	.. ..	June 8
		Peter Des Maizeaux, Esq.	.. ..	Nov. 10, 1720
		Mons. Heuricus Hoffman	Nov. 30	
	June 25, 1743	Mr. John Douglas, Surgeon	.. ..	Dec. 8
	April 9, 1771	{ Henry Pemberton, M. D.; afterwards Med. Prof. Gresh. }	.. ..	8
1675	Nov. 16, 1751	Mr. George Graham	March 9	March 16
	March 14, 1740	{ Conrad Sprengell, M. D. since Sir C. Sprengell, Kt. }	.. ..	23
		Mr. William East	.. ..	16
	Sept. 26, 1728	Richard Hale, M. D.		
	April 9, 1726	Sir Th. Hewett, Kt.	Nov. 2, 1721	Nov. 30, 1721
	Sept. 16, 1743	Sir George Savile, Bart.	.. ..	Jan. 11
		Paul Dudley, Esq.	Nov. 2	
1682	1758	William Barrowby, M. D.	.. ..	Nov. 9
		John Beale, M. L.	.. ..	23
	Jan. 1734	John Wolhouse, Esq.		
	Aug. 12, 1729	William Western, Esq.	Nov. 30	Feb. 1
	June 8, 1735	Mr. John Brown	.. ..	Dec. 7
	Oct. 27, 1731	Talbot, Earl of Sussex	Feb. 1	Feb. 15
	Dec. 9, 1767	Thomas Miles, M. A.	March 15	May 3, 1722
Dec. 9, 1684	Nov. 10, 1751	Mons. Abram Vater		
		Dr. Musgrave		
	1736	Thomas Slater Bacon, Esq.	.. ..	May 3
		Richard Lucas, M. A.	.. ..	March 29, 1721
	Oct. 13, 1758	Richard Molesworth, Esq.; afterwards Lord	.. ..	March 15
	Feb. 15, 1767	William Sloane, jun. Esq.	May 24, 1722	May 31, 1722
	Dec. 21, 1733	Samuel Harris, M. A.	Nov. 1	Feb. 28
		Samuel Morland, M. A.	May 24	Jan. 10
	July 6, 1766	Charles Taylor, Esq.	Nov. 1	Dec. 13
	March 17, 1764	{ George, Lord Parker, since Earl of Macclesfield, and P. R. S. }	Oct. 25	Nov. 8
April 25, 1695	1753	Richard, Earl of Burlington	25	
	Aug. 25, 1747	Mr. Ambrose Dikins, Sergeant Surgeon ..	Nov. 1	Nov. 8
		Robert Hucks, Esq.	1	8
		Mons. Philip Julius Bornemann	.. ..	Nov. 8, 1723
May 8, 1670	May 11, 1726	Charles, Duke of St. Albans	.. ..	30
Feb. 25, 1682	Dec. 5, 1771	Sign. Jean Baptista Morgagni	30	
1698	1778	Charles, Duke of Queensborough	8	
1654	Dec. 25, 1732	William, Earl of Yarmouth	30	
	July 14, 1763	Sir John Evelyn, Bart.	Jan. 17	Feb. 28, 1722
	Oct. 11, 1769	John White, Esq.	.. ..	Jan. 24
		Nicola Alerbe d'Aragona, Prince de Cassano	March 21	
		Or. Le Duc		
		Gebhard D'Anteny, Esq.	May 2, 1723	May 23, 1723
		West Fenton, Esq.	.. ..	Nov. 21
		Simon Degge, Esq.	.. ..	May 16
1673	1742	{ John Armstrong, Esq.; afterwards Surveyor general of the Ordnance }	.. ..	23
		Philip Glover, Esq.	.. ..	9
1677	May 13, 1734	Sir James Thornill, Kt.	May 9	April 2, 1724
	Nov. 6, 1728	Hewer Edgley Hewer, Esq.	June 27	Feb. 27, 1723
	Dec. 27, 1774	Francis Wollaston, Esq.	.. ..	July 4
Sept. $\frac{8}{19}$ 1697	July 10, 1767	{ Mr. Alexander Munro; afterwards M. D. Prof. Anat. Edinburgh }		
	1743	Isaac de Sequeyra Samuda, M. D.	.. ..	Oct. 24, 1724

e 2

BORN.	DIED.		ELECTED.	ADMITTED.
		Henry Jones, M. A.		June 18, 1724
	May 21, 1744	Dr. Dominico Ferrari	Nov. 14, 1723	
		Dr. Antonius Deidier, of Montpellier		
	1742 or 3	Sign. Conte Giulio Carlo de Fagnini	30	
		Ralph Ord, Esq.		Dec. 12, 1723
	1726	Gilbert Burnet, M. A.		19
	Oct. 17, 1758	Mr. John Ward, Rhet. Prof. Gresh. LL. D.		
		Benjamin Holloway, LL. B.		Nov. 4, 1725
	Nov. 28, 1724	Robert, Lord Romney	21	Dec. 19, 1723
		Mr. John Meres	March 12	June 18, 1724
		Mr. John Byron		March 19, 1723
	March 22, 1761	Edward Vernon, M. A.; afterwards D. D.		April 2, 1724
	Jan. 17, 1761	{ Anthony Ellys, M. A.; afterwards Bishop of St. Davids		2
	Feb. 12, 1778	{ Robert Ord, Esq.; afterwards Lord Chief Baron, Scotland		March 19, 1723
1701	Aug. 8, 1750	Charles, Duke of Richmond	Feb. 6	5
	Aug. 27, 1760	Smart Lethieullier, Esq.	March 12	19
		Mons. Nicolaus Cruquius		
April 4, 1688	Sept. 12, 1768	Mons. Joseph Nicolas de l'Isle		
	Jan. 1, 1735	John Kendall, Esq.		April 2, 1724
	Jan. 3, 1731	Mr. John Dobyns, Surgeon		June 25, 1723
	March 1750-1	{ Francis, Earl of Dalkeith, since Duke of Buccleugh		March 19
May 14, 1686	Sept. 16, 1786	Mons. Daniel Gabriel Fahrenheit	May 7, 1724	May 14, 1724
1702	April 10, 1729	John Gaspar Scheuchzer, M. D.		14
		Ralph Leicester, Esq.		
Dec. 22, 1684	April 5, 1747	{ Mons. Joannes Jacob. Dillenius, M.D. Prof. Bot. Oxon	June 25	June 25
		Mons. Joannes Adolphus Jacobeus		Oct. 22
	June 29, 1744	Mr. John Eames		Nov. 19
	Oct. 26, 1753	Joseph Danvers, Esq. since Sir J. Danvers, Bart.	Nov. 30	Dec. 10
	March 16, 1732	Sir Littleton Powis		
	Aug. 28, 1773	Mr. John Ranby, since Chr. Reg.		10
		Stephen Chase, M. D.		10
	May 23, 1727	John Diodate, M. D.		10
		Don Antonio Galvaon, Envoy from Portugal	April 15, 1725	April 29, 1725
	Sept. 20, 1758	Thomas Hill, Esq.	15	May 13
	May 27, 1761	Robert Nesbitt, M. D.		April 22
		Nathan Hickman, M. A.; afterwards M. D.		May 6
		Mr. Roby of New England		
		Richard Poley, Esq.		April 22, 1728
		Mr. Thomas Hunt		May 2, 1725
	Withdrawn } 1742 or 3 }	Mr. Edmond Stone		April 22
		Mons. Andreas Henry de Cronhelm		22
		Sign. Maurice Antonio Capeller		
1681	Sept. 16, 1769	Taylor White, Esq.	April 29	May 6
	March 26, 1742	George Lewis Teissier, since M. D.	Nov. 4	Nov. 18
	May 22, 1742	James Theobald, Esq.		25
	Feb. 20, 1759	Charles de la Faye, Esq.		25
	Dec. 11, 1762	Mons. Gaspar Newman	Dec. 9	
	May 15, 1734	Robert Houstoun, M. D.		Feb. 10
		Mr. Silvanus Bevan		Dec. 16
		{ Sidi Mohommed Ben Ali Abgali, Ambassador from the Emperor of Morocco	March 24, 1726	March 31, 1726
	March 16, 1770	Richard Hassell, Esq.	May 12	May 12
	1728	Sir Brook Bridges, Bart.		Nov. 30
		John Jeffreys, Esq.		May 19
		Edmund Allen, Esq.		12
	March 25, 1752	Temple Stanyan, Esq.		June 30
	March 20, 1742 or 3 }	Kingsmill Eyre, Esq.		May 26
	June 30, 1743	Robert Johnston Ketelby, Esq.		19
	1742 or 3	Henry Walther Gerdes, D. D. of Denmark	May 12	19
		Thomas Palmer, Esq.		
	Oct. 14, 1726	{ Sir Jeffrey Gilbert, Lord Chief Baron of the Exchequer		
		Richard Beard, M. D.	May 26	May 16, 1728
	March 17, 1760	Charles Stanhope, Esq.	July 7	Oct. 27, 1726
	Feb. 23, 1774	James Hargrave, M. A.		
		Dr. Zabdiel Boylston		July 7
	1763	Edward Rudge, Esq.	Nov. 3	Nov. 10
	April 1768	Edward Pawlet, Esq.		10
	Withdrawn 1754	Mr. James Stirling		Dec.

BORN.	DIED.		ELECTED.	ADMITTED.
1698	Sept. 10, 1761	Mons. Bernard Forrest de Belidor		
	May 8, 1749	Richard Graham, Jun. Esq.	Nov. 17, 1726	Nov. 24, 1726
	Nov. 14, 1754	Mr. Thomas White	.. ..	Jan. 12
		Richard Holland, M. D.	30	26
	March 4, 1761	Dr. Meyer Schamberg	.. ..	12
		{ William Billiers, Esq.; afterwards Sir W. } { Billiers, Kt.	.. ..	Dec. 22
	March 3, 1777	{ Thomas Robinson, Esq.; afterwards Sir T. } { Robinson, Bart.	.. ..	Feb. 2, 1727
	March 23, 1753	{ Hon. Thomas Trevor, Esq.; afterwards Ld. } { Trevor	March 9	April 13, 1726
		Joseph Andrews, Esq.	.. ..	March 16
		Mr. Cyriacus Ahlers	.. ..	16
	May 10, 1739	John Hollings, M. D.	.. ..	16
		Mons le Baron de Mamsbergh		
	Feb. 1, 1755	John Fuller, Jun. Esq.	.. ..	16
	Aug. 12, 1757	Benjamin Hoadley, Esq.; afterwards M. D.	.. ..	16
	July 2, 1772	James West, Esq.; afterwards P. R. S. ..	March 9	16
	Jan. 13, 1734	E. Hughes, Esq.	.. ..	Nov. 23, 1727
	April 9, 1773	William Ffolkes, Esq.	.. ..	April 20
Sept. 12, 1699	Jan. 29, 1768	{ Mr. John Martyn; afterwards Prof. Bot. } { Cambridge	March 30, 1727	March 30
	June 1735	Mr. John Harper	9, 1726	30
		Dr. Cyrillus, Prof. of Physic, at Naples		
		Charles Lamotte, M. A.	June 22, 1727	Feb. 15
		Philip Henry Zollman, Esq.	.. ..	Jan. 18
	1757	Wa ter Cary, Esq.		
Aug. 17, 1699	Nov. 6, 1777	Dr. Bernard de Jussieu		
	April 25, 1772	Theodore Jacobson, Esq.	.. ..	Oct. 26
	May 16, 1742	William Carr, Esq.		
		Francis Clifton, M. D.	.. ..	June 29
	April 28, 1783	Lord Charles Cavendish	June 8	22
		Philemon Lloyd, Esq.	Nov. 9	
	Feb. 12, 1752	Mr. Benjamin Robins	.. ..	Nov. 16
	May 23, 1734	Sir Robert Pye, Bart.	Jan. 11	Jan. 18
	Oct. 14, 1734	Hon. Henry Colepeper Fairfax, Esq.		
		Lord Hope, Son of the Earl of Hopetoun ..	Feb. 8	April 23, 1730
	Oct. 15, 1743	Erasmus Philips, Esq.; afterwards Bart. ..	15	Feb. 8, 1727
	1727	King George I., Patron	May 9	
		George, Prince of Wales	15	
Nov. 10, 1633	Oct. 25, 1760	King George II., Patron	July 11	
	Nov. 22, 1751	Mr. Stephen Horseman	Feb. 15	Feb. 29, 1728
		Mr. Thomas Pocock, M. A.	22	Nov. 21
	March 1768	Wm. Hanbury, Esq.	May 2, 1728	June 20
		Mr. William Dugood	.. ..	May 2
		Mr. Robert Gray	.. ..	9
	Jan. 14, 1778	Frank Nicholls, M. D. Anat. Prof. Oxon ..	.. ..	16
	May 9, 1767	Samuel Clarke, Esq.	.. ..	16
		Sign. Jac. Barthol. Beccari, of Bologna ..	June 27	
	May 16, 1770	Granville Wheller, Esq.	.. ..	Oct. 17
	1756	Mons. Chs. Freder. Weichman		
Sept. 8, 1694	Aug. 7, 1771	{ Mons. John Daniel Schoepflin, Prof. of His- } { tory in the University of Strasburg		
Sept. 28, 1698	July 27, 1759	Mons. Pierre Louis Moreau de Maupertuis		
	Dec. 30, 1740	Mr. John Senex	.. ..	July 4
1692	Nov. 3, 1729	{ Ericus Burman, Prof. Ast. in the University } { of Upsal		
	Jan. 7, 1752	Cromwell Mortimer, M. D. since Sec. R. S.	.. ..	July 4
1705	Feb. 28, 1779	{ Mr. Adrianus Van Royen, M. D. Prof. } { Bot. Leyd.	.. ..	June 27
		Mr. Paul de Saint Hyacinthe, of Paris	Oct. 24	Oct. 31
1665	May 1, 1736	Mr. Albertus Seba ..		
	June 13, 1761	Mr. Edward Nourse	.. ..	31
Sept. 20, 1674	Feb. 15, 1739	Sign. Eustachio Manfredi, of Bologna		
	July 22, 1734	Peter, Lord King, Lord Chancellor	Nov. 14	
	Sept. 27, 1764	John Trevor, Esq.; afterwards Lord Trevor	.. ..	Nov. 21
1694	Aug. 11, 1768	Mr. Peter Collinson	Dec. 5	Dec. 12
		Mr. Joh. Geo. Liebknecht, of Giessen		
		Mr. Samuel Palmer		12
		Cassem Aga Algiada, Ambassador from Tripoly	.. ..	Jan. 9
Jan. 31, 1707	March 31; 1751	Frederick, Prince of Wales	17	
	Ejected } June 9, 1757 }	John, Lord Delawar; afterwards Earl	19	Nov. 26, 1730
	Dec. 21, 1754	Thomas, Lord Viscount Gage	.. ..	Nov. 25, 1731
April 2, 1697	July 21, 1773	Mons. Sauveur Francois Morand	Jan. 9	May 8, 1729

BORN.	DIED.		ELECTED.	ADMITTED.
		{ Mons. René Jacques de Grangeot, Surgeon } at Paris		
	Jan. 16, 1763	Mr. David Durand	Jan 16, 1728	Jan. 23, 1728
1677	May 12, 1742	Abbe Joseph Privat de Molieres		
	June 1768	George Heathcote, Esq.	.. ..	Nov. 26, 1730
1679	May 19, 1763	Dr. Theophilus Lobb	March 6	March 13, 1728
	Dec. 29, 1783	Daniel Wray, Esq.	.. ..	13
Aug. 15, 1685	Feb. 27, 1759	Mons. Jacobus Theodorus Klein, of Danzig		
		Joseph Atwell, M. A.; afterwards D. D. ..	March 20, 1729	March 27, 1729
	Jan. 6, 1730	Robert Mather, Esq.	May 8	May 22
Sept. 14, 1698	July 16, 1739	Mons. Charles François de Cisternay du Fay	.. ..	March 31, 1737
		Mr. John Horsley	.. ..	April 23, 1730
1681	Dec. 16, 1770	Roger Long, D. D.	.. ..	May 15, 1729
		Hon. James d'Arcy, heir to Lord d'Arcy ..	June 5	June 12
		Geo. Carpenter, Esq. since Lord G. Carpenter	.. ..	Nov. 20
		Mr. John Swinton, M. A.; afterwards B. D.	Oct. 16	
1683	Aug. 5, 1754	Mr. James Gibbs, Architect	.. ..	Nov. 20, 1733
	Oct. 3, 1755	Sir John Clerk, Bart.	.. ..	May 24
		John Guilielm Abruz, M. D. of Hanover ..	Nov. 6	
		D—— Rolserius	Oct 16	
		Mr. Peter Kinch, M. A.		
	Nov. 7, 1756	Mr. John Freke	Nov. 6	
	March 12, 1737	Mr. Wm. Green, Surgeon	.. ..	Nov. 6, 1729
		{ Conr. D. Alev à Dehn, Count of the Roman } Empire	6	
		Frederic de Thom, Legatus Brunsvico Guelferibyt	.. ..	Nov. 20
		Father Joanne Baptista Carbone, at Lisbon		
March 30, 1664	April 20, 1750	D. Jean Lewis Petit		
	May 15, 1740	Mr. Ephraim Chambers		
		{ Joannes Henricus Heucherus, M. D. Prof. of Anat. and Bot. at the University of Wittenberg }		
		Andreas Michael Ramsay, Eques. Sti Lazari	Dec. 11	Dec 18
		Dr. Paul Antonio Rolli	.. ..	11
	April 17, 1732	Jonathan Gouldsmyth, M. D.	Jan. 15	Jan. 29
Oct. 29, 1693	Nov. 22, 1786	{ Edward Wilmot, M. D.; afterwards Bt. and Med. Reg. }	.. ..	29
		Littleton Brown, M. A.	.. ..	22
	Aug. 4, 1749	Henry, Lord. Colerane	Jan. 8	Feb. 19
	March 17, 1764	William Oliver, M. D.	22	
	Sept. 17, 1764	Thomas Walker, M. A.; afterwards LL. D.	.. ..	Jan. 29
	Dec. 18, 1771	Mr. Philip Miller	.. ..	29
		Mr. Ambrose Godfrey Hanckewitz	.. ..	Feb. 5
	Sept. 14, 1762	Jacob. de Castro Sarmento, M. D.	Feb. 5	12
		D. Joh. Sigism. August. Frobenius, M.D.	.. ..	12
		D. Joannes Henricus Hampe, M. D.	.. ..	12
1704	Jan. 5, 1780	Robert More, Esq.	.. ..	19
	Feb. 10, 1755	Mons. de Montesquieu	26	
	Aug. 7, 1754	Pierc Dod, M. D.	.. ..	March 19
		Wm. Bogdani, Esq.		
	April 2, 1783	Wm. Dixon, Esq.		
		Mr. Melchior de Ruischer		
Aug. 18, 1698	Oct. 26, 1765	{ Samuel Klingenstierna, Prof. Math. at the University of Upsal }	April 23, 1730	
	Jan. 11, 1757	Pere Ludovic. Castell, S. P.		May 21, 1730
		H. Berenger de Beaufarn	.. ..	7
		Henry Dry .. .. .. —	30	7
		Arch. Patoun	.. ..	
Dec. 31, 1668	Sept. 23, 1738	Herman Boerhaave, M. D.		
		Wm. Græme, M. D.	.. ..	May 7, 1731
		Simon Degge, Esq.	May 14	Dec. 9
		Thomas Anson, Esq.		
	Ejected June 30, 1757	James Justice Esq.	Oct. 22	
		John Allen, M. D.	.. ..	Oct. 26, 1732
		Laurent Garcin, M. D.	.. ..	May 27, 1731
	Nov. 23, 1731	Jeremiah Cray, Esq.	.. ..	13
		Joseph Banks, Esq.	Dec. 10	Feb. 25, 1730
		Colin Campbell, Esq.	.. ..	Nov. 7, 1734
		Mr. George Campbell	.. ..	Jan. 7, 1730
Oct. 18, 1705	Oct. 22, 1783	Mr. Ger. Fred. Muller		
Sept. 19, 1683	April 18, 1758	Dr. Lawrence Heister	17	
		Dr. John Amman	March 18	March 25, 1731
1698	Nov. 29, 1782	{ Hon. Coote Molesworth, M. D. Son of Viscount Molesworth }	.. ..	25

BORN.	DIED.		ELECTED.	ADMITTED.
	June 1, 1732	Hon. Benedict Leonard Calvert, Brother to Lord Baltimore	March 25, 1731	
	April 19, 1781	Sir Joseph Ayloffe, Bart.	May 27	
		Dr. Hieron Giuntini, of Florence		
		Hon. Wm. Count Bentinck		
Nov. 1691	Aug. 8, 1747	Mr. Martin Friewald	July 1	
	July 2, 1742	Robt. James Lord Petre	Oct. 28	Nov. 4, 1731
		Wm. Fellows, Esq.	Nov. 4	Dec. 9
Dec. 8, 1708	Aug. 18, 1765	His Royal Highness Francis Duke of Lorrain; afterwards Emperor of Germany	18	Nov. 25
1700	Jan. 12, 1749	Philip, Count Kinsky, his Imperial Majesty's Ambassador	.. ..	25
		Charles, Lord Baltimore	Dec. 9	Jan. 27
Aug. 12, 1711	Jan. 22, 1767	Hon Benjamin Bathurst, eldest Son of Lord Bathurst	.. ..	March 2
		Mons. Lewis de la Nauze, of the Academy des Belles Lettres, at Paris	Jan. 27	
		Edward Barry, M. D.; afterwards Bart. ..	Feb. 3	May 17, 1759
July 25, 1700	Sept. 7, 1776	Mr. Jacob Serenius; afterwards D. D. and Bishop of Strengnes, in Sweden	.. ..	Feb. 17, 1731
		Sir James Edwards, Bart.	.. ..	March 2, 1732
	Oct. 6, 1765	Robt. Barker, M. D.	March 9, 1732	March 30, 1731
	July 15, 1756	Thomas Lee Dummer, Esq.	.. ..	March 16
	July 17, 1769	John Robartes, Esq.; afterwards E. of Radnor	16	23
		Mr. John Gray	.. ..	23
		Baron Pfüitschner, Director of Experimental Philosophy, at Nancy, and in the University of Ponta Mousson	23	
Nov. 24, 1680	Nov. 30, 1742	Baron Joannes Baptista Bassand, his Imperial Majesty's Physician		
		Fayrer Hall, Esq.	30	April 6
	May 7, 1777	Rose Fuller, M. D.	April 20	May 4
		Jean Patrice Piers de Girardin	27	11
	Sept. 18, 1764	William, Earl Cowper	May 11	25
	1735	James, Earl of Stathmore	.. ..	25
	Dec. 24, 1749	John, Earl of Craufurd	June 15	June 29
	March 12, 1737	Wm. Fullerton, M. D.	22	29
		John Frederick Weidler, Prof. Math. at Wittenburg	Nov. 9	
	1739	Mr. Vincent Bacon, Surgeon	9	Nov. 16
		Conde de Montigo, Spanish Embassador ..	16	
May 2/13, 1706	Feb. 6, 1785	Mr. John Belchier, Surgeon	23	Dec. 7
Feb. 11, 1657	Jan. 9, 1757	Mons. Bernard le Bovier de Fontenelle	Jan. 18	
	July 14, 1733	Wm. Houstoun, M. D.		
		Mr. Stephen Gray	25	March 15
		Sign. Carol. Taglini, Prof. of Philosophy, in the University of Pisa	Feb. 1	
	1744	William, Lord Viscount Bateman	22	Dec. 20, 1733
	Nov. 7, 1759	Edward Hody, M. D.	March 8	March 22, 1732
		George Douglas, M. D.	15	22
		Mr. Wm. Maitland	April 12, 1733	April 19, 1733
1707	Oct. 12, 1768	James, Lord Aberdour; afterwards Earl of Morton, and P. R. S.	19	26
	1749	Mr. Mark Catesby	26	May 3
	June 25, 1767	Godfrey Sellius, of Danzig, LL. D.	May 3	
		Charles Frederick, Esq.; afterwards K. B. ..	24	31
		Christianus Ludovicus Gersten, from Giessen in Hessem Darmstadt	Oct. 25	Nov. 1
1678	Dec. 1, 1750	Dr. John Gabriel Doppelmayer, Prof. Ast. at Nürenberg	Dec. 6	
	Ejected June 9, 1757	Henry David, Lord Cardross; afterwards Earl of Buchan	Jan. 10	Jan. 24
	April 18, 1757	Mons. l'Abbe Joan de la Grive, a Paris ..	24	
		Sign. Ludovicus a Ripa, Prof. Ast. a Padua		
		Father Joannes Criv lli, at Venice		
Feb. 24, 1701	Dec. 10, 1742	Francis Joseph Hunauld, M. D.	Feb. 14	
Dec. 1, 1711	Oct. 22, 1751	His Highness Wm. Prince of Orange	March 7	March 7
	April 13, 1758	Mr. Alex. Ouchterlony, Merchant	7	21
		John Winthorp, Esq. of New England	April 4, 1734	April 25, 1734
	July 30, 1757	Edward Harrington, M. D.		
	Nov. 29, 1759	Mr. Browne Langrish, Surgeon; afterwards M. D.	May 16	
		Mons. Le Protti, first Physician to the Pope		
		Thomas Shaw, D. D.	June 13	Oct. 31
		Father Didacus de Revillas, at Rome		

1

BORN.	DIED.		ELECTED.	ADMITTED.
		Stephen Williams, M. D.		
	Dec. 8, 1761	Mynheer Hop, Envoy from the States General	Oct. 24, 1734	
	Nov. 19, 1763	James Spilman, Esq.	31	Nov. 14, 1734
		{ John Stevens, Esq. Surgeon to the Prince of } Wales		Oct. 31
		George Peter Domcke, of Hall in Saxony ..	Nov. 7	Nov. 21
Sept. 19, 1662	March 14, 1743	Abbe Jean Paul Bignon		
	Sept. 19, 1761	{ Petrus Van Muschenbrock, Prof. Math. in } the University of Utrecht	14	
		John Hamilton, Esq.	21	21
	Feb. 10, 1740	John Lord King	Jan. 9	May 1, 1735
1678	Feb. 20, 1771	Jean Jacques d'Outous de Mairan	23	
1700	Aug. 23, 1782	Henricus Lodovicus du Hamel du Monceau		
Nov. 19, 1700	April 24, 1770	Abbe Jean Antoine Nollet		
		{ Monsignor Celestina Galliani, Archbishop of } the Salonica and Cappelan Major of the kingdom of Naples		
		Mr. Peter Sainthill, Surgeon	Feb. 6	Feb. 20, 1734
	Dec. 12, 1781	Mr. John Chandler	.. ..	6
	Dec. 27, 1753	Sir Marmaduke Wyvill, Bart.		13
		George Hadley, Esq	Feb. 6	27
Nov. 23, 1705	Jan. 9, 1766	{ Thos. Birch, M. A. ; afterwards D. D. and } S. R. S.	20	27
	Nov. 18, 1766	Thomas Lord Southwell	March 13, 1735	March 27, 1735
		{ Thomas, Lord Lovell; afterwards Earl of } Leicester	27	
	Feb. 20, 1750	Wm. Freeman, Esq.	.. ..	April 17
Feb. 28, 1704	Sept. 11, 1760	Mons. Louis Godin		
	May 19, 1751	Martin Clare, M. A.		24
		Samuel Clarke, Esq.	April 17	24
	Feb. 27, 1754	Lord Viscount Tyrconnel	.. ..	17
		Mr. Moreton Gilkes, Surgeon	24	Jan. 26, 1743
		Alvaro Lopes Suasso, Esq.	.. ..	May 8, 1735
	Nov. 18, 1738	George Tilson, Esq	May 22	
July 2, 1698	Feb. 22, 1780	{ Francois Marie de Este, Prince of Modena ; } afterwards Duke of Modena	Nov. 6	Nov. 6
Aug. 15, 1714	March 7, 1786	Philip, Earl Stanhope	.. ..	Jan. 22
Nov. 27, 1701	April 25, 1744	{ Andreas Celsius, Prof. Ast. at the University } of Upsal	Jan. 29	29
	Withdrawn May 10, 1751	Mr. Jonathan Fawconer, Lapidary	Feb. 5	Feb. 12
	June 1, 1757	Antonio Cocchi, M. D. Prof. Anat. at Florence		
		{ Don Diego de Mendoca Corte Real, one of } his Portuguese Majesty's Privy Council		
		{ Jacobus Iattica, Physician to the Prince of } Modena	12	
		Sir Daniel Molyneaux, Bart	19	26
	Sept. 1747	Hon. Edw. Legge, Son to the Earl of Dartmouth	.. ..	March 4
	July 26, 1767	{ Paul Gottlieb Werlhof, M. D. Physician to } his Majesty's Household at Hanover		
	Jan. 28, 1771	Andrew Mitchell, Esq.; afterwards K. B.	.. ..	4
		Mr. Benjamin Cooke, Surgeon	March 11	
	July 7, 1755	Sir John de Lange, Kt.	.. ..	April 15, 1736
		Roger Jones, M. D.	March 11, 1736	1
	Aug. 30, 1757	David Hartley, M. A.	25	1
	March 2, 1742	James, Duke of Hamilton	April 8	
		{ Marco Antonio de Azevedo Coutinho, Ple- } nipotentiary from Portugal	May 6	July 1, 1737
	Feb. 10, 1762	Henry Hilsall, Esq.	27	Dec. 15, 1736
	{ Withdrawn 1769, died 1770 }	Mr. Francis Drake, Surgeon	June 10	July 1
	June 6, 1786	{ Sir Hugh Smithson, Bart.; afterwards Duke } of Northumberland	10	Jan. 27
Nov. 28, 1686		{ John Philip Seip, M. D. Physician to the } Prince of Waldeck	1	
		Sign. Francesco Algarotti, of Venice	8	April 19, 1739
		Robt. Bankes, Anatomy Prof. Cantab. ..	Nov. 11	
	Jan. 2, 1755	Sir James Lowther, Bart.	25	Jan 27, 1736
		{ Sign. Cervi, M. D. President of the Royal } Academy of Physic in Spain	Dec. 9	
	Feb. 11, 1755	The Marquis Scipio Maffei, of Verona	23	
	May 7, 1770	Moses da Costa, Esq.	Feb. 10	Feb. 17
	Dec. 5, 1783	George Lloyd, Esq.	10	
	June 14, 1768	Mr. James Short	March 24, 1737	April 27, 1738

5

BORN.	DIED.		ELECTED.	ADMITTED.
	Nov. 1782	{ James Barrow, Esq.; afterwards Sir James, and P. R. S. }	April 7, 1737	April 21, 1737
	Feb. 20, 1770	Captain Christopher Middleton	7	21
	Sept. 27, 1743	Henry Popple, Esq.	28	May 12
	Dec. 7, 1780	George Lewis Scott, Esq.	May 5	12
	May 15, 1762	Robert Taylor, M. D.	June 23	Dec. 7, 1752
May 13, 1713	May 17, 1765	Mons. Alexis Claude Claraut	Oct. 27	Nov. 23, 1737
	Jan. 8, 1742¾	William, Earl of Essex	Nov. 17	March 2
		{ Stephanus Evodius Assemanni, Archbishop of Apamea }	Jan. 19	
		Edward Smith, Esq.	19	
		John Peter Bernard, M. A. at Leyden	19	Jan. 26
		Thomas Stack, M. D.	26	26
April 26, 1704	June 2, 1780	{ Mr. Petrus Filenius, Prof. of the Oriental Languages in the University of Abo; afterwards D. D. and Bishop of Linkeping in Sweden }	26	April 20, 1738
	Nov. 24, 1753	Nicholas Mann, Esq.	26	
1704	Oct. 5, 1777	{ Dr. John Andrew Segner, Prof. Math. at the University of Gottingen }	26	
	July 22, 1747	Edward, Earl of Darnley	Feb. 9	Feb. 16, 1737
1705	April 27, 1782	John, Earl of Loudoun	9	16
		Capt. Wm. Walker	9	June 15, 1738
	March 30, 1742	Addison Hutton, M. D.	9	Feb. 23, 1737
May 16, 1698	Jan. 24, 1771	Lewis Way, Esq.	23	March 16
		{ Mr. Philip Naude, Prof. Math. in the Royal College of Joachimsthal (Berlin) }	23	
1680	Dec. 29, 1780	Gilbert Kennedy, M. D.	23	March 31, 1748
	Sept. 9, 1745	James Gambier, Esq.	March 23, 1738	April 20, 1738
	1782	Mr. John Maude	April 20	27
		Dr. Antonius Franciscus Gorius, Florentinus	May 4	
		Rev. Mr. Matthias Belius, of Hungary	June 1	
April 2, 1711	March 7, 1775	Job Baster, M. D.	1	
		Andrew Cantwell, M. D. of Paris	1	
		Mr. Richard Davies; afterwards M. D. ..	8	June 15
		Joseph Rogers, M. D.	15	
	1772	Mr. John Ellicott, Watchmaker	Oct. 26	Nov. 2
		Herman Bernard, M. D. of Berlin	Nov. 2	16
		{ Don Francisco Xavier de Menezes, Count da Erieeyra, Director of the Royal Academy of History in Portugal }	2	
1688	Oct. 17, 1757	Dr. Renalus Antonius Ferchault de Reaumur	9	
		{ Dr. Stephanus de Fourmont, Prof. Regius Linguæ Arabicæ at Paris }	9	
		{ Wm. Browne, M. D.; afterwards Sir Wm. Browne, Kt. }	Feb. 15	March 1
		{ Mons. Francois Xavier de Bon, Marquis de St. Hilaire, President of the Royal Society of Sciences at Montpellier }	15	
	Oct. 11, 1776	Samuel Mead, Esq.	March 8	15
	April 9, 1747	John Myddleton, Esq.	8	22
	Aug. 12, 1768	John Huxham, M. D.	April 5, 1739	
Nov. 20, 1715	April 2, 1799	Mons. le Monnier, of Paris (Astronomer) ..	5	June 22, 1748
		Henry Rowe, Esq.	12	
		{ Dr. Joseph de Montagny, Prof. of Metaphysics in the Academy of Lausanne }	June 7	
June 21, 1703	Dec. 11, 1780	Mons. Joseph Lieutaud, M. D.	21	
	April 30, 1762	John Sawbridge, Esq.	28	March 20, 1739
Oct. 16, 1708	Dec. 12, 1777	Dr. Albertus Haller; afterwards Baron ..	Nov. 1	
	July 2, 1753	John Peter Stehelin, M. A.	8	Nov. 22
March 1709	May 1, 1762	William, Duke of Portland	Dec. 6	20, 1740
Nov. 6, 1713	March 23, 1789	Thomas, Duke of Leeds	20	Feb. 12, 1739
		Mr. John Van Rixtel	20	Jan. 10
	May 1740	Mr. Thomas Haselden	Jan. 17	
Sept. 6, 1700	Aug. 21, 1768	Dr. Claude Nic. Le Cat	31	
	Sept. 19, 1756	Robert, Lord Raymond	Feb. 7	Feb. 14
	Oct. 1746	Westby Gill, Esq.	7	14
Sept. 7, 1707		{ Mons. George Louis le Clerc de Buffon; afterwards Count }	7	
	April 14, 1759	{ Thomas, Lord Lovell; afterwards Earl of Leicester }	March 6	Jan. 29, 1740
	Ejected June 9, 1757	John, Earl of Sandwich	20	May 22
	Dec. 17, 1777	James Hawley, M. D.	May 1, 1740	15
	Feb. 12, 1752	Charles Lockyer, Esq.	15	June 5

f

BORN.	DIED.		ELECTED.	ADMITTED.
	May 15, 1782	Sebastian Josepn de Carvalho e Mello, Envoy from Portugal; afterwards Conte de Oeyros, and at last Marchese de Pombal	May 15, 1740	June 5, 1740
		Josephus de Panicis, M. D. Prof. Physic, in the University of Rome	June 5	
		Samuel Skinner, Esq.	19	Oct. 23
		Henry Stuart Stevens, Esq.	Oct. 23	Feb. 19
	Jan. 8, 1763	Peter Davall, Esq.	23	Nov. 6
Nov. 20, 1685	Feb. 15, 1766	Mons. Jean Hellot	23	
		Sign. Eustachio Zanotti, Prof. Ast. a Bologna	Nov. 6	
		Rev. Monsigneur Giulio Sachetti, at Rome ..	6	
		Rev. Monsigneur Michael Angelo Giacometti, of Pistoria	6	
		Rev. Mr. Thos. Hunt; afterwards D. D. Heb. Prof. Oxon	13	March 19
May 3, 1695	Dec. 27, 1771	Dr. Henry Pitot	13	
		Jos. de Seystres, Marquis de Caumont	13	
		Rev. John Copping, Dean of Clogher in Ireland	13	Nov. 20
		Dr. Joannes Jallabert, Prof. of Experimental Philosophy and Mathematics, Geneva	13	
		Hon. Richard Arundell, son of Lord Arundell, of Trerice	20	27
	Jan. 8, 1777	Thomas, Lord Foley	27	
	1742	George Martini, M. D.	Dec. 11	
Sept. 5, 1711	Dec. 17, 1756	Ds. Johannes Nathaniel Lieberkulm, M. D. ..	18	
March 10, 1707	April 15, 1788	Ds. Joannes Paulus Grand jean de Fouchy; afterwards Secretary of the Royal Academy of Sciences of Paris	18	
Oct. 22, 1708	Sept. 22, 1742	Frederick Lewis Norden, Captain in the Danish Navy	Jan. 8	Nov. 12, 1741
	Dec. 25, 1777	Charles Chauncy, M. D.	29	Feb. 12, 1740
		Benito de Moura Portuga, of Lisbon	Feb. 5	
	May 16, 1790	Hon. Philip York; afterwards E. of Hardwicke	Jan. 29	5
		Daniel de Superville, M. D. Archiater to the Margrave of Bayreuth	Feb. 12	
Jan. 6, 1692	Dec. 25, 1777	Ds. Franciscus Maria Zanotti	26	
Sept. 14, 1713	March 21, 1742	Ds. Francois de Bremond	26	
	Nov. 25, 1774	Mr. Henry Baker	March 12	March 19
	Aug. 8, 1794	Count Jerom de Salis	19	26, 1741
		Shallet Turner, Esq.	26, 1741	April 9
	May 10, 1787	Mr. Wm. Watson; afterwards M. D. at last Kt.	April 9	16
	Withdrawn Feb. 15, 1749	Francis Philip Duval, M.D.	16	
	April 4, 1770	James Parsons, M. D.	May 7	May 14
		Ralph Knight, Esq.	28	
		John Welin, Philosophy Prof. in the University of Abo	28	
1602	Feb. 16, 1749	George Holmes, Esq.	Nov. 12	
		Father Thomas Le Seur, a Rome	Dec. 10	
		Father Francis Jacquier, a Rome	10	
		Mons. Pierre de Vigny, Member of the Royal Academy of Architecture at Paris	10	
	Dec. 8, 1756	Wm. Lord Harrington; afterwards Earl of	17	March 11
	Withdrawn Jan. 7, 1767	Mr. John Robertson	17	Dec. 17
	Withdrawn 1752	Wm. Battie, M. D.	Jan. 7	Jan. 14
		Edward Milward, M. D.	14	21
	March 6, 1751	Henry, Lord Viscount Lonsdale	21	March 11
	Nov. 26, 1767	John Lock, Esq.	Feb. 4	Feb. 11
	Sept. 15, 1765	Richard Pococke, LL. D; afterwards Bp. of Ossory, and since of Meath	11	18
Sept. 30, 1710	Jan. 1771	John, Duke of Bedford	March 11	
	Oct. 30, 1778	Sidney Stafford Smythe, Esq.; afterwards Lord Chief Baron and Kt.	11	April 1, 1742
	Nov. 18, 1797	muel Reynardson, Esq.	11	March 18, 1741
	April 10, 1782	Thomas Wilbraham, LL. D; afterwards M. D.	18	25
		Michael Russell, Jun. Esq.	March 25, 1742	May 6
		William Bristow, Esq.	25	Nov. 25
		Mr. Jerry Pierce, Surgeon at Bath	25	
	June 1, 1759	James Nihell, M. D.	April 1	
	Feb. 12, 1784	Jeremiah Milles, M. A.; afterwards D. D. and Dean of Exeter	1	April 8
		William Nicholas, Esq.	1	June 17
		John Thomas Batt, M. A.; afterwards M. D.	8	Dec. 9

3

BORN.	DIED.		ELECTED.	ADMITTED.
1714	April 18, 1794	Charles Pratt, Esq.; afterwards Lord Camden and Ld. Chancellor, at last El. Camden	April 8, 1742	April 29, 1742
Nov. 28, 1700	Sept. 2, 1764	Nathaniel Bliss, M. A.; afterwards Sav. P. G. and Ast. Royal	May 20	Jan. 13, 1743
		Mr. James Horne	20	May 27
	Jan. 1800	Wm. Brownrigg, M. D.	20	
	Oct. 3, 1749	Samuel Lisle, D. D. Archdeacon of Canterbury; afterwards Ld. Bp. of Norwich	June 17	Nov. 18
	Withdrawn June 4, 1767	John Lawry, M. A.	17	4
	Dec. 18, 1770	Charles Mason, B. D. Prof. Woodw. Cantab.; afterwards D. D.	24	Feb. 23, 1743
	July 15, 1757	Hon. Henry Finch, brother to the Earl of Aylesford	Oct. 28	June 2
		Mr. Thomas Bayes	Nov. 4	Nov. 11, 1742
	Withdrawn 1773	Walter Bowman, Esq.	4	25
	1746	Mons. Michael Fourmont, the younger, Prof. of the Oriental languages at Paris	4	
	1751	Mr. Peter Coste, of Paris	25	
	June 1743	Thomas Lediard, Esq.	Dec. 9	Dec. 16
	Withdrawn 1753	Wm. Talbot, Esq.	16	Jan. 13
		Dr. Chrisfrid Kirch, of Berlin	16	
	Jan 10, 1794	Clifton Wintringham, M. D.; afterwards Bart.	23	13
		John Phillips, Esq.; afterwards Bart.	23	
	Dec. 22, 1768	Charles Lyttelton, LL. D.; afterwards Bishop of Carlisle	Jan. 27	April 28, 1743
	Oct. 5, 1771	Thomas Rutherford, B. D.; afterwards D. D. and Prof. Theo. Cant.	27	May 5
		Wilhelmus Wilhelmius, M. A. of Rotterdam	27	
	Feb. 7, 1762	James Creed, Esq.; afterwards Sir	Feb. 10	March 3, 1742
	Jan. 25, 1749	James Lever, Esq.	10	Feb. 17
	-	Mr. Henry Henrickson, of Copenhagen ..	10	
	1779	Peter Newcome, M. A.	24	March 17
		Francis Hoskins Eyles Stiles, Esq.; afterwards Bart.	March 10	17
		Francis Chute, Esq.	10	May 12, 1743
	April 1745	George Shelvocke, Esq.	10	March 17, 1742
	1786	Abbe de Gua de Malves, Prof. Philosophy in the College Royal at Paris	10	
		Matthew Sarayva, M. D. Chev. of the order of Christ at Rio de Janeiro	April 21, 1743	
		Mr. Roger Paman	May 12	May 19, 1743
Sept. 3, 1710	May 12, 1784	Mons. Abraham Trembley	19	June 20, 1745
	Feb. 12, 1763	Rev. Henry Miles, D. D.	June 9	16, 1743
	1786	Solomon Dayrolle, Esq.	16	Nov. 17
	May 1768	Daniel Rich, Esq.	16	Jan. 19
	Feb. 28, 1751	William, Earl of Stafford	16	June 23
1694	May 30, 1778	Mons. Francois Marie Arouet de Voltaire ..	Nov. 3	
March 13, 1720	April 9, 1793	Mr. Charles Bonnet, of Geneva	17	
		Mons. Jean Masson, Chev. de Besse	Dec. 8	
		Mr. Joseph Ames	15	Dec. 22
	Jan. 9, 1750	Henry, Earl of Pembroke	15	
		Robt. Clayton, Lord Bishop of Corke; afterwards of Clogher	12	Oct. 30, 1746
Nov. 22, 1706	Oct. 20, 1758	Charles, Duke of Marlborough	12	March 8, 1743
		Mons. le Marquis de Lockmaria, of Paris ..	12	
		H. F. Baron de Solenthal, Ambassador from Denmark	19	Dec. 10, 1747
Jan. 17, 1709	Aug. 22, 1773	Geo. Lyttleton, Esq. since Baron Ld. Lyttleton	26	Jan. 31, 1754
	1778	Hon. Wm. Pitt, Esq. since Earl of Chatham	26	May 15, 1755
1706	May 4, 1789	Wm. Wyndham, Esq.	26	April 26, 1744
		Joseph Lawrentius Bruni, M. D. of Turin ..	Feb. 23	March 1, 1743
	Feb. 7, 1766	Col. Wm. Sotheby	March 8	15
		Mr. John Nickols Jun. Merchant	8	15
1720	June 20, 1793	Sir John Rawden, Bart.; afterwards Lord Rawden, afterwards Earl of Moira	April 12, 1744	May 24, 1744
	Dec. 30, 1791	Hon. Chas. Hop. Weir, brother to the Earl of Hopton	26	Jan. 24
	1761	Abbe Claude Sallier, of Paris, keeper of the King's Library	May 10	
	May 18, 1755	Mr. Roger Pickering	10	May 24
		Sign. Petro Andrea Capello, Ambassador from Venice	June 7	Feb. 7
1683	July 7, 1764	William, Earl of Bath	Nov. 15	Nov. 22
	May 1777	Rev. John Nixon, M. A.	15	

BORN.	DIED.		ELECTED.	ADMITTED.
		Joh. Benjamin Fischer, of Livonia	Nov. 15, 1744	
	1795	Mons. Secondat de Montesquieu	Dec. 6	
	Withdrawn 1771	Swithin Adee, M. D.	6	May 16, 1745
		{ Mons. de Boffrand, First Architect to the } King of France	Jan. 10	
	Nov. 13, 1770	{ Mons. Henry Francois Le Dran, Secretary } to the Royal Academy of Surg. at Paris	10	
March 1708	Aug. 24, 1764	Hon. Henry Bilson Legge, Esq.	10	June 10, 1746
	Dec. 15, 1757	Benjamin Keen, Esq. ; afterwards K. B. ..	31	Feb. 7, 1744
1717	Sept. 7, 1799	Mons. le Monnier, M. D.	Feb. 7	
	Jan. 21, 1755	Hugh, Lord Willoughby, of Parkham	14	Feb. 28, 1744
	Aug. 17, 1754	Samuel Hickman, Esq.	March 7	May 30, 1745
		John Merrill, Esq.	21	April 4
	June 9, 1772	Gowin Knight, M. B.	April 25, 1745	May 2
	Nov. 19, 1763	Peter Wyche, Esq.	May 2	9
	Withdrawn } Oct. 31, 1751	Wm. Saunders, M. D.	2	16
	Withdrawn } June 16, 1757	Wm. Mace, Esq. Prof. Rhet. Gresh. J. C. ..	23	30
		{ Wm. Bedford, M. D. ; afterwards Director } of the Mathemat. Class of the Royal Acad.	30	June 13
Jan. 15, 1708	Oct. 11, 1791	{ Sign. Giovan. Francisco Mauro Melchier Salvemini da Castiglione, of Lausanne ; afterwards Director of the Royal Academy of Sciences, Berlin	30	
		The Baron de Hardenberg, of Hanover ..	June 13	
1710	Nov. 17, 1756	Robt. Hoblyn, Esq.	13	Oct. 24
1667	July 20, 1752	John Christopher Vepusch, Mus. D.	13	June 20, 1748
	Feb. 2, 1802	Welbore Ellis, Esq. ; afterwards Lord Mendip	20	Nov. 10, 1745
	April 14, 1764	Mr. Noah Sherwood, Surgeon	20	June 27
	Jan. 21, 1781	Wm. Lewis, M. D.	Oct. 31	
April 10, 1707	Jan. 10, 1782	John Pringle ; afterwards Bart. and P. R. S.	31	Nov. 14, 1746
April 23, 1697	June 6, 1762	George Anson, Esq. ; afterwards Lord	Dec. 5	June 19, 1745
	May 22, 1761	Mr. Thomas Simpson	5	Dec.. 19, 1746
		Edward Montague, Esq.	12	April 24
		Mr. Wm. Arderon	12	
	Dec. 18, 1755	{ Hon. John Hamilton, Capt. son of the Earl } of Abercorn	March 13	17
	Jan. 10, 1756	{ Sign. Giovanni Giacomo Marinoni, Prof. } Ast. at Vienna	20	
	Oct. 1774	Rev. Patrick Murdoch ; afterwards D. D. ..	20	17
		Abbe Guyon, of Paris	April 10, 1746	
1694	March 6, 1754	{ Rt. Hon. Henry Pelham, brother to Hollis, } Duke of Newcastle	17	May 8
April 26, 1695	July 18, 1769	Christopherus Jacobus Trew, M. D.	17	
1714	May 7, 1766	{ Samuel Squire, M. A. Archdeacon of Bath ; afterwards D. D. and Bp. of St. David's	May 15	29
		Mr. Tycho Hoffman, of Denmark	29	
	Dec. 11, 1781	{ Sir Francis Dashwood, Bart. ; afterwards } Lord Le Despencer	June 19	Jan. 29, 1747
		Edward Milles, Esq.	19	Oct. 23
	March 29, 1780	Francis Blake, Esq. ; afterwards Bart. ..	19	Nov. 20
Jan. 2, 1706-7	Nov. 16, 1762	John, Earl of Cork and Orrery	Oct. 23	Jan. 24, 1750
		Mr. Lewis de Beaufort, of Leipzig	23	
	Oct. 31, 1770	Sir Peter Thompson, Kt.	Nov. 20	Nov. 27, 1746
		Matt. Robinson, Esq. ; afterwards Ld. Rokeby	27	April 30, 1747
	1784	{ John, Lord Viscount Castlemain, since Earl } of Tilney	Dec. 11	Dec. 18, 1746
1716	July 5, 1795	Don Antorio de Ulloa	11	
		Benjamin Prideaux, Esq.	11	18
1703	May 18, 1770	{ Dr. Johannes Henricus Winckler, Prof. Nat. } Phil. at the University of Leipzig	Jan. 8	
	Oct. 11, 1758	John Claphane, M. D.	8	May 4, 1710
Sept. 10, 1713	Dec. 30, 1781	Mr. Turberville Needham	22	May 25
	March 2, 1787	{ Mons. Johannes Nicolaus Sebastian Alle- } mand, of Leyden	22	
	1802	Mr. Daniel Peter Layard, Surgeon, since M. D.	22	Jan. 29, 1746
April 15, 1707	Sept. 1/16, 1783	Mr. Leonhard Euler	22	
		Wm. Faquier, Esq.	29	Feb. 5
Nov. 11, 1719	July 18, 1804	{ Peter Holford, Esq. Father of the Society, } since the death of Ld. Mendip, Feb. 2, 1802	Feb. 12	19
1707	March 2, 1797	Hon. Horatio Walpole	19	26
	July 22, 1802	Wm. Parker, M. A.	19	March 19
		Mr. Alexander Mac Farlane, of Jamaica ..	19	
	Dec. 12, 1797	Richard Brocklesby, M. D.	26	March 5

BORN.	DIED.		ELECTED.	ADMITTED.
	Oct. 3, 1769	The Marchese Anthony Nicolini, of Florence	March 26, 1747	March 26, 1747
	Sept. 18, 1786	{ Hon. Charles Hamilton, brother to the late Earl of Abercorn	26	April 2
		The Marchese Reinuccini, of Florence	26	
	Dec. 31, 1804	{ Joshua Iremonger, Father of the Society, since the death of Mr. Holford, July 18, 1804	April 30	May 14
		Benjamin Bosanquet, M. A.	May 21	June 18
		David Ravaud, M. A.	21	18
Dec. 3, 1696	Sept. 4, 1787	{ Samuel Christianus Hollman, Prof. at the University of Gottingen	21	
		Mons. Johanus Baptista, de la Chapelle of Paris	June 18	Dec. 12, 1776
April 27, 1724	Jan. 11, 1784	{ Ferdinand Philip, Duke of Sagon and Prince Lobkowitz	July 2	July 2, 1747
April 20, 1720	Feb. 22, 1794	{ Henry, Earl of Lincoln; afterwards Duke of Newcastle, and B. Soleuthal	Oct. 22	Dec. 10
	Nov. 8, 1789	John Silvester, M. D.; afterwards Sir John, Kt.	Nov. 12	Nov. 19
		Abbe Gurianni Baptista Passeri, of Pesara ..	12	
Oct. 1723	Feb. 9, 1795	{ Thomas, Lord Parker; afterwards Earl of Macclesfield	19	26
June 4, 1717	Withdrawn Feb. 10, 1763	Mr. Emanuel Mendes da Costa	26	26
Nov. 22, 1728		Charles Fred. Margrave of Baden Dourlack	Dec. 10	Dec. 17
	Jan. 9, 1784	Sir George Savil, Bart.	10	17
July 1711	May 25, 1776	Richard, Viscount Fitzwilliam	17	Jan. 7
	July 14, 1773	James, Lord Tyrawly	Jan. 7	14
1707	Jan. 10, 1789	Mr. Pierre Lyonet, of Hague	14	
1698	Dec. 5, 1755	William, Duke of Devonshire	21	28
	Dec. 13, 1762	Henry Reade, Esq.	21	Feb. 25
		Mynheer Munck, of Middelburg in Zealand	21	
	Sept. 2, 1780	George Scott, Esq.	March 10	May 5, 1745
	April 8, 1788	William Young, Esq. since Bart.	10	
		Count P. Czernicheu, Min. Plenip. from Russia	10	Jan. 19
	1749	{ Christopher Count de Manteufell, Minister of State to the King of Poland	24, 1748	
		Chevalier Ossorio, Envoy Ext. from Sardinia	April 21	Feb. 16
		Hugh Campbell, Esq.	May 19	May 26
		Wm. Lee, Esq.	26	Nov. 3
		Rev. Mr. Griffith Hughes	June 9	June 9
	Aug. 10, 1755	Rt. Hon. Sir Wm. Yonge, Bart. K. B.	23	30
	Withdrawn 1754	Rice Charleton, M. A.	Nov. 3	Nov. 17
	July 9, 1781	Mr. Robt. Dingley, Merchant	3	10
		Charles Joye, Esq. (not admitted)	17	
		Mr. James Simson, of Dublin, Merchant ..	17	
	July 3, 1753	John Hill, Esq.	Dec. 8	Dec. 22
	March 1768	John Mitchell, M. D.	15	22
Jan. 21, 1701	Feb. 4, 1774	Mons Charles Maria de la Condamine	15	May 12, 1763
		Sir Thomas Burnett, Kt.	22	Feb. 13, 1748
Nov. 16, 1717	1783	Mons. Jean le Roud D'Alembert	22	
		Wm. Windham Ashe, Esq.	Jan. 19	Nov. 2, 1749
	Aug. 2, 1762	Charles Feake, M. D.	26	Feb. 9, 1748
		Nicholas Munckley, M. D.	26	2
		Mr. Gabriel Cramer, Prof. Math. Geneva ..	Feb. 9, 1748	
	Jan. 5, 1782	Abraham Dixon, Esq.	9	
		Mons. de Mont Audouin, of Nantes	16	
	Oct. 30, 1770	Samuel Cornish, Esq. afterwards Bart.	March 9	March 16, 1748
1690	1754	{ Mr. Claudius Gros de Boze, of the Royal Academy of Sciences at Paris	April 6, 1749	
	March 24, 1778	Mr. Samuel Sharp, Surgeon	13	May 25, 1749
	Dec. 26, 1797	John Wilkes, Esq.	13	4
	May 24, 1787	William Mitford, Esq.	13	April 20
1694	1768	Mons. Jean Baptiste Boyer, M. D. of Paris ..	20	
		{ Peter Paul Molinelli, M. D. Prof. of Physic in the University of Bologna	20	
		{ Antonio Freyero d'Andrade, Envoy from Portugal	May 4	June 15
May 7, 1700	June 18, 1772	Gerard Van Sweiten, M. D.	4	
	Dec. 9, 1776	{ Jame Porter, Esq. Ambass. Constantinople, since Sir	11	Jan. 13, 1763
May 12, 1706	Feb. 19, 1767	Mons. Franciscus Boissier Sauvage de la Croix	25	
	Feb. 25, 1763	Rev. Mr. John Williamson; afterwards D. D.	June 15	May 27, 1756
July 21, 1694	Nov. 17, 1768	Thomas, Duke of Newcastle	Oct. 26	
1678	April 28, 1770	{ Sir John Ligonier, K. B.; afterwards Lord Ligonier	26	March 22, 1753
		Chevalier de Baillou, of Florence	26	
		The Duke de Medina Sidonia	Nov. 9	

BORN.	DIED.		ELECTED.	ADMITTED.
	Ejected June 9, 1757	Hon. James Oglethorpe, Lieut.-General ..	Nov. 9, 1749	Nov. 16, 1749
	June 21, 1770	Philip Carteret Webb, Esq.	9	16
Jan. 5, 1713	July 6, 1773	Don Jorge Juan	9	16
Oct. 25, 1683	May 6, 1757	Charles, Duke of Grafton	23	
1712	May 28, 1790	George, Earl of Cardigan ; afterwards Duke of Montague	Dec. 7	Dec. 14
	April 8, 1763	James, Earl of Waldegrave	14	Feb. 8
		Chevalier St. George, of Paris	Jan. 11	
	Dec. 9, 1779	Nathan Alcock, M. D.	25	
Aug. 1710	May 17, 1801	Wm. Heberden, M. D.	25	Feb. 1, 1749
		Mr. George Bell, Surgeon	25	11
Feb. 10, 1698	Aug. 15, 1758	Mons. Pierre Bouguer	25	
May 31, 1711	March 8, 1797	Mons. Samuel Formey, Secretary to the Royal Academy of Sciences, Berlin	25	
		Anthony Askew, M. B.	Feb. 1	8
	Jan. 21, 1806	Mr. Henry Ellis, Father of the Society since the death of Mr. Iremonger, Dec. 31, 1804, but the Elect. of Baden older Memb.	8	22
	Feb. 1752	Chevalier Charles de Folard, of Paris	8	
Nov. 4, 1705	Oct. 31, 1783	Louis Elizabeth de la Vergue, Compte de Tressan, Lieut.-General of the armies of France	8	
	Nov. 22, 1765	Isaac Townsend, Esq. Admiral of the White	22	March 1
Jan. 21, 170⅚	Feb. 14, 1760	Isaac Hawkins Brown, Esq.	22	April 5, 1750
	June 21, 1754	Charles Tough, M. A.	22	March 1, 1749
		Lewis Jean le Thieullier, M. D. of Paris ..	22	
		Mr. Nicholas Struych, of Amsterdam	22	
		James Mounsey, M. D.	March 8	Jan. 19, 1764
July 31, 1718	March 22, 1772	Mr. John Canton	22	Jan. 29, 1750
	Jan. 9, 1762	Thomas Hayter, Lord Bishop of Norwich; afterwards of London	22	May 10
July 4, 1680	Nov. 30, 1765	Mons. Antoine Joseph de Sallier d'Argenville ..	22	
Jan. 29, 1700	March 17, 1782	Mr. Daniel Bernoulli	May 3, 1750	
		Joannes Mendez Saquet Barbosa, Physician to the King of Portugal	10	
Feb. 2, 169⅚	Aug. 31, 1772	Rev. Mr. Wm. Borlase ; afterwards D. D. ..	17	Oct. 31, 1751
		Edw. Wortly Montague, Junr. Esq. ..	31	Feb. 7, 1750
	July 20, 1756	Richard Roderick, M. A.	June 21	Oct. 25
	Nov. 25, 1797	Charles Walmesley, Pere Benedict a Paris	Nov. 1	
Sept. 29, 1714	Aug. 21, 1773	Mons. François David Herrissant, M. D. ..	1	
		Abbe Guasco, of Piermont	1	
	Dec. 9, 1760	Henry, Lord Viscount Downe	Dec. 6	
		Mr. Joseph Warner, Surgeon	6	Dec. 13
	April 20, 1770	Hon. Charles Yorke ; (afterwards Lord Chancellor,) second Son to the Lord Chancellor	13	20
June 17, 1714	Sept. 4, 1784	Mons. Cæsar François Cassini de Thury ..	Jan. 17	
May 11, 1722	April 7, 1781	Peter Camper, M. D.	17	Nov. 16, 1785
		John Bacon, Esq.	24	April 18, 1751
	June 23, 1766	Isaac Pacatus Shard, Esq.	24	Jan. 31, 1750
	March 2, 1756	Robert Watson, M. D.	Feb. 7	Feb. 14, 1751
	May 2, 1779	Mr. William Mountaine	March 14	April 18
	June 27, 1787	Sir Thomas Heathcote, Bart.	May 16, 1751	June 13
		Sign. Marsilius Venturi, of Parma, First Physician to the Queen Dowager of Spain	June 6	
	June 14, 1787	Sir Israel Mauduit, Merchant	13	20
		Joseph Brookesbank, Esq.	13	20
		Joachim Tozé, Fidalgo de Silveria, Envoy from Portugal	Oct. 31	April 23
May 13, 1730	July 1, 1782	Charles, Marquis of Rockingham	Nov. 7	Dec. 5
	June 6, 1788	Mr. Benjamin Wilson	Dec. 5	12
May 17, 1718	Aug. 2, 1776	Matthew Maty, M. D.	19	Jan. 0, 1752
	Feb. 10, 1799	Charles Morton, M. D.	Jan. 16, 1752	23
		Justus. Jo. Torkos, M. D. of Presburg	16	
	March 18, 1774	Sir Matthew Fetherstonhaugh, Bart.	Feb. 13	Feb. 20
		Richard Russell, M. D.	13	Jan. 11, 1753
	March 22, 1800	Mons. Joseph de Guignes	20	
		Thomas Steavens, Esq.	March 5	
	April 1766	Robt. Whyte, M. D. Prof. Med. University of Edinburgh	April 16	April 23, 1752
	March 15, 1765	Peter Shaw, M. D.	16	June 18
		Sir Richard Hoare, Kt. Alderman of London	May 7	18
Sept. 9, 1707		Joannes Georgius Huber, M. D. of Basel ..	7	
June 4, 1694	Dec. 16, 1774	Mons. François Quesnay	28	

BORN.	DIED.		ELECTED.	ADMITTED.
		Mr. Wm. Mazeas, Librarian to the Duke of Noailles	May 28, 1752	
		John Hyde, Esq.	28	June 4, 1752
	Sept. 9, 1758	Mr. Arthur Pond, Portrait Painter	28	4
	1802	Charles Moss, D. D.; afterwards Bp. of St David's, and then Bp. of Bath and Wells	28	4
		Peter Burrell, Junr. Esq.	28	Nov. 9
	Withdrawn 1765	Wm. Cadogan, M. D. of Bristol	28	16
	Sept. 21, 1784	Richard Hazard, Esq.	June 4	16
		Henry Pacey, Esq.	4	Jun 11
		Wm. Brakenridge, D. D. ..	Nov. 9	Nov. 16
1702	1762	Mons. Jean Baptiste Faget, of the Academy of Surgery, Paris	Dec. 14	
	1784	Sir James Caldwell, Bart.	21	Feb. 8, 1753
		Wm. Allix, Esq. ..	Jan. 25, 1753	1
	May 17, 1792	Noah Thomas, M. A.; since M. D., since Sir N. Thomas, Kt.	25	1
	Jan. 10, 1794	Hugh, Earl of Marchmont	Feb. 1	8
	April 21, 1782	John, Earl of Clanricarde	8	8
Nov. 9, 1721	June 23, 1770	Mark Akenside, M. D.	8	15
	April 1768	Francis Faquier, Esq.	15	22
		Charles Petley, Esq.	22	Feb. 6, 1755
		Mons. Bellin, Engineer of the Navy of France	22	
1701	Jan. 3, 1778	Paul Jaques Malouin, Physician to the Queen of France	March 1	
		Don Joseph Hortega	8	
	Nov. 1775	Caleb Hardinge, M. D.	15	May 3, 1753
1724	Oct. 28, 1792	Mr. John Smeaton	15	March 29
	May 7, 1770	Peter Gabrey, M. D. of Hague	15	
	Sept. 16, 1777	Simon, Earl Harcourt	15	April 12
	Withdrawn 1771	Wm. Price, Esq.	22	March, 29
March $\frac{5}{15}$, 1693	March 23, 1754	Mr. John James Wetstein, Prof. Philo. Amster.	April 5	
May $1\frac{3}{25}$, 1707	Jan. 10, 1778	Charles Linnæus, M. D. Prof. Botany in the University of Upsala; afterwards Von Linne, Kt. of the Polar Star	May 3	
	1802	Joseph Hoare, B. D.; afterwards D. D. Princp. of Jesus Coll. Oxford, Prebendary, Westminster	24	Jan. 31, 1754
	1786	Richard Green, D. D.	31	June 7, 1753
Nov. $1\frac{7}{26}$, 1723	Dec. 19, 1802	Samuel Wegg, Esq.	31	7
		The Chevalier Richard Wall, Ambass. from Spain	31	7
	Dec. 24, 1770	Wm. Northey, Esq.	June 21	March 21, 1754
	June 19, 1763	Robt. Symmer, Esq.	Nov. 15	Dec. 6, 1753
1691	March 6, 1764	Philip, Lord Hardwicke, Lord Chancellor; afterwards Earl of Hardwicke	Dec. 6	
	Oct. 21, 1761	Wm. Lock, Esq.	Feb. 7, 1754	Feb. 14, 1754
	July 5, 1761	John Girle, Esq.	7	14
	Sept. 6, 1761	James Postlethwayt, Esq.	7	28
	Oct. 1776	Mr. John Ellis	14	28
	Sept. 27, 1774	Sholto Charles, Ld. Aberdour; since El. Morton	21	28
	Jan. 21, 1787	Gustavus Brander, Esq.	March 7	March 14
	May 23, 1775	Lewis Crusius, M. A. since D. D.	7	14
	1776	Bartholomew Hammond, Esq.	14	21
	May 21, 1771	Anthony, Earl of Shaftsbury	28	May 16
		John, Lord Viscount Castlecomer; afterwards Earl of Wandesford	April 4	
	Withdrawn 1770	John Cooksey, M. A.	4	April 25
		Richard Blacow, M. A.	4	25
		Mons. R. Caumont, M. D. Physician to the King of France	May 9	May 16
	Jan. 8, 1771	Gregory Sharpe, LL. D.	9	Dec. 5
		Father John Chevalier, Ord. Orat. Pres. at Lisbon	23	
	1778	Matthew Raper, Esq.	30	June 27, 1777
March 11, 1711	March 6, 1796	Guillaume Thomas Raynal	30	June 5, 1754
	Dec. 12, 1782	Charles Gray, Esq.	June 13	Nov. 21, 1755
		George Lewis, Esq.	13	April 24, 1754
	Aug. 1760	Rev. Caspar Wetstein, Chaplain to her R. H. the Princess Dowager of Wales	July 4	Nov. 7
	July 25, 1801	William, Earl of Dartmouth	Nov. 7	21
	Nov. 13, 1764	Rt. Hon. Sir Thomas Clarke, Kt. Master of the Rolls	14	Jan. 9, 1755
	May 8, 1766	Mr. Samuel Chandler; afterwards D. D. ..	Dec. 5	Dec. 12, 1754
		John Hudson, Esq.	5	Dec. 12
	June 24, 1782	John Blair, LL. D.	Jan. 9, 1755	Jan. 16, 1755

BORN.	DIED.		ELECTED.	ADMITTED.
		Mr. Lewis Jean Marie Danbenton	Jan. 9, 1755	
	Nov. 23, 1757	Mr. James Dodson	16	Jan. 23, 1755
	1757	James Dawkins, Esq.	23	
		Wm. Henry, D. D.	Feb. 20	June 6, 1765
	Dec. 22, 1760	Wm. Hirst, M. A.	20	
	Aug. 16, 1775	Mr. Josiah Colebrooke	March 13	March 20, 1755
	April 18, 1774	Roger Pettiward, D. D.	13	April 17
April 11, 1715	May 8, 1762	Charles Fred. Handertmark, M. D. Prof. of Physic at Leipzig	13	
		Mons. Martin Hubner, Prof. History, at the University of Copenhagen	20	
		Count Cyrille Rausaumousky, President of the Imperial Academy of Sciences, at St. Petersburgh	April 24	
July 18, 1685	July 17, 1755	Mous. Jean Claude Adrian Helvetius, M. D.	24	
		Mons. Ottavio Antonio Bayardi, of Naples	24	
		Sign. Camillo Paderni, Keeper of the Museum Herculaneum	24	
		Nathaniel Forster, D. D.	May 1	May 15
Oct. 2, 1716	May 27, 1781	Johannes Baptista Beccaria, Proff. Exp. Philosophy, in the University of Turin	29	
Aug. 22, 1728		James, Lord Viscount Charlemont ; since Earl	29	Feb. 1, 1759
	Withdrawn 1795	Rev. Henry Owen, M. D.	June 12	
	Jan. 27, 1764	Robt. Hunter Morris, Esq.	12	Feb. 15, 1753
		Rodolph de Valltravers, Esq.	12	June 19
Jan. 20, 1716	April 30, 1795	Jean Jaques Barthelemy, Keeper of the King of France's Cabinet of Medals	12	
Feb. 22, 1734-5	Dec. 29, 1806	Charles, Duke of Richmond	Dec. 11	Jan. 8, 1756
		Peter Ascanius, M. D. of Norway	11	
		Chevalier de Jacour	Jan. 8, 1756	
		Mons. de Bougainville	8	
		Father Guiseppe Maria Pancrazzi, of Naples	15	
1733	March 20, 1769	John Albert Schlosser, M. D. of Utrecht ..	22	
		Mons. John Andreas Peysonnel, M. D. ..	Feb. 5	
	Jan. 27, 1792	Shukburgh Ashby. Esq.	March 18	March 25
		Mons. Daviel Surgeon to the King of France	18	
	July 26, 1764	Charlton Wollaston, Esq. ; afterwards M. D.	25	April 1
	June 29, 1782	Keane Fitzgerald, Esq.	25	1
		James Jurin, Esq.	April 1	May 6
Jan. 17, 1706	April 17, 1790	Benjamin Franklin, Esq.; afterwards LL. D.	29	Nov. 24, 1757
	Nov. 28, 1768	Alexander Russell, M. D.	May 6	May 13, 1756
	Jan. 20, 1790	John Howard, Esq.	13	20
		Jacob Fred. Lantsheer, LL. D. of Middleburg	13	Feb. 5, 1761
	July 26, 1762	John Kidby, M. A.	June 3	June 17, 1756
	Sept. 19, 1804	Thomas Brand, Esq. ; afterwards Thomas Brand Hollis	3	July 8
		John Gregory, M. D.	Nov. 4	Jan. 17, 1765
		Xaverius Manetti, M. D.	11	
		Count Maffeo, of Udina in the Venetian State	11	
		Thomas Percival, Esq.	25	
		The Abbot Rudolphino Venuti, of Rome ..	March 17, 1757	
	March 24, 1795	Mons. Martinho de Mello e Castro, Envoy from Portugal; afterwards Secretary of State in Portugal	April 21	May 26, 1757
	Oct. 5, 1771	Alexander Thislethwayte, Esq.	21	April 28
	March 31, 1767	Thomas Lisle, D. D.	May 5	Dec. 8
	Dec. 18, 1759	Mr. Isaac Rommilly, Merchant	12	May 19
	Nov. 16, 1793	Robert, Lord Romney	19	26
		Charles Grave Hudson, Esq.; afterwards Bart..	19	26
	Dec. 24, 1779	Corbyn Morris, Esq.	19	26
Jan. 1708		Mr. George-Dionysius Ehret	19	26
Sept. 8, 1717		Dr. Vitaliano Donati, of Turin	19	
	Nov. 22, 1784	Father Paul Frisi, Prof. of Ethics, in the University of Pisa	June 9	
		Il Cavaliere Paulo Celesia, Minister from Genoa	16	June 30
		John Upton, Esq.	16	23
April 14, 1720	Dec. 1773	Thomas Hollis, Esq.	30	Nov. 10
	Sept. 1, 1761	George Matthias Bose, Prof. Nat. Philosophy in the University of Wittenburg	30	
April 3, 1694	July 23, 1773	Mr. George Edwards	Nov. 10	17
Aug. 2, 1711	Jan. 2, 1784	Charles Rogers, Esq.	17	24
March 7, 1719	Nov. 10, 1806	Don Joano, Duke de Bragança	17	Dec. 8, 1758
		Philip Barton, LL. D.	24	March 9, 1759
		Mr. Lorenz Natter, of Bibench, in Swabia ..	24	June 21, 1757
		Robt. Bootle, Esq.	Dec. 8	Dec. 15

i

BORN.	DIED.		ELECTED.	ADMITTED.
		George Wegg, Esq.	Feb. 9, 1758	
	Dec. 1802	Wm. Man Godschall, Esq.	16	May 31, 1759
	May 9, 1768	Philip Fred. Gmelin, M. D. Prof. Bot. and Chemistry, in the University of Tubingen	16	
	Aug. 14, 1792	John Ross, D. D.; afterwards Bp. of Exeter	23	March 16, 1758
March 13, 1728	Oct. 4, 1789	Francis, Earl of Huntingdon	March 2	16
	April 18, 1804	Father Theodore de Almeyda, Ord. Orator. at Lisbon	9	
	Nov. 1797	John de Schuvaloff, Grand Chamberlain and actual Privy Councellor to the Empress of Russia	16	
	Nov. 3, 1769	Robt. Lambe, LL. D. Dean of Peterborough; and afterwards Bishop of Peterborough	April 6	
	April 2, 1797	Mr. Richard Grindall, Surgeon	6	April 13
		Giovanin Joannes Marsili, M. D. of Venice ..	6	13
	July 28, 1804	Charles Allioni, M. D. of Turin	6	
	March 31, 1791	Ralph, Earl Verney	20	May 25
	Withdrawn 1774	George Forster Tufnell, Esq.	20	11
Oct. 6, 1732	1811	Nevil Maskelyne, M. A.; since Ast. Royal and D. D.	27	4
Oct. 27, 1714	Feb. 2, 1788	James Stuart, Esq.	27	25
	Nov. 5, 1764	John Hadley, M A. Chem. Prof. Cantab. ..	May 25	Dec. 14
		Amyas Bushe, Esq.	June 15	
	Jan. 20, 1759	Sir Thomas Drury, Bart.	Nov. 16	Nov. 23
	Feb. 24, 1806	Thomas Gisborne, M. D.	16	Dec. 21
		Joseph Ignatio de Torres, J. V. D. Physician to the Royal Family of Spain	Dec. 7	Jan. 18, 1759
	Jan. 13, 1796	John Anderson, Prof. Nat. Phil. Glascow	Feb. 1, 1759	June 28
Jan. 1696	March 30, 1763	Sign. Marco Antonio Foscarini, Nobile Venetian; afterwards Doge of Venice	8	
		John Carafa, Duke de Noia, of Naples	March 8	
		Joseph Salvador, Esq.	15	March 22
	Dec. 23, 1775	Erasmus Sauuders, D. D.	22	29
	Aug. 20, 1761	Edward Wright, M. D.	April 5	
	1795	Edward Hooper, Esq.	May 31	
		Philip Venuti, of Leghorn	June 14	
		John Lloyd, M. B.	Nov. 8	April 24, 1760
	May 27, 1780	John Lewis Petit, M. A.; afterwards M. D.	22	Jan. 24
		Edward Delaval, Esq.	Dec. 6	
		Wilkinson Blanchard; M. B; afterwards M. D.	6	Jan. 15, 1761
	April 29, 1799	David Van Royen, Prof. Bot. in the University of Leiden	6	
Nov. 1709	May 15, 1782	Sign. Eustachio Zanotti, of Bologna	Jan. 10, 1760	
March 15, 1713	March 21, 1762	M. Nicolas Louis de la Caille	17	
		Mons. Pereira, of Paris	Jan. 24	
		The Count de Saluce, of Turin	Feb. 21	
	April 3, 1783	David de Gorter, M. D.	21	
	Sept. 15, 1772	Samuel Dyer, Esq.	March 6	April 24, 1760
	1805	Ismel Wilkes, Esq.	6	March 13
		Dr. Tissot, of Lausanne	20	
May 14, 1702	1781	John Bernard, M. D. Prof. Physic, in the University of Dauay	27	
Dec. 6, 1722	Dec. 15, 1771	Gerard Meerman, LL. D.	April 24	
		James Comyn, Esq.	24	June 19
Oct. 10, 1731	Feb. 24, 1810	Hon. Henry Cavendish	May 1	May 8
	April 21, 1793	John Michell, B. D.	June 12	Feb. 5, 1761
	July 4, 1780	Samuel Musgrave, M. A.	12	
		King George III., Patron	Nov. 17	
	July 24, 1789	Thomas Ryves, Esq.	20	Nov. 26
	Sept. 17, 1767	Edward, Duke of York ··	27	
	1789	John Gideon Loten, Esq.	27	Dec. 11, 1760
		Nicholas de Himsel, M. D. of Riga	27	
	Oct. 31, 1765	William, Duke of Cumberland	Dec. 4	
		Giam. Baptista Albertini, Prince de San Saverino, Envoy from the King of the Two Sicilies	11	
	Oct. 1802	Benedict Ferner, Pr. As. Upsal	11	
	Aug. 28, 1801	Roger Baldwin, M. A.	18	Jan. 15, 1761
1710	Dec. 10, 1792	Mr. Jean Joseph Le Sue, of Paris, Surgeon	18	
May 18, 1711	Feb. 12, 1787	Roger Boscowick, S. J. S.	Jan. 15, 1761	
June 16, 1723	Feb. 28, 1792	Mr. Joshua Reynolds, since Kt.	15	22
April 7, 1725	Aug. 3, 1806	Mr. Michael Adanson, of Paris	22	
	Oct. 28, 1805	Daniel Dumaresq, D. D.	Feb. 5	May 12. 1763
		Mr. John Alex. Genevois, of Burtigny in Switzerland	19	

g

BORN.	DIED.		ELECTED.	ADMITTED.
	Dec. 1, 1805	Hugh Hamilton, Prof. Nat. Phil. Dublin; afterwards Dean of Armagh, since Lord Bp. of Clonfert; afterwards of Ossory	Feb. 19, 1761	Jan. 20, 1763
	May 31, 1765	Mr. Henry Van Haemstede, Minister of the Dutch Church in Austin Friars	March 12	April 2, 1761
	March 13, 1796	Richard Wilbraham Bootle, Esq.	April 2	30
	April 18, 1802	Erasmus Darwin, M. D.	9	
	Dec. 17, 1770	George Eckersall, Esq.	23	30
	March 3, 1792	Robert Adam, Esq.	May 7	June 11
1707	Nov. 30, 1761	Mr. John Dolland, Optician	28	11
	1779	Dom. Martin Panzano, F. S. A.	June 11	
	May 23, 1793	Mr. Wm. Hudson, Apothecary	Nov. 5	Nov. 12
	Oct. 2, 1764	William, Duke of Devonshire	12	April 29, 1762
	June 14, 1791	Joseph Geartner, M. D. of Stutgard	12	
		Thomas Wood, LL. D.	19	Nov. 26, 1761
		Thomas Heberden, M. D.	Dec. 10	
	Oct. 2, 1778	Washington, Earl Ferrers		May 13, 1762
	June 15, 1809	George Baker, M. D.; since Sir G. Baker, Bart.	Feb. 4, 1762	Feb. 11
	May 29, 1785	Andrew Coltee Ducarel, LL. D.	4	18
		Mr. Charles White, Surgeon, at Manchester	18	18
1709	Nov. 30, 1781	Theodore Tronchin, M. D. Prof. Med. Geneva	March 18	
		Johan Lulofs, Prof. Ast. Ludg. Bat.	18	
	Aug. 1793	Robt. Burrow, Esq.	18	March 25
	Feb. 6, 1795	Charles Dodgson, M. A.; afterwards Lord Bp. of Ossory, and then Ld. Bp. of Elphin	April 1	Dec. 16
		Fran. Hen. the Marquis de Turbilly	22	
		Ludov. Paul Abeille, Secretary of the Academy of Agriculture at Rennes	22	
1715	Sept. 9, 1798	Owen Salusbury Brereton, Esq.	June 17	Nov. 25
	Jan. 29, 1763	John Lewis Count Holstein, of Lethraborg, in Denmark	July 1	
	Jan. 2, 1772	Wm. Fitzherbert, Esq.	Nov. 11	June 27, 1765
	Sept. 9, 1765	Robt. Webb, Esq.	11	Jan. 13, 1763
	Sept. 7, 1766	Thomas Tyndall, Esq.	25	
Feb. 17, 1730	Oct. 13, 1801	Mr. Rd. Pulteney, Apothecary; afterwards M. D.	25	May 12
	1802	Samuel Felton, Esq.	Dec. 9	Dec. 16, 1762
	Dec. 26, 1776	Joshua Platt	23	
		John Baptist Bohadsch, M. D. of Prag. ..	23	
		John Anthony Helvetius, of Amsterdam ..	Feb. 10, 1763	
Dec. 16, 1716	Feb. 26, 1798	Ludov. Jul. Duke de Nivernois	Jan. 27	Feb. 10, 1763
		George Wollaston, M. A.; afterwards D. D.	Feb. 17	April 14
April 19, 1701	July 30, 1769	Andreas Elias Büchner, President of the Imperial. Academiæ Naturæ Curiosorum	17	
		Lorenzo Morosini, Venetian Ambassador ..	March 10	March 17
	June 21, 1779	Antonio Matani, M. D. of Pisa	10	
	1778	Lawrence Theodore Grouovius, of Leyden ..	10	
		Thomas Hornsby, M. A.. Sav. Prof. Ast. ..	April 21	June 16
	June 15, 1795	Anthony Shepherd, B. D.; since D. D. and Plum. Prof. Math.	21	16
	1784	Joseph Raulin, M. D. of Paris	May 12	
	Withdrawn 1795	Edward Waring, M. A.; since M. D. and Luc. Prof. Math.	June 2	
	1803	Jonathan Watson, Esq.	9	16
	Feb. 8, 1785	Matthew Duane, Esq.	9	16
	Dec. 21, 1780	James Harris, Esq.	23	Dec. 15
	Dec. 26, 1780	John Fothergill, M. D.	23	Nov. 24
		Charles Hyacinth Anthony, Duke de Gallean	Nov. 10	
Sept. 18, 1737	April 15, 1806	John Turton, M. B.; afterwards M. D. ..	17	March 5, 1767
Nov. 15, 1735	Nov. 16, 1776	Mr. James Ferguson	24	Dec. 8
1710	April 4, 1807	Mr. Joseph Jerome le François de la Lande	24	
July 11, 1732	1772	Mons. Charles Duclos, Secretary of the Academie François	Jan. 12, 1764	
1704	Jan. 24, 1781	Mr. Thomas Yeoman	12	Jan. 19, 1764
		Charles Wm. Ferdinand, Hereditary Prince of Brunswick; afterwards Duke of Brun.	19	19
Oct. 9, 1735		Count Simon Stratico, Prof. in the University of Padua	19	
		Mons. Charles Etienne Louis Camus	26	
Aug. 25, 1699	May 4, 1768	Benjamin Kennicott, D. D.	Feb. 16	Feb. 23
	Aug. 18, 1783	William Samuel Powell, D. D.	16	March 15
	Jan. 19, 1775	Mons. Ferd. Berthoud, of Paris, Clockmaker	16	
March 19, 1727	June 20, 1807	Sidney Swinney, D. D.	16	Feb. 23
	Nov. 1783	Bernard Siegfried Albinus, M. D. Prof. Anat. in the University of Leyden	23	
Feb. 24, 169	Sept. 9, 1770			

5

BORN.	DIED.		ELECTED.	ADMITTED.
1705	Nov. 29, 1780	Hieron. Dav. Gaubius, Prof. Chem. in the University of Leyden	Feb. 23, 1764	
1718	Jan. 5, 1790	Jacob Christ. Schæffer, D. D. Pastor at Ratisbon	23	
	Oct. 1793	Wills, Earl of Hillsborough; afterwards Marquis of Devonshire	March 8	July 5, 1764
	June 22, 1797	Rich. Warren, M. D. ; afterwards Med. Reg.	8	May 3
	Jan. 22, 1773	Charles Lloyd, Esq.	8	April 5
Nov. 18, 1711		Louis Marie Jos. d'Albert d'Ailly, Duke de Picquigny; afterwards Duc de Chaulnes	15	March 29
	Oct. 23, 1769	Eliah Harisey, Esq.	29	April 12
	Sept. 25, 1792	Adam Gottlob Moltke, Count de Bregentwed	April 5	
		Frederick Christian Meuschen, of Hague ..	5	
	May 29, 1791	Michael Morris, M. D.	5	12
	Dec. 22, 1788	Mr. Percivall Pott, Surgeon	5	May 10
	1796	Naphtali Franks, Esq.	May 3	10
Feb. 28, 1724	Dec. 2, 1805	Mr. Joseph Bernard, Marquis de Chabert ..	10	March 21, 1793
	Dec. 16, 1790	John Campbell, Esq.	17	May 24, 1764
Feb. 28, 1736	May 13, 1782	Daniel Charles Solander, M. D.	June 7	June 21
		Peter Simon Pallas, of Berlin, M. D.	7	
Feb. 24, 1711	Dec. 15, 1770	John, Earl of Egmont	21	28
	Jan. 23, 1785	Matthew Stewart, D. D. Prof. Math. University of Edinburgh	21	
	Jan. 13, 1795	Ralph Willett, Esq....	21	Dec. 6
		Mons. Jean Baptiste de Feronce, Ministr. Plenip. from the Duke of Brunsvig Lunen.	28	
		Sir John Webb, Bart.	28	6
		John Wilkinson, M. D.	28	Nov. 15
	Oct. 21, 1771	Fane Wm. Sharpe, Esq.	July 5	July 5
		Samuel Glasse, M. A.; afterwards D. D. ..	5	12
Sept. 22, 1717	Dec. 13, 1783	Peter Wargentin, Secretary to the Academy of Sciences at Stockholm, and Kt. of the Polar Star	12	
1718	July 16, 1794	John Roebuck, M. D.	12	March 24, 1791
July 2, 1734	July 16, 1790	John Francis Cigna, M. D. of Turin	Nov. 22	
June 26, 1730		Charles Messier, of Paris	Dec. 6	
1718	Aug. 22, 1782	Robert, Lord Trevor; afterwards Viscount Hampden	13	Dec. 20, 1764
	Nov. 13, 1787	Henry Stebbing, D. D.	Jan. 24, 1765	Jan. 31, 1765
Jan. 19, 1735	1790	John George Henry Count de Werthern	Feb. 14	
		John Nicoll, Esq.	28	March 7
	Aug. 30, 1804	Thomas Percivall, Esq. ; since M. D.	March 7	June 6
		Edward Stanley, Esq.	7	March 14
	Oct. 30, 1789	John Morgan, M. D. of Philadelphia	7	
		Compte Jo. Baptiste Carburi, Prof. Med. Turin	21	April 18
	July 4, 1787	Richard Jebb, M. D. ; since Sir R. Jebb, Bart.	28	18
Oct. 1732	Jan. 10, 1786	Mons. Jean Baptiste Jaques Elie de Beaumont, Avocat a Parlement de Paris	April 25	
March 9, 1735	July 9, 1784	Mr. Thorbern Bergman; afterwards Prof. Chemistry, at the University of Upsal, and Kt. of the Order of Wasa	25	
	Withdrawn 1804	Mr. William Harrison	May 9	May 16
		Peter Canvane, M. D. of Bath	16	May 15, 1766
	July 17, 1789	Marchio Dominicus Caraccioli, Envoy from the King of the Two Sicilies	23	June 6, 1765
	Dec. 15, 1782	John Cuthbert, Esq.	June 13	20
Oct. 22, 1728	March 9, 1795	Henry Houghton, Esq.; since Bart.	13	20
		John Tennent, M. D. of New York	13	20
	March 7, 1779	Sir John Mordaunt Cope, Bart.	20	27
		John Bentinck, Esq.	20	27
		John Lewin, Esq.	20	20
		David, Lord Cardross; afterwards El. of Buchan	27	Nov. 7
		Hon. Matthias Barnewall	27	Dec. 5
	Sept. 4, 1791	Mr. Daniel Hopkins, of Huntingdon, Surgeon; afterwards M. D.	27	Feb. 6, 1766
Dec. 20, 1736	Jan. 1809	John Mauritius Count de Bruhl	Nov. 7	Nov. 21, 1765
	1790	Count Sigesmond de Redern, Curator of the Academy of Sciences at Berlin	14	
	Nov. 26, 1771	John Bevis, M. D.	21	Dec. 5
Dec. 1710	Nov. 3, 1787	Robert Lowth, D. D.; afterwards Bp. of Oxford, and afterwards of London	21	
	April 1791	Mr. Richard Price; since D. D.	Dec. 5	Dec. 12
	Dec. 25, 1809	Richard Kaye, LL. B.; since LL. D. and Dean of Lincoln and Bart.	5	March 6, 1766
1720	April 7, 1783	Christian Mayer, S. J. Ast. Elect. Palat. ..	19	
	Jan. 16, 1790	Mr. John Landen	Jan. 16, 1766	April 24

BORN.	DIED.		ELECTED.	ADMITTED.
	June 28, 1797	George Keate, Esq.	Jan. 23, 1766	Feb. 6, 1766
		{ Mr. Charles Daniel Trudaine de Montigny, } of Paris	23	
		John Mills, Esq.	Feb. 13	April 24
	May 1779	{ Jno. Winthorpe, Esq. Prof. Math. and Nat. } Philo. in Harvard College at Cambridge, in New England	20	
Nov. 19, 1718	Nov. 4, 1785	{ Mr. Pierre Jean Grosley, of the Acad. des } Inser. at Paris	20	
	Feb. 7, 1768	Tyringham Stephens, Esq.	March 13	March 20
		{ Wm. Græme, Esq. Commander in Chief of } the Venetian Land Forces	13	
	Dec. 1, 1803	Thomas Altle, Esq.	20	20
	Sept. 17, 1781	John Letch, M. D.	20	April 10
	1782	Mons. l'Abbe Francois Gabriel Coyer	20	April 10, 1777
	March 19, 1799	John Strange, Esq.	April 10	April 24, 1766
	1802	Donald Monro, M. D.	24	May 1
	Withdrawn 1802	Samuel Harper, M. A.	24	1
		Thomas Anguish, Esq.	24	1
Feb. 2/13, 1743, O. S.		{ Joseph Banks, Esq.; afterwards P. R. S. } and Bart. K. B.	May 1	Feb. 12, 1767
Dec. 20, 1732		Edward Hasted, Esq.	8	May 8, 1766
		Mr. Dionysius Williams	8	
	Jan. 31, 1808	Leonard Morse, Esq.	15	June 12
	Nov. 5, 1790	Michael Lort, B. D.; since D. D.	15	May 15
		Arthur Lee, M. D.	29	Nov. 10, 1768
April 14, 1738	Oct. 30, 1809	Wm. Henry Duke of Portland	June 5	June 12, 1766
	1796	Wm. Webber, Esq.	5	Nov. 6
	Expelled May 1, 1783	Mr. David Riz	5	June 19
March 24, 1733	Feb. 6, 1804	Joseph Priestley, LL. D.	12	Jan. 28, 1768
	1785	Thomas Forster, M. A.	19	Nov. 13, 1766
	April 1803	{ Wm. Hamilton, Esq. Envoy Ext. to Naples; } afterwards K. B. and Privy Counsellor	Nov. 6	14, 1771
Jan. 17, 1732	Feb. 12, 1798	His Majesty Stanislaus Augustus King of Poland	Dec. 11	
	March 1, 1797	{ Mr. Henry Putman, Minister of the Dutch } Church in Austin Friars	Jan. 8, 1767	Jan. 29, 1767
Sept. 21, 1725	Dec. 26, 1793	Brownlow, Earl of Exeter	15	Feb. 12
	Feb. 19, 1803	Allan Pollok, M. D.	22	Jan. 29
	Nov. 5, 1798	John Zephan Holwell, Esq.	29	Feb. 5
		Anthony Tissington, Esq.	29	5
	1803	Mr. Peter Woulfe	Feb. 5	March 12
		Mr. John Hunter	5	Feb. 19
Feb. 14, 1730	Oct. 16, 1793	Sir James Naesmith, Bart.	12	
	1779	Sir Thomas Fludyer	12	March 5
	Dec. 1, 1786	John Hope, M. D.	12	Nov. 9, 1776
		Richard Eyre, D. D.	19	Feb. 26, 1767
	1771	Frederick, Lord Baltimore	26	March 19
June 15/26, 1726	Dec. 16, 1798	Thomas Pennant, Esq.	26	5
		John Mytton, Esq.	26	12
		John Martin Butt, M. D. of Jamaica	26	
	Oct. 14, 1786	Rd. Wright, M. A.; afterwards M. D.	March 19	April 2
		Joshua Kirby, Esq.	26	9
	June 30, 1790	Wm. Roy, Esq.; afterwards Major-General	26	9
	Aug. 11, 1788	Edward, Earl Winterton	April 2	9
	Sept. 1808	Steddy Grinfield, Esq.	2	9
	1786	John Malliet, Esq.	2	9
	Oct. 30, 1793	Mr. Henry Watson, Surgeon	2	9
	April 29, 1788	{ John Parker, Jun. Esq; afterwards Lord } Borringdon	9	June 4
	Dec. 12, 1786	William James, Esq.	9	April 30
	Oct. 4, 1806	{ Samuel Horsley, M. A.; afterwards LL. D. } and Lord Bp. of St. David's, afterwards of Rochester, since of St. Asaph	9	30
May 23, 1718	March 30, 1783	Wm. Hunter, M. D.	30	May 7
	Feb. 17, 1798	{ Mr. Francis Geach, of Plymouth, Surgeon; } afterwards M. D.	May 7	
	April 18, 1807	Edward King, Esq.	14	May 21
June 5, 1723	July 18, 1790	Adam Smith, LL. D.	21	May 27, 1773
		Robert Mylne, Esq.	21	June 4, 1767
	Oct. 11, 1780	Anthony Chamier, Esq.	28	4
	Jan 21, 1800	George Stevens, Esq.	28	18
		Hon. Chas. Dillon; afterwards C. Dillon Lee	28	July 2
		Robert Weston, Esq.	28	June 25
		Hon. Daines Barrington	June 4	18

1

BORN.	DIED.		ELECTED.	ADMITTED.
March 15, 1745		Charles Howard, Jun. Esq.; afterwards El. of Surry, at last Duke of Norfolk	June 18, 1767	April 14, 1768
	Feb. 1790	Daniel Minet, Esq.	18	Nov. 12, 1767
		Mr. Charles L'Epinasse	July 9	July 9
	1785	Mr. George Witchell	9	May 19, 1768
May 11, 1743		Rd. Henry Alex. Bennet, Esq.	Dec. 10	Dec. 17, 1767
Aug. 28 Sept. 8, 1744		Wm. Watson, Jun. M. B.; since M. D. and Kt (name in the Charter-book, between Bennett and Lord Greville)	10	May 19, 1768
Sept. 6, 1746	Sept. 9, 1801	Owen Manning, B. D.	10	Jan. 21
		George Lord Greville; afterwards El. of Warwick	17	7
	Feb. 1, 1776	William, Earl of Radnor	17	Feb. 18
	July 24, 1785	Rd Huck, M. D.; since Rd. Saunders, M. D.	Feb. 18, 1768	25
		Charles Moore, Esq.	25	May 12
		Mr. James Dargent	March 17	March 24
	Aug. 31, 1786	Chas. Howard, Sen. Esq.; since Duke of Norfolk	24	April 14
	Nov. 4, 1775	Mr. Daniel Harris	24	14
	April 25, 1785	Mr. James Horsfall	April 14	May 5
	Jan. 2, 1803	Richard Penneck, A. M.	14	April 28
	Nov. 12, 1775	Christopher Nugent, M. D.	14	28
		Mr. John Lodge Cowley	14	28
	Aug. 7, 1783	Mr. John James Majendie; since D. D. ..	21	May 19
	Feb. 8, 1784	John Darker, Esq.	May 5	12
		Thomas Dundas, Esq.; afterwards Bart. and Lord Dundas	5	Feb. 16, 1769
	July 29, 1795	John Heathcote, Esq.	12	
	Feb. 19, 1784	Thomas Morell, D. D.	June 16	June 23, 1768
Jan. 29, 1749	March 13, 1808	Christian VII., King of Denmark	Sept. 1	
	1797	Thomas Emlyn, Esq.	Nov. 10	Nov. 17
1725	1774	Robert Lord Clive	24	April 13, 1769
	Feb. 8, 1807	Thomas Lashley, M. D. of Barbadoes ..	24	Nov. 25, 1784
Jan. 10, 1729	Feb. 10, 1799	Mr. Lazarus Spallanzani, of Modena	June 2	
1710	Nov. 15, 1783	Joseph Etienne Bertier, Ord. Ora. of Paris, ..	2	
		Richard Watson, M. A.; afterwards D. D. and Prof. Div. Cant. afterwards Lord Bishop of Landaff	Feb. 2, 1769	April 27, 1769
	Sept. 10, 1781	Mr. John Cavernill; afterwards M. D.	9	March 16
		Richard Hill Waring, Esq.	16	23
		James Greive, M. D.	23	2
	Dec. 30, 1800	Thomas Dimsdale, M. D. Baron of the Empire of Russia	March 16	May 11
		John Corham Huxham, M. A.	16	April 13
		Francis Wollaston, LL. B.	April 13	20
	March 17, 1810	Mr. Wm Snarp, Surgeon	20	27
		Thaddæus Joseph Count de Burzynski, Senator of Poland	May 11	Jan. 25, 1770
	Sept. 7, 1799	John Ingenhousz, M. D.	25	March 21, 1771
	Ejected Dec. 7, 1775	Mr. Rudolph Erick Raspe	June 1	
April 29 1726	Aug. 13, 1784	John François Clement Morand, M. D. ..	1	
		Wm. Wyatt, M. A.	8	April 26, 1770
	Sept. 1804	John Ibbetson, Esq.	Nov. 16	Jan. 18
	Nov. 2, 1777	The Prince de Masserano	Jan. 11, 1770	May 10
June 1734	July 5, 1807	Mr. Timothy Lane	11	Jan. 18
	Aug. 23, 1809	Wm. Farr, M. D. of Plymouth	March 1	June 3, 1779
	March 12, 1780	Topham Beauclerk, Esq.	8	March 22, 1770
	1774	Mr. Wm. Hewson	8	15
	June 1778	James Welsh, M. D. of Winchester	15	March 4, 1773
		Andrew Joseph Planta, M. A.	15	March 22, 1770
	Sept. 1, 1795	Francis Russell, Esq.	15	Nov. 12, 1772
		Justin Macarthy, Esq.	April 5	
		Mr. David Bayford, Surgeon; afterwards M. D	26	May 10, 1770
		Henry Jerome de Salis, M. A.; afterwards D. D.	May 3	Nov. 26, 1772
	Withdrawn 1784	Charles Collignon, M. D.	3	
	Nov. 25, 1791	William Pitcairn, M. D.	10	May 17. 1770
	June 1, 1784	Thomas Dickson, M. D.	10	17
	1789	Charles Douglas, Esq.; afterwards. Bart and Rear Admiral of the Blue Squadron	17	24
	1790	Peter Jonas Bergius, M. D. of Stockholm ..	31	
Jan. 12, 1726	Dec. 10, 1779	Chas. Le Roy, M. D. Prof. Phys. at Montpeilier	31	
	1789	Thomas Healde, M. D.; afterwards Prof. Med. Gresh.	June 21	June 28
	1779	Edward Thomas, M. A.	21	July 5
		Isaac Hawkins Browne, Esq.	July 5	Nov. 15
	March 9, 1795	John Walsh, Esq.	Nov. 8	22

BORN	DIED		ELECTED	ADMITTED
	Aug. 1778	Samuel Wells Thomson, D. D.	Nov. 8, 1770	Nov. 22, 1770
		John Arbuthnot, Esq.	22	Dec. 20
		Mr. Robert Erskine, Engineer	Jan. 31, 1771	Feb. 7, 1771
July 24, 1737	June 19, 1808	Alexander Dalrymple, Esq.	Feb. 14	March 14
		John Wynn Baker, Esq.	14	
	Nov. 3, 1787	John Glen King, M. A.	21	Feb. 28
	Aug. 15, 1786	Thomas Tyrwhitt, Esq.	28	May 9
		Mr. Samuel Howard, Surgeon	March 14	March 21
	Oct. 11, 1790	Marmaduke Tunstall, Esq.	April 11	April 18
		Francis Maseres, Esq.	May 2	May 16
April 1743	Dec. 12, 1795	John Paradise, Esq.	2	9
Dec. 22, 1744	Jan. 16, 1787	Paul Henry Matty, M. A.	16	30
	April 21, 1807	Rev. Mr. George Walker	30	Dec. 5
		{ Father Martin Poezobut, Warsaw, Astron. } { to the King of Poland }	30	
		John Philip de Limbourg, M. D. of Spaa ..	30	
	Nov. 20, 1809	Philip Stevens, Esq.; afterwards Bart.	June 6	June 13
		James Petty, Esq.	6	Feb. 6, 1772
Aug. 10, 1740	July 11, 1807	John Frere, Esq.	6	June 20, 1771
Feb. 4/15, 1724	Jan. 13, 1800	His Serene Highness, Peter Duke of Courland	Nov. 7	
	Sept. 1774	Sir Wm. Duncan, Bart. Med. Reg.	14	
Sept. 18/29, 1740	Aug. 29 1808	Benjamin Way, Esq.	Dec. 5	Dec. 12
1746	Oct. 10, 1792	{ Hon. Constantine John Phipps; since Lord } { Mulgrave }	12	12
		Richard Cope Hopton, Esq.	12	Feb. 27, 1772
		Richard Paul Jodrell, Esq.	Jan. 9, 1772	Jan. 23
April 30 } May 11, 1730 }	Oct. 19, 1805	Alexander Aubert, Esq.	9	16
		Benjamin Booth, Esq.	9	16
		{ Cyril Jackson, Esq.; afterwards D. D. and } { Dean of Christ Church, Oxford }	16	Feb. 13
	1804	Nathaniel Pigott, Esq.	16	June 4
May 12, 1749	April 23, 1809	Hon. Charles Francis Greville	Feb. 13	Feb. 27
Nov. 8, 1782	Dec. 25, 1800	John Carnac, Esq.	20	Dec. 10
Oct. 22, 1729	Dec. 16, 1798	Mr. John Reinhold Forster; afterwards LL. D.	27	March 5
		Martin Folkes, Esq.; afterwards Bart. ..	April 2	May 13, 1790
	Expelled Nov. } 15, 1787 }	Philip Van Swinden, Jun. M. A.	2	May 14, 1772
		Thomas Pownall, Esq.	9	7
	April 27, 1794	William Jones, Esq.; afterwards Kt.	30	14
		{ Francis Milman, M. B.; afterwards M. D. } { and Bart. }	May 7	Dec. 5, 1776
		Richard Warburton Lytton, Esq.	28	June 4, 1772
May 9, 1742		Wm. Philip Perrin, Esq.	28	25
		Mr. Tesser Samuel Kuckhan	June 4	
		{ Marcus Antonius Leopoldus Caldanus, Prof. } { Physic in the University of Padua }	4	
		Isaac Gossett, M. A.; afterwards D. D. ..	18	25
April 17, 1748		Charles Blagden, M. D.; afterwards Kt. ..	25	Nov. 12
	1803	Edward Poore, Esq.	July 9	19
	June 12, 1795	John Greg, Esq.	9	Feb. 10, 1795
	Aug. 1781	Kenneth, Earl of Seaforth	Nov. 12	Nov. 19, 1772
	1801	Humphry Jackson, Esq.	19	26
Aug. 3, 1753		{ Charles, Lord Viscount Mahon; afterwards } { Earl of Stanhope }	19	Jan. 12, 1795
		John Lauder, Esq.	Dec. 24	Dec. 24, 1772
	Sept. 2, 1801	Hon. John Yorke	Feb. 18, 1773	April 29, 1773
		Dr. John Bethume	18	
	Jan. 31, 1788	Ashton Lever, Esq.; since Sir A. Lever, Kt.	18	May 1, 1777
	1802	Thomas Butterworth Bayley, Esq.	18	March 4, 1773
		{ Heneage, Lord Guernsey; afterwards Earl } { of Aylesford }	25	11
July 15, 1751		Wm. Benson Earle, Esq.	March 4	May 20
	March 21, 1796	Patrick Brydone, Esq.	4	March 11
		Wm. Falconer, M. D.	18	
		John Ives, Esq.	25	April 22
July 1750	Jan. 9, 1776	Alexander Lord Polwarth	April 1	29
Jan. 29, 1751	March 9, 1781	{ Francis, Marquis of Carmarthen; afterwards } { Duke of Leeds }	1	March 3, 1774
May 30, 1751	Jan. 31, 1799	Other, Earl of Plymouth	22	May 6, 1773
	June 12, 1799	Peter Livius, Esq.	29	July 1
	July 23, 1795	Richard Blyke, Esq.	29	May 13
		Mr. Wm. Henly, Linen Draper	May 20	June 10
		Edward Bancroft, M. B.; afterwards M. D.	20	10
	June 3, 1781	Thomas Dummer, Esq.	27	March 24, 1774

BORN.	DIED.		ELECTED.	ADMITTED.
	Nov. 1807	John Smith, Esq.; afterwards Sir J. Smith, Bart.	May 27, 1773	June 10, 1773
	Jan. 15, 1795	Sir Lucius O'Brien, Bart.	27	June 23, 1774
Sept. 18, 1750		Thomas Frankland, Esq.; afterwards Bart.	June 10	Jan. 27
Jan. 1730	April 15, 1791	Alexander Garden, M. D.	10	May 15, 1783
1710	July 6, 1785	Mr. Jacob de Stehelin, Secretary to the Imperial Acad. of Sciences at St. Petersburg	10	
		Mr. Jean Baptiste Le Roy, Parisiensis	10	
		Mr. Jean Andre de Luc, Genevensis	10	
	Oct. 25, 1787	Jacob Preston, Esq.	17	Jan. 27, 1774
	Never admitted	Sir Watkin Wm. Wynne, Bart.	17	
		John Coakley Lettsom, M. D.	Nov. 18	Nov. 25, 1773
		Jeremiah Dixon, Esq.	18	Jan. 9, 1777
	Jan. 12, 1781	John Lind, Esq.	25	Dec. 9, 1773
		Charles Burney, Mus. D.	Dec. 16	23
Feb. 21, 1744		Mr. Joseph Planta	Feb. 17, 1774	Feb. 24, 1774
	Feb. 5, 1807	General Pasquale de Paoli	March 3	March 17
		John Mervin Nooth, M. D.	3	24
	Aug. 30, 1784	Michael Teighe, M. D.	10	17
Oct. 1735	Jan. 3, 1800	Sir Wm. Musgrave, Bart.	17	April 14
Nov. 4, 1722	Feb. 1790	Mr. John Hyacinth de Magalhaens or Magellan	April 21	21
		Arthur Young, Esq.	28	28
	March 14, 1808	Philip Duval, M. A.; afterwards D. D. ..	May 12	Feb. 9, 1775
	May 11, 1789	Richard Michell, Esq.	12	May 19, 1774
	March 16, 1799	Wm. Gould, M. A.; afterwards D. D.	19	June 16
	Withdrawn 1796	Mr. Murdoch Mackenzie, Sen.	19	9
		John Ellis, Esq.	19	9
	1802	Granado Pigott, Esq.	June 2	Jan. 11, 1776
July 5, 1720	Sept. 22, 1798	Mr. Pierre Poissonnier, M. D. of Paris ..	2	
	July 1809	Mr. Anthony Geo. Eckhardt, at Hague	June 2	June 9, 1774
	July 24, 1791	Le Chevalier Ignace de Born	2	
	Feb. 22, 1789	James Walker, M. B.	9	Nov. 10
	Feb. 5, 1804	Patrick Geo. Crafurd, Esq.	9	10
	Withdrawn 1794	Richard Twiss, Esq.	9	June 16
		Mr. Charles Hutton; afterwards LL. D. ..	16	Nov. 10
	Withdrawn 1797	Francis Duroure, Esq.	Nov. 10	17
	1790	Robert Salusbury Cotton, Esq.	24	Dec. 8
	Withdrawn 1800	John Bagnall, Esq.	Dec. 8	March 16, 1775
		John Willett Adye, Esq.	8	Jan. 12
Jan. 14/25, 1750		John Lloyd, Esq.	15	May 11
Aug. 23, 1751	Aug. 11, 1804	Sir Geo. Shuckburgh, Bart.	22	Dec. 5, 1776
	1809	Alexander Hunter, M. D.	Feb. 2, 1775	
Oct. 12, 1723	Sept. 1809	Robt. Melville, Esq.; afterwards General ..	16	Feb. 23, 1775
	April 22, 1797	Jeremiah Milles, Esq.	23	March 9
June 21, 1733	Oct. 8, 1785	Sir John Cullum, Bart.	March 2	30
	Withdrawn 1795	Richard Gough, Esq.	9	June 29
	Sept. 15, 1789	Sir Robt. Barker, Kt.; afterwards Bart. ..	16	March 23
		Maxwell Garthshore, M. D.	23	30
	March 13, 1805	Sir Walter Rawlinson, Kt.	30	May 4
	Dec. 16, 1783	Wm. James, Esq.; since Sir Wm. James, Bart.	30	4
	Dec. 21, 1799	James Napier, Esq.; since Sir J. Napier, Kt.	30	April 6
		Mr. Wm. Hey, Surgeon of Leeds	30	June 18, 1778
		Lewis Dutens, M. A.	April 27	April 17, 1777
	May 1791	Wm. Constable, Esq.	May 4	May 11
		Mr. Thomas Henry, Apothecary at Manchester	18	
June 27, 1740		Mr. John Latham, Surgeon, of Dartford; afterwards M. D.	25	July 6
	Dec. 19, 1787	Mr. Robt. Sterling, Surgeon, of Colchester ..	June 1	June 15
June 13, 1724	Nov. 19, 1803	Mr. George Lewis Le Sage, of Geneva	1	
	Aug. 22, 1794	Chev. Achille Pierre Dionis Du Séjour	1	
	Jan. 6, 1800	Wm. Jones, A. B; afterwards A. M.	22	Jan. 9, 1777
	March 7, 1801	John Call, Esq.; afterwards Bart.	Nov. 9	Jan. 25, 1776
	1787	John Pitt, Esq.	9	Dec. 7, 1775
	March 27, 1786	Mr. John Obadiah Justamond, Surgeon ..	Dec. 14	Jan. 11, 1776
		George Finch Hatton, Esq.	7	Feb. 29
		Sir Abraham Hume, Bart.	14	March 14
	April 1794	James Bruce, Esq.	Jan. 11, 1776	14
Sept. 23, 1743		Mr. Charles Coombe, Apothecary; afterwards M. D.	11	Jan. 18
		John Elliot, Esq.; afterwards Admiral ..	18	Jan. 9, 1777
		Robert Ker, Esq.	25	April 25, 1776
	1802	George Fordyce, M. D.	Feb. 15	Feb. 22
Oct. 27, 1728	Feb. 14, 1779	James Cook, Esq. Captain in the Royal Navy	29	March 7
Sept. 28, 1744		Hon. Charles Marsham; afterwards Lord Romney, afterwards Earl	March 14	28

BORN.	DIED.		ELECTED.	ADMITTED.
		Samuel Prime, Esq.	March 21, 1776	March 28, 1776
June 23, 1716	Jan. 1, 1789	Rt. Hon. Sir Fletcher Norton, Kt.; afterwards Lord Grantley	April 18	
	March 8, 1796	Sir Wm. Chambers, Kt. of the Polar Star ..	25	May 2
		Wm. Cooper, D. D.	25	Nov. 7
	May 6, 1786	John Taylor, Esq.; afterwards Bart.	May 9	June 20
Sept. 26, 1752		Owen Putland Meyrick, Esq.	9	Feb. 6, 1777
		John Alleyne, Esq.	16	May 23, 1776
		Sir John Chetwode, Bart.	16	23
		John Stewart, Esq.	23	Nov. 21
	April 30, 1783	George Stinton, D.D.	23	June 6
	July 3, 1787	Wm. Calderwood, Esq.	June 6	13
Feb. 19, 1742		Prince Abondio Rezzonico, of Rome .. .	6	Nov. 9, 1786
	1805	M. Jean Baptiste Gasper Dausse de Villoeson, of Paris	6	
		Don Pedro Davila, Madrid	6	
	1807	George Atwood, M. A.	13	Nov. 13, 1776
	Nov. 8, 1781	Thomas Crofts, M. A.	13	20
	Aug. 1806	Mr. Edw. Nairne, Mathamat. Instrument-maker	20	27
	Dec. 6, 1785	Samuel Hemming, M. A.	20	27
Dec. 4, 1739	April 19, 1802	Henry, Viscount Palmerston	Nov. 7	Nov. 7
	Dec. 29, 1798	Mr. Wm. Wales	7	14
Nov. 10, 1755		Philip, Earl of Chesterfield	Dec. 19	Feb. 6, 1777
Oct. 12/23, 17—	Aug. 7, 1787	Wm. Russell, Esq.	Jan. 9, 1777	Jan. 16
Nov. 26, 1754	Jan. 11, 1794	Mr. George Forster; afterwards M. D. ..	9	Feb. 6
	Sept. 1791	Sir Herbert Mackworth, Bart.	9	March 6
Aug. 26, 1738	Dec. 22, 1789	George Nassau Clavering, Earl Cowper ..	Feb. 13	
	1802	Richard, Lord Grosvenor; afterwards Earl	13	6
	April 26, 1794	Wm. Brown, Esq.	20	Feb 27
	1798	Molyneux, Lord Shuldham	March 13	Dec. 11
	July 28, 1787	Edward Bridgen, Esq.	13	April 10
Jan. 26, 1724	Dec. 15, 1784	Nath. Matthew Wolf, M. D. of Dantzig ..	April 10	
		John Osborn, Esq.	17	24
Mar 29 / April 9, 1744		Robert Shuttleworth, Esq.	24	May 8
		Anthony Hamilton, D.D.	May 1	15
		Gustavus Adam Baron Nolcken, Envoy from the King of Sweden	8	June 19
	Feb. 1790	Wm. Cullen, M. D. Prof. Physic at Edinburgh	8	
March 13, 1749		Charles Anderson Pelham, Esq.; afterwards Lord Yarborough	8	Dec. 18
	Jan. 2, 1801	Edw. Ld. Amiens; afterwards El. of Aldborough	29	June 12
March 16, 1749		John Peachey, Esq.; afterwards Lord Selsey	29	12
	March 27, 1793	Mr. John Mudge, Surgeon, of Plymouth ..	29	
		Cassinier Gomez Ortega, M. D. of Madrid ..	June 5	
July 11, 1719		Mr. Joseph Toaldo, Prof. Ast. at Padua ..	5	
	Feb. 12, 1797	Thomas White, Esq.	19	Nov. 6
		George Samuel Wegg, Esq.	Nov. 13	20
Feb. 6/17, 1726	July 2, 1805	Patrick Russell, M. D.	27	Dec. 11
May 17, 1736		James Lind, M. D.	Dec. 18	Jan. 8, 1778
		Hon. Archibald Campbell Frazer	Jan. 8, 1778	15
	Aug. 1, 1784	Matthew Dobson, M. D. of Liverpool	Feb. 12	Feb. 19
		Mr. John Wyatt, Surgeon	12	19
		Henry Chas. Englefield, Esq.; afterwards Bart.	12	19
		Henry Partridge, Esq.	19	26
March 1736	1780	Hon. Robt. Boyle Walsingham; afterwards Lord Walsingham	March 5	Feb. 10, 1780
		Thomas de Grey, Jun. Esq.	5	Nov. 19, 1778
March 18, 1740		Wm. Wright, M. D.	12	Feb. 13, 1780
	May 2, 1796	James Watson, Esq.; afterwards Kt.	12	March 19, 1778
March 2, 1749	April 10, 1803	Rev Chas. Peter Layard, M. A.; afterwards D. D. and Dean of Bristol	26	May 7
		Wade Toby Caulfield, Esq.	April 2	April 9
	Oct. 17, 1782	Joseph Nash, Esq.	2	30
	May 1780	Thomas Cave, Esq.; afterwards Bart.	9	May 21
		Benjamin Heath, M. A.	30	21
	Nov. 11, 1792	Robt. Banks Hodgkinson, Esq.	30	21
	Oct. 20, 1800	Wm. Augustus Howard, M. D.	May 14	21
	March 9, 1780	Joseph Flse, Esq. Surgeon	14	21
		Lancelot Shadwell, Esq.	June 4	Nov. 5
	1805	Sir Rich. Worsley, Bart.; afterwards Rt. Hon.	18	26
		Alexander Hay, M. D.	18	June 25
	May 18, 1807	John Douglas, D.D.; afterwards Ld. Bp of Carlisle, afterwards of Salisbury	25	Nov. 19

BORN.	DIED.		ELECTED.	ADMITTED.
	1789	{ Rev. Wm. Preston, M. A. ; afterwards Ld. Bishop of Killala, afterwards of Leighlin and Ferns	June 25, 1778	July 9, 1778
	Withdrawn 1804	Rev. John Lockman, D. D.	25	Nov. 19
		Henry Dawkins, Esq.	Nov. 5	May 13, 1779
		Anthony Fothergill, M. D. of Northampton ..	12	Nov. 12, 1778
	Oct. 4, 1786	John Alstroemer, Esq.	Dec. 24	
Nov. 18, 1745	April 8, 1796	{ Hon. Thomas Francis Wenman, Brother to Viscount Wenman	Jan. 21, 1779	March 18, 1779
Aug. 29, 1751		John Joshua Lord Carysfort; afterwards Earl	Feb. 4	Feb. 4
	April 24, 1799	Wm. Seward, Esq. ..	11	18
March 21, 1748	Dec. 27, 1806	Edward Whitaker Gray, M. D.	11	March 4
	May 4, 1780	Rev. Michael Tyson, B. D.	11	4
		John Jebb, M. D.	18	Feb. 25
	Sept. 27, 1781	Rev. Robt. Richardson, D. D.	25	March 11
	Never admitted.	Samuel Farr, M. D.	25	
		Mr. Thomas Vage; afterwards M. D.	25	4
	Expelled } April 7, 1791 }	Henry Dagge, Esq.	March 11	April 15
		Lieut. James Glenie	18	Jan. 18, 1781
	March 24, 1786	Robt. Bromfield, M. D.	April 15	April 22, 1779
Jan. 6, 1752	Aug. 19, 1803	John Tonham, Esq.	15	May 13
	Jan. 1, 1805	George Buxton, M. D.	22	6
		{ Benjamin Thompson, Esq. ; afterwards Kt and Count of Rumford of the Holy Roman Empire	22	6
	Feb. 1, 1807	{ Sir Ralph Payne, K. B. ; afterwards Lord Lavington	29	April 29
		James Carmichael Smyth, M. D.	May 6	May 13
		Mr. Joseph Poli	6	13
		{ John Rogerson, M. D. Physician to the Empress of Russia	6	April 11, 1799
March 31, 1714	Feb. 18, 1788	Mr. John Whitehurst	13	May 13, 1779
Feb. 14, 1728	Withdrawn 1808	Major-Gen. Chas. Rainsford; afterwards Gen.	13	June 3
	Oct. 1, 1780	Josias Dupré, Esq.	June 3	10
March 28, 1725	Oct. 8, 1795	Andrew Kippis, D. D.	17	24
	Feb. 13, 1808	Wm. Fullarton, Esq.	17	Dec. 16
	March 19, 1794	Hon. James Murray	24	16
March 17, 1749 50		Samuel Foart Simmons, M. D.	Nov. 4	Nov. 11
	April 1803	{ John Henniker, Esq. ; afterwards Bart. and last Lord	11	18
		John Grant, Esq.	11	Dec. 16
		John Jennings, Esq.	18	23
		John Wilmot, Esq.	18	16
March 30, 1749	Dec. 21, 1809	Mr. Tiberius Cavallo	Dec. 9	Jan. 27, 1780
	Jan. 22, 1790	Hugh Hamersley, Esq.	16	Feb. 17
		Mr. Bernardo de Beluga	Jan. 20, 1780	
	Withdrawn 1781	Hon. Edw. Onslow, son of Lord Onslow ..	27	3
Nov. 25, 1743	Aug. 25, 1805	Wm. Henry, Duke of Gloucester	Feb. 10	
Feb. 24, 1736	Jan. 5, 1806	{ Christian Frederick Charles Alexander Margrave of Anspach and Bayreuth	10	10
May 1745		John, Earl of Upper Ossory	17	April 2, 1789
Aug. 1, 1735		Richard Kirwin, Esq.	24	March 2, 1780
	March 1, 1780	Lieut. Gen. Thomas Desaguliers	24	{ Died before admission.
		John Silvester, Esq.	March 2	March 2
	Sept. 4, 1797	Robt. Marsham, Sen. Esq.	9	May 31, 1781
	July 2, 1801	Robt. Edw. Lord Petre	April 6	March 8
Sept. 1, 1758		{ George John Viscount Althorpe; afterwards Earl Spencer	6	April 20, 1780
	1796	Mr. Chas. de Casaux, of Grenada	13	April 6, 1786
		Mr. Pierre Moultou, of Geneva	27	May 4, 1780
	Feb. 28, 1801	Lieut. Colonel John Duroure	May 25	June 1
		John Ord, Esq.	June 1	15
July 29, 1756		{ Joseph Lewis de Podmaniczky, Seigneur d'Aszod Földrár	8	
		{ Isaac Milner, M. A. ; afterwards D. D. and Dean of Carlisle	15	Dec. 14
June 30, 1755		John, Duke of Athol	Nov. 9	Feb. 8, 1781
		{ Lucas Pepys, M. D., Physician Extraordinary to his Majesty; afterwards Bart.	9	Nov. 16, 1780
		Mr. Philip Hurlock, Surgeon	16	23
June 16, 1713		Henry Penton, Esq.	16	Dec. 14
		Paul Prince Daschkaw, of the Russian Empire	Feb. 8, 1781	
		John Haygarth, M. B. of Chester	8	May 10, 1798

h

BORN.	DIED.		ELECTED.	ADMITTED.
		George Young, Esq. Captain in the Royal Navy; afterwards Kt.	Feb. 15, 1781	March 8, 1781
		John Lee, M. D. of Bath	15	Feb. 7, 1782
April 18, 1753		George, Lord De Ferrars; afterwards Earl of Leicester	15	March 15, 1781
May 22, 1751	1784	James King, Esq. Capt. in the Royal Navy, LL. D.	March 1	8
	Aug. 17, 1782	Thomas Pattinson Yeats, Esq.	8	22
Dec. 2, 1735	Feb. 21, 1783	The Rev. John Lightfoot, A. M.	8	29
Nov. 22 Dec. 7, 1742		James Rennell, Esq.	8	15
		Thomas Davies, Esq. Capt. in the Royal Artillery; afterwards Col. Lieut.-Maj. Gen.	8	22
	May 22, 1790	Wm. Franks, Esq.	8	22
		Mr. Rich. Brown Cheston, Surgeon, Gloucester	15	April 5
	Oct. 30, 1805	Welbore Ellis Agar, Esq.	22	5
	April 13, 1788	Peter Calvert, LL. D.	29	5
	Dec. 12, 1789	Anthony de Weselincboven, Esq. of Brussels	April 5	
Oct. 3, 1755		George, Viscount Lewisham; afterwards Earl of Dartmouth	May 3	June 14
Jan. 2, 1728		Lewis Charles Maria, Count of Barbiano and Belgisioso of the Holy Roman Empire, Envoy and Minster Plenipotentiary from the Emperor	3	May 10
May 28, 1738	May 14, 1782	Richard Parry Price, Esq.	3	10
		Thomas Bowdler, M. D.	10	24
	Jan. 3, 1809	Richard Shepherd, B. D.; afterwards D. D.	10	31
	July 31, 1783	James Price, A. M.; afterwards M. D. ..	10	17
		Henry Fly, A. M.; afterwards D. D.	10	17
		Wm. Vyse, LL. D.	17	31
		Henry Revell Reynolds, M. D.	17	24
	June 30, 1782	Thomas Blackburne, M. D.	June 14	
		Richard Lovell Edgeworth, Esq.	July 5	Feb. 28, 1782
	March 12, 1783	Patrick Dugud Leslie, M. D. of Durham ..	Nov. 8	April 25
		Joseph Windham, Esq.	8	Nov. 15, 1781
Sept. 1, 1739		Rev. Francis Henry Egerton	8	Jan. 24, 1792
	1803	Paul Jodrell, Esq.; afterwards M. D. and Kt.	15	10
Nov. 1738		Mr. Wm. Herschell; afterwards LL. D. ..	Dec. 6	May 30
		Theodore Forbes Leith, M. D.	20	Jan. 24
Sept. 11, 1757	April 20, 1802	Hon. Geo. Augustus North; afterwards Ld. North and Earl of Guilford	Jan. 17, 1782	March 7
		Rev. Wm. Coxe, M. A.	Feb. 14	Feb. 21
Jan. 19, 1761	1807	Peter Maria Augustus Broussonet, M. D. of Montpelier; afterwards Prof. Bot. at Montpelier, and Member of the Institute of France	14	21
	May 17, 1795	Henry Beaufoy, Esq.	21	28
March 1723	Feb. 1, 1808	Sir Jas. Peachey, Bart.; afterwards Ld. Selsey	28	March 14
		Roger Wilbraham, Esq.	28	21
Aug. 1730	1803	Frederick, Earl of Bristol, Lord Bp. of Derry	28	14
Nov. 30, 1735	1788	Samuel Greigg, Esq. Vice Adm. in the Russian Navy, Kt. of the Orders of Alex. Nevsky, St. George, and St. Ann	March 14	
	Aug. 10, 1793	Joseph Hurlock, Esq.	14	April 11
		Daniel Braithwaite, Esq.	14	11
	April 17, 1809	David Pitcairn, M. B.; afterwards M. D. ..	April 11	18
	Aug. 7, 1807	Matthew Guthrie, M. D. Physician to the Imperial Land Cadett Corps of Nobles at St. Petersburg	11	June 6
	Feb. 14, 1798	John Gunning, Esq. Surgeon Extraordinary to his Majesty	25	May 2
	1805	Barnard Christian Anker, Esq. of Christiana	Nov. 17	May 8, 1783
		Henry Penruddocke Wyndham, Esq.	Jan. 9, 1783	8
	Jan. 3, 1795	Mr. Josiah Wedgewood, Potter to her Majesty	16	Feb. 13
	Aug. 1, 1792	Thomas Hollingbery, D. D. Archdeacon of Chichester	23	13
		Wm Marsden, Esq.	23	6
Nov. 16, 1754		Matthew Raper, Esq.	March 6	March 13
Sept. 24, 1742	Oct. 18, 1793	John Wilson, Esq.; afterwards Kt. and one of the Justices of the Court of Com. Pleas	13	May 1
		John Law, Lord Bp. of Clonfert and Kilmacduah; afterwards of Killala, since of Elphin	20	March 27
		John, Lord Sheffield	April 3	May 1
	Dec. 25, 1807	Brownlow, Lord Brownlow	May 8	15
		Gideon Fournier, Esq.	15	22

1

BORN.	DIED.		ELECTED.	ADMITTED.
	June 8, 1806	Thomas Barnard, Lord Bp. of Killaloe and Kilfenora; afterwards of Limerick	May 29, 1783	June 5, 1783
	Jan. 6, 1795	Francis d'Aquino, Prince of Caramanico, Kt. of the Orders of St. Januarius and Malta, Envoy Ext. and Minister Plenip. from the King of the Two Sicilies	June 5	19
	Aug. 29, 1810	Christopher Wm. de Dreyer, Kt of the Order of Dannebrog, Envoy Extraordinary from the King of Denmark	5	19
Jan. 18, 1744		The Rev. Wm. Tooke, Minister of the English Church at St. Petersburg	5	Feb. 26, 1784
		Capt. Thomas Hyde Page, one of his Majesty's Engineers; afterwards Kt.	July 10	July 10, 1783
		John, Earl of Breadalbane	Feb. 19, 1784	Feb. 19, 1784
	Oct. 1805	George Lord Kinnaird	19	26
	1786	The Rev. Thomas Gresley, D. D.	April 22	May 6
July 6, 1754	Oct. 8, 1808	Mr. John Sheldon, Surgeon, Prof. Anat. to the Royal Academy	29	6
		James, Earl of Salisbury; afterwards Marquis	May 13	13
		Rt. Hon. Sir George Yonge, Bart.	13	13
		Busick Harwood, M. B.; afterwards Prof. Anat. in the University of Cambridge	27	Nov. 11
	Feb. 4, 1795	George, Viscount Mount Edgecumbe; afterwards Earl	June 10	June 17
		Henry Hugh Hoare, Esq.	17	Nov. 25
		Marchese Lewis Malaspina de Sannazaro, of Milan	24	
		John Sinclair, Esq.; afterwards Bart.	24	Nov. 18
	Feb. 4, 1810	Caleb Whitefoord, Esq.	24	July 1
		Alex. Duke of Gordon	24	Feb. 10, 1785
Dec. 10, 1724	Feb. 16, 1799	Charles Theodor, Elector Palatine, Duke of Bavaria	Aug. 12	
	Nov. 14, 1807	Thomas Potter, Esq.	Dec. 16	Dec. 23, 1784
		Gilbert Blane, M. D.	23	Jan. 13, 1785
		Smithson Tennant, Esq.	Jan. 13, 1785	Feb. 10
	Sept. 3, 1793	John, Earl of Buckinghamshire	Feb. 3	17
		Anthony, Earl of Shaftesbury	3	10
		George, Earl of Morton	24	March 3
		Rev. George Pretyman, D. D.; afterwards Lord Bishop of Lincoln	March 17	April 7
		Aaron Graham, Esq.	17	7
		Robert Halifax, M. D.	April 21	28
	May 9, 1790	The Rev. Charles Godfrey Woide, M. A. ..	21	28
	1793	Lieut.-Col. Robt. Pringle	28	May 12
	March 1790	Stayner Holford, Esq.	May 12	June 2
Dec. 2, 1759		James Edward Smith, Esq.; afterwards M. D.	26	May 26
	1789	Rev. Osmund Beauvoir, D. D.	June 9	Nov. 10
	Jan. 8, 1802	Robert Udny, Esq.	16	10
		Richard Neave, Esq. Governor of the Bank of England; afterwards Bart.	16	10
	May 8, 1790	The Rev. Henry Usher, D. D. Prof. Ast. in Dublin College	Nov. 24	
		James Watt, Esq.	24	Jan. 12, 1786
	1741	Wm. Withering, M. D.	24	
	Oct. 6, 1799	Matthew Boulton, Esq.	24	March 9
Sept. 14, 1728	Aug. 17, 1809	Samuel Galton, Jun. Esq.	Dec. 8	Jan. 12
		James Keir, Esq.	8	Jan. 26, 1791
		John Henniker, Major, Esq; afterwards Lord Henniker	15	Dec. 22, 1785
	April 5, 1799	Rev. Clayton Mordaunt Cracherode, M. A.	15	22
	July 1806	Rich. Joseph Sulivan, Esq.; afterwards Bart.	Dec. 22	Jan. 19, 1786
	Jan. 29, 1809	John Hunter, M. D.	Jan. 12, 1786	19
Oct. 6, 1730	Nov. 5, 1800	Mr. Jesse Ramsden, Mathm. Instrument-maker	12	19
		Charles George Lord Arden	19	26
	Jan. 1, 1809	James Bucknall Viscount Grimston	Feb. 2	Feb. 16
		Alex. Thomson, Esq. Accomptant General and Master in Chancery; afterwards Kt. and Baron of the Exchequer	9	23
		The Rev. Thomas Parkinson, M. A.	23	March 16
	March 9, 1801	John Holliday, Esq.	March 9	16
	1740	Rev. Henry Whitfield, D. D.	9	30
		Rev. John Berlow Seale, M. A.	9	23
	Withdrawn 1790	Wm. Thomson, M. B.; afterwards M. D. ..	16	April 3, 1788
		Rt. Hon. Wm. Eden; afterwards Ld. Auckland	23	
	April 20, 1786	John Goodricke, Esq.	April 6	Died before admission.

BORN.	DIED.		ELECTED.	ADMITTED
	Not admitted.	Col. Charles Vallancey	May 4, 1786	
	Sept. 29, 1799	Richard Molesworth, Esq.	4	May 11, 1786
		Philip d'Auvergne, Esq. Capt. in the Royal Navy, afterwards Prince successor of Bouillon	11	18
	July 29, 1795	Adair Crawford, M. D.	11	18
	June 3, 1800	Sir Godfrey Webster, Bart.	18	June 22
		Wm. Finch Palmer, Esq.	18	1
Sept 23, 1735		Rev. Thos. Martin, B. D. Prof. Bot. Camb.	18	15
		Rev. John Hewett, M. A.	25	May 25
		George, Duke of Marlborough	25	June 15
		Rev. Abram Rees, D. D.	June 1	15
		Edmund Turner, Esq.	15	22
		Wm. Young, Esq.; afterwards Bart.	15	22
		Rev. Samuel Vince, M. A.	22	29
	Withdrawn 1800	Forbes M'Bean, Esq. Col. in the Royal Regt. of Artillery; afterwards Lieut.-Gen.	Nov. 9	Nov. 23
		Rev. John Oldershaw, M. A.	16	Dec. 21
		Rev. Francis John Hyde Wollaston, M. A.	23	14
		Rev. Wm. Rose, M. A.	23	7
	Dec 4, 1792	Sir Wm. Fordyce, Kt.	Jan. 11, 1787	Jan. 25, 1787
		Arthur Piggott, Esq.; afterwards Kt.	18	25
		Wm. Morton Pitt, Esq.	25	Feb. 22
	June 14, 1800	Henry, Lord Middleton	21	1
May 6, 1756		Mr. Everard Home, Surgeon	Feb. 15	22
	Jan. 4, 1801	Sir George Leonard Staunton, Bart.	15	22
		Hon. Thomas Erskine; afterwards Lord	22	Jan. 10, 1788
May 2, 1762		Rich. Anthony Salisbury, Esq.	March 15	April 19, 1787
	Jan. 24, 1809	James, Earl of Fife	29	March 29
		Louis Pinto de Sousa Cautinho, Kt. of the Orders of Christ. and Malta, Envoy Ext. and Minister Plenip. from the Queen of Portugal	April 19	May 17
		Sir Thomas Gerry Cullum, Bart.	19	April 26
		James Louis Macie, Esq.; afterwards J. Smitnson	19	26
		Francis Ld. Rowden; afterwards Earl of Moira	May 3	May 17
		Craven Ord, Esq.	3	10
		Mr. Wm. Blizard, Surgeon; afterwards Kt.	3	10
		Wm. Bentinck, Esq. Capt. in the Royal Navy	17	24
Jan. 30, 1745	April 21, 1804	Ernestus, Duke of Saxe Gotha and Attenburg	June 14	
	Withdrawn 1797	John Ash, M. D.	Nov. 8	Nov. 15
		Baron Nicolas Vay de Vaja, of Hungary	22	Dec. 6
		Wm. Parsons, Esq.	22	6
	Jan. 3, 1805	Alexander, Lord Loughborough; afterwards Earl of Rosslyn	Dec. 6	13
		Thos. Boothby Parkyns, Esq.; afterwards Lord Rancliffe	6	13
		George Trenchard Goodenough, Esq.	6	20
Dec. 23, 1753		Rev. Richard Relham, M. A.	6	13
	Jan. 22, 1803	John Crisp, Esq.	Jan. 17, 1788	Jan. 24, 1788
		Robt. Waring Darwin, M. D.	Feb. 21	April 3
Aug. 25, 1742		Hugh, Duke of Northumberland	March 6	3
	Feb. 7, 1796	John Sibthorp, M. D. Prof. Botany, Oxford	6	3
		George Hardinge, Esq.	April 3	May 1
1744		Florens Laurentius Fridericus Crell, M. D. Prof. Medicine in the University of Helmstadt		
1708		Jean Rodolphe Perronet, Member of the Royal Academy of Sciences at Paris, Kt. of the Order of St. Michael		
		Louis Bernard Guyton de Morveau, Member of the Academy at Paris		
		Anton. Maria Lorgna, Coll. of Engineers in the Venetian Service, and Kt. of the Orders of St. Mauritius and St. Lazarus	On the Foreign List. April 3	Dec. 6, 1792
		Thos. Bugge, Prof. Ast. in the University of Copenhagen		
Feb. 16, 1727		Nicholas Joseph de Jacquin, M. D. Prof. Chemistry and Botany in the University of Vienna		
Oct. 8, 1730	Feb. 18, 1799	John Hedwig, M. D. of Leipzig		
		Eugenius Bulgarius, Archbp. of Cherson		
		Theodore Augustine Mann, Secretary of the Imperial and Royal Academy of Brussels		

BORN.	DIED.			ELECTED.	ADMITTED.
Aug. 7, 1726	Nov. 6, 1790	James Bowdoin, Esq. President of the American Acad. of Arts and Sciences	On the Foreign List.	April 3, 1788	Dec. 6, 1792
Nov. 11, 1743		Charles Peter Thumberg, M. D. Prof. Medicine and Botany, in the University of Upsala, and Kt. of the Order of Wasa			
	Jan. 21, 1799	Horace Benedict de Saussure, Member of the Great Council of Geneva			
Aug. 26, 1745	May 8, 1794	Antoine Laurent Lavoisier, Member of the Royal Academy of Sciences at Paris			
	July 21, 1798	James Adair, Esq		10	June 8, 1788
	1797	Robt. Augustus Johnson, Esq.		17	May 1
		Reginald Pole Carew, Esq.		24	22
		Rev. Wm. Pearce, D. D. Master of the Temple		May 1	8
		Richard Brooke Supple, Esq.; afterwards Sir R. Brooke de Capell Brooke, Bart.		1	8
		Martin Wall, M. D. Clinical Prof. Oxford		8	June 5
		Philip Rasleigh, Esq.		29	Dec. 18
	1802	Lieut. John Finlay, of the Corps of Royal Engineers		June 5	Nov. 20
		George, Earl of Glasgow		5	June 12
		Charles Wilkins, Esq.		12	12
	March 19, 1804	Rt. Hon. Sir Richard Pepper Arden, Kt. Master of the Rolls; afterwards Lord Alvanley, and Chief Justice in Com. Pleas		Nov. 13	Nov. 20
	Jan. 16, 1794	Edward Gibbon, Esq.		27	
Aug. 12, 1762		George, Prince of Wales		Jan. 26, 1789	
Aug. 16, 1763		Frederick, Duke of York		26	
Nov. 7, 1745	Sept 19, 1790	Henry Frederick, Duke of Cumberland		26	
		John Gillies, LL. D.		29	Feb. 5, 1789
		Mr. Edward Jenner, Surgeon, at Gloucester; afterwards M. D.		Feb. 26	May 7
		George Shaw, M. D.		26	March 5
		Richard, Viscount Fitzwilliam		March 5	19
	1799	Rev. Abraham Bennet		19	May 28
	1809	Rev. Jonathan Davies, D. D. Master of Eton College		April 2	Dec. 17
Sept. 2, 1725	May 27, 1795	Ewall Frederic Count de Hertsberg, Minister of State to the King of Prussia, and Member of the Academy of Sciences at Berlin	On the Foreign List.	April 30	
		Claude Louis Berthollet, Member of the Royal Academy of Sciences at Paris			
March 23, 1749		Pierre Simon de la Place, Member of the Royal Academy of Sciences at Paris			
June 30, 1748		Jean Dominique Compte de Cassini, Director of the Royal Observatory, and Member of the Royal Academy of Sciences, Paris			
		Adrien Marie Le Gendre, Member of the Royal Academy of Sciences, Paris			
	1804	Pierre Francois André Méchain, Member of the Royal Academy of Sciences at Paris, Astronome Hydrogrophe de la Marine			
Sept. 6, 1732	April 18, 1796	John Charles Wilcke, Secretary to the Royal Acad. of Sciences at Stockholm			
1747		John Elert Bode, Astronomer and Member of the Royal Academy of Sciences, Berlin			
1729		Christian Gottlieb Heyne, Prof. in the University, and Secretary of the Royal Society of Gottingen			
Sept. 27, 1719	June 20, 1800	Abraham Gotthelf Kestner, Prof. Math. in the University of Gottingen			
Feb. 27, 1727	Aug. 22, 1791	John David Michaeles, Prof. in the University of Gottingen, Kt. of the Order of the Polar Star			
		The Rev. Samuel Goodenough, LL. D.; afterwards Dean of Rochester, since Bishop of Carlisle		May 14	May 21, 1789
		Samson, Lord Eardley		Nov. 5	Dec. 24
		Col. Robt. Morse, of the Royal Corps of Engineers		12	17
		George Rogers, Esq. Commissioner of the Navy		19	Nov. 19
		Robert Wood, Esq.		19	Dec. 24

BORN.	DIED.		ELECTED.	ADMITTED.
Feb. 23, 1753	Sept. 7, 1801	Arthur, Earl of Hilsborough; afterwards Marquis of Downshire	Jan. 22, 1790	Feb. 11, 1790
	1803	The Rev. Richard Fisher, M. A.; afterwards Belward	Feb. 11	11
		Mark Beaufoy, Esq.	18	25
		John Reeves, Esq.	March 18	March 25
		Sir Wm. Green, Bart. Major General and Chief Royal Engineer of Great Britain	April 29	June 3
		John Thomas Stanley, Esq.; afterwards Bart.	29	May 13
		Wm. Elford, Esq.; afterwards Bart.	29	13
		Charles Warren, Esq.	29	6
		Wm. Morgan, Esq.	May 6	13
		Frederick Augustus Barnard, Esq.	13	June 3
		Matthew Baillie, M. D.	June 3	10
		Joseph Jekyll, Esq.	3	17
		Hon. George Keith Elphinstone, Capt. in the Royal Navy; afterwards Lord Keith	24	Nov. 4
		Philip Metcalfe, Esq.	Nov. 4	Dec. 16
		James Robertson, M. D.; afterwards Robertson Barclay.	11	Nov. 18
		Philip, Earl of Hardwicke	25	March 31, 1791
	Sept. 8, 1797	The Rev. Richard Farmer, D. D. Canon Residentiary of St. Paul's, and Master of Emanuel College, Cambridge	Feb. 17, 1791	Feb. 24
		Wm. Heberden, Jun. Esq.; afterwards M. D.	24	March 3
Oct. 1, 1737	Jan. 3, 1805	Charles Townley, Esq.	March 10	24
		George Pocock, Esq.	10	17
		Lewis Alexander Grant, Esq.	May 10	April 7
		George Best, Esq.	17	March 24
		Col. Norman Macleod; afterwards Major-Gen.	24	31
	July 14, 1804	John Spranger, Esq. Master in Chancery ..	24	May 5
Oct. 12, 1736	1794	Michael, Prince Poniatowsky, Archbishop of Gnesen, and Primate of Poland	31	April 7
		Cypriano Ribeiro Freire, Kt. of the Order of St. Jago, Chargé d'Affaires of Portugal	31	7
		Aylmer Bourke Lambert, Esq.	31	May 5
	May 20, 1804	Fowler Walker, Esq.	31	Nov. 10
		William Frazer, Esq.; afterwards Bart. ..	April 14	May 5
		George Chalmers, Esq.	May 5	12
		John Hawkins, Esq.	5	19
	1805	Abraham D'Aubant, Esq. Lieut.-Col in the Royal Corps of Engineers; afterwards Lieut.-Gen.	5	Nov. 10
	Jan. 29, 1802	Thomas Walker, Esq. Accomptant General and Master in Chancery	5	June 2
	July 6, 1799	Sir James Eyre, Kt. Lord Chief Baron of the Court of Exchequer	5	Dec. 15
		Louis de la Grange, Member of the Royal Academy of Sciences, Paris		
		Alessandro Volta, Prof. Nat. Philosophy in the University of Pavia		
		Antonio Scarpa, Prof. Anat. in the University of Pavia		
		Marc Auguste Pictet, Prof. Nat. Philosophy at Geneva	May 5	June 18, 1801
		Jean Baptiste Joseph de Lambre, of Paris; afterwards Member of the Royal Academy of Sciences, Paris		
		Simon L'Huilier, Citizen of Geneva		
		John Bruce, Esq. Prof. Philosophy in the University of Edinburgh	26	June 2, 1791
		Richard Stanley, Esq.	26	2
		George Pearson, M. D.	June 23	30
	1795	The Rev. John Scally, LL. D.	30	Nov. 17
		Davies Giddy, Esq.	Nov. 17	April 19, 1792
		Sir Cecil Bisshopp, Bart.; afterwards Lord Zouch	17	Nov. 24, 1791
		Joseph Huddart, Esq.	17	24
		John Turnbull, Esq.	17	24
		Henry Norton Willis, Esq.	Dec. 15	Jan. 12, 1792
	Aug. 25, 1793	James Six, Esq.	Jan. 19, 1792	Feb. 16
		Charles Long, Esq.	Feb. 2	16
	July 30, 1800	Rt. Hon. Frederick Montagu	16	March 8
		The Rev. Stephen Weston, B. D.	March 1	8
	July 1803	The Rev. Thomas Hussey, D. D.	8	15

On the Foreign List.

BORN.	DIED.		ELECTED.	ADMITTED.
		The Rev. Leonard Chappelow, M. A.	Feb. 15, 1792	Feb. 22, 1792
		The Rev. James Stanier Clarke	April 19	May 10
		Sir Richard Colt Hoare, Bart.	26	3
		John Komarzewski, Lieut.-Gen. in the Polish Army, Kt. of the Orders of St. Stanislaus, and St. Alexander Newski	May 10	17
	March 31, 1806	George, Lord Macartney; afterwards Viscount and Earl	June 7	June 21
		Wm. Bosville, Esq.	7	14
		Samuel Davies, Esq. of Bhagalpur, in the East Indies	28	Feb. 26, 1807
		The Rev. Richard Dickson Shackleford, D. D.	28	June 28, 1792
		Rev. Archibald Allison, B. LL. Preb. of Sarum	Nov. 15	
		David Pennant, Esq.	15	Dec. 6
	Jan. 1807	Samuel Solly, Esq.	22	6
	Aug. 31, 1805	James Currie, M. D.	Dec. 20	May 2, 1793
		Sir John Ingilby, Bart.	Feb. 7, 1793	March 7
		Sir John Scott, Kt. his Majesty's Attorney General; afterwards Lord Eldon, and Ld. High Chancellor	14	
		Sir Wm. Scott, Kt. his Majesty's Advocate-Gen.	14	March 20, 1794
Oct. 14, 1757		Charles Abbot, Esq.; afterwards Rt. Hon. and Speaker of the House of Commons	14	Feb. 21, 1793
		Richard Richards, Esq.	14	28
	June 1806	Andrew Douglas, Esq.	March 7	Jan. 12, 1797
Dec. 1739	Dec. 20, 1807	Francis Stephens, Esq. Commissioner of the Victualling Office	7	March 14
		Don Joseph Mendoza y Rios, Captain in the Spanish Navy	April 11	May 2
July 1, 1744	Feb. 24, 1799	Gottfried Charles Lichtenberg, Prof. of Natural Philosophy in the University of Gottingen	11	
1752		John Frederic Blumenback, M. D. Prof. Med. in the University of Gottingen		
		Robert Stearne Tighe, Esq.	18	May 2
		Wm. Saunders, M. D.	May 9	16
		Wm. Hyde Wollaston, M. D. of Huntingdon	9	March 6, 1794
		Mr. Samuel Bosanquet, Jun.	June 6	June 13, 1793
	June 14, 1808	Sir John Day, Kt.	6	Jan. 16, 1794
	Sept. 22, 1794	John Far Abbot, Esq.	20	
		Rt. Hon. Lord Frederick Campbell	Nov. 7	Nov. 14, 1793
		Wm. Charles Wells, M. D.	7	14
		Richard Wilson Greatheed, Esq.	21	March 6, 1794
	Aug. 18, 1797	Harvey, Viscount Mount Morres	Dec. 12	Dec. 19, 1793
		George Gostling, Esq.	12	19
		Rev. Thomas Watkins	Jan. 23, 1794	March 20, 1794
		Hon. Frederic North	Feb. 6	Feb. 13
		Sir John Mitford, Kt. his Majesty's Solicitor-Gen.; afterwards Ld. Reddesdale, and Lord Chancellor of Ireland	March 6	April 3
		James Earle, Esq. Surgeon Ext. to his Majesty's Household; afterwards Kt.	6	March 20
		Thomas Plumer, Esq.	6	April 10
		Rt. Hon. Sir Wm. Wynne, Kt. Dean of the Arches	6	March 20
	1794	Rt. Hon. John Hely Hutshinson, Secretary of State in Ireland	13	27
		Sir John Henslow, Kt. First Surveyor of the Royal Navy	20	27
		Mr. John Godfrey Schmeisser	20	27
		Thos Keate, Esq. Surgeon to her Majesty and to his R. H. the Prince of Wales	27	April 3
		John Walker, Esq.	27	3
		Hon. Robt. Fulk Greville	April 3	10
		John Gottlieb Walter, M. D. Professor of Anatomy, and Member of the Academy of Sciences, Berlin	May 1	
	July 16, 1800	Bryan Edwards, Esq.	22	June 26
		John Grieve, M. D.	22	5
		Hon. Robt. Banks Jenkinson; afterwards Lord Hawkesbury	29	19
		Mr. Thomas Young; afterwards M. D. ..	June 19	26
		Francis Humberston Mackenzie, Esq.; afterwards Lord Seaforth	26	May 14, 1795
	Dec. 1808	Peter Peirson, Esq.	July 3	Nov. 6, 1794

BORN.	DIED.		ELECTED.		ADMITTED.	
	1802	Robert Aldersey, Esq.	July	3, 1794	Jan.	8, 1795
	March 28, 1807	Nathanael Hulme, M. D.		3	July	10, 1794
		Lewis Majendie, Esq.		3	Nov.	13
		Matthew Martin, Esq.		10	Not admitted.	
		Major Alexander Dirom		10		
		John Symmons, Esq.		10	Nov.	13
		Wm. Sotheby, Esq.	Nov.	13	May	21, 1795
		John Blackburne, Esq.		20	Nov.	27, 1794
		Lieut.-Col. Patrick Ross	Dec.	11	Dec.	18
		Jacob, Earl of Radnor	Feb.	12, 1795	Feb.	26, 1795
	May 19, 1798	Hugh Gillan, M. D.		12		19
		George, Viscount Morpeth		26	March	5
		John, Lord Borringdon		26		5
		{ Rt. Hon. Sylvester Douglas; afterwards Lord Glenbervie }	March	5		12
		Matthew Montague, Esq.		12		26
		{ Christopher Pegge, M. D. Reader of Anat. at Christ Church, Oxford; afterwards Kt. }		19	April	16
		Thomas James Mathias, Esq.		19	March	26
	March 23, 1804	Rev. Benjamin Hutchinson, M. A.		19	April	23
		Samuel Young, Esq.		26		23
		Rev. Geo. Heath, D. D.		26	Dec.	17
		Wm. Blane, Esq.	April	16		24
April 8, 1732	June 20, 1796	Gregorio Fontana, Prof. Mathematics in the University of Pavia				
		Barnaba Oriani, Astronomer in the Royal Observatory at Milan				
		David Rittenhouse, Esq. President of the American Philosophical Society at Philadelphia				
	1811	Johann Christian Daniel von Schreber, M. D. President of the Imperial Acad. Naturæ Curiosorum, and Prof. Medicine in the University of Erlangen		16		
	1803	Alberto Fortis, Prof. Natural History in the University of Padua				
		Martin Heinrich Klaproth, Member of the Royal Academy of Sciences at Berlin				
	Sept. 11, 1801	August. Ferdinand von Veltheim, of Harbke; afterwards Count				
		Archibald Hamilton, Esq.		23	April	30
	Dec. 5, 1797	Harry Crathorne, Esq.		23		30
		Rev. Abraham Robertson, M. A.	May	21	June	4
		John Campbell, Esq.; afterwards Ld. Cawdor	June	4		25
		Wm. Petrie, Esq.	Nov.	19	Nov.	26
		Matt. Smith, Esq. Major of the Tower of London		19		26
		George Smith Gibbes, M. A.; afterwards M. D.	Feb.	18, 1796	Feb.	25, 1796
June 5, 1751		{ Joseph Correa de Serra, LL. D. Secretary of the Royal Acad. of Sciences, Lisbon }	March	3	March	10
		Rev. Wm. Langford, D. D.		17	Dec.	22
	April 21, 1800	Wm. Larkins, Esq.	April	14	April	21
	Feb. 1809	Lieut.-Gen. Thomas Osbert Mordaunt		14		21
		Mr. John Abernethy, Surgeon		14		21
		Glocester Wilson, Esq.		28	June	15, 1797
		{ Rev. Wm. Lax, M. A. Lawndes's Prof. of Astronomy, Cambridge }	May	5	June	9, 1796
	April 2, 1801	Edw. Riou, Esq. Captain in the Royal Navy		5	May	26
	Oct. 10, 1798	John Dalrymple, Esq. Admiral of the Blue ..		26	June	2
		Benjamin Hyett, Esq.	June	2	Jan.	26, 1797
		Charles Shaw Le Fevre, Esq.	Nov.	10	Nov.	17, 1796
	Feb. 4, 1807	Wm. Latham, Esq.		10		24
		George Holme Summer, Esq.		17		24
		Samuel Rogers, Esq.		17		24
		Rev. Wm. Hawley, M. A.		17	Dec.	8
		Robt. Smith, Esq.		24		8
		George, Viscount Valentia		24		8
		Rev. John Hellins	Dec.	22	Jan.	26, 1797
		Christopher Robt. Pemberton, M. D.		22		12
Jan. 15, 1776		His Highness Prince Wm. of Glocester	Jan.	14, 1797		
		Robt. Capper, Esq.		26	May	4
		Samuel Lysons, Esq.	Feb.	2	Feb.	9
	Withdrawn 1809	{ Sir Andrew Snape Hamond, Bart. Comptroller of the Royal Navy }	March	2	March	30
		Charles Hatchett, Esq.		9		16
		George Aust, Esq.		9		16

On the Foreign List.

BORN.	DIED.		ELECTED.	ADMITTED.
		Edward Adolphus, Duke of Somerset	March 9, 1797	March 23, 1797
		Bartholomew Parr, M. D. of Exeter	23	March 26, 1801
		Samuel Ferris, M. D.	30	April 1797
	Feb. 13, 1806	{ Rev. Stephen Eaton, M. A. Archdeacon of } Middlesex	April 27	May 4
June 1⅙, 1731		John Townley, Esq.	May 4	11
Nov. ⁷⁄₇, 1754		{ Frederick Charles William, Hereditary } Prince of Wirtemberg	11	
		Sir John St. Aubyn, Bart.	18	May 25
		George Ellis, Esq.	18	June 1
		Rev. Daniel Lysons, M. A.	25	1
		Henry Browne, Esq.	25	1
		Hon. Robt. Clifford	June 1	1
		Wm. Battine, LL. D.	1	22
	June 26, 1800	Mr. Wm. Cruikshank	1	15
		Charles Freeman, Esq.	15	July 6
		Isaak Titsingh, Esq.	22	Nov. 9
		James Brodie, Esq.	July 6	9
		John Spalding, Esq.	6	July 6
		Rev. George Whitmore, B. D.	Nov. 23	Jan. 11, 1798
		George, Earl of Egremont	Dec. 7	April 19
		{ John Heaviside, Esq. Surgeon Extraordinary } to his Majesty	14	Dec. 21, 1797
	Nov. 1805	{ Rev. Robt. Holmes, D. D. ; afterwards Dean } of Winchester	14	Feb. 15, 1798
		Thomas Greene, Esq.	March 22, 1798	March 29
		John Rennie, Esq.	29	April 19
	Jan. 1808	John Ryan, Esq.	April 19	26
		Mr. Stephen Lee	19	26
		Alexander Duncan, Esq.	19	26
Oct. ⁹⁄₁₀, 1750	March 17, 1803	{ Prince Demetri Gallitzin — Adam Afzelius, M. D. Demonstrator of Bot. in the University of Upsula — John Jerome Schroeter Oberamtmann, of Lilienthal in the Dutchy of Bremen — Martin Van Marum, M. D. Sec. of the Society of Sciences of Harlem } On the For. List.	19	May 3
Aug. 17, 1755		Lieut.-Col. Wm. Paterson	May 10	17
		Philip Hills, Esq.	June 7	{ Not admitted, re-elect. 1799
		Finlay Fergusson, Esq.	21	June 28
		Captain Wm. Mudge, of the Royal Artillery	28	Nov. 8
		Samuel Jackson, Esq.	Nov. 8	15
	Sept. 20, 1803	Nicholas Gay, Esq.	15	22
		Henry Gregg, Esq.	Dec. 6	Not admitted.
		Benjamin Hobhouse, Esq. M. P.	13	Feb. 21, 1799
		Edward Howard, Esq.	Jan. 17, 1799	Jan. 31, 1799
		William Drummond, Esq.	April 4	April 11
		Edward Hynde East, Esq.	11	25
		James Clark, M. D.	11	18
		Philip Hills, Esq.	18	25
		Home Popham, Esq. Captain in the Royal Navy	18	25
		Archibald Blair, Esq.	May 2	May 9
	Nov. 1805	Hon. Reginald Cocks	23	Feb. 6, 1800
		Abraham Mills, Esq.	30	June 6, 1799
		Joseph Sabine, Esq.	Nov. 7	Nov. 14
	1806	Count Apollos Mussin Puschkin	14	
	1805	Sir David Carnegie, Bart. M. P.	21	Jan. 15, 1801
		Edward Roberts, M. D.	Dec. 12	Dec. 19, 1799
		John, Marquis of Bute	12	Jan. 23, 1800
		Thomas Jones, Esq. M. P.	Jan. 9, 1800	16
		John Corse Scott, Esq.	16	April 3
		Frederick Morton, Lord Henley	Feb. 6	Feb. 13
		{ Rt. Hon. Thomas Pelham ; afterwards Lord } Pelham, since Earl of Chichester	April 24	May 8
		Alexander Crichton, M. D.	May 8	15
	May 1, 1804	Henry, Earl of Exeter ; afterwards Marquis	8	June 12
		Lieut.-Colonel John Macdonald	15	May 22
		Caleb Hillier Parry, M. D.	22	June 19
		Gibbs Walker Jordan, Esq.	29	12
		{ Charles Morice Pole, Esq. Rear-Admiral of } the Red ; afterwards Bart.	29	12
		Robt. Lord Carrington	29	12
		Sir John Cox Hippisley, Bart.	June 12	19
		James Meyrick, Esq.	19	26

i

BORN.	DIED.		ELECTED.	ADMITTED.
Jan. 31, 1774		Wm. George Maton, M. B. ; afterwards M. D.	June 26, 1800	July 3, 1800
		Charles Dickinson, Esq.	Nov. 27	Dec. 11
		Rev. Wm. Douglas, Canon Resid. of Salisbury	Dec. 18	April 16, 1801
		{ Codrington Edmund Carrington, Esq. ; afterwards Kt.	18	Jan. 15
	Jan. 22, 1809	Lieut.-Colonel Michael Symes	18	8
		Arthur, Earl of Mountnorris	18	8
		Rev. Herbert Marsh, B. D.	Jan. 8, 1801	29
		Samuel Turner, Esq.	15	22
		Matthew Smith, Esq. Capt. in the Royal Navy	Feb. 26	March 12
		Sir Walter Stirling, Bart.	March 5	26
		Richard Chenevix, Esq.	5	12
		John Ellis, Esq.	12	April 16
		Rev. Edward Balme, M. A.		March 26
		Edmund Antrobus, Esq.	26	April 16
		George Isted, Esq.	April 16	30
		Giffin Wilson, Esq.	16	May 7
		{ Wm. Long, Esq. Master of the Royal College of Surgeons	16	April 30
		Martin Davy, M. D.	23	June 18
		{ John Latham, M. D. Physician Extraordinary to his R H. the Prince of Wales	April 30	May 7
		{ Rev. John Hailstone, M. A. Woodward, Prof. Cambridge	May 7	June 4
Sept. 9, 1754		Wm. Bligh, Esq. Capt. in the Royal Navy ..	21	June 3, 1802
		John Lloyd Williams, Esq.	21	25, 1801
		Roger Elliot Roberts, Esq.	June 4	11
		Lieut.-Col. James Willoughby Gordon	11	Dec. 16, 1802
		Rev. Robt. Nixon, B. D.	11	June 18, 1801
		Edward Ash, M. D.	18	25
		Warren Hastings, Esq.	25	Nov. 5
		Rt. Hon. Charles Yorke	Nov. 12	26
		Rev. Edward Forster, M. A.	Dec. 10	Dec. 24
		Robert Wissett, Esq.	24	Jan. 14, 1802
		Mr. Astley Cooper, Surgeon	Feb. 18, 1802	Feb. 25
		Hon. George Knox	25	March 18
		Charles Burney, Jun. LL. D.	25	4
		James Louis Compte de Bournon	25	11
		{ Maximilian Joseph, Elector Palatine, Duke of Bavaria	March 4	
		John Liptrap, Esq.	4	18
		Mr. James Ware, Surgeon	11	18
		Richard Fowler, M. D	April 1	
		Sir Edward Knatchbull, Bart.	May 6	May 13
	1807	Langford Millington, Esq.	6	13
		Alexander, Marquis of Douglas	20	June 3
		William, Earl of Mansfield	20	3
		{ Wm. Cruickshank, Esq. Prof. Chemistry in the Royal Academy at Woolwich	June 24	Nov. 4
	Nov. 2, 1803	George Biggin, Esq.	July 1	July 8
		John Trotter, Esq.	8	8
		Lord Webb Seymour	Nov. 11	Nov. 25
		Henry Robt. Viscount Castlereagh	11	Nov. 10, 1803
		Dawson Turner, Esq.	Dec. 9	Feb. 24
		Robert Woodhouse, A. M.	16	Jan. 27
		Gilbert, Lord Minto	23	Feb. 3
		Edward Hilliard, Esq.	23	3
		John, Lord de Blaquiere	Jan. 13, 1803	Jan. 27
		Hon. Fulk Greville Upton	Feb. 10	March 3
		Rev. Matthew Raine, D. D.	17	Feb. 24
		Rev. Thomas Rackett, M. A.	17	March 10
		John, Earl of Glandore	24	24
		Henry Brougham, Jun. Esq.	March 3	March 8, 1804
		John Spencer Smith, Esq.	10	March 17, 1803
		Mr. Thomas Blizard, Surgeon	10	17
		Rev. John Brinkley, M. A.	17	May 26
		Mr. John Pearson, Surgeon	24	March 31
		James Forbes, Esq.	24	31
		{ Charles William Viscount Charleville ; afterwards Earl	31	March 20, 1806
		Sir George Thomas Staunton, Bart.	April 28	May 19, 1803
		Mr. James Wilson, Surgeon	May 19	26
		{ Humphry Davy, Esq. Prof. Chemistry to the Royal Institution of Great Britain	Nov. 17	Nov. 24
		Richard Gregory, Esq.	24	March 22, 1804

2

BORN.	DIED.		ELECTED.	ADMITTED.
Aug. 21, 1778		Lewis Weston Dillwyn, Esq.	Feb. 2, 1804	Feb. 26, 1804
		{ George Isaac Huntingford, Lord Bishop of Gloucester	23	March 15
		Carsten Anker, Esq.	March 1	8
		Thomas Bayly Howell, Esq.	8	April 12
Feb. 15, 1768		Mr. Anthony Carlisle, Surgeon	8	March 15
		Valentine Conolly, Esq.	15	22
Oct. 9, 1778		John Viscount Kirkwall	April 12	April 19
		{ Joseph Piazzi, Astronomer Royal at Palermo		
June 14, 1754		Francis Baren Von Zach, Col. and Director of the Observatory at Seeberg, near Gotha	12	
Oct. 11, 1758 1778		Wm. Olbers, M. D. of Bremen		
		Frederick Gauss, Ph. D. of Brunswick; afterwards Prof. Mathamatics in the University of Gottingen		
		Andrew Hutchinson, M. B.	26	May 3
		{ Rev. Robt. Nares, M. A. Canon Residentiary and Archdeacon of Litchfield	May 10	31
		Charles Short, Esq.	17	June 7
		{ Robert Robertson, Esq. Physician to Greenwich Hospital	31	14
		Thomas Harrison, Esq.	June 7	7
		Sir Thomas Hanmer, Bart.	21	Nov. 15
		Rev. Francis Wrangham, M. A.	Nov. 15	May 9, 1805
		Col. Tomkyns Hilgrove Turner	Dec. 6	Dec. 13, 1804
		Thomas Hope, Esq.	6	Jan. 10, 1805
		James Cockshutt, Esq.	6	April 4
	March 23, 1810	Thomas Finch, Esq.	13	Jan. 24
		{ Sir Edward Winnington, Bart. elected Jan. 10, but he had died the day before		
		{ Mr. Olaus Warberg, Prof. Ast. and Member of the Board of Longitude, Copenhagen	Jan. 10, 1805	Jan. 17
June 27, 1763		Edward Rudge, Esq.	31	Feb. 15
		George Paulett Morris, M. D.	Feb. 14	28
		Hon. Lieut.-Col. Wm. Blaquiere	21	28
		Robt. Fergusson, Esq.	March 7	April 25
		Hon. Lieut.-Col. Thomas Wm. Fermor	14	March 28
		Thomas Andrew Knight, Esq.	21	21
		Robert Holford, Esq.	28	May 2
		{ Wm. Smith, Esq. one of the Barons of the Exchequer in Ireland	April 25	
		Hon. John Cust; afterwards Lord Brownlow	May 2	May 30
		Frederick William, Earl of Bristol	23	June 13
		William Babington, M. D.	30	20
		Stephen Peter Rigaud, Esq.	June 13	Dec. 19
		Thomas Murdoch, Esq.	July 4	July 4
		John Barrow, Esq.	4	4
		Wilbraham, Earl of Dysart	Nov. 7	
		Edward Loveden Loveden, Esq.	14	Feb. 6, 1806
		Joseph Whidbey, Esq.	21	Dec. 12, 1805
		{ Nathaniel Dimsdale, Baron of the Russian Empire	21	12
		John Guillemard, Esq.	Jan. 16, 1806	Jan. 23, 1806
		Rev. Wm. Holwell Carr, B. D.	16	Feb. 6
		Mr. Honoratus Leigh Thomas, Surgeon	16	Jan. 23
		Sir Charles Warre Malet, Bart.	23	23
		William Smith, Esq. M. P.	Feb. 13	Feb. 20
		Rt. Hon. John Forster	20	27
		Robert Wigram, Esq.	27	March 6
		James Horsburgh, Esq.	March 13	20
		Sir Richard Clayton, Bart.	20	Not admitted.
		{ Sir John Nichol, Kt. his Majesty's Advocate General	20	March 27
		{ James Henry Arnold, Esq. Chancellor of the Diocese of Worcester	20	27
		{ George Cuvier, Member of the Institute of France		
		Bernard Germaine Etienne Lacepede, Member of the Institute of France	April 17	
		Pierre Prevost, Prof. Philos. at Geneva		
		Charles Harding, Prof. of Astronomy in the University of Gottingen		
		John Griffiths, Esq.	May 1	May 15

BORN.	DIED.		ELECTED.	ADMITTED.
		Major Edward Moor	May 1, 1806	June 5, 1806
		Francis Buchannan, M.D.	1	May 15
		John Kearney, Lord Bishop of Ossory	15	22
		Sir James Hall, Bart.	22	June 5
		Richard Sharp, Esq.	June 12	Nov. 6
		{ Wm. Higgins, Esq. Prof. of Chemistry to } the Dublin Society	12	
		Charles Stirling, Esq. Rear Adm. of the White	19	
		William Penn, Esq.	Nov. 13	Not admitted.
		Thomas Reid, Esq.	Dec. 18	Jan. 8, 1807
		Philip Henry, Viscount Mahon	Jan. 8, 1807	15
		Henry Cline, Esq.	15	Feb. 5
		Hon. Major-General John Leslie	22	19
		{ John Playfair, Esq. Prof. Nat. Philosophy } in the University of Edinburgh	Feb. 5	June 1, 1809
		George Frederick Stratton, Esq.	5	Feb. 26, 1807
		George Harrison, Esq.	5	March 5
		Thomas Burgess, Lord Bishop of St. David's	19	12
		John Pond, Esq.	26	5
		George Billas Greenough, Esq.	March 5	12
		William Garrow, Esq.	5	12
		Taylor Combe, Esq.	5	12
		John George Children, Esq.	12	19
		William Gell, Esq.	April 16	April 23
		William Hodgson, Esq.	23	May 14
		William Jacob, Esq.	23	28
		George, Earl of Winchilsea	May 7	28
		Richard Horsman Solly, Esq.	7	14
		William Blake, Esq.	14	28
		Major-General Robert Nicholson	June 4	June 11
		Col. David Humphries	11	
		Mr. Wm. Allen	Nov. 19	Nov. 26
		Lewis Hayes Petit, Esq.	Dec. 10	Dec. 17
		Charles Brandon Trye, Esq.	17	May 4, 1809
		James Peter Auriol, Esq.	Jan. 14, 1808	Jan. 21, 1808
		{ Alexander Hamilton, Esq. Prof. of Oriental Languages, in the East India Company's College at Hertford	14	21.
		John William, Earl of Bridgewater	28	
		Wm. Hasledine Pepys, Esq.	28	Feb. 4
		Robert Bree, M.D.	Feb. 11	18
		St. Andrew Lord St. John. of Bletso	18	25
Sept. 14, 1764		Richard, Earl of Mount Edgcombe	March 24	May 19
		{ Wm. Johnstone Hope, Esq. Captain in the } Royal Navy	24	April 7
		Mr. John Mason Good	31	7
		William Watson, Esq.	April 7	28
		George, Earl of Aberdeen	28	May 5
		John Goldingham, Esq.	May 26	June 2
		Alexander Marcet, M.D.	June 2	16
		Edward Astle, Esq.	2	16
		Thomas, Earl of Selkirk	July 7	June 15, 1809
		Wm. Henry White, Esq.	Nov. 10	Nov. 24, 1808
		Colin Chisholm, M.D.	24	
		George Ducket, Esq.	Dec. 8	Dec. 19, 1809
		Jerome de Salis, Esq.	15	19
		Alexander Macleay, Esq.	Jan. 19, 1809	Feb. 2
		Lieut.-Col. John Rowley, of the Royal Engineers	Feb. 9	16
		Henry Warburton, Esq.	16	March 2
		William Henry, M.D.	23	
		Robert Willan, M.D.	23	2
		Francis Augustus, Lord Heathfield	March 2	April 13
		Charles Frederick Barnwell, Esq.	9	13
		John Gillon, Esq.	23	{ Died before admission.
		John Anthony Noguier, Esq.	April 13	27
		Mr. William Thomas Brand	13	20
		Peter Leopold, Earl Cowper	May 11	June 1
		John Smith, Esq.	11	May 18
		James Burney, Esq. Capt. in the Royal Navy	June 8	Not admitted.
		{ Robert Bingley, Esq. King's Assay Master } in the Mint.	22	Nov. 9
		{ Lord Amelius Beauclerc, Captain in the } Royal Navy	Dec. 7	

BORN.	DIED.		ELECTED.	ADMITTED.
		Charles Hoare, Esq.	Dec. 21, 1809	March 8,1 10
		John, Earl of St. Vincent	21	Jan. 25
		Charles Konig, Esq.	Jan. 18, 1810	Feb. 1
		George Canning, Esq.	Feb. 1	8
		Mr. Benjamin Brodie, Surgeon	15	22
		{ Sir Richard Bickerton, Bart. Vice-Adm. of } the Red	22	March 15
		Sir Henry Halford, Bart.	March 8	15
		George Tuthill, A. M.	15	22
		Mr. Edward Troughton, Optician	15	22
		Joseph Cotton, Esq.	15	29
		{ Bowyer Edward Sparke, Lord Bishop of } Chester	15	April 5
		John, Earl of Darnley	22	5
		Sir George Shee, Bart.	May 10	June 21
		Lieut.-Col. Thomas Brisbane	10	
		{ Thomas Charles Hope, M. D. Prof. of Che- } mistry in the University of Edinburgh	30	
		Edward Stracey, Esq.	June 7	June 21
		Edward Thornton, Esq.	7	28
		Daniel Moore, Esq.	28	July 5
		William A. Cadell, Esq.	28	May 9, 1811
		William Viscount Lowther	July 5	July 12
		{ John Wilson Croker, Esq. Secretary of the } Admiralty	5	Nov. 22, 1810
		Rev. Robert Hodgson	5	Jan. 31, 1811
		George Ridge, Esq.	5	July 12, 1810
		William Wix, Esq.	12	Nov. 22, 1811
		Richard Wharton, Esq. M. P.	12	Feb. 14
		Sir Alexander Johnston, Kt.	Nov. 22	Dec. 20
		Hon. William Beauchamp Lyon, M. P.	Dec. 6	March 14
		James Robertson, Esq.	13	May 2
		John Baker, Esq.	13	Dec. 20
		Lord Viscount Milton	Jan. 17, 1811	April 25
		Rt. Hon. Isaac Corry	Feb. 21	March 28
		James Macartney, Esq.	21	14
		Rev. Wm. Dealtry	28	7
		Rev. John Kaye	March 7	April 4
		Sir Frederick Baker, Bart.	14	March 28
		John Carstairs, Esq.	14	28
		Walter Wade, M. D.	14	
		Rev. Richard Dixon	21	April 4
		Thomas Thomson, M. D.	28	Nov. 7
		William Congreve, Esq.	28	April 25
		Henry, Marquis of Lansdown	April 4	Dec. 5
		Robert Chalone, Esq. M. P.	4	
		Thomas Egan, M. D.	4	
		John Dent, Esq. M. P.	May 16	June 13
		John Elliot, Esq.	16	May 30
		John Proctor Anderdon, Esq.	23	30
		George Hibbert, Esq. M. P.	30	June 27
		Henry Ellis, Esq. LL. B.	30	13
		Rear Adm. Sir Wm. Sidney Smith, K. S. ..	June 13	
		Thomas Hoblyn, Esq.	27	July 4
		Rev. Thomas Sampson, D. D. F. S. A.	July 4	Nov. 7
		Rev. George Rowley, A. M.	Nov. 14	Dec. 19
		William Ford Stevenson, Esq.	21	5
		Edward Hawke Locker, Esq.	Dec. 5	
		Robert Brown, Esq.	12	19
		William Franks, Esq.	12	Jan. 23, 1812
		John Randolph, Lord Bishop of London ..	19	23
		Henry Richard Lord Holland	19	9
		Rev. Henry Hasted	Jan. 9, 1812	
		William Jackson Hooker	9	23
		Charles Henry Parry, M. D.	Feb. 20	

No. V.

Alphabetical List of the Fellows of the Royal Society, with the dates of their Election, as references to the Chronological List.

PATRONS.

KING CHARLES II. 1664
KING JAMES II. 1685
KING GEORGE I. 1727
KING GEORGE II. 1727
KING GEORGE III. 1766

The PRESIDENT, COUNCIL, and FELLOWS, elected at its first Institution in 1663.

Alleyn, J. Esq.
Annesley, James, Ld.
Ashmole, E. Esq.
Aubrey, J. Esq.
Austen, J. Esq.

Balle, P. M.D.
Balle, W. Esq.
Barrow, J. B.D.
Bate, G. M.D.
Bayne, T. M.D.
Berkeley, George, Ld.
Birkenhead, Sir J. Kt.
Boyle, Rob. Esq.
Boyle, Rich. Esq.
Brereton, W. Esq.
Brook, J. Esq.
Brouncker, William, Ld. Visc.
Bruce, Robert, Ld.
Bruce, D. M.D.
Buckingham, George, Duke of.
Bysshe, Sir E. Kt.

Cavendish, William, Ld.
Charleton, W. M.D.
Clarke, T. M.D.
Clayton, J. Esq.
Colwall, D. Esq.
Cotton, E. D.D.
Cox, T. M.D.
Crawford and Lindsay, John, Earl of.
Croone, W. M.D.

Denham, Sir J. K.B.
Devonshire, William, Earl of.
Digby, Sir K. Kt.
Dorchester, Henry, Marquis of.
Dryden, Mr. J.

Ellis, A. Esq.
Ent, G. M.D.
Erskine, W. Esq.
Evelyn, J. Esq.

Frane, Sir F. K.B.
Febure, Mons. Le.

Finch, Sir J. Kt.

Glisson, F. M.D.
Goddart, J. M.D.
Graunt, J. Esq.

Haake, T. Esq.
Hammond, W. Esq.
Hatton, Christopher, Ld
Harley, Sir R. Kt.
Hayes, J. Esq.
Henshaw, T. Esq.
Henshaw, N. M.D.
Hill, A. Esq.
Hoare, W. M.D.
Holder, W. D.D.
Hooke, R. M.A.
Hoskyns, J. Esq.
Howard, C. Esq.
Huygens, Mons. C.

Jones, R. Esq.

Kincardin, Alexander, Earl of.
King, Sir A. Kt.

Long, J. Esq.
Lowther, A. Esq.
Lucas, Ld. J.

Massarene, Ld. Visc. John.
Merret, C. M.D.
Moray, Sir R. Kt
Morgan, Sir A. Kt.

Needham, J. M.D.
Neile, Sir P. Kt.
Neile, W. Esq.
Northampton, James, Earl of.
Nott, Sir T. Kt.

Oldenberg, H. Esq.

Packer, P. Esq.
Palmer, D. Esq.
Paston, Sir R. K.B.

Pell, J. D.D.
Persal, Sir W. Kt.
Pett, Sir P. Kt.
Pett, P. Esq.
Petty, Sir W. Kt.
Pope, W. M.D.
Povry, T. Esq.
Powle, Sir R. K.B.
Powle, H. Esq.
Proby, H. Esq.

Quatremaine, W. M.D.

Sandwich, Edward, Earl of.
Scarburgh, E. M.D.
Schroter, W. Esq.
Seth, Bp.
Shaen, Sir J. Kt.
Slingesby, H. Esq.
Smyth, G. M.D.
Stanhope, A. Esq.
Stanley, T. Esq.
Sorbiere, Mons. S.
Southwell, R. Esq.
Spratt, T. M.A.

Talbot, Sir G. Kt.
Terne, C. M.D.
Tuke, S. Esq.

Vermuyden, C. Esq.

Waller, F. Esq.
Wallis, J. D.D.
Whistler, D. M.D.
Wilkins, J. M.D.
Williamson, J. Esq.
Willughby, F. Esq.
Winde, W. Esq.
Winthrop, J. Esq.
Wren, M. Esq.
Wren, T. M.D.
Wren, C. LL.D
Wyche, Sir C. Kt.
Wyche, Sir P. Kt.
Wylde, E. Esq.

After which the Society proceeded from time to time to elect Fellows by virtue of their Charter.

Abbot, G. Esq.	1793	Arthington, C. Esq.	1701	Barrington, Hon. Daines,	1767
Abbot, J. F. Esq.	1793	Arundell, Hon. Richard,	1740	Barrow, J. Esq.	1737
Abeille, L. Paul	1762	Ascanius, P. M.D.	1755	Barrow, J. Esq.	1805
Aberdeen, George, Earl of,	1808	Ash, St. G. A.M.	1686	Barrowby, W. M.D.	1721
Aberdour, James, Ld.	1733	Ash, J. M.D.	1787	Barthelemy, Mr. J. J.	1755
Aberdour, S. Charles, Ld.	1754	Ash, E. M.D.	1801	Barton, P. LL.D.	1757
Abernethy, Mr. J.	1796	Ashby, S. Esq.	1756	Barry, E. M.D.	1731
Ablancourt, Mons. F. d',	1684	Ashe, W. W. Esq.	1748	Bassand, Baron J. B.	1732
Abruz, J. G. M.D.	1729	Askew, A. M.B.	1749	Baster, J. M.D.	1738
Adair, J. Esq.	1788	Assemanni, S. E. Archbp.	1737	Bateman, W. Ld. Visc.	1732
Adair, ——, Esq.	1688	Astle, E. Esq.	1808	Bates, Mr. T.	1718
Adam, R. Esq.	1761	Aston, F. Esq.	1678	Bath, William, Earl of,	1744
Adanson, Mr. M.	1761	Athol, J. Duke of,	1780	Bathurst, R. M.D.	1663
Adee, S. M.D.	1744	Atkyns, Sir R. Kt.	1664	Bathurst, Hon. B.	1731
Adye, J. W. Esq.	1774	Atwell, J. M.A.	1729	Batt, J. T. M.A.	1742
Afzelius, A. M.D.	1798	Atwood, G. M.A.	1776	Battie, W. M.D.	1741
Agar, W. E. Esq.	1781	Aubant, Abraham, D', Esq.	1791	Battine, W. LL.D.	1797
Agliouby, W. M.D.	1667	Aubert, A. Esq.	1772	Bavaria, M. Joseph, Duke of,	1802
Agricola, Mr.	1698	Aubyn, Sir J. Saint, Bart.	1797	Bayardi, Mons. O. A.	1755
Ahlers, Mr. C.	1726	Audouin, Mons. de Mont,	1748	Bayes, Mr. T.	1742
Akenside, M. M.D.	1753	Aumont, Louis d' Aumont de		Bayford, Mr. D.	1770
Aland, J. F. Esq.	1712	Rochebaron, Duc d',	1713	Bayley, T. B. Esq.	1773
Albans, Charles, Duke of St.	1722	Auriol, J. P. Esq.	1808	Beal, Sir J.	1663
Albemarle, George, Duke of,	1664	Aust, G. Esq.	1797	Beale, J. M.L.	1721
Albinus, B. S. M.D.	1764	Auvergne, Philip d',	1786	Beard, R. M.D.	1726
Alcock, N. M.D.	1749	Auzout, Mons. A.	1666	Beauclerc, Lord Amelius,	1809
Aldersey, R. Esq.	1794	Averinus, Sig. J.	1712	Beauclerk, T. Esq.	1770
Alembert, Mons. J. le Roud d',	1748	Ayloffe, Sir J. Bart.	1731	Beaufarn, H. B. de,	1730
Algarotti, Sig. F.	1736	Ayres, Mr. T.	1707	Beaufort, Mr. Lewis de,	1746
Allemand, Mons. J. N. S.	1746	Ayres, Mr. C. N.	1708	Beaufort, Mons. V.	1663
Allen, T. M.D.	1667			Beaufoy, M. Esq.	1790
Allen, E. Esq.	1726	Babington, W. M.D.	1805	Beaufoy, H. Esq.	1782
Allen, J. M.D.	1736	Bacon, T. S. Esq.	1721	Beaumont, R. Esq.	1684
Allen, Mr. W.	1807	Bacon, Mr. V.	1731	Beaumont, Mr. J.	1685
Alleyne, J. Esq.	1776	Bacon, J. Esq.	1750	Beaumont, Mons. J. B. J.	
Allioni, C. M.D.	1758	Baden, Dourlack, Ch. Frede-		Elie de,	1765
Allison, Rev. A. B.LL.	1792	rick, Margrave of,	1747	Beauval, Mons. J. Basnage de,	1697
Allix, W. Esq.	1753	Baganell, N. Esq.	1664	Beauvoir, Rev. O. D.D.	1785
Allonville, Mons. J. E. de',	1715	Baglivi, Sig. G.	1698	Beccari, Sig. J. B.	1728
Almeyda, Father Theodore d',	1758	Bagnall, J. Esq.	1774	Beccaria, Mr. J. B.	1755
Alstroemer, J. Esq.	1778	Bailey, Mr. A.	1683	Becker, Mons. B. M.D.	1698
Althorpe, G. J. Visc.	1780	Baillie, M. M.D.	1790	Beckett, Mr. W.	1808
Altle, T. Esq.	1766	Baillon, Chev. de,	1749	Bedford, John, Duke of,	1741
Ames, Mr. J.	1743	Baker, Mr. T.	1684	Bedford, W. M.D.	1745
Amiens, Edward, Ld.	1777	Baker, Mr. H.	1740	Beighton, Mr. H.	1720
Amman, Dr. J.	1730	Baker, G. M.D.	1764	Belchier, Mr. J.	1732
Amyand, C. Esq.	1716	Baker, J. W. Esq.	1771	Belidor, Mons. B. Forrest de,	1726
Anderdon, J. P. Esq.	1811	Baker, J. Esq.	1810	Belius, Rev. M.	1738
Anderson, Mr. J.	1759	Baker, Sir F. Bart.	1811	Bell, Mr. G.	1749
Andrade, A. F. d',	1749	Baldini, Sig. J. A. Count,	1712	Bellers, Mr. F.	1711
Andrews, J. Esq.	1726	Baldwin, Dr. C. A.	1676	Bellers, Mr. J.	1718
Anglesey, Arthur, Earl of,	1668	Baldwin, R. M.A.	1760	Bellin, Mons.	1753
Anguish, T. Esq.	1766	Bale, C. M.D.	1709	Beluga, Mr. Bernardo de,	1780
Anker, B. C. Esq.	1782	Balle, R. Esq.	1708	Bemde, J. Esq.	1678
Anker, C. Esq.	1804	Balme, Rev. E. M.A.	1801	Bennet, R. H. A. Esq.	1767
Annesley, Mr.	1704	Baltimore, Charles, Ld.	1731	Bennet, Rev. A.	1789
Anson, T. Esq.	1730	Baltimore, Frederick, Ld.	1767	Bentinck, Hon. W. Count,	1731
Anson, G. Esq.	1745	Bamber, Mr. J.	1718	Bentinck, J. Esq.	1765
Anspach and Bayreuth, Alex-		Bancroft, E. M.B.	1773	Bentinck, Capt. W.	1787
ander, Margrave of,	1780	Bankes, Mr. R.	1736	Bentley, Mr. R.	1695
Anteny, Gebhard d', Esq.	1723	Banks, Sir J. Bart.	1668	Bergius, P. J. M.D.	1770
Antrobus, E. Esq.	1801	Banks, J. Esq.	1730	Bergman, Mr. T.	1765
Arbuthnot, J. M.D.	1704	Banks, J. Esq.	1766	Beringhen, Mons. Theodore	
Arbuthnot, J. Esq.	1770	Barbiano, Lewis, Count of,	1781	de,	1667
Arden, C. George, Ld.	1786	Barbosa, J. M. S. M.D.	1750	Berkley, Sir M. Kt.	1667
Arden, Rt. Hon. Sir R. Pep-		Barham, Mr. H.	1717	Berkley Sir C. K.B.	1667
per, Kt.	1788	Barker, R. M.D.	1732	Bernard, E. B.D.	1673
Arderne, Mr. J. A.M.	1668	Barker, Sir R. Kt.	1775	Bernard, Mr. C.	1696
Arderon, Mr. W.	1745	Barnard, Dr. T. Bp.	1783	Bernard, J. P. M.A.	1737
Areskyne, R. M.D.	1703	Barnard, F. A. Esq.	1790	Bernard, H. M.D.	1738
Argenville, Mons. A. J. de Sa-		Barnes, Mr. J.	1710	Bernard, J. M.D.	1760
liers d',	1750	Barnewall, Hon. M.	1765	Bernoulli, Mons. J.	1712
Argyle, Archibald, Earl of,	1663	Barnwell, C. F. Esq.	1809	Bernoulli, Mons. N. M.D.	1713
Armstrong, J. Esq.	1723	Barrett, R. Esq.	1713	Bernoulli, Mr. D.	1756
Arnold, J. H. Esq.	1806	Barrington, T. Esq.	1669	Berthier, Mr. J. E.	1763

Berthollet, Mons. C. L. 1789
Berthoud, Mons. F. 1764
Best, G. Esq. 1791
Bethune, Dr. J. 1773
Beuninghen, Mous. Conrad
    Van, ... 1682
Bevan, Mr. S. 1725
Bevis, J. M.D. 1765
Bianchi, Sig. V. 1710
Bianchini, Sig. F. 1712
Bickerton, Sir R. Bart. .. 1810
Bidloo, G. M.D. 1696
Biggin, G. Esq. 1802
Bignon, Abbe J. Paul, .. 1734
Billiers, W. Esq. 1726
Bingley, R. Esq. 1809
Birch, Mr. A. 1673
Birch, T. M.A. 1734
Bisse, Mr. P. 1706
Bisshopp, Sir C. Bart. .. 1791
Blackburne, Mr. S. 1681
Blackburne, T. M.D. .. 1781
Blackburne, J. Esq. .. 1794
Blackwell, J. Esq. 1692
Blacow, R. M.A. 1754
Blagden, C. M.D. 1772
Blair, Mr. P. 1712
Blair, J. LL.D. 1755
Blair, A. Esq. 1799
Blake, F. Esq. 1746
Blake, W. Esq. 1807
Blanchard, W. M.B. .. 1759
Blane, G. M.D. 1784
Blane, W. Esq. 1795
Blaquiere, John, Lord de, .. 1803
Blaquiere, Hon. Lieut.-Col. W. 1805
Bligh, Capt. W. 1801
Bliss, N. M.A. 1742
Blizard, Mr. W. 1787
Blizard, Mr. T. 1803
Blumenback, J. F. M.D. .. 1793
Blunt, Col. T. 1664
Blyke, R. Esq. 1773
Bode, J. E. 1789
Boerhaave, H. M.D. .. 1730
Boffrand, Mons. De, .. 1744
Bogdani, W. Esq. 1729
Bohadsch, J. B. M.D. .. 1762
Bois, Charles du, Esq. .. 1770
Bon, Mons. F. Xavier de, .. 1738
Bonet, Mons. L. F. .. 1711
Bon Figlilio, Sig. 1696
Bonnet, Mr. C. 1743
Booth, B. Esq. 1772
Boothe, Mr. P. 1703
Bootle, R. Esq. 1757
Bootle, R. W. Esq. 1761
Borghese, Mark Ant. Principe, 1682
Borlase, Rev. Mr. W. .. 1750
Born, Le Chev. Ignace de, 1774
Bornemann, Mons. P. J. .. 1722
Borringdon, John, Ld. .. 1795
Bosanquet, Mr. S. 1793
Bose, Mr. G. M. 1757
Bosville, W. Esq. 1792
Bottine, Sig. D. M.D. .. 1695
Bougainville, Mons. de, 1756
Bouguer, Mons. P. 1749
Boulton, M. Esq. 1785
Bournon, J. Lewis, Compte de, 1802
Bowdler, T. M.D. .. 1781
Bowdowin, J. Esq. .. 1788
Bower, T. M.D. 1712
Bowes, M. Esq. 1699
Bowman, W. Esq. 1742
Boyer, Mons. J. B. M.D. .. 1749

Boylston, Dr. Z. 1726
Bozé Mr. C. Gros de, .. 1749
Braddon, Mr. 1681
Bradley, Mr. R. 1712
Bradley, J. M.A. 1718
Bragança, Don Joano, Duke de, 1757
Braithwaite, D. Esq. .. 1782
Brakenridge, W. D.D. .. 1752
Brand, T. Esq. 1756
Brand, Mr. W. T. 1809
Brander, G. Esq. 1754
Brattle, Mr. W. 1713
Bredalbane, John, Earl of, 1784
Bree, R. M.D. 1808
Bregentwed, A. G. Molthe
    Count de, ... 1764
Bremond, Ds. François de, 1740
Brereton, O. S. Esq. .. 1762
Breynius, J. P. M.D. .. 1703
Briançon, Mons. le Compte de, 1706
Bridgeman, O. Esq. .. 1696
Bridgeman, O. Esq. .. 1697
Bridgen, E. Esq. 1777
Bridges, J. Esq. 1708
Bridges, Sir B. Bart. .. 1726
Bridgewater, J. Wm., Earl of 1808
Bridgman, W. Esq. .. 1679
Briggs, R. M.A. 1693
Brigstock, O. Esq. .. 1710
Brinkley, Rev. J. M.A. .. 1803
Brisbane, Lieut.-Col. T. .. 1810
Bristol, Frederick, Earl of, Bp. 1782
Bristol, F. William, Earl of, 1805
Bristow, W. Esq. 1742
Brocklesby, R. M.D. .. 1746
Brodie, J. Esq. 1797
Brodie, Mr. B. 1810
Bromfield, T. M.D. .. 1713
Bromfield, R. M.D. .. 1779
Brookesbank, J. Esq. .. 1751
Brougham, H. Esq. .. 1803
Brown, E. M.D. 1667
Brown, Mr. J. 1721
Brown, L. M.A. 1729
Brown, I. H. Esq. 1749
Brown, W. Esq. 1777
Brown, R. Esq 1811
Browne, T. M.D. 1699
Browne, W. M.D. 1738
Browne, J. H. Esq. .. 1770
Browne, H. Esq. 1797
Brownlow, Ld. 1783
Brownrigg, W. M.D. .. 1742
Broussonet, P. M. Augustus, 1782
Bruce, J. Esq. 1776
Bruce, J. Esq. 1791
Bruhl, J. Mauritius, Count de, 1765
Bruni, J. L. M.D. .. 1743
Brunswick, F. Albert, Duke of, 1664
Brunswick, C. W. Ferdinand,
    Hereditary Prince of, .. 1764
Brydges, J. Esq. 1694
Brydone, P. Esq. 1773
Buchannan, F. M.D. .. 1806
Büchner, Mr. A. E. .. 1767
Buckinghamshire, John, Earl
    of, ... 1785
Buffon, Mons. G. L. Le Clerc
    de, ... 1739
Bugge, Mr. T. 1788
Bulgarius, E. Archbp. .. 1788
Bulkley, Sir R. Kt. .. 1685
Bullialdus, Mons. J. .. 1667
Burgess, T. Bp. 1807
Burlington, Richard, Earl of, 1722
Burman, Mr. E. 1728

Burnet, Mr. G. 1665
Burnet, G. M.A. 1723
Burnett, W. Esq. 1706
Burnett, Sir T. Kt. 1748
Burney, C. Mus. D. .. 1773
Burney, C. LL.D. 1802
Burney, Capt. J. 1809
Burrell, P. Esq. 1752
Burrow, R. Esq. 1762
Bury, Sir T. 1781
Burzynski, T. Joseph, Count
    de, ... 1769
Bushby, Mr. J. 1719
Bushe, A. Esq. 1758
Bussiere, Mr. P. 1699
Bute, John, Marquis of, .. 1799
Butt, J. M. M.D. 1767
Buxton, G. M.D. 1779
Buys, Mr. 1706
Byrd, W. Esq. 1696
Byron, Mr. J. 1723

Cadell, W. A. Esq. 1810
Cadogan, C. Esq. 1718
Cadogan, W. M.D. .. 1752
Caille, M. N. Louis de la, 1760
Caldanus, M. A. L. M.D. .. 1772
Calderwood, W. Esq. .. 1776
Caldwell, Sir J. Bart. .. 1752
Call, J. Esq. 1775
Calvert, Hon. B. L. .. 1731
Calvert, P. LL.D. 1781
Campbell, J. M.D. 1718
Campbell, Mr. G. 1730
Campbell, C. Esq. 1730
Campbell, H. Esq. 1740
Campbell, J. Esq. 1760
Campbell, Rt. Hon. Frederick
    Ld. 1793
Campbell, J. Esq. 1795
Camper, P. M.D. 1750
Camus, Mons. C. E. Louis, .. 1764
Canning, G. Esq. 1810
Canton, Mr. J. 1749
Cantwell, A. M.D. 1738
Canvane, P. M.D. 1765
Capeller, Sig. M. A. .. 1725
Capello, Sig. P. A. .. 1744
Capper, R. Esq. 1797
Caraccioli, Mr. Marc. Dom. 1765
Caramanico, Francis, Prince
    of, ... 1783
Carbone, Father J. B. .. 1729
Carbura, Compte J. Baptiste, 1765
Cardigan, George, Earl of, .. 1749
Cardross, H. David, Ld. .. 1733
Cardross, David, Ld. .. 1765
Carew, R. P. Esq. 1788
Carkes, Mr. J. 1663
Carlisle, Charles, Earl of, .. 1665
Carlisle, Mr. A. 1804
Carmarthen, Francis, Marquis
    of, ... 1773
Carnac, J. Esq. 1772
Carnegio, Sir D. Bart. .. 1799
Carpenter, G. Esq. 1729
Carr, W. Esq. 1727
Carr, Rev. W. H. B.D. .. 1806
Carrington, Robert, Ld. .. 1800
Carrington, C. E. Esq. .. 1800
Carstairs, J. Esq. 1811
Carteret, P. Esq. 1664
Cartwright, Mr. 1716
Carvalho e Mello, S. Joseph de, 1740
Cary, W. Esq. 1727
Carysfort, J. Joshua, Ld. .. 1779

Davila, Don Pedro, ...	1776
Davy, M. M.D.	1801
Davy, H. Esq.	1803
Dawkins, J. Esq.	1755
Dawkins, H. Esq.	1778
Day, T. Esq.	1691
Day, Sir J. Kt.	1739
Dayrolle, S. Esq.	1743
Dealtry, Rev. W.	1811
Deane, Sir A. Kt.	1681
D'Arcy, Hon. James, ...	1729
Degge, S. Esq.	1723
Degge, S. Esq.	1730
Dehn, C. D. Alev à	1729
Deidier, Dr. A.	1723
Delaval, E. Esq.	1759
Delawar, John, Ld.	1726
Del Bene, Sig. T.	1695
Denmark, Prince George of,	1704
Denmark, Christian VII, King of, ...	1768
Dent, J. Esq.	1811
Dereham, Sir T. Bart. ..	1720
Derham, Mr. W.	1702
Desaguliers, J. T. M.A. ..	1714
Desaguliers, Lieut.-Gen. T.	1780
Devonshire, William, Duke of,	1747
Devonshire, William, Duke of,	1761
Dickenson, E. D.M.	1677
Dickinson, C. Esq.	1800
Dikins, Mr. A.	1722
Dickson, T. M.D.	1770
Dillenius, Mons. J. J. M.D.	1724
Dillon, Hon. Charles ..	1767
Dillwyn, L. W. Esq.	1804
Dimsdale, Baron T.	1769
Dimsdale, Baron N.	1805
Dingley, Mr. R.	1748
Diodate, J. M.D.	1724
Dirom, Major A.	1794
Dixon, W. Esq.	1729
Dixon, A. Esq.	1748
Dixon, J. Esq.	1773
Dixon, Rev. R.	1811
Dobson, M. M.D.	1778
Dobyns, Mr. J.	1723
Dod, P. M.D.	1729
Dodgson, C. M.A.	1762
Dodson, Mr. J.	1755
Dolben, J. D.D.	1665
Dolceus, J. M.D.	1692
Dolland, Mr. J.	1761
Domcke, Mr. G. P.	1734
Donati, Dr. V.	1757
Doody, Mr. S.	1695
Dopplemayer, Dr. J. G. ..	1733
Dorislaus, Mr. I.	1681
Dorset, Richard, Earl of, ..	1665
Dorset, Charles, Earl of, ..	1698
Douglas, Ld. G.	1692
Douglas, J. M.D.	1706
Douglas, C. Esq.	1770
Douglas, W. Esq.	1711
Douglas, Mr. J.	1720
Douglas, G. M.D.	1732
Douglas, J. D.D.	1778
Douglas, A. Esq.	1793
Douglas, Rt. Hon. S.	1795
Douglas, Rev. W.	1800
Douglas, Alexander, Marquis of, ...	1802
Downe, Henry, Ld. Visc. ..	1750
Downs, J. M.D.	1669
Drake, J. M.D.	1701
Drake, Mr. F.	1736
Drau, Mons. H. François Le,	1744

Dreyer, C. Wm. de, ...	1783
Drummond, W. Esq.	1799
Drury, Sir T. Bart.	1758
Dry, H.	1730
Duane, M. Esq.	1763
Duc, Dr. Le, ...	1722
Ducarel, A. C. LL.D. ..	1762
Ducket, G. Esq.	1808
Duclos, Mons.	1764
Dudley, Sir M. Bart. ..	1703
Dudley, P. Esq.	1721
Dugood, Mr. W.	1728
Duillier, Mr. N. F. de, ..	1587
Duillier, Mons. G. C. Facio de, ...	1706
Duliolo, Sig. R. de, ...	1712
Dumaresq, D. D.D.	1761
Dumas, Mons. V.	1665
Dummer, T. Lee, Esq. ..	1732
Dummer, T. Esq.	1773
Duncan, Sir W. Bart. ..	1771
Dundas, T. Esq.	1768
Dungarvan, Charles, Visc. ..	1663
Duplin, Ld. Visc.	1712
Dupré, J Esq.	1779
Durand, Mr. D.	1728
Duroure, F. Esq.	1774
Dutens, L. M.A.	1775
Duval, F. P. M.D.	1741
Duval, P. M.A.	1774
Dyer, S. Esq.	1760
Dysart, Wilbraham, Earl of,	1805
Eames, Mr. J.	1724
Eardley, Samson, Ld	1789
Earle, W. B. Esq.	1773
Earle, J. Esq.	1794
East, Mr. W.	1720
East, E. H. Esq.	1799
Eaton, Rev. S. M.A. ..	1797
Eckersall, G. Esq.	1761
Eckhardt, Mr. A. G.	1774
Eden, Rt. Hon. W.	1786
Edgcombe, Richard, Earl of Mount, ...	1808
Edgecumbe, Sir R. Bart. ..	1676
Edgecumbe, Geo. Visc. Mount,	1784
Edgeworth, R. L. Esq. ..	1781
Edwards, Sir J. Bart. ..	1731
Edwards, Mr. G.	1757
Edwards, B. Esq.	1794
Effen, Mons. J. Vau, ..	1715
Egan, T. M.D.	1811
Egerton, Rev. F. H.	1781
Egmont, John, Earl of, ..	1764
Egremont, George, Earl of,	1797
Ehret, Mr. G. D.	1757
Elford, W. Esq.	1790
Ellicott, Mr. J.	1738
Elliot, J. Esq.	1776
Elliot, J. Esq.	1811
Ellis, W. Esq.	1745
Ellis, Mr. H.	1749
Ellis, Mr. J.	1754
Ellis, J. Esq.	1774
Ellis, G. Esq.	1797
Ellis, J Esq.	1801
Ellis, H. Esq.	1811
Elphinstone, Hon. G. Keith,	1790
Ellys, A. M.A.	1723
Else, J. Esq.	1778
Emlyn, T Esq.	1768
Emmett, M. Esq.	1697
Englefield, H. C. Esq. ..	1778
Ent, G. Esq.	1676
Epinasse, Mr. Charles L', ..	1767

Erieeÿra, Don Francisco Xavier de Menezes, Count da,	1795
Erskine, Mr.	1771
Erskine, Hon. T.	1787
Essex, William, Earl of, ..	1737
Euler, Mr. L.	1746
Eustace, Sir M.	1667
Eve, H. Esq.	1681
Evelyn, Sir J. Bart.	1722
Exeter, Brownlow, Earl of,	1767
Exeter, Henry, Earl of, ..	1800
Eyre, K. Esq.	1726
Eyre, R. D.D.	1767
Eyre, Sir J. Kt.	1791
Faget, Mons. J. B.	1752
Fagnini, Sig. Con te iulio Carlo de, ...	1723
Fahrenheit, Mons. D. G. ..	1724
Fairfax, Hon. H. C. ..	1727
Falconer, W. M.D. ..	1773
Faquier, W. Esq.	1746
Faquier, F. Esq.	1753
Faria, Don Jos. de, ...	1682
Farmer, Rev. R. D.D. ..	1791
Farr, W. M.D.	1770
Farr, S. M.D.	1779
Fawconor, Mr. J.	1735
Fay, Mons. C. F. di Cisternay du, ...	1729
Faye, Charles, de la, Esq. ..	1725
Feake, C. M.D.	1748
Fellowes, W. Esq.	1704
Fellows, W. Esq.	1708
Fellows, W. Esq.	1731
Felton, S. Esq.	1762
Fenton, W. Esq.	1723
Ferguson, Mr. J.	1763
Fergusson, F. Esq.	1798
Fergusson, R. Esq.	1805
Fermor, Hon. Lieut.-Col. T. W.	1805
Ferner, Benedict, ...	1760
Feronce, Mons. J. Baptiste de,	1764
Ferrari, Dr. D.	1723
Ferrars, George, Ld. de, ..	1781
Ferrers, Washington, Earl,	1761
Ferris, S. M.D.	1797
Fetherstonhaugh, Sir M. Bart.	1752
Fevre, C. Shaw Le, Esq. ..	1796
Ffolkes, W. Esq.	1226
Fidalgo de Silveria, J. Toze,	1751
Fife, James, Earl of, ..	1787
Filenius, Mr. P.	1737
Finch, D. Esq.	1668
Finch, Hon. Henry, ...	1742
Finch, T. Esq.	1804
Finley, Lieut. J.	1788
Firmin, Mr. T.	1680
Fischer, Mr. J. B.	1744
Fisher, Rev. R. M.A. ..	1790
Fitzgerald, K. Esq.	1756
Fitzharding, Maurice, Ld. Visc.	1668
Fitzherbert, W. Esq. ..	1762
Fitzwilliam, Richard, Visc.	1747
Fitzwilliam, Richard, Visc.	1789
Flamsteed, Mr. J.	1676
Flatman, T. Esq.	1668
Flower, ——, Esq.	1667
Fludyer, Sir T.	1767
Fly, H. A.M.	1781
Folard, Chev. Charles de, ..	1749
Foley, T. Esq.	1696
Foley, R. Esq.	1708
Foley, Thomas, Ld.	1740

Hamilton, Hon. John,	1745	Hellot, Mons. J.	1740	Hollingberg, T. D.D.	1783
Hamilton, Hon. Charles,	1747	Helmfeld, Mons. G.	1670	Hollings, J. M.D.	1726
Hamilton, Mr. H.	1761	Helvetius, Mons. J. C. A. M.D.	1755	Hollis, Sir F.	1671
Hamilton, W. Esq.	1766	Helvetius, J. A.	1763	Hollis, T. Esq.	1757
Hamilton, A. D.D.	1777	Hemming, S. M.A.	1776	Hollman, M. S. C.	1747
Hamilton, A. Esq.	1795	Henchman, Dr. H. Bp.	1665	Holloway, B. LL.B.	1723
Hamilton, A. Esq.	1808	Henley, J. Esq.	1693	Holmes, G. Esq.	1741
Hammond, A. Esq.	1698	Henley, F. Morton, Ld.	1780	Holmes, Rev. R. D.D.	1797
Hammond, B. Esq.	1754	Henly, Mr. W.	1773	Holstein, J. Lewis, Count,	1762
Hammond, Sir A. Snape, Bart.	1797	Henrickson, Mr. H.	1742	Holt, R. Esq.	1707
Hampe, D. J. H. M.D.	1729	Henniker, J. Esq.	1779	Holwell, J. Z. Esq.	1767
Hanbury, W. Esq.	1728	Henniker, J. Esq.	1785	Home, Mr. E.	1787
Hanckewitz, Mr. A. G.	1729	Henry, W. D.D.	1755	Hooker, Mr. W. J.	1812
Hanmer, Sir T. Bart.	1804	Henry, Mr. T.	1775	Hooper, E. Esq.	1759
Hannisius, Mr. D.	1678	Henry, W. M.D.	1809	Hop, Mynheer,	1734
Harcourt, Simon, Earl of,	1753	Henshaw, Sir J. Kt.	1794	Hope, Ld.	1727
Hardenberg, The Baron de,	1745	Herbert, Charles, Ld.	1673	Hope, J. M.D.	1767
Harding, Mr. C.	1806	Herbert, J. Esq.	1677	Hope, T. Esq.	1804
Hardinge, C. M.D.	1743	Herfissant, Mons. F. D. M.D.	1750	Hope, Capt. W. J.	1808
Hardinge, G. Esq.	1788	Herschell, Mr. W.	1781	Hope, T. C. M.D.	1810
Hardwicke, Philip, Ld.	1753	Hertsberg, E. Frederic Count de,	1789	Hopkins, Mr. D.	1765
Hardwicke, P. Earl of,	1790	Heucherus, J. H. M.D.	1729	Hopton, R. C. Esq.	1771
Hardy, Sir E. Bart.	1668	Heusch, Mons. J. C. M.D.	1680	Horley, Sir E. K.B.	1663
Hargrave, J. M.A.	1726	Hevelius, Mons. J.	1664	Horne, Mr. J.	1742
Harisey, E. Esq.	1764	Hewer, H. E. Esq.	1723	Horneck, A. M.A.	1668
Harley, Robert, Ld.	1712	Hewett, Sir T. Kt.	1721	Hornsby, T. M.A.	1763
Harper, Mr. J.	1726	Hewett, Rev. J. M.A.	1736	Horsburgh, J. Esq.	1806
Harper, S. M.A.	1766	Hewson, Mr. W.	1770	Horseman, Mr. S.	1727
Harrington, W. Esq.	1665	Hey, Mr. W.	1775	Horsfall, Mr. J.	1768
Harrington, E. M.D.	1734	Heyne, C. G.	1789	Horsley, Mr. J.	1729
Harrington, William, Ld.	1741	Hiaerne, Dr. H.	1669	Horsley, S. M.A.	1767
Harris, S. M.A.	1722	Hibbert, G. Esq.	1811	Hortega, Don. J.	1753
Harris, J. Esq.	1763	Hickes, J. Esq.	1703	Hotham, C. Esq.	1667
Harris, Mr. D.	1768	Hickman, N. M.A.	1725	Houghton, Mr. J.	1680
Harrison, Mr. W.	1765	Hickman, S. Esq.	1744	Houghton, H. Esq.	1765
Harrison, T. Esq.	1804	Higgins, W. Esq.	1806	Houstoun, R. M.D.	1725
Harrison, G. Esq.	1807	Hill, Mr. O.	1677	Houstoun, W. M.D.	1732
Hartley, D. M.A.	1736	Hill, Mr. S.	1711	Howard, H. of Norfolk,	1666
Harvey, J. Esq.	1664	Hill, J. Esq.	1719	Howard, E. Esq.	1668
Harwood, J. A.M.	1686	Hill, T. Esq.	1725	Howard, Henry, Ld.	1672
Harwood, B. M.B.	1784	Hill, J. Esq.	1748	Howard, T. of Norfolk,	1672
Haselden, Mr. T.	1739	Hilliard, E. Esq.	1802	Howard, J. S. Esq.	1673
Hassell, R. Esq.	1726	Hills, P. Esq.	1799	Howard, H. Esq.	1696
Hasted, E. Esq.	1766	Hilsall, H. Esq.	1736	Howard, J. Esq.	1756
Hasted, Rev. H.	1812	Hilsborough, Wills, Earl of,	1764	Howard, C. Esq.	1767
Hastings, W. Esq.	1801	Hilsborough, Arthur, Earl of,	1790	Howard, C. Esq.	1768
Hatchett, C. Esq.	1797	Himsel, Nic. de, M.D.	1760	Howard, Mr. S.	1771
Hatton, G. F. Esq.	1775	Hippisley, Sir J. Cox, Bart.	1800	Howard, W. A. M.D.	1778
Havers, C. M.D.	1686	Hirst, W. M.A.	1755	Howard, E. Esq.	1799
Hawkins, J. Esq.	1791	Hoadley, B. Esq.	1726	Howell, T. B. Esq.	1804
Hawksbee, Mr. F.	1705	Hoare, J. Esq.	1664	Hoy, T. M.D.	1707
Hawley, J. M.D.	1740	Hoare, J. Esq.	1668	Huber, J. G. M.D.	1752
Hawley, Rev. W. M.A.	1796	Hoare, Sir R. Kt.	1752	Hubner, Mons. M.	1755
Hay, A. M.D.	1778	Hoare, J. B.D.	1753	Huck, R. M.D.	1765
Haygarth, J. M.B.	1780	Hoare, H. Hugh, Esq.	1784	Hucks, R. Esq.	1722
Haynes, E. Esq.	1683	Hoare, Sir R. C. Bart.	1792	Huddart, J. Esq.	1791
Hayter, Dr. T. Bp.	1749	Hoare, C. Esq.	1809	Hudson, J. Esq.	1754
Hayward, Sir W.	1665	Hobhouse, B. Esq.	1798	Hudson, C. G. Esq.	1757
Hazard, R. Esq.	1752	Hoblyn, R. Esq.	1745	Hudson, Mr. W.	1761
Healde, T. M.D.	1770	Hoblyn, T. Esq.	1810	Hughes, E. Esq.	1726
Heath, B. M.A.	1778	Hodges, T. Esq.	1715	Hughes, Rev. Mr. G.	1748
Heath, Rev. G. D.D.	1795	Hodges, Sir J. Bart.	1716	Hugo, Dr. J. A.	1717
Heathcote, Sir G. Kt.	1705	Hodgkinson, R. B. Esq.	1778	Hulme, M. M.D.	1794
Heathcote, H. Esq.	1720	Hodgson, Mr. J.	1703	Huilier, Mr Simon Le,	1791
Heathcote, G. Esq.	1728	Hodgson, W. Esq.	1807	Hume, Sir A. Bart.	1775
Heathcote, Sir T. Bart.	1751	Hodgson, Rev. R.	1810	Humphries, Col. D.	1807
Heathcote, J. Esq.	1768	Hody, E. M.D.	1732	Hunauld, F. J. M.D.	1733
Heathfield, F. Augustus, Ld.	1809	Hoffman, Mons. H.	1720	Hundertmark, C. F. M.D.	1755
Henviside, J. Esq.	1797	Hoffman, Mr. T.	1746	Hunt, William Le, Esq.	1667
Heberden, W. M.D.	1749	Holford, P. Esq.	1746	Hunt, Mr. T.	1725
Heberden, T. M.D.	1761	Holford, S. Esq.	1785	Hunt, Rev. T.	1740
Heberden, W. Esq.	1791	Holford, R. Esq.	1805	Hunter, R. Esq.	1709
Hedwig, J. M.D.	1788	Holland, R. M.D.	1726	Hunter, Mr. J.	1767
Heinson, Mons. J. T.	1692	Holland, H. Richard, Ld.	1811	Hunter, W. M.D.	1767
Heister, Dr. L.	1730	Holliday, J. Esq.	1786	Hunter, A. M.D.	1775
Hellins, Rev. J.	1796	Hollier, J. M.D.	1718	Hunter, J. M.D.	1786

l

Petit, J. L. M.A.	1759	Pott, Mr. P.	1764	Reyner, Col. B.	1667
Petit, L. H. Esq.	1807	Potter, F. B.D.	1663	Reynardson, S. Esq.	1741
Petiver, Mr. J.	1695	Potter, T. Esq.	1784	Reynolds, Mr. J.	1761
Petley, C. Esq.	1753	Pound, Mr. J. M.B.	1699	Reynolds, H. R. M.D.	1781
Petre, R. James, Ld.	1731	Powell, W. S. D.D.	1764	Rezzonico, Prince Abondio,	1776
Petre, Edward, Ld.	1780	Power, H. M.D.	1663	Rich, D. Esq.	1743
Petre, W. Esq.	1795	Powis, Sir L.	1724	Richards, R. Esq.	1793
Pettit, Mons. P.	1667	Powlet, John, Ld.	1706	Richardson, R. M.D.	1712
Pettiward, R. D.D.	1755	Pownall, T. Esq.	1772	Richardson, Rev. R. D.D.	1779
Pettus, Sir J. Bart.	1663	Pratt, B. D.D.	1708	Richmond, Charles, Duke of,	1723
Petty, H. Esq.	1696	Pratt, C. Esq.	1742	Richmond, Charles, Duke of,	1755
Petty, J. Esq.	1771	Preston, J. Esq.	1773	Ridge, G. Esq.	1810
Peysonnel, Mons. J. A.	1756	Preston, Rev. W. M.A.	1778	Rigaud, S. P. Esq.	1805
Pfütschner, Baron,	1732	Pretyman, Rev. G. D.D.	1785	Riou, Capt. E.	1796
Philips, E. Esq.	1727	Prevost, Mr. P.	1806	Ripa, Sig. Ludovicus a,	1718
Phillips, J. Esq.	1742	Price, W. Esq.	1753	Ripa, Sig. Ludovicus a,	1733
Phipps, Hon. C. J.	1771	Price, Mr. R.	1765	Rittenhouse, D. Esq.	1795
Piazzi, Mr. J.	1804	Price, R. P. Esq.	1781	Rivinus, A. Q. M.D.	1703
Pickering, Mr. R.	1744	Price, J. A.M.	1781	Rixtel, Mr. J. Van.	1739
Picquigny, L. M. Jos. d'Al-		Prideaux, B. Esq.	1746	Riz, Mr. D.	1766
bert d'Ailly, Duke de,	1764	Priestley, J. LL.D.	1766	Robartes, John, Ld.	1666
Pictet, Mr. M.A.	1791	Prime, S. Esq.	1776	Robartes, F. Esq.	1673
Pierce, Mr. J.	1742	Pringle, J. Esq.	1745	Robartes, R. Esq.	1703
Piggott, A. Esq.	1787	Pringle, Lieut.-Col. R.	1785	Robartes, J. Esq.	1732
Pighius, Dr. J.	1680	Prior, M. Esq.	1697	Roberts, E. M.D.	1799
Pigot, Mr. T.	1679	Protti, Mons. Le,	1734	Roberts, R. F. Esq.	1801
Pigott, N. Esq.	1772	Prymc, Mous. da la,	1701	Robertson, Mr. J.	1741
Pigott, G. Esq.	1774	Pujolas, Mr. M.	1695	Robertson, J. M.D.	1790
Pinto, Chev. Louis,	1787	Pulteney, Mr. R.	1762	Robertson, Rev. A. M.A.	1795
Pitcairn, W. M.D.	1770	Putman, Mr. H.	1767	Robertson, R. Esq.	1804
Pitcairn, D. M.B.	1782	Pye, Sir R. Bart.	1727	Bobertson, J. Esq.	1810
Pitfield, A. Esq.	1684			Robins, Mr. B.	1727
Pitot, Dr. H.	1740	Queensborough, Charles, Duke		Robinson, R. M.D.	1681
Pitt, R. M.D.	1682	of,	1722	Robinson, T. M.D.	1684
Pitt, Hon. W.	1743	Quesnay, Mons. F.	1752	Robinson, T. Esq.	1726
Pitt, J. Esq.	1775			Robinson, M. Esq.	1746
Pitt, W. M. Esq.	1787	Rackett, Rev. T. M.A.	1803	Roby, Mr.	1725
Place, P. Simon de la,	1789	Radnor, C. Bodvill, Earl of,	1693	Rockingham, Charles, Mar-	
Planta, A. J. M.A.	1770	Radnor, William, Earl of,	1767	quis of,	1751
Planta, Mr. J.	1774	Radnor, Jacob, Earl of,	1795	Roderick, R. M.A.	1750
Platt, J.	1762	Raine, Rev. M. D.D.	1803	Roebuck, J. M.D.	1764
Player, Sir T. Kt.	1673	Rainsford, Major Gen. C.	1779	Rogers, J. M.A.	1681
Playfair, J. Esq.	1807	Ramsay, And. Mic. E.S.L.	1729	Rogers, J. M.D.	1738
Plott, R. LL.D.	1677	Ramsden, Mr. J.	1786	Rogers, C. Esq.	1757
Plumer, T. Esq.	1794	Ranby, Mr. J.	1724	Rogers, G. Esq.	1789
Plumptre, H. M.D.	1707	Rand, Mr. I.	1719	Rogers, S. Esq.	1796
Plymouth, Other, Earl of,	1773	Randolph, Ld. John, Bp.	1811	Rogerson, J. M.D.	1779
Pocock, T. M.A.	1727	Raper, M. Esq.	1754	Rolli, Dr. P. A.	1729
Pocock, G. Esq.	1791	Raper, M. Esq.	1783	Rolserius, D.	1729
Pococke, R. LL.D.	1741	Raphson, J. M.A.	1689	Rolt, T. Esq.	1664
Poczobut, Father M.	1771	Rasleigh, P. Esq.	1788	Romilly, Mr. I.	1757
Podmaniczky, Joseph Lewis		Raspe, Mr. R. E.	1769	Romney, Robert, Ld.	1723
de,	1780	Raulin, J. M.D.	1763	Romney, Robert, Ld.	1757
Poissonnier, Mr. P. M.D.	1774	Rausaumousky, Count Cyrille,	1755	Rose, Rev. W. M.A.	1786
Poland, Stanislaus Augustus,		Ravaud, D. M.A.	1747	Rosenkrantz, Mons. I. Baron,	1713
King of,	1766	Rawden, Sir J. Bart.	1744	Ross, J. D.D.	1758
Pole, C. M. Esq.	1800	Rawlinson, T. Esq.	1712	Ross, Lieut.-Col. P.	1794
Poleni, Sig. J.	1710	Rawlinson, R. M.A.	1714	Rowden, Francis, Ld.	1787
Poley, R. Esq.	1725	Rawlinson, Sir W. Kt.	1775	Rowe, H. Esq.	1739
Poli, Mr. J.	1779	Raymond, Robert, Ld.	1739	Rowley, Lieut.-Col. J.	1809
Pollok, A. M.D.	1767	Raynal, Mr. G. T.	1754	Rowley, Rev. G.	1811
Polwarth, Alexander, Ld.	1773	Reade, H. Esq.	1747	Roxburgh, John, Duke of,	1707
Pond, Mr. A.	1752	Reaumer, Dr. R. A. Fer-		Roy, W. Esq.	1767
Pond, J. Esq.	1807	chault de,	1738	Roy, Mr. J. Baptiste Le,	1773
Poniatowsky, Michael, Prince,	1791	Reay, Ld. George,	1698	Royen, Adrianus Van,	1728
Poore, E. Esq.	1772	Recanati, Sig. J. B.	1720	Royen, David Van,	1759
Popham, Capt. H.	1799	Redding, R. Esq.	1671	Rudge, E. Esq.	1726
Popple, H. Esq.	1737	Redern, Count Sigesmund de,	1765	Rudge, E. Esq.	1805
Porter, J. Esq.	1749	Rees, Rev. A. D.D.	1786	Ruischer, Mr. Melchior de,	1729
Portland, William, Duke of,	1739	Reeves, J. Esq.	1790	Russell, Dr.	1681
Portland, Wm. Henry, Duke		Reid, T. Esq.	1806	Russell, M. Esq.	1742
of,	1766	Reinuccini, The Marchese,	1747	Russell, R. M.D.	1752
Portman, Sir W. Kt. Bart.		Relham, Rev. R. M.A.	1787	Russell, A. M.D.	1756
and K.B.	1664	Rennell, J. Esq.	1781	Russell, F. Esq.	1770
Portuga, Benito de Moura,	1740	Rennie, J. Esq.	1798	Russell, W. Esq.	1777
Postlethwayt, J. Esq.	1754	Revillas, Father Didacus de,	1734	Russell, P. M.D.	1777

Struych, Mr. N. 1749	Tough, C. M.A. 1749	Vigny, Mons. Pierre de, .. 1741
Stuart, A. Esq. 1714	Townley, C. Esq. 1791	Villermont, Mons. E. Carbat de, 1685
Stuart,.C. M.D. 1719	Townley, J. Esq. 1797	Villoeson, Mons. J. B. G.
Stuart, J. Esq. 1758	Townsend, I. Esq. 1749	Dansse de, ... 1776
Stubs, P. M.A. 1703	Townshend, Charles, Ld. Visc. 1706	Vince, Rev. S. M.A. 1786
Stukeley, W. M.D. 1717	Tozzi, B. 1715	Vincent, N. D.D. 1683
Suasso, A. L. Esq. 1735	Trembley, Mons. A. .. 1743	Vincent, Mons. L. 1715
Sue, Mr. J..Joseph Le, .. 1760	Trevor, Sir T. Kt. 1707	Vincent, John, Earl of St., 1809
Sulivan, R..J. Esq. 1785	Trevor, Hon. T. 1726	Viviani, Sig. V. 1696
Summer, G. H. Esq. 1796	Trevor, J. Esq. 1728	Volkra, O. C. Count, ... 1716
Superville, Daniel de, M.D. 1740	Trevor, Robert, Ld. .. 1764	Volta, Mr. A. 1791
Supple, R. B. Esq. 1788	Trew, C. J. M.D 1746	Voltaire, Mons. F. M. Arouet
Sussex,.Talbot, Earl of, 1721	Tronch.n, T. M.D. .. 1762	de, ... 1743
Sweiteu, G. Van, M.D. .. 1749	Trotter, J. Esq. 1802	Vrijberge, Mons. Van, 1706
Swinden, Philip Van, M.A. 1772	Trou, Sig. N. 1719	Vyse, W. LL.D. 1781
Swinney, S. D.D. 1764	Trouton, Mr. E. 1810	Wade, W. M.D. 1811
Swinton, Mr. J. M A. .. 1729	Trumball, Sir W. Kt. .. 1692	Wagstaffe, W. M.D. 1717
Sydenham, Sir P. Bart. .. 1700	Trye, C. B. Esq. 1807	Waldegrave, James, Earl of, 1749
Sylvius, ——, M.D. 1687	Tufnell, S. Esq. 1709	Wales, George, Prince of, .. 1727
Symes, Lieut.-Col. M. 1800	Tufnell, G. F. Esq. .. 1758	Wales, Frederick, Prince of, 1728
Symmer, R. Esq. 1753	Tunstall, M. Esq. 1771	Wales, Mr. W. 1776
Symmons, J..Esq. 1794	Turbilly, F. Hon. the Marquis	Wales, George, Prince of, 1789
	de, ... 1762	Walker, T. M.A. 1729
Taglini, Sig. C.. 1732	Turnbull, J Esq. 1791	Walker, Capt. W. 1737
Talbot, Sir J. Kt. 1663	Turner, J. Esq. 1682	Walker, Rev. G. 1771
Talbot, W. Esq. 1742	Turner, E. Esq. 1713	Walker, J. M.B. 1774
Tanner, Mr. J. 1710	Turner, S. Esq. 1741	Walker, F. Esq. 1791
Tarhat, Geo. Visc. 1692	Turner, E. Esq. 1786	Walker, T. Esq. 1791
Taylor,. B. LL.B. 1712	Turner, S. Esq. 1801	Walker, J. Esq. 1794
Taylor,. C. Esq. 1722	Turner, D. Esq. 1802	Wall, Chev. Richard, .. 1753
Taylor,.R. M.D. 1737	Turner, Col. T. H. .. 1804	Wall, M. M.D. 1783
Taylor,.J. Esq. 1776	Turton, J. M.B. 1762	Waller, Sir W. 1679
Teighe, M. M.D. .. 1774	Tuthill, G. A.M. 1810	Waller, R. Esq. 1681
Teissier, Mr. G. L. 1725	Tweedale, John, Earl of, 1663	Walmesley, Pere C. 1750
Tempest, W. Esq. 1712	Twiss, R. Esq. 1774	Walpole, Hon. Horatio, .. 1746
Tennant, S..Esq. 1785	Tyndall, T. Esq. 1762	Walsh, J. Esq. 1770
Tennent, J. M.D. 1765	Tyrawly, James, Ld. .. 1747	Walsingham, Hon. R. B. .. 1778
Theobald, J. Esq. 1725	Tyrconnel, Ld. Visc. .. 1735	Walsted, R. M.D. 1717
Thieullier, L. Jean le, M.D. 1749	Tyrwhitt, T. Esq. 1771	Walter, J. G. M.D. .. 1794
Thistlethwayte, A. Esq. .. 1757	Tyson, E. M.D. 1679	Wanley, Mr. H. 1706
Thom, Mr. Frederick de, 1729	Tyson, Rev. M. B.D. .. 1779	Warberg, Mr. O. 1805
Thomas, N. M.A. 1753		Warburton, J. Esq. 1719
Thomas, E. M.A. 1770	Ubaldini, Count Charles, of, 1667	Warburton H. Esq. 1809
Thomas, Mr. H. L. 1806	Udina, Maffeo, Count of, .. 1756	Ward, E. A.M. 1667
Thompson, R. Esq. 1702	Udny, R. Esq. 1785	Ward, Sir P. Kt. 1681
Thompson, Sir P. Kt. .. 1746	Ulloa, Don Antonio de, .. 1746	Ward, Mr. J. 1723
Thompson, B. Esq. 1779	Upper Ossory, John, Earl of, 1780	Ware, Mr. J. 1802
Thomson, S. W. D.D. .. 1770	Upton, J. Esq. 1757	Wargentin, P. Kt. 1764
Thomson, W. M.B. .. 1786	Upton, Hon. F. Greville, .. 1803	Waring, E. M.A. 1763
Thomson, A..Esq. .. 1786	Usher, Rev. H. D.D. .. 1785	Waring, R. H. Esq. 1769
Thomson, T. M.D. 1811		Warner, Mr. J. 1750
Thoresby, R. Esq. 1697	Vage, Mr. T. 1773	Warren, R. M.D. 1764
Thornhill, Sir J. Kt. .. 1723	Valentia, George, Visc. .. 1796	Warren, C. Esq. 1790
Thornton, E. Esq. 1810	Valentini, M. B. 1715	Waterhouse, F. Esq. .. 1663
Thorpe, J. M.B. 1705	Valisnieri, Sig. A. 1703	Watkins, Mr. T. 1714
Thruston, M. M.D. 1665	Vallancey, Col. C. 1786	Watkins, Rev. T. 1794
Thumberg, C. P. M.D. .. 1788	Valtravers, Radolph de, Esq. 1755	Watson, Mr. W. 1741
Thury, Mons. C. F. Cassini de, 1750	Valvasor, Mons. J. W. .. 1687	Watson, R. M.D. 1750
Thynne, T. Esq. 1664	Varignon, Mons. P. 1714	Watson, J. Esq. 1763
Tighe, Mr. R. 1708	Vater, Mons. A. 1721	Watson, Mr. H. 1767
Tighe, R. S. Esq. 1793	Vaughan, John, Ld. .. 1685	Watson, W. M.B. 1767
Tilli, Sig. M. A. M.D. .. 1708	Vaux, Sir T. de, ... 1665	Watson, R. M.A. 1769
Tillotson, J..D.D. 1671	Vay.de Vaja, Baron Nicolas, 1787	Watson, J. Esq. 1778
Tilson, G. Esq. 1735	Veltheim, A. F. Von, .. 1795	Watson, W. Esq. 1808
Timone, Sig. E. M.D. .. 1703	Venables, J. Esq. 1707	Watt, J. Esq. 1785
Tissington, A. Esq. 1767	Venturi, Sig. M. 1751	Way, L. Esq. 1737
Tissot, Dr. 1760	Venuti, The Abbot Rudolphino, 1757	Way, B. Esq. 1771
Titsingh, I. Esq. 1797	Venuti, Mr. P. 1759	Webb, P. C. Esq. 1749
Titus, Col. S. 1668	Vepusch, J. G. Mus. D. .. 1745	Webb, R. Esq. 1762
Toaldo, Mr. J. 1777	Vergue, L. Elizabeth de la,	Webb, Sir J. Bart. 1764
Tollett, M. G. 1713	Compte de Tressan, ... 1749	Wehber, W. Esq. 1766
Tooke, A. M.A. 1704	Verney, Ralph, Earl, .. 1758	Webster, Sir G. Bart. .. 1736
Tooke, Rev. W. 1783	Vernon, F. Esq. 1672	Wedgewood, Mr. J. 1783
Topham, J. Esq. 1779	Vernon, Mr. 1702	Wegg, S. Esq. 1759
Torkos, J. J. M.D. 1752	Vernon, E. M.A. 1723	Wegg, G. Esq. 1758
Torres, J. Ignatio de, .. 1758	Vessius, Mons. I. D.D. .. 1664	Wegg, G. S. Esq. 1777
Torriano, A. M.A. 1691	Vieussens, Mons. R. M.D. .. 1688	

INDEX.

n

INDEX.

THE END.

C. Baldwin, Printer,
New Bridge Street, London.